REINFORCED CONCRETE DESIGN TO EUROCODES

DESIGN THEORY AND EXAMPLES

FOURTH EDITION

REINFORCED CONCRETE DESIGN TO EUROCODES
DESIGN THEORY AND EXAMPLES
FOURTH EDITION

Prab Bhatt

Thomas J. MacGinley

Ban Seng Choo

CRC Press
Taylor & Francis Group
Boca Raton London New York

CRC Press is an imprint of the
Taylor & Francis Group, an **informa** business

A SPON PRESS BOOK

First published 1978 as Reinforced Concrete: Design Theory and Examples by E&FN Spon © 1978 T.J. MacGinley

Second edition published 1990 © 1990 T.J. MacGinley and B.S.Choo

Third edition published 2006 by Taylor & Francis © 2006 P. Bhatt, T.J. MacGinley and B.S. Choo

CRC Press
Taylor & Francis Group
6000 Broken Sound Parkway NW, Suite 300
Boca Raton, FL 33487-2742

© 2014 by P. Bhatt and the estates of T.J. MacGinley and B.S. Choo
CRC Press is an imprint of Taylor & Francis Group, an Informa business

No claim to original U.S. Government works

ISBN 13: 978-1-4665-5252-4 (pbk)

Library of Congress Cataloging-in-Publication Data

Bhatt, P.
    [Reinforced concrete]
    Reinforced concrete design to eurocodes : design theory and examples / Prab Bhatt, Thomas J. MacGinley, Ban Seng Choo. -- Fourth edition.
        pages cm
    "First published 1978 as Reinforced concrete : design theory and examples, by E&FN Spon ... 1978 [written by] T.J. MacGinley"--Title page verso.
    Includes bibliographical references and index.
    ISBN 978-1-4665-5252-4 (paperback : acid-free paper) 1. Reinforced concrete construction.
    2. Reinforced concrete construction--Standards--Europe. I. MacGinley, T. J. (Thomas Joseph) II. Choo, B. S. III. Title.

TA683.2.M33 2014
624.1'8341--dc23   2013028845

Visit the Taylor & Francis Web site at
http://www.taylorandfrancis.com

and the CRC Press Web site at
http://www.crcpress.com

Dedicated with love and affection to our grandsons

Veeraj Rohan Bhatt Verma

Devan Taran Bhatt

Kieron Arjun Bhatt

CONTENTS

Preface

The fourth edition of the book has been written to conform to Eurocode 2 covering structural use of concrete and related Eurocode 1. The aim remains as stated in the first edition: to set out design theory and illustrate the practical applications of code rules by the inclusion of as many useful examples as possible. The book is written primarily for students in civil engineering degree courses to assist them to understand the principles of element design and the procedures for the design of complete concrete buildings. The book will also be of assistance to new graduates starting their careers in structural design and to experienced engineers coming to grips with Eurocodes.

The book has been thoroughly revised to conform to the Eurocode rules. Many new examples and sections have been added. Apart from referring to the code clauses, reference to the full code has been made easier by using the equation numbers from the code.

Grateful acknowledgements are extended to:

- The British Standards Institution for permission to reproduce extracts from Eurocodes. Full copies of the standards can be obtained from BSI Customer Services, 389, Chiswick High Road, London W4 4AL, Tel: +44(0)20 8996 9001. e-mail: cservices@bsi-global.com
- Professor Christopher Pearce, Deputy Head, School of Science and Engineering , University of Glasgow, Scotland for use of the facilities.
- Mr. Ken McColl, computer manager of School of Engineering, Glasgow University for help with computational matters.
- Dr. Lee Cunningham, Lecturer in Engineering, University of Manchester for reviewing Chapter 19.
- Sheila, Arun, Sujaatha, Ranjana and Amit for moral support.

P. Bhatt
2 October 2013 (Mahatma Gandhi's birthday)

ABOUT THE AUTHORS

Prab Bhatt is Honorary Senior Research Fellow at Glasgow University, UK and author or editor of eight other books, including *Programming the Dynamic Analysis of Structures*, and *Design of Prestressed Concrete Structures*, both published by Taylor & Francis.

He has lectured on design of reinforced and prestressed concrete structures and also on structural mechanics to undergraduate and postgraduate classes in universities in India, Canada and Scotland. He has also carried out research, theoretical and experimental, in the area of behaviour of concrete structures, and has also been extensively involved in design office work.

Tom MacGinley and **Ban Seng Choo**, both deceased, were academics with extensive experience of teaching and research in Singapore, Newcastle, Nottingham and Edinburgh.

CHAPTER 1

INTRODUCTION

1.1 REINFORCED CONCRETE STRUCTURES

Concrete is arguably the most important building material, playing a part in all building structures. Its virtue is its versatility, i.e. its ability to be moulded to take up the shapes required for the various structural forms. It is also very durable and fire resistant when specification and construction procedures are correct.

Concrete can be used for all standard buildings both single-storey and multi-storey and for containment and retaining structures and bridges. Some of the common building structures are shown in Fig. 1.1 and are as follows:

1. The single-storey portal supported on isolated footings.
2. The medium-rise framed structure which may be braced by shear walls or unbraced. The building may be supported on isolated footings, strip foundations or a raft.
3. The tall multi-storey frame and core structure where the core and rigid frames together resist wind loads. The building is usually supported on a raft which in turn may bear directly on the ground or be carried on piles or caissons. These buildings usually include a basement.

Complete designs for types 1 and 2 are given. The analysis and design for type 3 is discussed. The design of all building elements and isolated foundations is described.

1.2 STRUCTURAL ELEMENTS AND FRAMES

The complete building structure can be broken down into the following elements:

- *Beams:* horizontal members carrying lateral loads
- *Slab:* horizontal plate elements carrying lateral loads
- *Columns:* vertical members carrying primarily axial load but generally subjected to axial load and moment
- *Walls:* vertical plate elements resisting vertical, lateral or in-plane loads
- *Bases and foundations:* pads or strips supported directly on the ground that spread the loads from columns or walls so that they can be supported by the ground without excessive settlement. Alternatively the bases may be supported on piles.

To learn about concrete design it is necessary to start by carrying out the design of separate elements. However, it is important to recognize the function of the element in the complete structure and that the complete structure or part of it needs to be analysed to obtain actions for design. The elements listed above are illustrated in Fig. 1.2 which shows typical *cast-in-situ* concrete building construction.

A *cast-in-situ* framed reinforced concrete building and the rigid frames and elements into which it is idealized for analysis and design are shown in Fig. 1.3. The design with regard to this building will cover

1. One-way continuous slabs
2. Transverse and longitudinal rigid frames
3. Foundations

Various types of floor are considered, two of which are shown in Fig. 1.4. A one-way floor slab supported on primary reinforced concrete frames and secondary continuous flanged beams is shown in Fig. 1.4(a). In Fig. 1.4(b) only primary reinforced concrete frames are constructed and the slab spans two ways. Flat slab construction, where the slab is supported by the columns without beams, is also described. Structural design for isolated pad, strip and combined and piled foundations and retaining walls (Fig. 1.5) is covered in this book.

1.3 STRUCTURAL DESIGN

The first function in design is the planning carried out by the architect to determine the arrangement and layout of the building to meet the client's requirements. The structural engineer then determines the best structural system or forms to bring the architect's concept into being. Construction in different materials and with different arrangements and systems may require investigation to determine the most economical answer. Architect and engineer should work together at this conceptual design stage.

Once the building form and structural arrangement have been finalized, the design problem consists of the following:

1. Idealization of the structure into load bearing frames and elements for analysis and design
2. Estimation of loads
3. Analysis to determine the maximum moments, thrusts and shears for design
4. Design of sections and reinforcement arrangements for slabs, beams, columns and walls using the results from 3
5. Production of arrangement and detail drawings and bar schedules

1.4 DESIGN STANDARDS

In Europe, design is generally to limit state theory in accordance with the following Eurocodes:

BS EN 1990:2002 Eurocode – *Basis of structural design*

BS EN 1992-1-1:2004: Eurocode 2: *Design of concrete structures Part-1: General rules and rules for buildings*

BS EN 1992-1-2:2004: Eurocode 2: *Design of concrete structures Part-1-2: General rules-Structural fire design*

The design of sections for strength is according to plastic theory based on behaviour at ultimate loads. Elastic analysis of sections is also covered because this is used in calculations for deflections and crack width.

The loading on structures conforms to:

BS EN 1991-1-1: 2002 Eurocode 1: Actions on Structures Part-1-1: General actions-Densities, self-weight, imposed loads on buildings

BS EN 1991-1-3: 2003 Eurocode 1: Actions on Structures. General actions. Snow loads

BS EN 1991-1-4: 2005 + A1:2010 Eurocode 1: Actions on structures. General actions. Wind actions

In addition to the above codes, although the code gives recommended values for certain parameters, each nation in Europe is allowed some leeway in terms of the values of certain parameters to be used in the codes. In U.K., the National Annex gives guidance on the specific parameters to be used.

The code also makes a clear distinction between principles and application rules. Principles are indicated by the letter P after the clause number. Principles comprise general statements, models, requirements for which no alternative is permitted. For example:

4.2 Environmental conditions
4.2(1)P Exposure conditions are chemical and physical conditions to which the structure is exposed.

4.3 Requirements for durability
4.3(1)P In order to achieve the required design working life of the structure, adequate measures shall be taken to protect each structural element against the relevant environmental actions.

Application rules are generally recognized to satisfy the principles. Application rules follow principle rules. For example under Section 5 Structural analysis:

5.1.1 (4) P Analysis shall be carried out using idealizations of both geometry and the behaviour of the structure. The idealizations selected shall be appropriate to the problem considered.
5.1.1 (7) Common idealizations of the behaviour used for the analysis are:
- Linear elastic behaviour
- Linear elastic behaviour with limited redistribution
- Plastic behaviour including strut and tie models
- Non-linear behaviour

In the above 5.1.1(4)P is a principle and 5.1.1 (7) is an application rule.

Note that different application rules can be used provided they are not in conflict with the principle rules. It is because of this, unlike say the British Standard BS 8110 for the design of reinforced concrete structures, the Eurocodes do not always give detailed equations for the design of an element or a structure.

The codes set out the design loads, load combinations and partial factors of safety, material strengths, design procedures and sound construction practice. A thorough knowledge of the codes is one of the essential requirements of a designer. Thus it is important that copies of these codes are obtained and read in conjunction with the book. Generally, only those parts of clauses and tables are quoted which are relevant to the particular problem, and the reader should consult the full text.

1.5 CALCULATIONS, DESIGN AIDS AND COMPUTING

Calculations form the major part of the design process. They are needed to determine the loading on the elements and structure and to carry out the analysis and design of the elements. The need for orderly and concise presentation of calculations cannot be emphasized too strongly. Very often in practice, projects are kept on hold after some preliminary work. Work should therefore be presented in a form such that persons other than those who did the initial design can follow what was done without too much looking back. A useful reference for the presentation of design office calculations is Higgins and Rogers (1999).

Design aids in the form of charts and tables are an important part of the designer's equipment. These aids make exact design methods easier to apply, shorten design time and lessen the possibility of making errors. Useful books are Reynolds et. al (2007) and Goodchild (1997).

The use of computers for the analysis and design of structures is standard practice. Familiarity with the use of spreadsheets is particularly useful. A useful reference is Goodchild and Webster (2000).

It is essential that students understand the design principles involved and are able to make manual design calculations before using computer programs. Manual calculations are also necessary to check that results from the computer are in the right 'ball park'. This ensures that no gross errors in terms of loads or structural idealizations have been committed.

1.6 DETAILING

The general arrangement drawings give the overall layout and principal dimensions of the structure. The structural requirements for the individual elements are presented in the detail drawings. The outputs of the design calculations are sketches giving sizes of members and the sizes, arrangement, spacing and cut-off points for reinforcing bars at various sections of the structure. Detailing translates this information into a suitable pattern of reinforcement for the structure as a whole. Useful references on detailing are by Institution of Structural Engineers, London (2006) and Calavera (2012).

It is essential for the student to know the conventions for making reinforced concrete drawings such as scales, methods for specifying steel bars, links, fabric, cut-off points etc. The main particulars for detailing are given for most of the worked exercises in the book. The bar schedule can be prepared on completion of the detail drawings. In the U.K., the form of the schedule and shape code for the bars conform to BS 8666: 2005, *Scheduling, Dimensioning, Bending and Cutting of Steel for Reinforcement for Concrete.*

It is essential that the student carry out practical work in detailing and preparation of bar schedules prior to and/or during a design course in reinforced concrete. Computer detailing suites are now in general use in design offices.

1.7 REFERENCES

Calavera, Jose. (2012). *Manual for Detailing Reinforced Concrete Structures to EC2.* Spon Press/Taylor & Francis.

Goodchild, C.H. and Webster, R.M. (2000). *Spresdsheets for Concrete Design to BS 8110 and EC2.* Reinforced Concrete Council.

Goodchild, C.H. (1997). *Economic Concrete Frame Elements.* Reinforced Concrete Council.

Higgins, J.B and Rogers, B.R. (1999). *Designed and Detailed,* 4th ed. British Cement Association.

Institution of Structural Engineers. (2006). *Standard Method of Detailing Structural Concrete: A Manual for Best Practice,* 3rd ed. London.

Reynolds, C.E., Steedman, J.C. and Threlfall, A.J. (2007). *Reinforced Concrete Designer's Handbook,* 11th ed. Routledge.

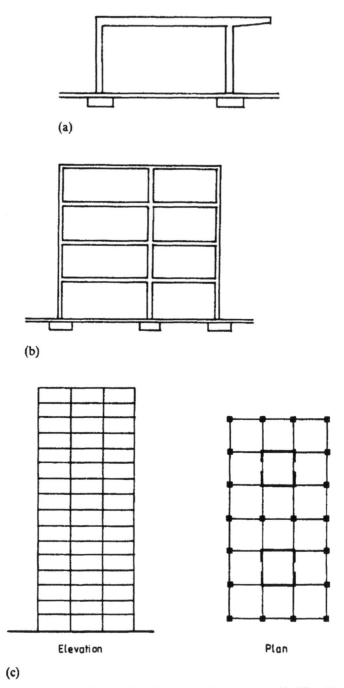

Fig. 1.1 (a) Single storey portal; (b) medium-rise reinforced concrete framed building; (c) reinforced concrete frame and core structure.

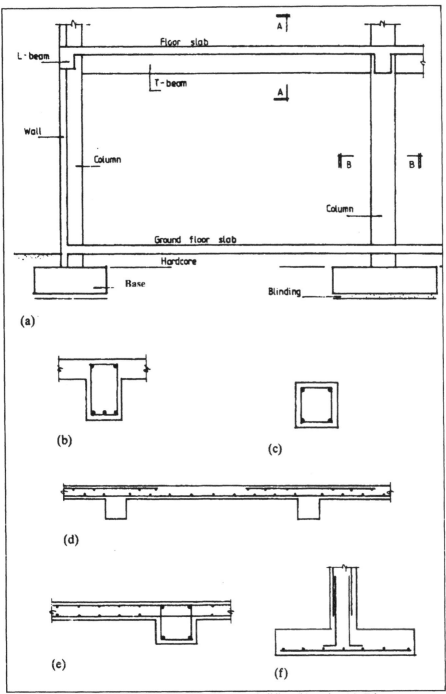

Fig. 1.2 (a) Part elevation of reinforced concrete building; (b) section AA, T-beam;
(c) section BB; (d) continuous slab; (e) wall; (f) column base.

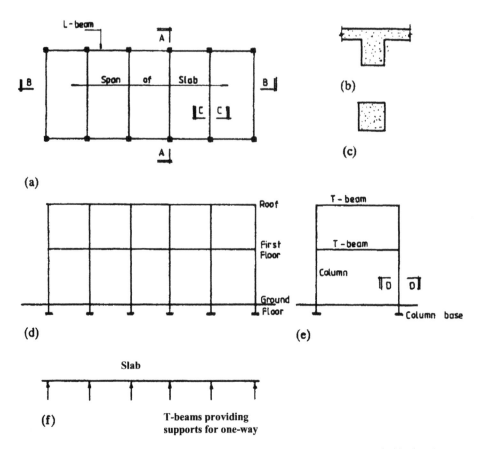

(a)

(b)

(c)

(d)

(e)

(f)

Fig. 1.3 (a) Plan of roof and floor; (b) section CC, T-beam; (c) section DD, column; (d) side elevation, longitudinal frame; (e) section AA, transverse frame; (f) continuous one-way slab.

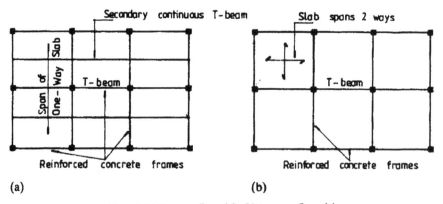

(a)

(b)

Fig. 1.4 (a) One-way floor slab; (b) two-way floor slab.

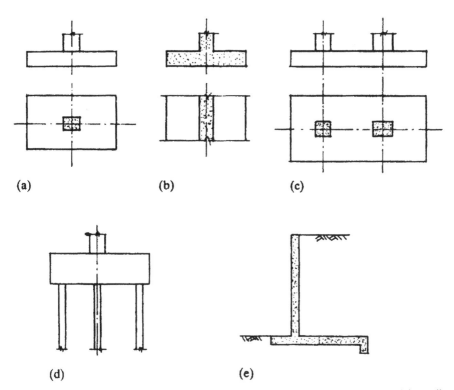

Fig. 1.5 (a) Isolated base; (b) wall footing; (c) combined base; (d) piled foundation; (e) retaining wall.

CHAPTER 2

MATERIALS, STRUCTURAL FAILURES AND DURABILITY

2.1 REINFORCED CONCRETE STRUCTURES

Reinforced concrete is a composite material of steel bars embedded in a hardened concrete matrix; concrete, assisted by the steel, carries the compressive forces, while steel resists tensile forces. Concrete itself is a composite material. The dry mix consists of cement and coarse and fine aggregates. Water is added and reacts with the cement which hardens and binds the aggregates into the concrete matrix; the concrete matrix sticks or bonds onto the reinforcing bars.

The properties of the constituents used in making concrete, mix design and the principal properties of concrete are discussed briefly. Knowledge of the properties and an understanding of the behaviour of concrete are important factors in the design process. The types and characteristics of reinforcing steels are noted.

Deterioration of and failures in concrete structures are now of widespread concern. This is reflected in the increased prominence given in the concrete codes to the durability of concrete structures. The types of failure that occur in concrete structures are listed and described. Finally the provisions regarding the durability of concrete structures noted in the code and the requirements for cover to prevent corrosion of the reinforcement and provide fire resistance are set out.

2.2 CONCRETE MATERIALS

2.2.1 Cement

The raw materials from which cement is made are lime, silica, alumina and iron oxide. These constituents are crushed and blended in the correct proportions and burnt in a rotary kiln. The resulting product is called clinker. The cooled clinker can be mixed with gypsum and various additional constituents and ground to a fine powder in order to produce different types of cements. The main chemical compounds in cement are calcium silicates and aluminates.

The Euro standard for cements is *BS EN 197-1:2011 Cement –Part 1: Composition, specifications and conformity criteria for common cements.*

When water is added to cement and the constituents are mixed to form cement paste, chemical reactions occur and the mix becomes stiffer with time and sets. The addition of gypsum mentioned above retards and controls the setting time.

This ensures that the concrete does not set too quickly before it can be placed in its final position or too slowly so as to hold up construction.

2.2.1.1 Types of Cements

The code gives five groups of cements, all of which are mixtures of different proportions of clinker and another major constituent. The five groups are:

1. CEM I Portland cement: This comprises mainly ground clinker and up to 5% of minor additional constituents.

2. CEM II Portland composite cement: This comprises of seven types which contain clinker and up to 35% of another single constituent.

i.   Portland slag cement (CEM II/A-S and CEM II/B-S). This comprises of clinker and blast furnace slag which originates from the rapid cooling of slag obtained by smelting iron ore in a blast furnace. The percentage of the slag varies between 6 and 35%.

ii. Portland silica fume cement (CEM II/A-D). This comprises of clinker and silica fume which originates from the reduction of high purity quartz with coal in an electric arc furnace in the production of silicon and ferrosilicon alloys.

iii. Portland-Pozzolana cement (CEM II/A-P, CEM II/B-P, CEM II/A-Q, CEM II/B-Q). This comprises clinker and natural pozzolana such as volcanic ashes or sedimentary rocks with suitable chemical and mineralogical composition or Natural calcined pozzolana such as materials of volcanic origin, clays, shales or sedimentary rocks activated by thermal treatment.

iv. Portland-fly ash cement (CEM II/A-V, CEM II/B-V, CEM II/A-W, CEM II/B-W). This mixture of clinker and fly ash dust-like particles precipitated from the flue gases from furnaces fired with pulverised coal.

v.   Portland burnt shale cement (CEM II/A-T, CEM II/B-T). This consists of clinker and burnt shale, specifically oil shale burnt in a special kiln at 800°C.

vi. Portland-limestone cement (CEM II/A-L, CEM II/B-L, CEM II/A-LL, CEM II/B-LL).

vii. Portland-composite cement (CEM II/A-M, CEM II/B-M).

3. CEM III blast furnace cement (CEM III/A, CEM III/B, CEM III/C): This comprises clinker and a higher percentage (36-95%) of blast furnace slag than that in CEM II/A-S and CEM II/B-S.

4. CEM IV pozzolanic cement (CEM IV/A, CEM IV/B): This comprises of clinker and a mixture of silica fume, pozzolanas and fly ash.

5. CEM V composite cement (CEM V/A, CEM V/B): This comprises clinker and a higher percentage of blast furnace slag and pozzolana or fly ash.

Table 2.1 Clinker content in cements

Cement type	Clinker content		
	A	B	C
CEM II	80-94%	65-79%	
CEM III	35-64%	20-34%	5-19%
CEM IV	65-89%	45-64%	-
CEM V	40-64%	20-38%	-

The letters A, B and C designate respectively higher, medium and lower proportion of clinker in the final mixture. However the percentage of clinker with the designations A, B, C can be different in different types of cement as shown in Table 2.1.

The second constituent in cement in addition to clinker is designated by the second letter as follows:

S = blast furnace slag
D = silica fume
P = natural pozzolana
Q = natural calcined pozzolana
V = siliceous fly ash
W = calcareous fly ash (e.g., high lime content fly ash)
L or LL = limestone
T = burnt shale
M = combination of two or more of the above components

2.2.1.2 Strength Class

There are three classes of strength as shown in Table 2.2. The strength class of cement classifies its compressive strength at 28 days.

Table 2.2 Strength class

Strength class	Compressive strength, MPa				Initial setting time
	Early strength		Standard strength		
	2 day	7 day	28 day		Minutes
32.5 N	-	≥ 16.0	≥ 32.5	≤ 52.5	≥ 75
32.5 R	≥ 10.0	-	≥ 32.5	≤ 52.5	
42.5 N	≥ 10.0	-	≥ 42.5	≤ 62.5	≥ 60
42.5 R	≥ 20.0	-	≥ 42.5	≤ 62.5	
52.5 N	≥ 20.0	-	≥ 52.5	-	≥ 45
52.5 R	≥ 30.0	-	≥ 52.5	-	

2.2.1.3 Sulfate-Resisting Cement

Sulfate resisting cements are used particularly in foundations where the presence of sulfates in the soil which can attack ordinary cements. The sulfate resisting cements have the designation SR and they are produced by controlling the amount of tricalcium aluminate (C_3A) in the clinker. The available types are:

    i.   Sulfate resisting Portland cements CEM I-SR0, CEM I-SR3, CEM I-SR5 which have the percentage of tricalcium aluminate in the clinker less than or equal to 0, 3 and 5% respectively.

    ii.   Sulfate resisting blast furnace cements CEM III/B-SR, CEM III/C-SR (no need for control of C3A content in the clinker).

    iii.   Sulphate resisting pozzolanic cements CEM IV/A-SR, CEM IV/B-SR (C3A content in the clinker should be less than 9%).

2.2.1.4 Low Early Strength Cement

These are CEM III blast furnace cements. Three classes of early strength are available with the designations N, R and L respectively signifying normal, ordinary, high and low early strength as shown in Table 2.2.

2.2.1.5 Standard Designation of Cements

CEM cement designation includes the following information:

    i.   Cement type (CEM I-CEM V)

    ii.   Strength class (32.5-52.5)

    iii.   Indication of early strength

    iv.   Additional designation SR for sulfate resisting cement

    v.   Additional designation LH for low heat cement

Examples:

1. CEM II/A-S 42.5 N

This indicates Portland composite cement (indicated by CEM II), with high proportion of clinker (indicated by letter A) and the second constituent is slag (indicated by letter S) and the strength class is 42.5 MPa (indicating that the characteristic strength at 28 days is a minimum of 42.5 MPa) and it gains normal early strength (indicated by letter N).

2. CEM III/B 32.5 N

This indicates blast furnace cement (indicated by CEM III); with medium proportion of clinker (indicated by letter B) and the strength class is 32.5 MPa (indicating that the characteristic strength at 28 days is a minimum of 32.5 MPa) and it gains normal early strength (indicated by letter N).

3. CEM I 42.5 R-SR3

This indicates Portland cement (indicated by CEM I), the strength class is 42.5 MPa (indicating that the characteristic strength at 28 days is a minimum of 42.5 MPa) and it gains high early strength (indicated by letter R) and is sulfate resisting with C_3A content in the clinker less than 3%.

4. CEM III-C 32.5 L – LH/SR

This indicates blast furnace cement (indicated by CEM III), the strength class is 32.5 MPa (indicating that the characteristic strength at 28 days is a minimum of 32.5 MPa) and it gains low early strength (indicated by letter L) and is sulfate resisting (indicated by letters SR) and is of low heat of hydration (indicated by LH).

2.2.1.6 Common Cements

Of twenty seven types of cement described in 2.2.1.1, the most common ones are mainly six:
   i.   CEM I
   ii. CEM II/B-S (containing 65-79% of clinker and 21-35% of blast furnace slag)
   iii. CEM II/B-V (containing 65-79% of clinker and 21-35% of siliceous fly ash)
   iv. CEM II/A-LL (containing 80-94% of clinker and 6-20% of limestone)
   v.   CEM III/A (containing 35-64% of clinker and 36-65% of other constituents)
   vi. CEM III/B (containing 20-34% of clinker and 66-80% of other constituents)

   The initial setting time must be a minimum of 75, 60 and 45 minutes for strength classes of 32.5, 42.5 and 52.5 respectively.

Useful references on aspects of cement are Neville (1996) and Popovics (1998).

2.2.2 Aggregates

The bulk of concrete is aggregate in the form of sand and gravel which is bound together by cement. Aggregate is classed into the following two sizes:
   1.   Coarse aggregate: gravel or crushed rock 5 mm or larger in size
   2.   Fine aggregate: sand less than 5 mm in size
   Natural aggregates are classified according to the rock type, e.g. basalt, granite, flint, limestone. Aggregates should be chemically inert, clean, hard and durable. Organic impurities can affect the hydration of cement and the bond between the cement and the aggregate. Some aggregates containing silica may react with alkali

in the cement causing the some of the larger aggregates to expand which may lead to concrete disintegration. This is the alkali–silica reaction. Presence of chlorides in aggregates, for example salt in marine sands, will cause corrosion of steel reinforcement. Excessive amounts of sulphate will also cause concrete to disintegrate.

To obtain a dense strong concrete with minimum use of cement, the cement paste should fill the voids in the fine aggregate while the fine aggregate and cement paste fill the voids in the coarse aggregate. Coarse and fine aggregates are graded by sieve analysis in which the percentage by weight passing a set of standard sieve sizes is determined. Grading limits for each size of coarse and fine aggregate are set out in *BS EN 12620:2002 + A1:2008: Aggregates for concrete.*

The grading affects the workability; a lower water-to-cement ratio can be used if the grading of the aggregate is good and therefore strength is also increased. Good grading saves cement content. It helps prevent segregation during placing and ensures a good finish.

2.2.3 Concrete Mix Design

Concrete mix design consists in selecting and proportioning the constituents to give the required strength, workability and durability. Three types of mixes are defined in *BS EN 206–1:2000: Concrete. Specification, performance, production and conformity* and *BS 8500-1:2006 Part 1: Method of specifying and guidance for the specifier.* This is Complementary British Standard to BS EN 206-1-1:2006.

The mixes are:

1. **Designed concrete**: This is concrete for which the required properties and additional characteristics are specified to the producer who is responsible for providing a concrete conforming to the specifications which shall contain:
   a.   Compressive strength class
   b.   Exposure classes (see Table 2.5)
   c.   Maximum nominal upper aggregate size
   d.   Chloride content class (maximum chloride content in cement is limited to 0.20-0.40% in the case of reinforced concrete and to 0.10-0.20% in the case of prestressed concrete).

2. **Prescribed concrete**: The composition of the concrete and the constituent materials to be used are specified to the producer who is responsible for providing a concrete with the specified composition. The specification shall contain:
   a.   Cement content
   b.   Cement type and strength class
   c.   Either w/c ratio or consistence in terms of slump or results of other test methods (see section 2.4.2)
   d.   Type, categories and maximum chloride content of aggregate

e. Maximum nominal upper size of aggregate and any limitations for grading

f. Type and quantity of admixture or other additives, if any

3. **Standardized prescribed concrete**: This is prescribed concrete for which the composition is given in a standard valid in the place of use of the concrete. Standardized prescribed concrete shall be specified by citing the standard valid in the place of use of the concrete giving the relevant requirements. Standardized prescribed concrete shall be used only for:

i. Normal-weight concrete for plain and reinforced concrete structures
ii. Compressive strength classes for design: Minimum characteristic cylinder strength of 16 MPa unless 20 MPa is permitted in provisions
iii. Exposure limited to concrete inside buildings with very low air humidity

The water-to-cement ratio is the single most important factor affecting concrete strength. For full hydration cement absorbs about 0.23 of its weight of water in normal conditions. This amount of water gives a very dry mix and extra water is added to give the required workability. The actual water-to-cement ratio used generally ranges from 0.45 to 0.6. The aggregate-to-cement ratio also affects workability through its influence on the water-to-cement ratio, as noted above. The mix is designed for the 'target mean strength' which is the characteristic strength required for design plus a specified number of times the standard deviation of the mean strength. In Eurocode 2, the mean value of cylinder compressive strength f_{cm} is taken as characteristic strength fck plus 8 MPa. Characteristic cylinder compressive strength f_{ck} is defined as not more than 5% of the results falling below the chosen strength.

Several methods of mix design are used in practice. Useful references are Day (2006) and Klett (2003).

2.2.4 Admixtures

Advice on admixtures is given in *BS EN 934–2: 2009 Admixtures for concrete, mortar and grout* and related standards.

The code defines admixtures as 'Materials added during the mixing process in a quantity not more than 5% by mass of the cement content of the concrete, to modify the properties of the mix in the fresh and/or hardened state'.

Admixtures covered by Euro Standards are as follows:

i. Set accelerating (retarding) admixture: admixture which decreases (increases) the time to commencement of transition of the mix from the plastic to the rigid state.

ii. Water resisting admixture: admixture which reduces the capillary absorption of hardened concrete.

iii. Water reducing/plasticizing admixture: admixture which, without affecting the consistence, permits a reduction in the water content of a given concrete mix, or which, without affecting the water content, increases the slump/flow or produces both effects simultaneously.

iv.   Air entraining admixture: admixture which allows a controlled quantity of small, uniformly distributed air bubbles to be incorporated during mixing which remain after hardening. This is used to increase resistance to freeze-thaw damage to concrete.

v.   High range water reducing agents/super plasticizers, which are more efficient than (3) above.

vi.   Hardening accelerating admixture: admixture which increases the rate of development of early strength in the concrete, with or without affecting the setting time.

vii.   Water retaining admixture: admixture which reduces the loss of water by a reduction of bleeding.

There are also admixtures which produce several different actions.

Some useful references on admixtures are Rixon et al. (1999), Paillere (1995) and Ramachandran (1995).

2.3 CONCRETE PROPERTIES

The main strength and deformational properties of concrete are discussed below.

2.3.1 Stress–Strain Relationship in Compression

Fig. 2.1 shows the stress–strain relationship for concrete in compression. The characteristic compressive strength of cylinder f_{ck} is defined as the strength below which not more than 5% of the results fall. The mean compressive strength f_{cm} is related to f_{ck} as $f_{cm} = f_{ck} + 8$ MPa.

For the design of cross sections, two simplified stress–strain relationships are proposed in Eurocode 2. The stress–strain relationship shown in Fig. 2.2 is a combination of a parabola and a straight line. The second simplified representation shown in Fig. 2.3 is bilinear.

The mathematical equation for the parabola–rectangle is given by

$$\text{stress}, \sigma_c = f_{ck}\left[1 - \left(1 - \frac{\varepsilon_c}{\varepsilon_{c2}}\right)^n\right] \text{ for } 0 \leq \varepsilon_c \leq \varepsilon_{c2}$$

$$\text{stress}, \sigma_c = f_{ck} \text{ for } \varepsilon_{c2} \leq \varepsilon_c \leq \varepsilon_{cu2}$$

Table 2.3 shows the several strength and strain properties of concrete. It also shows relationship between characteristic cylinder strength f_{ck} and cube strength f_{cu}. An approximate relationship between cylinder strength f_{ck} and cube strength $f_{ck, cube}$ is

$$f_{ck} \approx 0.8 \, f_{ck, cube}$$

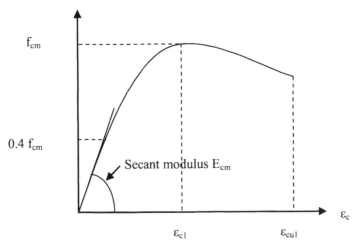

Fig. 2.1 Stress–strain curve for concrete in compression.

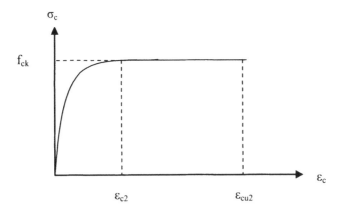

Fig. 2.2 Parabola–rectangle stress–strain relationship for concrete in compression.

2.3.2 Compressive Strength

The compressive strength is the most important property of concrete. The characteristic strength that is the concrete grade is measured by the 28-day cylinder/cube strength. Standard cylinders 150 mm diameter and 300 mm high or cubes of 150 or 100 mm for aggregate not exceeding 25 mm in size are crushed to determine the strength. The test procedure is given in *BS EN 12390:2: 2009: Testing Hardened Concrete: Making and curing specimens for strength tests* and *BS EN 12390:3: 2009: Testing Hardened Concrete: Compressive strength of test specimen.*

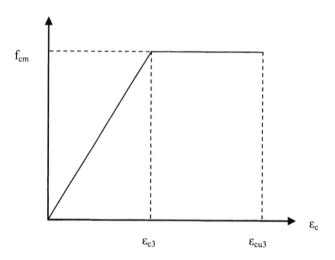

Fig. 2.3 Bilinear stress–strain relationship for concrete in compression.

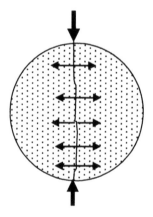

Fig. 2.4 Split cylinder test.

2.3.3 Tensile Strength

The tensile strength of concrete is about a tenth of the compressive strength. It is determined by loading a concrete cylinder across a diameter as shown in Fig. 2.4. The test procedure is given in *BS EN 12390:6: 2009: Testing Hardened Concrete: Tensile splitting strength of test specimens.*

The mean characteristic tensile strength f_{ctm} is related to mean cylinder compressive strength f_{cm} as follows.

$$f_{ctm} = 0.30 \times f_{ck}^{\frac{2}{3}} \text{ for } f_{ck} \leq 50 \text{ MPa}$$

$$f_{ctm} = 2.12 \times \ell n[1 + \frac{f_{cm}}{10}] \text{ for } f_{ck} > 50 \text{ MPa}$$

The 5% and 95% fractiles of the characteristic tensile strength of concrete are respectively $f_{ctk,\,0.05} = 0.7\,f_{ctk}$ and $f_{ctk,\,0.95} = 1.3\,f_{ctm}$.

2.3.4 Modulus of Elasticity

The short-term stress–strain curve for concrete in compression is shown in Fig. 2.1. The slope of the initial straight portion is the initial tangent modulus. At any point, the slope of the line joining the point to the origin is the secant modulus. The value of the secant modulus depends on the stress and rate of application of the load. The code giving details of the method of determining the elastic modulus is *BS 1881–121:1983 Testing concrete. Methods for determination of Static modulus of elasticity in compression. Note: A new Eurocode version is in preparation.*

The dynamic modulus is determined by subjecting a beam specimen to longitudinal vibration. The value obtained is unaffected by creep and is approximately equal to the initial tangent modulus. The code *BS 1881–209:1990 Testing concrete. Recommendations for the measurement of dynamic modulus of elasticity* gives the details.

BS EN 1002-1-1:2004 Eurocode 2 Design of concrete structures gives the following expression for the short term secant modulus of elasticity (see Fig. 2.1) between zero stress and 0.4 f_{cm} for concretes made with quartzite aggregates as

$$E_{cm} = 22[\frac{f_{cm}}{10}]^{0.3} \text{ GPa}$$

where $f_{cm} = f_{ck} + 8$ MPa, $f_{ck} =$ characteristic cylinder strength.

Because of the fact that the elastic modulus is greatly dependent on the stiffness of the aggregates, for limestone and sandstone aggregates the value from the equation should be *reduced* by 10% and 30% respectively. For basalt aggregates the value should be *increased* by 20%.

The tangent modulus $E_c = 1.05\,E_{cm}$

2.3.5 Creep

Creep in concrete is the gradual increase in strain with time in a member subjected to prolonged stress. The creep strain is much larger than the elastic strain on loading. If the specimen is unloaded there is an immediate elastic recovery and a slower recovery in the strain due to creep. Both amounts of recovery are much less than the original strains under load.

The main factors affecting creep strain are the concrete mix and strength, the type of aggregate, curing, ambient relative humidity, the magnitude and duration of sustained loading and the age of concrete at which load is first applied.   In clause 3.1.4(2), Eurocode 2 specifies that provided the concrete is not subjected to a stress greater than 45% of the compressive strength at the time of loading, long term creep strain $\varepsilon_{cc}(\infty, t_0)$ is calculated from the creep coefficient $\phi(\infty, t_0)$ by the equation

$$\varepsilon_{cc}(\infty, t_0) = \frac{\text{stress}}{E_c} \phi(\infty, t_0)$$

where E_c is the tangent modulus of elasticity of the concrete at the age of loading, t_0. The creep coefficient $\phi(\infty, t_0)$ depends on the effective section thickness, the age of loading and the relative ambient humidity. The creep coefficient is used in deflection calculations.   Clause 3.1.4 and Annex B of Eurocode 2 give the equations for determining the creep coefficient.   More details and examples are given in section 19.1.17, Chapter 19.

2.3.6 Shrinkage

The total shrinkage strain is composed of two parts, the drying shrinkage strain and the autogenous shrinkage strain. Drying shrinkage strain is the contraction that occurs in concrete when it dries and hardens. Drying shrinkage develops slowly due to migration of water and is irreversible but alternate wetting and drying causes expansion and contraction of concrete. The autogenous shrinkage strain develops during the hardening of concrete and develops quite fast during the early days after casting of concrete.

The aggregate type and content are the most important factors influencing shrinkage. The larger the size of the aggregate is, the lower is the shrinkage and the higher is the aggregate content; the lower the workability and water-to-cement ratio are, the lower is the shrinkage. Aggregates that change volume on wetting and drying, such as sandstone or basalt, produce concrete which experiences a large shrinkage strain, while concrete made with non-shrinking aggregates such as granite or gravel experience lower shrinkage strain. A decrease in the ambient relative humidity also increases shrinkage.

Eurocode 2 gives necessary data for calculating the drying shrinkage in equations (3.9)–(3.10) and in equations (3.11)–(3.13). Values of shrinkage strain are used in deflection calculations. More details and an example are given in section 19.1.18, Chapter 19.

2.4 TESTS ON WET CONCRETE

2.4.1 Workability

The workability of a concrete mix gives a measure of the ease with which fresh concrete can be placed and compacted. The concrete should flow readily into the forms and go around and cover the reinforcement, the mix should retain its consistency and the aggregates should not segregate. A mix with high workability is needed where sections are thin and/or reinforcement is complicated and congested.

For a given concrete, the main factor affecting workability is the water content of the mix. Plasticizing admixtures will increase workability. The size of aggregate, its grading and shape, the ratio of coarse to fine aggregate and the aggregate-to-cement ratio also affect workability to some degree.

2.4.2 Measurement of Workability

(a) Slump test
The fresh concrete is tamped into a standard cone which is lifted off after filling and the slump is measured. The slump is 25–50 mm for low workability, 50-100 mm for medium workability and 100–175 mm for high workability. Normal reinforced concrete requires fresh concrete of medium workability. The slump test is the usual workability test specified. The standard covering slump testing is *BS EN 12350–2:2009 Testing fresh concrete. Slump test.*

(b) Degree of compactability test
The fresh concrete is carefully placed in a container using a trowel, avoiding any compaction of the concrete. When the container is full, the top surface is struck off level with the top of the container. The concrete is compacted by vibration and the distance from the surface of the compacted concrete to the upper edge of the container is used to determine the degree of compactability. Details are given in *BS EN 12350–4: 2009 Testing fresh concrete–Part 4: Degree of compactability.*

(c) Flow table test
This test determines the consistency of fresh concrete by measuring the spread of concrete on a flat plate which is subjected to jolting.
Details are given in *BS EN 12350–5: 2009 Testing fresh concrete–part 5: Flow table test.*

2.5 TESTS ON HARDENED CONCRETE

2.5.1 Normal Tests

The main destructive tests on hardened concrete are as follows.
(a) Compression test: Refer to section 2.3.2 above.
(b) Tensile splitting test: Refer to section 2.3.3 above.

(c) Flexure test: A plain concrete specimen is tested to failure in pure bending. The British standard covering testing of flexural strength is *BS EN 12390:5: 2009: Testing hardened concrete: Flexural strength of test specimens.*

(d) Test cores: Cylindrical cores are cut from the finished structure with a rotary cutting tool. The core is soaked, capped and tested in compression to give a measure of the concrete strength in the actual structure. The ratio of core height to diameter and the location where the core is taken affect the strength. The strength is lowest at the top surface and increases with depth through the element. A ratio of core height-to-diameter of two gives a standard cylinder test. The Euro standard covering testing of cores is *BS EN 12504-1:2009 Testing concrete in structures. Cored specimens. Taking, examining and testing in compression.*

2.5.2 Non-destructive Tests

The main non-destructive tests for strength on hardened concrete are as follows.

(a) Rebound hardness test

The Schmidt hammer is used in the rebound hardness test in which a metal hammer held against the concrete is struck by another spring-driven metal mass and rebounds. The amount of rebound is recorded on a scale and this gives an indication of the concrete strength. The larger the rebound number is, the higher is the concrete strength. The standard covering testing by Rebound hammer is *BS EN 12504-2:2001 Testing concrete in structures. Non-destructive testing. Determination of rebound number.*

(b) Pullout force test

A small metal disc with a rod fixed centrally on one side is glued into concrete using adhesives, so that the rod protrudes from the surface of the concrete. The force required to pull the disc out of the concrete is measured. *BS EN 12504-3:2005 Testing concrete in structures. Determination of pull-out force* covers the test procedure.

(c) Ultrasonic pulse velocity test

In the ultrasonic pulse velocity test, the velocity of ultrasonic pulses that pass through a concrete section from a transmitter to a receiver is measured. The pulse velocity is correlated against strength. The higher the velocity, the stronger is the concrete.

(d) Other non-destructive tests

Equipment has been developed to measure
1. Crack widths and depths
2. Water permeability and the surface dampness of concrete
3. Depth of cover and the location of reinforcing bars

4. The electrochemical potential of reinforcing bars and hence the presence of corrosion

A useful reference on testing of concrete in structures is Bungey et al. (2006).

2.5.3 Chemical Tests

A complete range of chemical tests is available to measure
1. Depth of carbonation
2. The cement content of the original mix
3. The content of salts such as chlorides and sulphates that may react and cause the concrete to disintegrate or cause corrosion of the reinforcement

The reader should consult specialist literature.

2.6 REINFORCEMENT

Reinforcing bars are produced as hot rolled or cold worked high yield steel bars. They have characteristic yield strength f_{yk} of 400 to 600 MPa. Steel fabric is made from cold drawn steel wires welded to form a mesh. High yield bars are produced as deformed bars with transverse ribs to improve bond with concrete.

The stress–strain curves for reinforcing bars are shown in Fig. 2.5. Hot rolled bars have a definite yield point. A defined proof stress at a strain of 0.2% is recorded for the cold worked bars. The value of Young's modulus E_s for steel is 200 GPa. The maximum breaking stress is k times the characteristic stress f_{yk}. The design stress $f_{yd} = f_{yk}/\gamma_s$, where $\gamma_s = 1.15$.

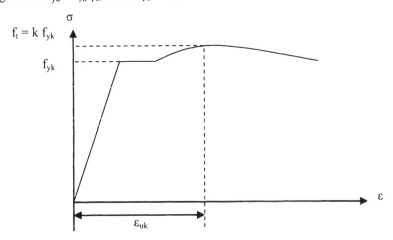

Fig. 2.5(a) Stress–strain curve for hot rolled steel reinforcing bars.

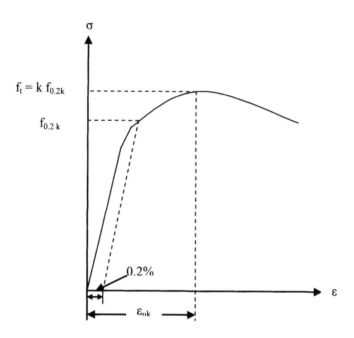

Fig. 2.5(b) Stress–strain curve for cold worked steel reinforcing bars.

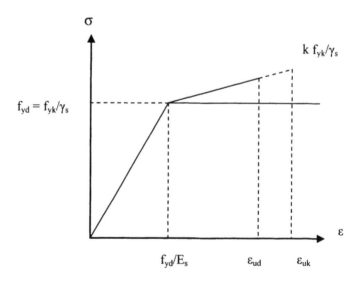

Fig. 2.6 Design stress–strain curve for steel.

The idealized design stress–strain curve for all reinforcing bars is shown in Fig. 2.6. In the first case the maximum design stress is greater than f_{yd} but the maximum strain is limited to ε_{ud} taken as equal to $0.9\ \varepsilon_{uk}$. An even more simplified option is to limit the maximum stress to f_{yd} with no limit on the maximum strain. For simplicity, the second option is preferred in all common design situations. The material safety factor for steel is taken as $\gamma_s = 1.15$. The behaviour in tension and compression is taken to be the same. Annex C in Eurocode 2 gives the properties of reinforcing bars shown in Table 2.3.

Table 2.3 Properties of reinforcing bars

Product form	Bars and de-coiled rods			Wire fabrics		
Class	A	B	C	A	B	C
f_{yk} or $f_{0.2k}$	400-600 MPa					
k, minimum	≥ 1.05	≥ 1.08	≥ 1.15	≥ 1.05	≥ 1.08	≥ 1.15
			< 1.35			< 1.35
$\varepsilon_{uk} \times 10^2$	≥ 2.5	≥ 5.0	≥ 7.5	≥ 2.5	≥ 5.0	≥ 7.5

As an example assuming:

$$f_{yk} = 500 \text{ MPa}, k = 1.2, \gamma_s = 1.15, f_{yd} = f_{yk}/\gamma_s = 435 \text{ MPa}, \varepsilon uk = 2.5 \times 10^{-2},$$

$$\varepsilon_{ud} = 0.9\varepsilon_{uk} = 2.25 \times 10^{-2}, E_s = 200 \text{ GPa}, fyd/E_s = 2.17 \times 10^{-2}$$

The maximum stress allowable at a strain of ε_{ud} is given by

$$\text{Max.stress} = f_{yd} + (k-1)f_{yd} \times \frac{(\varepsilon_{ud} - \dfrac{f_{yd}}{E_s})}{(\varepsilon_{uk} - \dfrac{f_{yd}}{E_s})} = 1.05 f_{yd}$$

This shows that there is only a 5% increase in the maximum allowable stress.

2.7 EXPOSURE CLASSES RELATED TO ENVIRONMENTAL CONDITIONS

BS EN 206-1:2000 Concrete -Part 1: Specification, performance, production and conformity defines six possible environmental conditions and the associated class designations which will determine the required cover to reinforcement to ensure durability. These are shown in Tables 2.4a to f. The main physical causes and attack by chemicals of concrete and reinforcement can be classified under the following headings. Data on limiting values for composition and properties of concrete are taken from Table F.1 of *EN 206-1-2000*.

Table 2.4a Exposure class X0, related environmental conditions.

Class designation	Description of the environment	Informative examples where exposure classes may occur
1. No risk of corrosion or attack		
X0	For concrete without reinforcement or embedded metal: All exposures except where there is freeze/thaw, abrasion or chemical attack. For concrete with reinforcement or embedded metal: Very dry	Concrete inside buildings with very low air humidity C12/15

a. Freezing and thawing

Concrete nearly always contains water which expands on freezing. The freezing–thawing cycle causes loss of strength, spalling and disintegration of the concrete. Resistance to damage is improved by using an air entraining agent.

b. Chlorides

High concentrations of chloride ions cause corrosion of reinforcement and the products of corrosion can disrupt the concrete. Chlorides can be introduced into the concrete either during or after construction as follows.

- i.   *Before construction* Chlorides can be admitted in admixtures containing calcium chloride, through using mixing water contaminated with salt water or improperly washed marine aggregates.
- ii.   *After construction* Chlorides in salt or sea water, in airborne sea spray and from deicing salts can attack permeable concrete causing corrosion of reinforcement.

c. Sulphates

Sulphates are present in most cements and in some aggregates. Sulphates may also be present in soils, groundwater and sea water, industrial wastes and acid rain. The products of sulphate attack on concrete occupy a larger space than the original material and this causes the concrete to disintegrate and permits corrosion of steel to begin. Sulphate-resisting cement should be used where sulphates are present in the soil, water or atmosphere and come into contact with the concrete.

d. Carbonation

Carbonation is the process by which carbon dioxide from the atmosphere slowly transforms calcium hydroxide into calcium carbonate in concrete. The concrete itself is not harmed and increases in strength, but the reinforcement can be seriously affected by corrosion as a result of this process.

Normally the high pH value of the concrete prevents corrosion of the reinforcing bars by keeping them in a highly alkaline environment due to the release of calcium hydroxide by the cement during its hydration. Carbonated concrete has a pH value of 8.3 while the passivation of steel starts at a pH value of 9.5. The depth of carbonation in good dense concrete is about 3 mm at an early stage and may increase to 6 to 10 mm after 30 to 40 years. Poor concrete may have a depth of carbonation of 50 mm after say 6-8 years. The rate of carbonation depends on time, cover, concrete density, cement content, water-to-cement ratio and the presence of cracks.

Table 2.4b Exposure classes XC, related environmental conditions, maximum w/c ratio, minimum strength class, minimum cement content in kg/m^3

Class designation	Description of the environment	Informative examples where exposure classes may occur
2. Corrosion induced by carbonation		
XC1	Dry or permanently wet	Concrete inside buildings with very low air humidity. Concrete permanently submerged in water. w/c = 0.65 C20/25 260
XC2	Wet, rarely dry	Concrete surfaces subjected to long term water contact. Many foundations. w/c = 0.60 C25/30 280
XC3	Moderate humidity	Concrete inside buildings with moderate or high air humidity. External concrete sheltered from rain. w/c = 0.55 C30/37 280
XC4	Cyclic wet and dry	Concrete surfaces subjected to water contact, not within exposure class XC2. w/c = 0.50 C30/37 300

e. Alkali–silica reaction

A chemical reaction can take place between alkali in cement and certain forms of silica in aggregate. The reaction produces a gel which absorbs water and expands in volume, resulting in cracking and disintegration of the concrete. The reaction only occurs when the following are present together:

i. A high moisture level in the concrete.
ii. Cement with a high alkali content or some other source of alkali
iii. Aggregate containing an alkali-reactive constituent

The following precautions should be taken if uncertainty exists:

i.   Reduce the saturation of the concrete.
ii.   Use low alkali Portland cement and limit the alkali content of the mix to a low level.
iii. Use replacement cementitious materials such as blast furnace slag or pulverized fuel ash. Most normal aggregates behave satisfactorily.

f. Acids

Portland cement is not acid resistant and acid attack may remove part of the set cement. Acids are formed by the dissolution in water of carbon dioxide or sulphur dioxide from the atmosphere. Acids can also come from industrial wastes. Good dense concrete with adequate cover is required and sulphate-resistant cements should be used if necessary.

Table 2.4c Exposure classes XD, related environmental conditions, maximum w/c ratio, minimum strength class, minimum cement content in kg/m^3

Class designation	Description of the environment	Informative examples where exposure classes may occur
3. Corrosion induced by chlorides		
XD1	Moderate humidity	Concrete surfaces exposed to airborne chlorides. w/c = 0.55 C30/37 300
XD2	Wet, rarely dry	Swimming pools Concrete components exposed to industrial waters containing chlorides. w/c = 0.55 C30/37 300
XD3	Cyclic wet and dry	Parts of bridges exposed to spray containing chlorides. Pavements Car park slabs w/c = 0.45 C35/45 320

Table 2.4d Exposure classes XS, related environmental conditions, maximum w/c ratio, minimum strength class, minimum cement content in kg/m^3

Class designation	Description of the environment	Informative examples where exposure classes may occur
4. Corrosion induced by chlorides from sea water		
XS1	Exposed to airborne salt but not in direct contact with sea water.	Structures near to or on the coast w/c = 0.50 C30/37 300
XS2	Permanently submerged	Parts of marine structures w/c = 0.45 C35/45 320
XS3	Tidal, splash and spray zones.	Parts of marine structures w/c = 0.45 C35/45 340

2.8 FAILURES IN CONCRETE STRUCTURES

2.8.1 Factors Affecting Failure

Failures in concrete structures can be due to any of the following factors:

i. Incorrect selection of materials
ii. Errors in design calculations and detailing
iii. Poor construction methods and inadequate quality control and supervision
iv. Chemical attack
v. External physical and/or mechanical factors including alterations made to the structure

The above items are discussed in more detail below.

2.8.1.1 Incorrect Selection of Materials

The concrete mix required should be selected to meet the environmental or soil conditions where the concrete is to be placed. The minimum grade that should be used for reinforced concrete is 25/30 class meaning that f_{ck} = 25 MPa and $f_{ck, cube}$ is 30 MPa. Higher grades should be used for some foundations and for structures near the sea or in an aggressive industrial environment. If sulphates are present in the soil or ground water, sulphate-resisting cement should be used. Where freezing and thawing occur, air entrainment should be adopted.

Table 2.4e Exposure classes XF, related environmental conditions, maximum w/c ratio, minimum strength class, minimum cement content in kg/m³ and minimum content of air entrained

Class designation	Description of the environment	Informative examples where exposure classes may occur
5. Freeze/Thaw attack		
XF1	Moderate water saturation, without deicing agent.	Vertical concrete surfaces exposed to rain and freezing. w/c = 0.55 C30/37 300
XF2	Moderate water saturation, with deicing agent.	Vertical concrete surfaces of road structures exposed to freezing and airborne de-icing agents. w/c = 0.55 C25/30 300 4% air
XF3	High water saturation, without deicing agent.	Horizontal concrete surfaces exposed to rain and freezing. w/c = 0.50 C30/37 320 4% Air
XF4	High water saturation, with deicing agents or sea water.	Road and bridge decks exposed to deicing agents. Concrete surfaces exposed to direct spray containing deicing agents and freezing. Splash zones of marine structures exposed to freezing. w/c = 0.45 C30/37 340 4%Air

Note: For XF2–XF4, use aggregates with sufficient freeze–thaw resistance.

2.8.1.2 Errors in Design Calculations and Detailing

An independent check should be made of all design calculations to ensure that the section sizes, slab thickness etc. and reinforcement sizes and spacing specified are adequate to carry the worst combination of design loads. The check should include overall stability, robustness and serviceability and foundation design.

Incorrect detailing is one of the most common causes of failure and cracking in concrete structures. First the overall arrangement of the structure should be correct, efficient and robust. Movement joints should be provided where required to reduce or eliminate cracking. The overall detail design should shed water.

Internal or element detailing must comply with the code requirements. The provisions specify the cover to reinforcement, minimum thicknesses for fire resistance, maximum and minimum steel areas, bar spacing limits and reinforcement to control cracking, lap lengths, anchorage of bars etc.

Table 2.4f Exposure classes XA, related environmental conditions, maximum w/c ratio, minimum strength class, minimum cement content in kg/m^3

Class designation	Description of the environment	Informative examples where exposure classes may occur
6. Chemical attack		
XA1	Slightly aggressive chemical environment	Natural soils and ground water w/c = 0.55 C30/37 300
XA2	Moderately aggressive chemical environment	Natural soils and ground water w/c = 0.50 C30/37 320 Sulfate-resisting cement
XA3	Highly aggressive chemical environment	Natural soils and ground water w/c = 0.45 C35/45 360 Sulfate-resisting cement

The limits on various chemicals in water and ground for the classes XA1 – XA3 are given in Table 2 of *BS EN 206-1:2000 Concrete -Part 1: Specification, performance, production and conformity.*
The reader should refer to the code for full details.

2.8.1.3 Poor Construction Methods

The main items that come under the heading of poor construction methods resulting from bad workmanship and inadequate quality control and supervision are as follows.

(a) Incorrect placement of steel
Incorrect placement of steel can result in insufficient cover, leading to corrosion of the reinforcement. If the bars are placed grossly out of position or in the wrong position, collapse can occur when the element is fully loaded.

(b) Inadequate cover to reinforcement
Inadequate cover to reinforcement permits ingress of moisture, gases and other substances and leads to corrosion of the reinforcement and cracking and spalling of the concrete.

(c) Incorrectly made construction joints
The main faults in construction joints are lack of preparation and poor compaction. The old concrete should be washed and a layer of rich concrete laid before pouring is continued. Poor joints allow ingress of moisture and staining of the concrete face.

(d) Grout leakage
Grout leakage occurs where formwork joints do not fit together properly. The result is a porous area of concrete that has little or no cement and fine aggregate. All formwork joints should be properly sealed.

(e) Poor compaction
If concrete is not properly compacted by ramming or vibration, the result is a portion of porous honeycomb concrete. This part must be hacked out and recast. Complete compaction is essential to give a dense, impermeable concrete.

(f) Segregation
Segregation occurs when the mix ingredients become separated. It is the result of
i. Dropping the mix from too great a height in placing. Chutes or pipes should be used in such cases.
ii. Using a harsh mix with high coarse aggregate content.
iii. Large aggregate sinking due to over-vibration or use of too much plasticizer.

Segregation results in uneven concrete texture, or porous concrete in some cases.

(g) Poor curing
A poor curing procedure can result in loss of water through evaporation. This can cause a reduction in strength if there is not sufficient water for complete hydration of the cement. Loss of water can cause shrinkage cracking. During curing the concrete should be kept damp and covered.

(h) Excessive water content
Excess water increases workability but decreases the strength and increases the porosity and permeability of the hardened concrete, which can lead to corrosion of the reinforcement. The correct water-to-cement ratio for the mix should be strictly enforced.

2.8.1.4 External Physical and/or Mechanical Factors

The main external factors causing concrete structures to fail are as follows.

(a) Restraint against movement
Restraint against movement causes cracking. Movement in concrete is due to elastic deformation and creep under constant load, shrinkage on drying and setting, temperature changes, changes in moisture content and the settlement of foundations. The design should include sufficient movement joints to prevent serious cracking. Cracking may only detract from the appearance rather than be of structural significance but cracks permit ingress of moisture and lead to corrosion of the steel. Various proprietary substances are available to seal cracks.

Movement joints should be clearly indicated for both members and structure as a whole. The joints are to permit relative movement to occur without impairing structural integrity. Diagrams of some movement joints are shown in Fig. 2.7. The location of movement joints is a matter of experience. Joints should be placed where cracks would probably develop, e.g., at abrupt changes of section, corners or locations where restraints from adjoining elements occur.

1. The **contraction joint** may be a complete or partial joint with reinforcement running through the joint. There is no initial gap and only contraction of the concrete is permitted.

2. The **expansion joint** is made with a complete discontinuity and gap between the concrete portions. Both expansion and contraction can occur in the same structure. The joint must be filled with a sealer.

3. There is complete discontinuity in a **sliding joint** and the design is such as to permit movement in the plane of the joint.

4. The **hinged joint** is specially designed to permit relative rotation of members meeting at the joint. The Freyssinet hinge has no reinforcement passing through the joint.

5. The **settlement joint** permits adjacent members to settle or displace vertically as a result of foundation or other movements relative to each other. Entire parts of the building can be separated to permit relative settlement, in which case the joint must run through the full height of the structure.

(b) Abrasion
Abrasion can be due to mechanical wear such as flow of grains in a silo, wave action etc. Abrasion reduces cover to reinforcement. Dense concrete with hard wearing aggregate and extra cover allowing for wear are required.

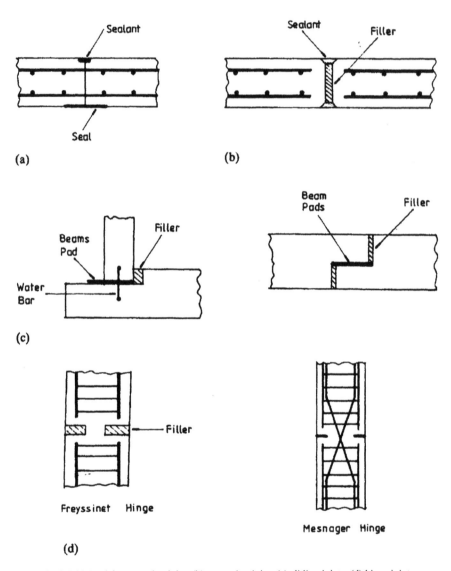

Fig. 2.7 (a) Partial contraction joint; (b) expansion joint; (c) sliding joints; (d) hinge joints.

(c) Wetting and drying

Wetting and drying leaches lime out of concrete and makes it more porous, which increases the risk of corrosion to the reinforcement. Wetting and drying also cause movement of the concrete which can cause cracking if restraint exists. Detail should be such as to shed water and the concrete may also be protected by impermeable membranes.

(e) Overloading
Extreme overloading will cause cracking and eventual collapse. Factors of safety in the original design allow for possible overloads but vigilance is always required to ensure that the structure is never grossly overloaded. A change in function of the building or room can lead to overloading, e.g., if a class room is changed to a library the imposed load can be greatly increased.

(f) Structural alterations
If major structural alterations are made to a building, the members affected and the overall integrity of the building should be rechecked. Common alterations are the removal of walls or columns to give a large clear space or provide additional doors or openings. Steel beams are inserted to carry loads from above. In such cases the bearing of the new beam on the original structure should be checked and if walls are removed the overall stability may be affected.

(g) Settlement
Differential settlement of foundations can cause cracking and failure in extreme cases. The foundation design must be adequate to carry the building loads without excessive settlement. Where a building with a large plan area is located on ground where subsidence may occur, the building should be constructed in sections on independent rafts with complete settlement joints between adjacent parts.

Many other factors can cause settlement and ground movement problems. Some problems are shrinkage of clays from ground dewatering or drying out in droughts, tree roots causing disruption, ground movement from nearby excavations, etc.

(h) Fire resistance
Concrete is a porous substance bound together by water-containing crystals. The binding material can decompose if heated to too high a temperature, with consequent loss of strength. The loss of moisture causes shrinkage and the temperature rise causes the aggregates to expand, leading to cracking and spalling of the concrete. High temperature also causes reinforcement to lose strength. At 550°C the yield stress of steel drops to about its normal working stress and failure occurs under service loads.

Concrete, however, is a material with very good fire resistance and protects the reinforcing steel. Fire resistance is a function of member thickness and cover. The code requirements regarding fire protection are set out in *BS EN 1992-1-2:2004 Eurocode 2: Design of concrete structures-part 1-2: General rules-Structural fire design.*

2.9 DURABILITY OF CONCRETE STRUCTURES

A durable structure should satisfy strength and serviceability requirements throughout its design working life. One of the main causes for poor durability is the corrosion of steel reinforcement. Good quality dense concrete and adequate cover are prime requirements in order to produce durable structures. Table 2.5 gives the details of water/cement ratio, minimum cement content for producing good quality concrete to satisfy various exposure classes and $C_{min, dur}$ which is the minimum cover to steel from durability considerations.

The minimum cover C_{min} to steel should satisfy the code equation (4.2):

$$C_{min} = \text{Maximum } \{C_{min, b}; C_{min,dur} ; 10 \text{ mm}\} \tag{4.2}$$

    i.   For safe transmission of bond forces, the required cover is $C_{min, b}$ as given in Table 4.2 of Eurocode 2.

    ii.   For separated bars, $C_{min, b} \geq$ bar diameter, φ.

    iii.   For bundled bars, $C_{min, b} \geq$ equivalent bar diameter, φ_n.

$\varphi_n = \varphi \sqrt{n_b} \leq 55$ mm. n_b = Number of bars in the bundle. $n_b \leq 4$ for vertical bars in compression and for bars in a lapped joint. $n_b \leq 3$ for all other cases.

If the nominal maximum size of the aggregate is greater than 32 mm, $C_{min, b}$ should be increased by 5 mm. The cover can be reduced if stainless steel bars are used.

For a design life of 50 years, minimum values of $C_{min, dur}$ for various classes of exposure are are given as follows in Table 4.4 N of Eurocode 2: X0 = 10 mm, XC1 = 15 mm, XC2/XC3 = 25 mm, XC4 = 30 mm, XD1/XS1 = 35 mm, XD2/XS2 = 40 mm, XD3/XS3 = 45 mm

2.10 FIRE PROTECTION

One of the prime requirements in terms of safety of a structure is that the structure gives enough protection to the occupants in case of fire. The code *BS EN 1992-1-2:2004 Eurocode 2: Design of concrete structures-part 1-2:General rules-Structural fire design* gives minimum dimensions for various types of members and also the necessary cover to reinforcement for different standards (periods in minutes) of fire resistance. Tables 2.6 to 2.9 give the necessary information. It is assumed that the ratio of flexural resistance in fire/resistance at normal temperature is 0.7.

Table 2.5 Recommended limiting values for composition and properties of concrete and minimum cover to steel for durability

Class designation	Maximum w/c ratio	Minimum strength class	Minimum cement content (kg/m³)	Minimum air content (%)	$C_{min, dur}$ mm *
X0	-	C12/15	-	-	10
XC1	0.65	C20/25	260	-	15(25)
XC2	0.60	C25/30	280	-	25(35)
XC3	0.55	C30/37	280	-	25(35)
XC4	0.50	C30/37	300	-	30(40)
XD1	0.55	C30/37	300	-	35(45)
XD2	0.55	C30/37	320	-	40(50)
XD3	0.45	C35/45	320	-	45(55)
XS1	0.50	C30/37	300	-	35(45)
XS2	0.45	C35/45	320	-	40(50)
XS3	0.45	C35/45	340	-	45(55)
XF1	0.55	C30/37	300	-	
XF2	0.55	C25/30	300	4.0	
XF3	0.50	C30/37	320	4.0	
XF4	0.45	C30/37	340	4.0	
XA1	0.55	C30/37	300	-	
XA2	0.50	C30/37	320	-	
XA3	0.45	C35/45	360	-	

*This column gives minimum cover to steel to ensure durability of reinforced concrete structures. Figures in parentheses refer to prestressed concrete structures. The values given are valid for a design life of 50 years.

Fig. 2.8 shows the width b for different types of members. If the web is of variable width, width b refers to the width at the centroid of tension steel.
In the case of I-section, $d_1 + 0.5 \times d_2 \geq$ (220, 380, 480 mm respectively for 120, 180 and 240 minutes of fire resistance).

Fig. 2.9 shows the nominal axis distance a, the distance from the surface to the centroid of steel and width b for different sections. Note that the value of a can be greater than that shown because of bond and durability requirements.

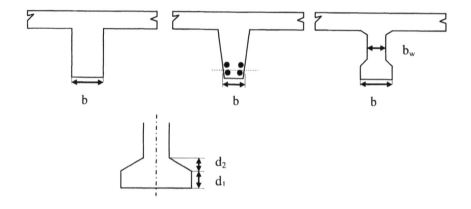

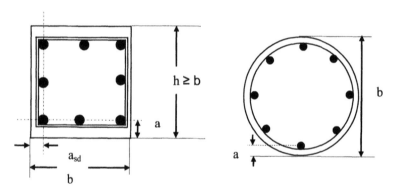

Fig. 2.8 Sections through members showing width b.

Fig. 2.9 Sections through members showing dimensions a and b.

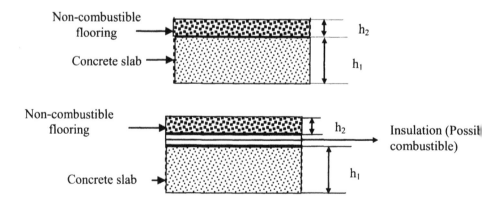

Fig. 2.10 Concrete slab with floor finishes.

Table 2.6 Dimensions a and b for simply supported beams (reinforced and prestressed)

Standard fire resistance (minutes)	Minimum distances (mm)				Web width, b_w (mm)
	Possible combinations of width b of beam and a, the average axis distance (mm)				
1	2	3	4	5	6
R 30	b = 80	120	160	200	80
	a = 25	20	15	15	
R 60	b = 120	160	200	300	80
	a = 40	35	30	25	
R 90	b = 150	200	300	400	100
	a = 55	45	40	35	
R 120	b = 200	240	300	500	120
	a = 65	60	55	50	
R 180	b = 240	300	400	600	150
	a = 80	70	65	60	
R 240	b = 280	350	500	700	170
	a = 90	80	75	70	
$a_{sd} = a + 10$ mm					

Note: Value of dimension a can be greater because of durability and bond considerations.

Table 2.7 Dimensions a and b for continuous beams (reinforced and prestressed)

Standard fire resistance (minutes)	Minimum distances (mm)				Web width, b_w (mm)
	Possible combinations of width b of beam and a, the average axis distance (mm)				
1	2	3	4	5	6
R 30	b = 80	160			80
	a = 15	12			
R 60	b = 120	200			100
	a = 25	12			
R 90	b = 150	250			110
	a = 35	25			
R 120	b = 200	300	450	500	130
	a = 45	35	35	30	
R 180	b = 240	400	550	600	150
	a = 60	50	50	40	
R 240	b = 280	500	650	700	170
	a = 75	60	60	50	
$a_{sd} = a + 10$ mm					

Note: Value of dimension a can be greater because of durability and bond considerations.

Table 2.8 Dimensions a and b for one-way and two-way slabs (reinforced and prestressed)

Standard fire resistance (minutes)	Minimum dimensions (mm)			
	Slab thickness, h_s	Axis distance, a (mm)		
	$h_s = h_1 + h_2$ (mm)	One way	Two way ($l_y \geq l_x$)	
	See Fig. 2.10		$l_y/l_x \leq 1.5$	$1.5 < l_y/l_x \leq 2$
R 30	60	10	10	10
R 60	80	20	10	15
R 90	100	30	15	20
R 120	120	40	20	25
R 180	150	55	30	40
R 240	175	65	40	50

Table 2.9 Dimensions a and b for rectangular and circular columns (reinforced and prestressed)

Standard fire resistance (minutes)	Minimum dimensions (mm) b and a			
	Column exposed on more than one side			Column exposed on one side
	$\mu_{fi} = 0.2$	$\mu_{fi} = 0.5$	$\mu_{fi} = 0.7$	$\mu_{fi} = 0.7$
R 30	b = 200	b = 200	b = 200 (300-)	b = 155
	a = 25	a = 25	a = 32(27)	a = 25
R 60	b = 200	b = 200 (300)	b = 250 (350)	b = 155
	a = 25	a = 36 (31)	a = 46 (40)	a = 25
R 90	b = 200 (300)	b = 300 (400)	b = 350 (450)	b = 155
	a = 31 (25)	a = 45 (38)	a = 53 (40)	a = 25
R 120	b = 250 (350)	b = 350 (450)	b = 350 (450)	b = 175
	a = 40 (35)	a = 45 (40)	a = 57 (51)	a = 35
R 180	b = 350	b = 350	b = 450	b = 230
	a = 45	a = 63	a = 70	a = 55
R 240	b = 350	b = 450	-	b = 295
	a = 61	a = 75		a = 70

μ_{fi} = Axial load in fire situation. Design resistance of column at normal temperature. A conservative value is $\mu_{fi} = 0.7$.

2.11 REFERENCES

Alexander, M. and Mindess, S. (2005). *Aggregates in Concrete.* CRC Press.

Blight, G.E., Alexander, M.G. (2011). *Alkali-Aggregate Reaction and Structural Damage to Concrete: Engineering Assessment, Repair and Management.* CRC Press.

Bungey, J.H., Millard, S.G. and Grantham, M. G. (2006). *Testing Concrete in Structures*, 4th ed. CRC Press/Taylor & Francis.

Day, K. (2006). *Concrete Mix Design: Quality Control and Specification,* 3rd ed. Spon/Taylor & Francis.

Dyer, T. (2011). *Concrete Durability*. Routledge.

Klett, I. (2000). *Engineered Concrete: Mix Design and Test Methods,* 2nd ed. CRC Press.

Malhotra, V.M. and Carino, N.J. (2003). *Handbook of Non-destructive Testing of Concrete,* 2nd ed. CRC Press.

Marchand, J., Odler, I. and Skalny, J. P. (2001). *Sulfate Attack on Concrete*. CRC Press.

Neville, A.M (1996). *Properties of Concrete,* 4th ed. Prentice–Hall.

Paillere, A.M. (1995). *Application of Admixtures in Concrete*. Taylor & Francis.

Popovics, S. (1998). *Strength and Related Properties of Concrete: A Quantitative Approach*. Wiley.

Poulsen, E. and Mejlbro, L. (2005). *Diffusion of Chloride in Concrete: Theory and Application*. CRC Press.

Ramachandran, V.S., Ed. (1995). *Concrete Admixtures Handbook,* 2nd ed. Noyes Publications.

Richardson, M.G. (2002). *Fundamentals of Durable Reinforced Concrete*. CRC Press.

Rixon, M.R. and Mailavaganam, R. (1999). *Chemical Admixtures for Concrete.* Taylor & Francis.

Schutter, G. (2012). *Damage to Concrete Structures*. CRC Press.

Tang, L., Nilsson, L. O. and Muhammed Bashir, P.A. (2011). *Resistance of Concrete to Chloride Ingress: Testing and Modelling*. CRC Press.

Whittle, R. (2012). *Failures in Concrete Structures: Case Studies of Reinforced and Prestressed Concrete Structures*. CRC Press.

CHAPTER 3

LIMIT STATE DESIGN AND STRUCTURAL ANALYSIS

3.1 STRUCTURAL DESIGN AND LIMIT STATES

3.1.1 Aims and Methods of Design

The Eurocode *BS EN 1990:2002 + A1:2005 Eurocode. Basis of structural design* states that a structure shall be designed and executed in such a way that it will, during its intended life (design life for building structures is generally taken as 50 years and for monumental structures like bridges as 100 years), with appropriate degrees of reliability and in an economical way sustain all actions and influences likely to occur during execution and use and be durable. In particular

a. A structure shall be designed to have adequate structural resistance, serviceability, and durability.
b. In the case of fire, the structural resistance shall be adequate for the required period of time.
c. A structure shall be designed and executed in such a way that it will not be damaged by events such as explosion, impact and the consequences of human errors to an extent disproportionate to the original cause.
d. The structure shall be designed such that deterioration over its design working life does not impair the performance of the structure below that intended, having due regard to its environment and the anticipated level of maintenance.

It is recognized that no structure can be made one hundred percent safe and that it is only possible to reduce the probability of failure to an acceptably low level.

The method recommended in the code is limit state design in conjunction with partial factor method. The loads (actions) will be according to the following standards:

- *BS EN 1991-1-1:2002 Eurocode 1: Actions on Structures part 1-1: General actions-Densities, self-weight, imposed loads on buildings*
- *BS EN 1991-1-3:2003 Eurocode 1: Actions on Structures. General actions. Snow loads*
- *BS EN 1991-1-4:2005 + A1:2010 Eurocode 1: Actions on Structures. General actions. Wind actions*

There is a U.K. National Annex to *BS EN 1990:2002 + A1:2005.*

It is recognized that calculations alone are not sufficient to produce a safe, serviceable and durable structure. The basic requirements should be met by:

- Choice of suitable materials
- Appropriate design and detailing
- Specifying control procedures for design, production, execution and appropriate use

3.1.2 Criteria for Safe Design: Limit States

One criterion for a safe design is that the structure should not become unfit for use, i.e., it should not reach a limit state during its design life. This is achieved, in particular, by designing the structure to ensure that it does not reach two important limit states.

1. **Ultimate limit state (ULS)**: This limit state is concerned with the safety of the people and of the structure. This requires that the whole structure or its elements should not collapse, overturn or buckle when subjected to the design loads.

2. **Serviceability limit states (SLS)**: This limit state is concerned with

- Comfort of the occupants: For example the structure should not suffer from excessive vibration or have large cracks or deflection so as to alarm the user of the building.
- Appearance of the structure. The structure should not become unfit for use due to excessive deflection or cracking.

For reinforced concrete structures, the normal practice is to design for the ultimate limit state, check for serviceability and take all necessary precautions to ensure durability.

3.1.3 Ultimate Limit State

The structure must be designed to carry the most severe combination of loads to which it is subjected. Each and every section of the elements must be capable of resisting the axial and shear forces, bending and twisting moments derived from the analysis. Overall stability of a structure is provided by shear walls, lift shafts, staircases and rigid frame action or a combination of these means. The structure should transmit all loads, dead, imposed, snow and wind, safely to the foundations.

The design is made for ultimate loads and design strengths of materials with partial safety factors applied to loads and material strengths. This permits uncertainties in the estimation of loads and in the performance of materials to be assessed separately. The section strength is determined using plastic analysis based on the short-term design stress–strain curves for concrete and reinforcing steel. As already noted in 3.1.1, the planning and design should be such that damage to a small area or failure of a single element should not cause collapse of a major part of a structure. This means that the design should be resistant to progressive collapse. The structure should resist the applied loads as a unit. This can be ensured by adequately tying the different parts of the structure using vertical and horizontal ties.

3.1.4 Serviceability Limit States

In checking for the serviceability limit states, account is to be taken of temperature, creep, shrinkage, sway and settlement and possibly other effects.

The main serviceability limit states are as follows.

(a) Deflection
The deformation of the structure should not adversely affect its efficiency or appearance. Deflections of beams may be calculated, but may tend to be complicated because of cracking, creep and shrinkage effects. In normal cases span-to-effective depth ratios can be used to check compliance with requirements.

(b) Cracking
Cracking should be kept within reasonable limits by correct detailing. Crack widths may be calculated, but may tend to be complicated and in normal cases cracking can be controlled by adhering to detailing rules with regard to bar spacing in zones where the concrete is in tension.

(c) Vibration
The structure should not under the action of wind loads or movement of the people vibrate so much as to make people uncomfortable or in worst cases even to alarm people.

In analysing a section for the serviceability limit states the behaviour is assessed assuming a linear elastic relationship for steel and concrete stresses. Allowance is made for the stiffening effect of concrete in the tension zone and for creep and shrinkage.

3.2 ACTIONS, CHARACTERISTIC AND DESIGN VALUES OF ACTIONS

Actions (loads) can be classified as
- Permanent actions (G): These are fixed values such as the self-weight of the structure and the weight of finishes, ceilings, services and partitions.
- Variable actions (Q): These are imposed loads due to people, furniture, and equipment etc. on floors, wind actions on the whole structure including roofs and snow loads on roofs.

- Accidental actions (A): These are loads due to crashing of vehicles against the building, bomb blasts and other forces.

The characteristic value of an action (load) is its main representative value defined by a nominal value which is normally expected to have a 95% probability of not being exceeded.

The characteristic loads used in design are as follows:

1. **The characteristic permanent action** G_k is given by a single value as its value does not vary significantly during the lifetime of the structure.

2. **The characteristic variable action** Q_k is represented as follows.
   - Combination value $\psi_0 Q_k$ is used for irreversible ultimate limit states.
   - Frequent value $\psi_1 Q_k$ is used for reversible limit states.
   - Quasi-permanent value $\psi_2 Q_k$ is used for calculating long term effects such as deflection due to creep and other aspects related to the appearance of the structure.

Note that combination factor ψ is a device for reducing the design value of variable loads when they act in combination.

Table 3.1 gives the ψ values for different imposed loads.

Table 3.1 Recommended values of ψ factors for imposed load on buildings

Imposed load on buildings				
Category	Description	Ψ_1	Ψ_2	Ψ_3
A	Domestic, residential areas	0.7	0.5	0.3
B	Office areas	0.7	0.5	0.3
C	Congregation areas	0.7	0.7	0.6
D	Shopping areas	0.7	0.7	0.6
E	Storage areas	1.0	0.9	0.8
F	Traffic area, Vehicle weight $\leq$ 30 kN	0.7	0.7	0.6
G	Traffic area, 30 kN < Vehicle weight $\leq$ 160 kN	0.7	0.5	0.3
H	Roofs	0	0	0
Snow loads for sites at an altitude > 1000 m		0.7	0.5	0.2
Snow loads for sites at an altitude $\leq$ 1000 m		0.5	0.2	0
Wind loads on buildings		0.6	0.2	0

The design value of an action is a product of the representative value and a load factor $\gamma_{F, i}$. Thus for permanent actions, design value is $\gamma_{F,i} G_k$. For variable actions, design value is $\gamma_{F, i} \psi_i Q_k$, where i = 0, 1, or 2 depending on whether it is a combination value, a frequent value or a quasi-permanent value. The value of $\gamma_{F, i}$ can be different for different Q_k and different from that for G_k.

The partial safety factor $\gamma_{F, i}$ takes account of
    a.   Possible increases in load
    b.   Inaccurate assessment of the effects of loads
    c.   Unforeseen stress distributions in members

     d.   Importance of the limit state being considered

Note: Uniformly distributed load will be represented by small letters. If for example concentrated permanent and imposed loads are G_k and Q_k respectively, their uniformly distributed values will be denoted by g_k and q_k respectively.

Permanent and variable actions are given in *BS EN 1991-1-1:2002 Eurocode 1: Actions on Structures Part-1-1: General actions-Densities, self-weight, imposed loads on buildings.* Snow loads are obtained from *BS EN 1991-1-3:2003 Eurocode 1: Actions on Structures. General actions. Snow loads.*

3. **The characteristic wind action** W_k depends on the location, shape and dimensions of the buildings. Wind loads are estimated using *BS EN 1991-1-4: 2005 + A1:2010 Eurocode 1: Actions on Structures. General actions. Wind actions.*

4. **The characteristic earth loads** E_n are to be obtained in accordance with *BS EN 1997-1:2004 Eurocode 7: Geotechnical design —Part 1: General rules.* Bond and Harris (2008) wrote a useful book covering the geotechnical aspects of design to Eurocode 7.

3.2.1 Load Combinations

In practice many different loads act together and this fact has to be considered in calculating the load for which the structure has to be designed.

Eurocode 1 gives the following load combinations depending on whether the overall equilibrium of the structure considered as a rigid body is being considered (EQU) or design of a structural element (STR) needs to be carried out.

3.2.2 Load Combination EQU

$$\text{Design load} = \sum_{j \geq i} \gamma_{G,j} G_{k,j} \ "+" \ \gamma_{Q,1} Q_{k,1} \ "+" \sum_{i>1} \gamma_{Q,i} \ \psi_{0,i} Q_{k,i} \qquad (6.10)$$

The load factors to be used are:

- $\gamma_{G,j} = 1.10$ (unfavourable), 0.90 (favourable)
- $Q_{k,1}$ is the leading variable action
- $\gamma_{Q,1} = 1.50$ (unfavourable), 1.00 (favourable)
- $Q_{k,i}$ are accompanying variable actions
- $\gamma_{Q,i} = 1.50$ (unfavourable), 1.00 (favourable)

In the equation (6.10) "+" implies "to be combined with".

For further information, see Table A1.2(A) in Design values of actions (EQU) (Set A) in *BS EN 1990:2002 Eurocode -Basis of Structural Design.*

3.2.3 Load Combination STR

The code gives the following alternative equations for design of structural elements. The equation numbers in the code are (6.10), (6.10a) and (6.10b). These equation numbers will be used in the rest of this book to make it convenient to refer to the code clauses.

$$\text{Design load} = \sum_{j\geq 1}\gamma_{G,j}\,G_{k,j}\ "+"\ \gamma_{Q,1}\,Q_{k,1}\ "+"\sum_{i>1}\gamma_{Q,i}\,\psi_{0,i}Q_{k,i} \tag{6.10}$$

or alternatively

$$\text{Design load} = \sum_{j\geq 1}\gamma_{G,j}\,G_{k,j}\ "+"\ \gamma_{Q,1}\,\psi_{0,1}\,Q_{k,1}\ "+"\sum_{i>1}\gamma_{Q,i}\,\psi_{0,i}Q_{k,i} \tag{6.10a}$$

$$\text{Design load} = \sum_{j\geq 1}\xi\gamma_{G,j}\,G_{k,j}\ "+"\ \gamma_{Q,1}\,Q_{k,1}\ "+"\sum_{i>1}\gamma_{Q,i}\,\psi_{0,i}Q_{k,i} \tag{6.10b}$$

In the above the symbols denote as follows

Σ = combined effect of

"+" = to be combined with

ξ = reduction factor for unfavourable permanent actions G

The load factors to be used are:

$\gamma_{G,j}$ = 1.35 (unfavourable), 1.00 (favourable)

$Q_{k,1}$ = leading variable action

$\gamma_{Q,1}$ = 1.50 (unfavourable), 1.00 (favourable)

$Q_{k,i}$ = accompanying variable actions

$\gamma_{Q,i}$ = 1.50 (unfavourable), 1.00 (favourable)

ξ = 0.85 (U.K. adopts 0.925)

For further information see Table A1.2(B) Design values of actions (STR/GEO) (Set B) in *BS EN 1990:2002 Eurocode -Basis of Structural Design.*

Table 3.2 Simplified equations for checking EQU and STR

Persistent and transient design situations	Permanent actions		Leading variable action		Accompanying variable action	
	Unfav.	Fav.	Unfav.	Fav.	Unfav.	Fav.
Equilibrium 6.10	$1.10\,G_{k,\,sup}$	$0.90\,G_{k,\,inf}$	$1.5\,Q_{k,1}$	0	$1.5\,\psi_{0,i}\,Q_{k,i}$	0
STR 6.10	$1.35\,G_{k,\,sup}$	$1.0\,G_{k,\,inf}$	$1.5\,Q_{k,1}$	0	$1.5\,\psi_{0,i}\,Q_{k,i}$	0
STR 6.10a	$1.35\,G_{k,\,sup}$	$1.0\,G_{k,\,inf}$	§$1.5\,\psi_{0,i}\,Q_{k,1}$	0	$1.5\,\psi_{0,i}\,Q_{k,i}$	0
STR 6.10b	*$1.15\,G_{k,\,sup}$	$1.0\,G_{k,\,inf}$	$1.5\,Q_{k,1}$	0	$1.5\,\psi_{0,i}\,Q_{k,i}$	0

* $\xi\,\gamma_{Gj,\,sup}$ = 0.85 × 1.35 = 1.15 (The U.K. National Annex value is 0.925 × 1.35= 1.25).

§ In case 6.10a, there is no leading variable.

The code in a note states that 'the characteristic values of all permanent actions from one source are multiplied by 1.35 if the total resulting action is unfavourable

and by 1.0 if the total resulting action effect is favourable. For example, all actions originating from the self weight of the structure may be considered as coming from one source; this also apples if different materials are involved.'
If only two variable loads are present, the above equations can be represented in a simplified form as shown in Table 3.2.

3.2.4 Examples

Example 1: Fig. 3.1 shows a simply supported beam with an overhang. It is subjected to the following loads.
- Permanent load due to self weight $g_k = 10.0$ kN/m
- Variable imposed load $q_k = 15$ kN/m
- Variable concentrated load $Q_k = 25$ kN/m

3.2.4.1 Checking for EQU (Stability)

The structure will become unstable if tension develops in the left hand support or if the right hand support sinks.

Fig. 3.1 A beam with an overhang.

1. Tension reaction at the first support

The influence line for tension at the left hand support is shown in Fig. 3.2. Using Muller-Breslau's principle, the ordinate at the left hand end is 1.0 and at the right hand end is $1.0 \times (1.5/12) = 0.125$. The area of the influence diagram in the simply supported section is $\frac{1}{2} \times 1.0 \times 12 = 6.0$ and in the overhang section is $\frac{1}{2} \times 0.125 \times 1.5 = 0.09375$. Load the structure so as to develop the maximum tension at the left hand support. The influence line shows that the maximum (unfavourable) load should be in the overhang and minimum (favourable) load in the simply supported section.

Fig. 3.2 Influence line for reaction at left hand support.

Depending on which variable (imposed) load is considered as the leading variable, there are two options for imposed loading.

- **Case 1**: Treat the distributed load q_k as the leading variable and the concentrated load Q_k as the accompanying variable.
- **Case 2**: Treat the concentrated load Q_k as the leading variable and the distributed load q_k as the accompanying variable.

Values of Ψ factors are taken from Table 3.1.

Using equation (6.10) for checking equilibrium, the loads in different parts of the beam are calculated as follows:

Case 1:

(1) In the simply supported section, all loads are favourable. Therefore

$\gamma_{G,1} = 0.90$, $g_k = 10$ kN/m, $\gamma_{Q,1} = 0$, $q_k = 0$ kN/m, $Q_k = 0$, $\psi_{0,i} = 0.7$

(2) In the overhang section, all loads are unfavourable. Therefore

$\gamma_{G,1} = 1.10$, $g_k = 10$ kN/m, $\gamma_{Q,1} = 1.5$, $q_k = 15$, $Q_k = 25$ kN, $\psi_{0,i} = 0.7$

Design reaction $= -\,[\{0.90 \times 10 + 0 \times 0\} \times 6.0 + 0 \times 0.7 \times 0]$ in simply supported
$+ [\{1.10 \times 10 + 1.5 \times 15\} \times 0.09375 + 1.50 \times 0.7 \times 25 \times 0.125]$
in overhang
$= -\,54.0 + 3.14 + 3.28 = -\,47.58$ kN (acting upwards)

No tension develops at the left hand support.

Case 2:

(1) In the simply supported section, all loads are favourable. Therefore

$\gamma_{G,1} = 0.90$, $g_k = 10$ kN/m, $\gamma_{Q,1} = 0$, $\psi_{0,i} = 0.7$, $q_k = 0$ kN/m, $Q_k = 0$,

(2) In the overhang section, all loads are unfavourable. Therefore

$\gamma_{G,1} = 1.10$, $g_k = 10$ kN/m, $\gamma_{Q,1} = 1.5$, $\psi_{0,i} = 0.7$, $q_k = 15$, $Q_k = 25$ kN

Design reaction $= -\,[\{0.90 \times 10 + 0 \times 0.7 \times 0\} \times 6.0 + 0 \times 0]$ in simply supported
$+ [\{1.10 \times 10 + 1.5 \times 0.7 \times 15\} \times 0.09375 + 1.50 \times 25 \times 0.125]$
in overhang
$= -\,54.0 + 2.51 + 4.69 = -\,46.85$ kN (acting upwards)

From the two cases considered, the minimum upward reaction is 46.85 kN and no tension develops at the left hand support.

2. Maximum reaction at the right hand support

If checking the adequacy of the foundation under the right hand support, then calculating the value of the reaction is a check for equilibrium. Fig. 3.3 shows the influence line for reaction at the right hand support. The ordinate at the right hand reaction is 1.0 and at the end of the overhang by proportion, it is $1 \times (12 + 1.5)/12 = 1.125$. The area of the influence diagram in the simply supported section is $\frac{1}{2} \times 1.0 \times 12 = 6.0$ and in the overhang section is $\frac{1}{2} \times (1 + 1.125) \times 1.5 = 1.59375$.

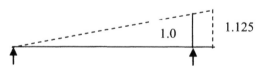

Fig. 3.3 Influence line for right hand reaction.

Case 1: In this case all loads are unfavourable. Therefore uniformly distributed loads occupy the entire length of the beam and the concentrated load acts at the tip of the overhang.

$\gamma_{G,1} = 1.10$, $g_k = 10$ kN/m, $\gamma_{Q,1} = 1.5$, $q_k = 15$, $Q_k = 25$ kN, $\psi_{0,i} = 0.7$

Design reaction $= [\{1.10 \times 10 + 1.50 \times 15.0\} \times 6.0 + 0 \times 0.7 \times 0] +$
$[\{1.10 \times 10 + 1.5 \times 15.0\} \times 1.59375 + 1.50 \times 0.7 \times 25 \times 1.125]$
$= 201.0 + 53.391 + 29.53 = 283.92$ kN (Acting upwards)

Case 2:
$\gamma_{G,1} = 1.10$, $gk = 10$ kN/m, $\gamma_{Q,1} = 1.50$, $\psi_{0,i} = 0.7$, $q_k = 15.0$ kN/m, $Q_k = 25$ kN

Design reaction $= [\{1.10 \times 10 + 1.50 \times 0.7 \times 15.0\} \times 6.0 + 0 \times 0.7 \times 0] +$
$[\{1.10 \times 10 + 1.5 \times 0.7 \times 15\} \times 1.59375 + 1.50 \times 25 \times 1.125]$
$= 160.5 + 42.63 + 42.19 = 245.32$ kN (acting upwards).

The maximum upward reaction is 283.92 kN.

Example 2: Fig. 3.4 shows an office building subjected to variable loading due to characteristic imposed loading Q_k and wind loading W_k, in addition to permanent gravity loading G_k. Check the loading patterns for maximum tension in the left hand column. From Table 3.1, for imposed office load take $\psi_0 = 0.7$ and for wind loading take $\psi_0 = 0.6$.

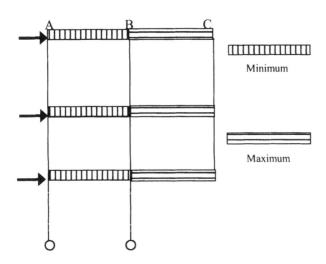

Fig. 3.4 An office building.

Maximum tension (and hence possible over topping) in the left hand column will occur when minimum gravity and imposed loading in the region AB, maximum gravity and imposed loading in the region BC and wind is blowing from left to right.

Case 1: Treating imposed load Q_k as the leading variable:

    a.   In the region AB the loading is 0.9 G_k

b. In the region BC the loading is $1.10\ G_k + 1.50\ Q_k$
c. The wind loading is $1.50 \times 0.6 \times W_k$

Case 2: Treating imposed load W_k as the leading variable:
    a. In the region AB the loading is $0.9\ G_k$
    b. In the region BC the loading is $1.10\ G_k + 1.50 \times 0.7 \times Q_k$
    c. The wind loading is $1.50\ W_k$

Note that the value of G_k, Q_k and W_k need not be same at all levels. For example, the loading on the roof is likely to be smaller than at other levels. Similarly the wind load at the roof level will be approximately half of that at lower levels.

3.2.4.2 Load Calculation for STR (Design)

Equations (6.10), (6.10a) and (6.10b) are used to determine the bending moment at the mid-span section and also moment over the right hand support

1. Design bending moment at mid-span section

The influence line bending moment at mid-span section is shown in Fig. 3.5. Using Muller-Breslau's principle, the ordinate at the mid-span is $L/4 = 12.0/4 = 3.0$ and at the right hand end the ordinate by proportion is, $3.0 \times ((1.5/6.0) = 0.75$. The area of the influence diagram in the simply supported section is $\frac{1}{2} \times 3.0 \times 12 = 18.0$ and in the overhang section is $\frac{1}{2} \times 0.75 \times 1.5 = 0.5625$.

It shows that for maximum positive bending moment (tension at the bottom face) the maximum (unfavourable) load should be in the simply supported section and minimum (favourable) load in the overhang section.

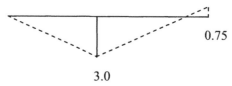

0.75

3.0

Fig. 3.5 Influence line for mid-span bending moment.

Case 1: Treat the distributed load q_k as the leading variable and the concentrated load Q_k as the accompanying variable.

In the simply supported section, all loads are unfavourable and in the overhang section all loads are favourable. In the simply supported section, place all uniformly distributed loads and a concentrated load at mid-span. In the overhang section only uniformly distributed permanent load.

    $\gamma_{G,1} = 1.35$, $g_k = 10$ kN/m, $\gamma_{Q,1} = 1.50$, $q_k = 15.0$ kN/m, $Q_k = 25$, $\psi_{0,i} = 0.7$

In the overhang section: Only permanent load
    $\gamma_{G,1} = 1.0$, $g_k = 10$ kN/m, $\gamma_{Q,1} = 0$, $q_k = 0$, $Q_k = 0$ kN, $\psi_{0,i} = 0.7$

Using code equation (6.10),

Mid-span BM $= (1.35 \times 10.0 + 1.50 \times 15.0) \times 18.0 + 1.50 \times 0.7 \times 25.0 \times 3.0$

$$- 1.0 \times 10.0 \times 0.5625$$
$$= 721.125 \text{ kNm}$$

Alternatively, using code equation (6.10a),

Mid-span BM = $(1.35 \times 10.0 + 1.50 \times 0.70 \times 15.0) \times 18.0 + 1.50 \times 0.7 \times 25.0 \times 3.0$
$$- 1.0 \times 10.0 \times 0.5625$$
$$= 599.625 \text{ kNm}$$

Using code equation (6.10b),

Mid-span BM = $(0.85 \times 1.35 \times 10.0 + 1.50 \times 15.0) \times 18.0 + 1.50 \times 0.7 \times 25.0 \times 3.0$
$$- 1.0 \times 10.0 \times 0.5625$$
$$= 684.675 \text{ kNm}$$

Note that in the above calculations, different values have been used for γ_G. If the same value of $\gamma_G = 1.35$ is used throughout, the value of the moment will reduce by 1.97 kNm.

Case 2: Treat the concentrated load Q_k as the leading variable and the distributed load q_k as the accompanying variable.

Using code equation (6.10),

Mid-span BM = $(1.35 \times 10.0 + 1.50 \times 0.7 \times 15.0) \times 18.0 + 1.50 \times 25.0 \times 3.0$
$$- 1.0 \times 10.0 \times 0.5625$$
$$= 633.375 \text{ kNm}$$

Using code equation (6.10a),

Mid-span BM = $(1.35 \times 10.0 + 1.50 \times 0.70 \times 15.0) \times 18.0 + 1.50 \times 0.7 \times 25.0 \times 3.0$
$$- 1.0 \times 10.0 \times 0.5625$$
$$= 599.625 \text{ kNm}$$

Using code equation (6.10b),

Mid-span BM = $(0.85 \times 1.35 \times 10.0 + 1.50 \times 0.7 \times 15.0) \times 18.0 + 1.50 \times 25.0 \times 3.0$
$$- 1.0 \times 10.0 \times 0.5625$$
$$= 596.925 \text{ kNm}$$

Note that in the above calculations, different values have been used for γ_G. If the same value of $\gamma_G = 1.35$ is used throughout, the value of the moment will reduce by 1.97 kNm.

The design bending moment is therefore 721.125 kNm from equation (6.10), Case 1.

2. Design bending moment over the support

The influence line bending moment at support section is shown in Fig. 3.6

1.5

Fig. 3.6 Influence line for bending moment over the support.

Using Muller-Breslau's principle, the ordinate at the end of overhang is 1.5. The area of the influence diagram in the overhang section is ½ × 1.5 × 1.5 = 1.125.

It shows that for maximum negative bending moment (tension at the top face) the maximum (unfavourable) load should be in the overhang section. Any load in the simply supported section does not produce any bending moment over the support. All loads in the overhang section are unfavourable.

Case 1: Treat the distributed load q_k as the leading variable and the concentrated load Q_k as the accompanying variable.
Using code equation (6.10),
Support BM $= (1.35 \times 10.0 + 1.50 \times 15.0) \times 1.125 + 1.50 \times 0.7 \times 25.0 \times 1.5$
$\qquad = 79.875$ kNm
Using code equation (6.10a),
Mid-span BM $= (1.35 \times 10.0 + 1.50 \times 0.70 \times 15.0) \times 1.125 + 1.50 \times 0.7 \times 25.0 \times 1.5$
$\qquad = 72.28$ kNm
Using code equation (6.10b)
Mid-span BM $= (0.85 \times 1.35 \times 10.0 + 1.50 \times 15.0) \times 1.125 + 1.50 \times 0.7 \times 25.0 \times 1.5$
$\qquad = 77.60$ kNm
Case 2: Treat the concentrated load Q_k as the leading variable and the distributed load q_k as the accompanying variable.
Using code equation (6.10),
Support BM $= (1.35 \times 10.0 + 1.50 \times 0.7 \times 15.0) \times 1.125 + 1.50 \times 25.0 \times 1.5$
$\qquad = 89.16$ kNm
Using code equation (6.10a),
Support BM $= (1.35 \times 10.0 + 1.50 \times 0.70 \times 15.0) \times 1.125 + 1.50 \times 0.7 \times 25.0 \times 1.5$
$\qquad = 72.28$ kNm
Using code equation (6.10b)
Support BM $= (0.85 \times 1.35 \times 10.0 + 1.50 \times 0.7 \times 15.0) \times 1.125 + 1.50 \times 25.0 \times 1.5$
$\qquad = 86.88$ kNm
The design bending moment is therefore 89.16 kNm from equation (6.10), Case 2.

3.2.5 Partial Factors for Serviceability Limit States

Table A1.4 of the code gives the following combination of actions for checking the serviceability limit state. All load factors γ_G and γ_Q are taken as unity. The combinations are shown in Table 3.3.

Table 3.3 Design values of actions for use in the combination of actions

Combination	Permanent actions G_d		Variable actions, Q_d	
	Unfavourable	Favourable	Leading	Others
Characteristic	$G_{kj, sup}$	$G_{kj, inf}$	$Q_{k,1}$	$\Psi_{0,i} Q_{k,i}$
Frequent	$G_{kj, sup}$	$G_{kj, inf}$	$\Psi_{1,1} Q_{k,1}$	$\Psi_{2,i} Q_{k,i}$
Quasi-permanent	$G_{kj, sup}$	$G_{kj, inf}$	$\Psi_{2,1} Q_{k,1}$	$\Psi_{2,i} Q_{k,i}$

3.3 PARTIAL FACTORS FOR MATERIALS

Eurocode 2 gives the following partial factors for materials.

For persistent and transient design situations, γ_c for concrete = 1.5, γ_s for reinforcing and prestressing steel = 1.15. For accidental design situations, γ_c = 1.3 and γ_s = 1.0.

Design strength for concrete is therefore $f_{cd} = f_{ck}/\gamma_c$ and for steel $f_{yd} = f_{yk}/\gamma_s$ where f_{ck} and f_{yk} are respectively the characteristic cylinder compressive strength of concrete and yield stress of steel.

The partial factor for materials takes account of
1. Uncertainties in the strength of materials in the structure
2. Uncertainties in the accuracy of the method used to predict the behaviour of members
3. Variations in member sizes and building dimensions

3.4 STRUCTURAL ANALYSIS

3.4.1 General Provisions

In clause 5.1.1, Eurocode 2 states that the purpose of structural analysis is to obtain distribution of internal stress resultants such as axial and shear forces, bending and twisting moments. The method normally used is frame analysis. However in some local areas such as

- In the vicinity of supports and concentrated loads
- Beam and column intersections
- Abrupt changes of cross section
- Anchorage zones in posttensioned members
- Deep beams where the span/depth ratio is less than about 3

In cases where stress resultants cannot give a true picture of the stress and strain distribution, methods such as finite element analysis will be required to carry out a detailed stress analysis.

In clause 5.1.1(7), the following four idealizations are stated. They are:

- Linear elastic behaviour.
- Linear elastic behaviour with limited redistribution. This applies only to statically indeterminate structures. See Chapter 13 and Chapter 14.
- Plastic behaviour including strut and tie models. See Chapter 8 on Yield line analysis of slabs and Chapter 18 on strut–tie and other models.
- Non-linear behaviour.

The most common method of analysis is linear elastic analysis. The analysis is carried out assuming

- Uncracked cross section
- Linear stress–strain relationship

- Mean value of modulus of elasticity

The complete structure may be analysed elastically by the matrix method of structural analysis using a computer program. It is normal practice to model beam elements using only the rectangular section of T-beam elements in the frame analysis.

It has to be remembered that an analysis of the complete statically indeterminate structure requires the cross section dimensions to be input as data. Therefore at the preliminary analysis/design stage, approximate methods of analysis which give sufficiently accurate values of the internal forces without having to analyse the entire structure so that one can decide on a preliminary design of the cross section are essential.

In some cases linear elastic analysis with limited redistribution or even a full plastic analysis can be used, provided certain limitations are observed. All these aspects will be discussed in later chapters.

3.5 REFERENCE

Bond, A. and Harris, A. (2008). *Decoding Eurocode 7*. Taylor & Francis.

CHAPTER 4

SECTION DESIGN FOR MOMENT

4.1 TYPES OF BEAM SECTION

The three common types of reinforced concrete beam section are:
a. Rectangular sections with tension steel only (this generally occurs when designing a given width of slab as a beam)
b. Rectangular sections with tension and compression steel
c. Flanged sections of either T or L shape with tension steel and rarely with or without compression steel

Beam sections are shown in Fig. 4.1. It will be established later that all beams of structural importance must have steel at top and at bottom to carry links to resist shear.

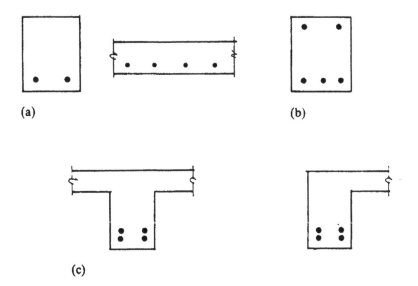

(a)   (b)

(c)

Fig. 4.1 (a) Rectangular beam and slab, tension steel only; (b) rectangular beam, tension and compression steel; (c) flanged beams.

4.2 REINFORCEMENT AND BAR SPACING

Before beginning section design, reinforcement data and code requirements with regard to minimum and maximum areas of bars in beams and bar spacing are set out. This is to enable sections to be designed with practical amounts and layout of steel. Requirements for cover were discussed in section 2.9, Chapter 2.

4.2.1 Reinforcement Data

In accordance with Eurocode 2, clause 8.9, bars may be placed singly or in pairs or in bundles provided all bars are of the same characteristic strength. In a bundle, bars of different diameters are allowed provided the ratio of diameters does not exceed 1.7. For design purposes the pair or bundle is treated as a single bar of equivalent diameter φ_n,

$$\varphi_n = \varphi \sqrt{n_b} \leq 55 \text{ mm}$$

where n_b = numbers of bars in the bundle with diameter φ, $n_b \leq 4$ for vertical bars in compression or bars at a lapped joint, $n_b \leq 3$ in all other cases.

Bars are **available** with diameters of 6, 8, 10, 12, 16, 20, 25, 32 and 40 mm and in grades with characteristic strengths f_{yk} = 400 to 600 MPa.

Preferred sizes of bars are 8, 10, 12, 16, 20, 25, 32 and 40 mm. For convenience in design calculations, areas of groups of bars are given in Table 4.1. Table 4.2 gives equivalent diameter of bundles of bars of same diameter.

Table 4.1 Areas of groups of bars

Size of bar (mm)	Numbers of bars in group							
	1	2	3	4	5	6	7	8
8	50	101	151	201	251	302	352	402
10	79	157	236	314	393	471	550	628
12	113	226	339	452	566	679	792	905
16	201	402	603	804	1005	1206	1407	1609
20	314	628	943	1257	1571	1885	2109	2513
25	491	982	1473	1964	2454	2945	3436	3927
32	804	1609	2413	3217	4021	4826	5630	6434

Table 4.2 Equivalent diameters of bars in groups

Size of bars in group (mm)	Number of bars in group			
	1	2	3	4
8	8	11.3	13.9	16
10	10	14.1	17.3	20
12	12	17.0	20.8	24
16	16	22.6	27.7	32
20	20	28.3	34.6	40
25	25	35.4	43.3	50
32	32	45.3	55.4	64

A useful publication for preparing detailed drawings was published in 2006 by the U.K. Institution of Structural Engineers.

Bars of size 12 mm and above are available from stock in lengths of 12 m. For sizes 8 mm and 10 mm available lengths are 8, 9 or 10 m.

As shown in Table 2.4, three grades of steel A, B and C are allowed, all with characteristic yield strength of 400 to 600 MPa but of different ductilities. The full details are shown in Table C.1 of Annex C in Eurocode 2. In the U.K. the corresponding grades of steel available are 500A, 500B and 500C with characteristic yield strength of 500 MPa.

4.2.2 Minimum and Maximum Areas of Reinforcement in Beams

According to clause 9.2.1.1, equation (9.1N) of Eurocode 2, the minimum area of tension reinforcement $A_{s, min}$ in a beam section to control cracking should be

$$A_{s,min} \geq 0.26 \frac{f_{ctm}}{f_{yk}} b_t d \text{ but not less than } 0.0013 \, b_t \, d \qquad (9.1N)$$

where
b_t = width of the tension zone. In a rectangular beam it is the width and in a T-beam it is the width of the web.
d = effective depth
f_{ctm} = mean axial tensile strength of concrete (see Chapter 2, section 2.3.3)
$\quad = 0.30 \times f_{ck}^{\ 0.667}$, $f_{ck} \leq 50$ MPa,
$\quad = 2.12 \times \ell n(1.8 + 0.1 \times f_{ck})$, $f_{ck} > 50$ MPa
For f_{yk} = 500 MPa, Table 4.3 gives the value of $A_{s,min}/(b_t \, d)$.

Table 4.3 Minimum value of tension reinforcement in beams

f_{ck} MPa	f_{ctm} MPa	$A_{s,min}/(b_t \, d)\%$
12	1.6	0.13
16	1.9	0.13
20	2.2	0.13
25	2.6	0.14
30	2.9	0.15
35	3.2	0.17
40	3.5	0.18
45	3.8	0.20
50	4.1	0.21
55	4.2	0.22
60	4.4	0.23
70	4.6	0.24
80	4.8	0.25
90	5.0	0.26

Maximum tension or compression steel area should not exceed 4% of the gross cross sectional area.

4.2.3 Minimum Spacing of Bars

Clause 8.2 of Eurocode 2, states that the spacing of bars should be such that concrete can be placed and compacted properly. The clear distance (horizontal and vertical) between individual parallel bars or horizontal layers of parallel bars should not be less than

Maximum {bar diameter; (maximum size of aggregate + 5 mm); 20mm}.

Where bars are positioned in separate horizontal layers, bars in each layer should be located vertically above each other.

4.3 BEHAVIOUR OF BEAMS IN BENDING

The behaviour of a cross section subjected to pure bending is studied by loading a beam at two points as shown in Fig. 4.2(a). Under this system of loading, sections between the loads are subjected to pure bending. Initially the beam behaves as a monolithic elastic beam till the stresses at the bottom fibre reach the tensile strength of concrete. Because of the very low tensile strength of concrete (about 10% of its compression strength), vertical cracks appear at a fairly low load. As the load is increased, cracks lengthen and penetrate deeper towards the compression face. Simultaneously, the strain in steel also increases. The final failure depends on the amount and yield stress of steel. The three possible modes of failure are:

1. **Steel yields first**: If the tensile force capacity of steel is 'low', then steel yields before the strain in the concrete at the compression face reaches the maximum permissible value of 0.0035 (see Fig. 2.2 and Fig. 2.3). Because steel is a ductile material, steel elongates while maintaining its strength. The beam continues to deform at constant load and the neutral axis moves up. The beam finally fails when the depth of the compression zone is too small to balance the tensile force in steel. This type of failure is the desired type because there is ample warning before failure. All beams, if overloaded, should be designed to fail in this manner. Fig. 4.2(b) shows the qualitative load versus deflection curve and Fig. 4.2(c) shows the stress distribution at elastic and ultimate stages.

2. **Simultaneous 'yielding' of steel and concrete**: If the tensile force capacity of steel is 'moderate', yielding of steel is simultaneously accompanied by the crushing of concrete. Unlike the failure mode where the steel yields first, there is little warning before failure. This is not a desirable mode of failure.

3. **Concrete crushes first**: If the tensile force capacity of steel is 'high', then steel does not yield at all before concrete crushes. Because concrete is a fairly brittle material, it fails in an explosive manner without any significant residual load bearing capacity. *This form of failure is to be avoided at all costs!*

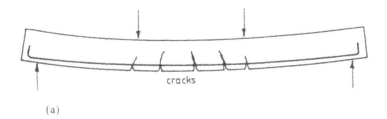

(a)

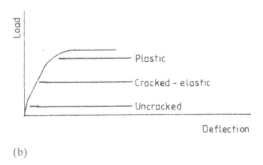

(b)

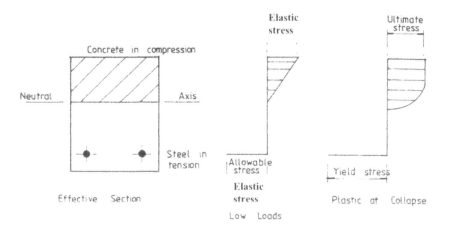

(c)

Fig. 4.2 (a) Flexural cracks at collapse; (b) load–deflection curve; (c) effective section and stress distribution.

4.4 SINGLY REINFORCED RECTANGULAR BEAMS

4.4.1 Assumptions and Stress–Strain Diagrams

The ultimate moment of resistance of a section is based on the assumptions set out in clause 3.1.7. These are as follows:

1. The strains in the concrete and reinforcement are derived assuming that plane sections remain plane;

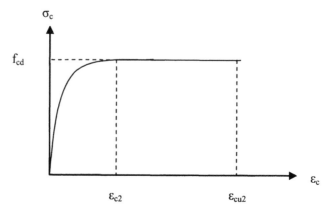

Fig. 4.3 Parabola–rectangle stress–strain relationship for concrete in compression.

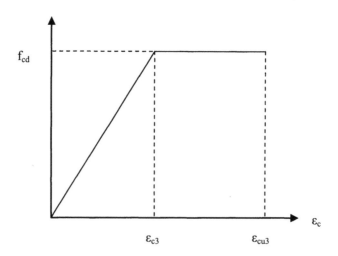

Fig. 4.4 Bilinear stress–strain relationship for concrete in compression.

2. The stresses in the concrete in compression are derived using either the design stress–strain curve given in Fig. 4.3 (parabolic–rectangular) or Fig. 4.4 (bilinear)

with $f_{cd} = f_{ck}/\gamma_c$, $\gamma_c = 1.5$. Note that in both cases the maximum strain in the concrete at failure (ε_{cu2} or ε_{cu3}) is 0.0035 for $f_{ck} \leq 50$ MPa and decreasing to 0.0026 for $f_{ck} = 90$ MPa.

Table 4.4 Values of λ, η, ε_{cu2} and ε_{cu3}

f_{ck} MPa	λ	η	$(\varepsilon_{cu2} = \varepsilon_{cu3}) \times 10^3$
≤ 50	0.80	1.0	3.5
55	0.7875	0.975	3.1
60	0.775	0.95	2.9
70	0.750	0.90	2.7
80	0.725	0.85	2.6
90	0.70	0.80	2.6

For convenience in calculation, as shown in Fig. 4.5, a drastic simplification is made by assuming rectangular stress distribution over a depth λ x, where the depth of the stress block is λ times the depth to the neutral axis denoted by x. The compressive stress in the stress block is a constant value equal to η f_{cd}. The maximum strain in the concrete at failure as ε_{cu3}.

3. The tensile strength of the concrete is ignored.

4. The stresses in the reinforcement are derived from the stress–strain curve shown in Fig. 2.6 (Chapter 2) where $\gamma_s = 1.15$.

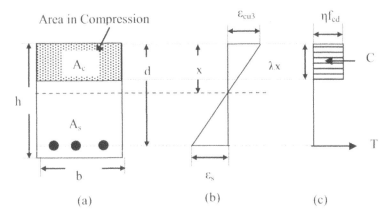

Fig. 4.5 (a) Section; (b) strain; (c) rectangular parabolic stress diagram; (d) simplified stress diagram.
h = overall depth of the section
d = effective depth, i.e. depth from the compression face to the *centroid* of tension steel
b = breadth of the section
x = depth to the neutral axis
f_s = stress in steel
A_s = area of tension reinforcement
ε_{cu3} = maximum strain in the concrete
ε_s = strain in steel

The values of λ and η depend on the value of f_{ck} as given in Eurocode equations (3.19) to (3.22).

$$\lambda = 0.8, \; f_{ck} \le 50 \text{ MPa} \qquad\qquad\qquad\qquad\qquad (3.19)$$
$$\lambda = 0.8 - (f_{ck} - 50)/400 \quad 50 < f_{ck} \le 90 \text{ MPa} \qquad (3.20)$$
$$\eta = 1.0, \; f_{ck} \le 50 \text{ MPa} \qquad\qquad\qquad\qquad\qquad (3.21)$$
$$\eta = 1.0 - (f_{ck} - 50)/200, \quad 50 < f_{ck} \le 90 \text{ MPa} \qquad (3.22)$$

Table 4.4 shows the values of λ and η for different values of f_{ck}.
The total compression force C in concrete $= (\lambda x) \times b \times (\eta \, f_{cd})$ and tension force T in steel $= A_s \, f_s$.

4.4.2 Moment of Resistance: Rectangular Stress Block

The total compressive force C is given by
$$C = (\lambda x) \times b \times (\eta \, f_{cd})$$
$$\text{Setting } f_{cd} = f_{ck}/\gamma_c, \; \gamma_c = 1.5, \; f_{cd} = f_{ck}/1.5 = 0.667 \, f_{ck}$$
$$k_c = C/(bd \, f_{ck}) = 0.667 \, \lambda \times \eta \times (x/d)$$

The lever arm z is
$$z = d - 0.5 \, \lambda x, \; z/d = 1 - 0.5 \, \lambda \, (x/d)$$

If M is the applied moment, then
$$M = C \times z = (\lambda x) \times b \times (\eta \, f_{cd}) \times (d - 0.5 \, \lambda x)$$
$$k = \frac{M}{bd^2 f_{ck}} = \frac{\eta}{1.5} \times \alpha \times (1 - 0.5\alpha), \; \alpha = \lambda \frac{x}{d}$$

$$\text{Rearranging, } \alpha^2 - 2\alpha + 3\frac{k}{\eta} = 0$$

Solving the quadratic equation for α,

$$\alpha = 1 - \sqrt{(1 - 3\frac{k}{\eta})}$$

$$\frac{z}{d} = 1 - 0.5\alpha = 0.5\{1.0 + \sqrt{(1 - 3\frac{k}{\eta})}\}$$

Total tensile force T in steel is
$$T = A_s \times f_s$$

For internal equilibrium, total tension T must be equal to total compression C. The forces T and C form a couple at a lever arm of z.
$$M = T \, z = A_s \, f_s \, z$$
$$A_s = M/(f_s \, z)$$

The stress f_s in steel depends on the strain ε_s in steel. As remarked in section 4.3, it is highly desirable that final failure is due to yielding of steel rather than due to crushing of concrete. It is useful therefore to calculate the maximum neutral axis depth in order to achieve this. Assuming that plane sections remain plane before and after bending, an assumption validated by experimental observations, if as shown in Fig. 4.5(b), the maximum permitted strain in concrete at the compression face is ε_{cu3}, then the strain ε_s in steel is calculated from the strain diagram by

$$\varepsilon_s = \frac{(d-x)}{x}\varepsilon_{cu3}$$

Strain ε_s in steel at a stress of f_{yk}/γ_s is given by

$$E_s\,\varepsilon_s = \frac{f_{yk}}{\gamma_m}$$

where f_{yk} = yield stress , γ_s = 1.15 and E_s is Young's modules for steel.
Taking f_{yk} = 500 MPa, γ_m = 1.15, f_{yk}/γ_s = 435 MPa, E_s = 200 GPa, ε_s = 0.0022
For ε_s = 0.0022, the depth of neutral axis x is given by

$$\varepsilon_s = 0.0022 = \frac{(d-x)}{x}\varepsilon_{cu3}$$

$$\frac{x}{d} = \frac{1}{\{1+\dfrac{0.0022}{\varepsilon_{cu3}}\}}$$

This is the value of x/d at which steel just reaches its 'yield' stress and concrete reaches its maximum compressive strain. This is called 'balanced design'. Table 4.5 shows the value of x/d for balanced design. However in order to ensure that failure is *preceded* by steel yielding well before the strain in concrete reaches ε_{cu3}, resulting in the desirable ductile form of failure, maximum value of x/d in practice is made much smaller than the value calculated for balanced failure.

In the case of statically indeterminate structures where linear elastic analysis with limited redistribution is used in the design at ULS (see Chapters 13 and Chapter 14), Eurocode 2 in clause 5.5 specifies that the following limits on the depth of neutral axis x as given by code equations (5.10a) and (5.10b)] have to be observed in order that the designed structure is reasonably ductile.

$$\delta \geq 0.44 + 1.25\, x_u/d \qquad f_{ck} \leq 50\ \text{MPa} \tag{5.10a}$$
$$\delta \geq 0.54 + 1.25(0.6 + 0.0014/\varepsilon_{cu2})\, x_u/d \quad f_{ck} > 50\ \text{MPa} \tag{5.10b}$$
$$\geq 0.7\ \text{if Class B and class C reinforcement is used.}$$
$$\geq 0.8\ \text{if Class A reinforcement is used as it is least ductile.}$$

δ = Redistributed bending moment/elastic bending moment
x_u = depth of neutral axis at the ULS
Table 4.5 shows the maximum values of x_u/d permitted for different ratios of redistribution. Note that all values are well below the value calculated for balanced design.
Similarly, Table 4.6 shows the values of $z_u/d = 1 - 0.5\,\lambda\,(x_u/d)$.
Table 4.7 shows the values of $k_c = 0.667\,\lambda \times \eta \times (x_u/d)$.

Table 4.8 shows the values of $k = 0.667 \times \eta \times \lambda \dfrac{x_u}{d} \times (1 - 0.5\lambda \dfrac{x_u}{d})$.

Reinforced concrete design to EC 2

Table 4.5 Maximum value of x_u/d for different ratios of redistribution

f_{ck} MPa	$\varepsilon_{cu2} \times 10^3$	x_u/d				
		$\delta = 1.0$	$\delta = 0.90$	$\delta = 0.80$	$\delta = 0.70*$	Balanced
≤ 50	3.5	0.448	0.368	0.288	*0.208*	0.6140
55	3.1	0.350	0.274	0.198	*0.122*	0.5849
60	2.9	0.340	0.266	0.192	*0.118*	0.5686
70	2.7	0.329	0.257	0.186	*0.114*	0.5510
80	2.6	0.323	0.253	0.183	*0.112*	0.5417
90	2.6	0.323	0.253	0.183	*0.112*	0.5417

Table 4.6 Maximum value of z_u/d for different ratios of redistribution

f_{ck} MPa	z_u/d			
	$\delta = 1.0$	$\delta = 0.90$	$\delta = 0.80$	$\delta = 0.70*$
≤ 50	0.8208	0.8528	0.8848	*0.9168*
55	0.8622	0.8922	0.9221	*0.9521*
60	0.8683	0.8969	0.9256	*0.9542*
70	0.8766	0.9034	0.9303	*0.9571*
80	0.8828	0.9083	0.9338	*0.9592*
90	0.8869	0.9115	0.9361	*0.9606*

Table 4.7 Maximum value of k_c for different f_{ck} and moment redistribution ratio

f_{ck}	$k_c = C/(bd\, f_{ck})$			
	$\delta = 1.0$	$\delta = 0.9$	$\delta = 0.8$	$\delta = 0.7*$
≤ 50	0.2401	0.1964	0.1544	0.1110
55	0.1800	0.1403	0.1018	0.0623
60	0.1677	0.1306	0.0948	0.0581
70	0.1488	0.1159	0.0841	0.0515
80	0.1335	0.1040	0.0754	0.0462
90	0.1213	0.0945	0.0686	0.0420

Table 4.8 Maximum value of k for different f_{ck} and moment redistribution ratio

f_{ck}	$k = M/(bd^2\, f_{ck})$				
	$\delta = 1.0$	$\delta = 0.9$	$\delta = 0.8$	$\delta = 0.7*$	Balanced
≤ 50	0.196	0.167	0.136	0.102	*0.247*
55	0.154	0.125	0.093	0.059	*0.227*
60	0.145	0.117	0.087	0.055	*0.213*
70	0.130	0.105	0.078	0.049	*0.191*
80	0.117	0.094	0.070	0.044	*0.173*
90	0.107	0.086	0.064	0.040	*0.159*

Note that in all the tables, for Class A steel, maximum value of $\delta = 0.80$.
For any value of δ and f_{ck}, $M = k\, bd^2\, f_{ck}$ is the maximum value of the applied moment that the section can resist utilizing fully the compression capacity $C = k_c\, bd\, f_{ck}$ of the cross section. This formula can be used to calculate the *minimum effective depth* required in a singly reinforced rectangular concrete section.

$$d_{min} = \sqrt{\frac{M}{k \ b \ f_{ck}}}$$

In practice the effective depth d is made larger than the required minimum consistent with the required headroom.

$$d \geq \sqrt{\frac{M}{k \ b \ f_{ck}}}$$

The reason for this is that with a larger depth, the neutral axis depth is smaller and hence the lever arm is larger leading for a given moment M, to a smaller amount of reinforcement. It has the additional advantage that in the event of unexpected overload, the beams will show large ductility before failure.

$$f_s = f_{yk}/1.15 = 0.87 \ f_{yk}$$
$$M = T \ z = A_s \ 0.87 \ f_{yk} \ z$$
$$A_s = M/(0.87 \ f_{yk} \ z)$$

4.4.2.1 U.K. National Annex Formula

According to the U.K. National Annex, the constant stress in the stress block is taken as $\alpha_{cc} \ \eta \ f_{cd}$, where $\alpha_{cc} = 0.85$

$$M = C \times z = (\lambda x) \times b \times (\alpha_{cc} \ \eta \ f_{cd}) \times (d - 0.5 \ \lambda x)$$

$$k = \frac{M}{bd^2 \ f_{ck}} = \frac{\alpha_{cc}\eta}{1.5} \times \alpha \times (1 - 0.5\alpha), \ \alpha = \lambda \frac{x}{d}$$

Rearranging, $\alpha^2 - 2\alpha + 3 \dfrac{k}{\alpha_{cc}\eta} = 0$.

Solving the quadratic equation for α,

$$\alpha = 1 - \sqrt{(1 - 3\frac{k}{\alpha_{cc}\eta})}$$

$$\frac{z}{d} = 1 - 0.5\alpha = 0.5 \{1.0 + \sqrt{(1 - 3\frac{k}{\alpha_{cc}\eta})}\}$$

The justification for assuming that the constant stress in the rectangular stress block is $0.85 \ \eta \ f_{cd}$ is given in

PD 6687-1, Background paper to the U.K. National Annex o Eurocode 2, Parts 1 and 3.

4.4.3 Procedure for the Design of Singly Reinforced Rectangular Beam

The steps to be followed in the design of singly reinforced rectangular beams can be summarised as follows.

- From the minimum requirements of span/depth ratio to control deflection (see Chapter 6), estimate a suitable effective depth d.
- Assuming the bar diameter for the main steel and links and the required cover as determined by exposure conditions, estimate an overall depth h.

  $h = d +$ bar diameter + Link diameter + Cover to links
- Assume breadth at about half the overall depth.
- Calculate the self–weight.
- Calculate the design live load and dead load moment using appropriate load factors. The load factors are normally 1.35 for dead loads and 1.5 for live loads.
- For the given value of f_{ck}, calculate λ and η from Table 4.4. Note that $f_{ck} \leq 50$ MPa, $\lambda = 0.8$ and $\eta = 1.0$
- As there is no redistribution possible in the case of statically determinate structures, $\delta = 1$
- Calculate value of k from Table 4.8
- In the case of singly reinforced sections, calculate the minimum effective depth using the formula

$$d_{min} = \sqrt{\frac{M}{k\ b\ f_{ck}}}$$

- Adopt an effective depth greater than the minimum depth in order to reduce the total tension reinforcement.
- Check that the new depth due to increased self–weight does not drastically affect the calculated design moment. If it does, calculate the revised ultimate moment required.
- Calculate $k = M/(b\ d^2\ f_{ck})$
- Calculate the lever arm z

$$\frac{z}{d} = 0.5\{1.0 + \sqrt{(1 - 3\frac{k}{\eta})}\}$$

- Calculate the required steel A_s

$$A_s = M/\{0.87\ f_{yk}\ z\}$$

- Check that the steel provided satisfies the minimum and maximum steel percentages specified in the code.

4.4.4 Examples of Design of Singly Reinforced Rectangular Sections

Example 1: A simply supported reinforced rectangular beam of 8 m span carries uniformly distributed characteristic dead load, which includes an allowance for self–weight of 7 kN/m and characteristic imposed load of 5 kN/m. The breadth $b = 250$ mm. Design the beam at mid-span section. Use strength class 25/30 concrete and $f_{yk} = 500$ MPa for steel reinforcement.

Solution: For 25/30 concrete, $f_{ck} = 25$ MPa, $\eta = 1.0$, $\lambda = 0.8$
Design load $= (1.35 \times 7) + (1.5 \times 5) = 16.95$ kN/m
Design ultimate moment M at mid–span:
$M = 16.95 \times 8^2/8 = 135.6$ kNm
For simply supported beam, $\delta = 1$ as there is no possibility of redistribution. From Table 4.8, for $f_{ck} \leq 50$ and $\delta = 1$, $k = 0.196$
Minimum effective depth to avoid any compression steel is given by

$$d_{min} = \sqrt{\frac{M}{0.196 \times b \times f_{ck}}} = \sqrt{\frac{135.6 \times 10^6}{0.196 \times 250 \times 25}} = 333\,mm$$

Using this value of d,

$$\frac{z}{d} = 0.5\{1.0 + \sqrt{(1-3\frac{k}{\eta})}\} = 0.5\{1.0 + \sqrt{(1-3\frac{0.196}{1.0})}\} = 0.82$$

The area of steel required is

$$A_s = \frac{M}{0.82\,d \times 0.87 f_{yk}} = \frac{135.6 \times 10^6}{0.82 \times 333 \times 0.87 \times 500} = 1142\ mm^2$$

However, if a value of d equal to say 400 mm, which is larger than the minimum value is used, then one can reduce the area of steel required.

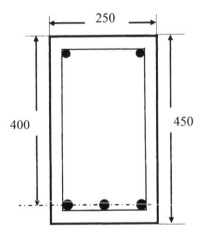

Fig. 4.6 Mid-span section of the beam.

Assuming d = 400 mm

$$k = \frac{M}{b\,d^2 f_{ck}} = \frac{135.6 \times 10^6}{250 \times 400^2 \times 25} = 0.136 < 0.196$$

$$\frac{z}{d} = 0.5[1.0 + \sqrt{(1-3\frac{k}{\eta})}] = 0.5\{1.0 + \sqrt{(1-3\frac{0.136}{1.0})}\} = 0.885$$

$$A_s = \frac{M}{0.885\,d \times 0.87 f_{yk}} = \frac{135.6 \times 10^6}{0.885 \times 400 \times 0.87 \times 500} = 881\,mm^2$$

Provide three 20 mm diameter bars as shown in Fig. 4.6. From Table 4.1, area of steel A = 943 mm^2. Assuming cover of 30 mm and link diameter of 8 mm, the overall depth h of the beam is

h = 400 + 30 + 8 + 20/2 = 448, say 450 mm.

From Table 4.3, for f_{ck} = 25 MPa that the minimum percentage steel required is 0.14. Check that the provided percentage is greater than the minimum of 0.14.

100 $A_s/(b_t\,d)$ = 100 × 943/(250 × 400) = 0.94 > 0.14.

Note that this is only one of several possible satisfactory solutions.

In simply supported beams bending moment decreases towards the supports. Therefore the amount of steel required towards the support region is much less than at mid-span. Therefore it is possible to reduce the area of steel away from the mid-span.

Example 2: Determination of tension steel cut-off. For the beam in Example 4.1, determine the position along the beam where *theoretically* the middle of the three 20 mm diameter bar may be cut off.

The section at cut-off has two 20 mm diameter bars continuing: A_s = 628 mm^2. The neutral axis depth can be determined by equating total compression in concrete to total tension in the beam.

$$T = 0.87\,f_{yk}\,A_s = 0.87 \times 500 \times 628 \times 10^{-3} = 273.2\,kN$$
$$C = \eta \times (f_{ck}/1.5) \times b \times \lambda x \times 10^{-3}, \lambda = 0.8, \eta = 1$$
$$C = (25/1.5 \times 250 \times 0.8x) \times 10^{-3} = 3.33x\,kN$$

Equating C = T:

$$x = 82\,mm$$
$$z = d - 0.5\,\lambda\,x = 367\,mm$$

Moment of resistance M_R:

$$M_R = T\,z = 273.2 \times 367 \times 10^{-3} = 100.3\,kNm$$

Determine the position 'a' from the support where M = 100.3 kN m.
Left hand reaction V is:

$$V = 16.95 \times 8/2 = 67.8\,kN$$
$$100.3 = 67.8\,a - 0.5 \times 16.95\,a^2$$

The solutions to this equation are a = 1.95 m and a = 6.04 m from end A.
The bars should NOT be stopped at the section but continued a certain length beyond the theoretical cut-off point because of anchorage considerations which will be explained in Chapter 6.

Example 3: Singly reinforced one-way slab section. Fig. 4.7 shows a slab section 1 m wide and 130 mm deep with an effective depth of 100 mm is subjected to a design ultimate moment of 10.5 kNm. Find the area of reinforcement required. f_{ck} = 25 MPa and f_{yk} = 500 MPa.

$$k = \frac{M}{bd^2 f_{ck}} = \frac{10.5 \times 10^6}{1000 \times 100^2 \times 25} = 0.042 < 0.196$$

$$\frac{z}{d} = 0.5\{1.0 + \sqrt{(1 - 3\frac{k}{\eta})}\} = 0.5\{1.0 + \sqrt{(1 - 3\frac{0.042}{1.0})}\} = 0.97$$

$$A_s = \frac{M}{0.97\,d \times 0.87 f_{yk}} = \frac{10.5 \times 10^6}{0.97 \times 100 \times 0.87 \times 500} = 249 \text{ mm}^2$$

In the case of slabs, because of the very large width, reinforcement is not usually specified as a fixed number of bars but in terms of the diameter of the bar and its spacing. Using Table 4.9, provide 8 mm diameter bars at 175 mm centres giving $A_s = 288$ mm^2.

Check the minimum percentage of steel.

The percentage steel = $100\, A_s/(b_t\, d) = 100 \times 288/(1000 \times 100) = 0.29 > 0.14$. The reinforcement for the slab is shown in Fig. 4.7.

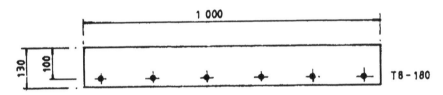

Fig. 4.7 Reinforcement in slab.

4.4.5 Design Graph

Using the equations developed in section 4.4.2, a graph for the design of singly reinforced rectangular beams can be constructed as follows.

- Choose strength class of concrete
- Calculate parameters η and λ
- Choose a value of $(x/d) \le 0.448$
- $(z/d) = \{1 - 0.5\,\lambda(x/d)\}$
- $C = \eta\,(f_{ck}/1.5) \times b \times \lambda x = bd\,f_{ck}\,\{0.667\,\eta\,\lambda\,(x/d)\}$
- $M = C\,z = bd^2\,f_{ck}\,[0.667\,\eta\,\lambda\,(x/d)\,(1 - 0.5\,\lambda x/d)$
- $A_s = M/(0.87\,f_{yk}\,z) = 0.766\,(x/d)\,\lambda\,\eta\,b\,d\,(f_{ck}/f_{yk})$
- $\dfrac{A_s\,f_{yk}}{bd\,f_{ck}} = 0.767\,\eta\lambda\,\dfrac{x}{d}$

Fig. 4.8 shows a plot of k = $M/(bd^2\,f_{ck})$ versus $A_s/(bd) \times (f_{yk}/f_{ck})$ for $f_{ck} \le 50$ MPa.

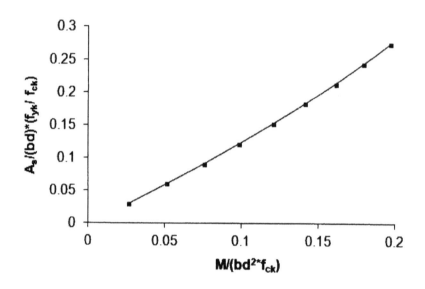

Fig. 4.8 Plot of $M/(bd^2 f_{ck})$ versus $A_s/(bd) * (f_{yk}/f_{ck})$ for $f_{ck} \leq 50$ MPa.

Table 4.9 (Table to be used for calculating steel areas in slabs, walls, etc.)

TOTAL REINFORCEMENT AREA (mm²/m)						
	Bar diameter (mm)					
Bar spacing (mm)	6	8	10	12	16	20	25
50	566	1010	1570	2260	4020	6280	9820
75	378	670	1050	1510	2680	4190	6550
100	283	503	785	1130	2010	3140	4910
125	226	402	628	904	1610	2510	3930
150	189	335	523	753	1340	2090	3270
175	162	288	448	646	1150	1790	2810
200	141	251	392	565	1010	1570	2460
250	113	201	314	452	804	1260	1960
300	94	167	261	376	670	1050	1640
350	81	144	224	323	574	897	1400
400	70	126	196	282	502	785	1230
450	63	112	174	251	447	697	1090
500	57	101	157	226	402	628	982

Note: $A = (\pi d^2/4) \{1000/(c/c \text{ spacing in mm}\}$.

4.5 DOUBLY REINFORCED BEAMS

The normal design practice is to use singly reinforced sections. However if for any reason, for example headroom considerations, it is necessary to restrict the overall depth and breadth of a beam, then it becomes necessary to use steel in the compression zone as well because concrete alone might not provide the necessary compression resistance.

4.5.1 Design Formulae Using the Rectangular Stress Block

The formulae for the design of a doubly reinforced beam are derived using the rectangular stress block.

Let M be the design ultimate moment. As shown in Table 4.7 and Table 4.8 respectively, a rectangular section as a *singly reinforced section* can resist a *maximum* value of the compressive force due to moment equal to
$$C_{sr} = k_c \, bd \, f_{ck}$$
and moment equal to
$$M_{sr} = k \, b \, d^2 \, f_{ck}$$
The corresponding neutral axis depth x_u / d can be determined from Table 4.5 and z_u / d from Table 4.6.

If the applied moment $M > M_{sr}$, then compression steel is required because concrete cannot provide any more compressive force than C_{sr}.
The compressive force C_s due to compression steel of area A_s' is
$$C_{sc} = A_s' \, f_s'$$
where f_s' is the stress in compression steel.

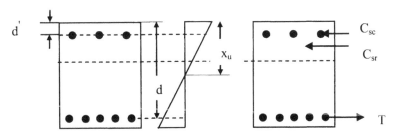

Fig. 4.9 Doubly reinforced beam.

As shown in Fig. 4.9, the lever arm z_{sc} for compression steel is
$$z_{sc} = (d - d')$$
The stress in the tensile steel is $f_{yk}/1.5 = 0.87 \, f_{yk}$ because the neutral axis depth of x_u/d is less than the value for balanced design. However the stress f_s' in the compressive steel depends on the corresponding strain ε_{sc} in concrete at compression steel level. ε_{sc} is given by

$$\varepsilon_{sc} = \varepsilon_{cu3} \frac{(x - d')}{x}$$

If the strain ε_{sc} is equal to or greater than the yield strain in steel, then steel yields and the stress f_s in compression steel is equal to 0.87 f_{yk}. Otherwise, the stress in compression steel is given by

$$f_s' = E_s \varepsilon_{sc}$$

If f_{yk} = 500 MPa and E_s = 200 GPa, then the yield strain in steel is equal to

$$\varepsilon_{yield} = \frac{0.87 f_{yk}}{E_s} = 0.0022$$

Therefore, steel will yield if

$$\varepsilon_{sc} = \varepsilon_{cu3} \frac{(x - d')}{x} \geq \{\frac{0.87 f_y}{E} = 0.0022\}$$

$$\therefore d' \leq \{1 - \frac{0.0022}{\varepsilon_{cu3}}\} x_u$$

$$\frac{d'}{d} \leq \{1 - \frac{0.0022}{\varepsilon_{cu3}}\} \frac{x_u}{d}$$

Table 4.10 shows the value of $\dfrac{d'}{d}$ for combinations of f_{ck} and δ.

Table 4.10 Maximum value of $\dfrac{d'}{d}$ for different ratios of redistribution

f_{ck} MPa	$\varepsilon_{cu3} \times 10^3$	$\dfrac{d'}{d}$			
		$\delta = 1.0$	$\delta = 0.90$	$\delta = 0.80$	$\delta = 0.70*$
≤ 50	3.5	0.1664	0.1367	0.1070	0.0773
55	3.1	0.1016	0.0795	0.0574	0.0353
60	2.9	0.0820	0.0642	0.0464	0.0285
70	2.7	0.0609	0.0477	0.0344	0.0212
80	2.6	0.0497	0.0389	0.0281	0.0173
90	2.6	0.0497	0.0389	0.0281	0.0173

* Note: For Class A steel, maximum value of δ = 0.80.

Taking moments about the tension steel,

$$M = C_{sr} z_u + C_{sc} z_{sc}$$

For various combinations of f_{ck} and δ, values of k_c and z_u/d are shown in Table 4.7 and Table 4.6 respectively.

$$M = \{k_c f_{ck} b d\} \times (z_u /d) + A_s' f_s' (d - d')$$

For various combinations of f_{ck} and δ, values of k are shown in Table 4.8.

$$M = k b d^2 f_{ck} + A_s' f_s' (d - d')$$

$$A_s' = (M - M_{sr})/\{f_s' (d - d')\}$$

For equilibrium, the tensile force T is equal to total compressive force.

$$T = A_s 0.87 f_{yk} = C_{sr} + C_{sc}$$

One important point to remember is that to prevent steel bars in compression from buckling, it is necessary to restrain the bars using links.

4.5.2 Examples of Rectangular Doubly Reinforced Concrete Beams

The use of the formulae developed in the previous section is illustrated by a few examples.

Example 1: A rectangular beam is simply supported over a span of 6 m and carries characteristic dead load including self-weight of 12.7 kN/m and characteristic imposed load of 6.0 kN/m. The beam is 200 mm wide by 300 mm effective depth and the inset d' of the compression steel is 40 mm. Design the steel for mid-span of the beam for $f_{ck} = 25$ MPa concrete and $f_{yk} = 500$ MPa reinforcement.

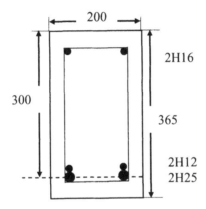

Fig. 4.10 Doubly reinforced beam.

$$\text{design load} = (12.7 \times 1.35) + (6 \times 1.5) = 26.15 \text{ kN/m}$$
Required ultimate moment M:
$$M = 26.15 \times 6^2/8 = 117.7 \text{ kN m}$$
Maximum moment that the beam section can resist as a singly reinforced section is
$$M_{sr} = 0.196 \times 25 \times 200 \times 300^2 \times 10^{-6} = 88.2 \text{ kNm}$$
Since $M > M_{sr}$, compression steel is required.
$d'/d = 40/300 = 0.13 < 0.1664$ (see Table 4.11). Therefore compression steel yields. The stress f_s' in the compression steel is $0.87 f_{yk}$.
$$A_s' = \{M - M_{sr}\}/[0.87 \ f_{yk} \ (d - d')]$$
$$A_s' = \{117.7 - 88.2\} \times 10^6/[0.87 \times 500 \times (300 - 40)] = 261 \text{ mm}^2$$
From equilibrium:
$$A_s \ 0.87 \ f_{yk} = C_{sr} + A_s' \ f_s'$$
$$A_s \ 0.87 \times 500 = k_c \times 200 \times 300 \times 25 + 261 \times 0.87 \times 500$$
$$k_c = 0.2401 \text{ from Table 4.7}$$
$$A_s = 1089 \text{ mm}^2$$

For the tension steel (2H25 + 2H12) give A_s = 1208 mm². For the compression steel 2H16 give $A_s^{'}$ = 402 mm². The beam section and flexural reinforcement steel are shown in Fig. 4.10.

Example 2: Design the beam in Example 4.6 but with $d^{'}$ = 60 mm.

$$d^{'}/d = 60/300 = 0.20 > 0.166 \text{ (Table 4.11)}$$

Compression steel does **not** yield.

Calculate the strain in compression steel: x_u/d = 0.448, x_u = 134 mm

$$\varepsilon_{sc} = (\varepsilon_{cu3} = 0.0035) \times \frac{(x - d^{'})}{x} = 0.0035\frac{(134 - 60)}{134} = 0.0019$$

Stress in compression steel is

$$f_s^{'} = E_s \, \varepsilon_{sc} = 200 \times 10^3 \times 0.0019 = 397 \text{ MPa}$$
$$A_s^{'} = \{M - M_{sr}\}/[397\,(d - d^{'})]$$
$$\{117.7 - 88.2\} \times 10^6/[397 \times (300 - 60)] = 310 \text{ mm}^2$$

From equilibrium:

$$A_s \, 0.87 \, f_{yk} = C_{sr} + A_s^{'} \, f_s^{'}$$
$$A_s \, 0.87 \times 500 = k_c \times 200 \times 300 \times 25 + 310 \times 397$$
$$k_c = 0.2401 \text{ from Table 4.7}$$
$$A_s \, 0.87 \times 500 = 0.2401 \times 200 \times 300 \times 25 + 310 \times 397, \, A_s = 1111 \text{ mm}^2$$

4.6 FLANGED BEAMS

4.6.1 General Considerations

In a simple slab–beam system shown in Fig. 4.11, the slab is designed to span between the beams. The beams span between external supports such as columns, walls, etc. The reactions from the slabs act as load on the beam.

When a series of beams are used to support a concrete slab, because of the monolithic nature of concrete construction, the slab acts as the flange of the beams. The end beams become L-beams while the intermediate beams become T-beams. In designing the intermediate beams, it is assumed that the loads acting on one half of the slab on the two sides of the beam are carried by the beam. Because of the comparatively small contact area at the junction of the flange and the rib of the beam, the distribution of the compressive stress in the flange is not uniform. It is higher at the junction and decreases away from the junction. This phenomenon is known as shear lag. For simplicity in design, it is assumed that only part of full physical flange width is considered to sustain compressive stress of uniform magnitude. This smaller width is known as effective breadth of the flange. Although the effective width actually varies even along the span as well, it is common to assume that the effective width remains constant over the entire span.

Fig. 4.12 shows typical bending moment distribution in continuous beams. The concept of effective width applies in the region between zero moments as shown in

Fig. 4.12 as here the flange will be in compression. Over the support region, the flange is in tension. Therefore the web of the beam resists compression and the beam behaves as a rectangular beam. Therefore the concept of effective width is irrelevant in this region.

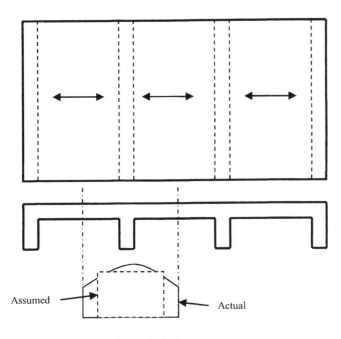

Fig. 4.11 Beam–slab system.

Clause 5.3.2.1 of Eurocode 2 gives the necessary information on effective width of flanges.

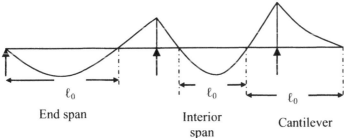

Fig. 4.12 Distance between points of zero moments ℓ_0 in the spans of a continuous beam.

Fig. 4.13 shows the simply supported lengths in end span and intermediate spans of a continuous beam.

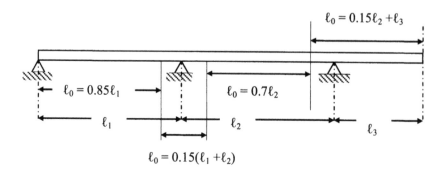

Fig. 4.13 Definition of ℓ_0 for calculation of effective flange width.

Note: The length of the cantilever ℓ_3 should be less than half the adjacent span and the ratio of the adjacent spans should lie between 2/3 and 1.5.

The effective width is given by Eurocode 2 equations (5.7), (5.7a) and (5.7b) as follows. The notation is shown in Fig. 4.14.

$$b_{effe,1} = 0.2\, b_1 + 0.1\, \ell_0 \leq 0.2\ell_0 \text{ and } b_{effe,1} \leq b_1$$
$$b_{effe,2} = 0.2\, b_2 + 0.1\, \ell_0 \leq 0.2\ell_0 \text{ and } b_{effe,2} \leq b_2$$
$$b_{eff} = b_{eff,1} + b_{eff,\,2} + b_w$$

The design procedure for flanged beams depends on the depth of the stress block. Two possibilities need to be considered.

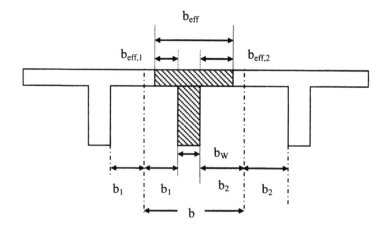

Fig. 4.14: Effective flange width parameters for flanged beams.

6.2 Stress Block within the Flange

If $\lambda x \leq h_f$, the depth of the flange (same as the total depth of the slab) then all the concrete below the flange is cracked and the beam may be treated as a rectangular beam of breadth b_{eff} and effective depth d and the method set out in sections 4.4.3 applies. The maximum moment of resistance when $\lambda x = h_f$ is equal to

$$M_{flange} = \eta \ f_{cd} \ b_{eff} \ h_f(d - h_f/2)$$

Thus if the design moment $M \leq M_{flange}$, then design the beam as singly reinforced rectangular section b × d.

4.6.3 Stress Block Extends into the Web

As shown in Fig. 4.15, the compression forces are as follows:
In the flange of width $(b_{eff} - b_w)$, the compression force C_1 is

$$C_1 = \eta \ f_{cd} \ (b_{eff} - b_w) \ h_f$$

In the web, the compression force C_2 is

$$C_2 = \eta \ f_{cd} \ b_w \ \lambda x$$

The corresponding lever arms about the tension steel are

$$z_1 = d - h_f/2$$
$$z_2 = (d - \lambda x \ /2)$$

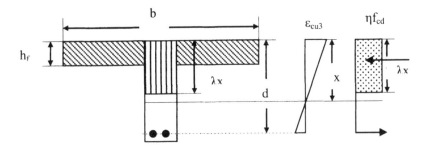

Fig. 4.15 T-beam with the stress block extending into the web.

Taking moments about tension steel, the moment of resistance M_R is given by

$$M_R = C_1 \ z_1 + C_2 \ z_2$$
$$M_R = \eta \ f_{cd} \ (b - b_w) \ h_f(d - h_f/2) + \eta \ f_{cd} \ b_w \ \lambda x \ (d - \lambda x \ /2)$$

From equilibrium,

$$T = A_s \ f_s = C_1 + C_2$$

If the amount of steel provided is sufficient to cause yielding of the steel, then $f_s = 0.87 f_{yk}$. The maximum moment of resistance without any compression steel is when $x = x_u$ as shown in Table 4.5. The maximum moment of resistance is

$$\frac{M_{max}}{b_{eff}d^2f_{cd}} = \eta\{1-\frac{b_w}{b_{eff}}\}\frac{h_f}{d}\{1-\frac{h_f}{2d}\}+\eta\lambda\frac{b_w}{b_{eff}}\frac{x_u}{d}\{1-0.5\lambda\frac{x_u}{d}\}$$

If $M_{flange} < M \le M_{max}$, then determine the value of x from

$$\frac{M}{b_{eff}d^2f_{cd}} = \eta\{1-\frac{b_w}{b_{eff}}\}\frac{h_f}{d}\{1-\frac{h_f}{2d}\}+\eta\lambda\frac{b_w}{b_{eff}}\frac{x}{d}\{1-0.5\lambda\frac{x}{d}\}$$

where $x \le x_u$ and the reinforcement required is obtained from the equilibrium condition,

$$A_s\, 0.87\, f_{yk} = C_1 + C_2$$

4.6.4 Steps in Reinforcement Calculation for a T-Beam or an L-Beam

- Calculate the total design load (including self–weight) and the corresponding design moment M using appropriate load factors.
- Calculate the maximum moment M_{flange} that can be resisted, when the entire flange is in compression.
$$M_{flange} = \eta\, f_{cd}\, b_{eff}\, h_f(d - h_f/2)$$
- Calculate the maximum moment that the section can withstand without requiring compression reinforcement.
$$\frac{M_{max}}{b_{eff}d^2f_{cd}} = \eta(1-\frac{b_w}{b_{eff}})\frac{h_f}{d}(1-\frac{h_f}{2d}) + \eta\lambda\frac{b_w}{b_{eff}}\frac{x_u}{d}(1-0.5\lambda\frac{x_u}{d})$$
- If $M \le M_{flange}$, then design as a rectangular beam of dimensions, $b \times d$.
  If $M_{flange} < M \le M_{max}$, then solve the following quadratic equation in (x/d)
$$\frac{M}{b_{eff}d^2f_{cd}} = \eta\{1-\frac{b_w}{b_{eff}}\}\frac{h_f}{d}\{1-\frac{h_f}{2d}\} + \eta\lambda\frac{b_w}{b_{eff}}\frac{x}{d}\{1-0.5\lambda\frac{x}{d}\}$$
- Check $x \le x_u$ and the reinforcement required is obtained from
$$A_s\, 0.87\, f_{yk} = C_1 + C_2$$
- If $M > M_{max}$, then compression steel is required or the section has to be revised. Compression steel is rarely required in the case of flanged beams because of the large compression area provided by the flange.

4.6.5 Examples of Design of Flanged Beams

Example 1: A continuous slab 100 mm thick is carried on T-beams at 2 m centres. The overall depth of the beam is 350 mm and the breadth b_w of the web is 250 mm. The 6 m span beams are simply supported. The characteristic dead load including self–weight and finishes is 7.4 kN/m² and the characteristic imposed load is 5 kN/m². Design the beam using the simplified stress block. The material strengths are $f_{ck} = 25$ MPa concrete and $f_{yk} = 500$ MPa reinforcement.
Since the beams are spaced at 2 m centres, the loads on the beam are:

Dead load = 7.4 × 2 = 14.8 kN/m
Live load = 5 × 2 = 10 kN/m

Design load = $(1.35 \times 14.8) + (1.5 \times 10) = 35.0$ kN/m
Ultimate moment at mid-span = $35.0 \times 6^2/8 = 157.4$ kN m
$b = b_1 + b_2 + b_w$ = spacing of beam = 2000 mm
$b_1 = b_2 = (2000 - 250)/2 = 875$ mm
ℓ_0 = Span of simply supported beam = 6000 mm
$b_{effe,1} = b_{effe,2} = 0.2\, b_1 + 0.1\, \ell_0 \le 0.2\ell_0$
$= 0.2 \times 875 + 0.1 \times 6000 = 775$ mm
$b_{effe,1} = b_{effe,2} = 775 \le (b_1 = b_2 = 875)$
Therefore $b_{effe,1} = b_{effe,2} = 775$
$b_{eff} = b_{eff,1} + b_{eff,2} + b_w = 775 + 775 + 250 = 1800 \le (b = 2000$ mm$)$
Effective width b_{eff} of flange = 1800 mm

Assuming a nominal cover on the links is 25 mm and if the links are H8 bars and the main bars are H25, then

$d = 350 - 25 - 8 - 25/2 = 304.5$ mm, say 300 mm.

First of all check if the beam can be designed as a rectangular beam by calculating M_{flange}.

$$M_{flange} = \eta\, f_{cd}\, b_{eff}\, h_f\, (d - h_f/2)$$

$f_{ck} = 25$ MPa, $f_{cd} = 25/1.5 = 16.7$ MPa, $\eta = 1$, $\lambda = 0.8, \delta = 1$ (simply supported beam)
$M_{flange} = 16.67 \times 1800 \times 100 \times (300 - 0.5 \times 100) \times 10^{-6} = 750.2$ kNm

The design moment of 165 kNm is less than M_{flange}. The beam can be designed as a rectangular beam of size 1800×300.

$$k = M/(b\, d^2\, f_{ck}) = 165 \times 10^6/(1800 \times 300^2 \times 25) = 0.041$$

$$\frac{z}{d} = 0.5[1.0 + \sqrt{(1 - 3\frac{k}{\eta})}] = 0.5[1.0 + \sqrt{(1 - 3\frac{0.041}{1.0})}] = 0.97$$

$$A_s = \frac{M}{0.97\, d \times 0.87\, f_{yk}} = \frac{165 \times 10^6}{0.97 \times 300 \times 0.87 \times 500} = 1303\, mm^2$$

Provide 3H25; $A_s = 1472\ mm^2$

Example 2: Determine the area of reinforcement required for the simply supported T-beam shown in Fig. 4.16. The dimensions of the beam are:
Effective width, $b = 600$ mm, $b_w = 250$ mm, $d = 340$ mm, $h_f = 100$ mm.
The beam is subjected to an ultimate moment of 305 kNm. The material strengths are $f_{ck} = 25$ MPa concrete and $f_{yk} = 500$ MPa reinforcement.
Note: $\eta = 1$, $\lambda = 0.8$, $\delta = 1$, $f_{cd} = 25/1.5 = 16.7$ MPa, $x_u/d = 0.448$ from Table 4.5.
Calculate M_{flange} to check if the stress block is inside the flange or not.

$$M_{flange} = \eta\, f_{cd}\, b_{eff}\, h_f\, (d - h_f/2)$$

$M_{flange} = 16.7 \times 600 \times 100 \times (340 - 0.5 \times 100) \times 10^{-6} = 290.6$ kNm

The design moment of 260 kNm is greater than M_{flange}. Therefore the stress block extends into the web.

Check if compression steel is required, although it is unlikely to be the case.

$$\frac{M_{max}}{b_{eff}\, d^2\, f_{cd}} = \eta\{1 - \frac{b_w}{b_{eff}}\}\frac{h_f}{d}\{1 - \frac{h_f}{2d}\} + \eta\lambda\frac{b_w}{b_{eff}}\frac{x_u}{d}\{1 - 0.5\lambda\frac{x_u}{d}\}$$

$$\frac{M_{max}}{b_{eff}d^2f_{cd}} = \{1-\frac{250}{600}\}\times\frac{100}{340}\times\{1-\frac{100}{2\times340}\} + 0.8\times\frac{250}{600}\times0.448\times\{1-0.5\times0.8\times0.448\}$$

$$= 0.583\times0.294\times0.853+0.8\times0.417\times0.448\times0.821$$

$$=0.269$$

$$M_{max} = 0.269\times600\times340^2\times16.67\times10^{-6}=311.0\,kNm$$

$$(M_{flange} = 290.6) < (M = 305) < (M_{max} = 311)$$

The beam can be designed without any need for compression steel.

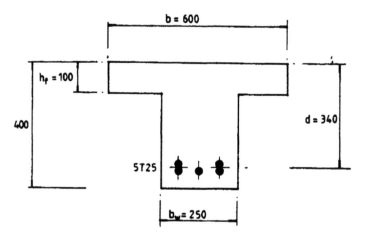

Fig. 4.16 Cross section of T-beam.

Determine the depth of the neutral axis from

$$\frac{M}{b_{eff}d^2f_{cd}} = \eta\{1-\frac{b_w}{b_{eff}}\}\frac{h_f}{d}\{1-\frac{h_f}{2d}\} + \eta\lambda\frac{b_w}{b_{eff}}\frac{x}{d}\{1-0.5\lambda\frac{x}{d}\}$$

$$\frac{305\times10^6}{600\times340^2\times16.67} = \{1-\frac{250}{600}\}\times\frac{100}{340}\times\{1-\frac{100}{2\times340}\} + 0.8\times\frac{250}{600}\frac{x}{d}\{1-0.5\times0.8\times\frac{x}{d}\}$$

$$0.264 = 0.1463 + 0.333\,(x/d) - 0.133\,(x/d)^2$$

Simplifying

$$(x/d)^2 - 2.50\,(x/d) + 0.885 = 0$$

Solving the quadratic in (x/d), $(x/d) = \dfrac{[2.50-\sqrt{(2.50^2-4\times0.885)}]}{2}$

$$x/d = (2.50 - 1.646)/2 = 0.427 < 0.448$$

$$x = 0.427 \times 340 = 145 \text{ mm}$$

$$C_1 = \eta\,f_{cd}\,(b - b_w)\,h_f = 16.67 \times (600 - 250) \times 100 \times 10^{-3} = 583.45 \text{ kN}$$

$$C_2 = \eta\,f_{cd}\,b_w\,\lambda x = 16.67 \times 250 \times 0.8 \times 145 \times 10^{-3} = 483.43 \text{ kN}$$

$$T = 0.87\,f_{yk}\,A_s = C_1 + C_2$$

$$0.87 \times 500 \times A_s = (583.45 + 483.43) \times 10^3$$

$$A_s = 2453 \text{ mm}^2$$

Provide 5H25, A = 2454 mm^2 as shown in Fig. 4.12.

4.7 CHECKING EXISTING SECTIONS

In the previous sections, methods have been described for designing rectangular and flanged sections for a given moment. In practice it may be necessary to calculate the ultimate moment capacity of a given section. This situation often occurs when there is change of use in a building and the owner wants to see if the structure will be suitable for the new purpose. Often moment capacity can be increased either by
- Increasing the effective depth. This can be done by adding a well bonded layer of concrete at the top of the beam/slab.
- Increasing the area of tension steel by bonding steel plates to the bottom of the beam.

4.7.1 Examples of Checking for Moment Capacity

Example 1: Calculate the moment of resistance of the singly reinforced beam section shown in Fig. 4.17. The material strengths are f_{ck} = 25 MPa concrete and f_{yk} = 500 MPa reinforcement. The tension reinforcement is 4H20 giving A_s = 1256 mm^2.

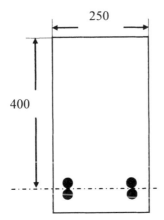

Fig. 4.17 Cross section of rectangular beam.

Solution: Assuming that tension steel yields, total tensile force T is given by
$$T = 0.87\, f_{yk}\, A_s = 0.87 \times 500 \times 1256 \times 10^{-3} = 546.4 \text{ kN}$$
If the neutral axis depth is x, then the compression force C is, taking λ = 0.8 and η = 1

$$C = \eta f_{cd}\,(\lambda x \times b) = \frac{25}{1.5} \times 0.8 \times x \times 250 \times 10^{-3} = 4.167\,x\;\text{kN}$$

For equilibrium, T = C. Solving for x

$$x = 131\;\text{mm} < (0.448\,d = 179\;\text{mm})$$

Check the strain in steel

$$\varepsilon_s = \frac{\varepsilon_{cu2} = 0.0035}{x}(d-x) = \frac{0.0035}{131}(400-131) = 0.007 > \{\text{yield strain} = \frac{0.87 \times 500}{200 \times 10^3}\}$$

Steel yields. Therefore the initial assumption is valid.

$$z = d - 0.5 \times \lambda \times x = 400 - 0.5 \times 0.8 \times 131 = 348\;\text{mm}$$

Moment of resistance M

$$M = T\,z = 546.4 \times 348 \times 10^{-3} = 190.2\;\text{kNm}$$

Example 2: Determine the ultimate moment capacity of the beam in Fig. 4.17, except, $A_s = 6T20 = 1885\;\text{mm}^2$.

Proceeding as in Example 1, assume that steel yields and calculate

$$T = 0.87\,f_{yk}\,A_s = 0.87 \times 500 \times 1885 \times 10^{-3} = 819.98\;\text{kN}$$

$$C = \eta f_{cd}\,(\lambda x \times b) = \frac{25}{1.5} \times 0.8 \times x \times 250 \times 10^{-3} = 4.167\,x\;\text{kN}$$

For equilibrium, T = C. Solving for x

$$x = 197\;\text{mm} > (0.448\,d = 179\;\text{mm})$$

Check the strain in steel

$$\varepsilon_s = \frac{\varepsilon_{cu2} = 0.0035}{x}(d-x) = \frac{0.0035}{197}(400-197) = 0.0036 > (\text{yield strain} = 0.0022)$$

Although the strain in steel is larger than the yield strain, in order to ensure sufficient ductility, the code limits the neutral axis depth to 0.448 d = 179 mm. Using this value of x,

$$C = \eta f_{cd}\,(\lambda x \times b) = \frac{25}{1.5} \times 0.8 \times 179 \times 250 \times 10^{-3} = 596.7\;\text{kN}$$

$$\text{Lever arm} = d - 0.5\,\lambda x = 400 - 0.5 \times 0.8 \times 179 = 328\;\text{mm}$$

$$M = C \times z = 596.7 \times 328 \times 10^{-3} = 195.7\;\text{kNm}$$

Example 3: Calculate the moment of resistance of the beam section shown in Fig. 4.18. The material strengths are $f_{ck} = 25$ MPa concrete and $f_{yk} = 500$ MPa reinforcement. $A_s = 4T25 = 1963\;\text{mm}^2$, $A_s = 2T20 + T16 = 829\;\text{mm}^2$.

Solution: Assume that both tension and compression steels yield and calculate the tension force T and compression force C_s in the steels.

$$T = 0.87\,f_{yk}\,A_s = 0.87 \times 500 \times 1963 \times 10^{-3} = 853.9\;\text{kN}$$

$$C_s = 0.87\,f_{yk}\,A_s = 0.87 \times 500 \times 829 \times 10^{-3} = 360.6\;\text{kN}$$

The compression force in concrete is

$$C_c = \eta f_{cd}\,b\,\lambda x = 1.0 \times \frac{25}{1.5} \times 250 \times 0.8 \times x \times 10^{-3} = 3.33\,x\;\text{kN}$$

For equilibrium, $C_c + C_s = T$.

$$3.33\,x + 360.6 = 853.9.$$

Solving x = 148 mm, x/d = 0.42 < 0.448

Calculate strain in tension and compression steels to verify the assumption.

$$\varepsilon_s = 0.0035\frac{(d-x)}{x} = 0.0035\frac{(350-148)}{148} = 0.0048$$

$$\varepsilon_s' = 0.0035\frac{(x-d')}{x} = 0.0035\frac{(148-50)}{148} = 0.0023$$

Both strains are larger than yield strain of 0.0022. Therefore both steels yield and the initial assumption is correct. The neutral axis depth is less than 0.448 d.

$$C_c = \eta f_{cd}\, b\, \lambda x = \frac{25}{1.5}\times 250\times 0.8\times 148\times 10^{-3} = 493.3\,\text{kN}$$

Taking moments about the tension steel, $M = C_c\,(d-0.5\,\lambda x) + C_s\,(d-d')$.

$$M = 493.3\times(350-0.5\times0.8\times148)\times10^{-3} + 360.6\times(350-50)\times10^{-3} = 251.6\ \text{kNm}$$

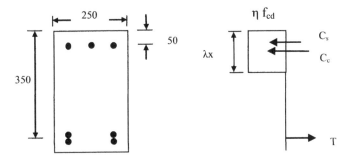

Fig. 4.18 Cross section of doubly reinforced beam.

4.7.2 Strain Compatibility Method

In the previous section, examples were given for calculating the moment of resistance of a given section. It required making initial assumptions about whether the compression and tension steels yield or not. After calculating the neutral axis depth from equilibrium considerations, strains in tension and compression steels are calculated to validate the assumptions. The problem can become complicated if say tension steel yields while the compression steel does not yield. A general approach in this case is the method of strain compatibility which has the advantage of avoiding the algebraic approach. The basic idea is to assume a neutral axis depth. From the assumed value of neutral axis depth, strains in steel in compression and tension are calculated. Thus

$$\varepsilon_s = \frac{\varepsilon_{cu3}}{x}(d-x),\ f_s = E\varepsilon_s \leq 0.97 f_{yk}$$

$$\varepsilon_s' = \frac{\varepsilon_{cu3}}{x}(x-d'),\ f_s' = E\varepsilon_s' \leq 0.87 f_{yk}$$

From the stresses, calculate the forces
$$T = A_s \, f_s, \quad C_s = A_s' \, f_s', \quad C_c = \eta \, f_{cd} \, b \, \lambda x, \quad C = C_s + C_c$$
For equilibrium, $T = C$. If equilibrium is not satisfied, then adjust the value of x and repeat until equilibrium is established. Normally only two sets of calculations for neutral axis depth are required. Linear interpolation can be used to find the appropriate value of x to satisfy equilibrium. The following example illustrates the method.

4.7.2.1 Example of Strain Compatibility Method

Example 1: Calculate the moment capacity of the section with $b = 250$ mm, $d = 350$ mm, $d' = 50$ mm, $f_{ck} = 25$ MPa and $f_{yk} = 500$ MPa,
$A_s' = 3H20 = 942.5$ mm^2, $A_s = 6H25 = 2945.2$ mm^2

Trial 1: Assume $x = 220$ mm, $\lambda = 0.8$, $\eta = 1$, $\delta = 1$, $f_{cd} = 25/1.5 = 16.67$ MPa
Strain ε_s' in compression steel is given by
$$\varepsilon_s' = 0.0035(x - d')/x = 0.0035 \times (220 - 50)/220 = 0.0027 > 0.0022$$
Therefore compression steel yields and the stress f_s' is equal to $0.87 \, f_{yk}$
Similarly, strain ε_s in tension steel is given by
$$\varepsilon_s = 0.0035(d - x)/x = 0.0035 \times (350 - 220)/220 = 0.00207 < 0.0022$$
Therefore tension steel does not yield and the stress f_s is equal to
$$f_s = \varepsilon_s \, E_s = 0.00207 \times 200 \times 10^3 = 413.6 \text{ MPa}$$
$$T = A_s \times f_s = 2945.2 \times 413.6 \times 10^{-3} = 1218.1 \text{ kN}$$
$$C = f_{cd} \times b \times 0.8 \, x + A_s' \times f_s'$$
$$C = \{16.67 \times 250 \times 0.8 \times 220 + 942.5 \times 0.87 \times 500\} \times 10^{-3}$$
$$C = (733.48 + 410.0) = 1043.5 \text{ kN}$$
$$T - C = 74.6 \text{ kN}$$
Total tensile force T is greater than the total compressive force C. Therefore increase the value of x in order to increase the compression area of concrete and also reduce the strain in tension steel but increase the strain in compression steel.

Trial 2: Assume $x = 240$ mm say
Strain ε_s' in compression steel is given by
$$\varepsilon_s' = 0.0035(x - d')/x = 0.00277 > 0.0022$$
Therefore compression steel yields and the stress f_s' is equal to $0.87 \, f_{yk}$.
Similarly, strain ε_s in tension steel is given by
$$\varepsilon_s = 0.0035(d - x)/x = 0.0016 < 0.0022$$
Therefore tension steel does not yield and the stress f_s is equal to
$$f_s = \varepsilon_s \, E = 0.001604 \times 200 \times 10^3 = 320.8 \text{ MPa}$$
$$T = A_s \times f_s = 2945.2 \times 320.8 \times 10^{-3} = 944.8 \text{ kN}$$
$$C = f_{cd} \times b \times 0.8 \, x + A_s' \times f_s'$$
$$C = \{16.67 \times 250 \times 0.8 \times 240 + 942.5 \times 0.87 \times 500\} \times 10^{-3}$$
$$C = (800.16 + 410.0) = 1210.2 \text{ kN}$$
$$T - C = -265.36 \text{ kN}$$

As shown in Fig. 4.19, linearly interpolate between x = 220 and 240 to obtain the value of x giving $T - C = 0$.

$$x = 220 + (240 - 220) \times (74.6)/(74.6 + 265.36) = 224 \text{ mm}$$
$$x/d = 224/350 = 0.64 > 0.448$$

As a check calculate T and C for x = 224 mm.

Strain ε_s' in compression steel is given by

$$\varepsilon_s' = 0.0035(x - d')/x = 0.0027 > 0.0022$$

Therefore compression steel yields and the stress f_s' is equal to $0.95 f_{yk}$.

Similarly, strain ε_s in tension steel is given by

$$\varepsilon_s = 0.0035(d - x)/x = 0.0018 < 0.0022$$

Therefore tension steel does not yield and the stress f_s is equal to

$$f_s = \varepsilon_s E = 0.0018 \times 200 \times 10^3 = 360 \text{ MPa}$$
$$T = A_s \times f_s = 2945.2 \times 360 \times 10^{-3} = 1160.3 \text{ kN}$$
$$C = f_{cd} \times b \times 0.8 x + A_s' \times f_s'$$
$$C = \{16.67 \times 250 \times 0.8 \times 224 + 942.5 \times 0.87 \times 500\} \times 10^{-3}$$
$$C = (746.8 + 410.0) = 1156.8 \text{ kN}$$
$$T - C = 3.5 \text{ kN}$$

This is close enough to be zero.

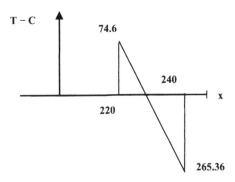

Fig. 4.19 Linear interpolation.

Taking moments about the tension steel, the lever arm for compression force in concrete is $(d - 0.5 \lambda x)$ and for the compression force in steel it is $(d - d')$.

$$M = \{746.8 \times (350 - 0.5 \times 0.8 \times 224) + 410.0 \times (350 - 50)\} \times 10^{-3} = 317.5 \text{ kNm}$$

Since x/d > 0.448, it is sensible to limit the permissible ultimate moment to a value less than 317.5 kNm. Limiting x to x = 0.448 d = 157 mm,

$$C_c = C = f_{cd} \times b \times 0.8 x = \{16.67 \times 250 \times 0.8 \times 157\} \times 10^{-3} = 523.4 \text{ kN}$$
$$\text{Lever arm } z_c = d - 0.5 \lambda x = 287 \text{ mm}$$

Strain ε_s' in compression steel is given by $\varepsilon_s' = 0.0035(x - d')/x = 0.0024 > 0.0022$.

Therefore compression steel yields and the stress f_s' is equal to $0.87 f_{yk}$.

$$C_s = \{942.5 \times 0.87 \times 500\} \times 10^{-3} = 410.0 \text{ kN}, \text{ Lever arm } z_s = d - d' = 300 \text{ mm}$$

Taking moments about the steel centroid, $M = C_c z_c + C_s z_s = 273.2 \text{ kNm}$.

4.8 REFERENCE

Institution of Structural Engineers. (2006). *Standard Method of Detailing Structural Concrete*, 3rd ed. London.

CHAPTER 5

SHEAR, BOND AND TORSION

5.1 SHEAR FORCES

In beams, a change in bending moment involves shear forces. Shear force at a section gives rise to diagonal tension in the concrete and leads to cracking. Shear failures are very brittle and therefore should be avoided. All beams should always be designed to fail in a ductile manner in flexure rather than in shear.

5.1.1 Shear in a Homogeneous Beam

According to engineers' theory of bending, in a beam a state of pure shear stress exists at the neutral axis. This causes principal tensile and compressive stresses of the same magnitude as the shear stress and inclined at 45° to the neutral axis. This is shown in Fig. 5.1(b) and Fig. 5.1(c) on an element at the neutral axis.
In an *elastic* rectangular beam shown in Fig. 5.1(a), the distribution of shear stress is parabolic as shown in Fig. 5.1(d). The maximum elastic shear stress at the neutral axis is given by

$$v_{max} = 1.5 \frac{V}{bh}$$

where V = shear force at the section.
In a T-beam or an L-beam, most of the shear force is resisted by the web and therefore for all practical purposes in shear calculations, flanged beams can be considered as rectangular beams of dimensions $b_w \times h$, where b_w = width of the web.

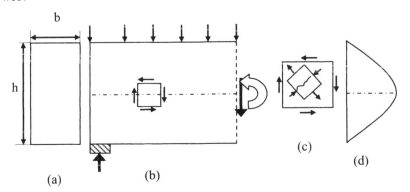

Fig. 5.1 (a) cross section; (b) beam; (c) enlarged element; (d) shear stress distribution.

5.1.2 Shear in a Reinforced Concrete Beam without Shear Reinforcement

(a) Shear failure
Shear in a reinforced concrete beam without shear reinforcement causes cracks on inclined planes near the support as shown in Fig. 5.2.

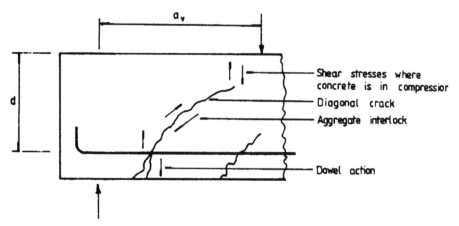

Fig. 5.2 Different actions contributing to shear strength.

The cracks are caused by the diagonal tensile stress mentioned above. The shear failure mechanism is complex and depends on the shear span a_v to effective depth d ratio (a_v/d). Shear span a_v is defined as the distance between the support and the major concentrated load acting on the span. When this ratio is large, the failure is as shown in Fig. 5.2.

The following actions form the three mechanisms resisting shear in the beam:
  a. Shear stresses in the compression zone resisted by uncracked concrete.
  b. Aggregate interlock along the cracks: Although cracks exist in the web due to tensile stresses caused by shear stresses, the width of the cracks is not large enough prevent frictional forces between cracked surfaces. These frictional forces exist along the cracked surfaces and contribute to resisting shear force.
  c. Dowel action in the bars where the concrete between the cracks transmits shear forces to the bars.

(b) Shear capacity
An accurate analysis for shear strength is not possible. The problem has been solved by testing beams of the type normally used in practice. Shear strength depends on several factors such as
  •   The percentage of flexural steel in the member. This affects the shear capacity by restraining the width of the cracks and thus enhancing the shear carried by the aggregate interlock along the cracks. It also naturally increases the shear capacity due to dowel action and increase the depth of the section in compression.

- Compression strength of concrete strength: It affects by increasing the aggregate interlock capacity and also the shear capacity of the uncracked portion of the beam.
- Type of aggregate: This affects the shear resisted by aggregate interlock. For example, lightweight aggregate concrete has approximately 20% lower shear capacity compared to normal weight concrete.
- Effective depth: Tests indicate that deeper beams have proportionally lower shear capacity compared to shallow beams. The reason for this is not clear but it is thought it might have some thing to do with lower aggregate interlock capacity.
- Restraining the tension steel separating from concrete by providing vertical links improves shear capacity by increasing dowel action.

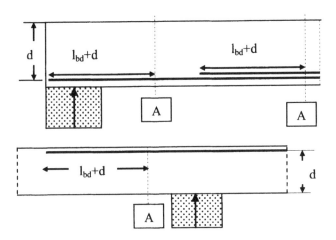

Fig. 5.3 Definition of A_{sl} at section A.

The shear capacity of a member without any shear reinforcement is given by code equations (6.2a), (6.2b) and (6.3N).

$$V_{Rd,c} = [C_{Rd,c}\, k\{100\rho_1\, f_{ck}\}^{1/3} + k_1\, \sigma_{cp}]b_w d \geq [v_{min} + k_1\, \sigma_{cp}]b_w d \quad (6.2a)$$

$$C_{Rd,c} = \frac{0.18}{(\gamma_c = 1.5)} = 0.12$$

$$k = 1 + \sqrt{\frac{200}{d}} \leq 2.0$$

$$\rho_1 = \frac{A_{sl}}{b_w\, d} \leq 0.02$$

A_{sl} = Area of tensile reinforcement which extends a length of (design anchorage length l_{bd} + effective depth) beyond the section where the shear capacity is being calculated as shown in Fig. 5.3.

$$v_{min} = 0.035\, k^{1.5}\, \sqrt{f_{ck}} \quad (6.3N)$$

$$k_1 = 0.15$$

σ_{cp} = Axial force/Area of concrete cross section

In the above equations, $V_{RD, c}$ is in Newtons, f_{ck} in MPa, all linear dimensions are in mm.

(c) Example

Calculate the shear capacity of a rectangular beam 250 × 450 mm, effective depth d = 400 mm reinforced at a section with 3H20 bars. f_{ck} = 25 MPa, σ_{cp} = 0. It may be assumed that the bars extend a length l_{bd} + d beyond the section under consideration. B_w = 250 mm, d = 400 mm, A_{sl} = 3 × 314 = 943 mm^2.

$$C_{Rd,c} = \frac{0.18}{(\gamma_c = 1.5)} = 0.12$$

$$k = 1 + \sqrt{\frac{200}{400}} = 1.71 \le 2.0$$

$$100\rho_1 = 100 \times \frac{943}{250 \times 400} = 0.94 \le 2.0$$

$$v_{min} = 0.035 \times 1.71^{1.5} \times \sqrt{25} = 0.39 MPa$$

$$V_{Rd,c} = [0.12 \times 1.71 \times \{0.94 \times 25\}^{1/3} + 0.15 \times 0] \times 250 \times 400 \times 10^{-3}$$

$$\ge [0.39 + 0.15 \times 0] \times 250 \times 400 \times 10^{-3}$$

$$V_{Rd,c} = 58.77 \ge 39.0 \text{ kN}$$

5.1.3 Shear Reinforcement in the Form of Links

Fig. 5.4 shows a reinforced concrete beam under third point loading. Due to the loading, bending and shear stresses act at all points in the beam. In areas where bending stresses dominate, vertical flexural tensile cracks develop. In areas where shear stresses dominate as at neutral axis, inclined tension cracks develop due to diagonal principal tensile stress caused by shear stress. At sections where both bending and shear are of equal importance, cracks which start as vertical tension cracks due to bending become inclined as they move up due to the action of shear stresses.

When sufficient inclined cracks form as shown in Fig. 5.5, concrete between cracks acts as concrete struts.

If vertical steel stirrup reinforcements are provided, the combination of vertical steel stirrups and inclined concrete struts together form the web of a composite truss whose tension chord is steel reinforcement and the compression chord is uncracked concrete as shown in Fig. 5.6. It is important to appreciate that this is a 'smeared truss' in the sense that both steel and concrete web members are not discrete members. If A_{sw} is the area of one steel stirrup and the spacing is s along the span, then the vertical force provided by the stirrups is $A_{sw} f_{yk}/s$ per unit length.

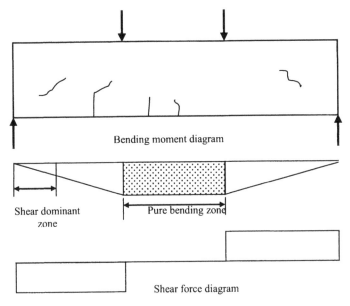

Bending moment diagram

Shear dominant zone

Pure bending zone

Shear force diagram

Fig. 5.4 Beam under third point loading.

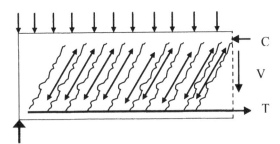

Fig. 5.5 A cracked reinforced concrete beam.

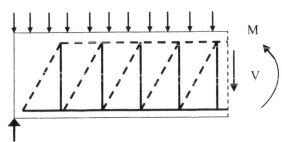

Fig. 5.6 A composite truss.

In this model of shear resistance, it is assumed that at ultimate loads, the steel stirrups yield and concrete struts **do not** crush.

Note that bending moment alone creates a force of M/z where z is the lever arm in the tension and compression chords. However, the smeared truss idealization introduces an additional force in the tension and compression chords due to shear force V. This can be seen from the force analysis of the truss shown in Fig. 5.7. The depth of the truss is z, the web members are inclined to the horizontal by θ and the panel width is z cot θ.

The force analysis of the truss shows that the tensile forces in the bottom chord are V cot θ, 2V cot θ, 3V cot θ and so on. The bending moments in the middle of each panel are 0.5V z cot θ, 1.5V z cot θ, 2.5V z cot θ and so on. Dividing the bending moment by the lever arm z, the force in the top and bottom chords due to bending moment is 0.5V cot θ, 1.5V cot θ, 2.5V cot θ and so on. The force in the top and bottom chords due to shear force alone is the difference between the total force and the force due to bending moment. This additional force is equal to 0.5V cot θ in the top and bottom chords. This has to be allowed for in the design of tension reinforcement.

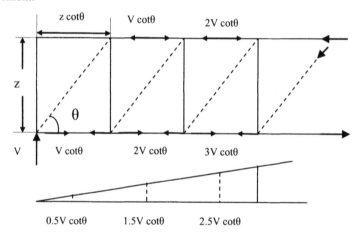

Bending moment distribution

Fig. 5.7 Additional tensile force due to shear force.

5.1.4 Derivation of Eurocode 2 Shear Design Equations

Fig. 5.8 shows the idealized truss with a cut section parallel to the concrete compression struts. Let z be the lever arm, the distance between the compression chord and longitudinal reinforcement acting as tension chord. If θ is the inclination of concrete struts to the horizontal and s is the spacing of the vertical shear links, the number of links in the distance z cot θ is z cot θ /s. If A_{sw} is the total area of shear link and f_{ywd} is the design yield stress of the link steel, the shear force V at the section is given by

$$V = \frac{z \cot \theta}{s} A_{sw} f_{ywd}$$

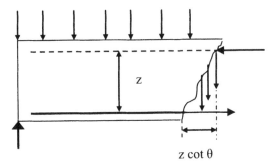

Fig. 5.8 Idealized truss with a cut section parallel to the struts.

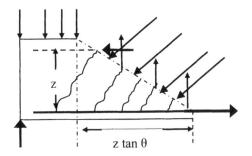

Fig. 5.9 Idealized truss with a cut section perpendicular to the struts.

Fig. 5.9 shows the same idealized truss but with a cut perpendicular to the struts. The length of the cut section is $z/\cos\theta$. If σ_c is the stress in the concrete struts, the total compressive force F_c parallel to the struts is

$$F_c = \sigma_c \, b_w \, \frac{z}{\cos\theta}$$

The total vertical tensile force F_s due to force in the stirrups is

$$F_s = \frac{z\tan\theta}{s} A_{sw} \, f_{ywd}$$

However, $V = \dfrac{z\cot\theta}{s} A_{sw} \, f_{ywd}$ can be expressed as

$$V\tan\theta = \frac{z}{s} A_{sw} \, f_{ywd}$$

F_s can now be expressed as

$$F_s = \frac{z\tan\theta}{s} A_{sw} \, f_{ywd} = V\tan^2\theta$$

From Fig. 5.8,

$$V = F_c \sin\theta - F_s$$

Replacing F_s by $V\tan^2\theta$

$$V = F_c \sin\theta - V\tan^2\theta$$

$$V (1 + \tan^2 \theta) = F_c \sin \theta$$

$$V = F_c \frac{\sin \theta}{(1 + \tan^2 \theta)}$$

$$= [\sigma_c b_w \frac{z}{\cos \theta}] \times \frac{\sin \theta}{(1 + \tan^2 \theta)}$$

$$= [\sigma_c b_w z] \times \frac{\tan \theta}{(1 + \tan^2 \theta)}$$

$$= \sigma_c b_w z \frac{1}{(\cot \theta + \tan \theta)}$$

The maximum shear force that can be resisted without crushing the concrete in the struts is given by

$$V_{Rd,max} = \sigma_c b_w z \frac{1}{(\cot \theta + \tan \theta)}$$

Code limits the concrete stress σ_c to

$$\sigma_c = \alpha_{cw} v_1 f_{cd}$$

where v_1 is an efficiency factor which allows for the effects of cracking as well for the actual distribution of stress in the struts. It is given by code equation (6.6N).

$$v_1 = 0.6 (1 - f_{ck}/250) \qquad (6.6N)$$
$$\alpha_{cw} = 1 \text{ for non-prestressed structures.} \qquad (6.10aN)$$

$V_{RD,\,max}$ is given by the code equation (6.9)

$$V_{RD,max} = \alpha_{cw} b_w z v_1 f_{cd} \frac{1}{(\cot \theta + \tan \theta)}$$

$$1 \leq \cot \theta \leq 2.5 \qquad (6.7N)$$

An approximate value for lever arm is z = 0.9 d. The value of b_w is the width of the web in T-beams. If the width of the web varies, then the minimum width between the tension and compression chords should be used as shown in Fig. 5.10.

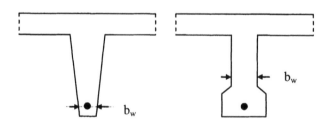

Fig. 5.10 Minimum web width b_w.

The shear reinforcement required is given by

$$V \tan \theta = \frac{z}{s} A_{sw} f_{ywd}$$

The expression for V can be written as

$$V = \frac{z}{s} A_{sw} f_{ywd} \cot \theta$$

In the code the above equation is the shear resistance due to stirrups alone and is given as equation (6.8).

$$V_{RD,s} = \frac{z}{s} A_{sw} f_{ywd} \cot \theta \qquad (6.8)$$

Table 5.1 Values of shear resistance

cot θ	$V_{RD,s}$	$V_{RD,max}$
1.0	$0.87(z/s) A_{sw} f_{ywk}$	$0.2(1 - f_{ck}/250) \, b_w \, z \, f_{ck}$
2.5	$2.175 \, (z/s) A_{sw} f_{ywk}$	$0.138(1 - f_{ck}/250) \, b_w \, z \, f_{ck}$

Table 5.1 shows the values of shear resistance of concrete and steel for the two extreme values of cot θ. Taking $f_{cd} = f_{ck}/ (\gamma_c = 1.5)$, $f_{ywd} = f_{yk}/ (\gamma_s = 1.15)$ and equating the value of shear resistance of concrete to the shear resistance of stirrups, the value of A_{sw}/s can be calculated. Clearly the minimum value of shear steel is needed when cot θ = 2.5 and the maximum value of shear steel is needed when cot θ = 1.0.

5.1.4.1 *Additional tension force due to shear in cracked concrete*

Fig. 5.11 shows a beam cracked in shear with concrete struts inclined at an angle to the beam axis.

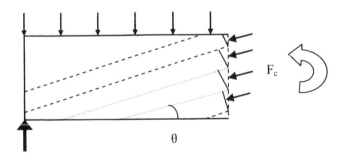

Fig. 5.11 Beam cracked in shear with concrete struts.

Taking a series of steps perpendicular to the struts, the total compressive across a vertical cut is

$$F_c = \sigma_c \, b \, h \cos \theta$$

where σ_c = compressive stress in concrete, b = width of the beam and h = depth of beam.
As shown in Fig. 5.12, the vertical component of F_c is equal to the shear force V_{Ed}.

$$V_{Ed} = F_c \sin \theta.$$

The horizontal component H is equal to $F_c \cos \theta$.

Expressing F_c in terms of V_{Ed},

$$H = F_c \cos\theta = (V_{Ed}/\sin\theta)\cos\theta = V_{Ed}\cot\theta.$$

The horizontal compressive force H is kept in equilibrium by tension forces in the top and bottom chords. The force in the bottom chord is approximately 0.5H. The additional tension force due to shear is therefore $0.5\,V_{Ed}\cot\theta$.

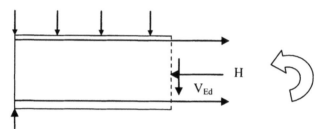

Fig. 5.12 Additional tension force in a beam cracked in shear with concrete struts.

In Eurocode 2 clause 6.2.3(7), the additional tensile force in the longitudinal reinforcement due to shear V_{Ed} in members with shear reinforcement is given by equation (6.18) as

$$\Delta V_{Ed} = 0.5\,V_{Ed}\,(\cot\theta - \cot\alpha) \qquad\qquad (6.18)$$

where θ = angle of the cracks and α = inclination of shear reinforcement to the vertical.

For members **with** shear reinforcement, the total tensile force F_s for which the reinforcement needs to be designed is given by

$$F_s = (M_{ED}/z + \Delta V_{Ed}) \le \text{Maximum moment along the beam } M_{ED,\,max}/z$$

In clause 9.2.1.3(2), Eurocode 2 suggests that this additional tensile force due to shear can be included by simply shifting the moment curve away from the section of maximum moment in the direction of decreasing moment a distance

$$a_1 = 0.5\,z\,(\cot\theta - \cot\alpha) \qquad\qquad (9.2)$$

The shifted bending moment diagram is used for design.

In the case of members **without** shear reinforcement, $a_1 = d$.

5.1.5 Minimum Shear Reinforcement

The code recommends in equations (9.4) and (9.5N) that minimum shear reinforcement must satisfy the condition

$$\frac{A_{sw}}{s\,b_w} \ge \frac{0.08\sqrt{f_{ck}}}{f_{yk}} \qquad\qquad (9.4) \text{ and } (9.5N)$$

In addition the maximum longitudinal spacing s is limited to 0.75 d according to equation (9.5N). The transverse spacing (across the width b_w) should be limited to $0.75\,d \le 600$ mm according to equation (9.8N).

5.1.6 Designing Shear Reinforcement

The steps in designing the shear reinforcement are as follows.
- Calculate the design shear force, V_{Ed}.
- Calculate $V_{RD, c}$. If $V_{Ed} < V_{RD, c}$, then no shear reinforcement is required but a minimum value should always be provided.
- If $V_{Ed} > V_{RD, c}$, shear reinforcement is required.
- Equate $V = V_{RD, max}$ and calculate the value of $\cot \theta$. Check that it is within the limits of 1.0 and 2.5. If it is outside the limits for minimum shear reinforcement, choose the maximum value within the limits and calculate the corresponding value of $V_{RD, max}$ and ensure that it is larger than V_{Ed}.
- Design the necessary shear reinforcement from the equation (6.8)

$$V_{RD,s} = \frac{z}{s} A_{sw} f_{ywd} \cot \theta \qquad (6.8)$$

- Check that the minimum reinforcement has been provided

$$\frac{A_{sw}}{s\, b_w} \geq \frac{0.08 \sqrt{f_{ck}}}{f_{yk}} \qquad (9.4) \text{ and } (9.5N)$$

Check the longitudinal and lateral spacings.

Example: At a cross section in a T-beam with flange width b = 600 mm, flange thickness h_f = 125 mm, effective depth d = 375 mm, width of web b_w = 200mm, flexural reinforcement is 2H32 mm bars, f_{ck} = 25 MPa, f_{ywk} = 500 MPa, design ultimate shear force V_{Ed} = 157.5 kN. Determine the spacing of 10 mm diameter links.

Solution:

i. Check if shear reinforcement is required, $V_{Ed} > V_{Rd, c}$.

V_{Ed} = 157.5 kN, b_w = 200 mm, d = 375 mm, A_{sl} = 2 × 804 = 1608 mm^2

$$C_{Rd,c} = \frac{0.18}{(\gamma_c = 1.5)} = 0.12$$

$$k = 1 + \sqrt{\frac{200}{375}} = 1.73 \leq 2.0$$

$$100\rho_1 = 100 \times \frac{1608}{200 \times 375} = 2.14 > 2.0$$

$$\text{Take } 100\rho_1 = 2.0$$

$$v_{min} = 0.035 \times 1.73^{1.5} \times \sqrt{25} = 0.39 \text{ MPa}$$

$$V_{Rd,c} = [0.12 \times 1.73 \times \{2.0 \times 25\}^{1/3} + 0.15 \times 0] \times 200 \times 375 \times 10^{-3}$$

$$\geq [0.39 + 0.15 \times 0] \times 200 \times 375 \times 10^{-3}$$

$$V_{Rd,c} = 57.35 \geq 29.3 \text{ kN}$$

$V_{Rd, c}$ = 57.35 kN < V_{Ed}. Therefore shear reinforcement is required.

ii. Check whether the section strength is adequate, $V_{Ed} < V_{Rd,max}$.

$$V_{Rd,max} = \alpha_{cw}\, b_w\, z\, v_1\, f_{cd}\, \frac{1}{(\cot\theta + \tan\theta)}$$

$$\cot\theta + \tan\theta = \frac{\cos\theta}{\sin\theta} + \frac{\sin\theta}{\cos\theta} = \frac{\cos^2\theta + \sin^2\theta}{\cos\theta\,\sin\theta} = \frac{1}{\cos\theta\,\sin\theta} = \frac{2}{\sin 2\theta}$$

$$V_{Rd,max} = \alpha_{cw}\, b_w\, z\, v_1\, f_{cd}\, \frac{\sin 2\theta}{2}$$

$$\sin 2\theta = \frac{2\,V_{Rd,max}}{\alpha_{cw}\, b_w\, z\, v_1\, f_{cd}}$$

$$\theta = 0.5\,\sin^{-1}\{\frac{2\,V_{Rd,max}}{\alpha_{cw}\, b_w\, z\, v_1\, f_{cd}}\}$$

Setting $V_{Rd,\ max} = V_{Ed}$,

$$\theta = 0.5\,\sin^{-1}\{\frac{2\,V_{Ed}}{\alpha_{cw}\, b_w\, z\, v_1\, f_{cd}}\}$$

$$\theta = 0.5\,\sin^{-1}\{\frac{2\times 157.5\times 10^3}{1.0\times 200\times (0.9\times 375)\times 0.6\times \dfrac{25}{1.5}}\}$$

$$= 0.5\sin^{-1}(0.467) = 0.5\times 27.8 = 13.9^0$$

$$\cot\theta = 4.04$$

The value of $\cot\theta$ is outside the limits of 1.0 and 2.5.
Choosing $\cot\theta = 2.5$ for minimum shear reinforcement, $V_{Rd,\ max} = 232.8$ kN $> V_{Ed}$
Section size is adequate.

iii. Design of shear reinforcement.
Ensure that $V_{Rd,\ s} \geq V_{Ed}$ and choosing 2-leg links of H8, $A_{sw} = 100.5$ mm^2, $\cot\theta = 2.5$, $z = 0.9d$, $f_{ywk} = 500$ MPa,

$$V_{Rd,s} = \frac{z}{s}\, A_{sw}\, f_{ywd}\, \cot\theta$$

$$= \frac{(0.9\times 375)}{s}\times 100.5\times \frac{500}{1.15}\times 2.5\times 10^{-3}$$

$$= \frac{36868}{s}\, kN \geq (V_{Ed} = 157.5\, kN)$$

$$s \leq 234\, mm$$

iv. Check minimum steel requirement.

$$\frac{A_{sw}}{s\, b_w} \geq \frac{0.08\sqrt{f_{ck}}}{f_{yk}} \qquad\qquad (9.4) \text{ and } (9.5N)$$

$$\frac{100.5}{s\times 200} \geq \frac{0.08\sqrt{25}}{500}$$

$$s \le \frac{100.5}{200} \times \frac{500}{0.08\sqrt{5}} = 628\,\text{mm}$$

Maximum spacing $s \le (0.75\,d = 0.75 \times 375 = 281$ mm).

Maximum spacing should be less than 280 mm. A spacing of 225 mm will be satisfactory. Note that for reasons of economy, it is always desirable to have the links at maximum spacing permitted as this will reduce the number of links.

5.1.7 Bent-up Bars as Shear Reinforcement

The most common method of providing shear reinforcement is in the form of links. A less common method is using bent–up bars. The reason bent-up bars are less popular is because of the increased cost of bending and fixing the reinforcement.

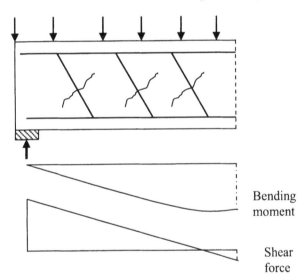

Fig. 5.13 Bent-up bars resisting shear force (simply supported end).

In the case of simply supported ends, the bending moment decreases but shear force increases towards the support. Normally the tension reinforcement is curtailed towards the supports. However, instead of curtailing the bottom tension reinforcement towards the supports, they can be bent *up* as shown in Fig. 5.13 to cross a potential shear crack and thus assist in resisting shear force.

A similar situation occurs at continuous supports as well. As shown in Fig. 5.14, both the bending moment as well as the shear force increase towards the supports. However as the bending moment causing tension at the top face decreases away from the support, the top tension steel can be bent *down* to act as shear reinforcement.

This is the main motivation for using bent-up bars as shear reinforcement. However, often there might not be sufficient bars to bend to maintain minimum spacing required.

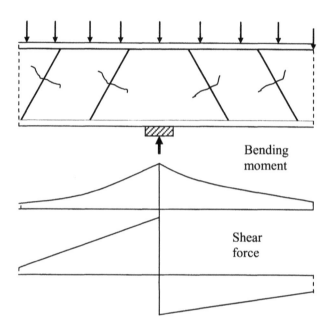

Fig. 5.14 Bent-up bars resisting shear force (continuous support).

Composite truss: As shown in Fig. 5.15, the composite truss is made up of bent-up bars and concrete struts. The concrete struts and the bent-up bars are inclined at angles θ and α respectively to the horizontal.

Taking a section parallel to the struts as shown in Fig. 5.16, the number of bent-up bars is $z(\cot\theta + \cot\alpha)/s$, where s is the spacing of the bent-up bars. The vertical component of the bars is therefore the shear resistance $V_{Rd,\,s}$ of the bent-up bars.

$$V_{Rd,s} = \frac{z(\cot\theta+\cot\alpha)}{s} A_{sw}\, f_{ywd}\sin\alpha \qquad (6.13)$$

Similarly taking a section perpendicular to the struts and proceeding as in section 5.1.4,

$$V = \sigma_c\, b_w\, z\frac{(\cot\theta+\cot\alpha)}{(1+\cot^2\theta)}$$

Replacing σ_c by $\alpha_{cw}\, \nu_1\, f_{cd}$ and V by $V_{Rd,\,max}$

$$V_{Rd,max} = \alpha_{cw}\, b_w\, z\nu_1\, f_{cd}\frac{(\cot\theta+\cot\alpha)}{(1+\cot^2\theta)} \qquad (6.14)$$

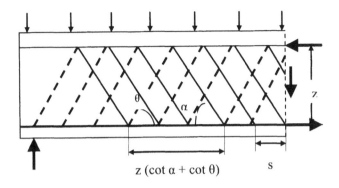

Fig. 5.15 Composite truss with bent-up bars and concrete struts.

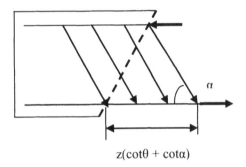

$$z(\cot\theta + \cot\alpha)$$

Fig. 5.16 A section parallel to the concrete struts.

5.1.7.1 Example of Design of Bent-up Bars and Link Reinforcement in Beams

Design shear reinforcement using a combination of shear links and bent-up bars for a rectangular beam b = 300 mm, d = 450 mm, f_{ck} = 25 MPa, f_{yk} = 500 MPa. The tension steel consists of 3H25 mm bars. Bent-up bars are H20 bars bent in pairs at an angle to the horizontal of 45^0 and at a spacing of 600 mm. The design shear force V_{Ed} = 320 kN.

Design shear reinforcement using one half of total shear force to be resisted by links and the other half by bent-up bars.

a. Link design

V_{Ed} = 0.5 ×320 = 160 kN, b_w = 300 mm, d = 450 mm, A_{sl} = 3 × 491 = 1473 mm^2

i. Check if shear reinforcement is required, $V_{Ed} > V_{Rd, c}$.

$$C_{Rd,c} = \frac{0.18}{(\gamma_c = 1.5)} = 0.12$$

$$k = 1 + \sqrt{\frac{200}{450}} = 1.67 \le 2.0$$

$$100\rho_1 = 100 \times \frac{1473}{300 \times 450} = 1.09 < 2.0$$

$$v_{min} = 0.035 \times 1.67^{1.5} \times \sqrt{25} = 0.38\,\text{MPa}$$

$$V_{Rd,c} = [0.12 \times 1.67 \times \{1.09 \times 25\}^{1/3} + 0.15 \times 0] \times 300 \times 450 \times 10^{-3}$$

$$= [0.60 + 0.15 \times 0] \times 300 \times 450 \times 10^{-3}$$

$$V_{Rd,c} = 81.4 \ge [0.38 + 0.15 \times 0] \times 300 \times 450 \times 10^{-3}\,\text{kN}$$

$$= 81.4 \ge 51.3$$

$V_{Rd,\,c} = 81.4\,\text{kN} < V_{Ed}$. Therefore shear reinforcement is required.

ii. Check the adequacy of the section, $V_{Rd,max} > V_{Ed}$.

$$\theta = 0.5\,\sin^{-1}\{\frac{2\,V_{Ed}}{\alpha_{cw}\,b_w\,z\,v_1\,f_{cd}}\}$$

$$\theta = 0.5\,\sin^{-1}\{\frac{2 \times 160 \times 10^3}{1.0 \times 300 \times (0.9 \times 450) \times 0.6 \times \dfrac{25}{1.5}}\}$$

$$= 0.5\,\sin^{-1}(0.263) = 0.5 \times 15.27 = 7.64^0$$

$$\cot\theta = 7.45$$

The value of $\cot\theta$ is outside the limits of 1.0 and 2.5.
Choosing $\cot\theta = 2.5$ for minimum shear reinforcement, $(V_{RD,\,max} = 419\,\text{kN}) > V_{Ed}$. Section size is adequate.

iii. Design of shear reinforcement

Ensure that $V_{Rd,\,s} \ge V_{Ed}$, and choose 2-leg links of 8 mm diameter, $A_{sw} = 100.5\,\text{mm}^2$, $\cot\theta = 2.5$, $z = 0.9d$, $f_{ywk} = 500\,\text{MPa}$,

$$V_{Rd,s} = \frac{z}{s}\,A_{sw}\,f_{ywd}\,\cot\theta$$

$$= \frac{(0.9 \times 450)}{s} \times 100.5 \times \frac{500}{1.15} \times 2.5 \times 10^{-3}$$

$$= \frac{17696}{s}\,\text{kN} \ge (V_{Ed} = 160\,\text{kN})$$

$$s \le 110\,\text{mm}$$

$$0.75\,d = 0.75 \times 450 = 338\,\text{mm}$$

Use a spacing of 100 mm.

iv. Check minimum area of links

$$\frac{A_{sw}}{s\,b_w} \geq \frac{0.08\sqrt{f_{ck}}}{f_{yk}}$$

$$\frac{A_{sw}}{s\,b_w} \geq [\frac{100.5}{100 \times 300} = 3.35\times10^{-3}] \geq [\frac{0.08\sqrt{25}}{500} = 0.83.35\times10^{-3}]$$

Minimum steel has been provided.

b. Bent-up bar design

i. Check if for bent-up bar case, $V_{Rd, max} > V_{Ed}$.

Assuming $\cot\theta = 2.5$ from link design and $\cot\alpha = 1.0$, $\alpha = 45^0$

$$V_{Rd,max} = \alpha_{cw}\, b_w\, z\, v_1\, f_{cd}\frac{(\cot\theta + \cot\alpha)}{(1+\cot^2\theta)}$$

$$= 1.0\times300\times(0.9\times450)\times0.6\times\frac{25}{1.5}\times\frac{(2.5+1)}{(1+2.5^2)}\times10^{-3}$$

$$= 1215\times\frac{(1+2.5)}{(1+2.5^2)}\,kN$$

$$= 586.6\ kN > (V_{Ed} = 160\,kN)$$

ii. Design of shear reinforcement

$A_{sw} = 2 \times 314 = 618\ mm^2$, $\cot\theta = 2.5$, $z = 0.9d$, $f_{ywk} = 500$ MPa, $s = 600$ mm, $\alpha = 45^0$

$$V_{Rd,s} = \frac{z}{s}A_{sw}\, f_{ywd}(\cot\theta + \cot\alpha)\times\sin\alpha$$

$$V_{Rd,s} = \frac{(0.9\times450)}{600}\times628\times\frac{500}{1.15}\times(2.5+1)\times0.7071\times10^{-3}$$

$$V_{Rd,s} = 456\,kN \geq (V_{Ed} = 160\,kN)$$

$s \leq 0.75$ d $(1+\cot\alpha) = 0.75 \times 450 \times (1+1) = 676$ mm which is greater than the spacing of 600 mm provided.

Note: The code is unclear about many aspects of combining vertical shear links and bent-up bars. For example there are two values for $V_{Rd,max}$, one for links steel and one for bent-up bars with the latter giving a much smaller value than the former. As it is clearly illogical to use two different values for $\cot\theta$; in the above example only one value of $\cot\theta$ calculated from $V_{RD, max}$ for link steel case is used.

5.1.8 Loads Applied Close to a Support

When loads are applied close to a support as shown in Fig. 5.17, a large proportion of the load is transferred to the support by strut action rather than by bending and shear action. Therefore one can take a reduced value of the load when designing shear reinforcement.

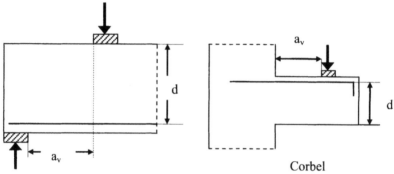

Fig. 5.17 Loads close to support.

Clause 6.2.2 (6) states that when loads are applied on the upper side within a distance $0.5 \leq a_v/d \leq 2.0$, where a_v is the distance from the edge of the support to the edge of the load as shown in Fig. 5.17 and d is the effective depth, the contribution of that load to shear force V_{Ed} may be reduced by a reduction factor β equal to 0.5 a_v/d. The reduction is applicable also when checking $V_{Rd, c}$.
If $a_v/d \leq 0.5$, then use $\beta = 0.25$.
In addition, V_{Ed} calculated without the reduction by β should always satisfy the condition given by code equation (6.5)

$$V_{Ed} \leq 0.5 b_w \, d \times \{0.6[1 - \frac{f_{ck}}{250}]\} \times f_{cd} \qquad (6.5)$$

In addition to the above requirement, V_{Ed} calculated with the reduction by β should always satisfy the condition given by code equation (6.19)

$$V_{Ed} \text{ reduced} \leq A_{sw} \, f_{ywd} \times \sin \alpha \qquad (6.19)$$

$\alpha = 90^0$ for links. A_{sw} is the total shear reinforcement crossing the inclined shear crack. Only the reinforcement in the central 0.75 a_v of the distance between the loaded areas should be taken into account.

5.1.8.1 Example

A corbel 350 mm wide and 500 mm deep is reinforced in tension by 2H25 mm bars to support a load of 400 kN. Assuming that $f_{yk} = 500$ MPa, $f_{ck} = 30$ MPa, effective depth d = 450 mm, $a_v = 600$ mm, design the necessary shear reinforcement.

$$a_v/d = 600/450 = 1.33$$
$$\beta = a_v/ (2d) = 1.33/2 = 0.67$$
$$V_{Ed} = \text{Reduced load for shear force} = \beta \times 400 = 268 \text{ kN}$$

i. Check that V_{Ed} satisfies equation (6.5).

$$V_{Ed} \leq 0.5 b_w \, d \times \{0.6[1 - \frac{f_{ck}}{250}]\} \times f_{cd}$$

$$269 \, \text{kN} \le 0.5 \times 350 \times 450 \times \{0.6[1 - \frac{30}{250}]\} \times \frac{30}{1.5} \times 10^{-3}$$

$$269 \text{ kN} \le 832 \, \text{kN}$$

ii. Check if shear reinforcement is needed, $V_{Rd,\, c} < V_{Ed}$.

$$A_{sl} = 2 \times 491 = 982 \text{ mm}^2$$

$$C_{Rd,c} = \frac{0.18}{(\gamma_c = 1.5)} = 0.12$$

$$k = 1 + \sqrt{\frac{200}{450}} = 1.67 \le 2.0$$

$$100\rho_1 = 100 \times \frac{982}{350 \times 450} = 0.62 < 2.0$$

$$v_{min} = 0.035 \times 1.67^{1.5} \times \sqrt{30} = 0.41 \, \text{MPa}$$

$$V_{Rd,c} = [0.12 \times 1.67 \times \{0.62 \times 30\}^{1/3} + 0.15 \times 0] \times 350 \times 450 \times 10^{-3}$$

$$= [0.53 + 0.15 \times 0] \times 350 \times 450 \times 10^{-3}$$

$$V_{Rd,c} = 83.5 \ge [0.41 + 0.15 \times 0] \times 300 \times 450 \times 10^{-3} \text{ kN}$$

$$= 83.5 \ge 64.6$$

$$V_{Rd,\, c} = 83.5 \text{ kN} < V_{Ed}.$$

Therefore shear reinforcement is required.

iii. Check adequacy of section, $V_{Rd,\, max} > V_{Ed}$.

$$\theta = 0.5 \sin^{-1}\{\frac{2 V_{Ed}}{\alpha_{cw} \, b_w \, z \, v_1 \, f_{cd}}\}$$

$$\theta = 0.5 \sin^{-1}\{\frac{2 \times 267 \times 10^3}{1.0 \times 350 \times (0.9 \times 450) \times 0.6 \times \dfrac{30}{1.5}}\}$$

$$= 0.5 \sin^{-1}(0.314) = 0.5 \times 18.3 = 9.15^0$$

$$\cot \theta = 6.21$$

The value of $\cot \theta$ is outside the limits of 1.0 and 2.5.
Choosing $\cot \theta = 2.5$ for minimum shear reinforcement, $V_{Rd,\, max} = 587 \text{ kN} > V_{Ed}$.
Section is adequate.

iv. Design of shear reinforcement $V_{Rd,\, s} > V_{Ed}$

Choosing 2-leg links of 8 mm diameter,
$A_{sw} = 100.5 \text{ mm}^2$, $\cot \theta = 2.5$, $z = 0.9d$ and $f_{ywk} = 500 \text{ MPa}$.

$$V_{RD,s} = \frac{z}{s} A_{sw} f_{ywd} \cot \theta$$

$$= \frac{(0.9 \times 450)}{s} \times 100.5 \times \frac{500}{1.15} \times 2.5 \times 10^{-3}$$

$$= \frac{44242}{s} \text{ kN} \geq (V_{Ed} = 267 \text{ kN})$$

$$s \leq 165 \text{ mm}$$

Maximum spacing = 0.75 d = 338 mm. Choose 150 mm spacing for links.

v. Check minimum shear steel:

$a_v = 600$ mm, $0.75\ a_v = 450$ mm, $\sin \alpha = 1$ and area of one 2-leg link = 100.5 mm^2.

$$V_{Ed} \text{ reduced} \leq A_{sw} f_{ywd} \times \sin \alpha$$

If N links are provided in the section 0.75 a_v,

$$268 \text{kN} \leq N \times 100.5 \times \frac{500}{1.15} \times 1.0 \times 10^{-3}$$

$$N \geq 6.1$$

Provide seven links in the 450 mm length. Provide 8 mm diameter 2-leg links at 450/ (7−1) = 75 mm centres.

Note: Annex J.3 of Eurocode 2 gives further information on the design of corbels using strut−tie method. See also Chapter 18.

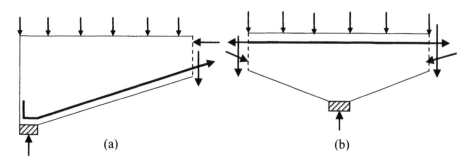

(a)   (b)

Fig. 5.18 Beams with sloping webs.

5.1.9 Beams with Sloping Webs

Very often the beam depth is increased towards the support to increase shear capacity. Fig. 5.18 shows the situation at a simply supported end and at a continuous support. Note that at a simply supported end, the total shear force is *increased* because of the vertical component of the tensile force. On the other hand, in the case of continuous beam, the total shear force is *reduced* by the vertical component of the compressive force. These changes need to be included when designing shear reinforcement.

5.1.10 Example of Complete Design of Shear Reinforcement for Beams

Example 1: A continuous slab 100 mm thick is carried on T-beams at 2 m centres. The overall depth of the beam is 450 mm and the breadth b_w of the web is 250 mm as shown in Fig. 5.19. The beams are 7 m clear span and are simply supported as shown in Fig. 5.20. The characteristic dead load including self-weight and finishes is 10 kN/m^2 and the characteristic imposed load is 6 kN/m^2. Design the beam using the simplified stress block. The materials are concrete f_{ck} =25 MPa and steel f_{yk} = 500 MPa.

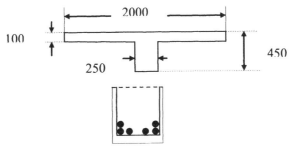

Fig. 5.19 Cross section of the T-beam.

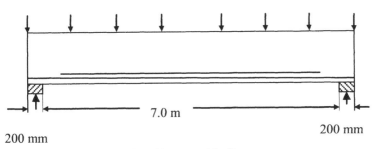

Fig. 5.20 Beam and loading.

Since the beams are spaced at 2 m centres, the loads on the beam are:
$$\text{Dead load} = 10.0 \times 2 = 20.0 \text{ kN/m}$$
$$\text{Live load} = 6 \times 2 = 12 \text{ kN/m}$$
$$\text{Design load} = (1.35 \times 20.0) + (1.5 \times 12) = 45.0 \text{ kN/m}$$
$$\text{Ultimate moment at mid-span} = 45.0 \times 7.2^2/8 = 291.6 \text{ kN m}$$

Design for bending:
Calculate the effective width:
$$b = b_1 + b_2 + b_w = \text{spacing of beam} = 2000 \text{ mm}$$
$$b_1 = b_2 = (2000 - 250)/2 = 875 \text{ mm}$$
$$\ell_0 = \text{Span of simply supported beam} = 7200 \text{ mm}$$
$$b_{effe,1} = b_{effe,2} = 0.2 \, b_1 + 0.1 \, \ell_0 \le 0.2\ell_0$$
$$= 0.2 \times 875 + 0.1 \times 7200 = 895 \text{ mm}$$
$$b_{effe,1} = b_{effe,2} = 895 > (b_1 = b_2 = 875)$$

Therefore $b_{effe, 1} = b_{effe, 2} = 875$

$b_{eff} = b_{eff,1} + b_{eff, 2} + b_w = 875 + 875 + 250 = 2000 \le (b = 2000 \text{ mm})$

Effective width b_{eff} of flange = 2000 mm

The beam section is shown in Fig. 5.17. Assuming a nominal cover on the links is 25 mm and if the links are 8 mm in diameter and the main bars are 20 mm in diameter, then

$$d = 450 - 25 - 8 - 20/2 = 407 \text{ mm, say } 400 \text{ mm.}$$

First of all check whether the beam can be designed as a rectangular beam by calculating M_{flange}.

$$M_{flange} = \eta \, f_{cd} \, b \, h_f \, (d - h_f/2)$$

$f_{ck} = 25$ MPa, $f_{cd} = 25/1.5 = 16.7$ MPa, $\eta = 1$, $\lambda = 0.8$, $\delta = 1$ (simply supported beam).

$$M_{flange} = 16.67 \times 2000 \times 100 \times (400 - 0.5 \times 100) \times 10^{-6} = 1166.9 \text{ kNm}$$

The design moment of 291.6 kNm is less than M_{flange}. The beam can be designed as a rectangular beam of size 1840×400.

$$k = M/ (b \, d^2 \, f_{ck}) = 291.6 \times 10^6/ (2000 \times 400^2 \times 25) = 0.037 < 0.196$$

$$\frac{z}{d} = 0.5\{1.0 + \sqrt{(1 - 3\frac{k}{\eta})}\} = 0.5\{1.0 + \sqrt{(1 - 3\frac{0.037}{1.0})}\} = 0.97$$

$$A_s = \frac{M}{0.97 \, d \times 0.87 \, f_{yk}} = \frac{291.6 \times 10^6}{0.97 \times 400 \times 0.87 \times 500} = 1728 \text{ mm}^2$$

Provide 6H20; $A_s = 1884 \text{ mm}^2$.

Check actual effective depth:

d_1 for bottom four bars = $450 - 25(\text{cover}) - 8 \text{ (links)} - 20/2 = 407$ mm.

d_2 for top two bars = $407 - 20 = 387$ mm.

$d = (4 \times 407 + 2 \times 387)/6 = 400$ mm. Assumption is valid.

Bar curtailment:

The top two bars can be curtailed, leaving the bottom four bars to run to full length. $A_s = 4H20 = 1256 \text{ mm}^2$.

Check minimum steel:

From equation (9.1N), $A_{s, min} = 0.26 \times (f_{ctm}/f_{yk}) \times b_t \, d$

$f_{ctm} = 0.30 \times (25)^{0.667} = 2.6$ MPa, $f_{yk} = 500$ MPa, $b_t = 250$ mm, $d = 400$ mm, $A_{s, min} = 135 \text{ mm} < 1256 \text{ mm}^2$ provided.

Ignoring the additional tensile force due to shear:

Equating total tension to total compression,

$A_s \times 0.87 \times f_{yk} = (2000 \times a) \times f_{cd}$, a = thickness of slab in compression

$1256 \times 0.87 \times 500 = (2000 \times a) \times 25/1.5$, a = 16.4 mm

lever arm $z = (d_1 - a/2) = 407 - 16.4/2 = 399$ mm

Moment of resistance = $A_s \times 0.87 \times f_{yk} \times z = 218$ kNm

$45 \times (7.2/2) \times x - 45 \times x^2/2 = 218$, x = 1.79 m and 5.40 m

Top two bars can be curtailed at 1.79 m from the centre of support at both ends. Including the additional tensile force due to shear: As a simple means of including the effect of shear force on tensile force in the reinforcement, using the shift rule from clause 9.2.1.3(2), the bending moment diagram is shifted by

$$a_1 = 0.5z \, (\cot\theta - \cot\alpha)$$

$z = 399$ mm, $\cot \theta = 2.5$, $\alpha = 90^0$, $\cot \alpha = 0$, $a_1 = 0.5$ m.
The moment of 218 kNm, instead of occurring at 1.79 m from the ends, is assumed to occur at $1.79 - a_1 = 1.29$ m from the ends.
Assume the anchorage length $l_{bd} \approx 41$ bar diameters $= 820$ mm (see section 5.2).
The top two bars can be stopped at $(1.79 - a_1 - l_{bd}) = (1.79 - 0.5 - 0.82) = 0.47$ m from the centre of support at both ends.

Design for shear:
In clause 6.2.1(8), the code states that for members subjected to predominantly uniformly distributed loading, the design shear force need not be checked at a distance less than d from the face of the support. Any shear reinforcement required should continue to the support. In addition it should be verified that that the shear at the support does not exceed $V_{Rd, max}$.
In clause 6.2.3 (5), the code states that where there is no discontinuity of V_{Ed}, (e.g., for uniformly distributed loading) the shear reinforcement in any length increment $\ell = z (\cot \theta + \cot \alpha)$ may be calculated using the smallest value of V_{Ed} in the increment.

Reaction $R = 45 \times 7.2/2 = 162.0$ kN. Width of support $= 200$ mm.
Shear force at d from face of support $= R - (100 + 400) \times 10^{-3} \times 45 = 139.5$ kN.
$$V_{Ed} = 139.5 \text{ kN.}$$

(a) Check if shear reinforcement is required, $V_{Ed} > V_{Rd, c}$
$V_{Ed} = 139.5$ kN, $b_w = 250$ mm, $d = 400$ mm, $A_{sl} = 4 \times 314 = 1256$ mm^2

$$C_{Rd,c} = \frac{0.18}{(\gamma_c = 1.5)} = 0.12$$

$$k = 1 + \sqrt{\frac{200}{400}} = 1.71 \le 2.0$$

$$100\rho_1 = 100 \times \frac{1256}{250 \times 400} = 1.26 < 2.0$$

$$v_{min} = 0.035 \times 1.71^{1.5} \times \sqrt{25} = 0.39 \text{ MPa}$$

$$V_{Rd,c} = [0.12 \times 1.71 \times \{1.26 \times 25\}^{1/3} + 0.15 \times 0] \times 250 \times 400 \times 10^{-3}$$

$$\ge [0.39 + 0.15 \times 0] \times 250 \times 400 \times 10^{-3}$$

$$V_{Rd,c} = 64.8 \ge 39.0 \text{ kN}$$

$V_{Rd, c} = 64.8$ kN < 139.5 kN. Therefore shear reinforcement is required.

(b) Check if the section strength is adequate, $V_{Ed} < V_{Rd, max}$.

$$\theta = 0.5 \sin^{-1} \{\frac{2 V_{Ed}}{\alpha_{cw} b_w z v_1 f_{cd}}\}$$

$$\theta = 0.5 \sin^{-1} \left\{ \frac{2 \times 139.5 \times 10^3}{1.0 \times 250 \times (0.9 \times 400) \times 0.6 \times \dfrac{25}{1.5}} \right\}$$

$$= 0.5 \sin^{-1} (0.31) = 0.5 \times 18.06 = 9.03^0$$

$$\cot \theta = 6.29$$

The value of $\cot \theta$ is outside the limits of 1.0 and 2.5.
Choosing $\cot \theta = 2.5$ for minimum shear reinforcement, $V_{Rd, max} = 310.3$ kN $> V_{Ed}$.
Check that shear force at support is less than $V_{Rd, max}$.
Shear force at support $= 162.0$ kN $< (V_{Rd, max} = 310.3$ kN).
Section size is adequate.

(c) Design of shear reinforcement:
Ensure that $V_{Rd, s} \geq V_{Ed}$ and choose 2-leg links of 8 mm diameter.
$A_{sw} = 100.5$ mm^2, $\cot \theta = 2.5$, $z = 0.9d$, $f_{ywk} = 500$ MPa, $V_{Ed} = 139.5$ kN.

$$V_{Rd,s} = \frac{z}{s} A_{sw} f_{ywd} \cot \theta$$

$$V_{Rd,s} = \frac{(0.9 \times 400)}{s} \times 100.5 \times \frac{500}{1.15} \times 2.5 \times 10^{-3}$$

$$V_{Rd,s} = \frac{39326}{s} \, \text{kN} \geq (V_{Ed} = 139.5 \, \text{kN})$$

$$s \leq 282 \, \text{mm}$$

Maximum spacing $s \leq (0.75 \ d = 0.75 \times 400 = 300$ mm).
Maximum spacing should be less than 300 mm. A spacing of 250 mm will be satisfactory.

(d) Check minimum steel requirement.

$$\frac{A_{sw}}{s b_w} \geq \frac{0.08 \sqrt{f_{ck}}}{f_{yk}}$$

$$\frac{100.5}{s \times 250} \geq \frac{0.08 \sqrt{25}}{500}$$

$$s \leq \frac{100.5}{250} \frac{500}{0.08 \sqrt{25}} = 503 \, \text{mm}$$

(e) Calculate the shear resistance with minimum shear steel.
$$s = 300 \text{ mm}.$$

$$V_{Rd,s} = \frac{z}{s} A_{sw} f_{ywd} \cot \theta$$

$$V_{Rd,s} = \frac{(0.9 \times 400)}{300} \times 100.5 \times \frac{500}{1.15} \times 2.5 \times 10^{-3} = 131.1 \, \text{kN}$$

This shear force occurs at $(7.2/2) \times [1.0 - 131.1/162] = 0.69$ m from the ends.

Therefore beyond 0.69 m from the ends, only minimum shear links at 300 mm are required.
Provide links at 250 mm centres for 0.69 m from the ends and links at 300 mm for the rest of the beam. 2H12 bars are provided at the top to hold the links. The arrangement is shown in Fig. 5.21.

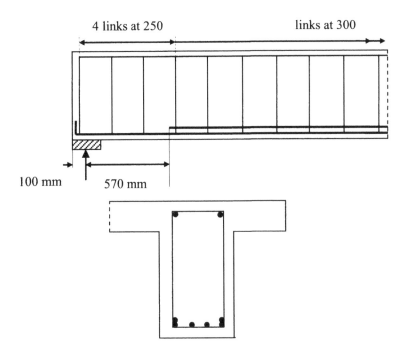

Fig. 5.21 Shear link arrangement.

In order to assist design calculations, Table 5.2 gives the values of $A_{sw}/(s \, b_w)$ for $f_{yk} = 500$ MPa and f_{ck} from 25–60 MPa.

Table 5.2 Minimum shear reinforcement: Value of $(A_{sw}/(s \, b_w)) \times 10^4$

f_{ck}, MPa	25	30	35	40	45	50	55	60
$(A_{sw}/(s \, b_w)) \times 10^4$	8.0	8.76	9.47	10.12	10.73	11.31	11.87	12.39

Table 5.3 gives values of A_{sw}/s for various link sizes and spacings assuming 2-leg links.

Table 5.3 A_{sw}/s for 2-leg links

Link size (mm)	Spacing s (mm)							
	75	100	125	150	175	200	250	300
H6	0.754	0.566	0.452	0.377	0.323	0.283	0.226	0.189
H8	1.340	1.005	0.804	0.670	0.575	0.503	0.402	0.335
H10	2.094	1.571	1.257	1.047	0.898	0.785	0.628	0.524
H12	3.016	2.262	1.810	1.508	1.293	1.131	0.905	0.754
H16	5.362	4.021	3.217	2.681	2.298	2.011	1.609	1.340

5.1.11 Shear Design of Slabs

Flexural design of slabs is treated in Chapter 8. One-way and two-way solid slabs are designed for shear like beams on the basis of a strip of unit width of 1 m. Slabs carrying moderate distributed loads such as floor slabs in office buildings and apartments do not normally require shear reinforcement. In clause 6.2.1(4), the Eurocode 2 suggests that minimum shear reinforcement may be omitted in members such as slabs (solid, ribbed or hollow core slabs) where transverse redistribution of loads is possible.

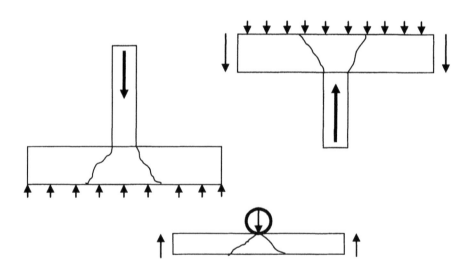

Fig. 5.22 Punching shear (a) pad footing; (b) flat slab–column junction; (c) wheel load on bride deck..

5.1.12 Shear Due to Concentrated Loads on Slabs

Fig. 5.22 shows situations where a slab is subjected to concentrated forces such as when the concentrated load is caused by a column reaction in a flat slab or in a pad footing or due to a concentrated wheel load on slabs in bridge decks. A concentrated load causes punching failure which occurs on inclined faces of a truncated cone or pyramid, depending on the shape of the loaded area as shown in Fig. 5.23.

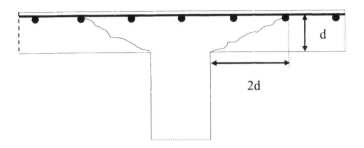

Fig. 5.23 Elevation of punching shear at a column.

Punching shear is considered in section 6.4 of Eurocode 2. Shear resistance is checked at the face of the column and at the basic control perimeter u_1 which is normally taken at a distance of 2d from the face of the column. Fig. 5.24 shows basic perimeters in the case of a circular and rectangular column.

    a.   Circular column of diameter D: Basic perimeter, $u_1 = \pi(D + 4d)$

    b.   Rectangular column b × h: Basic perimeter, $u_1 = 2(b + h) + 4\pi\,d$

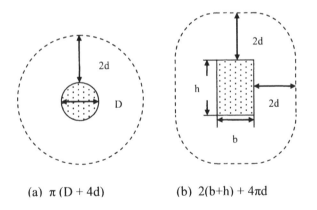

(a) $\pi\,(D + 4d)$ (b) $2(b+h) + 4\pi d$

Fig. 5.24 Control perimeters for circular and rectangular columns.

Fig. 5.25 shows basic perimeters in the case of a rectangular column close to an edge or at a corner at distance smaller than twice the effective depth d. Note that the unsupported edge is excluded in the perimeter calculation.

      (a) Column b × h, near to an edge: $u_1 = 2(a+b) + h + 2\pi d$, a < 2d

    (b) Column b × h, near to a corner: $u_1 = a_1 + b + a_2 + h + \pi d$, a_1 and a_2 < 2d

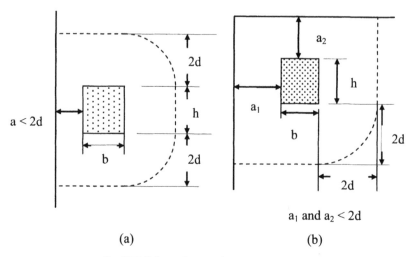

Fig. 5.25 Column close to edge or a corner.

5.1.13 Procedure for Designing Shear Reinforcement against Punching Shear

a. The effective depth d and the axial compressive stress σ_{cp} are taken as the average of the values in y- and z-directions.

$$d = 0.5\,(d_y + d_z),\ \sigma_{cp} = 0.5(\sigma_{cy} + \sigma_{cz})$$

b. At the column perimeter or perimeter of the loaded area u_0, ensure that

$$\frac{V_{Ed}}{u_0\,d} < [v_{Rd,max} = 0.5\,v\,f_{cd}]\ \text{where}\ v = 0.6\,(1 - f_{ck}/250),\ f_{cd} = f_{ck}/\,(\gamma_c = 1.5)$$

c. At the basic control perimeter u_1, calculate

$$v_{Rd,c} = [C_{Rd,c}\,k\{100\,\rho_1\,f_{ck}\}^{1/3} + k_1\,\sigma_{cp}] \ge (v_{min} + k_1\,\sigma_{cp})$$

$$C_{Rd,c} = \frac{0.18}{(\gamma_c = 1.5)} = 0.12$$

$$k = 1 + \sqrt{\frac{200}{d}} \le 2.0$$

$$\rho_1 = \sqrt{\rho_{ly}\,\rho_{lz}}\ \le 0.02$$

ρ_{ly} and ρ_{lz} refer to reinforcement ratio in y- and z-directions respectively calculated over a width of slab equal to width of the column plus 3d on each side.

$$v_{min} = 0.035\,k^{1.5}\,\sqrt{f_{ck}}$$

If $[v_{Ed} = \dfrac{V_{Ed}}{u_1\,d}] < v_{Rd,c}$, then no shear reinforcement is necessary.

If $v_{Ed} \ge v_{Rd,c}$, provide shear reinforcement.

d. Shear reinforcement is calculated according to the code equation (6.52)

$$V_{Rd,cs} = 0.75\, v_{Rd,c} + 1.5\{\frac{d}{s_r}\}\, A_{sw}\, f_{ywd,ef}\,\{\frac{1}{u_1 d}\}\,\sin\alpha \qquad (6.52)$$

where: A_{sw} = Total area of one perimeter of reinforcement
      s_r = Radial spacing of perimeters of reinforcement
      $f_{ywd,ef}$ = Effective design strength of punching shear reinforcement
         = $250 + 0.25\, d \le f_{ywd}$
      α = angle between the shear reinforcement and the plane of the slab.
If only a single line of bent-down bars is provided, take the ratio $(d/s_r) = 0.67$.

e. Determine the position of the outermost perimeter u_{out} where $v_{Ed} = v_{Rd,\,c}$.

f. Arrange the reinforcement.

5.1.13.1 Example of punching shear reinforcement design: Zero moment case

Design the shear reinforcement around the column of a flat slab. The flat slab is supported by 400 × 600 mm columns spaced at 7.5 m in both directions. The slab is 400 mm thick and is reinforced with H20 bars at 150 mm c/c in both directions with 30 mm cover. Assume $f_{ck} = 30$ MPa, $f_{ywk} = 500$ MPa and shear links are H8 *single* leg.
The characteristic loads on the slab are:
            Live load = 15.0 kN/m².
Dead load including self weight, screed, partitions, etc. = 13.5 kN/m².

(i) Effective depths
         In y-direction, $d_{ly} = 400 - 30 - 20/2 = 360$ mm
         In z-direction, $d_{lz} = 400 - 30 - 20 - 20/2 = 340$ mm
            $d = 0.5(d_{ly} + d_{lz}) = 350$ mm

(ii) Steel percentage
     A_s = H20 bars at 150 mm c/c = $\pi \times 20^2/4 \times (1000/150) = 2094$ mm²/m
    $100\, \rho_{1y} = 100\, \rho_{1z} = 100\, A_s/\,(bd) = 100 \times 2094/\,(1000 \times 350) = 0.60$
            $100\, \rho = 0.5(0.60 + 0.60) = 0.60$

(iii) Column reaction
Design load on slab:
         $q = 1.35 \times 13.5 + 1.5 \times 15 = 40.73$ kN/m²
Column reaction, $V_{Ed} = q \times$ spacing in y-direction × spacing in z-direction
         $V_{Ed} = 40.73 \times 7.5 \times 7.5 = 2291$ kN

(iv) Calculate $v_{Rd,\,max}$ and $v_{Rd,\,c}$.
    $f_{ck} = 30$ MPa, $v = 0.6\,(1 - f_{ck}/250) = 0.53$, $f_{cd} = f_{ck}/(\gamma_c = 1.5) = 20$ MPa
       $v_{Rd,max} = 0.5\, v\, f_{cd} = 0.5 \times 0.53 \times 20 = 5.3$ MPa
   $v_{Rd,c} = (C_{Rd,c}\, k\{100\rho_1\, f_{ck}\}^{1/3} + k_1\sigma_{cp}) \ge (v_{min} + k_1\sigma_{cp})$

$$C_{Rd,c} = \frac{0.18}{(\gamma_c = 1.5)} = 0.12, \ k = 1 + \sqrt{\frac{200}{350}} = 1.76 \leq 2.0, \ \sigma_{cp} = 0$$

$$100\,\rho_l = \sqrt{\rho_{ly}\,\rho_{lz}} = \sqrt{0.60 \times 0.60} = 0.60 \leq 2.0, \ v_{min} = 0.035\,k^{1.5}\,\sqrt{f_{ck}} = 0.45.$$

Substituting in the formula, $v_{Rd,c} = 0.55$ MPa.

(v) Check for maximum shear around the column perimeter
$$u_0 = \text{Column perimeter} = 2(400 + 600) = 2000 \text{ mm}$$
Load on slab acting downwards on the column $= 400 \times 600 \times 40.73 \times 10^{-6}$
$$= 9.8 \text{ kN}$$
$$V = 2291 - 9.8 = 2281.2 \text{ kN}$$
$$v = V/(u_0\,d) = 2281.2 \times 10^3 / (2000 \times 350) = 3.26 \text{ MPa} < (v_{Rd,\,max} = 5.3)$$
The slab thickness is therefore adequate.

(vi) Calculate the shear stress at the perimeter u_1 at 2d from the column face:
$$u_1 = 2(400 + 600) + 2\pi \times 2d = 6398 \text{ mm}$$
The load acting within the perimeter is equal to
$$[400 \times 600 + 2 \times (400 + 600) \times 2d + \pi\,(2d)^2] \times 40.73 \times 10^{-6} \text{ kN}$$
$$= 129.50 \text{ kN}$$
$$V_{Ed} = 2291 - 129.50 = 2161.5 \text{ kN}$$
$$v_{Ed} = V_{Ed}/(u_1 \times d) = 2161.5 \times 10^3 / (6398 \times 350) = 0.97 \text{ MPa} > (v_{Rd,\,c} = 0.55)$$
Shear reinforcement is needed.

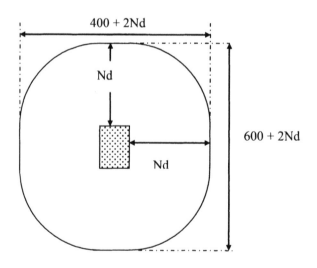

Fig. 5.26 Outer perimeter u_{out}.

(vii) Calculate the perimeter u_{out} where shear stress is equal to $v_{Rd,\,c}$
As shown in Fig. 5.26, let the perimeter be at a distance Nd from the face of the column.

$$u_{out} = 2(400 + 600) + 2\pi \text{ Nd mm}$$

The load acting within the perimeter is equal to

$$[400 \times 600 + 2 \times (400 + 600) \times Nd + \pi (Nd)^2] \times 40.73 \times 10^{-6} \text{ kN}$$
$$V_{Ed} = [2291 - \text{Load inside perimeter}] \text{ kN}$$
$$v_{Ed} = V_{Ed}/ (u_{out} \times d) = v_{Rd,c} \text{ MPa}$$

By trial and error, N = 3.72 and u_{out} = 2(400 + 600) + 2π × 3.72 ×350 = 10181 mm. At this perimeter no shear reinforcement is required.

(viii) Calculate the position of the outermost perimeter where shear reinforcement is required

According to clause 6.4.5(4) of Eurocode 2, the last ring of shear reinforcement must be within kd, where k = 1.5 from the u_{out}.
This perimeter lies at (Nd – kd) = (3.72 d – 1.5d) = 2.22 d from the face of the column.
Perimeter length = $u_{2.22 d}$ = 2(400 + 600) + 2π × 2.22 d = 6882 mm.

(ix) Calculate shear reinforcement using the code equation (6.52)

$$v_{Rd,cs} = 0.75\, v_{Rd,c} + 1.5 \left(\frac{d}{s_r}\right) A_{sw}\, f_{ywd,ef} \left(\frac{1}{u_1 d}\right) \sin \alpha$$

s_r = 0.75d, f_{ywk} = 500 MPa, γ_s = 1.15, d = 350 mm, f_{ywd} = 500/1.15 = 435 MPa
$f_{ywd,ef}$ = (250 + 0.25× 350 = 338) ≤ 435 MPa, $f_{ywd,ef}$ = 338 MPa, $v_{Rd,c}$ = 0.55

At basic control perimeter u_1 at 2d from column:
$$v_{Rd,cs} = v_{Ed} = 0.97 \text{ MPa}, u_1 = 6398 \text{ mm}$$
Substituting in code equation (6.52),

$$0.97 = 0.75 \times 0.55 + 1.5 \frac{d}{0.75d} A_{sw} \times 338 \times \frac{1}{6398 \times 350}$$
$$A_{sw} = 1847 \text{ mm}^2$$

(x) Calculate the minimum link leg area

Using code equation (9.11) to calculate the area of a single link leg

$$A_{sw,min} \times \frac{(1.5 \sin \alpha + \cos \alpha)}{s_r\, s_t} \geq 0.08 \frac{\sqrt{f_{ck}}}{f_{yk}} \qquad (9.11)$$

Substituting f_{ck} = 30 MPa, f_{yk} = 500 MPa, d = 350 mm, s_r = 0.75 d, s_t = 2d,
sin α = 1 for vertical links, $A_{sw, min}$ = 107 mm^2.
Choosing H12 bars, $A_{sw, min}$ = 113 mm^2.
No. of links required = A_{sw}/ Area of one link = 1847/113 = 17 links.
A minimum of 17 links should be provided at all perimeters with the spacing between the perimeters equal to or less than 0.75d. The first perimeter is at a distance > (0.3d = 105 mm) from the face of the column. The last perimeter is within 1.5d from the perimeter where shear reinforcement is no longer required. This is at 2.22d from the face of the column (see viii above).
(2.22d – 0.3d)/0.75d = 2.56.

Choosing the first perimeter at, say, (200 mm = 0.57d) from the face of the column, the number of perimeters at which reinforcement needs to be provided is (2.22d – 0.57d)/0.75d = 1.46, say 2.
Provide reinforcement on three perimeters.
Choose the perimeters as follows:

- Last perimeter at 2.22 d from the face of the column.
- Middle perimeter at (2.22d – 0.75 d) = 1.47 d from the face of the column.
- The first perimeter at 1.47 d – 0.75 d = 0.72 d from the face of the column.

(xi) Arrange link reinforcement

Arrange the perimeters as follows.
(i) First perimeter at a distance > (0.3d = 105 mm)
Choose first perimeter at 0.72d = 252 mm
Perimeter length = $u_{0.72\,d}$ = 2(400 + 600) + 2π × 0.72d = 3593 mm
Maximum spacing of links ≤ 1.5d = 525 mm
Spacing of links = perimeter length/Minimum no. of links = 3583/17 = 211 mm
Provide 17 links at say 210 mm.

(ii) Second perimeter at (252 + 0.75d) = 515 mm = 1.47d
Perimeter length = $u_{1.47\,d}$ = 2(400 + 600) + 2π × 1.47 d = 5233 mm
Maximum spacing of links ≤ 1.5d = 525 mm
Spacing of links = perimeter length/Minimum no. of links = 5233/17 = 308 mm.

(iii) Third perimeter at (515 + 0.75d) = 778 mm = 2.22d
Perimeter length = $u_{2.22\,d}$ = 2(400 + 600) + 2π × 2.22 d = 6486 mm
Maximum spacing of links ≤ 2d = 700 mm
Spacing of links = perimeter length/minimum no. of links = 6882/17 = 405 mm
Reinforcement is provided on three perimeters. Once the numbers are rounded up to practical dimensions, design will be satisfactory.

5.1.14 Shear Reinforcement Design: Shear and Moment Combined

Fig. 5.27 shows the shear stress distribution in the slab due to moment acting on the column or loaded area. It is generally assumed that the distribution of shear stress due to moment is 'plastic' in the sense that apart from the sign (i.e., up or down), the shear stress is constant around the perimeter.
The maximum shear stress v_{Ed} is taken, according to code equation (6.38) as

$$v_{Ed} = \beta \frac{V_{Ed}}{u_1 d} \qquad (6.38)$$

where the factor β accounts for the combined action of shear force V_{Ed} and moment M_{Ed}.

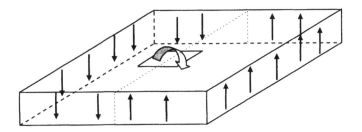

Fig. 5.27 Shear stress distribution due to bending moment on column.

5.1.14.1 Support reaction eccentric with regard to control perimeter for rectangular columns

Use code equations (6.39) and (6.41) to calculate β and W_1 in the case of rectangular columns subjected to a shear force V_{Ed} and a moment M_{Ed}.

$$\beta = 1 + k \frac{M_{Ed}}{V_{Ed}} \frac{u_1}{W_1} \qquad (6.39)$$

$$W_1 = 0.5c_1^2 + c_1\, c_2 + 4\, c_2\, d + 16\, d^2 + 2\, \pi\, d\, c_1 \qquad (6.41)$$

where
c_1 is the column dimension *parallel* to the eccentricity of the load.
c_2 is the column dimension *perpendicular* to the eccentricity of the load.
The value of k is given in Table 5.4.

Table 5.4 Values of k for rectangular loaded areas/columns

c_1/c_2	≤ 0.5	1.0	2.0	≥ 3.0
k	0.45	0.60	0.70	0.80

Example: Design shear reinforcement for a column 300×500 mm, $d = 200$ mm subjected to $V_{Ed} = 450$ kN and $M_{Ed} = 160$ kNm. The moment acts about an axis parallel to the longer side of the column.
As the moment acts about the longer side, eccentricity of the load will be parallel to the shorter side. Therefore $c_1 = 300$ mm, $c_2 = 500$ mm. Substituting in code equation (6.41),

$$W_1 = 0.5 \times 300^2 + 300 \times 500 + 4 \times 500 \times 200 + 16 \times 200^2 + 2\,\pi \times 200 \times 300$$
$$W_1 = 1.612 \times 10^6 \text{ mm}^2$$

Interpolating from Table 5.2 for k at $c_1/c_2 = 300/500 = 0.6$,

$$k = 0.45 + (0.60 - 0.45) \times (0.6 - 0.5)/ (1.0 - 0.5) = 0.48$$
$$u_1 = 2(300 + 500) + 2\pi \times 2d = 4113 \text{ mm}$$
$$M_{Ed}/V_{Ed} = 160 \times 10^6/ (450 \times 10^3) = 356 \text{ mm}$$

Substituting in code equation (6.39),

$$\beta = 1 + 0.48 \times 356 \times \frac{4113}{1.612 \times 10^6} = 1.44$$

$$v_{Ed} = \beta \frac{V_{Ed}}{u_1 d} = 1.44 \times \frac{450 \times 10^3}{4113 \times 200} = 0.79 \, \text{MPa}$$

5.1.14.2 Support reaction eccentric with regard to control perimeter for circular columns

For internal circular columns of diameter D, code equation (6.42) gives the value of β as

$$\beta = 1 + 0.6 \pi \frac{e}{D + 4d} \qquad (6.42)$$

where eccentricity, $e = M_{Ed}/V_{Ed}$.

5.1.14.3 Support reaction eccentric with regard to control perimeter about two axes for rectangular columns

For internal rectangular columns with moment $M_{Ed, y}$ and $M_{Ed, z}$ about y- and z-axes respectively, code equation (6.43) gives the value of β as

$$\beta = 1 + 1.8 \sqrt{\{(\frac{e_y}{b_z})^2 + (\frac{e_z}{b_y})^2\}} \qquad (6.43)$$

where eccentricities, $e_y = M_{Ed, z}/V_{Ed}$ and $e_z = M_{Ed, y}/V_{Ed}$. b_y and b_z are the overall widths of the critical perimeter in the y- and z-directions respectively as shown in Fig. 5.28.

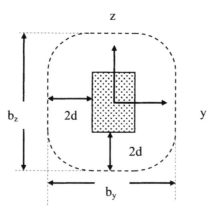

Fig. 5.28 Control perimeter dimensions.

Example: Calculate v_{Ed} for a column 300 × 500 mm, d = 200 mm, subjected to:
$$V_{Ed} = 450 \, \text{kN},$$
$M_{Ed, y}$ = 160 kNm, moment acts about the longer side of the column.
$M_{Ed, z}$ = 200 kNm, moment acts about the shorter side of the column.

$$b_y = 500 + 2 \times 2d = 1300 \text{ mm, } b_z = 300 + 2 \times 2d = 1100 \text{ mm}$$
$$e_y = M_{Ed, z}/V_{Ed} = 200 \times 10^6/ (450 \times 10^3) = 444 \text{ mm}$$
$$e_z = M_{Ed, y}/V_{Ed} = 160 \times 10^6/ (450 \times 10^3) = 356 \text{ mm}$$
$$u_1 = 2(300 + 500) + 2\pi \times 2d = 4113 \text{ mm}$$

$$\beta = 1 + 1.8\sqrt{\{(\frac{444}{1100})^2 + (\frac{356}{1300})^2\}} = 1.88$$

$$v_{Ed} = \beta \frac{V_{Ed}}{u_1 d} = 1.88 \times \frac{450 \times 10^3}{4113 \times 200} = 1.03 \text{ MPa}$$

5.1.14.4 Rectangular edge columns

Two cases are considered:
a. Eccentricity *perpendicular* to the free edge of the slab toward the interior and no eccentricity parallel to the edge.
The value of $\beta = 1$ (i.e., the shear stress v_{Ed} is constant) on the reduced control perimeter u_1* as shown in Fig. 5.29.

$$\leq \min(1.5 \text{ d}; 0.5c_1)$$

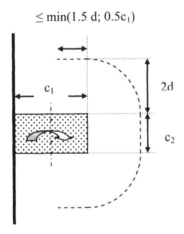

Fig. 5.29 Edge column: Reduced control perimeter.

Example: A 300×500 mm column orientated with the shorter side parallel to the free edge, d = 200 mm, $V_{Ed} = 450$ kN, $M_{Ed} = 160$ kNm, moment acts about the shorter side of the column and directed away from the free edge.
c_1 = dimension perpendicular to free edge = 500mm.
c_2 = dimension parallel to free edge = 300 mm.
1.5d = 300 mm, $0.5c_1 = 250$ mm, min (300; 250) = 250 mm.
Reduced perimeter $u_1* = c_2 + \pi \times 2d + 2 \times 250 = 2056$ mm.

Taking $\beta = 1$ and substituting in code equation (6.43),

$$v_{Ed} = \beta \frac{V_{Ed}}{u_{1*} \times d} = 1.0 \times \frac{450 \times 10^3}{2056 \times 200} = 1.09 \, \text{MPa}$$

b. Where eccentricity is with respect to both axes, β can be determined from code equation (6.44):

$$\beta = \frac{u_1}{u_{1*}} + k \frac{u_1}{W_1} e_{parallel} \qquad (6.44)$$

The moment acting about an axis parallel to the free edge acts toward the *interior* of the slab as shown in Fig. 5.28.
u_1 is the basic control perimeter as shown in Fig. 5.24.
u_{1*} is the reduced control perimeter as shown in Fig. 5.29.
k from Table 5.2 replacing the ratio c_1/c_2 by $0.5 \, c_1/c_2$.
$e_{parallel}$ = eccentricity parallel to the edge caused by a moment acting about an axis perpendicular to the edge.

$$W_1 = 0.25c_2^2 + c_1 \, c_2 + 4 \, c_1 \, d + 8 \, d^2 + \pi \, d \, c_2$$

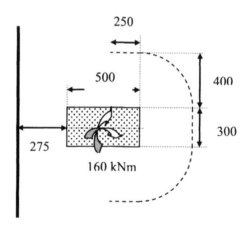

Fig. 5.30 Column near to a free edge.

Example: A 300 × 500 mm column orientated with the shorter side parallel to the free edge at a distance of 275 mm from the free edge is shown in Fig. 5.30.
$d = 200$ mm, $V_{Ed} = 450$ kN.
Assume that moments act about both the long side as well as the short side.
Moment about the longer side = 160 kNm.
$e_{parallel} = 160 \times 10^6 / (450 \times 10^3) = 356$ mm.
c_2 = Column dimension parallel to $e_{parallel}$ = 300 mm, c_1 = 500 mm.
$1.5d = 300$ mm, $0.5c_1 = 250$ mm.
Min $(1.5d; 0.5c_1)$ = min $(300; 250)$ = 250 mm.
Reduced perimeter u_{1*} from Fig. 5.28.
$u_{1*} = c_2 + \pi \times 2d + 2 \times 250 = 300 + \pi \times 2 \times 200 + 2 \times 250 = 2057$ mm.
Since the dimension 275 mm is less than $2d = 400$ mm, in calculating the value of basic control perimeter u_1, although the control perimeter extends right up to the free edge as shown in Fig. 5.23, the free edge is not taken into consideration.

$$u_1 = c_2 + \pi \times 2d + 2 \times (500 + 275) = 3107 \text{ mm}$$
$$W_1 = 0.25c_2^2 + c_1\,c_2 + 4\,c_1\,d + 8\,d^2 + \pi\,d\,c_2$$
$$W_1 = 0.25 \times 300^2 + 300 \times 500 + 4 \times 500 \times 200 + \pi \times 200 \times 300 = 0.761 \times 10^6 \text{ mm}^2$$
$$0.5\,c_1/c_2 = 0.5 \times 500/\,300 = 0.83$$

From Table 5.2, interpolating for $c_1/c_2 = 0.83$,
$$k = 0.45 + (0.60 - 0.45) \times (0.83 - 0.5)/\,(1.0 - 0.5) = 0.55$$

Substituting in code equation (6.44),
$$\beta = \frac{3107}{2057} + 0.55 \times \frac{3107}{0.761 \times 10^6} \times 356 = 2.31$$

$$v_{Ed} = \beta \frac{V_{Ed}}{u_1\,d} = 2.31 \times \frac{450 \times 10^3}{3107 \times 200} = 1.67 \text{ MPa}$$

5.1.14.5 Support Reaction Eccentric toward the Interior for Rectangular Corner Column

In the case of corner columns with eccentricity toward the *interior* of the slab, punching shear may be assumed to be uniformly distributed along the reduced control perimeter u_{1*} as shown in Fig. 5.31.
$$u_{1*} = \min(1.5d, 0.5c_1) + \min(1.5d, 0.5c_2) + \pi\,d.$$
β is given by the ratio $u_1/\,u_{1*}$ and u_1 is as given in Fig. 5.22.

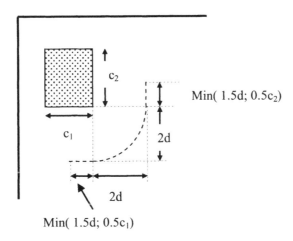

Fig. 5.31 Corner column.

Example: A 300×500 corner column located 275 mm from the edges is subjected to a shear force of 180 kN. The eccentricity of the load is toward the interior of the slab. Taking $d = 200$ mm, calculate the value of β and the shear stress v_{Ed}.
$c_1 = 300$ mm, $c_2 = 500$ mm. $1.5d = 300$ mm.
$$u_{1*} = \min(300, 150) + \min(300, 250) + \pi \times 200 = 1028 \text{ mm}$$

$$u_1 = c_1 + 275 + c_2 + 275 + \pi\,d$$
$$u_1 = 300 + 275 + 500 + 275 + \pi \times 200 = 1978 \text{ mm}$$
$$\beta = 1978/1028 = 1.92$$
$$v_{Ed} = \beta\,V_{Ed}/\,(u_1 \cdot d) = 1.92 \times 180 \times 10^3/\,(1028 \times 200) = 1.68 \text{ MPa}$$

5.1.14.6 Approximate values of β for columns of a flat slab

In cases where the lateral stability of the structure *does not* depend on the frame action of the slab and columns, the following approximate values of β may be used.
- Interior column: $\beta = 1.15$
- Edge column: $\beta = 1.4$
- Corner column: $\beta = 1.5$

5.2 BOND STRESS

Bond is the grip due to adhesion or mechanical interlock and bearing in ribbed bars between the reinforcement and the concrete as shown in Fig. 5.32(a). The bearing of the forces on the ribs causes radial cracks as shown in Fig. 5.32(b) and hoop reinforcement to resist the cracking greatly enhances the bond strength.

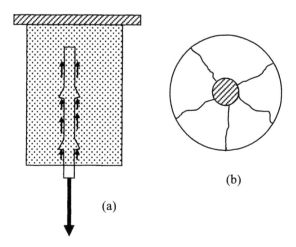

(b)

(a)

Fig. 5.32 (a) 'Bond' forces; (b) radial cracks.

Anchorage is the embedment of a bar in concrete so that it can carry loads through the bond between the steel and concrete. If the anchorage length is sufficient, then the full strength of the bar can be developed by bond. The area over which the bond stress acts is the product of the anchorage length $l_{bd,\,reqd}$ and the perimeter $\pi\,\phi$ of the bar.

If σ_{sd} is the design stress in the bar of diameter ϕ and constant bond stress is f_{bd}, equilibrium requires that

Force in the bar = Resistance due to bond

$$\frac{\pi}{4}\phi^2 \sigma_{sd} = \pi\phi \times l_{bd,reqd} \times f_{bd}$$

Simplifying the above equation leads to the code equation (8.3)

$$l_{bd,reqd} = \frac{\phi}{4}\frac{\sigma_{sd}}{f_{bd}} \tag{8.3}$$

If $\sigma_{sd} = f_{yd}$, $f_{yd} = f_{yk}/(\gamma_s = 1.15)$,

$$l_{bd,reqd} = \{\frac{1}{4.6}\frac{f_{yk}}{f_{bd}}\}\phi$$

Clause 8.4.2 of Eurocode 2 covers aspects of bond stress. The ultimate bond stress f_{bd} is given by code equation (8.2) as

$$f_{bd} = 2.25\,\eta_1\,\eta_2\,f_{ctd} \tag{8.2}$$

where
η_1 is a coefficient related to quality of bond conditions and the position of the bar during concreting.
$\eta_1 = 1.0$ when good conditions are obtained as in the following cases as shown in Fig. 5.33:

   a. Bars inclined at an angle α to the horizontal such that $45^0 \le \alpha \le 90^0$
   b. For all bars in a slab if the total depth of slab $h \le 250$ mm
Restricted good conditions occur in the following cases as shown in Fig. 5.34.

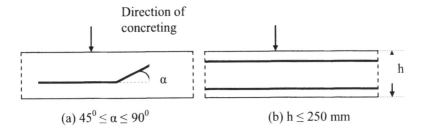

          (a) $45^0 \le \alpha \le 90^0$ (b) $h \le 250$ mm

Fig. 5.33 Cases of good bond conditions for all bars.

   a. For slabs $h > 250$ mm and concreted from above, only for bars in the bottom 250 mm of slab.
   b. For slabs $h > 600$ mm and concreted from above, only for bars in the zone beyond 300 mm from the top face.
$$\eta_1 = 0.7 \text{ in all other cases}$$
η_2 is related to bar diameter φ
$\eta_2 = 1.0$, $\varphi \le 32$ mm, $\eta_2 = (132 - \varphi)/100$, $\varphi > 32$ mm
$f_{ctd} = 0.7 \times f_{ctm}/\gamma_c$, $\gamma_c = 1.5$, (see code equation (3.16)
$$f_{ctm} = 0.3 \times f_{ck}^{0.67}, f_{ck} \le 50 \text{ MPa}$$
$$= 2.12 \times \ell n\,(1.8 + 0.1 \times f_{ck}), f_{ck} > 50 \text{ MPa}$$

In clause 8.4.2(2), it is suggested that the value of f_{ck} should be limited to 60 MPa as concrete above this strength tends to be brittle. Table 5.5 gives the values of bond stress and anchorage length $l_{bd, req}$ when $\eta_1 = 1.0$.

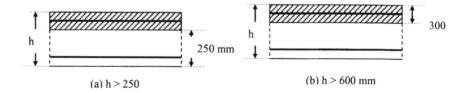

Fig. 5.34 Hatched areas indicate poor bonding.

For situations where the bond conditions are poor for which $\eta_1 = 0.7$, the values of bond stress in Table 5.5 must be multiplied by 0.7 and $l_{bd, req}$ divided by 0.7.

Table 5.5 Bond stress f_{bd} and basic anchorage length $l_{bd, reqd}$ for good bond conditions, $f_{yk}= 500$ MPa

f_{ck} MPa	f_{bd}, MPa		$l_{bd, reqd}/\varphi$	
	$\varphi \leq 32$ mm	$\varphi = 40$ mm	$\varphi \leq 32$ mm	$\varphi = 40$ mm
20	2.32	2.13	47	51
25	2.70	2.48	40	44
30	3.05	2.81	36	39
35	3.37	3.10	32	35
40	3.69	3.40	30	32
45	4.00	3.68	27	30
50	4.28	3.94	25	28
55	4.43	4.08	25	27
60	4.57	4.20	24	26

5.3 ANCHORAGE OF BARS

Bars are anchored by providing sufficient anchorage length in the case of a straight bar. The length required can be reduced by providing a standard bend (Fig. 5.35(a)) or a standard hook (Fig. 5.35(b)) or by a standard loop (Fig. 5.35(c)). In the case of the standard bend or hook, the bar has to have a straight length of at least five times the bar diameter beyond the end of the curved portion. Anchorage length can also be reduced by welding a transverse bar as in the case of welded mats. Fig. 5.35(d) shows the corresponding anchorage length. The diameter φ_1 of the transverse bar must be equal to or greater than 0.6 times the diameter of φ the main bar.

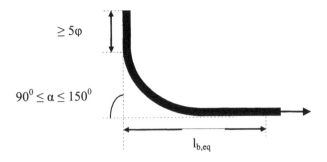

Fig. 5.35(a) A standard bend.

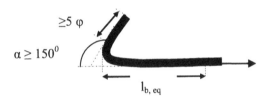

Fig. 5.35(b) A standard hook.

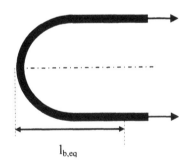

Fig. 5.35(c) A standard loop.

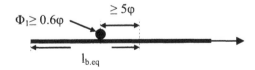

Fig. 5.35(d) Welded transverse bar.

5.3.1 Design Anchorage Length

The basic anchorage length $l_{b, reqd}$ given by equation (8.3) is modified as follows to obtain the design anchorage length l_{bd}.

$$l_{bd,reqd} = \frac{\phi}{4} \frac{\sigma_{sd}}{f_{bd}} = \{\frac{1}{4.6} \frac{f_{yk}}{f_{bd}}\}\phi \tag{8.3}$$

l_{bd} is given by code equation (8.4) as

$$l_{bd} = \alpha_1\,\alpha_2\,\alpha_3\,\alpha_4\,\alpha_5\,l_{b,\,reqd} \geq l_{b,\,min} \tag{8.4}$$

where: $\alpha_1,\,\alpha_2,\,\alpha_3,\,\alpha_4,\,\alpha_5$ are coefficients shown in Table 5.6.

$$(\alpha_2 \times \alpha_3 \times \alpha_5) \geq 0.7 \tag{8.5}$$

Table 5.6 shows the values of α_1 to α_5 for various situations that will influence 'bond length' and help to reduce the required bond length. Note that the smaller the value of these factors, lower will be the anchorage length required. However it has to be said that in practice, conditions generally limit the number of situations where one can reduce the bond length l_{bd} to less than $l_{b,\,reqd}$ to a small number of cases.

α_1 reflects the shape of bar (i.e. straight, hooked, looped, etc.) with adequate cover.
α_2 reflects the effect of minimum cover for the bar.
α_1 and α_2 are functions of the parameter c_d as shown in Fig. 5.36.

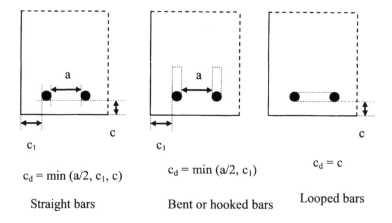

Fig. 5.36 Values of c_d for beams and slabs.

α_3 reflects the fact that confining transverse reinforcement resists cracking of concrete around the bars as explained in section 5.2 and Fig. 5.32(b). This is reflected by the parameter K shown in Fig. 5.37.

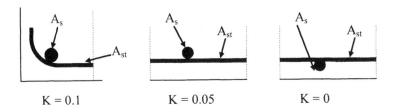

Fig. 5.37 Values of K for beams and slabs.

Table 5.6 Values of α_1 α_2 α_3 α_4 α_5 coefficients

Influencing factor	Type of anchorage	Reinforcing bar	
		Tension	Compression
Shape of bars	Straight	$\alpha_1 = 1$	$\alpha_1 = 1$
	Other than straight	$\alpha_1 = 0.7$, if $c_d > 3\varphi$ Otherwise $\alpha_1 = 1$ *See Fig. 5.36 for c_d values*	$\alpha_1 = 1$
Concrete cover	Straight	$\alpha_2 = 1 - 0.15(c_d - \varphi)/\varphi$ ≥ 0.7 ≤ 1.0 *See Fig. 5.36 for c_d values*	$\alpha_2 = 1$
	Other than straight	$\alpha_2 = 1 - 0.15(c_d - 3\varphi)/\varphi$ ≥ 0.7 ≤ 1.0 *See Fig. 5.36 for c_d values*	$\alpha_2 = 1$
Confinement by unwelded transverse reinforcement	All types	$\alpha_3 = 1 - k\lambda$ ≥ 0.7 ≤ 1.0 *See Fig. 5.37 for K value*	$\alpha_3 = 1$
Confinement by welded transverse reinforcement	*Fig. 5.32(d)*	$\alpha_4 = 0.7$	$\alpha_4 = 0.7$
Confinement by transverse pressure	All types	$\alpha_5 = 1 - 0.04\rho$ ≥ 0.7 ≤ 1.0	

$$\lambda = (\Sigma A_{st} - \Sigma A_{s,min})/A_s$$

ΣA_{st} = cross sectional area of transverse reinforcement along the design anchorage length.

$\Sigma A_{st, min}$ = Cross sectional area of the minimum transverse reinforcement
= 0.25 A_s for beams and 0 for slabs.

A_s = area of a single anchored bar with maximum bar diameter.

ρ = transverse pressure in MPa at ultimate limit state along l_{bd}.

α_4 takes account of the fact that welded transverse reinforcement naturally enhances the 'bond'.

α_5 accounts for the effect of transverse compression to the plane of splitting along the design anchorage length.

$l_{b,\,min}$ is defined by code equations (8.6) and (8.7) as follows.

$l_{b,\,min} > \max(0.3\, l_{b,\,reqd};\ 10\varphi;\ 100\ mm)$ for anchorage in tension   (8.6)

$l_{b,\,min} > \max(0.6\, l_{b,\,reqd};\ 10\varphi;\ 100\ mm)$ for anchorage in compression   (8.7)

Clause 8.4.2(2) of Eurocode 2 states that as a simplified alternative to code equation (8.4), in the case of standard bend or hook or loop

$$l_{bd} = l_{b,\,eq} = \alpha_1\, l_{b,\,reqd} \geq l_{b,\,min}$$

In the case of a welded transverse bar

$$l_{bd} = l_{b,\,eq} = \alpha_4\, l_{b,\,reqd} \geq l_{b,\,min}$$

5.3.2 Example of Calculation of Anchorage Length

Example : Calculate the full anchorage lengths in tension for a H32 bar, $f_{yk} = 500$ MPa in $f_{ck} = 25$ MPa concrete. Assume that the bar is fully stressed. Assume link diameter = 8 mm, clear cover to links 30 mm. A cross section of the beam is shown in Fig. 5.38.

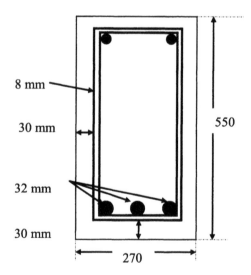

Fig. 5.38 Beam cross section.

Calculate the coefficients α_1 to α_5:

From Fig. 5.33, clear cover to the bars at the bottom as well as at sides is the sum of link diameter + cover = $c = c_1 = 30 + 8 = 38$ mm.

Clear distance between bars = $a = [270 - 2(30 + 8 + 32)]/2 - 32 = 33$ mm.

Calculate α_1:

$$c_d = \min (a/2; c_1; c) = \min (33/2, 38, 38) = 16.5 \text{ mm} < (3\varphi = 96 \text{ mm})$$

For a straight bar or one with a standard bend, $\alpha_1 = 1.0$.

Calculate α_2:

Straight bar: $\alpha_2 = 1 - 0.15(c_d - \varphi) /\varphi = 1.07 > 1.0, \alpha_2 = 1.0$.

Standard bend: $\alpha_2 = 1 - 0.15(c_d - 3\varphi) /\varphi = 1.37 > 1.0, \alpha_2 = 1.0$.

Calculate α_3: If links are provided at 375 mm c/c, then there are approximately three 8 mm links over a distance of 1125 mm.

$$\text{Link area} = \pi/4 \times 8^2 = 50.3 \text{ mm}^2$$
$$\Sigma A_{st} \approx 3 \times 50.3 = 151 \text{ mm}^2$$
$$A_s = 804 \text{ mm}^2 \text{ (cross sectional area of one 32 mm bar)}$$
$$\Sigma A_{min} = 0.25 \, A_s = 0.25 \times 804 = 201 \text{ mm}^2$$
$$\lambda = (151 - 201)/804 = -0.06$$
$$K = 0.05, \alpha_3 = 1.003 > 1.0, \alpha_3 = 1.0$$

Calculate α_4: No welded bar. $\alpha_4 = 1.0$.

Calculate α_5: No confining lateral pressure. $\alpha_5 = 1.0$.

$$\alpha_2 \times \alpha_3 \times \alpha_5 = 1.0 > 0.7$$
$$f_{ctm} = 0.30 \times f_{ck}^{0.667} = 0.30 \times (25)^{0.667} = 2.6 \text{ MPa}$$

From equation (3.16) in clause 3.1.6(2)P,

$$f_{ctd} = 0.7 \times f_{ctm}/ (\gamma_c = 1.5) = 1.2 \text{ MPa}$$

$\varphi = 32 \text{ mm}, \eta_1 = 1.0$ for good bond, $\eta_2 = 1.0$ as $\varphi \leq 32 \text{ mm}$

$$f_{bd} = 2.25 \eta_1 \, \eta_2 \, f_{ctd} = 2.25 \times 1.0 \times 1.0 \times 1.2 = 2.7 \text{ MPa}$$
$$f_{yk} = 500 \text{ MPa}, \sigma_{sd} = f_{yk}/ (\gamma_s = 1.15) = 435 \text{ MPa}$$
$$l_{bd,reqd} = \frac{\phi}{4} \frac{\sigma_{sd}}{f_{bd}} = \{\frac{32}{4.0} \times \frac{435}{2.7}\} = 1289 \text{ mm}$$

$l_{b, min} > \max (0.3 \, l_{b, reqd}; 10\varphi; 100 \text{ mm})$ for anchorage in tension
$$l_{b, min} > \max (0.3 \times 1289; 10 \times 32; 100 \text{ mm})$$
$$= \max (387; 320; 100) = 387 \text{ mm}$$
$$l_{bd} = \alpha_1 \, \alpha_2 \, \alpha_3 \, \alpha_4 \, \alpha_5 \, l_{b, reqd} \geq l_{b, min}$$

As all values of $\alpha_i = 1.0$, $l_{bd} = l_{b, reqd} \geq l_{b, min} = 1289 \text{ mm}$.

As a simplification, $l_{b, eq} = \alpha_1 \times l_{b, rqd}$.

$$l_{b, eq} = 1.0 \times l_{b, rqd} = 1289 > (l_{b, min} = 320) = 915 \text{ mm}$$
$$1289 \text{ mm} \approx 40 \, \varphi$$

In compression because $\alpha_1 = \alpha_2 = \alpha_3 = 1.0$, the anchorage length in compression in this case will be same as in tension viz. $1289 \text{ mm} \approx 40 \, \varphi$.

5.3.3 Curtailment and anchorage of bars

Sufficient reinforcement must be provided at all sections to resist the envelope of the acting tensile force including the additional tensile force due to the effect of inclined tensile cracks in the webs due to shear. The tensile force due to bending is given by M_{Ed}/z where M_{Ed} is the design bending moment at the section and z is the lever arm.

The additional tensile force ΔF_{td} due to shear V_{Ed} is given by code equation (6.18) as

$$\Delta F_{td} = 0.5 \, V_{Ed} \, (\cot\varphi - \cot\alpha)$$

where $2.5 \geq \cot\varphi \geq 1.0$ and $\alpha =$ Inclination of shear reinforcement to the beam axis. In the case of vertical shear links, $\cot\alpha = 0$ and $\cot\varphi$ is generally about 2.5. The total tensile force F_{td}

$$F_{td} = M_{Ed}/z + \Delta F_{td} \leq M_{Ed, \, max}/z$$

where $M_{Ed, \, max}$ is the maximum moment along the beam.
As a simplification, the effect of the shear force can be accounted for by shifting the bending moment in the direction of the decreasing bending moment (and therefore increasing shear force) by a value a_1 given by code equation (9.2)

$$a_1 = z \, (\cot\varphi - \cot\alpha)/2$$

where $z \approx 0.9d$.
Generally taking $\cot\varphi = 2.5$ and $\cot\alpha = o$ for vertical shear links,

$$a_1 = z \, (\cot\varphi - \cot\alpha)/2 = 1.125 \, d$$

5.3.4 Example of Moment Envelope

A three span continuous T-beam with spans of 8 m is used to support a 100 mm thick slab spanning 3m between the T-beams. The characteristic loads are:

Super imposed dead load due to partitions, ceiling, floor finishes = 3 kN/m².
Imposed live load = 3.5 kN/m².

T-beams have a total depth of 500 mm and a web width of 300 mm as shown in Fig. 5.39. $f_{ck} = 25$ MPa, $f_{yk} = 500$ MPa.

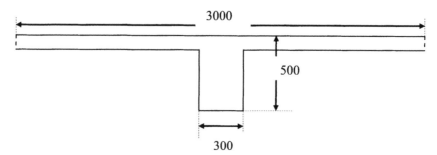

Fig. 5.39 Cross section of T-beam.

Cross sectional area = $0.5 \times 0.3 + (3.0 - 0.3) \times 0.1 = 0.42$ m²
Self weight = $0.42 \times 25 = 10.5$ kN/m
Super dead load = $3 \times$ (span of slab = 3) = 9.0 kN/m
$g_k = 10.5 + 9.0 = 19.5$ kN/m
$q_k = 3.5 \times$ (span of slab = 3) = 10.5 kN/m
$q_{max} = 1.35 \times 19.5 + 1.5 \times 10.5 = 42.08$ kN/m
$q_{min} = 1.0 \times 19.5 + 0.0 \times 10.5 = 19.5$ kN/m

Load cases: (see clause 5.1.3 of Eurocode 2).
Case 1: $q_{max}, q_{min}, q_{max}$ to give maximum span moment in spans 1–2 and 3–4 as shown in Fig. 5.40.

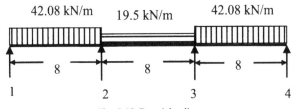

Fig. 5.40 Case 1 loading.

Support moment at 2 and 3 = 197.06 kNm.
Span 1–2:
$$\text{Reaction } V_1 = 42.08 \times 8/2 - 197.06/8 = 143.69 \text{ kN}$$
$$M = 143.69 \times x - 42.08 \times x^2/2$$
$$M_{max} = 245.2 \text{ at } x = 143.69/42.08 = 3.42 \text{ m}$$
Span 2–3:
$$\text{Reaction } V_2 = 42.08 \times 8/2 = 78.0 \text{ kN}$$
$$M = -197.06 + 78.0 \times x - 19.5 \times x^2/2$$
$$M_{max} = -41.1 \text{ at } x = 4.0$$

Case 2: q_{max}, q_{max}, q_{min} to give maximum support moment at 2 as shown in Fig. 5.41.

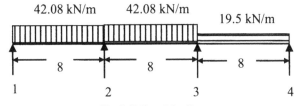

Fig. 5.41 Case 2 loading.

Support moment at 2 = 293.69 kNm and support moment at 3 = 172.87 kNm.
Span 1–2:
$$\text{Reaction } V_1 = 42.08 \times 8/2 - 293.69/8 = 131.61 \text{ kN}$$
$$M = 131.61 \times x - 42.08 \times x^2/2$$
$$M_{max} = 205.8 \text{ at } x = 3.13 \text{ m}$$
Span 2–3:
$$\text{Reaction } V_2 = 42.08 \times 8/2 + (293.69 - 172.87)/8 = 183.32 \text{ kN}$$
$$M = -293.69 + 183.32 \times x - 42.08 \times x^2/2$$
$$M_{max} = 105.6 \text{ at } x = 4.36 \text{ m}$$

Case 3: q_{min}, q_{max}, q_{min} to give maximum span moment in span 2–3 as shown in Fig. 5.42.

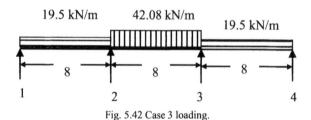

Fig. 5.42 Case 3 loading.

Support moment at 2 and 3 = 197.06 kNm.
Span 1-2:

$$\text{Reaction } V_1 = 19.5 \times 8/2 - 197.06/8 = 53.77 \text{ kN}$$
$$M = 53.77 \times x - 19.5 \times x^2/2$$
$$M_{max} = 74.1 \text{ at } x = 2.76 \text{ m}$$

Span 2-3:

$$\text{Reaction } V_2 = 42.08 \times 8/2 = 168.32 \text{ kN}$$
$$M = -197.06 + 168.32 \times x - 42.08 \times x^2/2$$
$$M_{max} = 139.58 \text{ at } x = 168.32/42.08 = 4.0 \text{ m}$$

Case 4: q_{min}, q_{max}, q_{max} to give maximum support moment at 3 as shown in Fig. 5.43.

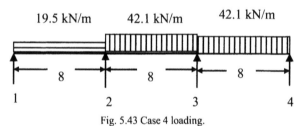

Fig. 5.43 Case 4 loading.

Support moment at 2 = 172.87 kNm and Support moment at 3 = 293.69 kNm.
Span 1-2:

$$\text{Reaction } V_1 = 19.5 \times 8/2 - 172.87/8 = 56.39 \text{ kN}$$
$$M = 56.39 \times x - 19.5 \times x^2/2$$
$$M_{max} = 81.53 \text{ at } x = 56.39/19.5 = 2.89 \text{ m}$$

Span 2-3:

$$\text{Reaction } V_2 = 42.08 \times 8/2 - (293.69 - 172.87)/8 = 153.22 \text{ kN}$$
$$M = -172.87 + 153.22 \times x - 42.08 \times x^2/2$$
$$M_{max} = 106.1 \text{ at } x = 3.64 \text{ m}$$

Bending moment diagrams are shown in Fig. 5.44. Bending moment envelope is shown in Fig. 5.45.

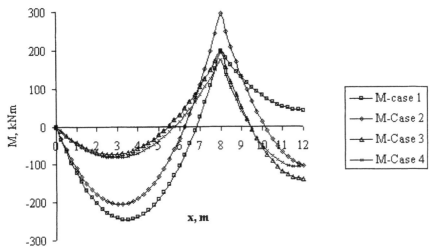

Fig. 5.44 Bending moment diagrams.

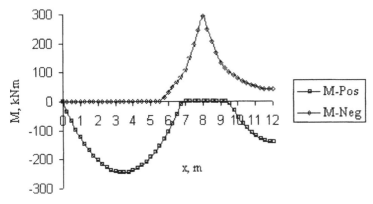

Fig. 5.45 Bending moment envelope.

Design of beam 1-2; span section: M = 245.2 kNm

Effective width:

$b = 3000$ mm, $b_w = 300$ mm, $b_1 = b_2 = 1350$ mm, $l_0 = 0.85 \times 8.0 = 6.8$ m

$b_{eff,1} = b_{eff,2} = 0.2 \times 1350 + 0.1 \times 6800 \leq 0.2 \times 6800$

$b_{eff,1} = b_{eff,2} = 1360$ mm $\leq b_1$

$b_{eff} = 2 \times 1350 + 300 = 3000$ mm

Effective depth, d: Assuming 30 mm cover, 8 mm links and 25 mm bar diameter

$d \approx 500 - 30 - 8 - 25/2 = 450$ mm

$f_{ck} = 25$ MPa, $f_{cd} = f_{ck}/(\gamma_c = 1.5) = 16.7$ MPa

$f_{yk} = 500$ MPa, $f_{yd} = f_{yk}/(\gamma_s = 1.15) = 435$ MPa

Maximum moment capacity M_{flange} if the entire flange is in compression:

$M_{flange} = f_{cd} \times b_{eff} \times h_f \times (d - h_f/2)$

$= 16.67 \times 3000 \times 100 \times (450 - 100/2) \times 10^{-6} = 20004$ kNm

Compression depth s is less than the slab depth.

$245.2 = 16.67 \times 3000 \times s \times (450 - s/2) \times 10^{-6}$

$s^2 - 900 s + 9789 = 0$

$s = 11$ mm

$A_s \times f_{yd} = (3000 \times s) \times f_{cd}$

$A_s = 1265$ mm $=$ Say 3H25 $= 1473$ mm^2

Check minimum steel (code equation (9.1N)) :

$f_{yk} = 500$ MPa, $f_{ctm} = 0.3 \times f_{ck}{}^{(2/3)} = 0.3 \times (25)^{(2/3)} = 2.6$ MPa,

$d = 450$ mm, $b_t = 300$ mm

$$A_{s,min} = 0.26 \frac{f_{ctm}}{f_{yk}} b_t d \geq 0.0013 b_t d$$

$$= 0.26 \times \frac{2.6}{500} \times 300 \times 450 \geq 0.0013 \times 300 \times 450$$

$$= 183 \geq 176 \, mm^2$$

$A_{s,\,provided} = 1473$ mm^2

Ratio $A_{s,\,Provided}/A_{s,\,required} = 0.86$

Maximum stress in bar $= f_{yd} \times 0.86 = 373$ MPa

$f_{ctd} = 0.7 \times f_{ctm}/1.5 = 1.21$ MPa

Assuming good bond, $f_{bd} = 2.25 \times f_{ctd} = 2.25 \times 1.21 = 2.7$ MPa

$$l_{bd,reqd} = \frac{\phi}{4} \frac{\sigma_{sd}}{f_{bd}} = \{\frac{25}{4} \times \frac{373}{2.7}\} = 863 \text{ mm}$$

Taking $\alpha_1 = 1.0$, $l_{bd} = l_{bd,\,reqd} = 863$ mm.

Bar curtailment at bottom in span 1–2:

If one bar is curtailed, A_s due to two bars $= 982$ mm$^2 > A_{s,\,min}$

Equating tension and compression forces, $A_s \times f_{yd} = (3000 \times s) \times f_{cd}$

$s = 8.5$ mm

$z = d - s/2 = 446$ mm

$M = A_s \times f_{yd} \times z = 190.3$ kNm

From Case 1 in Span 1–2: $M = 190.3 = 143.69 \times x - 42.08 \times x^2/2$

Moment of 190.3 kNm occurs in span 1–2 in Case 1 at 1.79m and 5.03 m from support. The maximum moment occurs at 3.42 m from support.

Taking $\alpha_1 = 1.0$, $l_{bd} = l_{bd,\,reqd} = 863$ mm.

Design support section: M = 293.7 kNm

As the flange is in tension, design as a rectangular beam 300 × d

$293.58 = 16.67 \times 300 \times s \times (450 - s/2) \times 10^{-6}$

$s^2 - 900 s + 117449 = 0$

$s = 158$ mm

$A_s \times f_{yd} = (3000 \times s) \times f_{cd}$

$A_s = 1817$ mm$^2 =$ Say 4H25 $= 1964$ mm^2

Ratio $A_{s,\,Provided}/A_{s,\,required} = 0.93$

Maximum stress in bar $= f_{yd} \times 0.93 = 402$ MPa

$$l_{bd,reqd} = \frac{\phi}{4} \frac{\sigma_{sd}}{f_{bd}} = [\frac{25}{4} \times \frac{402}{2.7}] = 931 \text{ mm}$$

Taking $\alpha_1 = 1.0$, $l_{bd} = l_{bd, reqd} = 931$ mm.

Bar curtailment at top at support 2 in spans 1−2 and 2−3:
If two bars are curtailed, A_s due to two bars $= 982$ mm^2
Equating tension and compression forces, $A_s \times f_{yd} = (300 \times s) \times f_{cd}$
$s = 85$ mm
$z = d − s/2 = 407$ mm
$M = A_s \times f_{yd} \times z = 174.0$ kNm
Span 1−2:
From Case 2, $M = −174.0 = 131.61 \text{ x} − 42.08 \text{ x}^2/2$
$x = 7.38$ m from left support or at $(8.0 − 7.38) = 0.62$ m from support in span 1−2.
From Case 3 in span 1-2, the negative moment is zero when
$M = 0 = 53.37 \text{ x} − 19.5 \text{ x}^2/2$ at $x = 5.47$ m from left hand support or at
$(8.0 − 5.47) = 2.53$ m from support 2.
Span 2−3:
From Case 2, $M = −174 = −293.69 + 183.32 \text{ x} − 42.08 \text{ x}^2/2$
$x = 0.71$ m from support in span 2−3.
In span 2-3, negative moment always exists as can be seen from the moment envelope shown in Fig. 5.45.

5.3.4.1 Anchorage of Curtailed Bars and Anchorage at Supports

As shown in section 5.1.4.1, the code in sections 9.2.1.3 gives rules for the curtailment of longitudinal tension reinforcement.

1. Sufficient reinforcement must be provided at all sections to resist the tension caused by the bending moment and additional tension caused by the effect of inclined cracks due to shear. The effect of the additional tension force can be accounted for by using the 'shift rule'.

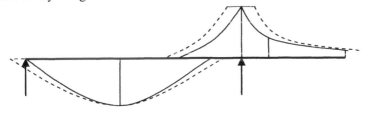

Fig. 5.46 Shift rule.

The modified bending moment diagram is obtained by shifting the basic bending moment diagram by a distance a_1 from the position of maximum moment (and zero shear force) towards the direction of decreasing moment. Therefore the theoretical cut-off point based on bending moment only is shifted a distance a_1 from the position of maximum moment towards the decreasing moment as shown in Fig. 5.46.

The equation for a_1 is given by code equation (9.2) as

$$a_1 = z\,(\cot\varphi - \cot\alpha)/2$$

where

$z \approx 0.9$ d,

α = inclination of shear reinforcement to beam axis.

φ = angle of inclination of concrete struts. $2.5 \geq \cot\varphi \geq 1.0$.

Note that if $\cot\varphi \approx 2.5$ and $\cot\alpha = 0$, $a_1 \approx 1.125$ d.

2. All bars should extend a length equal to l_{bd} beyond the point where they are no longer needed.

The cut off point from a support = Theoretical cut off $- a_1 - l_{bd}$.

Bottom steel: Taking $\cot\varphi \approx 2.5$, $\cot\alpha = 0$ because of vertical links only, d = 450 mm and z = 0.9 d = 405, then

$$a_1 = z\,(\cot\varphi - \cot\alpha)/2 = 506 \text{ mm}$$

The bending moment diagram is shifted by 0.506 m from the position of maximum moment towards the supports as shown in Fig. 5.46.

The bars can be curtailed as follows.

Bottom steel in span 1–2: Middle bar H25.

Adding l_{bd} = 0.863 m and a_1 to the end of the bar, the final position of the bar is $(1.79 - 0.506 - 0.863) = 0.421$ m to $(5.03 + 0.506 + 0.863 = 6.4$ m) from the left hand support.

Two bars extend all the way through with an anchorage length of l_{bd} = 0.0.863 m at the end supports. At the intermediate support, the bar should have an anchorage length of 10 φ = 250 mm.

Top steel: At the top, 4H25 extend from the support a length equal to $0.62 + a_1 + l_{bd} = 0.62 + 0.503 + 0.931 = 2.054$ m to the left of support 2 and $0.71 + a_1 + l_{bd} = 0.71 + 0.503 + 0.931 = 2.144$ m to the right of support 2.

From the moment envelope, the negative moment is zero at 2.53 m to the left of support 2. Therefore, 2H25 continue a length of $2.53 + a_1 + l_{bd} = 2.53 + 0.503 + 0.931 = 3.964$ m to the left of support 2.

On the right, the bars continue through the entire span 2-3 and extend 3.964 m beyond support 3.

5.3.4.2 Anchorage of Bottom Reinforcement at an End Support

In clause 9.2.1.4, the rules are given as follows.

1. If in design it is assumed that at an end support provides little or no end fixity, then the area of bottom reinforcement provided should be at least 25% of that provided in the span. The anchorage length is measured from the line of contact of the beam with support as shown in Fig. 5.47.

2. At intermediate supports, as shown in Fig. 5.48, the anchorage length should not be less than 10φ, where φ is the diameter of the bar.

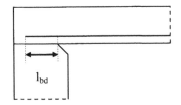

Fig. 5.47 Anchorage of bottom steel at end support.

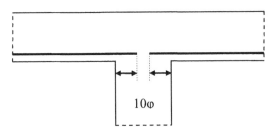

Fig. 5.48 Anchorage of bottom steel at intermediate support.

5.3.5 Laps

Lengths of reinforcing bars are joined by lapping, by mechanical couplers or by butt or lap welded joints. Only lapping which is the usual way of joining bars is discussed here. It is important that laps are not located in areas subjected to high moments. In order that sections are not weakened, laps should be staggered so that not all laps occur at the same section.

The lap length l_0 is given by code equations (8.10) and (8.11) as

$$l_0 = \alpha_1\,\alpha_2\,\alpha_3\,\alpha_4\,\alpha_5\,\alpha_6\,l_{b,reqd} \geq l_{0,min} \qquad (8.10)$$

$$l_{0,min} = max(0.3\alpha_6; 15\phi; 200 \text{ mm}) \qquad (8.11)$$

α_1 to α_5 are as given Table 5.6 but α_3 calculated taking

$$\Sigma A_{st,\,min} = \text{Area of one lapped bar} \times (\sigma_{sd}/\,f_{yd})$$

α_3 is calculated as

$$1.5 \geq [\alpha_6 = \sqrt{\frac{\rho_1}{25}}] \geq 1.0$$

$\rho_1 < 0.33$, $\alpha_6 = 1.0$ and $\rho_1 > 0.50$, $\alpha_6 = 1.5$.

ρ_1 = % of reinforcement lapped with in 0.65 l_0 from the centre of lap considered. As an example, consider the situation shown in Fig. 5.49.

In this example four bars are lapped. Counting only those bars whose centre lines of the lapped section are within the 0.65 l_0 on either side of the first lapped bar, one can count only the top and bottom bars. Therefore in this case $\rho_1 = 50\%$.

$$\alpha_6 = \sqrt{(50/25)} = 1.41.$$

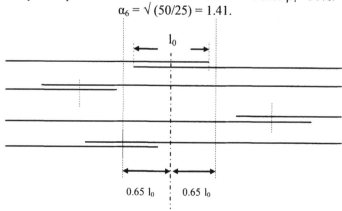

Fig. 5.49 Percentage of lapped bars at one section.

The rules for laps are as follows.

a. The clear distance between the lapped bars should be equal to or less than 4φ or 50 mm as shown in Fig. 5.50. If it is greater than these, then the lap length should be increased by the amount by which the limitation is exceeded.

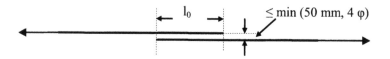

Fig. 5.50 Single lap.

b. In the case of adjacent laps, the clear distance between **adjacent bars** should be equal to greater than 2φ or 20 mm and the longitudinal distance between two **adjacent laps** should not be less than 0.3 l_0 as shown in Fig. 5.51.

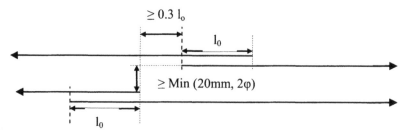

Fig. 5.51 Adjacent lap.

Where all the conditions stated above are complied with, the lapped bars in tension may be 100% provided all bars are in a single layer. Where the bars are in several layers, the percentage should be reduced to 50%

5.3.5.1 Transverse Reinforcement in the Lap Zone

Transverse reinforcement is required in the lap zone to resist transverse tension forces. Fig. 5.52 and Fig. 5.53 show the details of transverse reinforcement in tension and compression laps.

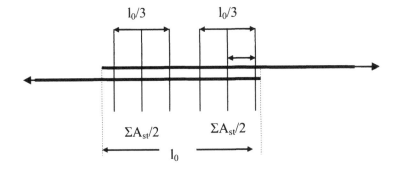

Fig. 5.52 Transverse reinforcement at a tension lap.

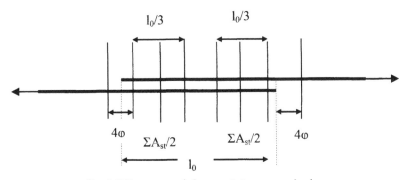

Fig. 5.53 Transverse reinforcement at a compression lap.

The requirements for tension laps are as follows.

*   If the diameter of the bars is less than 20 mm and the percentage of bars lapped in any section is less than 25%, then any transverse reinforcement or links necessary for other reasons such as shear may be assumed to be sufficient.
*   If the diameter of the bars is greater than or equal to 20 mm then total transverse reinforcement $\Sigma A_{st} \geq$ area of one lapped bar should be provided

as shown in Fig. 5.52. The spacing between the links must be less than 150 mm.

The requirements for compression laps are similar to that of tension laps, except that the last link should be placed at a distance of four diameters from the end of the lap.

5.3.5.2 Example of Transverse Reinforcement in the Lap Zone

The tension reinforcement in the web of a T-beam 250 mm wide consists of three 32 mm bars. The bars are lapped as shown in Fig. 5.54. Calculate the transverse reinforcement required. Concrete cover is 30 mm and link diameter is 12 mm. Assume $f_{ck} = 30$ MPa, $f_{yk} = 500$ MPa.

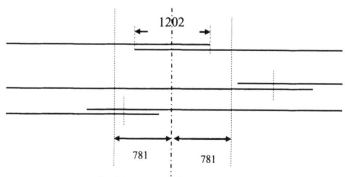

Fig. 5.54 Lapping of bars in T-beam.

Solution: Calculate factors α_1 to α_6:

$\alpha_1 = 1.0$ as the bars are straight

α_2: (See Fig. 5.36)

$c = c_1 = 30 + 12 = 42$ mm

a = clear space between bar = $[250 - 2 \times (30 + 12) - 3 \times 32]/2 = 35$ mm

$c_d = \min (35/2; 42; 42) = 18$ mm

$\alpha_2 = 1 - 0.15 \times (18 - 32)/32 = 1.44 > 1.0$

Take $\alpha_2 = 1.0$

Assume α_3 to $\alpha_5 = 1.0$

Calculate α_6:

Percentage of reinforcement lapped: As the centre lines of only two out of three laps are likely to be in the $0.65l_0$ region, $\rho_1 = (2/3) \times 100 = 67$

$\alpha_6 = \sqrt{(67/25)} = 1.64 > 1.5$

Take $\alpha_6 = 1.5$

Calculate $l_{b, reqd}$

$$f_{ctm} = 0.30 \times f_{ck}^{0.667} = 0.30 \times (30)^{0.667} = 2.9 \, \text{MPa}$$

$$f_{ctd} = 0.7 \times f_{ctm} / (\gamma_c = 1.5) = 1.35 \, \text{MPa}$$

$\varphi = 32$ mm, $\eta_1 = 1.0$ for good bond, $\eta_2 = 1.0$ as $\varphi \leq 32$ mm

$$f_{bd} = 2.25\eta_1 \, \eta_2 \, f_{ctd} = 2.25 \times 1.0 \times 1.0 \times 1.35 = 3.1 \, \text{MPa}$$

$$f_{yk} = 500 \text{ MPa}, \sigma_{sd} = f_{yk}/(\gamma_s = 1.15) = 435 \text{ MPa}$$

$$l_{bd,reqd} = \frac{\phi}{4} \frac{\sigma_{sd}}{f_{bd}} = \{\frac{32}{4.0} \times \frac{435}{3.1}\} = 1123 \text{ mm}$$

$$l_0 = (\alpha_1 \, \alpha_2 \, \alpha_3 \, \alpha_4 \, \alpha_5 \, \alpha_6) \times l_{b, reqd} = 1.5 \times 1123 = 1684 \text{ mm}$$

$$0.65 \, l_0 = 1095 \text{ mm}, \, l_0/3 = 505 \text{ mm}$$

Calculate A_{st}:

A_s = Area of 32 mm bar = 804 mm^2

$\Sigma A_{st} = A_s = 804$ mm^2, $\Sigma A_{st}/2 = 402$ mm^2

$l_0/3 = 505$ mm

$505/150 = 3.4$, say 4

Number of links required on each side of lap = 4 + 1 = 5

Area per link = 402/5 = 80 mm^2

Area of a 12 mm bar = 113 mm^2

Provide at laps a total of ten 12 mm diameter bars as shown in Fig. 5.52.

5.3.6 Bearing Stresses Inside Bends

Bars are anchored by providing a standard bend as shown in Fig. 5.35(c). It is often necessary to anchor a bar by extending it around a bend in a stressed state, as shown in Fig. 5.55.

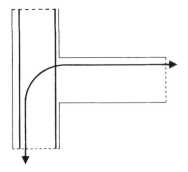

Fig. 5.55 Bar anchored around a bend.

In both cases concrete inside the bend is subjected to compressive stress and it is necessary to make the radius of the bend large enough to prevent the crushing of the concrete. It is also necessary that the radius of the bend is not so small that it might crack the bars during fabrication of the reinforcement.

In clause 8.3, the code prescribes the following limitations.

1. In order to prevent the bar from cracking during the bending of the bar, the minimum inside diameter of the bend should be 4φ for bars of 16 mm diameter and under. For bars of greater diameter than 16 mm, the radius of the bend should be 7 φ.

2. Explicit verification of concrete stress inside the bend is not necessary, if:

a. The anchorage does not require the bar to extend by more than 5 φ beyond the end of the bend
b. Inside radius complies with the requirement in (a) above.
c. The plane of the bend is not close to concrete face and there is a bar of a diameter at least equal to the diameter of the bar inside the bend as shown in Fig. 5.56 which will clearly reduce the stress in concrete inside the bend.
d. If the above conditions are not fulfilled, then the minimum diameter of the bar around which the bar should be bent (called diameter of the mandrel) is given by code equation (8.1)

$$\phi_{m,min} \geq \frac{F_{bt}}{f_{cd}} \times \{\frac{1}{a_b} + \frac{1}{2\phi}\} \qquad (8.1)$$

where

- F_{bt} = tensile force in the bar, φ = diameter of the bar, a_b = **half** the c/c distance between bars, $f_{cd} = f_{ck} / (\gamma_c = 1.5)$, $f_{ck} \leq 50$ MPa.
- For bars adjacent to the face of a member, a_b = cover + φ/2.
- Table 5.7 gives the ratio of $\phi_{m,\,min}/\phi$ for $f_{yk} = 500$ MPa, $\gamma_s = 1.15$, $F_{bt} = (\pi/4) \phi^2 f_{yd}$.

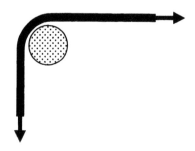

Fig. 5.56 Longitudinal bar inside a bend.

Table 5.7 Mandrel diameter in terms of bar diameter

f_{ck}	$\phi_{m,\,min}/\phi$				
	$a_b = 2\,\phi$	$a_b = 3\,\phi$	$a_b = 4\,\phi$	$a_b = 5\,\phi$	$a_b = 10\,\phi$
20	25.6	21.3	19.2	17.9	15.4
25	20.5	17.1	15.4	14.3	12.3
30	17.1	14.2	12.8	12.0	10.2
35	14.6	12.2	11.0	10.2	8.8
40	12.8	10.7	9.6	9.0	7.7
45	11.4	9.5	8.5	8.0	6.8
50	10.2	8.5	7.7	7.2	6.2
≥55	9.3	7.8	7.0	6.5	5.6

5.4 TORSION

5.4.1 Occurrence and Analysis of Torsion

In practice two types of torsion occur.

1. Equilibrium torsion: This is the case where torsional resistance must be provided in order to maintain equilibrium. Fig. 5.57 shows an example of a cantilever L-shaped in horizontal plane, carrying a vertical load at the tip. The longitudinal beam is subjected to a twisting moment equal to [W × a] and it is necessary that the beam can resist this twisting action as otherwise the beam will simply collapse. The twisting moment (W × a) is equilibrium torsion.

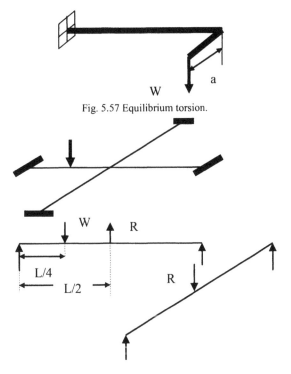

Fig. 5.57 Equilibrium torsion.

Fig. 5.58 Two interconnected beams.

2. Compatibility torsion: This is the case where members are subjected to twisting moment in order to preserve continuity of displacements, but torsional resistance is not required to maintain equilibrium. Fig. 5.58 shows a grid work of two intersecting beams. Only the longitudinal beam carries an external load W normal to the plane of the grid. An analysis assuming that the two beams are identical and are pin connected at the intersection shows that for the two beams to have the same deflection at the intersection, the load distribution is as shown. The vertical force R at the intersection is 0.344W, where W is the applied load. The longitudinal beam rotates in the clockwise direction at the intersection. If the two beams are rigidly connected, then the transverse beam, for compatibility reasons, will be forced to rotate in the same direction. This twisting of the transverse beam will create a twisting moment in the longitudinal beam. This is called compatibility torsion. This torsion is not needed to maintain equilibrium.

The code suggests in section 6.3.1 (2) that the minimum longitudinal and shear reinforcement normally provided can provide sufficient resistance against excessive cracking caused by compatibility torsion. Reinforcement needs to be designed to resist equilibrium torsion only.

5.4.2 Torsional Shear Stress in a Concrete Section

Fig. 5.59 shows a rectangular beam subjected to a torsional moment. Because concrete is weak in tension, the resulting shear stresses cause cracks which spiral around the axis.

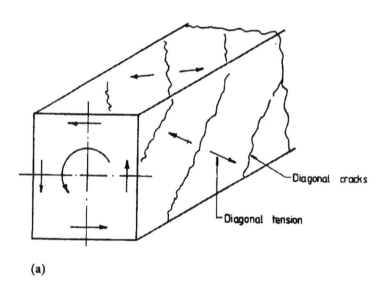

(a)

Fig. 5.59 Diagonal cracking pattern.

Fig. 5.60 shows a rectangular box beam whose wall thickness can be considered as small compared to other cross sectional dimensions. It is shown in books on strength of materials that when the box section is subjected to a torsional moment T_{Ed}, the shear flow q defined as the product of shear stress in the wall and its thickness is a constant in the walls of the box. The walls of the box are in a state of pure shear. The shear stress τ_i in the side of thickness t_i is given by

$$q = \tau_i \, t_i = \frac{T}{2 A_k}$$

where A_k is the area enclosed by the centre line dimensions of the sides of the box and t is wall thickness.

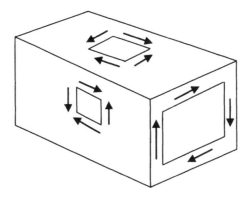

Fig. 5.60 Stresses in a thin walled box beam under torsion.

Fig. 5.61 shows the elastic stress distribution in a solid rectangular section subjected to a torsional moment. The shear stresses due to torsion are tangential to the sides and in an elastic material, the maximum shear stress occurs in the middle of the longer side of a rectangular section. The stress is zero at the centroid of the section and increases in a non-linear manner towards the edges. A very high proportion of the torsional resistance comes from the shear stresses acting over a short thickness near to the surface of the box. Therefore for all practical purposes the solid section can be treated as a thin-walled hollow section

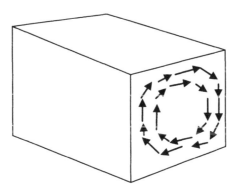

Fig. 5.61 Torsional shear stress distribution in a solid rectangular section.

The above ideas can be generalized and hold true for any hollow section. In clause 6.3.2, the code gives the following design procedure.

As shown in Fig. 5.62, the shear stress $\tau_{t, i}$ in the ith wall of thickness $t_{ef, i}$ of a section subjected to pure torsional moment T_{Ed} may be calculated from code equation (6.26) as

$$q = \tau_{t,i} \, t_{ef,i} = \frac{T_{Ed}}{2 A_k} \qquad (6.26)$$

where
q = constant shear flow the walls of the section.
$\tau_{t, i}$ = shear stress in the ith wall.
$t_{ef, i}$ = effective wall thickness of the ith wall.
A_k = area enclosed by the centrelines of the connecting walls including the inner
    hollow areas.
Note that
    1. $t_{ef, i}$ may be taken as A/u, but should not be taken as less than twice the distance between the edge and the centreline of the longitudinal reinforcement. For hollow sections, the real thickness is the upper limit.
    2. A = total area of the cross section within the outer circumference, including inner hollow areas.
    3. u = outer circumference of the cross section.
The shear force $V_{Ed, i}$ in the ith wall is given by code equation (6.27) as
$$V_{Ed, i} = t_{ef, i} \, t_{ef, i} \, z_i = q \, z_i \qquad (6.27)$$

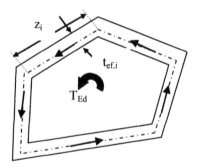

Fig. 5.62 Torsional stress distribution in a polygonal hollow section.

5.4.2.1 Example

Fig. 5.63 shows the cross section of a trapezoidal box girder with side cantilevers. The webs and bottom flange are 300 mm thick; the top flange is 400 mm. It is subjected to a torsional moment T_{Ed} = 5000 kNm. Calculate the shear forces in the walls of the girder.

Solution: As the thin cantilevers will not provide any significant torsional resistance, the cross section can be reduced to a simple trapezoidal box as shown in Fig. 5.64. The top width of the box:
    $1200 + 2 \times \{(3000 - 1200 - 2 \times 600)/2\} \times 2000/ (2000 - 400) = 1950$ mm

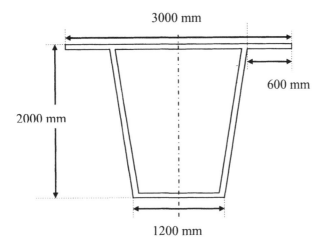

Fig. 5.63 Trapezoidal box girder with side cantilevers.

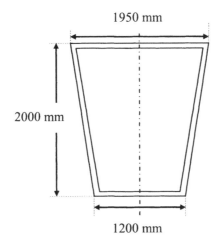

Fig. 5.64 Trapezoidal box girder.

(i) The centre-line dimensions:

Top flange = 1950 − 300 = 1650 mm

Bottom flange = 1200 − 300 = 900 mm

Height = 2000 − 400/2 − 300/2 = 1965 mm

Inclined length of web = $\sqrt{[1965^2 + \{(1650 - 900)/2\}^2]}$ = 2000 mm

A_k = 1965 × (1650 + 900)/2 = 2.505 × 10^6 mm^2

u_k = Perimeter of A_k = 1650 + 900 + 2 × 2000 = 6550 mm

$q = T_{Ed}/ (2 \times A_k)$ = 5000 × 10^6/ (2 × 2.505 × 10^6) = 998 N/mm

(ii) Shear stress in the walls:

Webs: τ = q/t = 998/300 = 3.3 MPa

Bottom flange: τ = q/t = 998/300 = 3.3 MPa

Top flange: $\tau = q/t = 998/400 = 2.5$ MPa
(iii) Shear forces in the walls:
Webs: $V = q \times$ length$= 998 \times 2000 \times 10^{-3} = 1996$ kN
Bottom flange: $V = q \times$ length$= 998 \times 900 \times 10^{-3} = 898$ kN
Top flange: $V = q \times$ length$= 998 \times 1650 \times 10^{-3} = 1647$ kN

5.4.3 Design for Torsion

As explained in the previous section, torsional moment induces shear stresses in the walls of the beam. Design of the walls for the shear stress induced by torsion follows the same lines as design for shear in beams. The walls are assumed to resist shear stresses by a combination of concrete struts and shear links. Unlike beams where there is compression force in the top chords due to bending moment, under pure shear induced by torsion, it is necessary to provide longitudinal reinforcement in the top chord also.
Fig. 5.65 shows the composite truss resisting the shear stress caused by torsion.

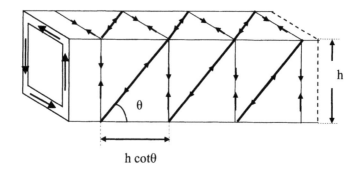

Fig. 5.65 Composite truss resisting torsional moment.

If $q = T_{Ed}/(2 A_k)$ is the shear flow per unit length, going around the perimeter u_k, the total force due to torsion is $q\, u_k$. In the same manner, if F_c is the force in the concrete compression strut, for equilibrium
$$F_c \sin \theta = q\, u_k$$
$$F_c = q\, u_k/\sin \theta$$
$$F_c \cos \theta = \text{Total horizontal force} = \Sigma A_{sl}\, f_{yd}$$
where ΣA_{sl} = total longitudinal steel area and f_{yd} = stress in the steel reinforcement. Expressing F_c in terms of u_k,
$$\Sigma A_{sl}\, f_{yd} = q\, u_k\, (\cos \theta/\sin \theta)$$
$$\frac{\Sigma A_{sl}\, f_{yd}}{u_k} = q \cot \theta$$
In the above equation substituting for q in terms of twisting moment as $T_{Ed}/(2 A_k)$ leads to the equation (6.28) of Eurocode 2.

$$\frac{\Sigma A_{sl} f_{yd}}{u_k} = \frac{T_{Ed}}{2 A_k} \cot \theta \qquad (6.28)$$

If A_{sw} is the area of one link spaced at s, the vertical force resisted by the link is

$$F_s = A_{sw} \frac{h \cot \theta}{s} f_{yd}$$

The vertical force from torsion is qh. Equating F_s to qh,

$$F_s = qh = A_{sw} \frac{h \cot \theta}{s} f_{yd}$$

Substituting for q in terms of T_{ED},

$$qh = \frac{T_{Ed}}{2 A_k} h = A_{sw} \frac{h \cot \theta}{s} f_{yd}$$

Simplifying, the equation to calculate the link reinforcement is given by

$$A_{sw} f_{yd} = \frac{T_{Ed}}{2 A_k} \frac{s}{\cot \theta}$$

It is necessary to put a limit on the compressive stress in the struts. Considering the truss in a single plane as shown in Fig. 5.66:

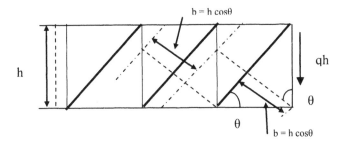

Fig. 5.66 Composite truss; width of struts.

- Compressive force $F_c = qh/\sin\theta$
- If σ_c is the stress in the strut, $F_c = t_{ef} b \sigma_c$
- The width covered by a strut is $b = h \cos\theta$

Equating the two expressions for Fc,

$$qh/\sin\theta = t_{ef} h \cos\theta \sigma_c$$

$$\sigma_c = \frac{q}{t_{ef} \sin\theta \cos\theta} = \frac{T_{Ed}}{2 A_k} \frac{1}{t_{ef}} \frac{1}{\sin\theta\cos\theta} = \frac{T_{Ed}}{A_k} \frac{1}{t_{ef}} \frac{1}{\sin 2\theta}$$

The code limits the concrete stress σ_c to

$$\sigma_c = v_1 f_{cd}$$

where $v_1 = v$ given by code equation (6.6N) is an efficiency factor which allows for the effects of cracking as well for the actual distribution of stress in the struts.

$$v_1 = v = 0.6 (1 - f_{ck}/250) \qquad (6.6N)$$

$$\sin 2\theta = \frac{T_{Ed}}{A_k} \frac{1}{t_{ef} \sigma_c}$$

$$2.5 \geq \cot \theta \geq 1.0$$

5.4.3.1 Example of Reinforcement Design for Torsion

Design the longitudinal and link reinforcement for the example in 5.3.3.1. Assume f_{ck} = 30 MPa, f_{cd} = $f_{ck}/1.5$ = 20 MPa, f_{yk} = 500 MPa, f_{yd} = 500/1.15 = 435 MP.

$$v_1 = 0.6(1 - 30/250) = 0.528$$
$$\sigma_c = v_1 f_{cd} = 0.528 \times 20 = 10.56 \text{ MPa}$$

As previously calculated, u_k = 6550 mm, q = 998 N/mm,

$$\sin 2\theta = \frac{T_{Ed}}{A_k} \frac{1}{t_{ef} \sigma_c} = \frac{5000 \times 10^6}{2.505 \times 10^6} \times \frac{1}{300 \times 10.56} = 0.63$$

$$\theta = 19.5^0, \cot \theta = 2.82 > 2.5$$

Take cot θ = 2.5

$$\Sigma A_{sl} \times 435 = \frac{T_{Ed}}{2 A_k} u_k \cot \theta = \frac{5000 \times 10^6}{2 \times 2.505 \times 10^6} \times 6550 \times 2.4 = 16.34 \times 10^6$$

ΣA_{sl} = 37568 mm², $\Sigma A_{sl}/u_k$ = 5.74 mm²/mm

Using H25 bars, this gives one H25 at every 86 mm. The above bars are distributed around the perimeter in a uniform manner. In order to tie the links, it is practical to put one bar at each corner.

Top flange, provide an additional 19 bars at 87 mm spacing.

Bottom flange, provide an additional 10 bars at 90 mm spacing.

In each web, provide 23 bars at 87 mm spacing.

Total number of bars provided = 4 + 19+ 10 + 2 × 23 = 79H25 = 38779 mm².

Link reinforcement: Using H12 links, A_{sw} = 113 mm²

$$A_{sw} f_{yd} = \frac{T_{Ed}}{2 A_k} \frac{s}{\cot \theta}$$

$$113 \times 435 = \frac{5000 \times 10^6}{2 \times 2.505 \times 10^6} \times \frac{s}{2.5}$$

$$s = 123 \text{mm}$$

Provide H12 links at 120 mm c/c.

5.4.4 Combined Shear and Torsion

As the mode of resistance for shear and torsion is by a composite truss consisting of concrete struts, shear links and longitudinal reinforcement, it is important to make sure that the concrete struts are not overstressed.

The code in equation (6.29) limits the combined effect of shear force and torsion as follows.

$$\frac{T_{Ed}}{T_{Rd,max}} + \frac{V_{Ed}}{V_{Rd,max}} \leq 1.0 \tag{6.29}$$

$$T_{Rd,max} = 2\nu\alpha_{cw}\, f_{cd}\, A_k\, t_{ef,i}\, \sin\theta\cos\theta \tag{6.30}$$

$$V_{Rd,max} = \alpha_{cw}\, b_w\, z\nu\, f_{cd}\frac{1}{\cot\theta + \tan\theta} \tag{6.9}$$

$\alpha_{cw} = 1$ for non-prestressed structures

5.4.4.1 Example of Design of Torsion Steel for a Rectangular Beam

A rectangular beam section has an overall depth of 500 mm and a breadth of 300 mm. It is subjected at ultimate to a vertical sagging moment of 320 kNm, a vertical shear of 230 kN and a torque of 30 kN m. Design the longitudinal steel and links required at the section. The material strengths are $f_{ck} = 30$ MPa, $f_{yk} = 500$ MPa. Use H25 bars for longitudinal steel and H8 bars for links.

Design for bending:
Using cover to steel of 30 mm,
$$d = 500 - 30 - 10 - 25/2 = 447 \text{ mm}$$
$$k = M/ (b\, d^2\, f_{ck}) = 320 \times 10^6/ (300 \times 447^2 \times 30) = 0.178 < 0.196$$
Design as a singly reinforced beam.
$$\eta = 1,\ z/d = 0.5[1.0 + \sqrt{(1 - 3 \times k)}] = 0.84$$
$$f_{yd} = 500/1.15 = 435 \text{ MPa}$$
$$A_s = M/ (z \times f_{yd}) = 320 \times 10^6/ (0.84 \times 447 \times 435) = 1959 \text{ mm}^2$$
Provide 4H25, $A_s = 1964$ mm$^2 > 1959$ mm^2
Check minimum steel:
$$A_{s,min} = 0.26 \times (f_{ctm}/f_{yk}) \times b_t\, d, \ f_{ctm} = 0.3 \times 30^{0.667} = 2.9 \text{ MPa}, f_{yk} = 500 \text{ MPa},$$
$$A_{s,min} = 202 \text{ mm}^2 < 1959 \text{ mm}^2$$

Design for shear:
$$V_{Ed} = 230 \text{ kN}$$

i. Check if shear reinforcement is required, $V_{Ed} > V_{Rd,c}$
$V_{Ed} = 230$ kN, $b_w = 300$ mm, $d = 447$ mm, $A_{sl} = 4H25 = 1964$ mm^2

$$C_{Rd,c} = \frac{0.18}{(\gamma_c = 1.5)} = 0.12$$

$$k = 1 + \sqrt{\frac{200}{447}} = 1.67 \leq 2.0$$

$$100\rho_1 = 100 \times \frac{1964}{300 \times 447} = 1.47 < 2.0$$

$$v_{min} = 0.035 \times 1.67^{1.5} \times \sqrt{30} = 0.41 \text{ MPa}$$

$$V_{Rd,c} = [0.12 \times 1.67 \times \{1.47 \times 30\}^{1/3} + 0.15 \times 0] \times 300 \times 447 \times 10^{-3}$$

$$\geq [0.41 + 0.15 \times 0] \times 300 \times 447 \times 10^{-3}$$

$$V_{Rd,c} = 94.9 \geq 55.0 \text{ kN}$$

$$V_{Rd, c} = 94.9 \text{ kN} < V_{Ed}.$$

Therefore shear reinforcement is required.

ii. Check if the section strength is adequate, $V_{Ed} < V_{Rd, max}$

$$\theta = 0.5 \sin^{-1} \{ \frac{2 V_{Ed}}{\alpha_{cw} \, b_w \, z \, v_1 \, f_{cd}} \}$$

$$\theta = 0.5 \sin^{-1} \{ \frac{2 \times 230 \times 10^3}{1.0 \times 300 \times (0.84 \times 447) \times 0.6 \times \dfrac{30}{1.5}} \}$$

$$= 0.5 \sin^{-1} (0.34) = 0.5 \times 19.9 = 9.95^0$$

$$\cot \theta = 5.7$$

The value of $\cot \theta$ is outside the limits of 1.0 and 2.5. Choose $\cot \theta = 2.5$.

$$v_1 = v = 0.6(1 - f_{ck}/250) = 0.6(1 - 30/250) = 0.528$$

$$V_{Rd,max} = \alpha_{cw} \, b_w \, z \, v_1 \, f_{cd} \frac{1}{\cot \theta + \tan \theta}$$

$$= 1.0 \times 300 \times (0.84 \times 447) \times 0.528 \times 20 \times \frac{1}{(2.5 + 0.4)} \times 10^{-3} = 410 \text{kN}$$

$$V_{Rd, max} = 410 \text{ kN} > V_{Ed}.$$

Section size is adequate.

iii. Design of shear reinforcement:

Ensure that $V_{Rd, s} \geq V_{Ed}$, and choose 2-leg H8 links,

$$A_{sw} = 101 \text{ mm}^2, \cot \theta = 2.5, z = 0.84d, f_{yk} = 500 \text{ MPa},$$

$$V_{Rd,s} = \frac{z}{s} A_{sw} f_{ywd} \cot \theta$$

$$= \frac{(0.84 \times 447)}{s} \times 101 \times \frac{500}{1.15} \times 2.5 \times 10^{-3}$$

$$= \frac{41221}{s} \text{kN} \geq (V_{Ed} = 230 \text{kN})$$

$$s \leq 179 \text{mm}$$

Maximum spacing $= 0.75d = 0.75 \times 447 = 335$ m.
Check minimum shear steel:

$$\rho_{w, min} = (0.08 \times \sqrt{f_{ck}})/ f_{yk} = 0.876 \times 10^{-3}$$

$$\rho_w = A_{sw}/(s \times b_w) = 101/(179 \times 300) = 1.881 \times 10^{-3} > 0.876 \times 10^{-3}$$

Design for torsion:

Calculate t_{ef}:

$$A = 300 \times 500, u = 2(300 + 500),$$

Effective wall thickness $t_{ef,i} = A/u = 94$ mm

Dimension of the 'hollow' section:

$$b = 300 - 94 = 206 \text{ mm}, h = 500 - 94 = 406 \text{ mm}$$
$$A_k = 206 \times 406 = 83636 \text{ mm}^2, u_k = 2(206 + 405) = 1224 \text{ mm}$$
$$T_{Ed} = 30 \text{ kNm}$$
$$q = T_{Ed}/ (2 A_k) = 179 \text{ N/mm}$$

$$\Sigma A_{sl} f_{yd} = \frac{T_{Ed}}{2 A_k} u_k \cot \theta$$

$$\Sigma A_{sl} \times 435 = \frac{30 \times 10^6}{2 \times 83636} \times 1224 \times 2.5$$

$$\Sigma A_{sl} = 1262 \text{ mm}^2, 3H25 = 1473 \text{ mm}^2$$

For symmetry, use 4H25. Since bending provides compression in the top chord, only 2H25 need be provided in the bottom chord.

Link design:

$$A_{sw} = \text{area of H8 bar} = 50 \text{ mm}^2, f_{yd} = 500/1.15 = 435 \text{ MPa}$$

$$A_{sw} f_{yd} \geq \frac{T_{Ed}}{2 A_k} \frac{s}{\cot \theta}$$

$$50 \times 435 \geq \frac{30 \times 10^6}{2 \times 83636} \times \frac{s}{2.5}$$

$$s \leq 303 \text{ mm}$$

Arrangement of reinforcement

Longitudinal bars: bending = 4H25, torsion = 2H25, total 6H25.
Provide 4H25 in the bottom and 1H25 in the middle of the sides.
Links: shear: 2-leg H8 links at 179 mm c/c, torsion: H8 links at 303 mm c/c.
Provide 8 mm links at 150 c/c but use 2-leg links alternatively.

Check shear-torsion interaction:

$$\cot\theta = 2.5, \theta = 21.8^0, \sin \theta = 0.37, \cos \theta = 0.929$$

$$T_{Rd,max} = 2 v_1 \alpha_{cw} f_{cd} A_k t_{ef,i} \sin\theta \cos \theta$$

$$= 2 \times 0.528 \times 20 \times 83636 \times 94 \times 0.37 \times 0.93 \times 10^{-6} = 57.1 \text{kNm}$$

$$V_{Ed} = 230 \text{ kN}, T_{Ed} = 30 \text{ kNm}$$

$$\frac{T_{Ed}}{T_{Rd,max}} + \frac{V_{Ed}}{V_{Rd,max}} \leq 1.0$$

$$\frac{30}{57.1} + \frac{230}{410} = 1.09 > 1.0$$

Design is unsatisfactory. Use a larger section.

5.5 SHEAR BETWEEN WEB AND FLANGE OF T-SECTIONS

Section 6.2.4 of Eurocode 2 deals with the reinforcement to resist the shear between web and flanges. Fig. 5.67 shows a T-beam between two sections. The difference in bending moment between the two sections causes differential bending forces in the flanges. In order to transmit this force to the web, shear stresses develop at the junction between the web and the flange. Reinforcement is needed in the flange to resist this shear stress.

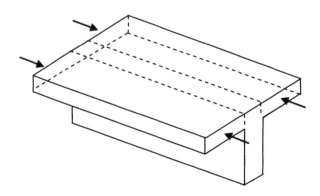

Fig. 5.67 Shear stresses at flange-web junction.

5.5.1 Example

Determine the shear reinforcement in the flanges for the T-beam example in section 5.1.10 and shown in Fig. 5.19 and Fig. 5.20.
Solution:

Mid-span bending moment = 291.6 kNm, $d = 400$ mm, $f_{ck} = 25$ MPa.
$$z/d = 0.97, z = 388 \text{ mm}.$$

Let s be the depth of flange in compression.
$$z = d - s/2, s = 24 \text{ mm}$$
Bending stress in the flange = $f_{cd} = 25/1.5 = 16.7$ MPa.
Width of one half of flange = $(2000 - 250)/2 = 875$ mm.
Compression force C in one half of flange at mid-span = $875 \times s \times f_{cd} \times 10^{-3}$
$$C = 875 \times 24 \times 16.7 \times 10^{-3} = 350.7 \text{ kN}$$
The compression force varies from zero where the bending moment is zero to a maximum value of 350.7 kN at position of maximum moment, i.e., mid-span.

The shear force $\Delta F_d = 350.7$ kN.

Assuming uniform distribution of shear stress at the flange-web junction, an average value v_{Ed} is given by code equation (6.20):
$$v_{Ed} = \Delta F_d / (h_f \times \Delta x) \tag{6.20}$$
h_f = thickness of the flange = 100 mm.
Δx = distance between the position of maximum moment and zero moment
$$\Delta x = \text{effective span}/2 = (7.2 / 2) = 3.6 \text{ m}$$

$$v_{Ed} = 350.7 \times 10^3 / (100 \times 3.6 \times 10^3) = 0.97 \text{ MPa}.$$

The transverse reinforcement A_{sf} at spacing s_f is given by code equation (6.21) as

$$\frac{A_{sf}}{s_f} f_{yd} \geq v_{Ed} \times \frac{h_f}{\cot \theta_1} \tag{6.21}$$

To prevent the crushing of the compression struts in the flange, the condition given by code equation (6.22) has to be satisfied

$$v_{Ed} \leq v f_{cd} \sin \theta_1 \cos \theta_1$$

The limits on $\cot \theta_1$ are:

$$1.0 \leq \cot \theta_1 \leq 2.0 \text{ for compression flanges}$$
$$1.0 \leq \cot \theta_1 \leq 1.25 \text{ for tension flanges}$$

$$v = 0.6 \ (1 - f_{ck}/250) = 0.528, \ f_{cd} = 25/1.5 = 16.7 \text{ MPa},$$
$$v_{Ed} \leq v f_{cd} \sin \theta_1 \cos \theta_1$$
$$0.97 \leq 0.528 \times 16.7 \times (0.5 \sin 2\theta_1)$$

$$\sin 2\theta_1 \geq 0.22, \theta_1 = 6.4^0$$
$$\cot \theta_1 = 9.0 > 2.0. \text{ Take } \cot \theta_1 = 2.0$$

Using H10 bars, $A_{sf} = 79 \text{ mm}^2$, $f_{yd} = 500/1.15 = 435 \text{ MPa}$,

$$\frac{A_{sf}}{s_f} f_{yd} \geq v_{Ed} \times \frac{h_f}{\cot \theta_1}$$

$$\frac{79}{s_f} \times 435 \geq 0.97 \times \frac{100}{2.0}$$

$$s_f \leq 704 \text{ mm}$$

Generally T-beams form a part of the monolithic slab and T-beam flooring with the slab spanning between the T-beams. The transverse bending of the flange caused by the slab spanning between the beams will need reinforcement which might very well exceed the reinforcement calculated to resist shear between flange and web.

CHAPTER 6

SERVICEABILITY LIMIT STATE CHECKS

6.1 SERVICEABILITY LIMIT STATE

In Chapter 4 and Chapter 5, design procedures for the ultimate limit state (ULS) in bending, shear and torsion were described. It is necessary in practice to ensure that the structure can not only withstand the forces at the ultimate limit state but also that it behaves satisfactorily at working loads. The main aspects to be satisfied at serviceability limit state (SLS) are deflection and cracking. In this chapter checks that are normally used to ensure satisfactory behaviour under SLS conditions without detailed calculations are considered. These are known as 'deemed to satisfy' clauses. Methods requiring detailed calculations are discussed in Chapter 17.

6.2 DEFLECTION

6.2.1 Deflection Limits and Checks

Limits for the serviceability limit state of deflection are set out in clause 7.4 of the code. It is stated in 7.4.1 (4) that the appearance and general utility of a structure could be impaired if the deflection exceeds $L/250$ where L is the span of a beam or length of a cantilever. Deflection due to dead load can be offset by pre-cambering. The code also states that deflections that could damage adjacent parts of the structure should be limited. For the deflection after construction, L/500 is normally appropriate under quasi-permanent loads.

Generally it is not necessary to calculate the deflections explicitly. Code allows limiting the span-to-effective depth ratio to ensure that deflection under SLS does not exceed the limits.

6.2.2 Span-to-Effective Depth Ratio

In a homogeneous elastic beam of span L, if the maximum stress is limited to an allowable value σ and the deflection Δ is limited to, say, span/250, then for a given load a unique value of span-to-effective depth ratio L/d can be determined to limit stress and deflection to their allowable values simultaneously. Thus for the simply supported beam with a uniformly distributed load,

$$\text{Maximum bending moment} = W\,L/8$$

where W = total load on the beam and

$$\text{Maximum stress } \sigma = M\,y/I = \frac{W\,L}{8}\frac{y}{I} = \frac{W\,L}{8}\frac{0.5d}{I}$$

where I is the second moment of area of the beam section, d is the depth of the beam and L is the span.

$$\Delta = \frac{L}{250} = \frac{5}{384}\frac{WL^3}{EI} = \frac{WL0.5d}{8I}\frac{5L^2}{24Ed} = \sigma\frac{5L^2}{24Ed}$$

$$\frac{L}{d} = \frac{4.8}{250}\times\frac{E}{\sigma}$$

where

E is Young's modulus, σ = allowable maximum stress

Similar reasoning may be used to establish span-to-effective depth ratios for reinforced concrete beams to control deflection. The method in the code is based on calculation and confirmed by tests. The main factors affecting the deflection of the beam are taken into account such as

- The basic span-to-effective depth ratio for rectangular or flanged beams and the support conditions
- The amount of tension and compression steel and the stress in them at SLS
- Concrete strength f_{ck}

The allowable value for the span-to-effective depth ratio can be calculated using the code equations (7.16a) and (7.16b) for normal cases.

The equations have been derived on the basis of the following assumptions:

- The maximum stress in steel σ_s at SLS is 310 MPa for $f_{yk} = 500$ MPa. If a different level of stress σ_s other than 310 MPa is used, then L/d from equation should be multiplied by $310/\sigma_s$ where

$$\frac{310}{\sigma_s} = \frac{500}{f_{yk}}\times\frac{A_{s,prov}}{A_{s,reqd}}$$

- For flanged sections, if b/b_w exceeds 3, then L/d from equation should be multiplied by 0.8.
- For beams and slabs other than flat slabs, where the effective span L_{eff} exceeds 7 m and the beam supports partitions liable to be damaged due to excessive deflections, L/d values from equation should be multiplied by $7/L_{eff}$.

$$\frac{L}{d} = K\left\{11 + 1.5\sqrt{f_{ck}}\frac{\rho_0}{\rho} + 3.2\sqrt{f_{ck}}\left(\frac{\rho_0}{\rho} - 1\right)^{\frac{3}{2}}\right\}\text{ if }\rho \le \rho_0 \qquad (7.16a)$$

$$\frac{L}{d} = K\left\{11 + 1.5\sqrt{f_{ck}}\frac{\rho_0}{\rho - \rho'} + \frac{1}{12}\sqrt{f_{ck}}\sqrt{\frac{\rho'}{\rho}}\right\}\text{ if }\rho > \rho_0 \qquad (7.16b)$$

where

L/d = limit of span/effective depth ratio.

K = factor to account for different structural systems.

$\rho_0 = 10^{-3} \times \sqrt{f_{ck}}$.

ρ = tension reinforcement ratio to resist the maximum moment due to design loads.

ρ' = compression reinforcement ratio to resist the maximum moment due to design loads.

Table 6.1 shows the basic L/d ratios for reinforced concrete members without axial restraint calculated from the code equations.

Table 6.1 Basic L/d ratios

Structural System	K	$\rho =$ 1.5%	$\rho =$ 0.5%
Simply supported beam, one-way or two-way spanning simply supported slab.	1.0	14	20
End span of continuous beam or one-way continuous slab or two-way spanning slab continuous over one long side.	1.3	18	26
Interior span of continuous beam or one-way or two-way spanning slab.	1.5	20	30
Flat slabs (based on longer span)	1.2	17	24
Cantilever	0.4	6	8

6.2.2.1 Examples of Deflection Check for Beams

Example 1: A continuous slab 100 mm thick is carried on T-beams at 2 m centres. The overall depth of the beam is 350 mm and the breadth b_w of the web is 250 mm. The beams are 6 m spans and are simply supported. The characteristic dead load including finishes is 4.5 kN/m^2 and the characteristic imposed load is 5 kN/m^2. The material strengths are f_{ck} =25 MPa and f_{yk} =500 MPa. Check if L/d is adequate. If inadequate, redesign the beam.

Since the beams are spaced at 2 m centres, the loads on the beam are:

100 mm thick slab + finishes = $(0.1 \times 25 + 4.5) \times 2 = 14.0$ kN/m

Weight of rib only = $(0.35 - 0.1) \times 0.25 \times 25 = 1.56$ kN/m

Live load = $5 \times 2 = 10$ kN/m

Design load = $1.35 \times (14.0 + 1.56)) + (1.5 \times 10) = 36.0$ kN/m

Ultimate moment at mid-span = $36.0 \times 6^2/8 = 162$ kN m

Effective width b_{eff} of flange can be shown to equal 1800 mm (See Chapter 4, section 4.6.5).

The beam section is shown in Fig. 4.17. Assuming a nominal cover on the links is 25 mm and if the links are H8 and the main bars are H25, then

$$d = 350 - 25 - 8 - 12.5 = 305 \text{ mm}$$

It can be shown that the beam can be designed as a rectangular beam of size 1800 × 300.

$$k = M/ (b \, d^2 \, f_{ck}) = 162 \times 10^6/ (1800 \times 305^2 \times 25) = 0.039 < 0.196$$

$$\frac{z}{d} = 0.5[1.0 + \sqrt{(1 - 3\frac{k}{\eta})}] = 0.5[1.0 + \sqrt{(1 - 3\frac{0.039}{1.0})}] = 0.97$$

$$A_s = \frac{M}{0.97\,d \times 0.87\,f_{yk}} = \frac{165 \times 10^6}{0.97 \times 305 \times 0.87 \times 500} = 1282\,\text{mm}^2$$

Provide 3H25; $A_s = 1472\,\text{mm}^2$

$A_{s,\,prov}/A_{s,\,reqd} = 1472/1282 = 1.15$

$\rho = A_s/(b_w\,d) = 1282/(250 \times 305) = 0.0168,\ \rho\% = 1.68$

$\rho_0 = 10^{-3} \times \sqrt{25} = 0.005,\ \rho_0\% = 0.5$

$\rho > \rho_0$

Simply supported beam, K = 1

No compression steel, $\rho' = 0$

$b/b_w = 1800/250 = 7.2 > 3.0$. Therefore L/d from equation is multiplies by 0.8.

$$\frac{L}{d} = K\left[11 + 1.5\,\sqrt{f_{ck}}\,\frac{\rho_0}{\rho - \rho'} + \frac{1}{12}\,\sqrt{f_{ck}}\,\sqrt{\frac{\rho'}{\rho}}\,\right] \text{ if } \rho > \rho_0 \qquad (7.16b)$$

As an approximation, the stress σ_s in the steel at SLS can be taken as

$\sigma_s = (\text{load at SLS/Load at ULS}) \times f_{yd}$

Load at ULS = $1.35 \times (14.0 + 1.56) + (1.5 \times 10) = 36.0$ kN/m

Load at SLS = $1.0 \times (14.0 + 1.56) + (1.0 \times 10) = 25.56$ kN/m

Note that load at SLS can be more appropriately taken as $g_k + \psi_2\,q_k$.

$\sigma_s = (25.56/\,36.0) \times (500/1.15) = 318$ MPa ≈ 310 MPa assumed in the formula.

L/d = $11 + 1.5 \times 5 \times 0.005/0.0168 = 13.2$

L/d = $13.2 \times 0.8 = 10.6$

Correcting for $A_{s,\,prov} > A_{s,\,reqd}$, L/d = $10.6 \times 1.15 = 12.1$

Actual L/d = $6000/305 = 19.7$

Depth is too small. Deflection will exceed the permitted L/250.

Redesign the beam:

L = 6 m, d = $6000/12 = 500$ mm

h = 500 + 25 (cover) + 8 (links) + 25/2 = 550 mm

d = 550 – 25 – 8 –25/2 = 504 mm

Increase in moment due to deeper web = $(0.55 – 0.35) \times 25 \times 1.35 \times 6^2/8$

= 30.38 kNm

$k = M/(b\,d^2\,f_{ck}) = (162 + 30.38) \times 10^6/(1800 \times 504^2 \times 25) = 0.02 < 0.196$

$$\frac{z}{d} = 0.5\left[1.0 + \sqrt{\left(1 - 3\frac{k}{\eta}\right)}\,\right] = 0.5\left[1.0 + \sqrt{\left(1 - 3\frac{0.02}{1.0}\right)}\,\right] = 0.99$$

$$A_s = \frac{M}{0.99\,d \times 0.87\,f_{yk}} = \frac{(165 + 30.38) \times 10^6}{0.99 \times 504 \times 0.87 \times 500} = 900\,\text{mm}^2$$

Provide 2H25; $A_s = 982\,\text{mm}^2$

$A_{s,\,prov}/A_{s,\,reqd} = 982/900 = 1.09$

$\rho = A_s/(b_w\,d) = 982/(250 \times 454) = 0.0087,\ \rho\% = 0.87$

$\rho_0 = 10^{-3} \times \sqrt{25} = 0.005,\ \rho_0\% = 0.5 < \rho\%$

Simply supported beam, K = 1

No compression steel, $\rho' = 0$

$b/b_w = 1800/250 = 7.2 > 3.0$. Therefore L/d from equation is multiplied by 0.8.

$$\frac{L}{d} = K[11 + 1.5 \sqrt{f_{ck}} \frac{\rho_0}{\rho - \rho'} + \frac{1}{12} \sqrt{f_{ck}} \sqrt{\frac{\rho'}{\rho}}] \text{ if } \rho > \rho_0 \qquad (7.16b)$$

L/d = 11 + 1.5 × 5 × 0.005/0.0087 = 15.3
L/d = 15.3×0.8 = 12.24
Correcting for $A_{s, prov} > A_{s, reqd}$, L/d = 12.24 × 1.09 = 13.34
Actual L/d = 6000/504 = 11.9

Deflection will not exceed the permitted L/250.

Example 2: A rectangular beam is simply supported over a span of 6 m and carries characteristic super dead load of 34 kN/m and characteristic imposed load of 17.0 kN/m. The beam is 300 mm wide by 500 mm deep and the inset of the compression steel is 40 mm. Design the steel for mid-span of the beam for the material strengths of f_{ck} =25 MPa and f_{yk} =500 MPa. Check the adequacy of L/d ratio.

Self weight = 0.3 × 0.500 × 25 = 3.75 kN/m
Design load = (34.0 + 3.75) × 1.35 + (17.0 × 1.5) = 76.46 kN/m
Required ultimate moment M:

$$M = 76.46 \times 6^2/8 = 344.1 \text{ kN m}$$

d ≈ 500 – 25(cover) – 8(link) – 25/2 = 454 mm

Maximum moment that the beam section can resist as a singly reinforced section is

$$M_{sr} = 0.196 \times 25 \times 300 \times 454^2 \times 10^{-6} = 303 \text{ kNm}$$

M > M_{sr}, Compression steel is required.

d'/d = 40/454 = 0.09 < 0.1664 (see Table 4.11)

The compression steel yields. The stress f_s' in the compression steel is $0.87f_{yk}$.

$$A_s' = \{M - M_{sr}\}/[0.87 f_{yk} (d - d')]$$

$A_s' = \{344.1 - 303.0\} \times 10^6/[0.87 \times 500 \times (454 - 40)] = 234 \text{ mm}^2$

From equilibrium:

$$A_s \, 0.87 \, f_{yk} = C_{sr} + A_s' \, f_s'$$

where $C_{sr} = k_c$ bd f_{ck}, compression force in concrete. k_c = 0.2401 from Table 4.7.

$A_s \times 0.87 \times 500 = k_c \times 300 \times 454 \times 25 + 234 \times 0.87 \times 500$

$$A_s = 2113 \text{ mm}^2$$

For the tension steel 5H25 give A_s = 2454 mm².
For the compression steel 2H12 give A_s' = 226 mm².

$A_{s, prov}/A_{s, reqd}$ = 2454/2113 = 1.16

$\rho = A_s/ (b_w \, d) = 2454/ (300 \times 454) = 0.018$, $\rho\%$ = 1.80

$\rho_0 = 10^{-3} \times \sqrt{25} = 0.005$, $\rho_0\%$ = 0.5

$$\rho > \rho_0$$

Simply supported beam, K = 1

Compression steel, ρ' = 234/ (300 × 454) = 0.0017, $\rho'\%$ = 0.17

Load at ULS = (34.0 + 3.75) × 1.35 + (17.0 × 1.5) = 76.46 kN/m
Load at SLS = (34.0 + 3.75) × 1.0 + (17.0 × 1.0) = 54.75 kN/m

Note that load at SLS can be more appropriately taken as $g_k + \psi_2 q_k$.

σ_s = (54.75/ 76.46) × (500/1.15) = 311 MPa ≈ 310 MPa assumed in the formula.

$$\frac{L}{d} = K[11 + 1.5 \sqrt{f_{ck}} \frac{\rho_0}{\rho - \rho'} + \frac{1}{12} \sqrt{f_{ck}} \sqrt{\frac{\rho'}{\rho}} \ ] \ \text{if} \ \rho > \rho_0 \qquad (7.16b)$$

L/d = 11 + 1.5 × 5 ×0.005/0.018 + (5/12) × √(0.0017/0.018) = 13.2.
L/d = 13.2.
Correcting for $A_{s, prov} > A_{s, reqd}$, L/d = 13.2 × 1.16 = 15.3).
Actual L/d = 6000/454 = 13.2.
Deflection will not exceed the permitted L/250.

6.3 CRACKING

6.3.1 Cracking Limits and Controls

Cracking is normal in reinforced concrete structures. However any prominent crack greatly detracts from the appearance and might even affect the proper functioning of the structure. Excessive cracking and wide deep cracks affect durability and can lead to corrosion of reinforcement. In section 7.3.1, code recommends the following limits on crack width.

For exposure classes X0 and XC1 (see Table 2.5), crack width is limited to 0.4 mm in order not to spoil the appearance of the structure. For exposure classes (XC2 to XC4, XD1, XD2, XS1 to XS3), crack width is limited to 0.3 mm.

The value given above is valid for ensuring appearance and durability under quasi-permanent loads but does not include any special requirement such as water tightness in structures retaining aqueous fluids.

Two methods are available to ensure crack widths limits are not violated. They are:

1. In normal cases control of cracking without direct calculation as given in section 7.3.3 of the code is followed. This is done by following the maximum bar diameter values given in Table 7.2N of the code reproduced here as Table 6.2 or the values for maximum spacing of bars given in Table 7.3N reproduced here as Table 6.3.
2. In special cases crack widths of an already designed structure can be calculated using the equations given section 7.3.4 of the code.

In this Chapter only rules for the normal case are considered. Rules for special cases are discussed in Chapter 19.

6.3.2 Bar Spacing Controls in Beams

Cracking is controlled by specifying the maximum distance between bars in tension as given in Table 6.2 *or* by restricting the maximum bar diameter as given in Table 6.3.

Table 6.2 Maximum bar spacing for crack control

Steel stress, MPa	Maximum bar spacing (mm) for maximum crack width, w_k		
	$w_k = 0.4$ mm	$w_k = 0.3$ mm	$w_k = 0.2$ mm
160	300	300	200
200	300	250	150
240	250	200	100
280	200	150	50
320	150	100	
360	100	50	

In the above table, the steel stress corresponds to stress in steel at SLS. It is reasonable to assume that the steel stress at SLS can be calculated as

Steel stress at SLS = Steel stress at ULS × [quasi-permanent load/load at ULS]
The stress in steel at ULS is normally taken as $f_{yd} = f_{yk}/ (\gamma_s = 1.15)$.

Table 6.3 Maximum bar diameter φ'_s for crack control

Steel stress, MPa	Maximum bar size (mm) for maximum crack width, w_k		
	$w_k = 0.4$ mm	$w_k = 0.3$ mm	$w_k = 0.2$ mm
160	30	32	25
200	32	25	16
240	20	16	12
280	16	12	8
320	12	10	6
360	10	8	5
400	8	6	4
450	6	5	

The above value of φ'_s has been calculated assuming $f_{ct, eff} = 2.9$ MPa, coefficient $k_c = 0.4$, depth of the tensile zone immediately prior to cracking, $h_{cr} = 0.5h$, $(h - d) = 0.1h$.
The above value needs to be modified as follows as given by code equations (7.6N) and ((7.7N) respectively.

$$\varphi_s = \varphi'_s \times \frac{f_{ct,eff}}{2.9} \times \frac{k_c \, h_{cr}}{2(h-d)} \quad \text{for cases of bending} \qquad (7.6N)$$

$$\varphi_s = \varphi'_s \times \frac{f_{ct,eff}}{2.9} \times \frac{h_{cr}}{8(h-d)} \quad \text{for cases of uniform axial tension} \qquad (7.7N)$$

6.3.3 Minimum Steel Areas

If crack control is required, a minimum amount of steel reinforcement is required in areas where tension is expected. In beams, a minimum reinforcement is required in individual parts of the beam such as webs, flanges, etc. The minimum reinforcement is calculated using the code formula (7.1).

$$A_{s,\,min}\,\sigma_s = k_c\,k\,f_{ct,\,eff}\,A_{ct} \qquad (7.1)$$

where

$A_{s,\,min}$ = the minimum area of steel in the tensile zone.

A_{ct} = Area of concrete in the tensile zone (See Fig. 6.1).

σ_s = Permitted steel stress. A lower value than f_{yk} is required so as not to violate crack width limitations (see Table 6.2).

$f_{ct,\,eff} = f_{ctm}$ of concrete at the time when cracks are expected to occur.

k = 1.0 for webs h ≤ 300 mm deep, flanges with widths less than 300 mm.

   0.65 for webs h ≥ 800 mm deep, flanges with widths greater than 800 mm.

   Interpolation for intermediate values is allowed.

k_c = 1.0 for pure tension.

For bending or bending with axial force,

k = 0.4 for webs of box and T-sections

k = 0.9 [F_{cr}/ ($A_{ct} \times f_{ct,\,eff}$) ≥ 0.5 where absolute value of the tensile force F_{cr} within the flange immediately prior to cracking due to the cracking moment calculated with $f_{ct,\,eff}$.

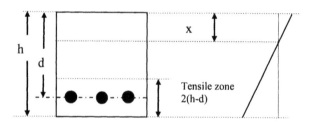

Fig. 6.1 Tensile area in beams.

6.3.3.1 Example of Minimum Steel Areas

Example 1: A rectangular beam 300 × 650 is required to span 7 m. The ends are simply supported. The characteristic dead load g_k = 20 kN/m and imposed load q_k = 30 kN/m. f_{ck} = 30 MPa, f_{yk} = 500 MPa.

$$q_{ult} = 1.35 \times 20 + 1.5 \times 30 = 72 \text{ kN/m}$$
$$M_{ult} = 72 \times 7^2/8 = 441 \text{ kNm}$$
$$d = 650 - 30 \text{ (cover)} - 8 \text{ (link)} - 25/2 \text{ (bar radius)} = 599 \text{ mm}$$
$$k = M/ (b\,d^2\,f_{ck}) = 441 \times 10^6/ (300 \times 599^2 \times 30) = 0.137 < 0.196$$
$$\eta = 1, \ z/d = 0.5\{1+ \surd\,(1 - 3 \times k/\,\eta)\} = 0.884$$
$$z = 0.884 \times 599 = 530 \text{ mm}$$
$$f_{yd} = 500/1.15 = 435 \text{ MPa}$$
$$A_s = M/ (z\,f_{yd}) = 441 \times 10^6/ (530 \times 435) = 1915 \text{ mm}^2$$
$$A_s = 4H25 = 1964 \text{ mm}^2$$

1. Check L/d ratio using code equation (7.16b)
$$f_{ck} = 30 \text{ MPa } \rho_0\% = 0.1 \times \sqrt{f_{ck}} = 0.55$$
$$\rho\% = 100 \times 1964/ (300 \times 599) = 1.09$$
$$\rho'\% = 0$$
Load at ULS = $1.35 \times 20 + 1.5 \times 30 = 72$ kN/m
Load at SLS = $1.0 \times 20 + 1.0 \times 30 = 50.0$ kN/m
$\sigma_s = (50.0/ 72.0) \times (500/1.15) = 302$ MPa ≈ 310 MPa assumed in the formula.

$$\frac{L}{d} = K[11 + 1.5 \sqrt{f_{ck}} \frac{\rho_0}{\rho - \rho'} + \frac{1}{12} \sqrt{f_{ck}} \sqrt{\frac{\rho'}{\rho}}] \text{ if } \rho > \rho_0 \qquad (7.16b)$$

$L/d = 11 + 1.5 \times 5.5 \times 0.55/1.09 = 15.2$
$A_{s, prov}/A_{s, reqd} = 1964/1915 = 1.026$
L/d corrected $= 15.2 \times 1.026 = 15.6$
Actual $L/d = 7000/599 = 11.7 < 15.6$
Deflection check is satisfactory.

2. Check minimum longitudinal steel using code equation (9.1N)
$f_{ctm} = 0.3 \times f_{ck}^{0.67} = 0.3 \times 30^{0.67} = 2.9$ MPa
$A_{s, min} = 0.26 (f_{ctm}/ f_{yk}) b_t d = 0.26 \times (2.9/500) \times 300 \times 599 = 271$ mm^2

3. Check maximum spacing of bars using Table 6.2
$q_{SLS} = g_k + q_k = 20 + 30 = 50$ kNm
$q_{SLS}/ q_{ULT} = 50/72 = 0.69$
$\sigma_s = f_{yd} \times 0.69 = 300$ MPa
From Table 6.2, for $\sigma_s = 300$ and $w_k = 0.3$ mm, maximum spacing = 125 mm
Bar spacing $= [300 - 2 \times (30 + 8) - 25]/3 = 66$ mm < 125 mm

4. Check minimum reinforcement areas using code equation (7.1)
Stress in steel at SLS = 302 MPa (see (1) above.
From code equation (7.2), as there is no axial load, $\sigma_c = 0$, $k_c = 0.4$.
$k = 1.0$ for $h \leq 300$ and $k = 0.65$ $h \geq 800$.
Interpolating, $k = 0.65 + \{(800 - 650)/ (800 - 300)\} \times (1.0 - 0.65) = 0.76$
$A_{ct} = \{2(h - d) = 102\} \times \{width = 300\} = 30600$ mm^2.
Assuming that cracks might appear at about 3 days, calculate $f_{ck}(t)$.
$$f_{ck}(3) \approx f_{cm}(t) - 8$$
$$f_{cm}(t) = \beta_{cc}(t) f_{cm} \qquad (3.1)$$
$$\beta_{cc}(t) = \exp\{s \times [1 - \sqrt{(28/t)]}\} \qquad (3.2)$$
Taking $s = 0.25$ for class N cement and $t = 3$, $\beta_{cc}(t) = 0.6$
$f_{cm} = f_{ck} + 8 = 38$ MPa, $f_{cm}(t) = 0.6 \times (38) = 22.8$ MPa
$f_{ctm} = 0.3 \times f_{ck}^{0.67} = 0.3 \times 22.8^{0.67} = 2.4$ MPa
Substituting in code equation (7.1),
$$A_{s, min} \times \sigma_s = k_c k f_{ct, eff} A_{ct} \qquad (7.1)$$
$$A_{s, min} = 0.4 \times 0.76 \times 2.4 \times 30600/ 302 = 74 \text{ mm}^2$$
All checks are satisfactory.

6.3.4 Bar Spacing Controls in Slabs

The maximum clear spacing between bars in slabs is given in clause 9.3.1.1(3). If h is the overall depth of the slab, for main steel, the spacing of the bars should not exceed the lesser of 3 h or 400 mm. Secondary steel spacing of the bars should not exceed the lesser of 3.5 h or 450 mm.

Example 1: A 350 mm deep slab has been designed for an ultimate moment of 200 kNm/m using f_{ck} = 30 MPa and f_{yk} = 500 MPa. Apply 'Deemed to satisfy' rule to check the maximum bar spacing.
Dimensions of slab: b = 1000 mm, h = 350 mm, cover = 30 mm, main steel H16 bars.

$$\text{Effective depth, d} = 350 - 30 - 16/2 = 312 \text{ mm.}$$
$$k = 200 \times 10^6 / (1000 \times 312^2 \times 30) = 0.069 < 0.196$$
$$\eta = 1, \ z/d = 0.5(1.0 + \sqrt{(1 - 3 \times 0.069)} = 0.95$$
$$A_{s\ required} = 200 \times 10^6 / (0.95 \times 312 \times 0.87 \times 500) = 1552 \text{ mm}^2/\text{m}$$
$$A_{s\ Provided} = \text{H16 at 125 mm} = 1609 \text{ mm}^2$$

Maximum spacing allowed: clause 9.3.1 (3) states that spacing 3 h ≤ 400 mm.
Spacing = 125 mm.
Actual spacing of 125 mm is less than permitted maximum spacing. Design is satisfactory.

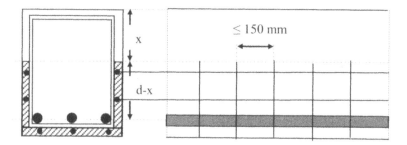

Fig. 6.2 Surface reinforcement.

6.3.5 Surface Reinforcement

Appendix J of the code gives details regarding surface reinforcement to resist spalling. It should be used where the main reinforcement is made up of bars of diameter greater than 32 mm or bundled bars with an equivalent diameter greater than 32 mm. The surface reinforcement consists of small diameter bars or of welded mesh and is placed outside the links parallel to and perpendicular to the main tension reinforcement as shown in Fig. 6.2. The area of surface reinforcement should not be less than 0.01 times the area of concrete external to the links. The spacing of the bars should not exceed 150 mm.

CHAPTER 7

SIMPLY SUPPORTED BEAMS

The aim in this chapter is to put together the design procedures developed in Chapters 4, 5 and 6 to make a complete design of a reinforced concrete beam. Beams carry lateral loads in roofs, floors etc. and resist the loading in bending, shear and bond. The design must comply with the ultimate and serviceability limit states.

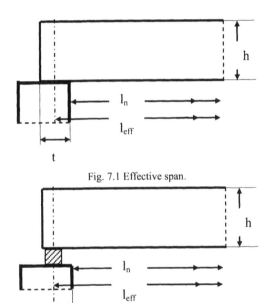

Fig. 7.1 Effective span.

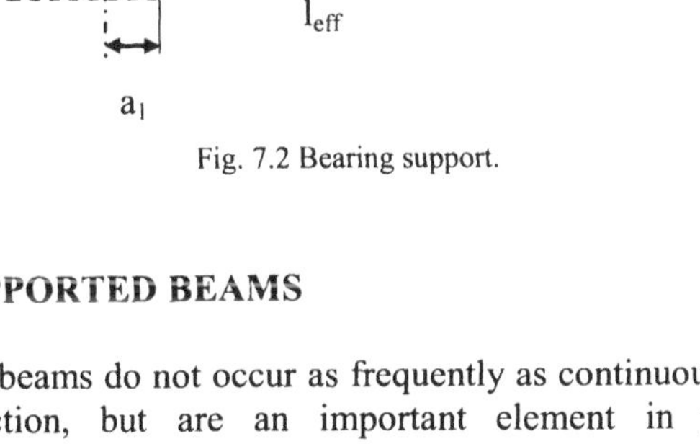

Fig. 7.2 Bearing support.

7.1 SIMPLY SUPPORTED BEAMS

Simply supported beams do not occur as frequently as continuous beams in *in-situ* concrete construction, but are an important element in pre-cast concrete construction.

The effective span l_{eff} of a simply supported beam is defined in the code in clause 5.3.2.2, equation (5.8) as

$$l_{eff} = l_n + a_1 + a_2 \tag{5.8}$$

l_n = distance between the faces of supports

a_1 and a_2 refer to support width at ends 1 and 2 of the beam

$a_1 = \min (0.5h, 0.5t)$
h = overall depth of the beam
t = width of support as shown in Fig. 7.1
If a bearing is provided as shown in Fig. 7.2, then a_1 (or a_2) is the distance from the face of the support to the centre of the bearing. In other words, the effective span is the distance between the centres of bearings.

7.1.1 Steps in Beam Design

Although the steps in beam design as shown in (a) to (j) below are presented in a sequential order, it is important to appreciate that design is an iterative process. Initial assumptions about size of the member, diameters of reinforcement bars are made and after calculations it might be necessary to revise the initial assumptions and start from the beginning. Experience built over some years helps to speed up the time taken to arrive at the final design. The two examples in this chapter *do not* show this iterative aspect of design.

(a) Preliminary size of beam
The size of beam required depends on the moment and shear that the beam carries. Clause 9.2.1.1(3) gives the maximum reinforcement as 4% of the cross sectional area of concrete. The minimum percentage must comply with the code equation (9.1N)

$$A_{s,\min} = 0.26 \times \frac{f_{ctm}}{f_{yk}} \times b_t\, d \qquad\qquad (9.1N)$$

where f_{ctm} = mean tensile strength of concrete, b_t = width of the web, d = effective depth, f_{yk} = characteristic strength of reinforcement.

A general guide to the size of beam required may be obtained from the basic span-to-effective depth ratio from Table 7.4N of the code.
The following values are generally found to be suitable.
Overall depth ≈ span/15.
Breadth ≈ (0.4 to 0.6) × depth.
The breadth may have to be very much greater in some cases. The size is generally chosen from experience. Many design guides are available which assist in design.

(b) Estimation of loads:
The loads should include an allowance for self-weight which will be based on experience or calculated from the assumed dimensions for the beam. The original estimate may require checking after the final design is complete. The estimation of loads should also include the weight of screed, finish, partitions, ceiling and services if applicable. The following values are often used:
Screed: 1.8 kN/m^2
Ceiling and service load: 0.5 kN/m^2
Demountable light weight partitions: 1.0 kN/m^2

Block-work partitions: 2.5 kN/m^2
The imposed loading, depending on the type of occupancy, is taken from *Eurocode 1: Actions on Structures-Part 1-1: General actions-Densities, self-weight, imposed loads for buildings.*

(c) Analysis

The ultimate design loads are calculated using appropriate partial factors of safety from *BS EN 1990-2002 Eurocode-Basis of Structural Design*. The load combinations to be considered are as explained in section 3.2.3, Chapter 3. The details are summarised in Table 3.2, repeated here for convenience.

Table 3.2 Simplified equations for checking EQU and STR

Persistent and transient design situations	Permanent actions		Leading variable action		Accompanying variable action	
	Unfav.	Fav.	Unfav.	Fav.	Unfav.	Fav.
6.10	$1.35 \, G_{k, \, sup}$	$1.0 \, G_{k, \, inf}$	$1.5 \, Q_{k,1}$	0	$1.5 \, \psi_{0,i} \, Q_{k,i}$	0
6.10a	$1.35 \, G_{k, \, sup}$	$1.0 \, G_{k, \, inf}$	$\S 1.5 \, \psi_{0,i} \, Q_{k,1}$	0	$1.5 \, \psi_{0,i} \, Q_{k,i}$	0
6.10b	$*1.15 \, G_{k, \, sup}$	$1.0 \, G_{k, \, inf}$	$1.5 \, Q_{k,1}$	0	$1.5 \, \psi_{0,i} \, Q_{k,i}$	0

Note: In equation (6.10b), The U.K National Annex uses the multiplier as 1.25 instead of 1.15 for unfavourable permanent actions.
The ultimate reactions, shears and moments are determined and the corresponding shear force and bending moment diagrams are drawn.

(d). Cover

Choose cover to suit the environment and fire resistance. See Chapter 2, sections 2.9 and 2.10 for details.

(e) Design of moment reinforcement

The flexural reinforcement is designed at the point of maximum moment. Refer to Chapter 4 for the steps involved.

(f) Curtailment and end anchorage

A sketch of the beam in elevation is made and the cut-off point for part of the tension reinforcement is determined. The end anchorage for bars continuing to the end of the beam is set out to comply with code requirements in clause 9.2.1.4. See Chapter 5 for details. It requires that the bars should continue for a length equal to anchorage length measured from the line of contact of the support with the beam. It is necessary to include the additional tensile force F_E due to shear equal to

$$F_E = 0.5 \, V_{ED} \cot\theta, \quad 1.0 \leq \cot\theta \leq 2.5$$

(g) Design for shear

Design ultimate shear stresses are checked and shear reinforcement is designed using the procedures set out in section 6.2. Refer to this and Chapter 5.

Note that except for minor beams such as lintels, all beams must be provided with at least minimum links as shear reinforcement. Small diameter bars are required in the top of the beam to carry and anchor the links.

(h) Deflection
Deflection is checked using the rules from section 7.4.2. Refer to Chapter 6.

(i) Cracking
The maximum clear distance between bars on the tension face is checked against the limits given in the code in section 7.3.3 and Table 7.3N. Refer to Chapter 6 for more details.

(j) Design sketch
Design sketches of the beam with elevation and sections are needed to show all information for the draft sperson.

7.1.2 Example of Design of a Simply Supported L-Beam in a Footbridge

(a) Specification
The section through a simply supported reinforced concrete footbridge of 7 m effective span is shown in Fig. 7.3. The characteristic imposed load is 5 kN/m^2 and the materials to be used are f_{ck} = 25 MPa concrete and f_{yk} = 500 MPa reinforcement. Design the L-beams that support the bridge. Concrete weighs 25 kN/m^3, and the unit mass of the handrails is 16 kg/m per side. The beam will be in XD1 environment.

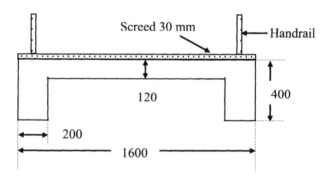

Fig. 7.3 Cross section of a foot bridge.

(b) Loads, shear force and bending moment diagram
The total load is carried by two L-beams. All the load acting on 0.8 m width acts on an L-beam.

The dead load carried by each L-beam is
$$[(0.12 \text{ slab} + 0.03 \text{ screed}) \times 0.8 + 0.2 \times (0.4 - 0.12) \text{ rib}] \times 25$$
$$+ 16 \times 9.81 \text{x } 10^{-3} \text{ hand rails} = 4.6 \text{ kN/m}$$
The total live load acting on each beam is $0.8 \times 5 = 4.0$ kN/m.
As there is only one variable load, when using equation (6.10) the design load at ultimate limit state is $(1.35 \times 4.6) + (1.5 \times 4) = 12.21$ kN/m
The ultimate moment at the centre of the beam is $12.21 \times 7^2/8 = 74.8$ kN m
Support reaction $= 12.21 \times 7.0/2 = 42.74$ kN.
The load, shear force and bending moment diagrams are shown in Fig. 7.4.

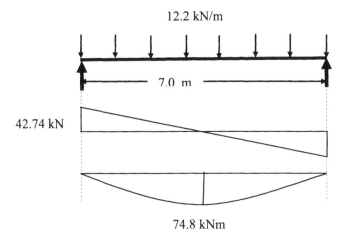

12.2 kN/m

7.0 m

42.74 kN

74.8 kNm

Fig. 7.4 Loading, shear force, and bending moment diagrams.

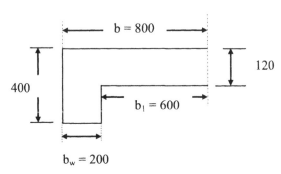

b = 800

120

400

$b_1 = 600$

$b_w = 200$

Fig. 7.5 The L-beam.

(c) Design of moment reinforcement
The effective width b of the flange of the L-beam shown in Fig. 7.5 is given by code equations (5.7), (5.7a) and (5.7b). See Fig. 4.11, Chapter 4.
$$b_1 = 600 \text{ mm, } l_0 = 7 \text{ m,}$$
$$b_{\text{eff, 1}} = 0.2 \text{ } b_1 + 0.1 \text{ } l_0 \leq 0.2 \text{ } l_0$$

$$b_{eff,\,1} = 0.2 \times 600 + 0.1 \times 7000 \le \min(0.2 \times 7000, b_1)$$
$$= 820 \le \min(1400, 600) = 600 \text{ mm}$$
$$b_{eff} = b_{eff,\,1} + b_w = 800 \text{ mm}$$
$$b = 800 \text{ mm}$$

Assume exposure class as XD1, from Table 2.26, Chapter 2, $c_{min,\,dur} = 35$ mm for structural class S4. Assume 25 mm bars for reinforcement.

Cover, $C_{min} = \max$ (diameter of bar; $c_{min,\,dur}$; 10 mm) = max (25; 35; 10) = 35 mm.

The effective depth d is estimated as

$$d = 400 - 35 \text{ (cover)} - 8 \text{ (link diameter)} - 25/2 = 344.5 \text{ mm, say } 345 \text{ mm}$$

Check for the depth of the stress block:

The moment of resistance of the section when the stress block is equal to the slab depth $h_f = 120$ mm is

$$M_{Flange} = f_{cd} \times b \times h_f \times (d - h_f/2)$$
$$M_{Flange} = (25/1.5) \times 800 \times 120 \times (345 - 0.5 \times 120) \times 10^{-6} = 456 \text{ kNm}$$
$$(M = 74.8) < (M_{Flange} = 456)$$

The stress block is inside the slab and the beam can be designed as a rectangular section.

$$k = M/(bd^2 f_{ck}) = 74.8 \times 10^6/(800 \times 345^2 \times 25) = 0.031 < 0.196$$
$$\eta = 1,\ z/d = 0.5[1.0 + \sqrt{(1 - 3k)} = 0.98$$
$$z = 0.98 \times 345 = 338 \text{ mm}$$
$$A_s = M/(0.87 f_{yk} z) = 74.8 \times 10^6/(0.87 \times 500 \times 338) = 509 \text{ mm}^2$$

Provide 2H25, $A_s = 982$ mm^2.

Check for minimum steel area using the code equation (9.1N).

$$f_{ctm} = 0.3 \times f_{ck}^{0.67} = 0.3 \times 25^{0.67} = 2.6 \text{ MPa}$$

$$A_{s,min} = 0.26 \times \frac{f_{ctm}}{f_{yk}} \times b_t \times d = 0.26 \times \frac{2.6}{500} \times 200 \times 345 = 93 \text{ mm}^2$$

$$A_{s \text{ provided}} > A_{s,\,min}$$

Curtailment of bars: As there are only two bars, all bars will be taken right to the end.

(d) Design of shear reinforcement

Maximum shear force V_{Ed} at d from the face of support

$$V_{Ed} = 42.74 - 12.2 \times 0.345 = 38.5 \text{ kN}$$

i. Check if shear reinforcement is required, $V_{Ed} > V_{Rd,\,c}$

$$V_{Ed} = 37.4 \text{ kN, } b_w = 200 \text{ mm, } d = 345 \text{ mm, } A_{sl} = 2H25 = 982 \text{ mm}^2$$

$$C_{Rd,c} = \frac{0.18}{(\gamma_c = 1.5)} = 0.12$$

$$k = 1 + \sqrt{\frac{200}{345}} = 1.76 \le 2.0$$

$$100\rho_1 = 100 \times \frac{982}{200 \times 345} = 1.42 < 2.0$$

$$v_{min} = 0.035 \times 1.42^{1.5} \times \sqrt{25} = 0.30 \, \text{MPa}$$

$$V_{Rd,c} = [0.12 \times 1.76 \times \{1.42 \times 25\}^{1/3} \geq 0.30] \times 200 \times 345 \times 10^{-3} = 47.9 \, \text{kN}$$

$V_{Rd, c} = 47.9$ kN $> V_{Ed}$. Therefore no shear reinforcement is required but nominal minimum reinforcement will be provided.

Check minimum shear steel requirement

$$\frac{A_{sw}}{s \, b_w} \geq \frac{0.08 \sqrt{f_{ck}}}{f_{yk}}$$

$$\frac{100.5}{s \times 200} \geq \frac{0.08 \sqrt{25}}{500}$$

$$s \leq \frac{100.5}{200} \times \frac{500}{0.08 \sqrt{25}} = 628 \, \text{mm}$$

Maximum spacing $s \leq (0.75 \, d = 0.75 \times 345 = 259 \, \text{mm})$.
Maximum spacing should be less than 259 mm. A spacing of 250 mm will be satisfactory. 2H12 bars are provided to carry the links at the top of the beam. The shear reinforcement is shown in Fig. 7.6.

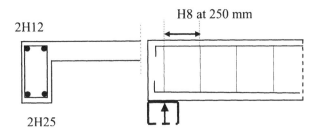

Fig. 7.6 Shear reinforcement.

(e) End anchorage
Clause 9.2.1.4 (1) of Eurocode 2 states that
- The area of bottom reinforcement provided at supports with little or no end fixity should be at least 25% of the area of steel provided in the span.
- The tensile force F_E to be anchored is given by code equation (9.3) as

$$F_E = |V_{ED}| \frac{a_1}{z} \qquad (9.3)$$

a_1 is given in code equation (9.2) as

$$a_1 = \frac{z}{2}(\cot \theta - \cot \alpha) \qquad (9.2)$$

α = angle of shear reinforcement and for vertical links, $\alpha = 90^0$ and $\cot \alpha = 0$.
$$\cot \theta \approx 2.5, \, a_1/z = 1.25$$

$$F_E = 1.25\ V_{Ed}$$

The bond length needed (see clause 9.2.1.4(3)) is measured from the line of contact between the beam and the support.

At support, $V_{Ed} = 38.5$ kN. $F_E = 1.25 \times 38.5 = 48.13$ kN.

Area of steel required to resist this force $= F_E/f_{yd}$
$$= 48.13 \times 10^3 / (500/1.15) = 111\ mm^2.$$

As all the steel is carried over to the support, area of steel provided is that of 2H25 equal to 982 mm^2 > 111 mm^2.

From Table 5.5, Chapter 5, for $f_{ck} = 25$ MPa, the bond length is 40 bar diameters.

The bond length ℓ_{bd} needed $= (A_s$ required$/A_s$ provided$) \times (40 \times 25)$.

$\ell_{bd} = (111/982) \times (40 \times 25) = 113$ mm.

(f) Deflection check

The deflection of the beam is checked using the equation (7.16a and b) of the code.

$A_{s,\ prov} / A_{s,\ reqd} = 982/509 = 1.93$.

$\rho = A_s / (b_w\ d) = 982/ (200 \times 345) = 0.014$, $\rho\% = 1.42$.

$\rho_0 = 10^{-3} \times \sqrt{25} = 0.005$, $\rho_0\% = 0.5$. $\rho > \rho_0$.

Simply supported beam, $K = 1$.

No compression steel, $\rho' = 0$.

$b/b_w = 800/200 = 4.0 > 3.0$. Therefore L/d from equation is multiplies by 0.8.

$$\text{Load at ULS} = (1.35 \times 4.6) + (1.5 \times 4) = 12.21\ kN/m$$
$$\text{Load at SLS} = (1.0 \times 4.6) + (1.0 \times 4) = 8.6\ kN/m$$

Stress σ_s in steel at SLS $= (8.6/12.21) \times f_{yd} = 306$ MPa

$$\frac{L}{d} = K\left[11 + 1.5\ \sqrt{f_{ck}}\ \frac{\rho_0}{\rho - \rho'} + \frac{1}{12}\ \sqrt{f_{ck}}\ \sqrt{\frac{\rho'}{\rho}}\ \right] \text{if}\ \rho > \rho_0 \qquad (7.16b)$$

L/d $= 13.6 \times 0.8 = 10.9$.

L/d ratio corrected for $A_{s,\ prov} / A_{s,\ reqd}$ is $10.9 \times 1.93 = 21.1$.

Actual L/d $= 7000/345 = 20.3$.

Deflection will not exceed the permitted L/250.

(g) Check for cracking

The clear distance between bars on the tension face is
$$200 - 2 \times (35 + 8) - 25 = 89\ mm$$

Sustained load at SLS $= g_k + q_k = 4.6 + 4.0 = 8.6$ kN/m.

Load at ULS $= 1.35\ g_k + 1.5\ q_k = 1.35 \times 4.6 + 1.5 \times 4.0 = 12.21$ kN/m.

Stress σ_s in steel at SLS $= 306$ MPa.

From Table 6.2, for 0.3 mm crack width, maximum spacing is approximately 120 mm. The actual spacing of 89 mm does not exceed 120 mm. The beam is satisfactory with regard to cracking.

(i) Beam reinforcement

The reinforcement for each L-beam is shown in Fig. 7.6. Note that the slab reinforcement also provides reinforcement across the flange of the L-beam.

7.1.3 Example of Design of Simply Supported Doubly Reinforced Rectangular Beam

(a) Specification
A rectangular beam is 300 mm wide by 575 mm overall depth with inset to the compression steel of 55 mm. The beam is simply supported and spans 9 m. The characteristic dead load including an allowance for self-weight is 20 kN/m and the characteristic imposed load is 11 kN/m. The materials to be used are f_{ck} = 25 MPa and f_{yk} = 500 MPa. Design the beam.

(b) Loads and shear force and bending moment diagrams
$$\text{Design load} = (1.35 \times 20) + (1.5 \times 11) = 43.5 \text{ kN/m}$$
$$\text{Ultimate moment} = 43.5 \times 9^2/8 = 440.4 \text{ kN m}$$
$$\text{Shear force at support} = 43.5 \times 9/2 = 195.8 \text{ kN}$$

(c) Design of the moment reinforcement
Calculate the effective depth d:

Assuming 25 mm bars for reinforcement in two layers, 10 mm diameter for links and cover to the reinforcement is taken as 35 mm for XD1 exposure, effective depth d is
$$d = 575 - 35 - 10 - 25 = 505 \text{ mm}$$
The maximum moment of resistance of a singly reinforced rectangular beam is
$$0.196 \text{ b d}^2 \text{ f}_{ck} = 0.196 \times 300 \times 505^2 \times 25 \times 10^{-6} = 374.9 \text{ kNm} < 440.4 \text{ kNm}$$
Compression reinforcement is required.

From Table 4.5, d'/ d = 55/450 = 0.12 < 0.1664.

The compression steel yields. Stress in the compression steel is f_{yd}.

The area of compression steel is
$$A'_s = (M - 0.196 \text{ bd}^2 \text{ f}_{ck}) / [(d - d') \times f_{yd}]$$
$$= (440.4 - 374.9) \times 10^6/ [(505 - 55) \times 435] = 335 \text{ mm}^2$$
Provide 2H16, A'_s = 402 mm^2.

Equate total tensile and compressive forces. Neutral axis depth with maximum moment for singly reinforced beam from Table 4.5 is 0.448 d. The stress block depth is 0.8x = 0.358 d
$$f_{cd} \times b \times 0.358d + A'_s \times f_{yd} = A_s \times f_{yd}$$
$$16.67 \times 300 \times 0.358 \times 505 + 335 \times 435 = A_s \times 435$$
$$A_s = 2414 \text{ mm}^2$$
Provide 5H25, A_s = 2454 mm^2.

Note: The value of d is slightly changed.

d_1 for the top 2H25 = 575 – (35+10) – 25 – 25/2 = 492.5 mm.

d_2 for the bottom 3H25 = d_1 + 25 = 517.5 mm.

$d = (2 \times d_1 + 3 \times d_2) / (2+3) = 508 \text{ mm}$.

The revised values are:
$$0.196 \text{ b d}^2 \text{ f}_{ck} = 0.196 \times 300 \times 508^2 \times 25 \times 10^{-6} = 379.4 \text{ kNm} < 440.4 \text{ kNm}$$
The area of compression steel is
$$A'_s = (M - 0.196 \text{ bd}^2 \text{ f}_{ck}) / [(d - d') \times f_{yd}]$$
$$= (440.4 - 379.4) \times 10^6/ [(508 - 55) \times 435] = 310 \text{ mm}^2$$
Provide 2T16, A'_s = 402 mm^2.

$$f_{cd} \times b \times 0.358d + A'_s \times f_{yd} = A_s \times f_{yd}$$
$$16.67 \times 300 \times 0.358 \times 508 + 310 \times 435 = A_s \times 435$$
$$A_s = 2401 \text{ mm}^2$$

Provide 5T25, $A_s = 2454$ mm².
There is a very slight reduction in tension and compression steel.
The top layer 2H25 can be curtailed.
Stress block depth s: $A_s = 3H25 = 1473$ mm², $d = d_2 = 473$ mm.

$$f_{cd} \times b \times s = A_s \times f_{yd}, \ s = 128 \text{ mm}$$
$$z = d - s/2 = 409 \text{ mm}$$
$$M = A_s \times f_{yd} \times (d - s/2) \times 10^{-6} = 262.1 \text{ kNm}$$
$$262.1 = 195.8 \ x - 0.5 \times 43.5 \times x^2$$
$$x = 1.64 \text{ and } 7.36 \text{ m}$$

(d) Design of shear reinforcement

V_{Ed} = Shear force at d from face of support = $195.8 - 43.5 \times 508 \times 10^{-3}$
$$= 173.7 \text{ kN}$$

i. Check whether shear reinforcement is required, $V_{Ed} > V_{Rd, c}$

$V_{Ed} = 173.7$ kN, $b_w = 300$ mm, $d = 508$ mm, $A_{sl} = 5H25 = 2454$ mm².

$$C_{Rd,c} = \frac{0.18}{(\gamma_c = 1.5)} = 0.12$$

$$k = 1 + \sqrt{\frac{200}{508}} = 1.63 \le 2.0$$

$$100\rho_1 = 100 \times \frac{2454}{300 \times 508} = 1.61 < 2.0$$

$$v_{min} = 0.035 \times 1.63^{1.5} \times \sqrt{25} = 0.36 \text{ MPa}$$

$$V_{Rd,c} = [0.12 \times 1.63 \times \{1.61 \times 25\}^{1/3} + 0.15 \times 0] \times 300 \times 508 \times 10^{-3}$$

$$\ge [0.36 + 0.15 \times 0] \times 300 \times 508 \times 10^{-3}$$

$$V_{Rd,c} = 102.2 \ge 54.9 \text{ kN}$$

$V_{Rd, c} = 102.2$ kN $< V_{Ed}$. Therefore shear reinforcement is required.

ii. Check whether the section strength is adequate, $V_{Ed} < V_{Rd, max}$

$$\theta = 0.5 \sin^{-1}\{\frac{2V_{Ed}}{\alpha_{cw} \ b_w \ z \ v_1 \ f_{cd}}\}$$

$$= 0.5 \sin^{-1}\{\frac{2 \times 173.5 \times 10^3}{1.0 \times 300 \times (0.9 \times 508) \times 0.6 \times (\frac{25}{1.5})}\}$$

$$= 0.5 \sin^{-1}(0.25) = 0.5 \times 14.65 = 7.33^0$$

cot θ = 7.8, which is outside the limits of 1.0 and 2.5. Choosing cot θ = 2.5 for minimum shear reinforcement, $V_{Rd, max} = 473.1$ kN > (V_{Ed} at support = 195.8 kN). Section size is adequate.

iii. Design of shear reinforcement

Ensuring that $V_{Rd,s} \geq V_{Ed}$, and choosing 2-leg links of 10 mm diameter, $A_{sw} = 157.1$ mm^2, cot $\theta = 2.5$, $z = 0.9d$, $f_{ywk} = 500$ MPa,

$$V_{Rd,s} = \frac{z}{s} A_{sw} f_{ywd} \cot \theta$$

$$= \frac{(0.9 \times 508)}{s} \times 157.1 \times \frac{500}{1.15} \times 2.5 \times 10^{-3}$$

$$= \frac{31229}{s} \text{kN} \geq (V_{Ed} = 173.7 \text{ kN})$$

$$s \leq 180 \text{ mm}$$

iv. Check minimum steel requirement

$$\frac{A_{sw}}{s\,b_w} \geq \frac{0.08\sqrt{f_{ck}}}{f_{yk}}$$

$$\frac{157.1}{s \times 300} \geq \frac{0.08\sqrt{25}}{500}$$

$$s \leq 655 \text{ mm}$$

Maximum spacing $s \leq (0.75\ d = 0.75 \times 508 = 381$ mm).
Calculate $V_{Rd,s}$ for $s = 375$ mm.

$$V_{Rd,s} = \frac{z}{s} A_{sw} f_{ywd} \cot \theta$$

$$= \frac{(0.9 \times 508)}{375} \times 157.1 \times \frac{500}{1.15} \times 2.5 \times 10^{-3} = 83.3 \text{ kN}$$

This shear force occurs at 2.59 m from the supports. This means that in the middle portion of the beam for a distance of $(9.0 - 2 \times 2.59) = 3.83$ m links at a spacing of 375 mm will be satisfactory. In the rest of the beam, a spacing of 175 mm will be satisfactory.

(e) Bar curtailment and end anchorage

The additional tensile force generated by shear that can be accommodated by shifting the bending moment diagram by a distance a_1 is given by code equation (9.2) as

$$a_1 = z\ (\cot\varphi - \cot\alpha)/2 \qquad (9.2)$$

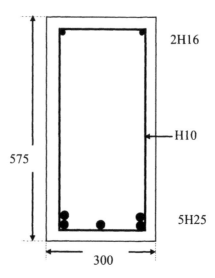

Fig. 7.7 Beam cross section.

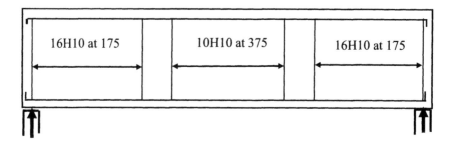

Fig. 7.8 Beam reinforcement.

Using z = 409 mm, cot θ = 2.5, cot α = as only vertical links are used as shear reinforcement, a_1 = 511 mm.

From Table 5.5, l_{bd} = 40 φ = 1.0 m, the top 2H25 can be stopped at $(1.64 - l_{bd} - a_1)$ = 0.14 m from the ends. This is too short a length to bother about. It is simply convenient to carry all bars to the ends. The compression bars will be carried through to the ends of the beam to anchor the links.

The amount of bottom reinforcement at the support is clearly greater than 25% of the steel at mid-span as required by the clause 9.2.1.4(3).

The tensile force F_E at the support to be anchored is

$$F_E = V_{ED} \times (a_1/z) \approx 1.25 \, V_{ED}$$

At support, V_{ED} = 195.8 kN, F_E = 245 kN.

Area of steel = 2H25 = 982 mm², stress in the bar = 199 MPa.

$$\text{Stress in the bar } /f_{yd} = 0.46$$

From Table 5.5, l_{bd} = 40 bar diameters.

l_{bd} required $= 0.46 \times 40 \times 25 = 460$ mm.
The anchorage of the bars at the supports will be provided by a 90° bend with an internal diameter of seven bar diameters equal to 175 mm. The diameter of the bend is determined on the basis that the bar is not damaged (see section 5.2.6).

(f) Deflection check
The deflection of the beam is checked using the equation (7.16 b) of the code.
$$A_{s, prov} / A_{s, reqd} = 2454/2091 = 1.17$$
$$\rho = A_s/ (b_w\, d) = 2454/ (300 \times 508) = 0.016, \rho\% = 1.61$$
$$\rho_0 = 10^{-3} \times \sqrt{25} = 0.005, \rho_0\% = 0.5$$
$$\rho' = A'_s/ (b_w\, d) = 402/ (300 \times 508) = 0.0026, \rho'\% = 0.26$$
$$\rho > \rho_0$$
Simply supported beam, $K = 1$.
$$\text{Load at ULS} = (1.35 \times 20) + (1.5 \times 11) = 43.5 \text{ kN/m}$$
$$\text{Load at SLS} = (1.0 \times 20) + (1.0 \times 11) = 31.0 \text{ kN/m}$$
Stress σ_s in steel at SLS $= (31.0/ 43.5) \times f_{yd} = 310$ MPa

$$\frac{L}{d} = K\left[11 + 1.5 \sqrt{f_{ck}}\, \frac{\rho_0}{\rho - \rho'} + \frac{1}{12} \sqrt{f_{ck}} \sqrt{\frac{\rho'}{\rho}}\, \right] \text{ if } \rho > \rho_0 \qquad (7.16b)$$

$$L/d = 13.9$$
L/d ratio corrected for $A_{s, prov} / A_{s, reqd}$ is $13.9 \times 1.17 = 16.3$.
Actual L/d $= 9000/508 = 17.7$.
Deflection will exceed the permitted L/250. **The beam needs to be redesigned with a deeper beam.**

(g) Check for cracking
The clear distance between bars on the tension face is
$$[300 - 2 \times (35 + 10) - 3 \times 25]/2 = 68 \text{ mm}$$
Stress σ_s in steel at SLS $= 310$ MPa.
From Table 6.2, for 0.3 mm crack width, maximum spacing is approximately 120 mm. The actual spacing of 68 mm does not exceed 300 mm.
The beam is satisfactory with regard to cracking.

(i) Beam reinforcement
The reinforcement for the beam is shown in Fig. 7.7 and Fig. 7.8.

7.2 REFERENCES

Institution of Structural Engineers, London. (2006). *Standard Method of Detailing Structural Concrete: A Manual for Best Practice.* 3rd ed.

Ray, S.S. (1995). *Reinforced Concrete.* Blackwell Science.

Goodchild, C.H. (1997). *Economic Concrete Frame Elements.* Reinforced Concrete Council.

CHAPTER 8

REINFORCED CONCRETE SLABS

8.1 DESIGN METHODS FOR SLABS

Slabs are plate elements forming floors and roofs in buildings, which normally carry uniformly distributed loads acting normal to the plane of the slab. In many ways the behaviour of a beam and a slab are similar but there are also some fundamental differences. A beam is essentially a one-dimensional element subjected on a face to bending moment and shear force as shown in Fig. 8.1(a).

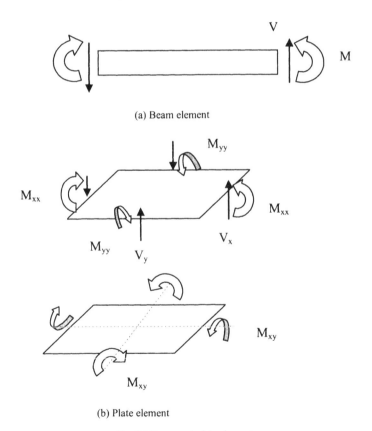

(a) Beam element

(b) Plate element

Fig. 8.1 Beam and plate elements.

A slab is similar to a beam element except that it is a two-dimensional element as shown in Fig. 8.1(b). It can be visualized as representing two beams at right angles. On the faces normal to the x–axis, bending moment M_{xx} and shear force V_x

act. Similarly, on the faces normal to the y-axis, bending moment M_{yy} and shear force V_y act. However, because of the two-dimensional nature of the slab, it is subjected not only to bending moments M_{xx} and M_{yy} and shear forces V_x and V_y but also to twisting moments M_{xy} on all the four faces.

The slab is essentially a statically indeterminate structure and for a slab of given shape and support conditions, the distribution of bending and twisting moments and shear forces in the slab subjected to loads normal to the plane of the slab cannot be determined easily. Elastic analysis can be done for simple shapes such as rectangular simply supported plates by analytical methods but for practical problems, the finite element method is used.

The object of determining the distribution of moments and shear forces is to obtain a set of stresses which are in equilibrium with the applied ultimate loads. Provided the slab cross section behaves in a ductile manner, elastic distribution of stresses or some variations within limits of the elastic stresses can be used for designing the slab.

The code gives very little guidance on the equations for bending moments and shear forces for which the slabs need to be designed. However clause 5.5 lays down a few basic principles which one needs to adhere to. The basic principles are:

- Linear analysis with limited redistribution may be applied to the analysis of structural members for the verification of ULS.
- The moments at ULS calculated using a linear elastic analysis may be redistributed provided that the resulting distribution of moments remains in equilibrium with the applied loads.
- In continuous beams and slabs which are predominantly subjected to flexure and which have the ratios of adjacent spans in the range of 0.5 to 2.0, redistribution of bending moments may be carried out without explicit check on the rotation capacity, provided that the following limitation given by code equations (5.10a) and (5.10b) on the neutral axis dept x_u at the ultimate is satisfied.

$$\delta \geq 0.44 + 1.25(0.6 + \frac{0.0014}{\varepsilon_{cu2}}) \times \frac{x_u}{d} \qquad f_{ck} \leq 50 \text{ MPa}$$

$$\text{(5.10a)}$$

$$\delta \geq 0.54 + 1.25 \times (0.6 + \frac{0.0014}{\varepsilon_{cu2}}) \times \frac{x_u}{d} \qquad f_{ck} > 50 \text{ MPa}$$

$$\text{(5.10b)}$$

where

δ = Ratio of redistributed moment to elastic bending moment.

x_u = Depth of neutral axis at ULS after redistribution.

$\varepsilon_{cu2} = 3.5 \times 10^{-3} \quad f_{ck} \leq 50 \text{ MPa}$

$\qquad = \{2.6 + 35 \times [(90 - f_{ck})/100]^4\} \times 10^{-3} \quad f_{ck} > 50 \text{ MPa}$

Note in equation (5.10b), $\delta \geq 0.7$ if Class B and Class C reinforcement is used and $\delta \geq 0.8$ if Class A reinforcement is used. Note that the **higher** the value of δ, the **lower** the amount of redistribution.

See Annex C of Eurocode 2 for properties of Class A, B and C reinforcements.

Equations (5.10a) and (5.10b) can be simplified as

$$\delta \geq 0.44 + 1.25 \times \frac{x_u}{d} \qquad f_{ck} \leq 50 \text{ MPa}$$

$$\delta \geq 0.54 + C_1 \times \frac{x_u}{d} \qquad f_{ck} > 50 \text{ MPa}$$

for f_{ck} = (55, 60, 70, 80 and 90), the corresponding values of C_1 = (1.32, 1.35, 1.40, 1.42, 1.42).
The following example illustrates the above ideas.

Example: Table 8.1 shows the results for the three span continuous beam in section 5.3.4. Fig. 8.1 shows the bending moment distribution for the four load cases.

Table 8.1 Bending moments from elastic analysis

Loading	Span 1–2	Support 2	Span 2–3	Support 3	Span 3–4
MAX, min, MAX	238.3$ 245.2*	197.1	–41.1	197.1	238.3$ 245.2*
MAX, MAX, min	190.0$ 205.8*	293.7	103.5$ 106.1*	172.9	69.6$ 81.5*
min, MAX, min	57.5$ 74.1*	197.1	139.7	197.1	57.5$ 74.1*
min, MAX, MAX	69.6$ 81.5*	172.9	103.5$ 106.1*	293.7	190.0

Note: If two values are shown for span moment, the figure with * is the maximum bending moment in the span and the figure with $ is the bending moment at mid-span. If only one value is shown then the maximum bending moment occurs at mid-span.

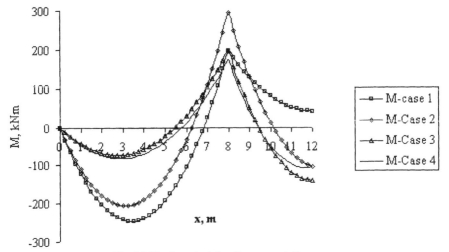

Fig. 8.2 Elastic analysis bending moment diagrams.

Because the support moments tend to peak over a short distance only, one can reduce the support moment to a more manageable value. For example, assuming δ 0.7 for reduction of maximum support moment, Table 8.2 shows the modified moments and Fig. 8.3 shows the modified bending moment diagrams. As can be seen, the change applies only to loading cases (MAX, MAX, min) and (min, MAX, MAX). Note that once the support moment is reduced, then the span moment increases in order to maintain equilibrium with the applied loads. The maximum support moments are changed from 293.7 kNm from elastic analysis to 70 percent of the elastic value. The new support moment value is 0.7 × 293.7 = 205.6 kNm.

Table 8.2 Modified elastic bending moments

Loading	Span 1–2	Support 2	Span 2–3	Support 3	Span 3–4
MAX, min, MAX	238.3$	197.1	–41.1	197.1	238.3$
	245.2*				245.2*
MAX, MAX, min	234.0$	205.6	*147.6$*	172.9	69.6$
	241.8*		*147.8**		81.5*
min, MAX, min	57.5$	197.1	139.7	197.1	57.5$
	74.1*				74.1*
min, MAX, MAX	69.6$	172.9	*144.6$*	205.6	234.0$
	81.5*		*147.8**		241.8*

Note: If two values are shown for span moment, then the figure with * is the maximum bending moment in the span and the figure with $ is the bending moment at mid-span. If only one value is shown then the maximum bending moment occurs at mid-span.

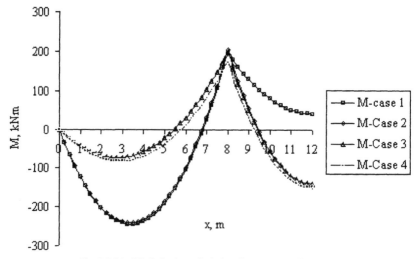

Fig. 8.3 Modified elastic analysis bending moment diagrams.

The continuous beam can be designed as shown in section 5.3.4, using the moment values shown in Table 8.1. In this case as there has been no change from the elastic analysis values, therefore $\delta = 1$ and the maximum depth of neutral axis and maximum moment capacity are given by Table 4.5 and Table 4.8.

For $f_{ck} \leq 50$ MPa, $\delta = 1.0$:

$$x_u/d = 0.448 \text{ and } M_{max} = 0.196 \text{ b } d^2 f_{ck}$$

Assuming a breadth of b = 300 mm, f_{ck} = 25 MPa, for the maximum support moment of 293.7 kNm, the minimum effective depth needed is

$$293.7 \times 10^6 = 0.196 \times 300 \times d^2 \times 25, d = 447 \text{ mm}$$

For $f_{ck} \leq 50$ MPa, $\delta = 0.7$:

$$x_u/d = 0.208 \text{ and } M_{max} = 0.102 \text{ b } d^2 f_{ck}$$

Assuming a breadth of b = 300 mm, f_{ck} = 25 MPa, for the maximum support moment of 205.6 kNm, the minimum effective depth needed is

$$205.6 \times 10^6 = 0.102 \times 300 \times d^2 \times 25, d = 518 \text{ mm}$$

Therefore, although the maximum moment has reduced, in order to ensure sufficient ductility, one needs to adopt a deeper section if one wants to use a singly reinforced section.

Because of the difficulty of obtaining the elastic moment distribution, historically many simplified methods of determining the distribution of bending moments and shear forces which intuitively reflect the distribution of the loads to the supports have been used in practice. Slabs designed by these methods have behaved satisfactorily and are widely used in design practices.

In practice, apart from the finite element method which is generally used for non-standard design situations, slabs are designed using the following methods.

1. Simplified elastic analysis which uses the idealization of a slab into strips or beams spanning one way or a grid with the strips spanning two ways
2. Using design coefficients for moment and shear coefficients which have been obtained from yield line analysis
3. The yield line and Hillerborg strip methods

These methods will be illustrated by several examples.

8.2 TYPES OF SLABS

Slabs may be simply supported or continuous over one or more supports and are classified according to the method of support as follows:

1. spanning one way between beams or walls
2. spanning two ways between the support beams or walls
3. flat slabs carried on columns and edge beams or walls with no interior beams

Slabs may be solid of uniform thickness or ribbed with ribs running in one or two directions. Slabs with varying depth are generally not used. Stairs with various support conditions form a special case of sloping slabs.

8.3 ONE-WAY SPANNING SOLID SLABS

8.3.1 Idealization for Design

Uniformly loaded slabs

In section 5.3.1(5), the code defines that a slab subjected dominantly to uniformly distributed loads may be considered as one-way spanning if either:

- It possesses two free (unsupported) and sensibly parallel edges.
- It is the central part of a sensibly rectangular slab supported on four edges with a ratio of longer to shorter span greater than 2.

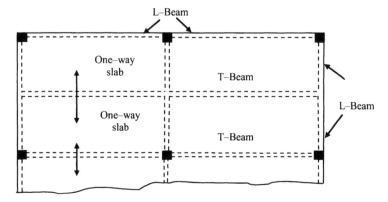

Fig. 8.4 Plan of a typical one-way slab spanning between beams.

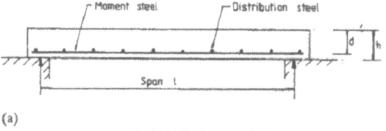

(a)

Fig. 8.5 (a) Simply supported slab.

Fig. 8.4 shows the plan of a typical one-way slab spanning between beams and the beams supported on columns. The beams could be either T-beams or L-Beams. L-beams occur at the edges and T-beams occur in the interior.

One-way slabs carrying predominantly uniform loadS are designed on the assumption that they consist of a series of rectangular beams 1 m wide spanning between supporting beams or walls. The sections through a simply supported slab and a continuous one-way slab are shown in Fig. 8.5 (a) and Fig. 8.5 (b) respectively.

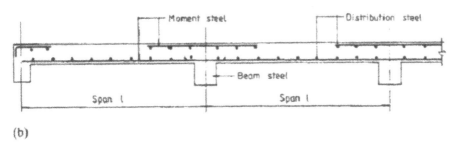

(b)

Fig. 8.5 (b) Continuous one-way slab.

8.3.2 Effective Span, Loading And Analysis

(a) Effective span
Clause 5.3.2.2 of the code gives the rules for calculating the effective spans. The effective span for one-way slabs is the same as that set out for beams in section 7.1. If l_n is the clear span (distance between faces of supports), the effective span l_{eff} is given by

$$l_{eff} = l_n + a_1 + a_2$$

The effective spans for non-continuous (simply supported), continuous and fully constrained situations are shown in Fig. 8.6.

(b) Arrangement of loads
The slab should be designed to resist the most unfavourable arrangement of loads. In clause 5.1.3 of Eurocode 2, the following two loading arrangements are recommended for buildings.

1. Alternate spans carrying $(\gamma_G G_k + \gamma_Q Q_k)$ other spans carrying only $\gamma_G G_k$.
2. Any two adjacent spans carrying $(\gamma_G G_k + \gamma_Q Q_k)$. All other spans carrying only $\gamma_G G_k$.

In note 3 to the Table A1.2 (B) of *BS EN 1990:2002 Eurocode –Basis of structural design*, it is stated that characteristic values of all permanent actions from one source such as that from self weight of the structure, are multiplied by $\gamma_{G, sup}$ if the total action effect is unfavourable and by $\gamma_{G, inf}$ if the total action effect is favourable.

$\gamma_{G, sup} = 1.35$, $\gamma_{G, inf} = 1.0$, $\gamma_Q = 1.5$ if unfavourable otherwise 0.

Fig. 8.7 shows the loading arrangements to cause the maximum bending moment in the chosen span and at the chosen support of a continuous beam. In order to cause the maximum bending moment in a chosen span, place the maximum load on that span and in all alternate spans. In the remaining spans, only minimum load is applied.

In order to cause the maximum bending moment in a chosen support, place the maximum load on spans on either side of the chosen support. On the remaining spans only minimum load is applied.

Maximum load is equal to $(1.35\ g_k + 1.5\ q_k)$ and minimum load is equal to $1.35g_k$.

Once all loading patterns are analysed, envelopes of maximum and minimum moments can be drawn and the slab is designed.

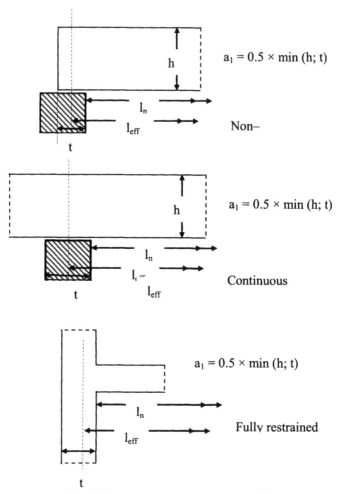

Fig. 8.6 Effective spans for various support conditions.

Fig. 8.8 to Fig. 8.11 show the bending moment diagrams for continuous beams of uniform constant cross section subjected to a uniformly distributed load on one span at a time. Table 8.3 gives the values of bending moments at support sections. The sign convention is positive values show hogging moments and negative values indicate sagging moments. These tabular values are very suitable for calculating the support moments using spreadsheets.

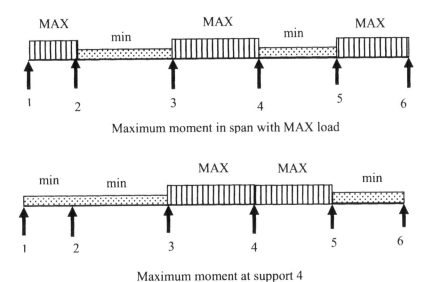

Maximum moment in span with MAX load

Maximum moment at support 4

Fig. 8.7 Eurocode 2 suggests loading to cause maximum bending moments in a span and at a support.

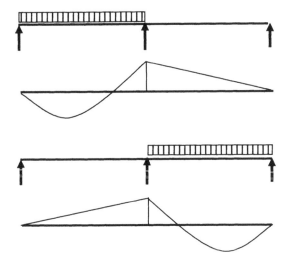

Fig. 8.8 Two-span continuous beam.

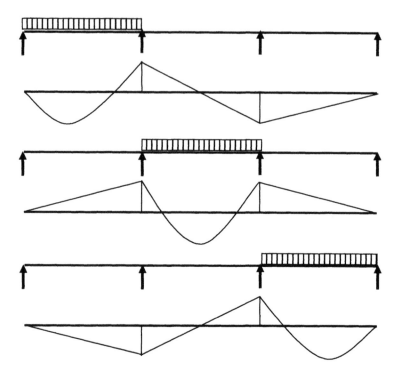

Fig. 8.9 Three-span continuous beam.

Once the support moments are known, then the maximum moment in the span of an isolated member can be calculated as shown in Fig. 8.12.

$$V_L = 0.5qL - \frac{[M_R - M_L]}{L}$$

Maximum span moment M_{Max} occurs at $x = V_L/q$.

$$M_{max} = -M_L + VL \times x - 0.5 \times q \times x^2$$

Note the signs of the moments at supports. M_L is positive clockwise but M_R is positive anticlockwise.

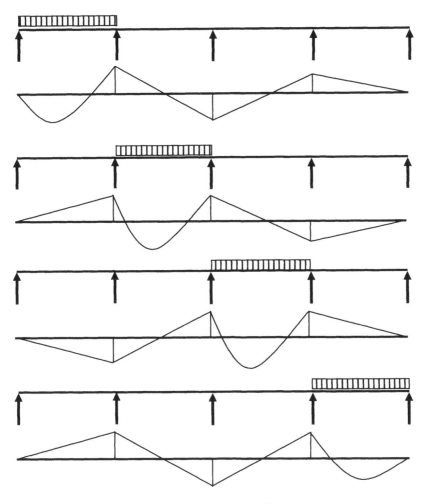

Fig. 8.10 Four-span continuous beam.

8.3.3 Section Design, Slab Reinforcement Curtailment and Cover

(a) Cover
The amount of cover required for durability is given in Table 2.5 and the cover for fire protection is given in Table 2.9 in Chapter 2.

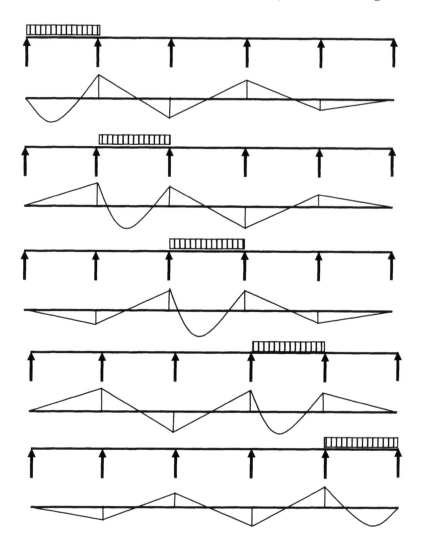

Fig. 8.11 Five-span continuous beam.

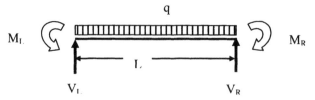

Fig. 8.12 Loading on an isolated beam.

(b) Minimum tension steel

The main moment steel spans between supports and over the interior supports of continuous slab. The slab sections are designed as rectangular beam sections 1 m wide.

The minimum area of main reinforcement has to satisfy clause 9.2.1.1(1).

$$A_{s,min} \geq 0.26 \frac{f_{ctm}}{f_{yk}} b_t \, d \text{ but } \geq 0.0013 \, b_t \, d$$

where b_t = width (for slab design 1000 mm), d = effective depth.

Table 4.3, Chapter 4 shows the value of $A_{s,\,min}$ calculated from the code equation (9.1N).

(c) Distribution steel

The distribution, transverse or secondary steel runs at right angles to the main moment steel and serves the purpose of tying the slab together and distributing non-uniform loads through the slab. Clause 9.3.1.1(2) states that in the case of one-way slabs, secondary reinforcement of not less than 20 percent of principal reinforcement should be provided.

Note that distribution steel is required at the top parallel to the supports of continuous slabs. The main steel is placed nearest to the surface to give the greatest effective depth.

Table 8.3 Support moment coefficients

No. of spans	Load on span	Support moment coefficients			
		Support 2	Support 3	Support 4	Support 5
2	1–2	6.25			
	2–3	6.25			
3	1–2	6.67	−1.67		
	2–3	5.00	5.00		
	3–4	−1.67	6.67		
4	1–2	6.6970	−1.7860	0.4463	
	2–3	4.9066	5.3568	−1.3352	
	3–4	−1.3352	5.3568	4.9066	
	4–5	0.4463	−1.7860	6.6970	
5	1–2	6.7003	−1.7945	0.4778	−0.1190
	2–3	4.9023	5.3836	−1.4346	0.3567
	3–4	−1.3157	5.2634	5.2634	−1.3157
	4–5	0.3567	−1.4346	5.3836	4.9023
	5–6	−0.1190	0.4778	−1.7945	6.7003

Note: Support moment = Coefficient $\times (qL^2/100)$.
q = load per unit length on span, L = constant span.
Sign convention: Positive values show hogging moments.

(d) Slab reinforcement

Slab reinforcement is a mesh and may be formed from two sets of bars placed at right angles. Table 4.9, Chapter 4 gives bar spacing data in the form of areas of steel per metre width for various bar diameters and spacings. Reinforcement in slabs consist of a large number bars both ways which need to be tied together to form a mat. This is an expensive operation. Although more steel might be used than strictly required, it is often economical to use cross-welded wire fabric. Table 8.4 shows the particulars of fabric produced from cold reduced steel wire with f_{yk} = 460 MPa as given in BS 4483:1985. The fabric is available in 4.8 m × 2.4 m sheets.

Table 8.4 Fabric types

Fabric reference	Longitudinal wire			Cross wire		
	Wire size (mm)	Pitch (mm)	Area (mm²/m)	Wire size (mm)	Pitch (mm)	Area (mm²/m)
Square mesh						
A393	10	200	393	10	200	393
A252	8	200	252	8	200	252
A193	7	200	193	7	200	193
A142	6	200	142	6	200	142
A98	5	200	98	5	200	98
Structural mesh						
B1131	12	100	1131	8	200	252
B785	10	100	785	8	200	252
B503	8	100	503	8	200	252
B385	7	100	385	7	200	193
B285	6	100	285	7	200	193
B196	5	100	196	7	200	193
Long mesh						
C785	10	100	785	6	400	70.8
C636	9	100	636	6	400	70.8
C503	8	100	503	5	400	49.1
C385	7	100	385	5	400	49.1
C283	6	100	283	5	400	49.1
Wrapping mesh						
D98	5	200	98	5	200	98
D49	2.5	100	49	2.5	100	49

(e) Crack Control

Maximum spacing of bars is given in Clause 9.3.1.1(3) as follows. If h is the total depth of slab, then maximum spacing is normally restricted to

$$3h \leq 400 \text{ mm for principal reinforcement}$$
$$3.5 \, h \leq 450 \text{ for secondary reinforcement}$$

However in areas of maximum moment, maximum spacing is restricted to

$$2h \leq 250 \text{ mm for principal reinforcement}$$
$$3 \, h \leq 400 \text{ mm for secondary reinforcement}$$

(f) Curtailment of bars in slabs

Curtailment of bars is done according to the moment envelope. However, clause 9.3.1.2(1) requires that half the calculated span reinforcement must continue up to support.

It is further stated that in monolithic construction, where partial fixity occurs along an edge of a slab but is not taken into account, the top reinforcement should be capable of resisting at least 25 percent of the maximum moment in the adjacent span and this reinforcement should extend at least 0.2 times the length of the adjacent span measured from the face of the support.

The above situation occurs in the case of simply supported slabs or the end support of a continuous slab cast integral with an L-beam which has been taken as a simple support for analysis but the end of the slab might not be permitted to rotate freely as assumed. Hence negative moments may arise and cause cracking.

(g) Shear

Under normal loads shear stresses are not critical and shear reinforcement is not required. Shear reinforcement is provided in heavily loaded thick slabs but should not be used in slabs less than 200 mm thick (clause 9.3.2 (1)).

(h) Deflection

The check for deflection is a very important consideration in slab design and usually controls the slab depth. In normal cases a strip of slab 1 m wide is checked against span-to-effective depth ratios.

8.4 EXAMPLE OF DESIGN OF CONTINUOUS ONE-WAY SLAB

(a) Specification

A continuous one-way slab has four equal spans of 4.0 m each. The slab depth is assumed to be 160 mm. The loading is as follows:

Dead loads due to self–weight, screed, finish, partitions, ceiling: 5.2 kN/m^2

Imposed load: 3.0 kN/m^2

The materials strengths are: concrete, f_{ck} = 25 MP, reinforcement, f_{yk} = 500 MPa. The condition of exposure is XC1. Fire resistance = 2 hours. Design the slab and show the reinforcement on a sketch of the cross section.

(b) Design loads

Consider a strip 1 m wide.

Design maximum ultimate load = $(1.35 \times 5.2) + (1.5 \times 3) = 11.52$ kN/m
Design minimum ultimate load = $(1.35 \times 5.2) + (0 \times 3) = 7.02$ kN/m
Five load cases are analysed as shown in Table 8.5. The analysis is done using the bending moment coefficients given in Table 8.3.

Table 8.5 Load cases analysed

Maximum moment at	Loads on spans			
	Span 1–2	Span 2–3	Span 3–4	Span 4–5
Support 2	11.52	11.52	7.02	7.02
Support 3	7.02	11.52	11.52	7.02
Support 4	7.02	7.02	11.52	11.52
Spans 1–2 & 3–4	11.52	7.02	11.52	7.02
Spans 2–3 & 4–5	7.02	11.52	7.02	11.52

(i) Maximum moment at support 2

q_{max} on spans 1–2 and 2–3.
q_{min} load on span 3–4 and 4–5.
Table 8.6 shows the results of calculations.

Table 8.6 Loading for maximum moment at support 2

Load on span	q	$qL^2/100$	Moment at		
			Support 2	Support 3	Support 4
1–2	11.52	1.8432	12.34	−3.29	0.82
2–3	11.52	1.8432	9.04	9.87	−2.46
3–4	7.02	1.123	−1.50	6.02	5.51
4–5	7.02	1.123	0.50	−2.01	7.52
SUM			Σ 20.38	Σ 10.59	Σ 11.39

Note: Loading for maximum moment at support 4 will be mirror image of the results for maximum moment at support 2. $M_4 = M_2$, $M_2 = M_4$ and $M_3 = M_3$.
Maximum moment at support 2 = 20.38 kNm.

Table 8.7 Loading for maximum moment at support 3

Load on span	q	$qL^2/100$	Moment at		
			Support 2	Support 3	Support 4
1–2	7.02	1.123	7.52	−2.01	0.50
2–3	11.52	1.8432	9.04	9.87	−2.46
3–4	11.52	1.8432	−2.46	9.87	9.04
4–5	7.02	1.123	0.50	−2.01	7.52
SUM			Σ 14.60	Σ 15.72	Σ 14.60

(ii) Maximum moment at support 3

q_{max} on spans 2–3 and 3–4.
q_{min} load on spans 1–2 and 4–5.
Table 8.7 shows the results of calculations.
Maximum moment at support 3 = 15.72 kNm.

(iii) Maximum moment in spans 1–2 and 3–4

q_{max} on spans 1–2 and 3–4.
q_{min} load on spans 2–3 and 4–5.
Table 8.8 shows the results of calculations.

Table 8.8 Loading for maximum moment in spans 1–2 and 3–4

Load on span	q	$qL^2/100$	Moment at		
			Support 2	Support 3	Support 4
1–2	11.52	1.8432	12.34	−3.29	0.82
2–3	7.02	1.123	5.51	6.02	−1.50
3–4	11.52	1.8432	−2.46	9.87	9.04
4–5	7.02	1.123	0.50	−2.01	7.52
	SUM		Σ 15.89	Σ 10.59	Σ 15.88

From the support moment values in Table 8.8:

span 1–2

$M_2 = 15.89$ and left hand reaction, $V_1 = 19.07$ kN/m and the maximum moment is 15.78 kNm/m at 1.66 m from support 1. Moment reduces to half the maximum at 1.2 m on either side of the maximum.

span 3–4

$M_3 = 10.59$ kNm/m, $M_4 = 15.88$ kNm/m, right hand reaction, $V_4 = 24.36$ kN/m and the maximum moment is 9.88 kNm/m at 2.12 m from support 4. Moment reduces to half the maximum at 0.92 m on either side of the maximum.

Table 8.9 Loading for maximum moment in spans 2–3 and 4–5

Load on span	q	$qL^2/100$	Moment at		
			Support 2	Support 3	Support 4
1–2	7.02	1.123	7.52	−2.01	0.50
2–3	11.52	1.8432	9.04	9.87	−2.46
3–4	7.02	1.123	−1.50	6.02	5.51
4–5	11.52	1.8432	0.82	−3.29	12.34
	SUM		Σ15.88	Σ 10.59	Σ 15.89

(iv) Maximum moment in spans 2–3 and 4–5

q_{max} load on spans 2–3 and 4–5.
q_{min} on spans 1–2 and 3–4.
Table 8.9 shows the results of calculations.
From the support moment values in Table 8.9:

span 2–3

$M_2 = 15.88$ kNm/m, $M_3 = 10.59$ kNm/m, left hand reaction, $V_2 = 24.36$ kN/m and the maximum moment is 9.88 kNm/m at 2.12 m from support 2. Moment reduces to half the maximum at 0.93 m on either side of the maximum.

span 4–5
M_4 = 15.89 kNm/m, right hand reaction, V_5 = 19.07 kN/m and the maximum moment is 15.78 kNm/m at 1.66 m from support 5. Moment reduces to half the maximum at 1.2m on either side of the maximum.

Fig. 8.13 shows the resulting bending moment diagrams.

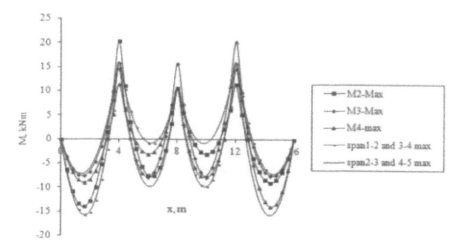

Fig 8.13 Bending moment distribution in a four-span continuous one-way slab.

Fig. 8.14 shows the symmetrical moment envelope which will be used for designing the reinforcement and also to decide on the bar curtailment.

The maximum bending moments are:

(i) Hogging
Supports 2 and 4: 20.38 kNm/m.
Support 3: 15.72 kNm/m.
In both cases, the moment reduces to half the peak value at approximately 0.4 m on either side of the support.

(ii) Sagging
Span 1–2 and span 4–5: 15.78 kNm/m.
Moment reduces to half the peak value at 1.2 m on either side of the peak value which occurs at 1.7 m from the simply supported end.
Span 2–3 and 3–4: 9.88 kNm/m.
Moment reduces to half the peak value at 0.92 m on either side of the peak value which occurs at 1.89 m from the end with a moment of 10.59 kNm/m.

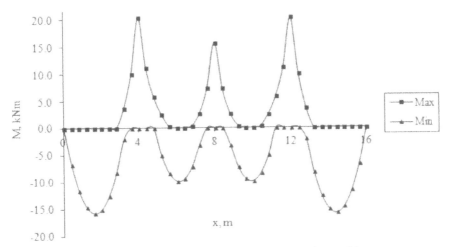

Fig.8.14 Moment envelope for four-span one-way continuous slab.

(c) Design of moment steel
The minimum cover required is 40 mm on the basis of fire protection and 15 mm on the basis of XC1 exposure. Assume 10 mm diameter bars. The effective depth: $d = 160 - 40 - 10/2 = 115$ mm. Width, b = 1000 mm.

(i) Hogging moment
Supports 2 and 4:
M = 20.38 kNm/m.
$k = M/ (bd^2 f_{ck}) = 20.38 \times 10^6/ (1000 \times 115^2 \times 25) = 0.062 < 0.196$.
Singly reinforced section can be designed
$$\frac{z}{d} = 0.5[1.0 + \sqrt{(1-3k)}] = 0.95$$
$A_s = M/ (0.87 f_{yk} z) = 20.38 \times 10^6/ (0.87 \times 500 \times 0.95 \times 115) = 429$ mm^2/m
10 mm bars at 175 mm spacing gives $A_s = 448$ mm^2/m (see Table 4.9, Chapter 4).
Support 3:
M = 15.72 kNm/m.
$k = M/ (bd^2 f_{ck}) = 15.72 \times 10^6/ (1000 \times 115^2 \times 25) = 0.048 < 0.196$.
Singly reinforced section can be designed
$$\frac{z}{d} = 0.5[1.0 + \sqrt{(1-3k)}] = 0.96$$
$A_s = M/ (0.87 f_{yk} z) = 15.72 \times 10^6/ (0.87 \times 500 \times 0.96 \times 115) = 327$ mm^2/m.
10 mm bars at 200 mm spacing gives $A_s = 392$ mm^2/m (see Table 4.9, Chapter 4).

(ii) Sagging moment
Spans 1–2 and 4–5:
M = 15.78 kNm/m.
Moment reduces to half the maximum at 1.2 m on either side of the maximum.
$k = M/ (bd^2 f_{ck}) = 15.78 \times 10^6/ (1000 \times 115^2 \times 25) = 0.048 < 0.196$.

Singly reinforced section can be designed

$$\frac{z}{d} = 0.5[1.0 + \sqrt{(1-3k)}] = 0.96$$

$A_s = M/ (0.87\ f_{yk}\ z) = 15.78 \times 10^6/ (0.87 \times 500 \times 0.96 \times 115) = 329\ mm^2/m$.
10 mm bars at 200 mm spacing gives $A_s = 392\ mm^2/m$ (see Table 4.9, Chapter 4).
Spans 2–3 and 3–4:
M = 9.88 kNm/m.
Moment reduces to half the maximum at 0.93m on either side of the maximum.
$k = M/ (bd^2\ f_{ck}) = 9.88 \times 10^6/ (1000 \times 115^2 \times 25) = 0.03 < 0.196$.
Singly reinforced section can be designed

$$\frac{z}{d} = 0.5[1.0 + \sqrt{(1-3k)}] = 0.98$$

$A_s = M/ (0.87\ f_{yk}\ z) = 9.88 \times 10^6/ (0.87 \times 500 \times 0.98 \times 115) = 202\ mm^2/m$.
10 mm bars at 300 mm spacing gives $A_s = 261\ mm^2/m$ (see Table 4.9, Chapter 4).

Fig. 8.15 shows the calculated steel at different locations. Taking into account the maximum spacing and also the minimum steel requirement, the above calculated value of steel is adjusted to simplify the layout and also minimize the number of variations in order to minimize errors during construction.

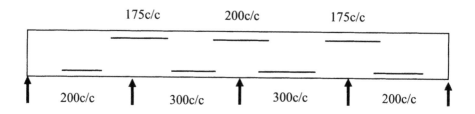

Fig. 8.15 Calculated H10 steel at supports and in spans.

Minimum steel

$f_{ctm} = 0.30 \times f_{ck}^{0.667} = 0.30 \times 25^{0.667} = 2.6\ MPa$.
$A_{s,\ min} = 0.26\ (f_{ctm}/f_{yk})\ b_t\ d \geq 0.0013\ b_t\ d$.
$A_{s,\ min} = 0.26 \times (2.6/500) \times 1000 \times 115 \geq 0.0013 \times 1000 \times 115$.
$A_{s,\ min} = 156\ mm^2/m$.
In this case h = 160 mm.
In areas of maximum moment, spacing ≤ 250 mm for main steel. For 10 mm diameter bars at 250 mm c/c, area of steel is 314 mm²/m.
In other areas, spacing ≤ 400 mm for main steel. For H10 bars at 400 mm c/c, area of steel is 196 mm²/m.
Both the above values are greater than the minimum steel of 156 mm²/m

Main steel
Bottom steel: The steel reinforcement can be rationalized as follows.

In spans 2–3 and 3–4, calculated steel is 10 mm bars at 300 c/c. However in maximum moment areas, the bar spacing is limited to 250 mm. In addition, away from the areas of maximum moment, the maximum spacing is limited to 400 mm. Therefore all bottom steel can be at 200 mm spacing and alternate bars can be curtailed to give a spacing of 400 mm c/c.

From Table 5.5 for f_{ck} = 25 MPa and $\varphi \leq 32$ mm, the anchorage length is 40 bar diameters which is equal to 400 mm.

In the end spans, sagging moment reduces to half the peak value at 1.2 m to the left and right of peak value. As there is generally no shear reinforcement in the slabs, the shift in bending moment to accommodate the tensile stress caused by shear is equal to the effective depth. Adding 400 mm of anchorage length plus (d = 115 mm) to these lengths, the length over which the bars at 200 mm c/c are required is $2 \times 1.2 + 2 \times 0.4 + 2 \times d$ = 3.43 m. The length of steel saved by curtailment in a span length of 4.0 m is only 0.57 m. To keep the layout simple, it is better to have the steel at 200 mm c/c in the bottom of the slab over the entire four spans.

Top steel: Maximum steel over the internal supports can be a constant value of 175 mm c/c. In the end spans, the moment reduces to half the peak value approximately 0.4 m from the support. Adding an anchorage length of 400 mm and a shift value of (d = 115 mm), the length is 0.0.915 m. The negative moment reduces to zero at 1.4 m from the support. Adding an anchorage length of 400 mm and a shift value of (d = 115 mm), this comes to 1.915 m. The steel can be provided as follows.

End spans: Provide steel over the supports at 175 mm c/c. Stop alternate bars 0.92 m from support. Continue the rest of the bars 1.92 m from support.

Middle spans: Provide steel over the supports at 175 mm c/c. Stop alternate bars 0.92 m from support. Continue the rest of the bars over the whole length.

Secondary steel: Secondary steel to be not less than 20 percent of main steel. Spacing restricted to 3h ≤ 400 mm in areas of maximum moment and 3.5h ≤ 450 mm.

Minimum steel as calculated is 156 mm²/m. Using 8 mm bars at 300 mm spacing, A_s = 168 mm²/m.

Use H8 at 300 as secondary steel.

Fig. 8.16 shows the final arrangement of flexural steel.

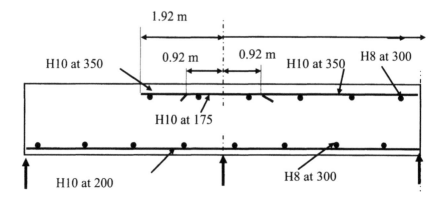

Fig. 8.16 Flexural steel.

(c) Shear force distribution in the slab

Fig. 8.17 shows the shear force diagrams for the five loading cases analysed.
Fig. 8.18 shows the corresponding shear force envelope.
The maximum shear force is 28.14 kN/m at support 2 for span 1–2.

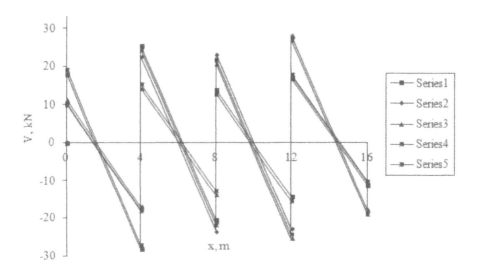

Fig. 8.17 Shear force diagrams.

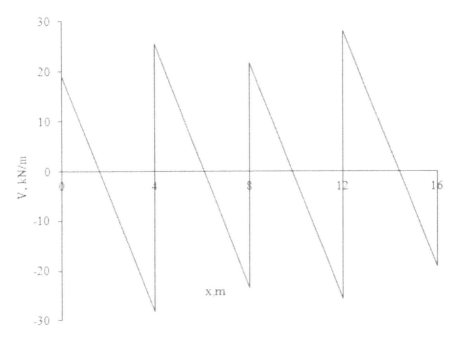

Fig. 8.18 Shear force envelope.

(d) Shear resistance
$b_w = 1000$ mm, $d = 115$ mm, $A_s = $ H10 at $200 = 393$ mm^2/m

$$C_{Rd,c} = \frac{0.18}{(\gamma_c = 1.5)} = 0.12$$

$$k = 1 + \sqrt{\frac{200}{115}} = 2.3 > 2.0$$

$$k = 2.0$$

$$100\rho_1 = 100 \times \frac{393}{115 \times 1000} = 0.34 \leq 2.0$$

$$v_{min} = 0.035 \times 2.0^{1.5} \times \sqrt{25} = 0.50 \text{ MPa}$$

$$V_{Rd,c} = [0.12 \times 2.0 \times \{0.34 \times 25\}^{1/3} + 0.15 \times 0] \times 115 \times 1000 \times 10^{-3}$$

$$\geq [0.50 + 0.15 \times 0] \times 115 \times 1000 \times 10^{-3}$$

$$V_{Rd,c} = 56.30 \geq 39.1 \text{ kN}$$

$$V_{Rd,c} = 58.30 \text{ kN/m} > (V_{Ed} = 28.14 \text{ kN/m})$$

No shear reinforcement is required.

(e) Crack control

As all the reinforcements satisfy the maximum spacing requirements, the slab will not suffer excessive cracking.

(f) Deflection

Use the code equation (7.16a).

$\rho_0\% = 0.1 \times \sqrt{f_{ck}} = 0.5$.

$b_w = 1000$ mm, $d = 115$ mm, $A_s = $ H10 at $200 = 393$ mm^2/m.

$\rho\% = 100 \times 393 / (1000 \times 115) = 0.34$.

$K = 1.3$.

$$\frac{L}{d} = K[11 + 1.5 \sqrt{f_{ck}} \frac{\rho_0}{\rho} + 3.2 \sqrt{f_{ck}} (\frac{\rho_0}{\rho} - 1)^{\frac{3}{2}}] \text{ if } \rho \leq \rho_0$$

(7.16a)

$$\frac{L}{d} = 1.3 \times [11 + 1.5 \sqrt{25} \frac{0.50}{0.34} + 3.2 \sqrt{25} (\frac{0.50}{0.34} - 1)^{\frac{3}{2}}] = 35.4$$

Actual $L/d = 4000/115 = 34.8 < 35.4$

L/d ratio is slightly higher than desirable but the slab will not deflect more than L/250.

8.5 ONE-WAY SPANNING RIBBED OR WAFFLE SLABS

8.5.1 Design Considerations

Solid slabs are uneconomic in spans over 4 m due to self weight. When spans are long (perhaps over 5 m) but the live loads are relatively moderate or light, it is advantageous to reduce the dead weight of the slab. By having a series of ribs (beams) connected by structural topping as shown in Fig. 8.19, the weight of the slab between the ribs is considerably reduced.

Ribbed slabs may be constructed in a variety of ways. Two principal methods of construction are:

    a.   Ribbed slabs without permanent blocks. The space between the beams is created using square or rectangular plastic formers during casting. Reinforcement is laid between the formers.

    b.   Ribbed slabs with permanent hollow or solid blocks to obtain a flat ceiling.

8.5.2 Ribbed Slab Proportions

Clause 5.3.1(6) states that ribbed or waffle slabs need not be treated as discrete elements for the purpose of analysis, provided that the flange or the structural topping and transverse ribs have sufficient torsional stiffness. This may be assumed provided that:

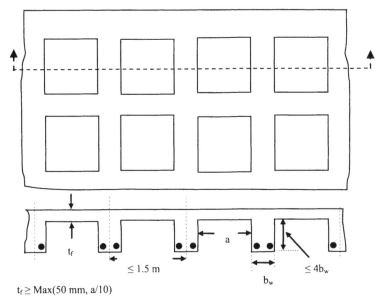

$t_f \geq$ Max(50 mm, a/10)

Fig. 8.19 Ribbed slab.

1. The centres of ribs should not exceed 1.5 m.
2. The depth of rib below the flange does not exceed four times its width.
3. The depth of flange is at least one-tenth of the clear distance between ribs or 50 mm, whichever is greater.
4. Transverse ribs are provided at a clear spacing not exceeding 10 times the overall depth of the slab.
5. The minimum flange thickness of 50 mm may be reduced to 40 mm where permanent blocks are incorporated between the ribs.

Note that to meet a specified fire resistance period, non-combustible finish, e.g., screed on top or sprayed protection can be included to give the minimum thickness.

8.5.3 Design Procedure and Reinforcement

(a) Shear forces and moments
Shear forces and moments for continuous slabs can be obtained by analysis as shown in section 8.3.

(b) Design for moment and moment reinforcement
Design consists of determining the reinforcement required in the ribs. The mid-span section is designed as a T-beam with an effective flange width. The support section is designed as a rectangular beam. The slab may be made solid near the support to increase shear resistance.

Moment reinforcement consisting of one or more bars is provided at the top and the bottom of the ribs. If appropriate, bars can be curtailed in a similar way to bars in solid slabs.

(c) Shear resistance and shear reinforcement
The shear resistance is checked as for beams and any necessary shear links are designed.

(d) Reinforcement in the topping
Minimum percentage of reinforcement is provided in the topping in each direction. The code states in clause 9.2.1.1(1) that minimum cross sectional area of not less than 0.13 percent of the area of the topping should be provided in each direction. The reinforcement normally consists of a mesh which is placed in the centre of the topping. If the ribs are widely spaced the topping may need to be designed for moment and shear as a continuous one-way slab between ribs.

8.5.4 Deflection

The deflection can be checked using the span-to-effective depth rules given in section 7.4.2 of the code.

8.5.5 Example of One-Way Ribbed Slab

(a) Specification
A ribbed slab is continuous over four equal spans each of 6 m. The characteristic dead loading including self-weight, finishes, partitions etc. is 4.7 kN/m^2 and the characteristic imposed load is 2.5 kN/m^2. The construction materials are concrete, f_{ck} = 25 MPa and reinforcement, f_{yk} = 500 MPa. Design the end span of the slab.

(b) Trial section
A cross section through the floor and a trial section for the slab are shown in Fig. 8.20. The thickness of topping is 60 mm and the minimum width of a rib is 125 mm. The deflection check will show whether the depth selected is satisfactory. The cover for mild exposure is 25 mm. For H12 bar, effective depth d = 350 − 25 − 12/2 = 319, say, 320 mm.
Note: Minimum thickness of topping = 50 mm or clear distance between ribs/10 whichever is greater. In this case topping is 60 mm which is greater than 50 mm or (450 − 125)/10 = 32 mm.

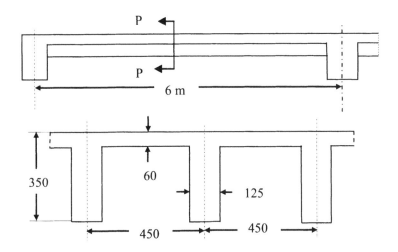

Fig. 8.20 Top: Section through floor. Bottom: Section PP through slab.

(c) Design loads

Consider a typical T-beam. Load over a width of 450 mm (equal to the spacing of beams) acts on the beam.

$$g_k = 4.7 \times 0.45 = 2.12 \text{ kN/m}$$
$$q_k = 2.5 \times 0.45 = 1.13 \text{ kN/m}$$

Design maximum ultimate load = $(1.35 \times 2.12) + (1.5 \times 1.13) = 4.56$ kN/m

Design minimum ultimate load = $(1.35 \times 2.12) + (0 \times 1.13) = 2.86$ kN/m

Five load cases are analysed as shown in Table 8.10. The analysis is done using the bending moment coefficients given in Table 8.3.

Table 8.10 Load cases analysed

Maximum	Loads on spans			
moment at	Span 1–2	Span 2–3	Span 3–4	Span 4–5
Support 2	4.56	4.56	2.86	2.86
Support 3	2.86	4.56	4.56	2.86
Support 4	2.86	2.86	4.56	4.56
Spans 1–2 & 3–4	4.56	2.86	4.56	2.86
Spans 2–3 & 4–5	2.86	4.56	2.86	4.56

Table 8.11 Loading for maximum moment at support 2

Load on	q	$qL^2/100$	Moment at		
span			Support 2	Support 3	Support 4
1–2	4.56	1.6416	10.99	–2.93	0.73
2–3	4.56	1.6416	8.06	8.79	–2.19
3–4	2.86	1.03	–1.38	5.52	5.06
4–5	4.56	1.03	0.46	–1.84	6.90
SUM			Σ 18.13	Σ 9.54	Σ 10.50

(i) Maximum moment at support 2
q_{max} on spans 1–2 and 2–3 and 4–5.
q_{min} load on span 3–4.
Table 8.11 shows the results of calculations.
Note: For loading for maximum moment at support 4 will be mirror image of the results for maximum moment at support 2. $M_4 = M_2$, $M_2 = M_4$ and $M_3 = M_3$.

(ii) Maximum moment at support 3
q_{max} on spans 2–3 and 3–4.
q_{min} load on spans 1–2 and 4–5.
Table 8.12 shows the results of calculations.

Table 8.12 Loading for maximum moment at support 3

Load on span	q	$qL^2/100$	Moment at		
			Support 2	Support 3	Support 4
1–2	2.86	1.03	6.89	−1.84	0.46
2–3	4.56	1.6416	8.06	8.79	−2.19
3–4	4.56	1.6416	−2.19	8.79	8.06
4–5	2.86	1.03	0.46	−1.84	6.89
SUM			Σ 13.22	Σ 13.90	Σ 13.22

(iii) Maximum moment in spans 1–2 and 3–4
q_{max} on spans 1–2 and 3–4.
q_{min} load on spans 2–3 and 4–5.
Table 8.13 shows the results of calculations.

Table 8.13 Loading for maximum moment in spans 1–2 and 3–4

Load on span	q	$qL^2/100$	Moment at		
			Support 2	Support 3	Support 4
1–2	4.56	1.6416	10.99	−2.93	0.73
2–3	2.86	1.03	5.06	5.52	−1.38
3–4	4.56	1.6416	−2.19	8.79	8.06
4–5	2.86	1.03	0.62	−1.84	6.89
SUM			Σ 14.48	Σ 9.54	Σ 14.30

From the support moment values in Table 8.13:
span 1–2
$M_2 = 14.48$ kNm and left hand reaction, $V_1 = 11.27$ kN/m and the maximum moment is 13.92 kNm at 2.47 m from support 1.
span 3–4
$M_3 = 9.54$ kNm, $M_4 = 14.30$ kNm, right hand reaction, $V_4 = 14.47$ kN/m and the maximum moment is 8.67 kNm at 3.17 m from support 4.

(iv) Maximum moment in spans 2–3 and 4–5
q_{max} load on spans 2–3 and 4–5.
q_{min} on spans 1–2 and 3–4.

Table 8.14 shows the results of calculations.

Table 8.14 Loading for maximum moment in spans 2–3 and 4–5

Load on span	q	qL^2/100	Moment at		
			Support 2	Support 3	Support 4
1–2	2.86	1.03	6.89	−1.84	0.62
2–3	4.56	1.6416	8.06	8.79	−2.19
3–4	2.86	1.03	−1.36	5.52	5.06
4–5	4.56	1.6416	0.73	−2.93	10.99
SUM			Σ14.32	Σ 9.54	Σ 14.48

From the support moment values in Table 814:
span 2–3
M_2 = 14.32 kNm, M_3 = 9.54 kNm, left hand reaction, V_2 = 14.48 kN and the maximum moment is 8.66 kNm/m at 3.17 m from support 2.
span 4–5
M_4 = 14.48 kNm, right hand reaction, V_5 = 11.27 kN/m and the maximum moment is 13.92 kNm/m at 2.47 m from support 5.
Fig. 8.21 shows the resulting bending moment diagrams. Fig. 8.22 shows the symmetrical moment envelope which will be used for designing the reinforcement and also to decide on the bar curtailment.
The maximum bending moments are:

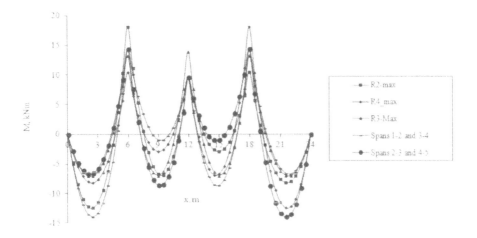

Fig. 8.21 Bending moment distribution in the four-span continuous ribbed slab.

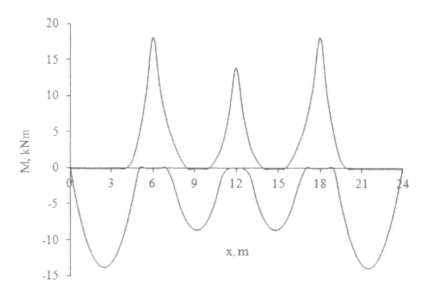

Fig. 8.22 Moment envelope for the four span one-way ribbed slab.

(i) Hogging
Supports 2 and 4: 18.13 kNm.
Support 3: 13.90 kNm/m.
In both cases, the moment reduces to half the peak value at approximately 0.6 m on either side of the support.

(ii) Sagging
Span 1–2 and span 4–5: 13.92 kNm.
Moment reduces to half the peak value at 1.75 m to the left and to the right of the peak value which occurs at 2.47 m from the simply supported end.
Span 2–3 and 3–4: 8.66 kNm.
Moment reduces to half the peak value at 1.4 m to the left and to the right of the peak value which occurs approximately at mid-span.

(d) Design of moment steel

(i) Hogging moment
Because the flange is in tension, design this section as a rectangular section 125 mm wide and d = 320 mm.
Supports 2 and 4:
$$M = 18.13 \text{ kNm/m}.$$
$$k = M/ (bd^2 \, f_{ck}) = 18.13 \times 10^6/ (125 \times 320^2 \times 25) = 0.057 < 0.196.$$
Section can be designed as a singly reinforced section.

$$\frac{z}{d} = 0.5[1.0 + \sqrt{(1-3k)}] = 0.96$$

$A_s = M/ (0.87 \, f_{yk} \, z) = 18.13 \times 10^6/ (0.87 \times 500 \times 0.96 \times 320) = 136 \text{ mm}^2$
2H10 gives $A_s = 157 \text{ mm}^2$.
Support 3:

$$M = 13.90 \text{ kNm/m}$$

$k = M/ (bd^2 \, f_{ck}) = 13.90 \times 10^6/ (125 \times 320^2 \times 25) = 0.043 < 0.196$
Section can be designed as a singly reinforced section.

$$\frac{z}{d} = 0.5[1.0 + \sqrt{(1-3k)}] = 0.97$$

$A_s = M/ (0.87 \, f_{yk} \, z) = 13.90 \times 10^6/ (0.87 \times 500 \times 0.97 \times 320) = 103 \text{ mm}^2$
Provide 2H10. $A_s = 157 \text{ mm}^2$.

(ii) Sagging moment
The flange breadth is 450 mm, $h_f = 60$ mm, d = 320 mm.
Effective width: (see Fig. 4.11).

$$b_1 = b_2 = (450-125)/2 = 163 \text{ mm}, \, \ell_0 \approx 0.7 \times 6 \text{ m} = 4200 \text{ mm}$$
$$b_{effe,\,1} = b_{effe,\,2} = 0.2 \, b_1 + 0.1 \, \ell_0 \le 0.2\ell_0 \text{ and } b_{effe,\,1} \le b_1$$
$$b_{effe,\,1} = 163 \text{ mm}$$
$$b_{eff} = b_{eff,\,1} + b_{eff,\,2} + b_w$$
$$b = 450 \text{ mm}$$

Spans 1–2 and 4–5
M = 13.92 kNm/m.
Check if the stress block is inside the flange.

$$M_{flange} = \eta \, f_{cd} \, b \, h_f \, (d - h_f/2)$$
$$f_{cd} = 25/1.5 = 16.7 \text{ MPa}, \, \eta = 1$$
$$M_{flange} = 1.0 \times 16.7 \times 450 \times 60 \times (320 - 60/2) \times 10^{-6} = 130.8 > 14.57 \text{ kNm}.$$

The neutral axis lies in the flange. The beam is designed as a rectangular beam.

$$k = 13.92 \times 10^6/ [450 \times 320^2 \times 25] = 0.012 < 0.196$$

$$\frac{z}{d} = 0.5[1.0 + \sqrt{(1-3k)}] = 0.99$$

$A_s = 13.92 \times 10^6/ (0.87 \times 500 \times 0.99 \times 320) = 101 \text{ mm}^2$
Provide 2H10. $A_s = 157 \text{ mm}^2$.

Spans 2–3 and 3–4
M = 8.66 kNm.

$$k = 8.66 \times 10^6/ [450 \times 320^2 \times 25] = 0.008 < 0.196$$

$$\frac{z}{d} = 0.5[1.0 + \sqrt{(1-3k)}] = 0.99$$

$A_s = 8.66 \times 10^6/ (0.87 \times 500 \times 0.99 \times 320) = 63 \text{ mm}^2$
Provide 2H8. $A_s = 101 \text{ mm}^2$.

Fig. 8.23 shows the calculated steel at different locations. Taking into account the maximum spacing and also the minimum steel requirement, the above calculated value of steel is adjusted to simplify the layout and also minimize the number variations in order to minimize errors during construction. As all steel is carried to the supports, there is no need to check on the possibility of bar curtailment.

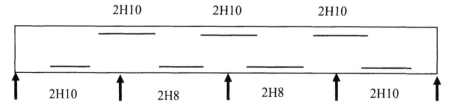

Fig. 8.23 Calculated steel at supports and in spans.

(iii) Minimum steel

$f_{ctm} = 0.30 \times f_{ck}^{0.667} = 0.30 \times 25^{0.667} = 2.6 \text{ MPa}$.

$A_{s, min} = 0.26 \ (f_{ctm}/f_{yk}) \ b_t \ d \geq 0.0013 \ b_t \ d$.

b_t = width of web = 125 mm, d = 320 mm.

$A_{s, min} = 0.26 \times (2.6/500) \times 125 \times 320 \geq 0.0013 \times 125 \times 320$.

$A_{s, min} = 54 \text{ mm}^2$. The areas of steel calculated are higher than the minimum values.

(e) Shear forces in the rib

Fig. 8.24 shows the shear force diagrams for the loading considered in Table 8.9. The 'shear force envelope' is shown in Fig. 8.25.

At support 2 for span 1–2: V_{Ed} = 16.70 kN.

Check if shear reinforcement is required, $V_{Ed} > V_{Rd, c}$.

V_{Ed} = 16.70 kN, b_w = 125 mm, d = 320 mm, A_{sl} = 2H10 = 157 mm^2.

$$C_{Rd,c} = \frac{0.18}{(\gamma_c = 1.5)} = 0.12$$

$$k = 1 + \sqrt{\frac{200}{320}} = 1.79 \leq 2.0$$

$$100\rho_1 = 100 \times \frac{157}{125 \times 320} = 0.39 < 2.0$$

$$v_{min} = 0.035 \times 1.79^{1.5} \times \sqrt{25} = 0.42 \text{ MPa}$$

$$V_{Rd,c} = [0.12 \times 1.79 \times \{0.39 \times 25\}^{1/3} + 0.15 \times 0] \times 125 \times 320 \times 10^{-3}$$

$$\geq [0.42 + 0.15 \times 0] \times 125 \times 320 \times 10^{-3}$$

$$V_{Rd,c} = 18.35 \geq 16.7 \text{ kN}$$

$V_{Rd, c}$ = 18.35 kN > (V_{Ed} = 16.70). Therefore no shear reinforcement is required.

H6 links at 1000 mm c/c will be used to make a reinforcement cage. The arrangement of moment and shear reinforcement in the rib are shown in Fig. 8.26.

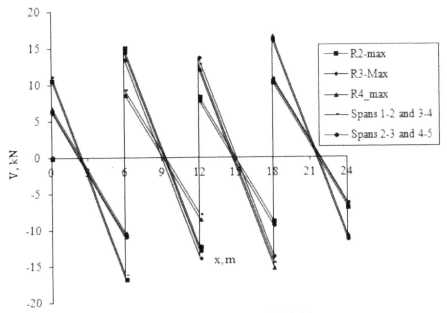

Fig. 8.24 Shear force diagram: Ribbed slab.

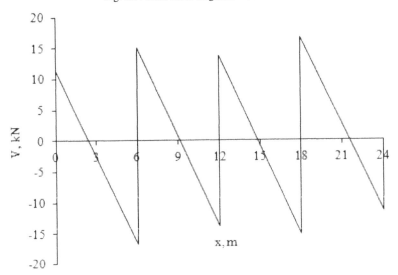

Fig. 8.25 Shear force envelope: Ribbed slab.

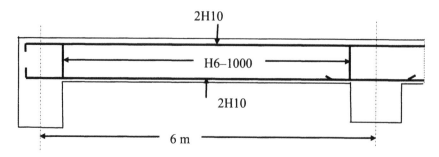

Fig. 8.26 Reinforcement detail in the ribs of a ribbed slab.

(f) Deflection

Using the code equation (7.16a):

$$\rho_0\% = 0.1 \times \sqrt{f_{ck}} = 0.5$$
$$b_w = 125 \text{ mm}, \, d = 320 \text{ mm}, \, A_s = 2H10 = 157 \text{ mm}^2$$
$$\rho\% = 100 \times 157/(125 \times 320) = 0.39 < \rho_0\%$$
$$\rho'\% = 0$$
$$K = 1.3 \text{ for end spans}$$

$$\frac{L}{d} = K[11 + 1.5\sqrt{f_{ck}}\frac{\rho_0}{\rho} + 3.2\sqrt{f_{ck}}(\frac{\rho_0}{\rho} - 1)^{\frac{3}{2}}] \text{ if } \rho \leq \rho_0$$

$$\frac{L}{d} = 1.3 \times [11 + 1.5\sqrt{25}\frac{0.50}{0.39} + 3.2\sqrt{25}(\frac{0.50}{0.39} - 1)^{\frac{3}{2}}] = 29.9$$

Correction for flange width to web width ratio:

$$b/b_w = 450/125 = 3.6 > 3$$
$$\text{Correction factor} = 0.8$$
$$L/d = 29.9 \times 0.8 = 23.9$$
$$\text{Actual } L/d = 6000/320 = 18.75 < 23.9$$

The slab is satisfactory with respect to deflection.

(g) Reinforcement in topping

Minimum steel

$$f_{ctm} = 0.30 \times f_{ck}^{0.667} = 0.30 \times 25^{0.667} = 2.6 \text{ MPa}$$
$$A_{s,\,min} = 0.26 \, (f_{ctm}/f_{yk}) \, b_t \, d \geq 0.0013 \, b_t \, d$$
$$A_{s,\,min} = 0.26 \times (2.6/500) \times 1000 \times 60 \geq 0.0013 \times 1000 \times 60$$
$$A_{s,\,min} = 81 \text{ mm}^2/\text{m}$$

Maximum spacing is normally restricted to

$$3h \leq 400 \text{ mm for principal reinforcement}$$
$$3.5 \, h \leq 450 \text{ for secondary reinforcement}$$

In this case h = 60 mm. Taking the maximum spacing as 3.5 h = 210 mm and the minimum steel as 81 mm²/m, 5 mm bars at 200 mm c/c both ways gives $A_s = 98$ mm²/m. From Table 8.3, wrapping mesh D98 fulfils the requirements.

8.6 TWO-WAY SPANNING SOLID SLABS

8.6.1 Slab Action, Analysis and Design

When floor slabs are supported on four sides, two-way spanning action occurs as shown in Fig. 8.29. In a square slab the action is equal in each direction. In long narrow slabs where the length is greater than twice the breadth, the action is effectively one way. However, the end beams always carry some slab load.

Slabs may be classified according to the edge conditions. In the following continuous over supports also includes the case where the slab is built in at the supports. They can be defined as follows:

1. Simply supported one–panel slabs where the corners can lift away from the supports.
2. A one panel slab held down on four sides by integral edge beams (the stiffness of the edge beam affects the slab design).
3. Slabs with all edges continuous over supports.
4. Slab with one, two or three edges continuous over supports. The discontinuous edge(s) may be simply supported or held down by integral edge beams.

Elastic analytical solutions of rectangular and circular slabs for standard cases are given in textbooks on the theory of plates. Irregularly shaped slabs, slabs with openings or slabs carrying non-uniform or concentrated loads, slabs with edge beams can be analysed using computer programs based on finite element analysis.

Commonly occurring cases in slab construction in buildings are discussed. The design is based on shear and moment coefficients based on intuitive understanding of load distribution in two orthogonal directions. The slabs are square or rectangular in shape and predominantly support uniformly distributed loads.

8.6.2 Rectangular Slabs Simply Supported on All Four Edges: Corners Free to Lift

A typical example of a slab based on an intuitive understanding of load distribution is the design of simply supported slabs that do not have adequate provision either to resist torsion at the corners or to prevent the corners from lifting. Fig. 8.27 shows a slab simply resting on a wall or on a steel beam which illustrates this situation.

If the corners are not held down, under loading, the slab curls up at the corners and is therefore not supported along its entire length. The portion of the slab not in contact with the support depends on the load as well as the stiffness of the slab. Even when using finite element programs, the exact portion of the slab not in contact with the support can be determined only by trial and error. The following is a sensible common sense approach. It is based on the design procedure given originally by Rankine and Grashoff and was included in British Standards BS8110 which has been used for many years leading to satisfactory designs.

In an elastic simply supported beam subjected to uniformly distributed load q per unit length as shown in Fig. 8.28, the deflection at mid-span is given by

$$\Delta = \frac{5}{384\,EI} q L^4$$

where EI = flexural rigidity.

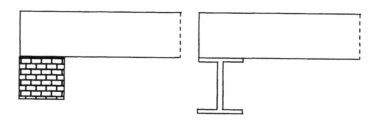

Fig. 8.27 Slab resting on a wall or on a steel beam.

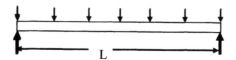

Fig. 8.28 Elastic simply supported beam.

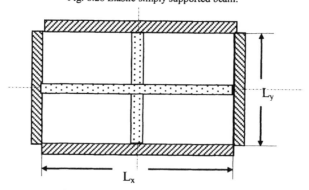

Fig. 8.29 Rectangular simply supported slab.

Fig. 8.29 shows a simply supported rectangular slab subjected to a uniformly distributed load q per unit area. Taking at the middle of the slab a strip of slab in the x-direction and another in the y-direction, if q_x is the load carried by the strip ion the x-direction and q_y is the load carried by the strip in the y-direction, then for compatibility of deflection at the centre,

$$\frac{5}{384\,EI} q_x L_x^4 = \frac{5}{384\,EI} q_y L_y^4$$

$$q_y = \alpha^4 \, q_x \,, \alpha = (\frac{L_x}{L_y}) \geq 1.0$$

Total load, $q = q_x + q_y$. Solving for q_x and q_y,

$$q_x = q\frac{1}{(1+\alpha^4)}, \quad q_y = q\frac{\alpha^4}{(1+\alpha^4)}$$

For a square slab: $\alpha = 1$. $q_x = q_y = 0.5q$.
For a rectangular slab:

$\alpha = 1.25$, $q_x = 0.29\,q$, $q_y = 0.71q$
$\alpha = 1.50$, $q_x = 0.17\,q$, $q_y = 0.83q$
$\alpha = 2.00$, $q_x = 0.06\,q$, $q_y = 0.94q$

As is to be expected, as the shape of the slab becomes more rectangular, a greater proportion of the load is carried in the short (L_y) direction. For values α greater than 2.0, the slab behaves essentially as a one-way spanning slab.
The maximum bending moments in the strips are:

Short (L_y-direction): $\quad q_y = q\dfrac{\alpha^4}{(1+\alpha^4)}$, $m_y = q_y\dfrac{L_y^2}{8} = q\dfrac{L_y^2}{8}\dfrac{\alpha^4}{(1+\alpha^4)}$

Long (L_x-direction): $\quad q_x = q\dfrac{1}{(1+\alpha^4)}$, $m_x = q_x\dfrac{L_x^2}{8} = q\dfrac{L_x^2}{8}\dfrac{1}{(1+\alpha^4)}$

The design of reinforcement is done for the above moments. 60 percent of the mid-span steel is carried over to the supports and fully anchored. The remaining 40 percent can be stopped at 0.1 of the span from the supports.

8.6.3 Example of a Simply Supported Two-Way Slab: Corners Free to Lift

(a) Specification
A slab in an office building measuring 5 m × 7.5 m is simply supported at the edges with no provision to resist torsion at the corners or to hold the corners down. The slab is assumed to be 200 mm thick. The total characteristic dead load including self–weight, screed, finishes, partitions, services etc. is 6.2 kN/m². The characteristic imposed load is 2.5 kN/m². Design the slab using $f_{ck} = 25$ MPa concrete and $f_{yk} = 500$ MPa reinforcement.

(b) Design of the moment reinforcement
Consider centre strips in each direction 1 m wide. The design load is
$$q = (1.35 \times 6.2) + (1.5 \times 2.5) = 12.12 \text{ kN/m}^2$$
$$L_x/L_y = 7.5/5 = 1.5$$
For cover of 25 mm and 12 mm diameter bars the effective depths are as follows:
For short-span bars in the bottom layer: $d_y = 200 - 25 - 12/2 = 169$ mm
For long-span bars in the top layer: $d_x = 200 - 25 - 12 - 12/2 = 157$mm
Minimum steel:

$f_{ctm} = 0.30 \times f_{ck}^{0.667} = 0.30 \times 25^{0.667} = 2.6$ MPa.

$A_{s, min} = 0.26 \, (f_{ctm}/f_{yk}) \, b_t \, d \geq 0.0013 \, b_t \, d$.

Short direction: $A_{s, min} = 0.26 \times (2.6/500) \times 1000 \times 169 \geq 0.0013 \times 1000 \times 169$.

$A_{s, min} = 229 \, mm^2/m$.

Long direction: $A_{s, min} = 0.26 \times (2.6/500) \times 1000 \times 157 \geq 0.0013 \times 1000 \times 169$.

$A_{s, min} = 212 \, mm^2/m$.

Slab depth, $h = 200$ mm. Spacing of steel $\leq$ min (3h; 400 mm), spacing ≤ 400 mm.

Short span

$$q_y = q \frac{\alpha^4}{(1+\alpha^4)} = 10.12 \, kN/m^2, \; m_y = q_y \frac{L_y^2}{8} = 31.63 \, kNm/m$$

$$k = 31.63 \times 10^6 / (1000 \times 169^2 \times 25) = 0.044 < 0.196$$

$$\frac{z}{d} = 0.5[1.0 + \sqrt{(1-3k)}] = 0.996$$

$A_s = M/ (0.87 \, f_{yk} \, z) = 31.63 \times 10^6 / (0.87 \times 500 \times 0.996 \times 169) = 432 \, mm^2/m$

Provide H12 bars at 250 mm centres to give an area of 452 mm^2/m. Curtailing 50 percent of bars gives a steel area of 226 mm^2/m which is approximately equal to the minimum value. However the spacing increases to 500 mm which is greater than the maximum value permitted. Therefore continue all the bars to the supports.

Long span

$$q_x = q \frac{1}{(1+\alpha^4)} = 2.0 \, kN/m^2, \; m_y = q_x \frac{L_x^2}{8} = 14.06 \, kNm/m$$

$$k = 14.06 \times 10^6 / (1000 \times 157^2 \times 25) = 0.023 < 0.196$$

$$\frac{z}{d} = 0.5[1.0 + \sqrt{(1-3k)}] = 0.998$$

$A_s = M/ (0.87 \, f_{yk} \, z) = 14.06 \times 10^6 / (0.87 \times 500 \times 0.998 \times 157) = 205 \, mm^2/m$

Provide H8 bars at 225 mm centres to give an area of 223 mm^2/m which is almost equal to minimum steel area. Therefore no curtailment is possible. All bars must be carried over to the support.

Fig. 8.30 shows the reinforcement details.

(c) Shear resistance:

Although for flexural design it was assumed that the total load is shared between the strips in the short and long directions, for calculating the shear force in the strips, it is assumed that the all the load in the portions of the slab as shown in Fig. 8.31 go to the supporting beams. This shows that the maximum shear force in the strip spanning in the short direction is almost equal to q $L_y/2$ for a meter-wide strip. This is obviously an over-estimation of the actual shear force but errs on the safe side.

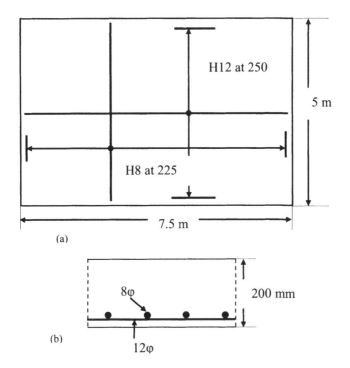

Fig. 8.30 Slab steel: (a) plan; (b) part section.

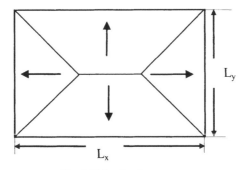

Fig. 8.31 Load to the supports.

$V_{Ed} = 12.12 \times 5.0/2 = 30.3$ kN/m.

Check if shear reinforcement is required, $V_{Ed} > V_{Rd, c}$.

$V_{Ed} = 30.3$ kN/m, $b_w = 1000$ mm, $d = 169$ mm, $A_{sl} = 452$ mm^2/m.

$$C_{Rd,c} = \frac{0.18}{(\gamma_c = 1.5)} = 0.12$$

$$k = 1 + \sqrt{\frac{200}{169}} = 1.09 \le 2.0$$

$$100\rho_1 = 100 \times \frac{452}{1000 \times 169} = 0.27 < 2.0$$

$$v_{min} = 0.035 \times 1.09^{1.5} \times \sqrt{25} = 0.20 \,\text{MPa}$$

$$V_{Rd,c} = [0.12 \times 1.09 \times \{0.27 \times 25\}^{1/3} + 0.15 \times 0] \times 1000 \times 169 \times 10^{-3}$$

$$\geq [0.20 + 0.15 \times 0] \times 1000 \times 169 \times 10^{-3}$$

$$V_{Rd,c} = 41.78 \geq 33.8 \text{ kN}$$

$V_{Rd,\,c} = 41.78$ kN $> (V_{Ed} = 30.3)$. Therefore no shear reinforcement is required.

(d) Deflection
Use the code equation (7.16a).
$$\rho_0\% = 0.1 \times \sqrt{f_{ck}} = 0.5$$
$$b_w = 1000 \text{ mm}, \, d = 169 \text{ mm}, \, A_{sl} = 452 \text{ mm}^2/\text{m}$$
$$\rho\% = 100 \times 452 / (1000 \times 169) = 0.27 < \rho_0\%$$
$$\rho'\% = 0$$
K = 1 for two-way spanning simply supported slab

$$\frac{L}{d} = K\{11 + 1.5 \sqrt{f_{ck}} \frac{\rho_0}{\rho} + 3.2 \sqrt{f_{ck}} (\frac{\rho_0}{\rho} - 1)^{\frac{3}{2}}\} \text{ if } \rho \leq \rho_0$$

$$\frac{L}{d} = \{11 + 1.5 \sqrt{25} \frac{0.50}{0.27} + 3.2 \sqrt{25} (\frac{0.50}{0.27} - 1)^{\frac{3}{2}}\} = 37.5$$

Actual L/d = 5000/169 = 29.6 < 37.5
The slab is satisfactory with respect to deflection.

(e) Cracking
Maximum spacing is normally restricted to
3h ≤ 400 mm for principal reinforcement
3.5 h ≤ 450 for secondary reinforcement
In this case h = 200 mm. Taking the maximum spacing as 3 h = 600 mm, the actual spacing of bars is 250 mm which is less than permitted. Therefore the design is safe against unacceptable crack widths.

(f) Finite element analysis
A finite element analysis of the slab was carried out by assuming that a meter length of the slab from each corner is not in contact with the support. The results are shown in Fig. 8.32 to Fig. 8.35. Fig. 8.32 and Fig. 8.33 show respectively the contour of bending moments in the long and short directions.
Fig. 8.34 shows that the bending moment in the long direction is sensibly constant over 60 percent of the span. The maximum value is about 13.25 kNm/m compared with 14.06 kNm/m from the approximate calculation. Similarly Fig. 8.35 shows that the maximum moment in the short direction is 25.4 kNm/m compared with 31.63 kNm/m from the approximate calculation. This shows that the assumptions made in the intuitive approach are reasonable.

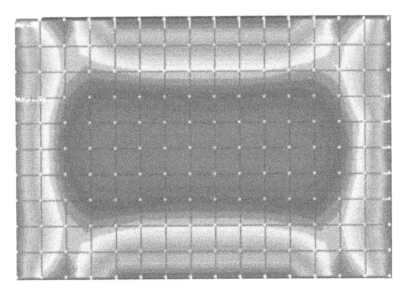

Fig. 8.32 Contour of bending moment in the slab in the long direction (L_x).

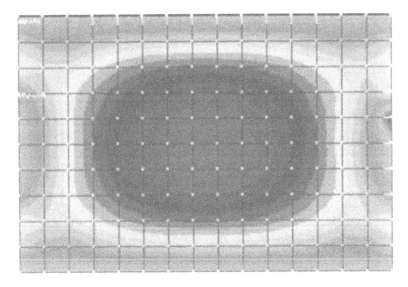

Fig. 8.33 Contour of bending moment distribution in the slab in the short direction (L_y).

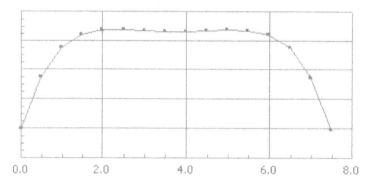

Fig. 8.34 Bending moment distribution in the slab in the long direction (L$_x$).

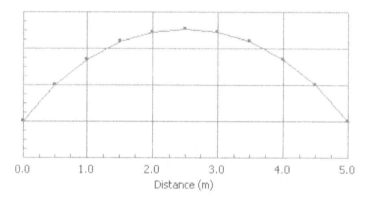

Distance (m)

Fig. 8.35 Bending moment distribution in the slab in the short direction (L$_y$).

8.7 RESTRAINED SOLID SLABS

In the previous section, design of slabs not restrained from lifting up at the corners was considered. In many cases, if the slabs are monolithic with the support beams or are continuous over supports, the slabs cannot freely lift. In such cases, along the supported edges not only restraining bending moments but also twisting moments and shear forces act.

The presence of twisting moments has two important consequences.

First, the twisting moment M_{xy} on an edge over an infinitesimal distance δy can be replaced by two forces $M_{xy}\,\delta y$ apart. Similarly over an adjacent point, the twisting moment will be $M_{xy}+\dfrac{\partial M_{xy}}{\partial y}\delta y$ which over a distance δy can also be replaced by two forces $F=M_{xy}+\dfrac{\partial M_{xy}}{\partial y}\delta y$, δy apart. As can be seen, the net result is a distributed

shear force equal to $\dfrac{\partial M_{xy}}{\partial y}$ except at the corner where the force will be M_{xy}.
Similarly from the adjacent edge, there is another force equal to M_{xy}. The net result is at the corner there is a concentrated force equal to $2\ M_{xy}$ acting in the same direction as the load as shown in Fig. 8.36. Note that the unit for M_{xy} is kNm/m which is the same thing as kN. The units for $2\ M_{xy}$ is kN. These concentrated forces act in the same direction at the diagonally opposite corners. On two corners the forces act down and at the other two corners they act up. These upward forces, if not opposed, lift the corners up.

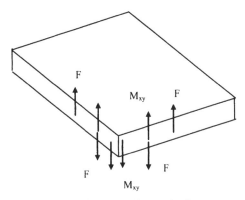

Fig. 8.36 Twisting moments at a simply supported corner leading to concentrated corner force.

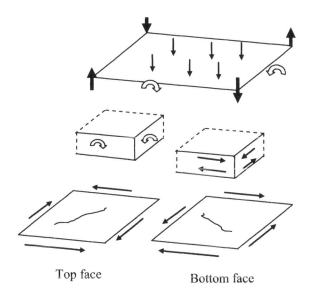

Top face   Bottom face

Fig. 8.37 Twisting moments at a simply supported corner.

Second, as shows in Fig. 8.37, the twisting moment can be thought of as two shear forces at the top and bottom edges. The shear force due to twisting moment produces tensile stresses in the top face causing a crack at 45^0 to the horizontal. Similarly, the shear force due to twisting moment produces tensile stresses in the bottom face causing a crack at 135^0 to the horizontal. If orthogonal reinforcement is used, then shear reinforcement will be needed in both x- and y-directions both at the top as well at the bottom face. In contrast to reinforcing for bending moment, reinforcing for twisting moment requires four layers of reinforcement: two in the top face and two in the bottom face.

8.7.1 Design and Arrangement of Reinforcement

Restrained slabs can be designed using moments calculated from finite element analysis. However for routine design, this approach is unnecessarily complicated. In this section, the design procedures as stated in the British Standards BS8110 are used. As stated before in section 8.6.2, this design has been used for many years and produced satisfactory designs.

Moment coefficients given here have been derived from yield line analysis. The derivation of the coefficients will be given in section 8.9.16. Table 8.15 shows the moment coefficients for the design two-way spanning slabs supported on all the four edges but only some of which are continuous. The moment coefficients are given both for support moment (hogging, negative) and span moment (sagging, positive) in both the short span and long span directions. In this method the corners are assumed to be prevented from lifting and provision is made for resisting torsion near the corners. The maximum moments at mid-span on strips of unit width for spans L_x and L_y are given by

$$m_{sx} = \beta_{sx}\, q\, L_y^{\,2}$$
$$m_{sy} = \beta_{sy}\, q\, L_y^{\,2}$$

where L_y = short span and L_x is the long span. q = applied load per unit area.

These equations may be used for continuous slabs when the following provisions are satisfied:

1. The characteristic dead and imposed loads are approximately the same on adjacent panels as on the panel being considered;
2. The spans of adjacent panels in the direction perpendicular to the line of the common support are approximately the same as that of the panel considered in that direction.

The design rules for slabs are as follows.

1. The slabs are divided in each direction into middle and edge strips as shown in Fig. 8.38.
2. The maximum moments defined above apply to the middle strips. The moment reinforcement is designed for 1 m wide strips using formulae in Chapter 4. The amount of reinforcement provided must not be less than the minimum area as given in section 9.2.1.1 of the Eurocode. The bars are spaced at the calculated spacing uniformly across the middle strip.

3. The minimum tension reinforcement specified is to be provided in the edge strips. The edge strips occupy a width equal to total width/8 parallel to the supports as shown in Fig. 8.38.

Table 8.15 Bending moment coefficients for rectangular panels supported on all sides with provision for torsion and corners prevented from lifting.

	Short span L_y coefficients: $\beta_{sy} \times 10^3$								Long span L_x coefficients $\beta_{sx} \times 10^3$
	Side ratio L_x/L_y								For all ratios
	1.0	1.1	1.2	1.3	1.4	1.5	1.75	2.0	
Case 1: Interior panel									
Edge	−31	−37	−42	−46	−50	−53	−59	−63	−32
Mid-span	24	28	32	35	37	40	44	48	24
Case 2: One short edge discontinuous									
Edge	−39	−44	−48	−52	−55	−58	−63	−67	−37
Mid-span	29	33	36	39	41	43	47	50	28
Case 3: One short edge discontinuous									
Edge	−39	−49	−56	−62	−68	−73	−82	−89	−37
Mid-span	30	36	42	47	51	55	62	67	28
Case 4: Two adjacent edges discontinuous									
Edge	−47	−56	−63	−69	−74	−78	−87	−93	−45
Mid-span	36	42	47	51	55	59	65	70	34
Case 5: Two short edges discontinuous									
Edge	−46	−50	−54	−57	−60	−62	−67	−70	−
Mid-span	34	38	40	43	45	47	50	53	34
Case 6: Two long edges discontinuous									
Edge	−	−	−	−	−	−	−	−	−45
Mid-span	34	46	56	65	72	78	91	100	34
Case 7: Two short edges and one long edge discontinuous									
Edge	−57	−65	−71	−76	−81	−84	−92	−98	−
Mid–span	43	48	53	57	60	63	69	74	44
Case 8: Two long edges and one short edge discontinuous									
Edge	−	−	−	−	−	−	−	−	−58
Mid-span	42	54	63	71	78	84	96	105	44
Case 9: All edges simply supported									
Mid-span	55	65	74	81	87	92	103	111	56

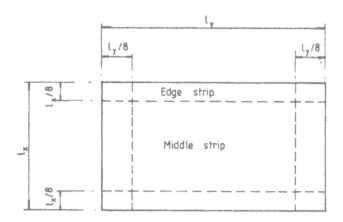

Fig. 8.38 Middle and edge strips.

4. The reinforcement is to be detailed in accordance with the simplified rules
    for curtailment of bars in slabs as shown in Fig. 8.39 and Fig. 8.40 at the
    continuous end and simply supported end respectively. At the
    discontinuous edge, top steel of one-half the area of the bottom steel at
    mid-span is to be provided to control cracking.

Continuous member (Fig. 8.39):
*   Sagging moment: 100 percent of steel calculated for maximum sagging
    moment is placed over 60 percent of effective span and 40 percent
    maximum steel is placed over the 20 percent of effective span towards the
    supports.
*   Hogging moment: 100 percent of steel calculated for maximum hogging
    moment to extend to 0.15 of the effective span from the face of the
    support or 45 bar diameters. 50 percent of maximum steel to extend to
    0.30 of the effective span from the face of the support.

Simply supported end (Fig. 8.40):
*   Sagging moment: 100 percent of steel calculated for maximum sagging
    moment to extend up to 0.1 of the effective span from the simply
    supported end and 40 percent of maximum steel to extend over (0.1
    effective span + 12 times the bar diameter or equivalent anchorage).
*   Torsion reinforcement is to be provided at corners where the slab is
    simply supported on *both edges meeting at the corners*. Corners X and Y
    shown in Fig. 8.41 require torsion reinforcement. This is to consist of a
    top and bottom mesh with bars parallel to the sides of the slab and
    extending from the edges a distance of one-fifth of the shorter span. The
    area of bars in each of the four layers should be, at X, three-quarters of the
    area of bars required for the maximum mid-span moment and at Y, one-

half of the area of the bars required at corner X. Note that no torsion reinforcement is required at the internal corners Z shown in Fig. 8.41.

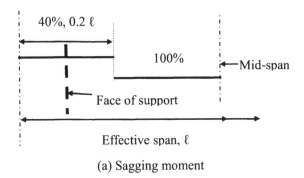

(a) Sagging moment

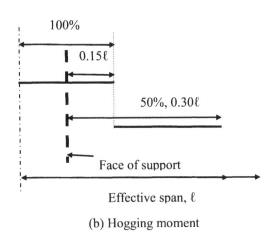

(b) Hogging moment

Fig. 8.39 Reinforcement arrangement at continuous end.

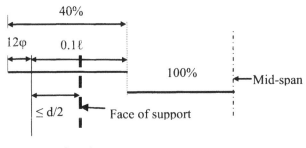

Sagging moment

Fig. 8.40 Reinforcement arrangement at simply supported end.

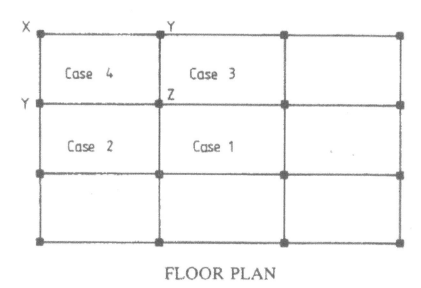

FLOOR PLAN

Fig. 8.41 Slab arrangement, floor plan: case 1: interior panel; case 2: one short edge discontinuous; case 3: one long edge discontinuous; case 4: two adjacent edges discontinuous.

8.7.2 Shear Forces and Shear Resistance

Shear forces

Shear force coefficients β_{vx} and β_{vy} for various support cases for continuous slab strips are given in Table 8.16. The maximum shear forces per unit width in the slab are given by

$$V_{sx} = \beta_{vx} \, q \, L_y$$
$$V_{sy} = \beta_{vy} \, q \, L_y$$

These are numerically the same as the design loads on supporting beams per unit length over the middle strip.

The coefficients have been derived using yield line analysis and will be discussed in section 8.9.16.

Shear resistance is checked as detailed in section 6.2 of the Eurocode 2.

8.7.3 Deflection

Deflection is checked in accordance with section 7.4.2 of the code by comparing the actual span-to-effective depth ratio with the corresponding allowable ratio. Calculations are done for the short span as it carries the maximum load.

Table 8.16 Shear force coefficients for rectangular panels supported on all sides with provision for torsion and corners prevented from lifting

Type of edge	Short span L_y coefficients: $\beta_{vy} \times 10^2$								Long span L_x coefficients: $\beta_{vx} \times 10^3$
	Side ratio L_x/L_y								For all
	1.0	1.1	1.2	1.3	1.4	1.5	1.75	2.0	ratios
Case 1: Interior panel									
Continuous	33	36	39	41	43	45	48	50	33
Case 2: One short edge discontinuous									
Continuous	36	39	42	44	45	47	50	52	36
Discontinuous	–	–	–	–	–	–	–	–	24
Case 3: One short edge discontinuous									
Continuous	36	40	44	47	49	51	55	59	36
Discontinuous	24	27	29	31	32	34	36	38	–
Case 4: Two adjacent edges discontinuous									
Continuous	40	44	47	50	52	54	57	60	40
Discontinuous	26	29	31	33	34	35	38	40	26
Case 5: Two short edges discontinuous									
Continuous	40	43	45	47	48	49	52	54	–
Discontinuous	–	–	–	–	–	–	–	–	26
Case 6: Two long edges discontinuous									
Continuous	–	–	–	–	–	–	–	–	40
Discontinuous	26	30	33	36	38	40	44	47	–
Case 7: Two short edges and one long edge discontinuous									
Continuous	45	48	51	53	55	57	60	63	–
Discontinuous	30	32	34	35	36	37	39	41	29
Case 8: Two long edges and one short edge discontinuous									
Continuous	–	–	–	–	–	–	–	–	45
Discontinuous	29	33	36	38	40	42	45	48	30
Case 9: All edges simply supported									
Discontinuous	33	36	39	41	43	45	48	50	33

8.7.4 Cracking

Crack control is attained by adhering to the maximum spacing of bars as detailed in section 9.3.a of the code.

8.7.5 Example of Design of Two-Way Restrained Solid Slab

(a) Specification
The part floor plan for an office building is shown in Fig. 8.42. It consists of restrained slabs poured monolithically with the downstand edge beams. The slab is 180 mm thick and the loading is as follows:
Total characteristic dead load $g_k = 6.2$ kN/m^2
Characteristic imposed load $q_k = 2.5$ kN/m^2
Design the comer slab using grade $f_{ck} = 30$ MPa concrete and $f_{yk} = 500$ MPa reinforcement. Show the reinforcement on sketches.

(b) Slab division, moments and reinforcement
The corner slab is divided into middle and edge strips as shown in Fig. 8.42.

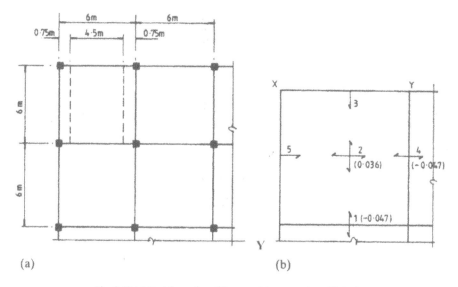

(a)   (b)

Fig. 8.42 (a) Part floor plan; (b) symmetric moment coefficients.

The moment coefficients are taken from Table 8.14 for a square slab for the case with two adjacent discontinuous edges. The values of the coefficients and locations of moments are shown in Fig. 8.42 (b).

Design ultimate load = $(1.35 \times 6.2) + (1.5 \times 2.5) = 12.12$ kN/m^2

Assuming 10 mm diameter bars and 20 mm cover, the effective depth of the outer layer to be used in the design for moments in the short span direction is
$$d = 180 - 20 - 10/2 = 155 \text{ mm}$$
The effective depth of the inner layer to be used in the design for moments in the long span direction is
$$d = 180 - 20 - 10 - 10/2 = 145 \text{ mm}$$
The moments and steel areas for the middle strips are calculated. Because the slab is square, only one direction need be considered.

(i) Positions 1 and 4 (over supports) short direction:

$$d = 155 \text{ mm}$$
$$m_{sy} = -0.047 \times 12.12 \times 6^2 = -20.51 \text{ kN m/m}$$
$$k = 20.51 \times 10^6 / (30 \times 1000 \times 155^2) = 0.029 < 0.196$$
$$z/d = 0.5[1.0 + \surd (1.0 - 3 \times 0.029)] = 0.98$$
$$A_s = 20.51 \times 10^6 / (0.87 \times 500 \times 0.98 \times 155) = 310 \text{ mm}^2/\text{m}$$

Provide H10 bars at 250 mm centres to give an area of 314 mm²/m.

(ii) Position 2 (mid-span) short direction:

Use the smaller value of d.

$$d = 145 \text{ mm}$$
$$m_{sx} = 0.036 \times 12.12 \times 6^2 = 15.71 \text{ kN m/m}$$
$$k = 15.71 \times 10^6 / (30 \times 1000 \times 145^2) = 0.025 < 0.196$$
$$z/d = 0.5[1.0 + \surd (1.0 - 3 \times 0.025)] = 0.98$$
$$A_s = 15.71 \times 10^6 / (0.87 \times 500 \times 0.98 \times 145) = 254 \text{ mm}^2/\text{m}$$

Provide H8 bars at 175 mm centres to give an area of 287 mm²/m.

(iii) Minimum steel:

$$f_{ctm} = 0.30 \times f_{ck}^{0.667} = 0.30 \times 30^{0.667} = 2.9 \text{ MPa}.$$
$$A_{s, \text{ min}} = 0.26 \, (f_{ctm}/f_{yk}) \, b_t \, d \geq 0.0013 \, b_t \, d.$$
b_t = width of web = 1000 mm, d = 155 mm.
$$A_{s, \text{ min}} = 0.26 \times (2.9/500) \times 1000 \times 155 \geq 0.0013 \times 1000 \times 155.$$
$A_{s, \text{ min}} = 233$ mm²/m. The areas of steel calculated are higher than the minimum values.

Note that H8 bars at 200 mm centres give an area of 251 mm²/m.

(iv) Positions 3 and 5 (discontinuous edges):

Top steel one half of the area of steel at mid-span is to be provided.
$$A_s = 0.5 \times 310 = 155 \text{ mm}^2/\text{m} < 210 \text{ mm}^2/\text{m} \text{ (minimum steel)}$$

Provide 8 mm diameter bars at 200 mm centres to give an area of 251 mm²/m. In detailing, the moment steel will not be curtailed because both negative and positive steel would fall below the minimum area if 50 percent of the bars were stopped off.

Fig. 8.43 shows the reinforcement arrangement.

(c) Shear forces and shear resistance

(i) Positions 1 and 4 (continuous edge):

$$\beta_{vy} = \beta_{vx} = 0.40$$
$$V_{Ed} = V_{sy} = 0.4 \times 12.12 \times 6 = 29.01 \text{ kN/m}$$

Check if shear reinforcement is required, $V_{Ed} > V_{Rd, c}$.

$$V_{Ed} = 29.01 \text{ kN/m}, b_w = 1000 \text{ mm}, d = 155 \text{ mm}, A_{sl} = 314 \text{ mm}^2/\text{m}$$

$$C_{Rd,c} = \frac{0.18}{(\gamma_c = 1.5)} = 0.12$$

$$k = 1 + \sqrt{\frac{200}{155}} = 2.15 > 2.0$$

$$\text{Take} \, k = 2.0$$

$$100\rho_1 = 100 \times \frac{314}{1000 \times 155} = 0.20 < 2.0$$

$$v_{min} = 0.035 \times 2.0^{1.5} \times \sqrt{30} = 0.54 \, \text{MPa}$$

$$V_{Rd,c} = [0.12 \times 2.0 \times \{0.20 \times 30\}^{1/3} + 0.15 \times 0] \times 1000 \times 155 \times 10^{-3}$$

$$\geq [0.54 + 0.15 \times 0] \times 1000 \times 155 \times 10^{-3}$$

$$V_{Rd,c} = 67.6 \geq 83.7 \, \text{kN}$$

$V_{Rd,c}$ = 83.7 kN > (V_{Ed} = 29.01). Therefore no shear reinforcement is required.

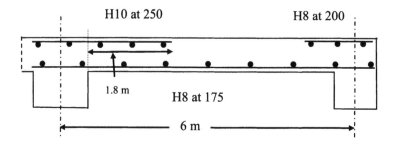

Fig. 8.43 Reinforcement arrangement (cross section).

(ii) Positions 3 and 5 (discontinuous edge):

$$\beta_{vy} = \beta_{vx} = 0.26$$

The bottom tension bars are to be stopped at the centre of the support. The shear resistance is based on the top steel with $A_s = 251 \, \text{mm}^2/\text{m}$.

$$V_{Ed} = V_{sx} = 0.26 \times 12.12 \times 6 = 18.91 \, \text{kN/m}$$

$$d = 145 \, \text{mm}$$

$$C_{Rd,c} = \frac{0.18}{(\gamma_c = 1.5)} = 0.12$$

$$k = 1 + \sqrt{\frac{200}{145}} = 2.17 > 2.0$$

$$\text{Take} \, k = 2.0$$

$$100\rho_1 = 100 \times \frac{251}{1000 \times 145} = 0.17 < 2.0$$

$$v_{min} = 0.035 \times 2.0^{1.5} \times \sqrt{30} = 0.54 \, \text{MPa}$$

$$V_{Rd,c} = [0.12 \times 2.0 \times \{0.17 \times 30\}^{1/3} + 0.15 \times 0] \times 1000 \times 145 \times 10^{-3}$$

$$\geq [0.54 + 0.15 \times 0] \times 1000 \times 145 \times 10^{-3}$$

$$V_{Rd,c} = 59.9 \geq 78.3 \text{ kN}$$

$V_{Rd, c} = 78.3$ kN $> (V_{Ed} = 18.91)$. Therefore no shear reinforcement is required.

(d) Torsion steel
Torsion steel of length equal to $1/5^{th}$ of shorter span $= 6/5 = 1.2$ m is to be provided in the top and bottom of the slab at the three external corners marked X and Y in Fig. 8.42 (b).

(i) Corner X
The area of torsion steel is 0.75 × (Required steel at maximum *mid-span* moment):
$$A_s = 0.75 \times 254 = 191 \text{ mm}^2/\text{m}$$
This will be provided by the minimum steel of 8 mm diameter bars at 200 mm centres giving a steel area of 251 mm²/m.

(ii) Corner Y
The area of torsion steel is one half of that at corner X.
$$A_s = 0.5 \times 191 = 96 \text{ mm}^2/\text{m}.$$
Again provide minimum H8 bars at 200 mm centres giving a steel area of 251 mm²/m.

(e) Edge strips
Provide minimum reinforcement, 8 mm diameter bars at 200 mm centres, in the edge strips both at top and bottom.

(f) Deflection
Check using steel at mid-span with $d = 145$ mm.
Using the code equation (7.16a),
$$\rho_0\% = 0.1 \times \sqrt{f_{ck}} = 0.55$$
$$b_w = 1000 \text{ mm}, d = 145 \text{ mm}, A_{sl} = 287 \text{ mm}^2/\text{m}$$
$$\rho\% = 100 \times 287/ (1000 \times 145) = 0.20 < \rho_0\%$$
$$\rho'\% = 0$$
$K = 1.3$ for two-way spanning slab continuous over one long edge

$$\frac{L}{d} = K[11+1.5 \sqrt{f_{ck}} \frac{\rho_0}{\rho} + 3.2 \sqrt{f_{ck}} (\frac{\rho_0}{\rho} - 1)^{\frac{3}{2}}] \text{ if } \rho \leq \rho_0$$
(7.16a)

$$\frac{L}{d} = 1.3 \times [11+1.5 \sqrt{30} \frac{0.55}{0.20} + 3.2 \sqrt{30} (\frac{0.55}{0.20} - 1)^{\frac{3}{2}}] = 96.4$$
$$\text{Actual } L/d = 6000/145 = 41.4 < 96.4$$
The percentage of reinforcement is quite low. The L/d ratio is rather higher than desirable but calculations indicate that the slab is satisfactory with respect to deflection.

(g) Cracking

The bar spacing does not exceed 3 h = 3 × 180 = 540 mm. The slab will not suffer excessive cracking.

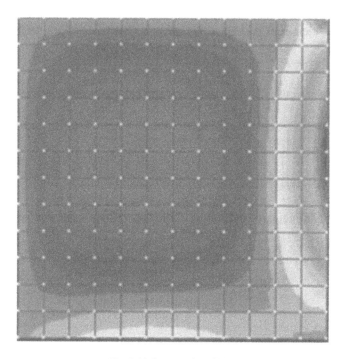

Fig. 8.44 Contour plot of M_{xx}.

8.7.6 Finite Element Analysis

In the previous section, the design was done using the moment coefficients obtained from yield line analysis. As a comparison, an elastic analysis of the slab was carried out using a finite element analysis program. The results are shown in Fig. 8.44 to Fig. 8.47.

Fig. 8.44 and Fig. 8.45 show respectively a contour plot of moment in the x-direction and twisting moment M_{xy}. Fig. 8.46 shows a plot of the variation of the moment along the centre line of the slab. The maximum hogging and sagging moments in kNm/m are 29.3 and 12.6 respectively giving a ratio of 2.3. The corresponding values from the yield line analysis are 20.5 and 15.7 respectively giving a ratio of 1.3. Clearly the support moment has decreased and the span moment has increased due to redistribution of the moment caused by yielding.

Fig. 8.47 shows the variation of M_{xy} along the diagonal. As is to be expected, the maximum twisting moment is at the corner where two simply supported edges meet. In section 8.12, design using the elastic values of (M_{xx}, M_{yy}, M_{xy}) will be discussed. This is more versatile than the design based on yield line analysis.

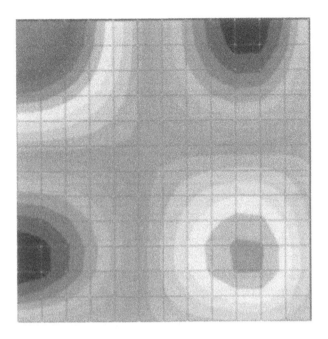

Fig. 8.45 Contour plot of M_{xy}.

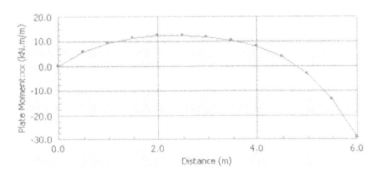

Fig. 8.46 Variation of M_{xx} along the middle section of the slab.

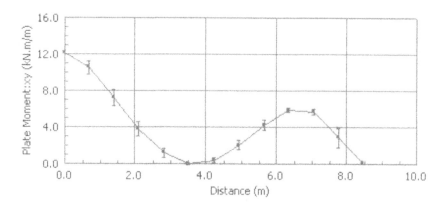

Fig. 8.47 Variation of M_{xy} along the diagonal section of the slab.

8.8 WAFFLE SLABS

8.8.1 Design Procedure

Two-way spanning *ribbed* slabs are termed waffle slabs. Waffle slabs can be designed using the method detailed in section 8.6 provided the slab satisfies the criteria detailed in section 8.5.2.

Slabs may be made solid near supports to increase moment and shear resistance and provide flanges for support beams. In edge slabs, solid areas are required to contain the torsion steel.

8.8.2 Example of Design of a Waffle Slab

(a) Specification

Design a waffle slab for an internal panel of a floor system that is constructed on an 8 m square module. The total characteristic dead load is 6.5 kN/m^2 and the characteristic imposed load is 2.5 kN/m^2. The materials for construction are $f_{ck} = 35$ MPa concrete and $f_{yk} = 500$ MPa reinforcement.

(b) Arrangement of slab

A plan of the slab arrangement is shown in Fig. 8.48 (top). The slab is made solid for 500 mm from each support. The proposed section through the slab is shown in Fig. 8.48 (bottom). The proportions chosen for rib width, rib depth, depth of topping and rib spacing meet various requirements set out in section 8.5.2.

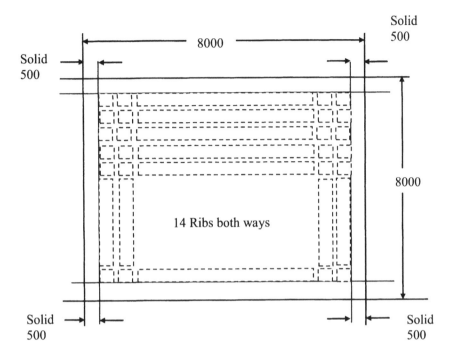

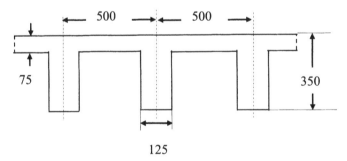

Fig. 8.48 Top: Plan of waffle slab. Bottom: Section through the slab.

(c) Reinforcement

Design ultimate load = $(1.35 \times 6.5) + (1.5 \times 2.5) = 12.53$ kN/m^2

The middle strip moments for an interior square panel are, from Table 3.14,

Support $msx = -0.031 \times 12.53 \times 8^2 = -24.86$ kNm/m

Mid-span $m_{sx} = 0.024 \times 12.53 \times 8^2 = 19.25$ kNm/m

Slab width supported by one rib = 500 mm.

The moment per rib is therefore

Support $msx = -24.86 \times 0.5 = -12.43$ kNm

Mid-span $m_{sx} = 19.25 \times 0.5 = 9.63$ kNm

The effective depths assuming 12 mm diameter main bars and 6 mm diameter links and 25 mm cover are as follows:

$$\text{Outer layer } d = 350 - 25 - 6 - 12/2 = 313 \text{ mm}$$
$$\text{Inner layer } d = 350 - 25 - 6 - 12 - 12/2 = 301 \text{ mm}$$

Support: solid section 500 mm wide

As the flooring dimensions and supports are symmetrical, it is convenient to have the steel arrangement also symmetrical. Section design is based on the smaller value of d equal to 301 mm.

$$k = 12.43 \times 10^6 / (500 \times 301^2 \times 35) = 0.008 < 0.196$$
$$z/d = 0.5[1.0 + \sqrt{(1 - 3 \times 0.008)}] = 0.99$$
$$A_s = 12.43 \times 10^6 / (0.87 \times 500 \times 0.99 \times 301) = 96 \text{ mm}^2$$

Check minimum steel

$$f_{ctm} = 0.30 \times f_{ck}^{0.667} = 0.30 \times 35^{0.667} = 3.2 \text{ MPa}$$
$$A_{s,\,min} = 0.26 \, (f_{ctm}/f_{yk}) \, b_t \, d \geq 0.0013 \, b_t \, d$$
$$b_t = \text{width of web} = 500 \text{ mm}, \, d = 313 \text{ mm}$$
$$A_{s,\,min} = 0.26 \times (3.2/500) \times 500 \times 313 \geq 0.0013 \times 500 \times 226$$
$$A_{s,\,min} = 260 \text{ mm}^2.$$

The areas of steel calculated are less than the minimum values. Provide 2H16 in the ribs giving an area of 402 mm². At the end of the solid section, the maximum moment of resistance of the concrete ribs with width 125 mm is given by

$$M = 0.196 \times 125 \times 301^2 \times 35 \times 10^{-6} = 77.7 \text{ kNm}$$

This exceeds the applied moment at the support and so the ribs are able to resist the applied moment without compression steel. The applied moment at 500 mm from the support will be somewhat less than the support moment.

Centre of span. T-beam, d = 301 mm

The flange breadth b is 500 mm and $h_f = 75$ mm. $f_{cd} = 35/1.5 = 23.3$ MPa.

$$M_{flange} = f_{cd} \times b \times h_f \times (d - h_f/2) \times 10^{-6}$$
$$M_{flange} = 23.3 \times 500 \times 75 \times (301 - 75/2) \times 10^{-6} = 230 \text{ kNm} > 10.06 \text{ kNm}$$

Hence the neutral axis lies in the flange and the beam is designed as a rectangular beam.

$$k = 10.06 \times 10^6 / (500 \times 301^2 \times 35) = 0.006 < 0.196$$
$$z/d = 0.5 \, [1.0 + \sqrt{(1 - 3 \times 0.006)}] = 0.995$$
$$z = 0.995 \, d = 299 \text{ mm}$$
$$A_s = 10.06 \times 10^6 / (0.87 \times 500 \times 299) = 77 \text{ mm}^2$$

Provide two 8 mm diameter bars with area 101 mm².

Check minimum steel

$$f_{ctm} = 0.30 \times f_{ck}^{0.667} = 0.30 \times 35^{0.667} = 3.2 \text{ MPa}$$
$$A_{s,\,min} = 0.26 \, (f_{ctm}/f_{yk}) \, b_t \, d \geq 0.0013 \, b_t \, d$$
$$b_t = \text{width of web} = 125 \text{ mm}, \, d = 301 \text{ mm}$$

$$A_{s,\,min} = 0.26 \times (3.2/500) \times 125 \times 301 \geq 0.0013 \times 125 \times 301$$
$$A_{s,\,min} = 63 \text{ mm}^2.$$

The areas of steel calculated are higher than the minimum values.

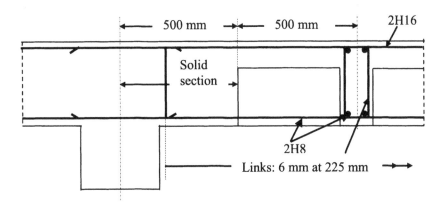

Fig. 8.49 Reinforcement detail in the rib including shear reinforcement.

(d) Shear resistance
The shear force coefficient is taken from Table 8.15. The shear at the support is
$$V_{sy} = 0.33 \times 12.53 \times 8 = 33.41 \text{ kN/m}$$
The shear at the support for the width of 500 mm supported by one rib is
$$V_{sy} = 33.41 \times 0.5 = 16.71 \text{ kN}$$
Loading over 500 mm width $= 12.53 \times 0.5 = 6.27$ kN/m
The shear on the ribs at 500 mm from support is
$$V_{Ed} = 16.71 - 6.27 \times 0.5 = 13.57 \text{ kN}$$
$$d = 301 \text{ mm}$$
$$C_{Rd,c} = \frac{0.18}{(\gamma_c = 1.5)} = 0.12$$
$$k = 1 + \sqrt{\frac{200}{301}} = 1.82 < 2.0$$
$$100\rho_1 = 100 \times \frac{101}{125 \times 301} = 0.27 < 2.0$$
$$v_{min} = 0.035 \times 1.82^{1.5} \times \sqrt{35} = 0.51 \text{ MPa}$$
$$V_{Rd,c} = [0.12 \times 1.82 \times \{0.27 \times 35\}^{1/3} + 0.15 \times 0] \times 125 \times 301 \times 10^{-3}$$
$$\geq [0.51 + 0.15 \times 0] \times 125 \times 301 \times 10^{-3}$$
$$V_{Rd,c} = 17.36 \geq 19.19 \text{ kN}$$
$$V_{Rd,\,c} = 19.19 \text{ kN} > (V_{Ed} = 13.57).$$
Therefore no shear reinforcement is required.

Provide 6 mm diameter 2-leg links to form a cage. The arrangement of the reinforcement and shear reinforcement in the rib is shown in Fig. 8.49.

(e) Deflection

$$b/b_w = 500/125 = 4.0 > 3$$
$$\text{Correction factor} = 0.8$$

Check using steel at mid-span with $d = 301$ mm.

Using the code equation (7.16a), $\rho_0\% = 0.1 \times \sqrt{f_{ck}} = 0.59$, $b_w = 125$ mm, $d = 301$ mm, $A_{sl} = 101$ mm^2

$\rho\% = 100 \times 101/(125 \times 301) = 0.27 < \rho_0\%$, $\rho'\% = 0$

$K = 1.5$ for interior span of two-way spanning slab

$$\frac{L}{d} = K\{11 + 1.5\sqrt{f_{ck}}\frac{\rho_0}{\rho} + 3.2\sqrt{f_{ck}}(\frac{\rho_0}{\rho} - 1)^{\frac{3}{2}}\} \text{ if } \rho \le \rho_0$$

(7.16a)

$$\frac{L}{d} = 1.5 \times \{11 + 1.5\sqrt{35}\frac{0.59}{0.27} + 3.2\sqrt{35}(\frac{0.59}{0.27} - 1)^{\frac{3}{2}}\} = 82.2$$

Correction factor for $b/b_w > 3$ is 0.8.

$L/d = 82.2 \times 0.8 = 65.6$.

Actual $L/d = 8000/226 = 35.3 < 65.6$.

L/d ratio is a bit high but the slab is satisfactory with respect to deflection.

(f) Reinforcement in topping

For a topping 75 mm thick the minimum area required per metre width is

$$f_{ctm} = 0.30 \times f_{ck}^{0.667} = 0.30 \times 35^{0.667} = 3.2 \text{ MPa}$$

$A_{s, min} = 0.26 \, (f_{ctm}/f_{yk}) \, b_t \, d \ge 0.0013 \, b_t \, d$

b_t = width of web = 1000 mm, h = 75 mm

$A_{s, min} = 0.26 \times (3.2/500) \times 1000 \times 75 \ge 0.0013 \times 1000 \times 75$

$A_{s, min} = 125$ mm^2/m.

Provide H6 bars at 225 mm spacing giving an area of 126 mm^2/m.   Provide in the centre of the topping A142 square structural mesh with H6 bars at 200 c/c both ways, giving a cross sectional area of 141 mm^2/m.

8.9 FLAT SLABS

8.9.1 Definition and Construction

The flat slab is a slab with or without drops, supported generally without beams by columns with or without column heads. The slab may be solid or have recesses formed on the soffit to give a waffle slab. Here only solid slabs will be discussed.

Flat slab construction is shown in Fig. 8.50 for a building with circular internal columns, square edge columns and drop panels. The slab is thicker than that required in T-beam floor slab construction but the omission of beams gives a smaller storey height for a given clear height and simplification in construction and formwork.

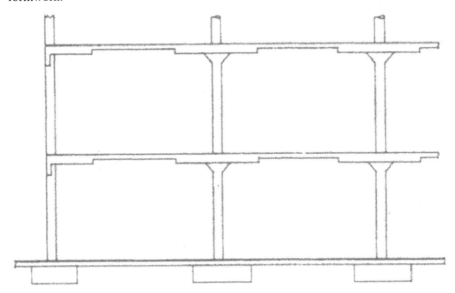

Fig. 8.50 Flat slab construction.

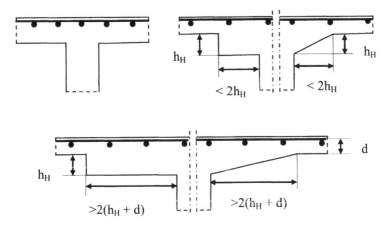

Fig. 8.51 Slab without drop panel, with a rectangular drop panel and flared column head.

Various column supports for the slab either without or with drop panels are shown in Fig. 8.51. As can be seen, the total width of the drop can vary over a wide range. Drop panels only influence the distribution of moments if the smaller dimension of the drop is at least equal to one-third of the smaller panel dimension.

Smaller drops provide resistance to punching shear. The panel thickness is generally controlled by deflection.

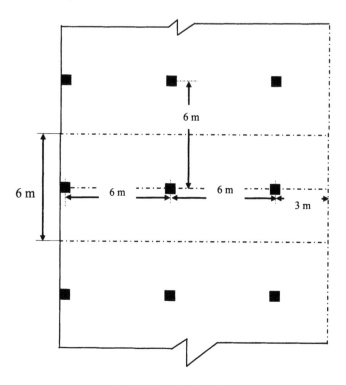

Fig. 8.52 Plan of a flat slab.

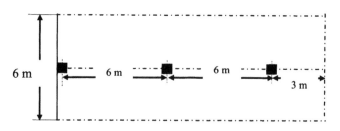

Fig. 8.53 Part of the flat slab used for analysis.

8.9.2 Analysis

The bending moment distribution in a flat slab is quite complex. Fig. 8.52 shows part plan of a flat slab 30 × 24 m with columns spaced at 6 m both ways. Fig. 8.53 shows a symmetrical half of a part of the slab 6 m wide lying between the centre lines of the panel. The slab is 300 mm thick and is loaded by a uniformly distributed load of 18.75 kN/m². Fig. 8.54 shows the contours of moments in the x-directions for a symmetrical half of a slab from finite element analysis.

Fig. 8.55 shows the distribution of bending moment along the line of columns. As can be observed it is similar to bending moment distribution in an equivalent continuous beam.

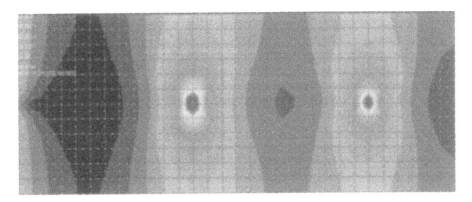

Fig. 8.54 Contour of bending moment in the x-direction.

Fig. 8.56 shows the variation of the bending moment across the width at a section at the second column and Fig. 8.57 shows the variation of the bending moment across the width at a section midway between the first and second columns.

The main point to notice is that the hogging moment is highly concentrated over a short width near the column line but the sagging moment is less concentrated than the hogging moment.

The results from the finite element analysis indicate that

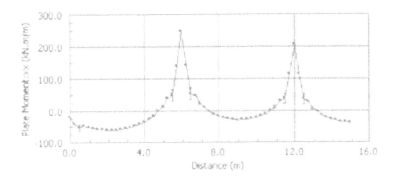

Fig. 8.55 Distribution of bending moment along the line of columns.

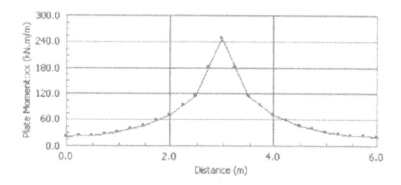

Fig. 8.56 Variation of bending moment M_{xx} across the width at the second column.

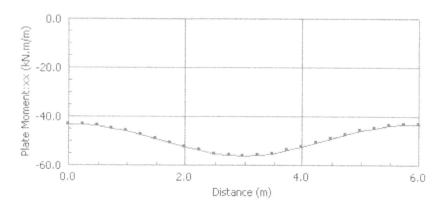

Fig. 8.57 Variation of bending moment M_{xx} across the width between first and second columns.

- Continuous beam analysis of a part of the slab lying between the centre lines of the panel will yield reasonable distribution of the bending moment in the flat slab.
- The moment tends to concentrate over a width near the column.

These observations are taken into account in the Eurocode 2 recommendations for flat slab analysis and design.

8.9.3 General Eurocode 2 Provisions

The design of slabs is covered in the Eurocode 2 in two places:
- Section 6.4 which gives the procedure to design against punching shear failure.
- Annex I which gives the procedure for the analysis of flat slabs.

In Annex I, the code simply states that design moments may be obtained by using a proven method of analysis such as

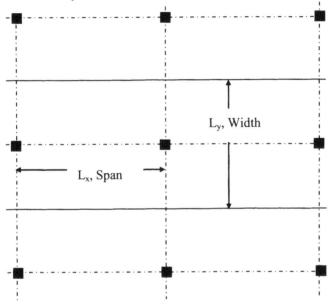

Fig. 8.58 Division of the structure into frames.

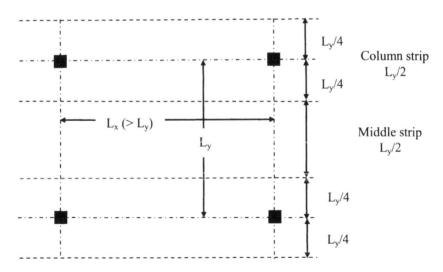

Fig. 8.59 Division of a panel into column and middle strips.

(a) Equivalent frame method
(b) Grillage method
(c) Finite element analysis
(d) Yield line analysis

Annex I also gives some guidance on the use of the equivalent frame method.
The code states that normally it is sufficient to consider only the single load case of maximum design load (1.35 × dead load + 1.5 × imposed load) on all spans. The following method of analysis is used to obtain the moments and shears for design.

8.9.4 Equivalent Frame Analysis Method

The structure is divided longitudinally and transversely into frames consisting of columns and strips of slab contained between the centre lines of adjacent panels as shown in Fig. 8.58. The stiffness of the members may be calculated from their gross cross sectional dimensions. For vertical loading, the stiffness may be based on the full width of the panels. For horizontal loading, 40 percent of the gross value should be used to reflect the increased flexibility of the column–slab joints in flat slab structures compared with that of column–beam joints.

The total bending moment obtained from the analysis is should be distributed across the width of the slab. The panel should be divided into column and middle strips as shown in Fig. 8.59. The bending moment should be apportioned to the strips as shown in Table 8.17.

Table 8.17 Distribution of moments in flat slabs

	Distribution between column and middle strip as percentage of total negative or positive moment	
	Column strip	*Middle strip*
Negative	60 to 80 percent	40 to 20 percent
Positive	50 to 70 percent	50 to 30 percent

Section 9.4.1 of Eurocode 2 states that at internal columns, 50 percent of the total reinforcement to resist the negative (hogging) moment should be placed within $L_y/8$ on either side of the column.

8.9.5 Shear Force and Shear Resistance

The punching shear around the column is the critical consideration in flat slabs. Rules are given in section 6.4 of the Eurocode 2 for calculating the ultimate design shear resistance capacity force and designing the required shear reinforcement. Chapter 5, sections 5.1.11 to 5.1.14 gives many examples of designing against punching shear. Section 6.4.2 of Eurocode 2 gives details about the control perimeter as shown in Fig. 8.60 for circular and rectangular columns. The control perimeter is at a distance of 2 d from the column perimeter.

Section 6.4.3 of Eurocode 2 gives equations for calculating punching shear capacity. The following checks are required:
- At the column perimeter, $v_{Ed} < v_{Rd,\ max}$
- Punching shear reinforcement is not required if $v_{Ed} < v_{Rd,\ c}$
- Where punching shear reinforcement is required, it is provided as detailed in section 6.4.5

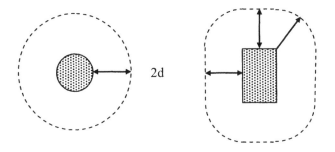

Fig. 8.60 Control perimeters for circular and rectangular columns.

Where the support reaction is eccentric, the calculated value of column reaction should be multiplied by a factor β. In the case of structures where lateral stability is not dependent on the frame action (braced structures), the values of $\beta = 1.15$, 1.4 and 1.5 respectively for internal, edge and corner columns. For other cases where the value of moment transferred to the column is known explicitly, β can be calculated using Eurocode 2 equations (6.30) to (6.46).

As conventional shear reinforcement in the form of links greatly complicates and slows down the steel fixing process, it is not desirable to have shear reinforcement in light or moderately loaded slabs. However some prefabricated proprietary shear reinforcements are available which considerably simplify the provision of shear reinforcement. Another form of shear reinforcement used is the stud rail which consists of headed shear studs welded to a steel plate.

8.9.6 Deflection

The check is to be carried out for the most critical direction, i.e., for the longest span using Eurocode 2 equations (7.16a) and (7.16b.)

8.9.7 Crack Control

The bar spacing rules for slabs given in Tables 7.2N and 7.3N of Eurocode 2 apply.

8.9.8 Example of Design for an Internal Panel of a Flat Slab Floor

(a) Specification
The floor of a building constructed of flat slabs is 30 m × 24 m. The column centres are 6 m in both directions and the building is braced with shear walls. The slab is 300 mm deep. The internal columns are 450 mm square.
The loading is as follows:
Screed, floor finishes, partitions and ceiling = 2.5 kN/m^2
Imposed load = 3.5 kN/m^2
The materials are f_{ck} = 30 MPa concrete and f_{yk} = 500MPa reinforcement.
Design the edge panel on two sides and show the reinforcement on a sketch.

(b) Slab and column details and design dimensions
A part floor plan and slab details are shown in Fig. 8.61.

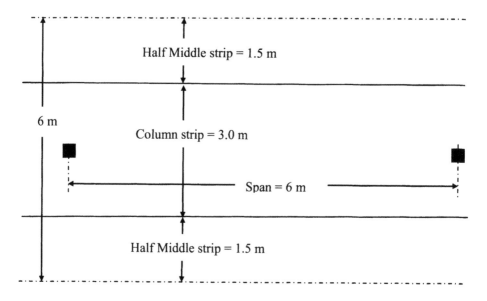

Fig. 8.61 Plan of continuous beam.

(c) Design loads and moments
Considering a 6 m width of the slab and taking unit weight of concrete as 25.0 kN/m^3.
Slab weight = 6 × 0.3× 25 = 45 kN/m
Screed weight = 6 × 2.5 = 15 kN/m
g_k = 45 + 15 = 60 kN/m
Imposed load, q_k = 6 × 3.5 = 21 kN/m
Design load = 1.35 × 60 + 1.5 × 21 = 112.5 kN/m

As shown in Fig. 8.62, the equivalent frame consists of five 6 m span continuous beams loaded by 112.5 kN/m. The continuous beam may be analysed by any suitable method.

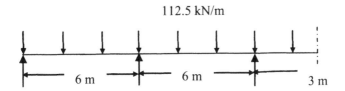

Fig. 8.62 'Equivalent frame' continuous beam.

The final moment is shown in Fig. 8.63.

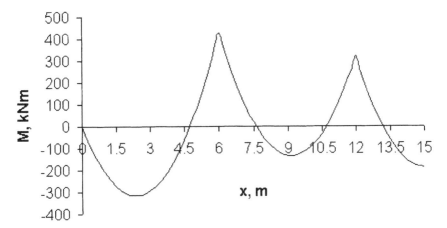

Fig. 8.63 Moment distribution in the continuous beam.

The key values are:

Total Negative moments:
Moment at second and fifth support = –426.4 kNm
Moment at third and fourth internal support = –319.9 kNm

Total Positive moments:
Span 1–2 and 5–6: M = 315.5 kNm
Span 2–3 and 4–5: M = 134.5 kNm

(d) Design of moment reinforcement
The cover is 25 mm and 16mm diameter bars in two layers and 8 mm links are assumed. The effective depth for the inner layer is
$$d = 300 - 25 - 8 - 25 - 16/2 = 234 \text{ mm}$$

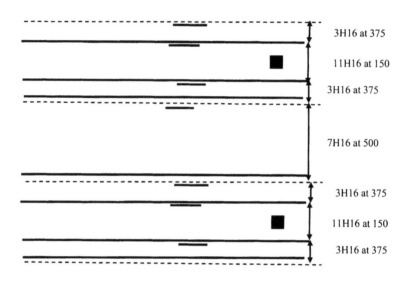

Fig. 8.64 Top steel in the x-direction.

Negative moment reinforcement (steel at top), end panel
$$M = 426.4 \text{ kNm}$$
$$k = 426.4 \times 10^6 / (6000 \times 234^2 \times 30) = 0.043 < 0.196$$
$$z/d = 0.5 \, [1.0 + \sqrt{(1 - 3 \, 0 \times 0.043)}] = 0.97$$
$$z = 0.97 \, d = 226 \text{ mm}$$
$$A_s = 426.4 \times 10^6 / (0.87 \times 500 \times 226) = 4337 \text{ mm}^2$$

Column strip reinforcement
50 percent of 4337 mm² equal to 2169 mm² is placed in the column strip of width of $L_y/4 = 1500$ mm. Provide 11H16 at 150 mm c/c giving steel area of 2211 mm². The remaining 20 percent of 4337 mm² equal to 867 mm² is placed in the column strip of width of $L_y/4 = 1500$ mm. Provide 3H16 at 375 mm c/c on 750 mm width on either side of column giving steel area of 2211 mm².

Middle strip reinforcement
The remaining one 30 percent of 4337 mm² equal to 1301 mm² is placed in the column strip of width of 3000 mm. Provide 7H16 at 500 mm.

Fig. 8.64 shows the reinforcement arrangement at the top of the slab.

Positive moment reinforcement (steel at bottom), end panel
$$M = 315.5 \text{ kNm}$$
$$k = 315.5 \times 10^6 / (6000 \times 234^2 \times 30) = 0.032 < 0.196$$
$$z/d = 0.5 \, [1.0 + \sqrt{(1 - 3 \, 0 \times 0.032)}] = 0.98$$
$$z = 0.98 \, d = 228 \text{ mm}$$
$$A_s = 315.5 \times 10^6 / (0.87 \times 500 \times 228) = 3181 \text{ mm}^2$$

Column strip reinforcement
60 percent of 3181 mm^2 equal to 1909 mm^2 is placed in the column strip of width of 3000 mm. Provide 10H16 at 300 mm c/c giving steel area of 2011 mm^2.
Note that in section 9.4.1 (3), the code recommends that bottom reinforcement ($\geq$ 2 bars) in each orthogonal direction should be provided at internal columns and this reinforcement should pass through the column.

Middle strip reinforcement
The remaining 40 percent of 1272 mm^2 is placed in the 3000 mm wide column strip. Provide 7H16 at 500 mm.
Fig. 8.65 shows the reinforcement arrangement at the bottom of the slab.

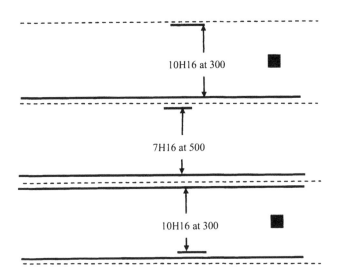

Fig. 8.65 Bottom steel in the x-direction.

(e) Shear resistance

Fig. 8.66 shows the shear force distribution. The reaction at the second support is the sum of right reaction of 408.57 kN from span 1–2 + left reaction of 355.25 kN from span 2–3 giving a total of 763.8 kN.

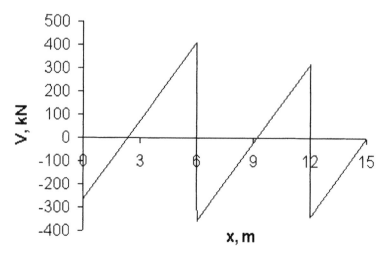

Fig. 8.66 Shear force distribution in the continuous beam.

i. At the column face

$$\text{Reaction} = 763.8 \text{ kN}$$
Internal column: Unbalnced moment factor, $\beta = 1.15$
$$V_{Ed} = 763.8 \times 1.11 = 847.8 \text{ kN}$$
Column: 450 mm square, $u_0 = 4 \times 450 = 1800$ mm, d = 234 mm
$$V_{Ed} = 847.8 \times 10^3 / (1800 \times 234) = 2.0 \text{ MPa}$$
$$v = 0.6 \ (1 - f_{ck}/250) = 0.6 \times (1 - 30/250) = 0.53, \ f_{cd} = 30/ \ (\gamma_c = 1.5) = 20 \text{ MPa}$$
$$V_{Rd, max} = 0.5 \ (v_1 = v) \times f_{cd} = 0.5 \times 0.53 \times 20 = 5.28 \text{ MPa}$$
$V_{Ed} < V_{Rd, max}$. The maximum shear stress is satisfactory.

ii. At 2.0 d from the column face

Shear perimeter, $u_1 = 4 \times 450 + 2 \times \pi \times (2 \times d) = 1800 + 2941 = 4741$ mm
$$v = 847.8 \times 10^3 / (4741 \times 234) = 0.76 \text{ MPa}$$
Note: It is more accurate to deduct the downward load inside the shear perimeter in calculating v.
Area inside shear perimeter = $\{450 \times 450 + 4 \times 450 \times (2d) + \pi \times (2d)^2\} \times 10^{-6}$
$$= 1.73 \text{ m}^2$$
Applied load = 18.75 kN/m^2
Load inside the perimeter = $1.73 \times 18.75 = 32.49$ kN
$$v = (847.8 - 32.49) \times 10^3 / (4741 \times 234) = 0.74 \text{ MPa}$$
The improvement is insignificant.
Calculate $v_{Rd, c}$:
In the centre half of the column strip 16 mm diameter bars are spaced at 150 mm centres giving an area of 1340 mm²/m.
$$100 \ \rho_1 = 100 \ A_s/ \ (bd) = 100 \times 1340 / (1000 \times 234) = 0.57 < 2.0$$
$$C_{Rd, c} = 0.18/ \ (\gamma_c = 1.5) = 0.12, \ k = 1 + \sqrt{(200/d)} = 1.93 < 2.0$$
$$v_{Rd,c} = C_{Rd,c} \ k \ (100 \rho_1 \ f_{ck})^{0.33} = 0.12 \times 1.93 \times (0.57 \times 30)^{0.33} = 0.60 \text{ MPa}$$

$$v_{min} = 0.035 \times k^{1.5} \times \sqrt{f_{ck}} = 0.035 \times 1.93^{1.5} \times \sqrt{30} = 0.51 \text{ MPa}$$
$$(v = 0.76) > (v_{Rd, c} = 0.60)$$

Shear reinforcement is needed.

iii. Calculate the perimeter u_{out} where shear stress is equal to $v_{Rd, c}$

As shown in Fig. 5.23, let the perimeter be at a distance Nd from the face of the column.

$$u_{out} = 2(450 + 450) + 2\pi \text{ Nd mm}$$

The load acting within the perimeter is equal to

$$[450 \times 450 + 2 \times (450 + 450) \times \text{Nd} + \pi \,(\text{Nd})^2] \times 18.75 \times 10^{-6} \text{ kN}$$
$$V_{Ed} = [847.8 - \text{Load inside perimeter}] \text{ kN}$$
$$v_{Ed} = V_{Ed} / (u_{out} \times d) = v_{Rd, c} \text{ MPa}$$

By trial and error, N = 2.65.

$$u_{out} = 2(450 + 450) + 2\pi \times 2.65 \times 234 = 5696 \text{ mm}$$

At this perimeter no shear reinforcement is required.

iv. Calculate the position of the outermost perimeter where shear reinforcement is required

The last ring of shear reinforcement must be within kd, where k = 1.5 from the u_{out}. This perimeter lies at (Nd – kd) = (2.65 d – 1.5d) =1.15 d from the face of the column.

Perimeter length = $u_{1.15\,d}$ = 2(450 + 450) + 2π × 1.15 d = 3491 mm.

v. Calculate shear reinforcement using the code equation (6.52)

$$v_{Rd,cs} = 0.75\, v_{Rd,c} + 1.5 \{\frac{d}{s_r}\}\, A_{sw}\, f_{ywd,ef}\, \{\frac{1}{u_1\, d}\}\, \sin\alpha \qquad (6.52)$$

s_r = 0.75d, f_{ywk} = 500 MPa, γ_s = 1.15, d = 234 mm, f_{ywd} = 500/1.15 = 435 MPa
$f_{ywd,\,ef}$ = (250 + 0.25× 234 = 309) ≤ 435 MPa, $f_{ywd,\,ef}$ = 309 MPa, $v_{Rd,\,c}$ = 0.60

$$v_{Rd,cs} = 0.75\, v_{Rd,c} + 1.5(\frac{d}{s_r})\, A_{sw}\, f_{ywd\,ef}\, (\frac{1}{u_1\, d})\, \sin\alpha$$

At basic control perimeter u_1 at 2d from column:

$$v_{Rd,\, cs} = v_{Ed} = 0.76 \text{ MPa}, u_1 = 4741 \text{ mm}$$

Substituting in code equation (6.52),

$$v_{Rd,cs} = 0.75\, v_{Rd,c} + 1.5(\frac{d}{s_r})\, A_{sw}\, f_{ywd\,ef}\, (\frac{1}{u_1\, d})\, \sin\alpha$$

$$0.76 = 0.75 \times 0.60 + 1.5 \frac{d}{0.75d} A_{sw} \times 309 \times [\frac{1}{4741 \times 234}]$$

$$A_{sw} = 557 \text{ mm}^2$$

vi. Calculate the minimum link leg area

Using code equation (9.11) to calculate the area of a single link leg.

$$A_{sw,min} \times \frac{(1.5\sin\alpha + \cos\alpha)}{s_r \, s_t} \geq 0.08 \frac{\sqrt{f_{ck}}}{f_{yk}}$$

Substituting $f_{ck} = 30$ MPa, $f_{yk} = 500$ MPa, $d = 234$ mm, $s_r = 0.75$ d, $s_t = 2d$, $\sin\alpha = 1$ for vertical links,

$A_{sw,\,min} = 48$ mm^2. Choosing 8 mm diameter bars, $A_{sw,\,min} = 50$ mm^2.

No. of links required $= A_{sw}/$ Area of one link $= 557/50 = 11.1$ links.

A minimum of 12 links should be provided at all perimeters with the spacing between the perimeters ≤ 0.75 d.

vii. Arrange link reinforcement

Arrange the perimeters as follows.

(i) First perimeter at 100 mm $= 0.43d > 0.3d$

Perimeter length $= u_{0.43\,d} = 2(450 + 450) + 2\pi \times 0.43$ d $= 2432$ mm

Maximum spacing of links $\leq 1.5d = 350$ mm

Spacing of links $=$ perimeter length/Minimum no. of links

$\qquad\qquad = 2432/12 = 203$ mm < 350 mm

Provide 12 links at say 200 mm

(ii) Second perimeter at $(100 + 0.72d) = 269$ mm say $\approx 1.15d$

Perimeter length $= u_{1.15\,d} = 2(450 + 450) + 2\pi \times 1.15$ d $= 3491$ mm

Spacing of links $\leq 1.5d = 350$ mm

Spacing of links $=$ perimeter length/Minimum no. of links

$\qquad\qquad = 3491/12 = 291$ mm < 350 mm

viii. Summary

Reinforcement is provided on two perimeters. Once the numbers are rounded up to practical dimensions, design will be satisfactory.

(f) Deflection

The calculations are made for the middle strip using the average of the column and middle strip tension steel.

10H16 at 300 mm c/c giving steel area of 2011 mm^2 is placed in the column strip of width of 3000 mm.

7H16 at 500 mm giving steel area of 1407 mm^2 is placed in the column strip of width of 3000 mm.

Permissible span depth ratio can be calculated using Eurocode 2 equation (7.16a).

$\rho\% = 100 \times 0.5 \times (2011 + 1407)/ (3000 \times 234) = 0.24$

$\rho_0\% = 0.1\sqrt{f_{ck}} = 0.1\sqrt{30} = 0.55$

$k = 1.2$ for flat slabs

$$\frac{L}{d} = K \{11 + 1.5 \sqrt{f_{ck}} \frac{\rho_0}{\rho} + 3.2 \sqrt{f_{ck}} (\frac{\rho_0}{\rho} - 1)^{\frac{3}{2}} \} \text{ if } \rho \leq \rho_0 \qquad (7.16a)$$

$$\frac{L}{d} = 1.2 \times \{11 + 1.5 \sqrt{30} \frac{0.55}{0.24} + 3.2 \sqrt{30} (\frac{0.55}{0.24} - 1)^{\frac{3}{2}} \} = 66.7$$

Allowable span/d ratio = 66.7
Actual span/d ratio = 6000/234 = 25.6
Hence the slab is satisfactory with respect to deflection.

(g) Cracking
The bar spacing does not exceed 3 h, i.e. 900 mm.

(h) Arrangement of reinforcement
The arrangement of the reinforcement is shown in Fig. 8.66 and Fig. 8.67. For clarity, only the steel in x-direction is shown separately for top and bottom steel. The steel arrangement is identical in both the x- and y-directions. Note that although in the diagrams steel shown arrangement is shown confined to the individual panel, in reality steel extends into adjacent panels.

8.10 YIELD LINE METHOD

8.10.1 Outline of Theory

The yield line method developed by Johansen is applicable to calculation of collapse load caused by yielding of under-reinforced concrete slab. It is based on the upper bound theorem (also known as the kinematic theorem) of the classical theory of plasticity. According to this theorem, for any assumed collapse mechanism, if the collapse load is calculated by equating the energy dissipation at the plastic 'hinges' to the work done by the external load, then the load so calculated is equal to or greater than the true collapse load. The yield line method applied to slabs is analogous to the calculation of ultimate load of frames by the formation of plastic hinges in the members of the frame. The collapse mechanism of a frame consists of a set of rigid members connected at plastic hinges. The only difference between a frame and a slab at collapse is that instead of discrete plastic hinges forming at several locations in a frame, in the case of the slab, yielding takes place along several lines of hinges, referred to as yield lines. All deformations are assumed to take place at the yield lines and the fractured slab at collapse consists of rigid portions held together by the yielded reinforcement at the yield lines. It is important to appreciate that the method assumes ductile behaviour at yield lines and does not consider the possibility of shear failure. Another important point to bear in mind is that because the method gives an upper bound solution to the true collapse load, it is important to investigate all possible collapse modes in order to determine the smallest collapse load.

In the one-way continuous slab shown in Fig. 8.67(a), straight yield lines form with a sagging yield line at the bottom of the slab near mid-span and hogging yield lines over the supports. The yield line patterns for a square and a rectangular simply supported two-way slab subjected to a uniform load are shown in Fig. 8.67(b) and Fig. 8.67(c) respectively. The deformed shape of the square slab is an inverted pyramid and that of the rectangular slab is an inverted roof shape.

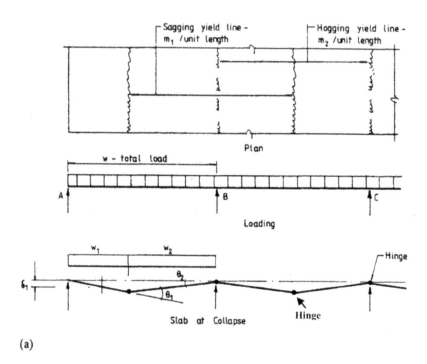

(a)

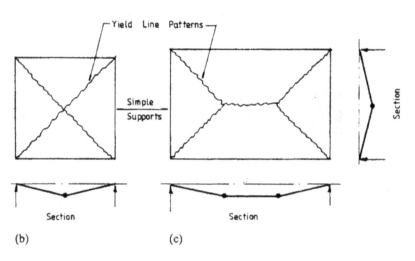

(b)   (c)

Fig. 8.67 (a) Continuous one-way slab; (b) square slab; (c) rectangular slab.

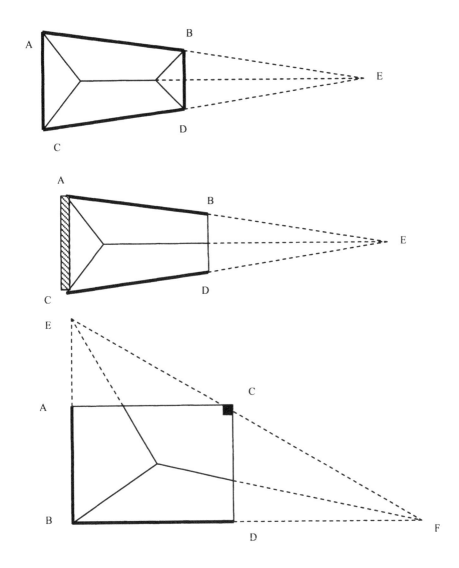

Fig. 8.68 Top: Simply supported trapezoidal slab. Middle: Trapezoidal slab with a free edge.
Bottom: Rectangular slab with a column support.

8.10.1.1 Properties of Yield Lines

The following properties of the yield lines will be found useful in proposing possible collapse mechanisms.

(i)   Yield lines are generally straight and they must end at a slab boundary.

(ii)   A yield line between two rigid regions must pass through the intersection of the axes of rotation of the two rigid regions. Edge supports act as axes of rotation.

(iii)   Axes of rotation lie along the line of supports. They can pass over a column at any angle.

Fig. 8.68 shows some yield line patterns, which illustrate the above properties.

(i)   Fig. 8.68(a) shows a trapezoidal slab simply supported on all four edges. The yield line between the two trapezoidal rigid regions passes through E where the axes of rotations AB and CD meet. The yield line between the trapezoidal rigid region rotating about AB and the triangular region rotating about BD meets at B, the intersection point of the two axes of rotation.

(ii)   Fig. 8.68(b) shows a trapezoidal slab simply supported on two opposite edges AB and CD, while edge AC is fixed against rotation while edge BD is free. The yield line between the two trapezoidal rigid regions passes through E where the axes of rotations AB and CD meet. The yield line ends at the free edge. The yield line between the trapezoidal rigid region rotating about AB and the triangular region rotating about AC meets at A, the intersection point of the two axes of rotation.

(iii)   Fig. 8.68(c) shows a rectangular slab simply supported on edges AB and BD and supported on a column at C. The axes of rotations are AB, BD and EF. The yield lines terminating at a free edge intersect the intersection of the two axes of rotations. The axis of rotation ECF passes over the column.

8.10.2 Johansen's Stepped Yield Criterion

As remarked earlier, the slab yields only at yield lines. Yielding is governed by Johansen's stepped yield criterion which assumes that yielding takes place when the applied moment normal to the yield line is equal to the moment of resistance provided by the reinforcement crossing the yield line. It assumes that all reinforcement crossing a yield line yields and that the reinforcement bars stay in their original directions.

As shown in Fig. 8.69, let the two sets of reinforcement in the x and y directions respectively have ultimate moment of resistance such that for a yield line parallel to the x-axis the normal moment of resistance is m_x and this resistance is provided by flexural steel in the *y-direction*. Similarly for a yield line parallel to the y-axis, the normal moment of resistance is m_y and this is provided by flexural steel in the *x-direction*.

If a yield line forms at an angle θ to the x-axis, then as shown in Fig. 8.70, the yield line can be imagined to be made up of a series of steps parallel to the reinforcement directions. For a unit length of yield line, the lengths of the horizontal and vertical steps are respectively cos θ and sin θ. The moment of resistance on the horizontal step is m_x cos θ and on the vertical step it is m_y sin θ. The components of these moments of resistance parallel to the yield line are m_x cos^2 θ and m_y sin^2 θ respectively. Thus the normal moment of resistance along the yield line is

$$m_n = m_x \cos^2 \theta + m_y \sin^2 \theta$$

Note that if $\theta = 0$, then the yield line is perpendicular to the reinforcement in the y-direction and hence $m_n = m_x$. Similarly if $\theta = 90^0$, then the yield line is perpendicular to the reinforcement in the x-direction and hence $m_n = m_y$. If $m_x = m_y = m$, a case of isotropic reinforcement, then $m_n = m$ irrespective of the direction of the yield line.

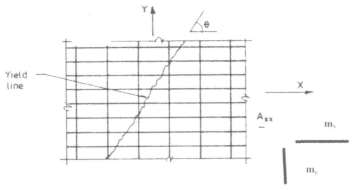

Fig. 8.69 Yield line in an orthogonally reinforced slab.

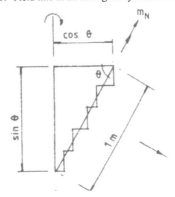

Fig. 8.70 Resistant moments on an inclined yield line.

8.10.3 Energy Dissipated in a Yield Line

Consider a slab ABCD, simply supported on the two adjacent edges AB and BC and free on the other two edges AD and CD as shown in Fig. 8.71 (a). Let a yield line BD form between the two rigid regions ABD and CBD as shown in Fig. 8.71 (b). Let the dimensions of the slab be as follows:

$$AF = 1.5, \quad BF = 6.0, \quad FD = 2.0, \quad BG = 4.0, \quad CG = 0.5$$

From geometry, the values for the following angles can be calculated.

Angle $ABF = \tan^{-1}(1.5/6.0) = 14.04°$, Angle $FBD = \tan^{-1}(2.0/6.0) = 18.44°$

Angle $CBG = \tan^{-1}(0.5/4.0) = 7.13°$, Angle $DBG = 90° -$ Angle $FBD = 71.56°$

The length L of the yield line $BD = \sqrt{(2^2 + 6^2)} = 6.325$

The energy dissipated at a yield line is given by the equation

$$E = m_n \, L \, \theta_n$$

where m_n = normal moment on the yield line, L = length of the yield line, θ_n = rotation at the yield line.

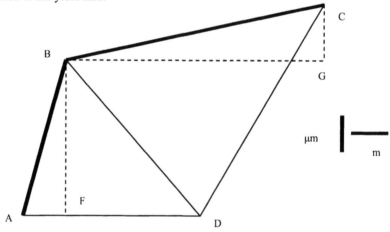

Fig. 8.71 (a) Slab supported on two edges only.

If the moments of resistance due to steel in the x- and y-directions respectively are μm and m, then the value of m_n on a yield line inclined at an angle φ to the x-axis is given by

$$m_n = m \cos^2\varphi + \mu m \sin^2\varphi$$

The energy dissipation at a yield line can be calculated by any of the three methods as follows.

Method 1: This is the most general and direct method but is not always the most convenient method to use. The inclination of the yield line to the horizontal is

$$\varphi = \text{Angle } FDB = 90° - 18.44° = 71.56°.$$

The length L of the yield lines is L = 6.325 and the moment of resistance normal to the yield line is

$$m_n = m \cos^2\varphi + \mu m \sin^2\varphi = 0.1m + 0.9\mu m$$

In order to calculate θ_n, draw a line JK perpendicular to the yield line BD as shown in Fig. 8.71 (b). From geometry,

Angle (JBD) $= 14.04° + 18.11° = 32.48°$, BD $= 6.325$, JD $=$ BD tan (JBD) $= 4.063$

Angle (DBK) $= 71.56° + 7.13° = 78.69°$, KD $=$ BD tan (DBK) $= 31.625$

If point D deflects vertically by Δ, then

$$\theta_n = \Delta/JD + \Delta/KD = 0.2777 \, \Delta$$

Energy dissipated in the yield line is

$$m_n \, L \, \theta_n = (0.1m + 0.9 \, \mu m) \, (6.325) \, (0.2777 \, \Delta)$$
$$= (1.5812 \, \mu m + 0.1757 \, m) \, \Delta$$

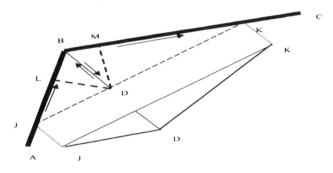

Fig. 8.71 (b) Deformation along the yield line BD.

Method 2: If the yield line is inclined at an angle φ to the x-axis, then the energy dissipated can be expressed as

$$m_n \, L \, \theta_n = (m_x \cos^2 \varphi + m_y \sin^2 \varphi) \, L \, \theta_n$$
$$m_n \, L \, \theta_n = m_x \, (L \cos \varphi) \, (\theta_n \cos \varphi) + m_y \, (L \sin \varphi) \, (\theta n \sin \varphi)$$
$$L_x = L \cos \varphi, \quad L_y = L \sin \varphi, \, \theta_x = \theta_n \cos \varphi, \quad \theta_y = \theta_n \sin \varphi$$
$$m_n \, L \, \theta_n = m_x \, L_x \, \theta_x + m_y \, L_y \, \theta_y$$

This formulation avoids having to calculate m_n and L and can be useful in some instances. Referring to Fig. 8.71 (a),

$$L_x = FD = 2.0, \, L_y = FB = 6.00,$$
$$m_x = m, \, m_y = \mu m,$$

From Method 1,

$$\theta_n = 0.2777 \, \Delta,$$
$$\varphi = \text{Angle FDB} = 71.56°$$
$$\theta_x = \theta_n \cos \varphi = 0.0879 \, \Delta, \, \theta_y = \theta_n \sin \varphi = 0.2635 \, \Delta$$
$$m_n \, L \, \theta_n = m \, (2.0) \, (0.0879 \, \Delta) + \mu m \, (6.0) \, (0.2635 \, \Delta)$$
$$= (1.5812 \, \mu m + 0.1757 \, m) \, \Delta$$

Method 3: This is the best approach if the axes of rotation of the rigid regions lie along the coordinate axes and the steel is orthogonal and the *reinforcement directions coincide with the coordinate axes*. The method is based on the fact that θ_n is the sum of the rotation of the two rigid regions. In Fig. 8.71(b),

$$\theta_n = \text{Angle DJK} + \text{Angle DKJ}.$$
$$m_n \, L \, \theta_n = m_x \, L_x \, \theta_x + m_y \, L_y \, \theta_y$$

$$= (m_x\ L_x\ \theta_{x1} + m_y\ L_y\ \theta_{y1}) + (mx\ L_x\ \theta_{x2} + m_y\ L_y\ \theta_{y2})$$

where $(\theta_{x1}, \theta_{y1})$ and $(\theta_{x2}, \theta_{y2})$ refer respectively to the x and y components of the rotation at the yield line due to rigid regions ABD (i.e., Angle DJK) and CBD (i.e., Angle DKJ).

In order to use this method, it is important to use a consistent notation for the moment and rotation vectors. For the **rotation vector**, it is assumed that it is positive if the right hand's thumb points along the positive direction of the rotation vector, then the slab rotates in the **clockwise** direction. The **moment vector** is assumed positive if the right hand's thumb points along the positive direction of the moment vector. The moment then acts in the **anticlockwise** direction.

In the example considered, the rigid portion ABD rotates about the support AB in a clockwise direction. Therefore the rotation vector points in the direction from A to B. Similarly, the rigid portion DBC rotates about the support BC in a clockwise direction. Therefore the rotation vector points in the direction from B to C.

The normal moment on the yield line BD causes tension on the bottom side. Therefore in the rigid portion ABD, the moment vector points in the direction from D to B while in the rigid portion DBC the moment vector points in the opposite direction from B to D.

Rotation of the rigid region ABD about the axis AB is

$$\theta_1 = \Delta/DL, \text{ where DL is perpendicular to AB.}$$
$$DL = BD \sin ABD = 6.325 \sin (14.04 + 18.44 = 32.48) = 3.397$$
$$\theta_1 = \Delta/3.397 = 0.2944\ \Delta$$
$$\text{Angle FAB} = 90 - 14.04° = 75.96°$$
$$\theta_{x1} = \theta_1 \cos (FAB) = 0.2944\ \Delta \times 0.2425 = 0.0714\ \Delta$$
$$\theta_{y1} = \theta_1 \sin (FAB) = 0.2944\ \Delta \times 0.9701 = 0.2856\ \Delta$$
$$m_x = -m, \quad m_y = \mu\ m, \quad L_x = 2.0, \quad L_y = 6.0$$

Note that the sign of m_x is negative because the horizontal component of the moment vector points in a direction opposite to that of the corresponding rotation component.

Rotation of the rigid region DBC about the axis BC is

$$\theta_2 = \Delta/DM, \text{ where DM is perpendicular to AB}$$
$$DM = BD \sin DBM = 6.325 \sin (90 - 18.44 + 7.13 = 78.69) = 6.202$$
$$\theta_2 = \Delta/6.202 = 0.1612\ \Delta$$
$$\theta_{x2} = \theta_2 \cos (GBC) = 0.1612\ \Delta \cos (7.13) = 0.16\ \Delta$$
$$\theta_{y2} = \theta_2 \sin (GBC) = 0.1612\ \Delta \sin (7.13) = 0.02\ \Delta$$
$$m_x = m, \quad m_y = -\mu\ m, \quad L_x = 2.0, \quad L_y = 6.0$$

Note that the sign of m_y is negative because the vertical component of the moment vector points in the direction opposite to that of the corresponding rotation component.

Energy E dissipated on the yield line is

$$E = - m\ (2.0)\ (0.0714\ \Delta) + \mu\ m\ (6.0)\ (0.2856\ \Delta)$$
$$+ m\ (2.0)\ (0.16\ \Delta) - \mu\ m\ (6.0)\ (0.02\Delta)$$
$$E = (1.5936\ \mu\ m + 0.1772\ m)\ \Delta$$

Although Method 3 appears to be more complicated than Method 1, in most cases of rectangular slabs where the axes of rotation coincide with the coordinate axes, Method 3 will be found to be the ideal method to use.

8.10.4 Work Done by External Loads

If a rigid region carries a uniformly distributed load q and rotates by θ about an axis AB as shown in Fig. 8.72, then the work done by q is given by

$$\text{Work done} = \int q\, dA\, r\, \theta$$

where dA = an element of area, r = perpendicular distance to the element of area from the axis of rotation.
Since q and θ are constant, work done $= q\, \theta \int r\, dA$.
But $\int r\, dA$ = first moment of area about the axis of rotation.

$$W = q\, \theta \,\{\text{First moment of area about the axis of rotation}\}$$
$$= q\, \theta \times \text{Area} \times \text{Distance to the centroid of area from the axis of rotation}$$
$$= q \times \text{Area} \times \text{Deflection at the centroid}$$

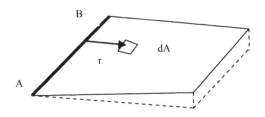

Fig. 8.72 External work done by loads on a slab.

8.10.5 Example of a Continuous One-Way Slab

Consider a strip of slab 1 m wide where the mid-span positive reinforcement has a moment of resistance of *m* per metre and the support negative reinforcement has a moment of resistance of *m'* per metre. The slab with ultimate load *W* per span is shown in Fig. 8.73(a).

(a) End span AB
The yield line in the span forms at point C at *x* from A. The rotation at A is θ. The deflection Δ at the hinge in the span is $\theta\, x$. If the rotation at the hinge over the support B is φ, then

$$\varphi\, (l - x) = \Delta = \theta\, x$$
$$\varphi = \theta\, x / (l - x)$$

The net rotation ψ at the hinge C is

$$\psi = (\theta + \varphi) = \theta\, l / (l - x)$$

The rotations at A, C and B are shown in Fig. 8.73 (b).

The work done by the loads is W (x θ)/2.
The energy dissipated E in the yield lines is

$$E = (m \, \psi + m' \, \varphi)$$
$$E = m \, \theta \, l/ \, (l-x) + m' \, \theta \, x / (l-x)$$
$$E = \theta/ \, (l-x) \, \{m \, l + m' \, x\}$$

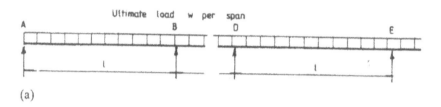

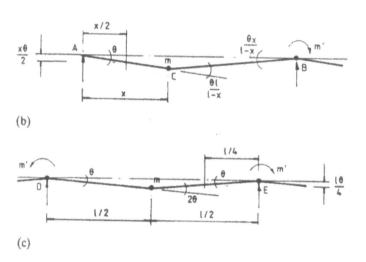

Fig. 8.73 (a) Continuous one-way slab; (b) end span; (c) internal span.

Equating the work done by the loads to energy dissipated at the hinges,

$$W (x \, \theta)/2 = \theta/ \, (l-x) \, \{m \, l + m' \, x\}$$
$$W = 2 \, \{m \, l + m' \, x\} / [x \, (l-x)]$$

The position x of the yield line in the span is determined so that the collapse load is minimum. Differentiating W with respect to x,

$$dW/dx = x \, (l-x) \, m' - (m \, l + m' \, x) \, (l-2x) = 0$$
$$m' \, x^2 - m \, l \, (l-2x) = 0$$

This equation can be solved for x for a given value of the ratio m'/m.
Under the section 5.6.2 for plastic analysis for beams, frames and slabs, clause 5.6.2(2) of the code states that the required ductility may be deemed to be satisfied

without explicit verification of rotation capacity if **all** the following conditions are fulfilled.

    i. $x_u/d \leq 0.25$, $f_{ck} \leq 50$ MPa and $x_u/d \leq 0.15$, $f_{ck} \geq 55$ MPa.

    ii. the ratio of moments at intermediate supports to the moments in the span should be between 0.5 and 2.0.

    iii. reinforcing steel is either Class B or C. (See Table C.1 in the Annex C of the code. This is reproduced in Table 2.4, Chapter 2.)

In Eurocode 2 code equations (5.10a) and (5.10b) given below, the ratio δ of redistributed/elastic moments is gives as

$$\delta \geq 0.44 + 1.25\{0.6 + \frac{0.0014}{\varepsilon_{cu2}}\}\frac{x_u}{d} \text{ for } f_{ck} \leq 50 \text{ MPa} \qquad (5.10a)$$

$$\delta \geq 0.54 + 1.25\{0.6 + \frac{0.0014}{\varepsilon_{cu2}}\}\frac{x_u}{d} \text{ for } f_{ck} > 50 \text{ MPa} \qquad (5.10b)$$

Table 8.18 Variation of δ with f_{ck} for plastic analysis

f_{ck}, MPa	$\varepsilon_{cu2} \times 10^3$	x_u/d	δ
≤ 50	3.5	0.25	0.7525
55	3.1	0.15	0.7372
60	2.9	0.15	0.7430
70	2.7	0.15	0.7497
80	2.6	0.15	0.7535
90	2.6	0.15	0.7535

For the limitations on x_u/d stated above, the corresponding values of redistribution ratio δ are shown in Table 8.18. The maximum reduction of the moment is about 25 percent of the elastic values.

Note that if $x_u/d = 0.25$, $M_u = 0.12 \, bd^2 \, f_{ck}$, $k = 0.12$.

When designing using yield line analysis, the depth of the slab should be so chosen that $k \leq 0.12$.

For the special case where $m = m'$, the equation $dW/dx = 0$ reduces to

$$x^2 + 2\,l\,x - l^2 = 0, \quad x = 0.414 \, l$$

Substitute in the work equation $x = 0.414 \, l$ to obtain the value of m:

$$m = m' = 0.086 \, Wl$$

Since the maximum moment in span is at $x = 0.414 \, l$, the contra-flexure point is at a distance of 2x from support A. Therefore the theoretical cut-off point for the top reinforcement is at $2x = 0.828 \, l$ from the support A or at $0.172 \, l$ from support B.

(b) Internal span DE
The hinge is at mid-span and the rotations are shown in Fig. 8.73(c). The work equation is

$$W \, (0.5 \, l \, \theta)/2 = m \, \theta + m \, 2 \, \theta + m \, \theta$$

For the case where $m = m'$

$$m = Wl/16 = 0.063Wl$$

Note that the moment value 0.063 W*l* is same that in Table 8.14 of the code. The theoretical cut-off points for the top bars are at 0.147 *l* from each support.

8.10.6 Simply Supported Rectangular Two-Way Slab

The slab and yield line pattern are shown in Fig. 8.76. The ultimate loading is *w* per square metre. As shown in Fig. 8.74, steel in the shorter y-direction provides a moment of resistance of m per unit length and the steel in the longer x-direction provides a moment of resistance of μm per unit length. The yield line pattern is defined by one parameter, *β*.

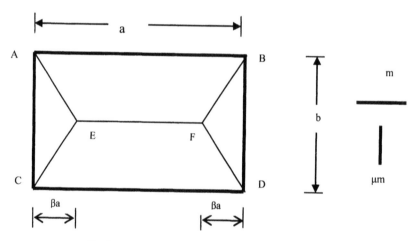

Fig. 8.74 Collapse mode for a simply supported slab.

1. Work done by external loads
The work done by the loads can be calculated by assuming that points E and F deflect by Δ.

(a) Triangles ACE and BFD
Area = 0.5 b βa, deflection at the centroid = Δ/3.
Work done by external loads is

$$W_1 = 2[q\ 0.5\,b\,\beta a\frac{\Delta}{3}] = qab\frac{\beta}{3}\Delta$$

(b) Trapeziums CEFD and AEFB
Dividing the trapezium into two triangles and a rectangle,

Triangle: area = 0.5 b/2 βa, deflection at the centroid = Δ/3
Rectangle: area = b/2 (a – 2 βa), deflection at the centroid = Δ/2
Work done by external loads is

$$W_2 = 2[2\{q\ 0.5\frac{b}{2}\beta a\frac{\Delta}{3}\} + q\frac{b}{2}(a-2\beta a)\frac{\Delta}{2}] = qab\{\frac{(3-4\beta)}{6}\}\Delta$$

Total work done by the loads $W = W_1 + W_2 = q\dfrac{ab}{6}(3 - 2\beta)\Delta$.

2. Energy dissipated at the yield lines
The energy dissipated at the yield lines can be calculated using Method 3.

(a) Yield line in triangles ACE and BFD
The triangles rotate only about y-axis.

$$l_y = b, \quad m_y = \mu m, \quad \theta_y = \dfrac{\Delta}{\beta a}$$

Hence the energy dissipated E_1 on the yield lines in triangles ACE and BFC is

$$E_1 = 2[\ell_x m_x \theta_x + \ell_y m_y \theta_y]$$

$$E_1 = 2[2\beta a \; m \; 0 + b \; \mu m \; \dfrac{\Delta}{\beta a}] = 2m\dfrac{b}{a}\dfrac{\mu}{\beta}\Delta$$

(b) Yield line in trapeziums AEFB and CEFD
The trapeziums rotate only about x-axis.

$$l_x = a, \quad m_x = m, \quad \theta_x = \dfrac{\Delta}{0.5b}$$

Hence the energy dissipated E_2 on the yield lines in trapeziums AEFB and CEFD is

$$E_2 = 2[a \; m \; \dfrac{\Delta}{0.5b} + b \; \mu m \; 0] = 4m\dfrac{a}{b}\Delta$$

The total energy dissipated E is therefore

$$E = E_1 + E_2 = \{2m\dfrac{b}{a}\dfrac{\mu}{\beta} + 4m\dfrac{a}{b}\}\Delta$$

3. Calculation of moment of resistance:
Equating the work done by the external loads to the energy dissipated at the yield lines,

$$E = 2m\dfrac{b}{a}\dfrac{\mu}{\beta} + 4m\dfrac{a}{b} = W = q\dfrac{ab}{6}(3 - 2\beta)$$

Solving for m,

$$m = q\dfrac{b^2}{12}\dfrac{(3\beta - 2\beta^2)}{(\mu\dfrac{b^2}{a^2} + 2\beta)}$$

In order to calculate the maximum value of m required, set dm/dβ = 0

$$(\mu\dfrac{b^2}{a^2} + 2\beta)(3 - 4\beta) - (3\beta - 2\beta^2)(2) = 0$$

Simplifying

$$4\beta^2 + 4\mu\dfrac{b^2}{a^2}\beta - 3\mu\dfrac{b^2}{a^2} = 0$$

Solving the quadratic in β,

$$\beta = \frac{1}{2}\frac{b^2}{a^2}\{-\mu + \sqrt{(\mu^2 + 3\mu\frac{a^2}{b^2})}\}$$

8.10.6.1 Example of Yield Line Analysis of a Simply Supported Rectangular Slab

A simply supported rectangular slab 4.5 m long by 3 m wide carries an ultimate load of 16 kN/m². Determine the design moments for the case when the value of μ = 0.5.

Substituting a = 4.5 m, b = 3.0 m, b/a = 0.667, μ = 0.5, in the formula for β,

$$\beta = 0.312$$

Substituting β = 0.312 and q = 16.0 in the equation for m,

$$m = 10.51 \text{ kNm/m}, \ \mu m = 5.26 \text{ kNm/m}$$

It is usual in designs based on the yield line analysis for the reinforcement to remain uniform in each direction. It is evident from the collapse mechanism that yield line analysis provides no information on where the reinforcement can be curtailed, nor does it give any information on the shear force distribution in the slab.

8.10.7 Rectangular Two-Way Clamped Slab

The solution derived in section 8.9.6 can be extended to the case of a continuous or clamped slab. The slab shown in Fig. 8.75 has a continuous hogging yield line around the supports. The negative moment of resistance of the slab at the supports has a value of γ m' per unit length in the shorter direction and m' per unit length in the longer direction.

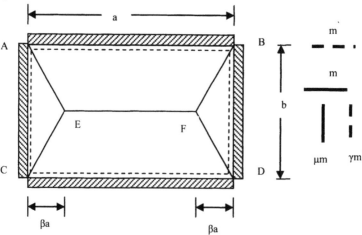

Fig. 8.75 Collapse mode for a clamped slab.

The basic yield line pattern will be as for the simply supported slab shown in Fig. 8.74 except that negative yield lines (tension at the top face) form parallel and close to the supports as shown in Fig. 8.75 by dotted lines. The work done by external loads and the energy dissipated at the positive yield lines (tension at the bottom face) remain as for the simply supported slab. The extra aspect to be considered is the energy dissipated at the negative yield lines.

(a) For the *two* negative yield lines parallel to the *shorter* sides

$$l_y = b, \ m_y = \gamma m', \ \theta_y = \frac{\Delta}{\beta a}$$

Hence the energy dissipated by the *two* negative yield lines parallel to the *shorter* side is

$$E_3 = 2\{b \ \gamma \ m' \frac{\Delta}{\beta a}\} = 2m' \frac{b}{a} \frac{\gamma}{\beta} \Delta$$

(b) For the *two* negative yield lines parallel to the *longer* sides:

$$l_x = a, \ m_x = m', \ \theta_x = \frac{\Delta}{0.5b}$$

Hence the energy dissipated by the *two* negative yield lines parallel to the longer sides is

$$E_4 = 2\{a \ m' \frac{\Delta}{0.5b}\} = 4m' \frac{a}{b} \Delta$$

Total energy dissipated at the negative yield lines is

$$E_{negative yield lines} = E_3 + E_4 = 2m' \{\frac{b}{a} \frac{\gamma}{\beta} + 2\frac{a}{b}\} \Delta$$

Total energy dissipated at the positive and negative yield lines is

$$E = [2m' \{\frac{b}{a} \frac{\gamma}{\beta} + 2\frac{a}{b}\} + 2m \{\frac{b}{a} \frac{\mu}{\beta} + 2\frac{a}{b}\}] \Delta$$

Equating the work done by the external loads to the energy dissipated at all the yield lines,

$$m = q \frac{b^2}{12} \frac{(3\beta - 2\beta^2)}{(\mu \frac{b^2}{a^2} + 2\beta) + \frac{m'}{m} (\gamma \frac{b^2}{a^2} + 2\beta)}$$

In order to calculate the maximum value of m required, set dm/dβ = 0

$$\{(\mu + \gamma \frac{m'}{m}) \frac{b^2}{a^2} + 2\beta(1 + \frac{m'}{m})\}(3 - 4\beta) - (3\beta - 2\beta^2)(2 + 2\frac{m'}{m}) = 0$$

Simplifying

$$4(1 + \frac{m'}{m})\beta^2 + 4(\mu + \gamma \frac{m'}{m})\frac{b^2}{a^2}\beta - 3(\mu + \gamma \frac{m'}{m})\frac{b^2}{a^2} = 0$$

Solving the quadratic in β,

$$\beta = \frac{1}{2}\frac{b^2}{a^2}\left[\frac{-(\mu+\gamma\frac{m'}{m})+\sqrt{\{(\mu+\gamma\frac{m'}{m})^2+3(\mu+\gamma\frac{m'}{m})(1+\frac{m'}{m})\frac{a^2}{b^2}\}}}{(1+\frac{m'}{m})}\right]$$

8.10.7.1 Example of Yield Line Analysis of a Clamped Rectangular Slab

A clamped rectangular slab 4.5 m long by 3 m wide carries an ultimate load of 16 kN/m^2. Determine the design moments for the case where

$$m'/m = 1.3, \quad \mu = 0.6, \quad \gamma = 0.6$$

as obtained from *average moment ratios* from elastic analysis.
Substituting in the formula for β,
$a = 4.5$ m, $b = 3.0$ m, $b/a = 0.667$, $m'/m = 1.3$, $\mu = 0.6$, $\gamma = 0.6$, $\beta = 0.33 < 0.5$
Substituting $\beta = 0.33$ and $q = 16.0$ kN/m^2, $m = 4.34$ kNm/m, $\mu m = 2.60$ kNm/m, $m' = 5.64$ kNm/m, $\gamma m' = 3.4$ kNm/m.

It is usual in designs based on the yield line analysis for the reinforcement to remain uniform in each direction.   However, the extent of the negative reinforcement required can be determined by finding the dimensions of a simply supported central rectangular region of dimensions $\alpha a \times \alpha b$ which has a collapse moment in the yield lines of $m = 4.1$ kNm/m, $\mu m = 2.47$ kNm/m and $q = 16.0$.

Since at the cut-offs of top bars, the hogging yield line has zero strength, this simulates simple support. The bars must be anchored beyond the theoretical cut-off lines.

Substituting $\mu = 0.6$, $b/a = 3/(4.5) = 0.67$ in the equation for β in section 8.9.6 gives $\beta = 0.33$.

Substituting for $m = 4.34$ kN/m, $b = 3\,\alpha$, $a = 4.5\,\alpha$, $\mu = 0.6$, $\beta = 0.33$, $q = 16$kN/m^2 in the equation for m in section 8.9.6, where

$$m = q\frac{(\alpha b)^2}{12}\frac{(3\beta-2\beta^2)}{(\mu\frac{b^2}{a^2}+2\beta)}$$

gives $\alpha = 0.66$. The theoretical cut-off lengths are therefore $(1 - \alpha)/2 = 0.17$ of the side dimensions.

8.10.8 Clamped Rectangular Slab with One Long Edge Free

A rectangular slab continuous on three edges and free on a long edge has two distinct modes of collapse as shown in Fig. 8.76 and Fig. 8.77. The slab shown in the figures has a continuous hogging yield line around the supports in addition to positive yield lines. As shown in Fig. 8.77, the positive and negative moments of resistance of the slab have values of μm and $\gamma m'$ per unit length respectively in yield lines parallel to the y-direction and m and m' per unit length respectively in yield lines parallel to the x-direction.

In the case of simply supported and clamped rectangular slabs, there was only one mode of collapse defined by a single parameter β. However when one of the edges is free, there are two different modes of collapse possible as shown in Fig. 8.76 and Fig. 8.77. Calculations have to be done for both modes of collapse to determine either the minimum collapse load or the maximum moment of resistance required.

8.10.8.1 Calculations for Collapse Mode 1

The mode of collapse is shown in Fig. 8.76. Assume that EF deflects by Δ.

(1) Energy dissipated at the yield lines

(a) Trapeziums ACFE and BDFE: The trapeziums rotate only about the y-axis.

(i) For the negative yield line

$l_y = b$, $m_y = \gamma m'$, $\theta_y = \dfrac{\Delta}{0.5a}$. Total energy dissipation for the two regions is

$$E_1 = 2\{b\,\gamma\,m'\,\frac{\Delta}{0.5a}\} = 4m'\gamma\frac{b}{a}\Delta$$

(ii) For the positive yield line

$l_y = b$, $m_y = \mu m$, $\theta_y = \dfrac{\Delta}{0.5a}$. Total energy dissipation for the two regions is

$$E_2 = 2\{b\,\mu m\,\frac{\Delta}{0.5a}\} = 4m\frac{b}{a}\mu\Delta$$

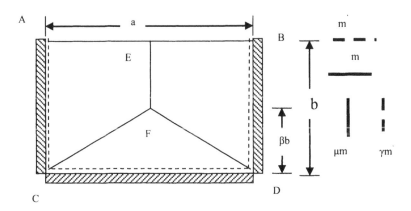

Fig. 8.76 Clamped slab with a free edge, collapse mode 1.

(b) Triangle CFD: The triangle rotates only about the x-axis.

(i) For the negative yield line

$l_x = a$, $m_x = m'$, $\theta_x = \dfrac{\Delta}{\beta b}$. Energy dissipation is

$$E_3 = a\, m' \frac{\Delta}{\beta b} = m' \frac{a}{b}\frac{1}{\beta}\Delta$$

(ii) For the positive yield line

$l_x = a$, $m_x = m$, $\theta_x = \dfrac{\Delta}{\beta b}$. Energy dissipation is

$$E_4 = a\, m \frac{\Delta}{\beta b} = m \frac{a}{b}\frac{1}{\beta}\Delta$$

Total energy dissipated at the positive and negative yield lines is

$$E = E_1 + E_2 + E_3 + E_4 = \{4\frac{b}{a}(\gamma m' + \mu m) + \frac{a}{b}(m' + m)\frac{1}{\beta}\}\Delta$$

Simplifying

$$E = \frac{1}{\beta ab}\{4(\gamma m' + \mu m)\beta b^2 + (m' + m)a^2\}\Delta$$

(2) Work done by external loads

(a) Triangle CFD: Area $= 0.5\, a\, \beta b$, deflection at the centroid $= \Delta/3$
Work done by external loads is

$$W_1 = q\ 0.5\ a\,\beta b\frac{\Delta}{3}$$

(b) Trapeziums ACFE and BDFE: Dividing it into a triangle and a rectangle
Triangle: area $= 0.5\ a/2\ \beta b$, deflection at the centroid $= \Delta/3$
Rectangle: area $= a/2\ (b - \beta b)$, deflection at the centroid $= \Delta/2$
Work done by the external loads is

$$W_2 = 2[\{q\ 0.5\frac{a}{2}\beta b\frac{\Delta}{3}\} + q\frac{a}{2}(b - \beta b)\frac{\Delta}{2}] = q\frac{ab}{6}(3 - 2\beta)\Delta$$

Total work W done by the loads $= W_1 + W_2$

$$W = q\frac{ab}{6}(3 - \beta)\Delta$$

(3) Calculation of m
Equating the work done by the external loads to the energy dissipated at all the yield lines,

$$m = q\frac{b^2}{6}\frac{(3\beta - \beta^2)}{\{4(\gamma\frac{m'}{m} + \mu)\beta\frac{b^2}{a^2} + (1 + \frac{m'}{m})\}}$$

In order to calculate the maximum value of m required, set dm/dβ = 0,

$$\{4(\gamma\frac{m'}{m}+\mu)\beta\frac{b^2}{a^2}+(1+\frac{m'}{m})\}(3-2\beta)-(3\beta-2\beta^2)4(\gamma\frac{m'}{m}+\mu)\frac{b^2}{a^2}=0$$

Simplifying

$$4(\gamma\frac{m'}{m}+\mu)\frac{b^2}{a^2}\beta^2+2(1+\frac{m'}{m})\beta-3(1+\frac{m'}{m})=0$$

8.10.8.2 Calculations for Collapse Mode 2

The mode of collapse is shown in Fig. 8.77. Assume that E F deflects by Δ.

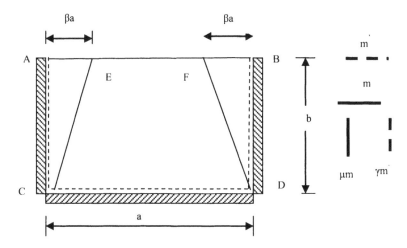

Fig. 8.77 Clamped slab with a free edge, collapse mode 2.

(1) Energy dissipated at the yield lines

(a) Triangles ACE and BDF: Rotation of the triangles is about y-axis only.

(i) For the negative yield line

$l_y = b$, $m_y = \gamma m'$, $\theta_y = \dfrac{\Delta}{\beta a}$. Total energy dissipation for the two triangles is

$$E_1 = 2\{b\,\gamma\,m'\,\frac{\Delta}{\beta a}\} = 2m'\gamma\frac{b}{a}\frac{1}{\beta}\Delta$$

(ii) For the positive yield line

$l_y = b$, $m_y = \mu m$, $\theta_y = \dfrac{\Delta}{\beta a}$, Total energy dissipation for the two triangles is

$$E_2 = 2\{b\,\mu\,m\,\frac{\Delta}{\beta a}\} = 2m\mu\frac{b}{a}\frac{\Delta}{\beta}$$

(b) Trapezium CEFD: The trapezium rotates only about the x-axis.

(i) For the negative yield line

$l_x = a$, $m_x = m'$, $\theta_x = \dfrac{\Delta}{b}$. Energy dissipation is

$$E_3 = a\,m'\,\frac{\Delta}{b}$$

(ii) For the positive yield line

$l_x = 2(\beta a)$, $m_x = m$, $\theta_x = \dfrac{\Delta}{b}$, Energy dissipation is

$$E_4 = 2\beta a\,m\,\frac{\Delta}{b}$$

Total energy dissipated at the positive and negative yield lines is

$$E = E_1 + E_2 + E_3 + E_4 = \{2\frac{b}{\beta a}(\gamma m' + \mu m) + \frac{a}{b}(m' + 2\beta m)\}\Delta$$

Simplifying

$$E = \frac{1}{\beta a b}\{2(\gamma m' + \mu m)b^2 + (m'\beta + 2\beta^2 m)a^2\}\Delta$$

(2) Work done by external loads

(a) Triangles ACE and BDF:
Area = 0.5 b βa, deflection at the centroid = $\Delta/3$

$$W_1 = 2\{q\frac{1}{2}b\,\beta a\frac{\Delta}{3}\} = q\frac{ab}{3}\beta\Delta$$

(b) Trapezium CEFD
Dividing it into two triangles and a rectangle,

   Two triangles: area = 0.5 b βa, deflection at the centroid = $\Delta/3$
   Rectangle: area = b (a − 2βa), deflection at the centroid = $\Delta/2$

$$W_2 = 2\{q\frac{1}{2}b\,\beta a\frac{\Delta}{3}\} + qb(a - 2\beta a)\frac{\Delta}{2} = q\frac{ab}{6}(3 - 4\beta)\Delta$$

Total work W done by the loads = $W_1 + W_2$

$$W = q\frac{ab}{6}(3 - 2\beta)\Delta$$

Equating the work done by the external loads to the energy dissipated at all the yield lines,

$$m = q\frac{b^2}{6}\frac{(3\beta - 2\beta^2)}{\{2(\gamma\frac{m'}{m} + \mu)\frac{b^2}{a^2} + (2\beta^2 + \frac{m'}{m}\beta)\}}$$

In order to calculate the maximum value of m required, set dm/dβ = 0,

$$\{2(\gamma\frac{m'}{m}+\mu)\frac{b^2}{a^2}+(2\beta^2+\frac{m'}{m}\beta)\}(3-4\beta)-\{3\beta-2\beta^2\}(4\beta+\frac{m'}{m})=0$$

Simplifying

$$(\frac{m'}{m}+3)\beta^2+\{4(\gamma\frac{m'}{m}+\mu)\frac{b^2}{a^2}\}\beta-3(\gamma\frac{m'}{m}+\mu)\frac{b^2}{a^2}=0$$

Solving the quadratic in β,

$$\beta=\frac{\{4(\gamma\frac{m'}{m}+\mu)\frac{b^2}{a^2}\}+\sqrt{[\{4(\gamma\frac{m'}{m}+\mu)\frac{b^2}{a^2}\}^2+12(\gamma\frac{m'}{m}+\mu)\frac{b^2}{a^2}(\frac{m'}{m}+3)]}}{2(\frac{m'}{m}+3)}$$

8.10.8.3 Example of Yield Line Analysis of a Clamped Rectangular Slab with One Long Edge Free

A clamped rectangular slab with one long edge free is 4.5 m long by 3 m wide and carries an ultimate load of 16 kN/m². Determine the design moments for the case when

$$m'/m = 5.0, \quad \mu = 3.0, \quad \gamma = 1.5$$

as obtained from maximum moment ratios from elastic finite element analysis.

When a slab has more than one distinct mode of failure, it is necessary to investigate both modes of failure and accept the *larger* of the two moments as the design moment.

Mode 1: Substituting the values of the parameters in the formula for β,
$$\beta = 0.712 < 1.0$$
Using this value of β,
$$m = 2.03 \text{ kNm/m}$$

Mode 2: Using the same parameters as for mode 1, calculate the value of β. The smaller root for β is
$$\beta = 0.597 > 0.5$$
Using this value of β = 0.5,
$$m = 1.95 \text{ kNm/m}.$$

For design the larger value for m is obtained from mode 1. Therefore
m = 2.03 kNm/m, μ m = 6.1 kNm/m, m' = 10.2 kNm/m, γm' = 15.2 kNm/m.

8.10.9 Trapezoidal Slab Continuous over Three Supports and Free on a Long Edge

Fig. 8.78 shows a uniformly loaded trapezoidal slab with three edges clamped and one edge free. Normal moment of resistance per unit length on positive and negative yield lines parallel to x- and y-axes are respectively (m, μm) and (m', γm') respectively.

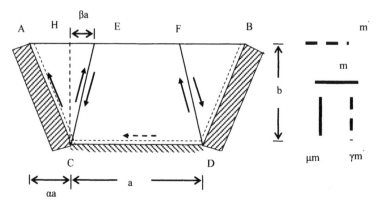

Fig. 8.78 Trapezoidal clamped slab with a free edge.

In the previous examples, the rotations of the rigid regions took place about edges which were parallel to x- or y-axis. In this example only one axis of rotation is parallel to the coordinate axes. One possible mode of collapse is shown by the positive yield lines CE and DF and negative yield lines parallel to the supports.

Assume that EF deflects by Δ. As shown in Fig. 8.79, let the yield line CE be inclined to the vertical by φ and the support CA is inclined to the vertical by ψ. Let EG be perpendicular to support AC. From geometry,

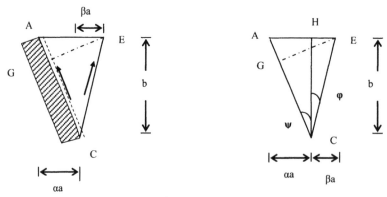

Fig. 8.79 Rotation and moment vectors.

$$\text{Angle ACH} = \psi, \quad \tan\psi = \alpha\frac{a}{b}, \quad \cos\psi = \frac{h}{AE} = \frac{h}{a(\alpha+\beta)} = \frac{1}{a(\alpha+\beta)}$$

Angle ECH $= \varphi$, $\tan\varphi = \beta\dfrac{a}{b}$, Angle CAH $= 90 - \psi$, GE $= h$, AE $= (\alpha+\beta)\, a$

Triangle ACE rotates clockwise about the support AC by θ. The rotational components of θ are

$$\theta = \frac{\Delta}{h}, \quad \theta_x = \theta\sin\psi, \quad \theta_y = \theta\cos\psi$$

Substituting for $\tan\psi$ and $\cos\psi$

$$\theta_x = \frac{\alpha}{(\alpha+\beta)}\frac{\Delta}{b}, \quad \theta_y = \frac{1}{(\alpha+\beta)}\frac{\Delta}{a}$$

(a) Energy dissipated in yield lines

(i) Negative yield line in triangles ACE and BDF
The triangles rotate about an axis inclined to both x- and y-axes.

$$l_x = \alpha\, a, \quad m_x = m\,, \quad \theta_x = \frac{\alpha}{(\alpha+\beta)}\frac{\Delta}{b}$$

$$l_y = b, \quad m_y = \gamma m\,, \quad \theta_y = \frac{1}{(\alpha+\beta)}\frac{\Delta}{a}$$

Hence the energy dissipated on the negative yield line in the two triangles is

$$E_1 = 2\{l_x\, m_x\, \theta_x + l_y\, m_y\, \theta_y\}$$

$$E_1 = 2\{\alpha a\, m'\,\frac{\alpha}{(\alpha+\beta)}\frac{\Delta}{b} + b\,\gamma\, m'\,\frac{1}{(\alpha+\beta)}\frac{\Delta}{a}\}$$

$$E_1 = 2\frac{m'}{(\alpha+\beta)}[\alpha^2\frac{a}{b} + \gamma\frac{b}{a}]\Delta$$

Note that because both the moment vector and the rotation vector act in the same direction, the energies dissipated by both the x- and y-components are positive.

(ii) Positive yield line in triangles ACE and BDF

$$l_x = \beta\, a, \quad m_x = m, \quad \theta_x = \frac{\alpha}{(\alpha+\beta)}\frac{\Delta}{b}$$

$$l_y = b, \quad m_y = \mu m, \quad \theta_y = \frac{1}{(\alpha+\beta)}\frac{\Delta}{a}$$

Note that x-components of the moment vector and rotation vector point in opposite directions. Therefore the energy dissipated on the positive yield line in triangle ACE is

$$E_2 = 2\{l_x\, m_x\, \theta_x + l_y\, m_y\, \theta_y\}$$

$$E_2 = 2\{-\beta a\, m\,\frac{\alpha}{(\alpha+\beta)}\frac{\Delta}{b} + b\,\mu m\,\frac{1}{(\alpha+\beta)}\frac{\Delta}{a}\}$$

$$E_2 = 2\frac{m}{(\alpha+\beta)}[-\alpha\beta\frac{a}{b}+\mu\frac{b}{a}]\Delta$$

(iii) Negative yield line in the trapezium ECDF
The trapezium rotates only about the x-axis.

$$l_x = a, \quad m_x = m', \quad \theta_x = \frac{\Delta}{b}$$

$$E_3 = l_x\, m_x\, \theta_x = a\, m'\, \frac{\Delta}{b}$$

(iv) Positive yield line in the trapezium ECDF

$$l_x = 2(\beta\, a), \quad m_x = m, \quad \theta_x = \frac{\Delta}{b}$$

$$E_4 = l_x\, m_x\, \theta_x = 2\beta a\, m\, \frac{\Delta}{b}$$

Therefore total energy dissipation is

$$E = E_1 + E_2 + E_3 + E_4$$

$$E = \frac{1}{(\alpha+\beta)}\frac{1}{ab}[m'\{(2\alpha^2+\alpha+\beta)a^2 + 2\gamma b^2\} + 2m\{\beta^2 a^2 + \mu b^2\}]\Delta$$

(b) Work done by external loads

(i) Triangles ACE and BDF
Area = 0.5 (α + β) a b,   deflection of the centroid = Δ/3
$W_1 = 2\, q\, \{0.5\, (\alpha+\beta)\, a\, b\, \Delta/3\}$

(ii) Trapezium ECDF

Divide into two triangles and a rectangle.
For each triangle:
area = 0.5 β a b,   deflection of the centroid = Δ/3
For the rectangle:
Area = (1 – 2β) a b,   deflection of the centroid = Δ/2
Work done is
$W_2 = 2\, q\, \{0.5\, a\, b\, \beta\}\, \Delta/3 + q\, a\, b\, (1 - 2\,\beta)\, \Delta/2$
Total work W done is $W = W_1 + W_2$
$W = q\, ab\, \{3 + 2\, (\alpha - \beta)\}\, \Delta/6$
Equating W = E and simplifying,

$$m = q\, b^2\, \frac{1}{6}\, \frac{3(\alpha+\beta)+2(\alpha^2-\beta^2)}{[\frac{m'}{m}\{(2\alpha^2+\alpha+\beta)+2\gamma\frac{b^2}{a^2}\} + 2\{\beta^2 + \mu\frac{b^2}{a^2}\}]}$$

If α = 0, then the equation will be same as for Mode 2 collapse of a clamped rectangular slab with a free edge.
Assuming:

a = 4.5 m, b = 3.0 m, α a = 1.0, m´/m = 5.0, μ = 3.0, γ= 1.5, q = 16 kN/m
β = 0.5, m = 3.04 kNm/m.

8.10.10 Slab with a Symmetrical Hole

Fig. 8.80 shows a simply supported rectangular slab of dimensions a × b with a central rectangular hole of dimensions αa × αb.
There are three distinct modes of collapse which have to be analysed in calculating the minimum collapse load.

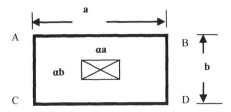

Fig. 8.80 Slab with a hole.

8.10.10.1 Calculations for Collapse Mode 1

Fig. 8.81 shows the collapse mode 1. Let the deflection at the apex of the triangle be Δ.

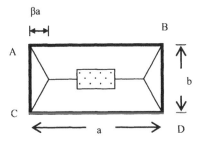

Fig. 8.81 Collapse mode 1.

(a) Work done by external loads
(i) Two side triangles

Area = 0.5 b βa, deflection at the centroid = Δ/3

$$W_1 = 2\{q\,0.5b\,\beta a\frac{\Delta}{3}\} = qab\frac{\beta}{3}\Delta$$

(ii) Two trapeziums
Dividing each into two triangles, two rectangles and a rectangle adjoining the side
of the hole,

Triangle
$$\text{Area} = 0.5\ b/2\ \beta a, \quad \text{deflection at the centroid} = \Delta/3$$
Rectangle
$$\text{Area} = b/2\ (a - \alpha a - 2\ \beta a), \quad \text{deflection at the centroid} = \Delta/2$$

Rectangle adjoining the hole
Note that the deflection at the edge of the hole is not Δ but $(1 - \alpha)\ \Delta$
$$\text{Area} = \{b\ (1 - \alpha)/2\}\ \alpha a, \quad \text{deflection at the centroid} = (1 - \alpha)\ \Delta/2$$

$$W_2 = 2[2q\{\frac{1}{2}\frac{b}{2}\beta a\frac{\Delta}{3}\} + q\frac{b}{2}(a - \alpha a - 2\beta a)\frac{\Delta}{2} + qb\frac{(1-\alpha)}{2}\alpha a(1-\alpha)\frac{\Delta}{2}]$$

$$W_2 = q\frac{ab}{6}\{3 - 4\beta - 6\alpha^2 + 3\alpha^3\}\Delta$$

Total work W done by the loads = $W_1 + W_2$

$$W = q\frac{ab}{6}(3 - 2\beta - 6\alpha^2 + 3\alpha^3)\Delta$$

(b) Energy dissipated at the yield lines

(i) Yield lines in the two side triangles
$$l_y = b, \quad m_y = \mu m, \quad \theta_y = \frac{\Delta}{\beta a}$$

$$E_1 = 2\{b\ \mu m\ \frac{\Delta}{\beta a}\} = 2m\frac{b}{a}\frac{\mu}{\beta}\Delta$$

(ii) Yield lines in the two trapeziums
$$l_x = (a - \alpha a), \quad m_x = m, \quad \theta_x = \frac{\Delta}{0.5b}$$

$$E_2 = 2\{a(1 - \alpha)\ m\ \frac{\Delta}{0.5b}\} = 4m(1 - \alpha)\frac{a}{b}\Delta$$

The total energy dissipated is therefore
$$E_1 + E_2 = E = \{2m\frac{b}{a}\frac{\mu}{\beta} + 4m(1 - \alpha)\frac{a}{b}\}\Delta$$

Equating the work done by the external loads to the energy dissipated at the yield
lines,

$$E = 2m\frac{b}{a}\frac{\mu}{\beta} + 4m(1 - \alpha)\frac{a}{b}\Delta = W = q\frac{ab}{6}(3 - 2\beta - 6\alpha^2 + 3\alpha^3)\Delta$$

Solving for m,

$$m = q\,\frac{b^2}{12}\,\frac{\{(3-6\alpha^2+3\alpha^3)\beta-2\beta^2\}}{\{\mu\dfrac{b^2}{a^2}+2(1-\alpha)\beta\}}$$

In order to calculate the maximum value of m required, set $dm/d\beta = 0$,

$$\{\mu\frac{b^2}{a^2}+2(1-\alpha)\beta\}(3-6\alpha^2+3\alpha^3-4\beta)-\{(3-6\alpha^2+3\alpha^3)\beta-2\beta^2\}2(1-\alpha)=0$$

The resulting quadratic equation in β can be solved numerically for specific values of the parameters α and μ and the corresponding value of m can be determined. Note that from geometry the above equations are valid for $0 \le \beta \le (1-\alpha)/2$.

8.10.10.2 Calculations for Collapse Mode 2

Fig. 8.82 shows the collapse mode 2. Assume that the deflection at the apex of side triangles is Δ.

(a) Work done by external loads

(i) Side trapeziums
Divide each into two triangles and a rectangle.

Triangle

$$\text{Area} = 0.5\ \{(1-\alpha)\ a/2\}\ \beta b, \quad \text{deflection at the centroid} = \Delta/3$$

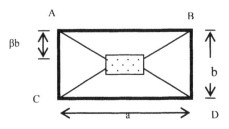

Fig. 8.82 Collapse mode 2.

Rectangle

$$\text{Area} = (1-2\beta)\ b\ (1-\alpha)\ a/2, \quad \text{deflection at the centroid} = \Delta/2$$

$$W_1 = 2[2q\{\frac{1}{2}\frac{(1-\alpha)a}{2}\beta b\frac{\Delta}{3}\}+q(1-2\beta)b\frac{(1-\alpha)a}{2}\frac{\Delta}{2}\ ]$$

$$W_1 = q\,\frac{ab}{6}\{3-3\alpha-4\beta(1-\alpha)\}\Delta$$

(ii) Top and bottom trapeziums
Divide it into two triangles and a rectangle which is part of the hole.

Triangle

Area = 0.5 {(1 − α) a /2} βb, deflection at the centroid = Δ/3

Rectangle

Note that the deflection at the edge of the hole is not Δ but (1 − α) Δ/ (2β)

Area = αa (1 − α) b/2, deflection at the centroid = (1 − α) Δ/ (4 β)

$$W_2 = 2[2q\{\frac{1}{2}\frac{(1-\alpha)}{2}a\,\beta b\frac{\Delta}{3}\} + q\{\alpha a\frac{(1-\alpha)b}{2}(1-\alpha)\frac{\Delta}{4\beta}\}]$$

$$W_2 = q\frac{ab}{6}(1-\alpha)\{2\beta - \frac{3\alpha^2}{2\beta} + \frac{3\alpha}{2\beta}\}\Delta$$

Total work W done by the loads

$$W = W_1 + W_2 = q\frac{ab}{6}(1-\alpha)\{3-2\beta - \frac{3(\alpha^2-\alpha)}{2\beta}\}\Delta$$

(b) Energy dissipated at the yield lines

(i) Yield lines in the two side trapeziums

$$l_y = 2(\beta b), \quad m_y = \mu m, \quad \theta_y = \frac{\Delta}{0.5(1-\alpha)a}$$

$$E_1 = 2\{2\beta b\,\mu m\,\frac{\Delta}{0.5(1-\alpha)a}\} = 8m\frac{b}{a}\frac{\beta\mu}{(1-\alpha)}\Delta$$

(ii) Yield line in the top and bottom trapeziums

$$l_x = 2(1-\alpha)\,a/2, \quad m_x = m, \quad \theta_x = \frac{\Delta}{\beta b}$$

$$E_2 = 2\{(1-\alpha)a\,m\,\frac{\Delta}{\beta b}\} = 2m(1-\alpha)\frac{1}{\beta}\frac{a}{b}\Delta$$

The total energy dissipated is therefore

$$E = E_1 + E_2 = \{8m\frac{b}{a}\frac{\beta}{(1-\alpha)}\mu + 2m\frac{(1-\alpha)}{\beta}\frac{a}{b}\}\Delta$$

Equating the work done by the external loads to the energy dissipated at the yield lines,

$$\{8m\frac{b}{a}\frac{\beta}{(1-\alpha)}\mu + 2m\frac{(1-\alpha)}{\beta}\frac{a}{b}\}\Delta = q\frac{ab}{6}(1-\alpha)\{3-2\beta - \frac{3(\alpha^2-\alpha)}{2\beta}\}\Delta$$

Solving for m,

$$m = q\frac{b^2}{24}\frac{(1-\alpha)^2\{(3(\alpha^2-\alpha)+6\beta-4\beta^2\}}{\{4\mu\frac{b^2}{a^2}\beta^2+(1-\alpha)^2\}}$$

In order to calculate the maximum value of m required, set dm/dβ = 0.

$$\{4\mu\frac{b^2}{a^2}\beta^2 + (1-\alpha)^2\}(6-8\beta) - \{(3(\alpha^2-\alpha)+6\beta-4\beta^2\}8\mu\frac{b^2}{a^2}\beta = 0$$

The resulting quadratic equation in β can be solved numerically for specific value of the parameters and the corresponding value of m can be determined. The above equations are valid only for $0.5(1-\alpha) \le \beta \le 0.5$.

8.10.10.3 Calculations for Collapse Mode 3

Fig. 8.83 shows the collapse mode 3. Assume that the deflection at the longer sides of the hole is Δ.

(a) Work done by external loads

(i) Side trapeziums
Divide each into two triangles and a rectangle.

Triangle
$$\text{Area} = 0.5 \{(1-\alpha)\, b\,/2\}\,\beta a, \quad \text{deflection at the centroid} = \Delta/3$$

Rectangle
Note that the deflection at the edge of the hole is not Δ but $\{0.5(1-\alpha)/\beta\}\,\Delta$
$$\text{Area} = \alpha b\,(1-\alpha)\,a/2, \quad \text{deflection at the centroid} = \{0.5(1-\alpha)/\beta\}\,\Delta/2$$

$$W_1 = 2[2q\{\frac{1}{2}\frac{(1-\alpha)b}{2}\beta a\frac{\Delta}{3}\} + q\,\alpha b\,(1-\alpha)\frac{a}{2}\frac{(1-\alpha)}{2\beta}\frac{\Delta}{2}]$$

$$W_1 = q\frac{ab}{12}\frac{(1-\alpha)}{\beta}\{4\beta^2 + 3\alpha(1-\alpha)\}\Delta$$

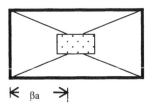

$\Kappa\,\beta a\,\rightarrow$

Fig. 8.83 Collapse mode 3.

(ii) Top and bottom trapeziums
Divide it into two triangles and a rectangle which is part of the hole.

Triangle
$$\text{Area} = 0.5 \{(1-\alpha)\, b\,/2\}\,\beta a, \quad \text{deflection at the centroid} = \Delta/3$$

Rectangle

Area $= (1 - 2\beta)$ a $(1 - \alpha)$ b/2,   deflection at the centroid $= \Delta/2$

$$W_2 = 2[2q\{\frac{1}{2}\frac{(1-\alpha)}{2}b\,\beta a\frac{\Delta}{3}\} + q\{(1-2\beta)a\frac{(1-\alpha)b}{2}\frac{\Delta}{2}\}]$$

$$W_2 = q\frac{ab}{6}(1-\alpha)\{3-4\beta\}\Delta$$

Total work W done by the loads $= W_1 + W_2$

$$W = q\frac{ab}{12\beta}(1-\alpha)(6\beta-4\beta^2-3\alpha^2+3\alpha)\Delta$$

(b) Energy dissipated at the yield lines

(i) Yield lines in the two side trapeziums

$$l_y = 2(1-\alpha)\,b/2, \quad m_y = \mu m, \quad \theta_y = \frac{\Delta}{\beta a}$$

$$E_1 = 2\{2(1-\alpha)\frac{b}{2}\,\mu m\,\frac{\Delta}{\beta a}\} = 2m\frac{b}{a}\frac{(1-\alpha)\mu}{\beta}\Delta$$

(ii) Yield line in the top and bottom trapeziums

$$l_x = 2(\beta a), \quad m_x = m, \quad \theta_x = \frac{\Delta}{0.5(1-\alpha)b}$$

$$E_2 = 2\{2\beta a\,m\,\frac{\Delta}{0.5(1-\alpha)b}\} = 8m\frac{\beta}{(1-\alpha)}\frac{a}{b}\Delta$$

The total energy dissipated is $E = E_1 + E_2$,

$$E = \{2m\frac{(1-\alpha)}{\beta}\mu\frac{b}{a} + 8m\frac{a}{b}\frac{\beta}{(1-\alpha)}\}\Delta$$

Equating the work done by the external loads to the energy dissipated at the yield lines,

$$\{2m\frac{(1-\alpha)}{\beta}\mu\frac{b}{a} + 8m\frac{a}{b}\frac{\beta}{(1-\alpha)}\}\Delta = q\frac{ab}{12\beta}(1-\alpha)(6\beta-4\beta^2-3\alpha^2+3\alpha)\Delta$$

Solving for m,

$$m = q\frac{b^2}{24}\frac{(1-\alpha)^2\{3\alpha(1-\alpha)\beta+6\beta^2-4\beta^3\}}{\{4\beta^2+(1-\alpha)^2\mu\frac{b^2}{a^2}\}}$$

In order to calculate the maximum value of m required, set dm/dβ = 0,

$$\{4\beta^2+(1-\alpha)^2\mu\frac{b^2}{a^2}\}\{3\alpha(1-\alpha)+12\beta-12\beta^2\}-\{3\alpha(1-\alpha)\beta+6\beta^2-4\beta^3\}8\beta = 0$$

The resulting quadratic equation in β can be solved numerically for specific values of the parameters and the corresponding value of m can be determined. The above equations are valid only for $0.5(1 - \alpha) \le \beta \le 0.5$.

Table 8.18 Collapse load for a simply supported slab with a hole

α	Mode 1		Mode 2		Mode 3	
	β	m/ (qb²)	β	m/ (qb²)	β	m/ (qb²)
0	0.312	**0.0730**	N/A	–	N/A	–
0.05	0.317	**0.0753**	0.5*	0.0715	0.5*	N/A
0.10	0.320	**0.0770**	0.5*	0.0741	0.5*	0.0336
0.20	0.322	**0.0779**	0.4567	0.0769	0.5*	0.0325
0.30	0.317	0.0760	0.4076	**0.0770**	0.5*	0.0290
0.40	0.300*	0.0701	0.3564	**0.0744**	0.5*	0.0242
0.50	0.250*	0.0607	0.3031	**0.0690**	0.5*	0.0189
0.6	0.20*	0.0474	0.2474	**0.0608**	0.2*	0.0136
0.7	0.15*	0.0316	0.1895	**0.0498**	0.15*	0.0120
0.8	0.10*	0.0158	0.1290	**0.0360**	0.10*	0.0074
0.9	0.05*	0.0041	0.0659	**0.0194**	0.05*	0.0036

*Not a stationary minimum.

8.10.10.4 Calculation of Moment of Resistance

For calculating the required ultimate moment, ultimate moments from all the three modes are calculated and the largest value is chosen. Results of calculations for a/b = 1.5, $\mu = 0.5$ for a range of $0 \le \alpha \le 0.9$ are shown in Table 8.19.

In some cases the stationary minimum value of β is obtained in the non-valid region. In such cases the minimum value of $m/(qb^2)$ has been calculated by limiting the value of β to the valid region. This is indicated in the table by *. It is noticed that up to $\alpha \approx 0.25$, mode 1 governs and afterwards mode 2 governs. It appears that mode 3 never governs.

8.10.11 Slab-and-Beam Systems

Combined beam-slab systems are commonly met in practice. Fig. 8.84 shows a typical case of a slab supported on beams cast integral with slabs, which in turn are supported on columns at the corners of the rectangle.

In considering this type of system, it is important to investigate yield line collapse modes involving independent collapse of the slab only and *combined* slab–beam collapse.

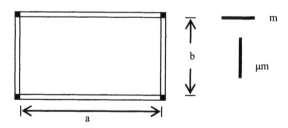

Fig. 8.84 Integral beam–slab systems.

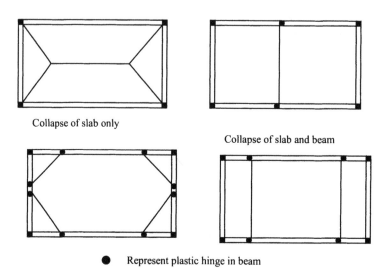

Collapse of slab only

Collapse of slab and beam

● Represent plastic hinge in beam

Fig. 8.85 Collapse of beam–slab systems.

If the torsional strength of the supporting beams is ignored, then the slab can be assumed to be simply supported on beams and the collapse of the slab only is treated as the collapse of a simply supported slab as discussed in section 8.9.6. The calculated moment of resistance is the minimum that should be provided in the slab.

In considering the combined slab–beam collapse, if the deflection at the plastic hinge is Δ, then the total rotation at the plastic hinge is

$$\theta_n = 2(\Delta/0.5a) = 4\Delta/a$$

The work done by external loads is

$$W = qb\,(\Delta a/2) = 0.5q\ a\ b\ \Delta$$

If the moment capacity of the beams is M_b, then the energy dissipated at the plastic hinge due to slab and beam is

$$E = (\mu m\ b + 2M_b)\,\theta_n = (m\ b + 2M_b)\,4\Delta/a$$

Equating $W = E$,

$$qa^2b/8 = \mu m\ b + 2M_b$$

Knowing m,

$$M_b = qa^2b/16 - \mu m\, b/2$$

It is possible to increase the value of m to a value larger than the minimum for collapse of slab only and provide lighter beams. For example using the slab designed in section 8.8.6.1, q = 16 kN/m², a = 4.5 m, b = 3.0 m, μ = 0.5, m = 10.51 kNm/m, μ m = 5.26 kNm/m, M_b = 52.86 kNm.

If it is decided to decrease the moment capacity of the beams towards the supports, then other possible collapse modes such as that shown in Fig. 8.85 need to be investigated.

8.10.12 Corner Levers

In sections 8.9.6 and 8.9.7 the yield lines for both simply supported and continuous slabs were assumed to run directly into the corners (Fig. 8.84 and Fig. 8.85). This situation will develop only if there is sufficient top steel at the corner region and the corner is held down. However if the corners are not held down then the yield line will divide to form a corner lever as shown in Fig. 8.86. Two possible situations occur.

1. Simply supported corner not held down: In this case the slab lifts off the corner and the sagging yield line divides as shown in Fig. 8.86(a) and the triangular portion rotates about the chain dotted line.

2. Simply supported corner held down: In this case the sagging yield line divides and a hogging yield line forms as shown in Fig. 8.86(b).

Solutions have been obtained for these cases which show that for a 90° corner, the corner lever mechanism decreases the overall strength of the slab by about 10 percent. In the case of slabs with acute corners the reduction in the calculated ultimate load due to corner levers is much larger. The reinforcement should be increased accordingly when the simplified solution is used. The top reinforcement commonly known as torsional reinforcement will prevent cracking in continuous slabs on the corner lever hogging yield line.

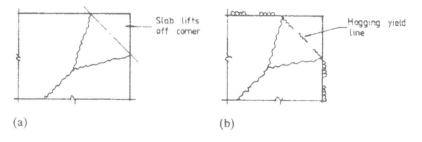

(a)   (b)

Fig. 8.86 (a) Corner not held down; (b) corner held down.

8.10.13 Collapse Mechanisms with More Than One Independent Variable

The collapse mechanisms considered previously were governed by a single variable β. Unfortunately this is not always the case. Fig. 8.87 shows a case of a slab clamped on two adjacent edges and the other two edges simply supported. In this case the collapse mechanism is defined by three independent variables β_1, β_2 and β_3.

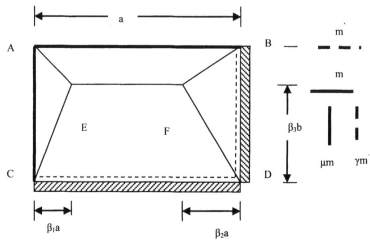

Fig. 8.87 Collapse mode for a slab with two adjacent discontinuous edges.

The problem of finding the maximum (or minimum) value of a function of several variables is not a trivial task. Fortunately computer programs are available for solving such problems.

8.10.14 Circular Fans

When concentrated loads act, flexural failure modes are likely to involve concentration of yield lines around the loaded area. This generally involves curved negative moment yield lines with radial positive moment yield lines as shown in Fig. 8.88.

If the moment of resistance is same in both directions and the radius of the fan is r, then the energy dissipated can be calculated by assuming that the deflection at the centre is Δ.

(a) Energy dissipated at yield line

(i) Negative yield line
The total length L of the negative yield line is $2\pi r$. Rotation θ_n at the yield line is Δ/r. Moment of resistance is m. Therefore energy dissipated is
$$E_1 = m\,(2\pi r)\,\Delta/r = 2\pi\,m\,\Delta$$

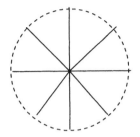

Fig. 8.88 Yield lines in a circular fan.

(ii) Positive yield lines

As the reinforcement in each direction is the same, the slab is isotropically reinforced. Therefore the x- and y-axes for each triangular segment can be different. Assuming that the x- and y-axes coincide with the radial and tangential directions of the circle, each segment rotates about the tangent only. The projection of the yield lines of each segment on the tangent is equal to the arc length corresponding to that segment.

The total projected length L of all positive yield line is $2\pi r$. The tangential rotation θ_n at the yield line is Δ/r. Moment of resistance is m.

Therefore energy dissipated is

$$E_2 = m(2\pi r) \Delta/r = 2\pi \, m \, \Delta$$

The total energy dissipation is

$$E = E_1 + E_2 = 2\pi (m' + m) \Delta$$

(b) Work done by external loads

Let q be the uniformly distributed load due to self weight and other externally applied loads and P is the concentrated load at the centre of the circle. The concentrated load could be an external load or a reaction from a column as in flat slab construction. The work done by the external uniformly distributed load q is calculated by noting that at the centroid of each triangular segment, deflection is $\Delta/3$ and the total load is $q(\pi r^2)$. Therefore

$$W = q(\pi r^2) \Delta/3 + P \Delta$$

Equating E and W,

$$(m' + m) = q \, r^2 /6 + P/(2\pi)$$

8.10.14.1 Collapse Mechanism for a Flat Slab Floor

Fig. 8.89 shows a flat slab floor with columns spaced at L_x and L_y in the x- and y-directions respectively. Postulate a collapse mechanism where the entire floor deflects by Δ with circular fans around columns as shown in Fig. 8.91. In any one panel:

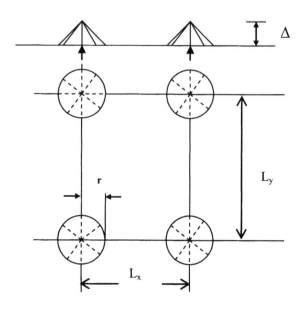

Fig. 8.89 Collapse of a flat slab floor.

(a) Energy dissipated at yield lines
From section 8.9.14 above,
$$E = 2\pi \, (m' + m) \, \Delta$$
(b) Work done by external load

(i) Uniformly distributed load outside the circular fans
$$\text{Area} = L_x \, L_y - \pi \, r^2, \quad \text{deflection} = \Delta$$
(ii) Uniformly distributed load inside the circular fans
$$\text{Area} = \pi \, r^2, \quad \text{deflection} = 2\Delta/3$$

Total work done is
$$W = q \, (L_x \, L_y - \pi \, r^2) \, \Delta + q \, \pi \, r^2 \, (2\Delta/3)$$
$$W = q \, (L_x \, L_y - \pi \, r^2/3) \, \Delta$$

Equating W and E,

$$m + m' = \frac{q}{2\pi} \{ L_x L_y - \frac{\pi r^2}{3} \}$$

8.10.15 Design of a Corner Panel of Floor Slab Using Yield Line Analysis

A square corner panel of a floor slab simply supported on the outer edges on steel beams and continuous over the interior beams is shown in Fig. 8.90.
$$E = 2m \{ 2/(1 - \beta) + 1/\beta \} \, \Delta$$

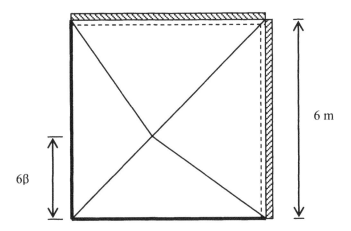

Fig. 8.90 Corner slab.

The design ultimate load is 12.4 kN/m². The slab is to be 175 mm thick and reinforced equally in both directions. The moment of resistance in the hogging and sagging yield lines is to be the same. The materials are $f_{ck} = 30$ MPa concrete and $f_{yk} = 500$ MPa reinforcement. Design the slab using the yield line method.

The yield line pattern, which is symmetric about the diagonal, depends on one variable β. Assuming the deflection at the meeting point of the sagging yield lines as Δ,

(a) Energy dissipated at yield lines

(i) Positive yield lines in bottom triangle
Rotates about x-axis only.
$$\theta_x = \Delta/ (6\beta), \ m_x = m, \ l_x = 6$$
(ii) Positive yield lines in left triangle
Rotates about y-axis only.
$$\theta_y = \Delta/ (6\beta), \quad m_y = m, \ l_y = 6$$
Total energy dissipation E_1 is
$$E_1 = 2\{m \ 6 \ \Delta/ (6\beta)\} = 2m \ \Delta/\beta$$

(iii) Positive and negative yield lines in the right triangle
Note that $m_y = (m+m')$ accounts for both positive and negative yield lines in the triangle. Rotates about y-axis only.
$$\theta_y = \Delta/ (6 - 6 \ \beta), \quad m_y = (m + m'), \ l_y = 6$$
(iv) Positive and negative yield lines in the top triangle
Rotates about x-axis only.
$$\theta_x = \Delta/ (6 - 6 \ \beta), \quad m_x = (m + m'), \ l_x = 6$$
Total energy dissipation E_2 is
$$E_2 = 2(m + m') \ \Delta/ (1 - \beta)$$

The total energy dissipated E by all yield lines is
$$E = E_1 + E_2 = 2\{(m + m')/ (1 - \beta) + m/ \beta\} \, \Delta$$
If $m = m'$, $E = 2m\{2/ (1 - \beta) + 1/ \beta\} \, \Delta$

(b) External work done by the loads

(i) Bottom and left triangles
$$\text{Area} = 0.5 \times 6 \times (6 \, \beta), \text{ deflection} = \Delta/3$$
$$W_1 = 2[q \, \{0.5 \times 6 \times (6 \, \beta)\} \, \Delta/3] = 12q \, \beta \, \Delta$$

(ii) Top and right triangles
$$\text{Area} = 0.5 \times 6 \times (6 - 6 \, \beta), \text{ deflection} = \Delta/3$$
$$W_2 = 2[q \, \{0.5 \times 6 \times (6 - 6 \, \beta)\} \, \Delta/3 = 12q \, (1 - \beta) \, \Delta$$
The total work W done is
$$W = W_1 + W_2 = 12 \, q \, \Delta$$

(c) Calculation of moment capacity required
Equating E and W, and solving for m

$$m = 6q \frac{\beta - \beta^2}{(1 + \beta)}$$

For maximum m, $dm/d\beta = 0$,
$$(1 + \beta) (1 - 2 \, \beta) - (\beta - \beta^2) = 0$$
Simplifying,
$$\beta^2 + 2\beta - 1 = 0, \beta = (\sqrt{2} - 1) = 0.4142$$
Substituting for β, $m = 1.03q$. If $q = 12.4 \text{ kN/m}^2$, $m = 12.76 \text{ kNm/m}$.

(d) Design for flexure
Assuming 10 mm diameter bars and 25 mm cover, the effective depth d of the inner layer is
$$d = 175 - 25 - 10 - 5 = 135 \text{ mm}$$
According to the code equation (5.10a), the ratio of redistributed /elastic moment is
$$\delta \geq 0.44 + 1.25 \, (x_u/d) \tag{5.10a}$$
If Class A steel is used, $\delta \geq 0.8$, $x_u/d \leq 0.29$.
Keeping $x_u/d \leq 0.25$ for approximately 25 percent redistribution,
$$z/d = 1 - 0.4 \, x_u/d = 1 - 0.1 = 0.9$$
$$M_u = f_{cd} \times (0.8x \times b) \times z = 0.12 \, bd^2 \, f_{ck}$$
$$k = m/ (bd^2 \, f_{cu}) = 12.76 \times 10^6/ (1000 \times 135^2 \times 30) = 0.023 < 0.12$$
$$z/d = 0.5(1.0 + \sqrt{(1.0 - 3 \times 0.023)} = 0.98$$
$$x_u/d = 0.05 < 0.25$$
$$z = 0.98 \times 135 = 132 \text{ mm}$$
$$A_s = 12.76 \times 10^6/ (0.87 \times 500 \times 132) = 222 \text{ mm}^2/m$$
Increase the steel area by 10 percent to 245 mm²/m to allow for the formation of corner levers.
Check minimum steel area:

$$f_{ctm} = 0.30 \times f_{ck}^{0.667} = 0.30 \times 25^{0.667} = 2.6 \text{ MPa}$$
$$A_{s, min} = 0.26 \ (f_{ctm}/f_{yk}) \ b_t \ d \geq 0.0013 \ b_t \ d$$
$$A_{s, min} = 0.26 \times (2.6/500) \times 1000 \times 135 \geq 0.0013 \times 1000 \times 135$$
$$A_{s, min} = 183 \text{ mm}^2/\text{m}$$

Maximum spacing is normally restricted to

3 h ≤ 400 mm for principal reinforcement
3.5 h ≤ 450 for secondary reinforcement

Provide 10 mm diameter bars at 300 mm centres to give a steel area of 262 mm^2/m which is greater than the minimum area of reinforcement = 183 mm^2/m.

As a comparison, using the moment coefficients from Table 8.14 for two adjacent edges discontinuous, the moments are:

$$\text{Support moment} = m' = -0.047 \times 12.4 \times 6^2 = -20.98 \text{ kNm/m}$$
$$\text{Span moment} = m = 0.036 \times 12.4 \times 6^2 = 16.07 \text{ kNm/m}$$
$$m'/m = 1.31$$

If this ratio is used in the yield line analysis, then the total energy dissipated by all yield lines is

$$E = E_1 + E_2 = 2\{(m + m')/(1 - \beta) + m/\beta\} \ \Delta$$

If m' = 1.31 m,

$$E = 2 \ m \ \{2.31/(1 - \beta) + 1/\beta\} \ \Delta$$

The total work W done is

$$W = W_1 + W_2 = 12 \ q \ \Delta$$

Equating E and W, and solving for m

$$m = 6q \frac{\beta - \beta^2}{(1 + 1.31\beta)}$$

For maximum m, dm/dβ = 0,

$$(1 + 1.31 \ \beta)(1 - 2 \ \beta) - 1.31 \ (\beta - \beta^2) = 0$$

Simplifying,

$$1.31 \ \beta^2 + 2\beta - 1 = 0, \ \beta = (\sqrt{2.31} - 1) = 0.52$$

Substituting for β, m = 0.89 q. If q = 12.4 kN/m^2, m = 11.05 kNm/m and m' = 14.5 kNm/m.

Compared with the yield line solution which gave m = 12.76 kNm/m, using the moment coefficients gives m = 11.05 kNm.

The cut-off for the top steel can be determined by considering a slab of length L as shown in Fig. 8.91 with no top steel i.e. m' = 0.

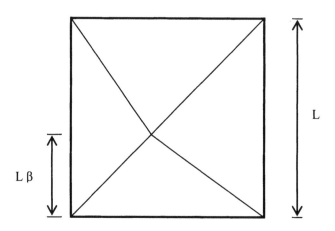

Fig. 8.91 Corner slab with simple supports.

Proceeding as before,

1. The total energy dissipated E by all yield lines is
$$E = 2\{(m + m')/ (1 - \beta) + m/ \beta\}\ \Delta$$
Setting m' = 0,
$$E = 2m\ \{1 / (1 - \beta) + 1/ \beta\}\ \Delta$$
$$= 2m\ \Delta / (\beta - \beta^2)$$

2. External work done by the loads
(i) Bottom and left triangles
$$\text{Area} = 0.5 \times L \times (L\ \beta), \quad \text{deflection} = \Delta/3$$
$$W_1 = 2[q\ \{0.5 \times L \times (L\ \beta)\}\ \Delta/3] = L^2 q\ \beta\ \Delta/3$$

(ii) Top and right triangles
$$\text{Area} = 0.5 \times L\times (L - L\ \beta), \quad \text{deflection} = \Delta/3$$
$$W_2 = 2[q\ \{0.5 \times L \times (L - L\ \beta)\}\ \Delta/3 = L^2 q\ (1 - \beta)\ \Delta/3$$
The total work W done is
$$W = W_1 + W_2 = L^2 q\ \Delta/3$$

3. Calculation of moment capacity required
Equating E and W, and solving for m

$$m = \frac{qL^2}{6}(\beta - \beta^2)$$

For maximum m, $dm/d\beta = 0$, $1 - 2\beta = 0$, $\beta = 0.5$, $m = qL^2/24$.
If m = 12.76 kNm/m and q = 12.4 kN/m^2, then L = 4.97 m.
The top bars can be cut off at (6.0 – 4.97) = 1.03 m from the support. However allowing for an anchorage length of 36 bar diameters, the anchorage length for a H10 bar is 36 × 10 = 360 mm. The top bars can be curtailed at 1.03 + 0.36 = 1.4 m

from the support. Area of steel at top is 10H at 300 mm. Over a length of 1.4 m provide 6H10 at 300 c/c.
The bottom steel is to run into and be continuous through the supports.

At the corner where two simply supported edges meet, torsional steel has to be provided. The steel is provided in four layers, two at top and two at bottom at right angles. The area of steel provided is 75 percent of the steel at mid-span and the bars extend a distance of 20 percent of the shorter span.
Area of steel at mid-span is 262 mm²/m. 75 percent of 262 = 197 mm²/m ≈ 10 mm bars at 400 c/c. The bars extend a distance from the edge of L/5 = 1.2 m. Provide 4H10 spaced at 400 mm.

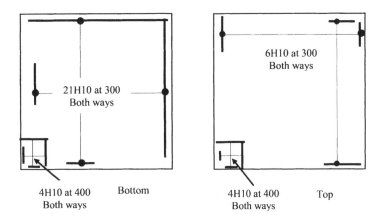

21H10 at 300
Both ways

6H10 at 300
Both ways

4H10 at 400   Bottom   4H10 at 400   Top
Both ways   Both ways

Fig. 8.92 Steel layout for corner slab.

4. Check shear capacity
The shear at the continuous edge is
$$V = 12.4 \times (6/2) + 12.76/6 = 39.3 \text{ kN /m}$$
$$v = 39.3 \times 10^3 / (1000 \times 135) = 0.29 \text{ MPa}$$
$$100 \, \rho_1 = 100 \, A_s/ (bd) = 100 \times 262/ (1000 \times 135) = 0.19 < 2.0$$
$$C_{Rd, c} = 0.18/ (\gamma_c = 1.5) = 0.12, \ k = 1 + (200/d)^{0.5} = 2.2 > 2.0$$
$$v_{Rd,c} = C_{Rd,c} \, k \, (100 \rho_1 \, f_{ck})^{0.33} = 0.12 \times 2.0 \times (0.19 \times 30)^{0.33} = 0.43 \text{ MPa}$$
$$v_{min} = 0.035 \times k^{1.5} \times \sqrt{f_{ck}} = 0.035 \times 2.0^{1.5} \times \sqrt{30} = 0.54 \text{ MPa}$$
$$(v = 0.29) < (v_{Rd, c} = 0.54)$$
No shear reinforcement is needed. The shear stress is satisfactory.

5. Deflection

Bottom steel $A_s = 10H$ at $300 = 262$ mm²/m.
$\rho\% = 100 \times 262/ (1000 \times 135) = 0.19$.
$\rho_0\% = 0.1\sqrt{f_{ck}} = 0.1\sqrt{30} = 0.55$.

k = 1.3 for end span of two-way spanning slab continuous over one long edge.

$$\frac{L}{d} = K\{11 + 1.5\sqrt{f_{ck}}\frac{\rho_0}{\rho} + 3.2\sqrt{f_{ck}}(\frac{\rho_0}{\rho} - 1)^{\frac{3}{2}}\} \text{ if } \rho \le \rho_0$$

(7.16a)

$$\frac{L}{d} = 1.2 \times \{11 + 1.5\sqrt{30}\frac{0.55}{0.19} + 3.2\sqrt{30}(\frac{0.55}{0.19} - 1)^{\frac{3}{2}}\} = 96.6$$

Allowable span/d ratio = 96.6
Actual span/d ratio = 6000/135 = 44.4
Hence the slab is satisfactory with respect to deflection.

6. Cracking
The minimum clear distance between bars is not to exceed 3h = 525 mm.

7. Reinforcement details
The reinforcement is shown in Fig. 8.92.   Note that at the corners torsion reinforcement is better provided as U-bars rather than as individual bars as indicated. Note that if the slab was supported on reinforced concrete L-beams on the outer edges as opposed to say steel I-beams, a value for the ultimate negative resistance moment at these edges could be assumed and used in the analysis.

8.10.16 Derivation of Moment and Shear Coefficients for the Design of Restrained Slabs

Bending moment and shear force coefficients in Tables 3.14 and 3.15 of the code for the design of two-way restrained slabs with corners held down and with provision for resisting torsion are derived on the basis of yield line analysis. The ratio of negative moment to positive moment is kept constant at 1.33. The 'long span' moments derived for a square slab are assumed to hold good for other values of the aspect ratio. Yield line analysis assumes that reinforcement in each direction is uniformly distributed over the width but the code recommends that the main steel is provided only in the middle strip which is 3/4 times the relevant width and only minimum steel in the edge strips. Therefore the value obtained from the yield line analysis is multiplied by 4/3. The shear in the slab is calculated by assuming that the total load on the support is uniformly distributed over the middle three quarters of the beam span.

8.10.16.1 Simply Supported Slab

Using the formulae derived in section 8.9.6,

$$2m\frac{b}{a}\frac{\mu}{\beta} + 4m\frac{a}{b} = q\frac{ab}{6}(3 - 2\beta)$$

Multiplying through by (b/a) and simplifying, the above equation becomes

$$m[\frac{b}{a}]^2 \frac{\mu}{\beta} + 2m = q\frac{b^2}{12}(3-2\beta)$$

(a) Square slab
a/b = 1, $\mu m = m$, $\beta = 0.5$, $m = 0.0417 \, qb^2$.
Multiplying this value for m by 4/3, $m = 0.056 \, qb^2$, $\beta_{sx} = 0.056$.

(b) Rectangular slab
a/b > 1.0.
Keep $\mu m = 0.0417 \, qb^2$ as constant for all values of a/b. Substituting this value in the above equation for m and simplifying,

$$m = q\,b^2\,\{0.125 - 0.0833\beta - 0.0209(\frac{b}{a})^2\,\frac{1}{\beta}\}$$

For a maximum value of m, $dm/d\beta = 0$.

$$-0.0833 + 0.0209(\frac{b}{a})^2\,\frac{1}{\beta^2} = 0, \; \beta = 0.5\frac{b}{a}$$

Substituting this value of β in the equation for m

$$m = q\,b^2\,\{0.125 - 0.0833(\frac{b}{a})\}$$

Multiplying this value for m by 4/3,

$$4/3 \, m = q\,b^2\,\{0.1667 - 0.1111(\frac{b}{a})\}, \beta_{sx} = \{0.1667 - 0.1111(\frac{b}{a})\}$$

$$\mu m = 0.0417 \, qb^2$$

Multiplying this value for μm by 4/3,

$$\mu m = 0.0556 \, qb^2, \; \beta_{sy} = 0.056$$

(c) Shear coefficients

$$\beta = 0.50 \, b/a$$

Short beam:

$$\text{Load} = 0.5 \times q \times b \times \beta a.$$

Spreading this uniformly over a length of 0.75 b,

$$v_x = qb\{0.6667\beta\frac{a}{b}\} = 0.333qb, \beta_{vy} = 0.3333$$

Long beam:

$$\text{Load} = 0.5 \times q \times 0.5b \times (2 - 2\beta) \, a.$$

Spreading this uniformly over a length of 0.75 a,

$$v_y = qb\{0.6667(1-\beta)\} = qb\{0.6667(1-0.5\frac{b}{a})\}, \; \beta_{vx} = \{0.6667(1-0.5\frac{b}{a})$$

8.10.16.2 Clamped Slab

Using the formula derived in section 8.9.7,

$$2(\frac{b}{a}\frac{\mu m}{\beta} + 2\frac{a}{b}m) + 2m'(\frac{b}{a}\frac{\gamma m'}{\beta} + 2\frac{a}{b}m') = q\frac{ab}{6}(3-2\beta)$$

Multiplying through by b/a and simplifying, the above equation becomes

$$[\frac{b}{a}]^2\frac{(\mu m + \gamma m')}{\beta} + 2(m+m') = q\frac{b^2}{12}(3-2\beta)$$

(a) Square slab

a/b = 1, μm = m, γm' = m' = 1.33 m, β = 0.5
m = 0.0179 qb², m' = 0.024 qb²

Multiplying these value for μm by 4/3,
m = 0.024 qb², β_{sx} = 0.024 and m' = 0.032 qb², β_{sy} = 0.032

(b) Rectangular slab

a/b > 1.0. Keep μm = 0.0179 qb², γm' = 0.024 qb² and m' = 1.33m as constant for all values of a/b. Substituting these values in the above equation for m and simplifying,

$$m = q b^2 \{0.0536 - 0.0357\beta - 0.009(\frac{b}{a})^2\frac{1}{\beta}\}$$

For a maximum value of m, dm/dβ = 0.

$$-0.0357 + 0.009(\frac{b}{a})^2\frac{1}{\beta^2} = 0, \ \beta = 0.502\frac{b}{a}$$

Substituting this value of β in the equation for m

$$m = q b^2 \{0.0536 - 0.0359(\frac{b}{a})\}$$

Multiplying this value for μm by 4/3,

$$m = q b^2 \{0.0715 - 0.0479(\frac{b}{a})\}$$

$$m' = 1.33m = q b^2 \{0.0953 - 0.0639(\frac{b}{a})\}$$

For positive moment at mid-span in short span

$$\beta_{sx} = \{0.0715 - 0.0479(\frac{b}{a})\}$$

Negative moment at short edge

$$\beta_{sx} = \{0.0953 - 0.0639(\frac{b}{a})\}$$

μm = 0.0179 qb², γm' = 0.024 qb²

Multiplying the above values by 4/3
μm = 0.0239 qb², γm' = 0.032 qb²

In the long span direction, coefficients for positive and negative moments are respectively

$$\beta_{sy} = 0.0239 \text{ and } \beta_{sy} = 0.032$$

(c) Shear coefficients
Follow the procedure for the simply supported slab in section 8.9.16.1.

8.10.16.3 Slab with Two Discontinuous Short Edges

Fig. 8.93 shows a slab with two discontinuous short edges.

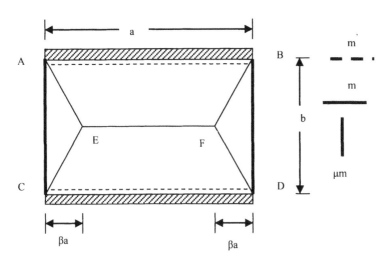

Fig. 8.93 Collapse mode for a slab with two discontinuous short edges.

Using the formulae derived in section 8.8.16.2 but with $\gamma m' = 0$,

$$[\frac{b}{a}]^2 \frac{\mu m}{\beta} + 2(m+m') = q\frac{b^2}{12}(3-2\beta)$$

(a) Square slab
$$a/b = 1, \quad \mu m = m, \quad m' = 1.33\ m.$$
For maximum m, $dm/d\beta = 0$
$$\beta^2 + 0.4286\beta - 0.3214 = 0, \quad \beta = 0.3918$$
Using $\beta = 0.3918$, $m = 0.026\ qb^2$, $m' = 0.034\ qb^2$.
Multiplying these values by 4/3,
$$m = 0.034\ qb^2, \quad \beta_{sx} = 0.034 \text{ and } m' = 0.046\ qb^2, \quad \beta_{sy} = 0.046$$

(b) Rectangular slab
$a/b > 1.0$. Keep $\mu m = 0.026\ qb^2$ and $m' = 1.33\ m$ as constant for all values of a/b.
Substituting these values in the equation for m and simplifying,
$$m = qb^2\{0.0536 - 0.0357\beta - 0.0056(\frac{b}{a})^2\frac{1}{\beta}\}$$

For a maximum value of m, $dm/d\beta = 0$.

$$-0.0357 + 0.0056(\frac{b}{a})^2\frac{1}{\beta^2} = 0, \ \beta = 0.3950\frac{b}{a}$$

Substituting this value of β in the equation for m

$$m = q\,b^2\{0.0536 - 0.0282(\frac{b}{a})\}$$

Multiplying the above value by 4/3,

$$m = q\,b^2\{0.0715 - 0.0376(\frac{b}{a})\}\,, \quad m'=1.333\,m = q\,b^2\{0.0953 - 0.0501(\frac{b}{a})\}$$

Short span

For positive moment: $\beta_{sx} = \{0.0715 - 0.0376(\frac{b}{a})\}$

Negative moment at short edge: $\beta_{sx} = \{0.0953 - 0.0501(\frac{b}{a})\}$

$$\mu m = 0.0260 \ qb^2$$

Multiplying the above values by 4/3, $\mu m = 0.0347 \ qb^2$

Long span

For positive moment at mid-span: $\beta_{sy} = 0.0347$

(c) Shear coefficients

Proceed as for the simply supported slab but use:

$$\beta = 0.3918, \ a/b = 1$$
$$\beta = 0.3950 \ b/a, \ a/b > 1.0.$$

$$v_x = qb\{0.6667\beta\frac{a}{b}\}$$

$$v_x = 0.2633\,qb, \beta_{vy} = 0.2633, b/a < 1.0$$

$$v_x = 0.2612\,qb, \beta_{vy} = 0.2612, b/a = 1.0$$

$$v_y = qb\{0.6667(1-\beta)\}$$

$$v_y = qb\{0.6667(1-0.3950\frac{b}{a})\}, \ \beta_{vy} = \{0.6667(1-0.3950\frac{b}{a}), b/a < 1.0$$

$$v_y = qb\{0.6667(1-0.3918)\}, \ \beta_{vy} = 0.4055, b/a = 1.0$$

8.10.16.4 Slab with Two Discontinuous Long Edges

Fig. 8.94 shows a slab with two discontinuous long edges. Using the formulae derived in section 8.9.16.2 and substituting m' = 0,

$$[\frac{b}{a}]^2\frac{(\mu m + \gamma m')}{\beta} + 2m = q\frac{b^2}{12}(3-2\beta)$$

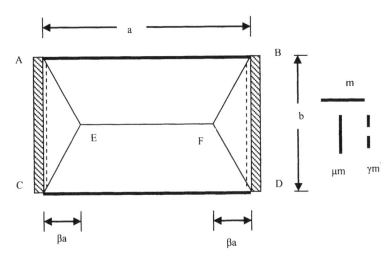

Fig. 8.94 Collapse mode for slab with two discontinuous long edges.

(a) Square slab

$a/b = 1$, $\mu m = m$, $\gamma m' = 1.33$ m.

For maximum, $dm/\beta = 0$.

$$\beta^2 + 2.333\beta - 1.75 = 0, \beta = 0.5972 > 0.5$$

Restrict $\beta = 0.5$ as this is the maximum value permissible. Using $\beta = 0.5$,
$m = 0.025$ qb^2, $\gamma m' = 0.033$ qb^2.

Multiplying these values by 4/3,

$$m = \mu m = 0.033 \text{ qb}^2, \quad \beta_{sx} = 0.033 \text{ and } \gamma m' = 0.044 \text{ qb}^2, \quad \beta_{sy} = 0.044$$

(b) Rectangular slab

$a/b > 1.0$. Keep $\mu m = 0.025$ qb^2 and $\gamma m' = 0.033$ qb^2 as constant for all values of
a/b. Substituting these values in the equation for m and simplifying,

$$m = q b^2 \{0.125 - 0.0833\beta - 0.029(\frac{b}{a})^2 \frac{1}{\beta}\}$$

For a maximum value of m, $dm/d\beta = 0$.

$$-0.0833 + 0.029(\frac{b}{a})^2 \frac{1}{\beta^2} = 0, \beta = 0.59\frac{b}{a}$$

Substituting this value of β in the equation for m and simplifying,

$$m = q b^2 \{0.125 - 0.098(\frac{b}{a})\}$$

Multiplying the above value by 4/3, $m = q b^2 \{0.1667 - 0.1307(\frac{b}{a})\}$

For positive moment at mid-span in short span

$$\beta_{sx} = \{0.1667 - 0.1307(\frac{b}{a})\}, b/a < 1.0, \quad \beta_{sx} = 0.033, b/a = 1.0$$

Long span direction

$\mu m = 0.033$ qb^2, $\gamma m' = 0.044$ qb^2

Positive moment: $\mu m = 0.033\, qb^2$, $\beta_{sy} = 0.033$.

Negative moment: $\gamma m' = 0.044\, qb^2$, $\beta_{sy} = 0.044$.

(c) Shear coefficients

Proceed as for the simply supported slab but use:

$\beta = 0.5$ for $a/b = 1$ and for $a/b > 1.0$, $\beta = 0.59\, b/a$

$$v_x = qb\{0.6667\beta\frac{a}{b}\}$$

$$v_x = 0.3933\, qb, \beta_{vy} = 0.3933, b/a < 1.0$$

$$v_x = 0.3333\, qb, \beta_{vy} = 0.3333, b/a = 1.0$$

$$v_y = qb\{0.6667(1-\beta)\}$$

$$v_y = qb\{0.6667(1-0.59\frac{b}{a})\},\ \beta_{vy} = \{0.6667(1-0.59\frac{b}{a}), b/a < 1.0$$

$$v_y = qb\{0.6667(1-0.5)\},\ \beta_{vy} = 0.3333, b/a = 1.0$$

8.10.16.5 Slab with One Discontinuous Long Edge

Fig. 8.95 shows the collapse mode which is governed by two parameters β_1 and β_2.

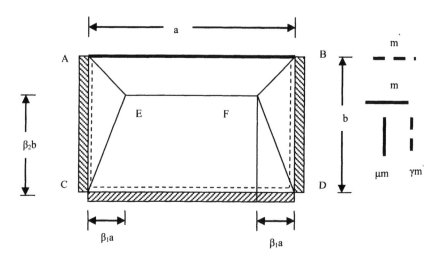

Fig. 8.95 Collapse mode for a slab with one discontinuous long edge.

It can be shown that the basic equation for solving the problem is

$$2[\frac{b}{a}]^2\frac{(\mu m + \gamma m')}{\beta_1} + \frac{(m+m')}{\beta_2} + \frac{m}{(1-\beta_2)} = q\frac{b^2}{6}(3-2\beta_1)$$

(a) Square slab

$a/b = 1$, $\mu m = m$, $\gamma m' = m' = 1.33$ m.

$$m\{\frac{4.6667}{\beta_1} + \frac{2.333}{\beta_2} + \frac{1}{(1-\beta_2)}\} = q\frac{b^2}{6}(3-2\beta_1)$$

For maximum m, $dm/d\beta_1 = 0$ and $dm/d\beta_2 = 0$.

$\quad\quad dm/d\beta_2 = 0$ leads to $\beta_2^2 - 3.5\beta_2 + 1.75 = 0$, $\beta_2 = 0.604 < 1.0$

Value of $\beta_2 = 0.604$ is independent of β_1. Using this value of β_2,

$\quad\quad dm/d\beta_1 = 0$ leads to $\beta_1^2 + 1.4612\beta_1 - 1.0959 = 0$, $\beta_1 = 0.546 > 0.5$

Using $\beta_1 = 0.5$ and $\beta_2 = 0.604$, $m = \mu m = 0.020$ qb^2, $m' = \gamma m' = 0.027$ qb^2.
Multiplying these values by 4/3,

$\quad\quad m = \gamma m' = 0.027$ qb^2, $\beta_{sx} = 0.027$ and $m' = \gamma m' = 0.036$ qb^2, $\beta_{sy} = 0.036$.

(b) Rectangular slab

$a/b > 1.0$. Keep $\mu m = 0.020$ qb^2 and $\gamma m' = 0.027$ qb^2 as constant for all values of a/b and $m' = 1.33$ m. Substituting these values in the equation for m and simplifying,

$$m\{\frac{2.333}{\beta_2} + \frac{1}{(1-\beta_2)}\} = qb^2\{0.5 - 0.333\beta_1 - \frac{0.094}{\beta_1}(\frac{b}{a})^2\}$$

$\quad\quad dm/d\beta_2 = 0$ leads to $\beta_2^2 - 3.5\beta_2 + 1.75 = 0$, $\beta_2 = 0.604 < 1.0$

$\quad\quad dm/d\beta_1 = 0$ leads to $-0.333 + 0.094(\frac{b}{a})^2\frac{1}{\beta_1^2} = 0$, $\beta_1 = 0.531\frac{b}{a}$

Substituting these values of β_1 and β_2 in the equation for m and simplifying,

$$m = qb^2\{0.0783 - 0.0554(\frac{b}{a})\}$$

Multiplying the above value by 4/3,

$$m = qb^2\{0.1044 - 0.0739(\frac{b}{a})\}, \quad m' = 1.33m = qb^2\{0.1392 - 0.0985(\frac{b}{a})\}$$

Short span

For positive moment

$$m = qb^2\{0.1044 - 0.0739(\frac{b}{a})\}, \quad \beta_{sx} = \{0.1044 - 0.0739(\frac{b}{a})\}, b/a < 1$$

$$m = qb^2 0.027, \quad \beta_{sx} = 0.027, b/a = 1$$

Negative moment at short edge

$$m' = 1.33m = qb^2\{0.1392 - 0.0985(\frac{b}{a})\}, \quad \beta_{sx} = \{0.1392 - 0.0985(\frac{b}{a})\}, b/a < 1$$

$$m' = qb^2 0.036, \quad \beta_{sx} = 0.036, b/a = 1$$

Long span

$\mu m = 0.027$ qb^2 and $\gamma m' = 0.036$ qb^2

For positive and negative moments, the moment coefficients β_{sy} are 0.027 and 0.036 respectively.

(c) Shear coefficients

Use and $\beta_1 = 0.5$ for a/b = 1 and for a/b > 1.0, use $\beta_1 = 0.531$ b/a. $\beta_2 = 0.604$ for all aspect ratios.

Short beam: Spread the load uniformly over a length of 0.75 b.

$$\text{Load} = 0.5 \times q \times b \times \beta_1 a.$$

$$v_x = qb\{0.6667\beta_1 \frac{a}{b}\}$$

$$v_x = 0.354qb, \beta_{vy} = 0.354, b/a < 1.0$$

$$v_x = 0.3333\,qb, \beta_{vy} = 0.3333, b/a = 1.0$$

Load on the longer beam: Spread the load on the beam uniformly over a length of 0.75 a and use $\beta_2 = 0.604$.

Continuous end

$$v_y = 0.5 \times q \times \beta_2 b \times (2 - 2\beta_1)\, a$$

$$v_y = qb\{0.805(1 - \beta_1)\}$$

$$v_y = qb\{0.805(1 - 0.531\frac{b}{a})\},\ \beta_{vy} = \{0.805(1 - 0.531\frac{b}{a}),\ b/a < 1.0$$

$$v_y = qb\{0.805(1 - 0.5)\},\ \beta_{vy} = 0.403, b/a = 1.0$$

Simply supported end

$$v_y = 0.5 \times q \times (1 - \beta_2)\, b \times (2 - 2\beta_1)\, a$$

$$v_y = qb\{0.528(1 - \beta_1)\}$$

$$v_y = qb\{0.528(1 - 0.531\frac{b}{a})\},\ \beta_{vy} = \{0.528(1 - 0.531\frac{b}{a}),\ b/a < 1.0$$

$$v_y = qb\{0.528(1 - 0.5)\},\ \beta_{vy} = 0.264, b/a = 1.0$$

8.10.16.6 Slab with One Discontinuous Short Edge

Fig. 8.96 shows the collapse mode which is governed by two parameters β_1 and β_2. It can be shown that the basic equation for solving the problem is

$$[\frac{b}{a}]^2 \{\frac{(\mu m + \gamma m')}{\beta_2} + \frac{\mu m}{\beta_1}\} + 4(m + m') = q\frac{b^2}{6}(3 - \beta_1 - \beta_2)$$

(a) Square slab

$$a/b = 1,\ \mu m = m,\ \gamma m' = m' = 1.33\ m.$$

$$m\{9.333 + \frac{1}{\beta_1} + \frac{2.333}{\beta_2}\} = q\frac{b^2}{6}(3 - \beta_1 - \beta_2)$$

Simplifying

$$m = \frac{qb^2}{6}\beta_1\beta_2 \frac{(3 - \beta_1 - \beta_2)}{(9.333\beta_1\beta_2 + 2.333\beta_1 + \beta_2)}$$

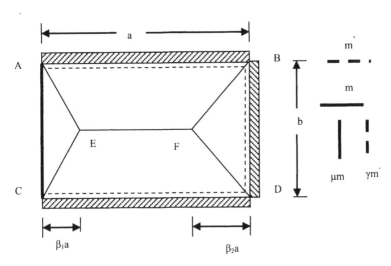

Fig. 8.96 Collapse mode for a slab with one discontinuous short edge.

For maximum m, $dm/d\beta_1 = 0$ and $dm/d\beta_2 = 0$.
$dm/d\beta_1 = 0$ leads to

$$A_1 \beta_1^2 + B_1 \beta_1 + C_1 = 0,$$
$$A_1 = 1 + 8\beta_2,$$
$$B_1 = -4\beta_2^2 + 0.85714\beta_2,$$
$$C_1 = 0.4286\beta_2^2 - 1.2857\beta_2$$

$dm/d\beta_2 = 0$ leads to

$$A_2 \beta_2^2 + B_2 \beta_2 + C_2 = 0,$$
$$A_2 = 1.0 + 9.333\beta_1,$$
$$B_2 = 4.6667\beta_1,$$
$$C_2 = 2.33\beta_1^2 - 7\beta_1$$

The values of β_1 and β_2 can be calculated as follows:
- Assume a value for β_2
- Calculate A_1, B_1 and C_1
- Solve the quadratic in β_1
- Using the calculated value of β_1, calculate A_2, B_2 and C_2
- Solve the quadratic in β_2

- Compare the assumed and calculated values of β_2
- Repeat calculations until the assumed and calculated values of β_2 differ by a very small value
- Calculate the value of m/ (qb^2)

Using this procedure, in this case
$$\beta_1 = 0.4013 \text{ and } \beta_2 = 0.5455, \text{ m} = 0.0213 \text{ qb}^2, \text{ m}' = 0.0284 \text{ qb}^2.$$
Multiplying these values by 4/3,
$$\text{m} = \mu\text{m} = 0.028 \text{ qb}^2, \ \beta_{sx} = 0.028 \text{ and m}' = \gamma\text{m}' = 0.038 \text{ qb}^2, \ \beta_{sx} = 0.038.$$

(b) Rectangular slab
a/b > 1.0. Keep $\overset{'}{m} = 1.33$ m, $\mu\text{m} = 0.021 \text{ qb}^2$ and $\gamma\overset{'}{m} = 0.028 \text{ qb}^2$ as constant for all values of a/b. Substituting these values in the equation for m and simplifying,
$$m = qb^2 \{0.0536 - 0.01786\beta_1 - \frac{0.00228}{\beta_1}(\frac{b}{a})^2 - 0.01786\beta_2 - \frac{0.00533}{\beta_2}(\frac{b}{a})^2\}$$

$dm/d\beta_1 = 0$ leads to
$$-0.01786 + \frac{0.00228}{\beta_1^2}(\frac{b}{a})^2 = 0, \ \beta_1 = 0.3575\frac{b}{a}$$

$dm/d\beta_2 = 0$ leads to
$$-0.01786 + \frac{0.00533}{\beta_2^2}(\frac{b}{a})^2 = 0, \ \beta_2 = 0.5460\frac{b}{a}$$

Substituting these values of β_1 and β_2 in the equation for m and simplifying,
$$m = qb^2 \{0.0536 - 0.0323(\frac{b}{a})\}$$

Multiplying the above value by 4/3, $m = qb^2 \{0.0715 - 0.0431(\frac{b}{a})\}$

Short span
Positive moment
$$m = qb^2 \{0.0715 - 0.0431(\frac{b}{a})\}, \ \beta_{sx} = \{0.0715 - 0.0431(\frac{b}{a})\}, b/a < 1$$
$$m = qb^2 0.028, \ \beta_{sx} = 0.028, b/a = 1$$

Negative moment at short edge
$$\overset{'}{m} = 1.33 m = qb^2 \{0.0953 - 0.0575(\frac{b}{a})\}, \ \beta_{sx} = \{0.0953 - 0.0575(\frac{b}{a})\}, b/a < 1$$
$$\overset{'}{m} = qb^2 0.038, \ \beta_{sx} = 0.038, b/a = 1$$

Long span
$$\mu\text{m} = 0.028 \text{ qb}^2 \text{ and } \gamma\text{m}' = 0.038 \text{ qb}^2$$
For positive and negative moments, the moment coefficients are respectively.
$$\beta_{sy} = 0.028 \text{ and } \beta_{sy} = 0.038$$

(c) Shear coefficients
Spread the load uniformly over a length of 0.75 b.
$$a/b = 1: \beta_1 = 0.4013 \text{ and } \beta_2 = 0.5455$$
$$a/b > 1.0: \beta_1 = 0.3575 \, b/a \text{ and } \beta_2 = 0.546 \, b/a$$

Load on the shorter beam
Continuous end
$$\text{Load} = 0.5 \times q \times b \times \beta_2 \, a.$$
$$v_x = qb\{0.6667\beta_2 \frac{a}{b}\} = qb\{0.364\}, \beta_{vx} = 0.364$$

Simply supported end
$$\text{Load} = 0.5 \times q \times b \times \beta_1 \, a.$$
$$v_x = qb\{0.6667\beta_1 \frac{a}{b}\} = qb\{0.1589\}, \beta_{vx} = 0.1589, b/a < 1$$

$$v_x = qb\{0.6667\beta_1 \frac{a}{b}\} = qb\{0.2675\}, \beta_{vx} = 0.2675, b/a = 1$$

Load on the longer beam
Spread the load uniformly over a length of 0.75 a.
$$\text{Load} = 0.5 \times q \times 0.5b \times (2 - \beta_1 - \beta_2) a$$
$$v_y = qb\{0.3333(2 - \beta_1 - \beta_2\}$$

$$v_y = qb\{0.333 \times (2 - 0.9035\frac{b}{a})\}, \; \beta_{vy} = \{(0.667 - 0.301\frac{b}{a}), b/a < 1.0$$
$$v_y = qb\{0.333(2 - 0.3575 - 0.546)\}, \; \beta_{vy} = 0.3655, b/a = 1.0$$

8.10.16.7 Slab with Two Adjacent Discontinuous Edges

Fig. 8.97 shows the collapse mode which is governed by three parameters β_1, β_2 and β_3.
It can be shown that the basic equation for solving the problem is

$$[\frac{b}{a}]^2 \{\frac{(\mu m + \gamma m')}{\beta_2} + \frac{\mu m}{\beta_1}\} + (m + m')\frac{1}{\beta_3} + m\frac{1}{(1 - \beta_3)} = q\frac{b^2}{6}(3 - \beta_1 - \beta_2)$$

(a) Square slab
a/b = 1, $\mu m = m$, $\gamma m' = m' = 1.33 \, m$.
$$m\{\frac{1}{\beta_1} + \frac{2.333}{\beta_2} + \frac{2.333}{\beta_3} + \frac{1}{(1 - \beta_3)}\} = q\frac{b^2}{6}(3 - \beta_1 - \beta_2)$$
For maximum m, $dm/d\beta_1 = 0$, $dm/d\beta_2 = 0$ and $dm/d\beta_3 = 0$.
$dm/d\beta_3 = 0$ leads to
$$-\frac{2.333}{\beta_3^2} + \frac{1}{(1 - \beta_3)^2} = 0, \; \beta_3^2 - 3.5\beta_3 + 1.75 = 0, \beta_3 = 0.604$$
Substituting this value of β_3 in the expression for m and simplifying,

$$m = \frac{qb^2}{6} \beta_1 \beta_2 \frac{(3 - \beta_1 - \beta_2)}{(6.3884\beta_1\beta_2 + 2.333\beta_1 + \beta_2)}$$

For maximum m, $dm/d\beta_1 = 0$ and $dm/d\beta_2 = 0$.
$dm/d\beta_1 = 0$ leads to

$$A_1 \beta_1^2 + B_1 \beta_1 + C_1 = 0,$$
$$A_1 = 6.3384 \beta_2 + 2.3333,$$
$$B_1 = 2\beta_2,$$
$$C_1 = \beta_2^2 - 3\beta_2$$

$dm/d\beta_2 = 0$ leads to

$$A_2 \beta_2^2 + B_2 \beta_2 + C_2 = 0,$$
$$A_2 = 6.3384 \beta_1 + 1,$$
$$B_2 = 4.6667\beta_1,$$
$$C_2 = 2.33\beta_1^2 - 7\beta_1$$

The values of β_1 and β_2 can be calculated following the same procedure as in section 8.9.16.7.
In this case $\beta_1 = 0.39565$ and $\beta_2 = 0.64356$, m = 0.0261 qb², m' = 0.0348 qb².
Multiplying these values by 4/3,

m = μm = 0.035 qb², β_{sx} = 0.035 and m' = γm' = 0.046 qb², β_{sx} = 0.046

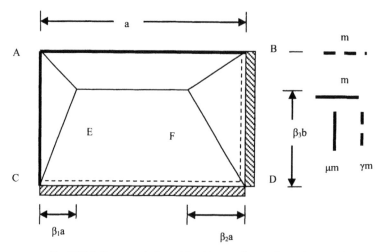

Fig. 8.97 Collapse mode for a slab with two adjacent discontinuous edges.

(b) Rectangular slab

a/b > 1.0. Keep μm = 0.0261 qb² and γm' = 0.0348 qb² as constant for all values of a/b. Substituting these values in the equation for m and simplifying,

$$m = qb^2 \{0.0783 - 0.0261\beta_1 - \frac{0.00407}{\beta_1}(\frac{b}{a})^2 - 0.0261\beta_2 - \frac{0.00955}{\beta_2}(\frac{b}{a})^2\}$$

$dm/d\beta_1 = 0$ leads to

$$-0.0261 + \frac{0.00407}{\beta_1^2}(\frac{b}{a})^2 = 0, \quad \beta_1 = 0.3949\frac{b}{a}$$

$dm/d\beta_1 = 0$ leads to

$$-0.0261 + \frac{0.00955}{\beta_2^2}(\frac{b}{a})^2 = 0, \quad \beta_2 = 0.6049\frac{b}{a}$$

Substituting these values of β_1 and β_2 in the equation for m and simplifying,

$$m = q b^2 \{0.0783 - 0.0522(\frac{b}{a})\}$$

Multiplying the above value by 4/3,

$$m = q b^2 \{0.1044 - 0.0696(\frac{b}{a})\}$$

Short span

Positive moment

$$m = q b^2 \{0.1044 - 0.0696(\frac{b}{a})\}, \quad \beta_{sx} = \{0.1044 - 0.0696(\frac{b}{a})\}, b/a < 1$$

$$m = q b^2 0.035, \quad \beta_{sx} = 0.035, b/a = 1$$

Negative moment at short edge

$$m' = 1.33 m = q b^2 \{0.1392 - 0.0928(\frac{b}{a})\}, \quad \beta_{sx} = \{0.1392 - 0.0928(\frac{b}{a})\}, b/a < 1$$

$$m' = 1.333 \times q b^2 0.035, \quad \beta_{sx} = 0.047, b/a = 1$$

Long span

$$\mu m = 0.035\ qb^2 \text{ and } \gamma m' = 0.046\ qb^2$$

For positive and negative moments, the moment coefficients β_{sy} are 0.035 and 0.046 respectively.

(c) Shear coefficients

$$a/b = 1, \beta_1 = 0.3957, \beta_2 = 0.6436$$
$$a/b > 1.0, \beta_1 = 0.395\ b/a, \beta_2 = 0.605\ b/a.$$

Shorter beam: Spread the load uniformly over a length of 0.75 b.
Continuous end

$$\text{Load} = 0.5 \times q \times b \times \beta_2\ a.$$

$$v_x = qb\{0.6667\beta_2\frac{a}{b}\}$$

$$v_x = qb\{0.6667\beta_2\frac{a}{b}\} = qb(0.403), \beta_{vx} = 0.403, b/a < 1.0$$

$$v_x = qb\{0.6667\beta_2\frac{a}{b}\} = qb(0.4291), \beta_{vx} = 0.4291, b/a = 1.0$$

Simply supported end

$$\text{Load} = 0.5 \times q \times b \times \beta_1 \, a.$$

$$v_x = qb\{0.6667\beta_1\frac{a}{b}\} = qb\{0.264\}, \beta_{vx} = 0.264$$

Longer beam: Spread the load uniformly over a length of 0.75 a, and $\beta_3 = 0.604$.

Continuous end

$$\text{Load} = 0.5 \times q \times \beta_3 \, b \times (2 - \beta_1 - \beta_2) \, a$$
$$v_y = qb\{0.4027(2 - \beta_1 - \beta_2)\}$$

$$v_y = qb\{0.4027(2 - 1.0007\frac{b}{a}\}, \beta_{vy} = 0.4027(2 - 1.0007\frac{b}{a}), b/a < 1.0$$
$$v_y = qb\{0.4027(2 - 1.0393)\}, \beta_{vy} = 0.3869, b/a = 1.0$$

Simply supported end

$$\text{Load} = 0.5 \times q \times (1 - \beta_3) \, b \times (2 - \beta_1 - \beta_2) \, a$$
$$v_y = qb\{0.2640(2 - \beta_1 - \beta_2)\}$$

$$v_y = qb\{0.2640(2 - 1.0007\frac{b}{a}\}, \beta_{vy} = 0.2640(2 - 1.0007\frac{b}{a}), b/a < 1.0$$
$$v_y = qb\{0.4027(2 - 1.0393)\}, \beta_{vy} = 0.3869, b/a = 1.0$$

8.10.16.8 Slab with Only a Continuous Short Edge

Fig. 8.98 shows the collapse mode which is governed by two parameters β_1 and β_2. It can be shown that the basic equation for solving the problem is

$$[\frac{b}{a}]^2\{\frac{(\mu m + \gamma m')}{\beta_2} + \frac{\mu m}{\beta_1}\} + 4m = q\frac{b^2}{6}(3 - \beta_1 - \beta_2)$$

(a) Square slab

a/b = 1, μm = m, γm' = 1.33 m.

$$m\{\frac{1}{\beta_1} + \frac{2.333}{\beta_2} + 4\} = q\frac{b^2}{6}(3 - \beta_1 - \beta_2)$$

Simplifying,

$$m = \frac{qb^2}{6}\beta_1\beta_2\frac{(3 - \beta_1 - \beta_2)}{(4\beta_1\beta_2 + 2.333\beta_1 + \beta_2)}$$

For maximum m, dm/dβ₁ = 0 and dm/dβ₂ = 0.
dm/dβ₁ = 0 leads to

$$A_1 \beta_1^2 + B_1 \beta_1 + C_1 = 0,$$
$$A_1 = 4\beta_2 + 2.3333,$$
$$B_1 = 6\beta_2,$$
$$C_1 = \beta_2^2 - 3\beta_2$$

dm/dβ₂ = 0 leads to

$$A_2 \beta_2^2 + B_2 \beta_2 + C_2 = 0,$$
$$A_2 = 4\beta_1 + 1,$$
$$B_2 = 4.6667\beta_1,$$
$$C_2 = 2.33\beta_1^2 - 7\beta_1$$

The values of β_1 and β_2 can be calculated following the same procedure as in section 8.9.16.6. In this case $\beta_1 = 0.2868$ and $\beta_2 = 0.65934$, m = 0.0311 qb², m' = 0.0414 qb². Multiplying these values by 4/3,
m = μm = 0.042 qb², β_{sx} = 0.042 and γm' = 0.055 qb², β_{sx} = 0.055.

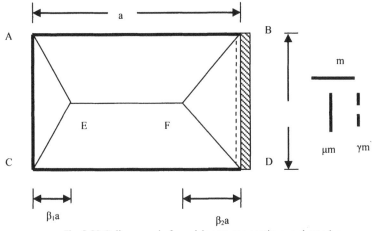

Fig. 8.98 Collapse mode for a slab with one continuous short edge.

(b) Rectangular slab
a/b > 1.0. Keep μm = 0.031 qb² and γm' = 0.041 qb² as constant for all values of a/b. Substituting these values in the equation for m and simplifying,

$$m = qb^2 \{0.125 - 0.0417\beta_1 - \frac{0.00776}{\beta_1}(\frac{b}{a})^2 - 0.0417\beta_2 - \frac{0.01811}{\beta_2}(\frac{b}{a})^2\}$$

dm/dβ₁ = 0 leads to

$$-0.0417 + \frac{0.00776}{\beta_1^2}(\frac{b}{a})^2 = 0, \quad \beta_1 = 0.4314\frac{b}{a}$$

dm/dβ₂ = 0 leads to

$$-0.0417 + \frac{0.01811}{\beta_2^2}(\frac{b}{a})^2 = 0, \ \ \beta_2 = 0.659\frac{b}{a}$$

Substituting these values of β_1 and β_2 in the equation for m and simplifying,

$$m = qb^2\{0.125 - 0.0909(\frac{b}{a})\}$$

Multiplying the above value by 4/3,

$$m = qb^2\{0.1667 - 0.1212(\frac{b}{a})\}$$

Short span
Positive moment

$$m = qb^2\{0.1667 - 0.1212(\frac{b}{a})\}, \beta_{sx} = \{0.1667 - 0.1212(\frac{b}{a})\}, b/a < 1,$$

$$m = qb^2 0.041, \beta_{sx} = 0.041, a/b = 1$$

Long span
$\mu m = 0.041 \ qb^2$ and $\gamma m' = 0.055 \ qb^2$
For positive and negative moments, the moment coefficients β_{sy} are 0.041 and 0.055 respectively.

(c) Shear coefficients

$$a/b = 1: \beta_1 = 0.2868 \text{ and } \beta_2 = 0.6593$$
$$a/b > 1.0: \beta_1 = 0.4314 \ b/a \text{ and } \beta_2 = 0.6590 \ b/a.$$

Short beam: Spread the load uniformly over a length of 0.75 b.

Continuous end

$$\text{Load} = 0.5 \times q \times b \times \beta_2 \ a$$

$$v_x = qb\{0.6667\beta_2 \frac{a}{b}\}, \beta_{vx} = 0.44$$

Simply supported end

$$\text{Load} = 0.5 \times q \times b \times \beta_1 \ a$$

$$v_x = qb\{0.6667\beta_1 \frac{a}{b}\}, \beta_{vx} = 0.2876, b/a < 1$$

$$v_x = qb\{0.6667\beta_1 \frac{a}{b}\}, \beta_{vx} = 0.1912, b/a = 1$$

Longer beam: Spread the load uniformly over a length of 0.75 a.
$$\text{Load:} = 0.5 \times q \times 0.5 \ b \times (2 - \beta_1 - \beta_2) \ a$$
$$v_y = qb\{0.3333(2 - \beta_1 - \beta_2\}$$

$$\beta_{vy} = (0.6667 - 0.3634\frac{b}{a}), b/a > 1, \ \ \beta_{vy} = 0.3513, b/a = 1$$

8.10.16.9 Slab with Only a Continuous Long Edge

Fig. 8.99 shows the collapse mode which is governed by two parameters β_1 and β_2.

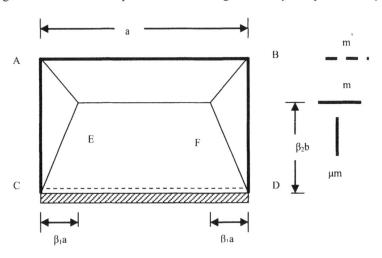

Fig. 8.99 Collapse mode for a slab with one continuous long edge.

It can be shown that the basic equation for solving the problem is

$$[\frac{b}{a}]^2\{ 2\frac{\mu m}{\beta_1}\}+\frac{(m+m')}{\beta_2} +\frac{m}{(1-\beta_2)} = q\frac{b^2}{6}(3-2\beta_1)$$

(a) Square slab
$a/b = 1$, $\mu m = m$, $m' = 1.33\ m$.

$$m\{\frac{2}{\beta_1} +\frac{2.333}{\beta_2}+\frac{1.0}{(1-\beta_2)} \} = q\frac{b^2}{6}(3-2\beta_1)$$

Simplifying:

$$m = \frac{qb^2}{6}\beta_1\beta_2(1-\beta_2)\frac{(3-2\beta_1)}{\{2\beta_2(1-\beta_2)-1.333\beta_1\beta_2 +2.333\beta_1 \}}$$

$dm/d\beta_2 = 0$ leads to

$$\beta_2^2 -3.5\beta_2 +1.75 = 0,\quad \beta_2 =0.604$$

Using this value of β_2, the expression for m is

$$m = \frac{qb^2}{6}\beta_1 \frac{(3-2\beta_1)}{\{6.3885\beta_1 +2.0\}}$$

$dm/d\beta_1 = 0$ leads to

$$\beta_1^2 +0.6261\beta_1 -0.4696 = 0,\quad \beta_1 =0.4404$$

Using $\beta_1 = 0.4404$ and $\beta_2 = 0.604$ in the expression for m, $m = 0.032\ qb^2$, $m' = 0.043\ qb^2$.

Multiplying these values by 4/3,

$$m = \mu m = 0.043 \ qb^2, \ \beta_{sx} = 0.043 \ \text{and} \ m' = 0.058 \ qb^2, \ \beta_{sx} = 0.058.$$

(b) Rectangular slab

$a/b > 1.0$. Keep $\mu m = 0.032 \ qb^2$, $m' = 1.333m$ and $\beta_2 = 0.604$ as constant for all values of a/b. Substituting these values in the equation for m and simplifying,

$$m = qb^2 \{0.0783 - 0.0522\beta_1 - \frac{0.010}{\beta_1}(\frac{b}{a})^2\}$$

$dm/d\beta_1 = 0$ leads to

$$-0.0522 + \frac{0.010}{\beta_1^2}(\frac{b}{a})^2 = 0, \ \beta_1 = 0.4377\frac{b}{a}$$

Substituting for β_1 in the equation for m and simplifying,

$$m = qb^2 \{0.0783 - 0.0457(\frac{b}{a})\}$$

Dividing the above value by 0.75,

$$m = qb^2 \{0.1044 - 0.0609(\frac{b}{a})\}$$

Short span
Positive moment

$$m = qb^2 \{0.1064 - 0.0609(\frac{b}{a})\}, \ \beta_{sx} = \{0.1064 - 0.0609(\frac{b}{a})\}, b/a < 1,$$

$$m = qb^2 0.043, \beta_{sx} = 0.041, a/b = 1$$

Negative moment

$$m' = 1.333m = qb^2 \{0.1419 - 0.0812(\frac{b}{a})\}, \ \beta_{sx} = \{0.1419 - 0.0812(\frac{b}{a})\}, b/a < 1$$

$$m' = m = qb^2 0.0573, \beta_{sx} = 0.0573, a/b = 1$$

Long span: $\mu m = 0.043 \ qb^2$, $\beta_{sy} = 0.058$

(c) Shear coefficients
Spread the load uniformly over a length of 0.75 b.

$$\beta_2 = 0.604 \ \text{and} \ \beta_1 = 0.4404, \ a/b = 1$$
$$\beta_2 = 0.604 \ \text{and} \ \beta_1 = 0.4382 \ b/a, \ a/b > 1.0$$

Shorter beam

$$\text{Load} = 0.5 \times q \times b \times \beta 1 \ a.$$

$$v_x = qb\{0.6667\beta_1 \frac{a}{b}\}$$

$$v_x = qb\{0.6667\beta_1 \frac{a}{b}\}, \beta_{vx} = 0.2921$$

Longer beam: Spread the load uniformly over a length of 0.75 a.

Continuous end

$$\text{Load} = 0.5 \times q \times \beta_2 b \times (2 - 2\beta_1) a$$

$$v_y = qb\{0.8053(1 - \beta_1)\}$$

$$v_y = qb\{0.8053 \times (1 - 0.4382\frac{b}{a}\}, \beta_{vy} = (0.8053 - 0.3529\frac{b}{a})$$

Simply supported end

$$\text{Load} = 0.5 \times q \times (1 - \beta_2) \; b \times (2 - 2\beta_1) \; a$$

$$v_y = qb\{0.5280(1 - \beta_1)\}$$

$$v_y = qb\{0.5280 \times (1 - 0.4382\frac{b}{a}\}, \beta_{vy} = (0.5280 - 0.2314\frac{b}{a})$$

8.11 HILLERBORG'S STRIP METHOD

This method of designing slabs is based on the lower bound theorem of plasticity. The basic idea is to find a distribution of moments, which fulfils the equilibrium equations, and design the slab for these moments. Normally in a slab not only moments about two axes but also torsional moments exist. Analysis is complicated because of the presence of these torsional moments. Strip method simplifies analysis by ignoring torsional moments and assuming that the load is carried by a set of strips in bending only.

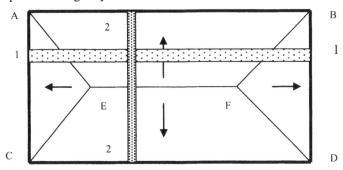

Fig. 8.100 Load distribution in a simply supported slab.

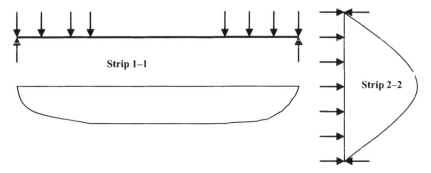

Fig. 8.101 Bending moments in horizontal (1–1) and vertical (2–2) strips.

8.11.1 Simply Supported Rectangular Slab

As an example, consider the rectangular slab simply supported on four sides and subjected to a uniformly distributed load q as shown in Fig. 8.100.

As shown in Fig. 8.101, for a typical strip 1–1 in the horizontal direction, the loading on the strip consists of uniformly distributed loading on the end portions only and for typical strip 2–2 in the vertical direction, with the loading on the strip consisting of uniformly distributed loading covering the entire span. The bending moments in these individual simply supported strips can be easily calculated and the slab may be reinforced accordingly.

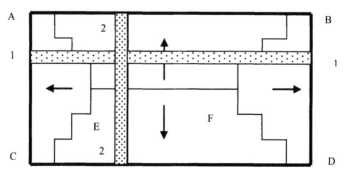

Fig. 8.102 Step-wise load distribution to the supports.

The main difficulty in assuming the load distribution as shown in Fig. 8.100 is that the loading on the strip across its width is not uniform. This difficulty can be avoided by assuming load distribution to the supports in a step-wise fashion as shown in Fig. 8.102.

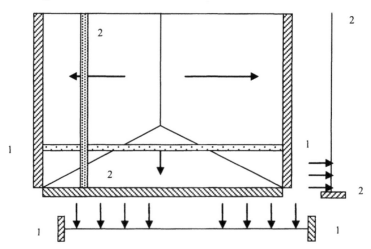

Fig. 8.103 Load distribution in a slab with one free edge.

8.11.2 Clamped Rectangular Slab with a Free Edge

Fig. 8.103 shows a slab clamped on three sides and free on one side. The load distribution to the supports is as indicated. If desired the step-wise load distribution can also be adopted. The strip 1–1 is a beam clamped at both ends while the strip 2–2 is a cantilever.

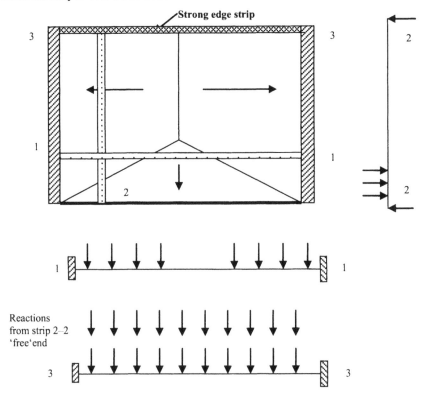

Fig. 8.104 Slab with opposite edges simply supported and free.

8.11.3 Slab Clamped on Two Opposite Sides, One Side Simply Supported and One Edge Free

Fig. 8.104 shows a slab clamped on two opposite sides and one side is simply supported while the opposite edge is free. The load distribution to the supports is as indicated.

The strip 1–1 is a beam clamped at both ends. In Fig. 8.103, strip 2–2 was clamped at one end and could therefore act as a cantilever. In Fig. 8.104, for the strip 2–2 to transmit any load to the simply supported end, it is necessary that there is a support at the 'free' end. Edge strip 3–3 provides this support. Therefore while designing strip 3–3, it is necessary to include not only the load applied directly onto the strip but also the reactions from strip 2–2. Strip 3–3 acts like an edge beam by being more heavily reinforced than the rest of strips 1–1. Strip 3–3 could be thickened in order to allow sufficient depth of lever arm to the reinforcement.

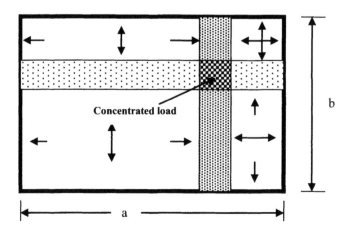

Fig. 8.105 Strong band reinforcement.

8.11.4 Strong Bands

Fig. 8.105 shows a rectangular slab simply supported on all sides and carrying a concentrated load W. The load is transmitted to the supports mainly through heavily reinforced strips in two directions. These strips are known as 'strong bands'. The strong bands act as beams and are more heavily reinforced compared to the rest of the slab. It is often convenient to increase the thickness in order to accommodate steel reinforcement and also to increase its lever arm. Distributed load on the rest of the slab can be distributed between the edge supports and strong bands. The load carried by the strong bands will be approximately in inverse proportion to the fourth power of the spans. Thus if the spans are a and b with $a \ge b$, then the loads W_a and W_b carried by the strips in the a and b direction are

$$W_b = \frac{\alpha}{(1+\alpha)}W, \quad W_a = \frac{1}{(1+\alpha)}W, \quad \alpha = [\frac{a}{b}]^4$$

$$W_b = \frac{\alpha}{(1+\alpha)}W, \quad W_a = \frac{1}{(1+\alpha)}W, \quad \alpha = [\frac{a}{b}]^4$$

The concept of the strong band is also useful when designing slabs with holes or slabs with re-entrant corners.

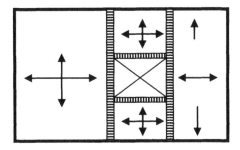

Fig. 8.106 Slab with a hole.

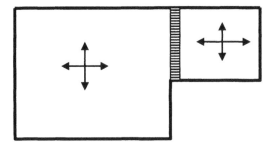

Fig. 8.107 Slab with a re-entrant corner.

Fig. 8.106 shows a slab with a rectangular hole. By providing strong bands around the hole, edge beams are created and the loads can be distributed between the edge supports and strong bands. The two strong bands running between the supports also provide support for the short edge beams around the hole.

Fig. 8.107 shows a slab with a re-entrant corner. By providing a strong band, the slab is conveniently divided into two rectangular slabs which can be effectively designed separately. The strong band acts as an additional support to the two slabs and allows the above simplification compared with the relatively complex distribution of moments obtained from an elastic analysis.

8.11.5 Comments on the Strip Method

One of the main attractions of the strip method as compared with the yield line method is that apart from the fact that it is a lower bound method and therefore there is no need to investigate alternative mechanisms, the method not only gives the bending moments and shear forces at every point in the structure but also gives information on the loads and their distribution acting on the supporting beams. This is of great attraction to designers.

It is important to remember that the method ensures safety against bending failure only. It does not take account of the possibility of shear failure. Because of the fact that the emphasis is on safety at ultimate limit state, additional considerations are necessary to ensure that the design meets serviceability limit state conditions as well. For any given structure, it is possible to choose an infinite number of possible distributions of loads to the supports and the corresponding moments.

As an example consider the load distribution on the rectangular slab simply supported on all edges shown in Fig. 8.108. The proportion of the uniformly distributed load q against the arrows indicates the value of the load carried to the support in the direction indicated. This load distribution is different from the one shown in Fig. 8.100. However from a serviceability limit state point of view, it is important to ensure that the chosen distribution of moments does not depart too far from the elastic distribution of moments.

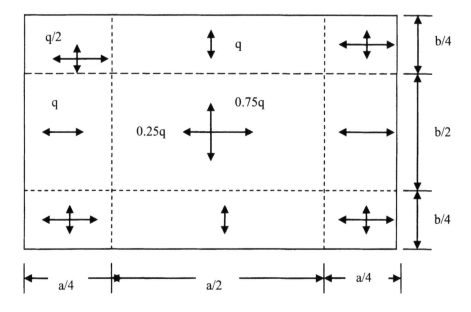

Fig. 8.108 A feasible load distribution to the four edge supports.

The following points should be borne in mind when deciding on the load distribution to the supports.

With fixed edges, the ratio between the proportion of load carried to the fixed edge and that carried to the simply supported edge should be increased by a factor of 1.6 to 1.8 compared to the case where both edges have same support conditions. If, for example, of the two opposite edges, one is fixed and the other is simply supported, then an appropriate load distribution should be as shown in Fig. 8.110. The dividing lines between the regions can be treated as zero shear lines. For example, for the strip 1–1 shown in Fig. 8.109, if the line of zero shear is at 1.1 from the simply supported end, then the reaction at the simply supported end is 1.1 q and at the fixed end is 1.9 q. The maximum bending moment in the span is at the point of zero shear and is equal to 0.605 q and at the fixed end is 1.2 q. Thus the ratio of reactions is 1.9/1.1 = 1.72 and the ratio of moments is 1.2/0.605 = 2.0. Thus by choosing the position of lines of zero shear, it is possible to control the moment distribution to correspond to the elastic values.

Although the strip method assumes that torsional moments are zero, however where two simply supported edges meet, torsional moments do exist. In the absence of proper reinforcement, this will lead to cracking which is best limited by providing torsional reinforcement as suggested in codes of practice. The ratio between the support and span design moments in a strip fixed at both ends and subjected to uniformly distributed loading should be about 2.

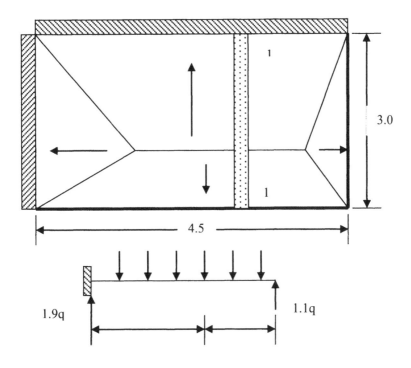

Fig. 8.109 Zero shear lines to control load distribution to supports.

8.12 DESIGN OF REINFORCEMENT FOR SLABS USING ELASTIC ANALYSIS MOMENTS

With the widespread availability of finite element programs to carry out elastic analysis of plates, it is necessary to have rules for designing reinforcement for a given set of bending and twisting moments in slabs. Fig. 8.110 shows the bending moments M_x and M_y and twisting moment M_{xy} acting on an element of slab. The convention used in representing a moment by a double-headed arrow is that if the right hand thumb is pointed in the direction of the arrow head, then the direction of the moment is given by the direction the fingers of the right hand bend. Bending moments as shown in Fig. 8.110, cause tension on the bottom face.

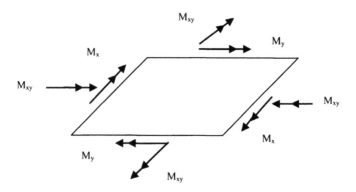

Fig. 8.110 Bending and twisting moments on an element of slab.

As shown in Fig. 8.111, on a section inclined at an angle α to the y-axis, normal bending moment M_n and twisting moment M_{nt} act. It can be shown that

$$M_n = M_x \cos^2 \alpha + M_y \sin^2 \alpha + 2 M_{xy} \sin \alpha \cos \alpha$$

If the ultimate sagging moment of resistance provided by steel in x- and y-directions are M^b_{xu} and M^b_{yu} respectively, then from Johansen's yield criterion (section 8.9.2), the normal moment of resistance on a section inclined at an angle α to the y-axis is given by

$$M^b_{nu} = M^b_{xu} \cos^2 \alpha + M^b_{yu} \sin^2 \alpha$$

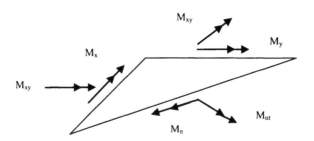

Fig. 8.111 Normal bending moment and twisting moments on an element of slab.

Since it is desirable that the applied M_n must not be greater than the resistance M_{nu},

$$[\{M^b_{xu} \cos^2 \alpha + M^b_{yu} \sin^2 \alpha\} - \{M_x \cos^2 \alpha + M_y \sin^2 \alpha + 2M_{xy} \sin \alpha \cos \alpha\}] \geq 0$$

Dividing throughout by $\cos^2 \alpha$ and setting $t = \tan \alpha$, the above equation simplifies to

$$\{(M^b_{xu} - M_x) + (M^b_{yu} - M_y) \, t^2 - 2M_{xy} \, t\} \geq 0$$

Yielding will take place when the difference between M_{nu} and M_n is a minimum. Differentiating with respect to t,

$$(M^b{}_{yu} - M_y)\, t - M_{xy} = 0$$

For the difference to be a minimum, the second derivative with respect to t must be positive. Therefore

$$(M^b{}_{yu} - M_y) \geq 0$$

Substituting the value of t and simplifying

$$(M^b{}_{xu} - M_x)(M^b{}_{yu} - M_y) - M_{xy}{}^2 = 0$$

This equation shows for what combination of bending and twisting moments a slab with a known moment of resistance in x- and y-directions yields. This equation is known as the yield criterion for a slab. Note that the twisting moment term appears as a square indicating that the sign of M_{xy} is irrelevant.

From the yield criterion, the following special cases can be noted.

Case 1:

$$\text{If } M^b{}_{xu} = 0, \text{ then } M^b{}_{yu} = M_y - M_{xy}{}^2/M_x$$

Case 2:

$$\text{If } M^b{}_{yu} = 0, \text{ then } M^b{}_{xu} = M_x - M_{xy}{}^2/M_y$$

Case 3:

If $M^b{}_{xu} \neq 0$ and $M^b{}_{yu} \neq 0$, then for economy $(M^b{}_{xu} + M^b{}_{yu})$ must be made a minimum. From the yield criterion

$$M^b{}_{xu} = M_x + M_{xy}{}^2 / (M^b{}_{yu} - M_y)$$
$$(M^b{}_{xu} + M^b{}_{yu}) = M_x + M_{xy}{}^2 / (M^b{}_{yu} - M_y) + M^b{}_{yu}$$

Minimizing the above expression with respect to $M^b{}_{yu}$,

$$-M_{xy}{}^2 / (M^b{}_{yu} - M_y)^2 + 1 = 0$$
$$(M^b{}_{yu} - M_y) = \pm M_{xy}$$

Since $(M^b{}_{yu} - M_y) \geq 0$, choosing the positive sign,

$$M^b_{xu} = M_x + |M_{xy}|, \quad M^b_{yu} = M_y + |M_{xy}|$$

Note that only the numerical value of M_{xy} is used.

8.12.1 Rules for Designing Bottom Steel

In the following, positive bending moments are sagging moments which cause tension on the bottom face. The rules for calculating the moment of resistance required for flexural steel at bottom are as follows.

(a) If $\dfrac{M_x}{|M_{xy}|} \geq -1.0$ and $\dfrac{M_y}{|M_{xy}|} \geq -1.0$, then $M^b_{xu} = M_x + |M_{xy}|, \quad M^b_{yu} = M_y + |M_{xy}|$

(b) If $\dfrac{M_x}{|M_{xy}|} < -1.0$ and $M_y - \dfrac{M_{xy}^2}{M_x} > 1.0$ then $M^b_{xu} = 0, \quad M^b_{yu} = M_y - \dfrac{M_{xy}^2}{M_x}$

(c) If $\dfrac{M_y}{|M_{xy}|} < -1.0$ and $M_x - \dfrac{M_{xy}^2}{M_y} > 1.0$ then $M^b_{yu} = 0, \quad M^b_{xu} = M_x - \dfrac{M_{xy}^2}{M_y}$

(d) If none of the above conditions are valid, then $M^b_{yu} = M^b_{xu} = 0$

8.12.1.1 Examples of Design of Bottom Steel

The following examples illustrate the use of equations derived in section 8.11.1. The four criteria are checked to see which is the valid one for a specific combination of M_x, M_y and M_{xy}.

Example 1: $M_x = 30$ kNm/m, $M_y = 15$ kNm/m, $M_{xy} = 20$ kNm/m

Check criterion (a): $\dfrac{M_x}{|M_{xy}|} = 1.5 > -1.0$, $\dfrac{M_y}{|M_{xy}|} = 0.75 > -1.0$, Therefore criterion (a) applies.

$$M^b_{xu} = 30+20 = 50 \text{ kNm/m, } M^b_{yu} = 15+20 = 35 \text{ kNm/m}$$

Example 2: $M_x = -35$ kNm/m, $M_y = 15$ kNm/m, $M_{xy} = 20$ kNm/m

(a) $\dfrac{M_x}{|M_{xy}|} = -1.75 < -1.0$, $\dfrac{M_y}{|M_{xy}|} = 0.75 > -1.0$.

(b) $\dfrac{M_x}{|M_{xy}|} = -1.75 < -1.0$, $M_y - \dfrac{M^2_{xy}}{M_x} = 26.43 > 1.0$. Therefore criterion (b) applies.

$$M^b_{xu} = 0, \ M^b_{yu} = M_y - \dfrac{M^2_{xy}}{M_x} = 26.43 \text{ kNm/m}$$

Example 3: $M_x = -15$ kNm/m, $M_y = -25$ kNm/m, $M_{xy} = 20$ kNm/m

(a) $\dfrac{M_x}{|M_{xy}|} = -0.75 > -1.0$, $\dfrac{M_y}{|M_{xy}|} = -1.25 < -1.0$

(b) $\dfrac{M_x}{|M_{xy}|} = -0.75 > -1.0$, $M_y - \dfrac{M^2_{xy}}{M_x} = 1.67 > 1.0$

(c) $\dfrac{M_y}{|M_{xy}|} = -1.25 < -1.0$, $M_x - \dfrac{M^2_{xy}}{M_y} = 1.0 > 0$. Therefore criterion (c) applies.

$$M^b_{yu} = 0, \ M^b_{xu} = M_x - \dfrac{M^2_{xy}}{M_y} = 1.0 \text{ kNm/m}$$

Example 4: $M_x = -30$ kNm/m, $M_y = -40$ kNm/m, $M_{xy} = 20$ kNm/m

(a) $\dfrac{M_x}{|M_{xy}|} = -1.5 < -1.0$, $\dfrac{M_y}{|M_{xy}|} = -2.0 < -1.0$

(b) $\dfrac{M_x}{|M_{xy}|} = -1.5 < -1.0$, $M_y - \dfrac{M^2_{xy}}{M_x} = -26.67 < 0$

(c) $\dfrac{M_y}{|M_{xy}|} = -2.0 < -1.0$, $M_x - \dfrac{M_{xy}^2}{M_y} = -20.0 < 0$

Since none of the criteria (a) to (c) apply, $M_{yu}^b = 0$, $M_{xu}^b = 0$. No steel is required at the bottom of the slab.

8.12.2 Rules for Designing Top Steel

In a manner similar to the determination of sagging moment of resistance, if the ultimate *hogging* moments of resistance provided by steel in x- and y-directions are M_{xu}^t and M_{yu}^t respectively, then the rules for calculating the moment of resistance required for flexural steel at top are as follows. Note that the value of M_{xu}^t and M_{yu}^t are both negative, indicating that they correspond to a hogging bending moment requiring steel at the top of the slab.

(a) If $\dfrac{M_x}{|M_{xy}|} \le 1.0$ and $\dfrac{M_y}{|M_{xy}|} \le 1.0$, then $M_{xu}^t = M_x - |M_{xy}|$, $M_{yu}^t = M_y - |M_{xy}|$

(b) If $\dfrac{M_y}{|M_{xy}|} > 1.0$ and $M_x - \dfrac{M_{xy}^2}{M_y} < 0$ then $M_{yu}^t = 0$, $M_{xu}^t = M_x - \dfrac{M_{xy}^2}{M_y}$

(c) If $\dfrac{M_x}{|M_{xy}|} > 1.0$ and $M_y - \dfrac{M_{xy}^2}{M_x} < 0$ then $M_{xu}^t = 0$, $M_{yu}^t = M_y - \dfrac{M_{xy}^2}{M_x}$

(d) If none of the above conditions are true, then $M_{yu}^t = M_{xu}^t = 0$

8.12.2.1 Examples of Design of Top Steel

Example 1: $M_x = -30$, $M_y = -40$, $M_{xy} = 20$ kNm/m

(a) $\dfrac{M_x}{|M_{xy}|} = -1.5 < 1.0$ and $\dfrac{M_y}{|M_{xy}|} = -2.0 < 1.0$. Criterion (a) applies.

$M_{xu}^t = M_x - |M_{xy}| = -50$ kNm/m, $M_{yu}^t = M_y - |M_{xy}| = -60$ kNm/m

Example 2: $M_x = -30$, $M_y = 35$, $M_{xy} = 20$ kNm/m

(a) $\dfrac{M_x}{|M_{xy}|} = -1.5 < 1.0$ and $\dfrac{M_y}{|M_{xy}|} = 1.75 > 1.0$

(b) $\dfrac{M_y}{|M_{xy}|} = 1.75 > 1.0$ and $M_x - \dfrac{M_{xy}^2}{M_y} = -41.43 < 0$. Criterion (b) applies.

$M_{yu}^t = 0$, $M_{xu}^t = M_x - \dfrac{M_{xy}^2}{M_y} = -41.43$

Example 3: $M_x = 25$, $M_y = -45$, $M_{xy} = 20$ kNm/m

(a) $\dfrac{M_x}{|M_{xy}|} = 1.25 > 1.0$ and $\dfrac{M_y}{|M_{xy}|} = -2.25 < 1.0$

(b) $\dfrac{M_y}{|M_{xy}|} = -2.25 < 1.0$ and $M_x - \dfrac{M_{xy}^2}{M_y} = 33.89 > 0.$

(c) $\dfrac{M_x}{|M_{xy}|} = 1.25 > 1.0$, $M_y - \dfrac{M_{xy}^2}{M_x} = -61.0 < 0$. Criterion (c) applies.

$$M'_{xu} = 0, \ \ M'_{yu} = M_y - \frac{M_{xy}^2}{M_x} = -61.0 \text{ kNm/m}$$

8.12.3 Examples of Design of Top and Bottom Steel

In sections 8.11.1.1 and 8.11.2.1 examples were concerned with determining the required moment of resistance either at the top or the bottom face of the slab. However cases do arise where for a given combination of bending and twisting moments there is need to provide steel at both the faces. This case generally arises when twisting moments larger than bending moments are present.

Example 1: $M_x = 15$, $M_y = -18$, $M_{xy} = 20$ kNm/m

Bottom steel: $\dfrac{M_x}{|M_{xy}|} = 0.75 > -1.0$ and $\dfrac{M_y}{|M_{xy}|} = -0.9 > -1.0$

$M_{xu}^b = M_x + |M_{xy}| = 35$ kNm/m, $M_{yu}^b = M_y + |M_{xy}| = 2$ kNm/m

Top steel: $\dfrac{M_x}{|M_{xy}|} = 0.75 < 1.0$ and $\dfrac{M_y}{|M_{xy}|} = -0.9 < 1.0$

$M'_{xu} = M_x - |M_{xy}| = -5$ kNm/m, $M'_{yu} = M_y - |M_{xy}| = -38$ kNm/m

This example shows that steel is required in both directions top and bottom.

Example 2: $M_x = 20$, $M_y = -20$, $M_{xy} = 20$ kNm/m

Bottom steel: $\dfrac{M_x}{|M_{xy}|} = 1.0 > -1$ and $\dfrac{M_y}{|M_{xy}|} = -1.0$

$M_{xu}^b = M_x + |M_{xy}| = 40$ kNm/m, $M_{yu}^b = M_y + |M_{xy}| = 0$

Top steel: $\dfrac{M_x}{|M_{xy}|} = 1.0$ and $\dfrac{M_y}{|M_{xy}|} = -1.0$

$M'_{xu} = M_x - |M_{xy}| = 0, \ M'_{yu} = M_y - |M_{xy}| = -40$ kNm/m

This example shows that steel is required in only y-direction at top and in only x-direction at bottom.

8.12.4 Comments on the Design Method Using Elastic Analysis

The Eurocode in clause 5.6.1 permits the design for ultimate limit state of slabs using Upper bound also known as kinematic method (Johansen's yield line method) or the Lower bound also known as static method (Hillerborg's strip method). These methods can be used without any direct check of rotation capacity provided conditions of code clause 5.6.1(2) P repeated in section 8.9.5 are adhered to. Any problems with serviceability limit state requirements can be minimized by ensuring recommended span/depth ratios and spacing of reinforcement are observed.

Using bending and twisting moments from elastic analysis to design slabs using the rules developed in sections 8.11.1 to 8.11.3 avoids this problem and leads to a very economical design. The main disadvantage is that the designed reinforcement will vary from point to point and some form of averaging is needed to convert the variable reinforcement into bands with constant reinforcement. The method is also highly amenable to computer-aided design of general slab structures.

8.13 STAIR SLABS

8.13.1 Building Regulations

In U.K., statutory requirements are laid down in *Building Regulations and Associated Approved Documents,* Part H that defines private and common stairways. The private stairway is for use with one dwelling and the common stairway is used for more than one dwelling. Requirements from the building regulations are shown in Fig. 8.112.

8.13.2 Types of Stair Slabs

Stairways are sloping one-way spanning slabs. Two methods of construction are used.

(a) Transverse spanning stair slabs

Transverse spanning stair slabs span between walls, a wall and stringer (an edge beam), or between two stringers. The stair slab may also be cantilevered from a wall. A stair slab spanning between a wall and a stringer is shown in Fig. 8.113(a). The stair slab is designed as a series of beams consisting of one step with assumed breadth and effective depth shown in Fig. 8.113(c). The moment reinforcement is generally one bar per step. Secondary reinforcement is placed longitudinally along the flight.

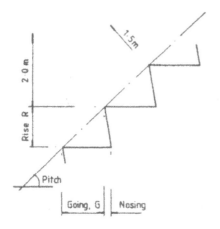

	Private	Common
Rise R	≯ 220 mm	≯ 190 mm
Going G	≮ 220 mm	≮ 230 mm
Pitch	≯ 42°	≯ 38°
No. of steps in flight	-	≯ 16

700 mm > G + 2R > 550 mm

Fig. 8.112 Building regulation for dimensions of stairs.

(b) Longitudinal spanning stair slabs

The stair slab spans between supports at the top and bottom of the flight. The supports may be beams, walls or landing slabs. A common type of staircase is shown in Fig. 8.114.

The effective span l lies between the top landing beam and the centre of support in the wall. If the total design load on the stair is W the positive design moment at mid-span and the negative moment over top beam B are both taken as $Wl/10$. The arrangement of moment reinforcement is shown in Fig. 8.114. Secondary reinforcement runs transversely across the stair.

A staircase around a lift well is shown in Fig. 8.115. The effective span l of the stair is normally taken as between the centres of landing. The maximum moment near mid-span and over supports is taken as $Wl/10$, where W is the total design load on the span.

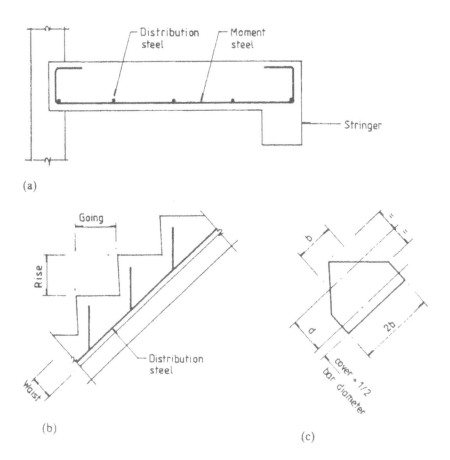

Fig. 8.113 (a) Transverse section; (b) longitudinal section; (c) assumptions for design.

8.13.3 Design Requirements

(a) Imposed loading
The imposed loading on stairs is given in Eurocode 1: Actions on Structures Part 1-1: General Actions, clause 6.3.1.2, Table 6.2. From this table the distributed loading is between 2 kN/m^2 and 4 kN/m^2 with the former value recommended.

(b) Design provisions
The stress analysis of a stair slab is complex. The slab is attached to the wall on one side and is connected to the landings at top and bottom. Because of the depth of the stairs, cantilever moment is unlikely to be important. The usual assumptions for design of staircases are as follows.

1. *Staircase* may be taken to include a section of the landing spanning in the same direction and continuous with the stair flight.
2. The design ultimate load is to be taken as uniform over the plan area. When two spans intersect at right angles as shown in Fig. 8.116 the load on the common area can be divided equally between the two spans.
3. When as shown in Fig. 8.116 the staircase is built monolithically at its ends into structural members spanning at right angles to its span, the effective span is given by $l_n + 0.5(a_1 + a_2)$, where l_n is the clear horizontal distance between supporting members. a_1 is the breadth of a supporting member at one end or the depth of the slab whichever is the smaller and a_2 is the dimension similar to a1 at the second end of the slab.
4. The effective span of simply supported staircases without stringer beams should be taken as the horizontal distance between centrelines of supports or the clear distance between faces of supports plus the effective depth whichever is less.
5. The depth of the section is to be taken as the minimum thickness perpendicular to the soffit of the stair slab.
6. The design procedure is the same as for beams and slabs.

8.13 4 Example of Design of Stair Slab

(a) Specification
Design the side flight of a staircase surrounding an open stair well. A section through the stairs is shown in Fig. 8.116(a).
The stair slab is supported on a beam at the top and on the landing of the flight at right angles at the bottom. The imposed loading is 2 kN/m². The stair is built 110 mm into the sidewall of the stair well. The clear width of the stairs is 1.25 m and the flight consists of steps with risers at 180 mm and goings of 220 mm with 20 mm nosing. The stair treads and landings have 15 mm granolithic finish and the underside of the stair and landing slab has 15 mm of plaster finish. The materials are $f_{ck} = 30$ MPa concrete and $f_{yk} = 500$ A reinforcement.

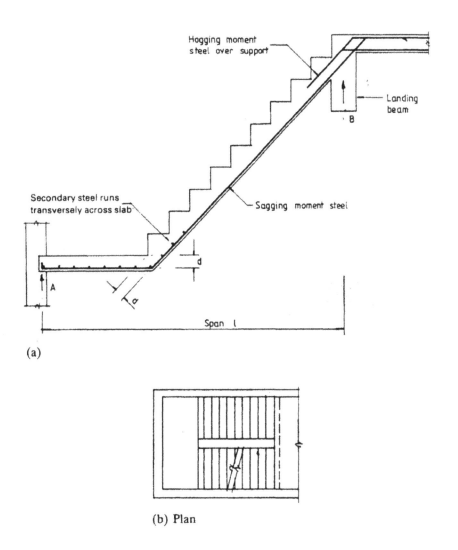

Hogging moment
steel over support

Landing
beam

· B

Secondary steel runs
transversely across slab

Sagging moment steel

d

A

a

Span   l

(a)

(b) Plan

Fig. 8.114 A common type of stair case.

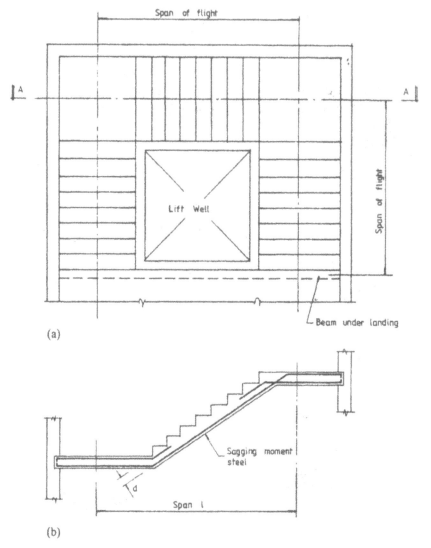

Fig. 8.115 (a) Plan; (b) section AA.

(b) Loading and moment
Assume the waist thickness of structural concrete is 100 mm, the cover is 25 mm
and the bar diameter is 10 mm. The loaded width is 1.10 m and effective breadth
of the stair slab is 1.25 m as shown in section AA in Fig. 8.116(a). The effective
span of the stair slab is taken as the clear horizontal distance (1540 mm) plus the
distance of the stair to the centre of the top beam (235 mm) plus *one-half of the
breadth of the landing* (625 mm), i.e., 2400 mm. The design ultimate loading on
the stairs is calculated first.

(i) Landing slab
The overall thickness including the top and underside finish is 130 mm.
$$\text{Dead load} = 0.13 \times 25 = 3.25 \text{ kN/m}^2$$
$$\text{Imposed load} = 2 \text{ kN/m}^2$$
$$\text{Total design ultimate load} = 1.35 \times 3.25 + 1.5 \times 2.0 = 7.4 \text{ kN/m}^2$$
Taking the loaded length of the landing as 1100 mm and assuming that only 50 percent of the landing load goes to each flight,
$$\text{Landing load} = 0.5 \times 7.4 \times (\text{landing breadth/2} = 0.625) \times (\text{loaded length} = 1.1)$$
$$= 2.54 \text{ kN}$$

(ii) Stair slab
The slope length is $\sqrt{(1.775^2 + 1.44^2)} = 2.29$ m and the steps project 152 mm perpendicularly to the top surface of the waist. The average thickness including finishes is $100 + 152/2 + 30 = 206$ mm.
$$\text{Dead load} = 0.206 \times 2.29 \times 1.1 \times 25 = 12.97 \text{ kN}$$
$$\text{Imposed load} = 1.775 \times 1.1 \times 2.0 = 3.92 \text{ kN}$$
$$\text{Total design load} = 1.35 \times 12.97 + 1.5 \times 3.92 = 23.39 \text{ kN}$$
The total load on the span is
$$2.54 + 23.39 = 25.93 \text{ kN}$$
Assuming that the design moment for sagging moment near mid-span and the hogging moment over the supports are both $Wl/10$, design moment M is
$$M = 25.93 \times 2.4/10 = 6.22 \text{ kNm}$$

(c) Moment reinforcement
The effective depth
$$d = 100 - 25 - 5 = 70 \text{mm}$$
The effective width b will be taken as the width of the stair slab, 1250 mm.
$$k = M/ (bd^2 f_{ck}) = 6.22 \times 10^6/ (1250 \times 70^2 \times 30) = 0.034 < 0.196$$
$$z/d = 0.5[1.0 + \sqrt{(1.0 - 3 \times 0.0349)}] = 0.97$$
$$z = 0.97 \times 70 = 68 \text{ mm}$$
$$A_s = 6.22 \times 10^6/ (0.87 \times 500 \times 68) = 210 \text{ mm}^2 \text{ for the full 1250 mm width}$$
$$= 210/1.25 = 168 \text{ mm}^2/\text{m}$$

Minimum steel
$$f_{ctm} = 0.30 \times f_{ck}^{0.667} = 0.30 \times 25^{0.667} = 2.6 \text{ MPa}$$
$A_{s, min} = 0.26 (f_{ctm}/f_{yk}) b_t d \geq 0.0013 b_t d$
$b_t = 1$ m width of slab $= 1000$ mm, $d = 700$ mm
$A_{s, min} = 0.26 \times (2.6/500) \times 1000 \times 70 \geq 0.0013 \times 1000 \times 70$
$A_{s, min} = 97 \text{ mm}^2/\text{m}$.
The area of steel calculated (210) is higher than the minimum value.
Provide five 8 mm diameter bars to give a total area of 251 mm². Space the bars at 300 mm centres. The same steel is provided in the top of the slab over both supports.
According to clause 9.3.1.1(2), secondary reinforcement should be not less than 20 percent of principal reinforcement.

Main steel required is 168 mm²/m. 20 percent of 168 mm²/m = 34 mm²/m, which is below the minimum value of 97 mm²/m. Provide minimum steel of 8 mm diameter bars at 300 mm centres (maximum spacing = 3h = 300 mm) to give 101 mm²/m transversely as distribution steel.

(d) Shear resistance
From the loading shown in Fig. 8.116(c), reaction at the right support is the maximum.

$$\text{Shear} = \{2.54 \times 0.625/2 + 23.39 \times (0.625+1.775/2)\}/2.4 = 15.1 \text{ kN}$$
$$v = 15.1 \times 10^3/(1250 \times 70) = 0.17 \text{ N/mm}^2$$
$$100\ \rho_1 = 100\ A_s/(bd) = 100 \times 251/(1250 \times 70) = 0.29 < 2.0$$
$$C_{Rd,\,c} = 0.18/(\gamma_c = 1.5) = 0.12,\ k = 1 + (200/70)^{0.5} = 2.69 > 2.0$$
$$v_{Rd,c} = C_{Rd,c}\ k\ (100\rho_1\ f_{ck})^{0.33} = 0.12 \times 2.0 \times (0.29 \times 30)^{0.33} = 0.49 \text{ MPa}$$
$$v_{min} = 0.035 \times k^{1.5} \times \sqrt{f_{ck}} = 0.035 \times 2.0^{1.5} \times \sqrt{30} = 0.54 \text{ MPa}$$
$$(v = 0.17) < (v_{Rd,\,c} = 0.54)$$

No shear reinforcement is needed.
The slab is satisfactory with respect to shear.

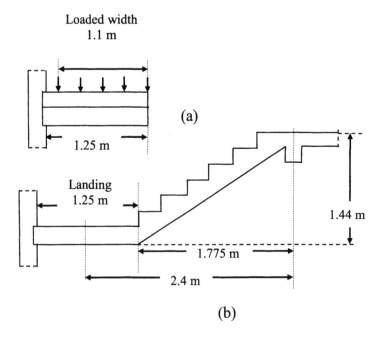

Fig. 8.116 (a) Section through the stairs; (b) loading diagram.

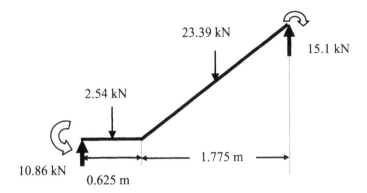

Fig. 8.116 (c) Loads and reactions.

(e) Deflection

The slab is checked for deflection using code equation (7.16a).

$$100 \, \rho_1\% = 100 \times 251/ (1250 \times 70) = 0.29 < 2.0$$
$$\rho_0\% = 0.1\sqrt{f_{ck}} = 0.1\sqrt{30} = 0.55$$

k = 1.5 for interior span of one-way slabs

$$\frac{L}{d} = K\{11+1.5 \sqrt{f_{ck}} \frac{\rho_0}{\rho} + 3.2 \sqrt{f_{ck}} (\frac{\rho_0}{\rho}-1)^{\frac{3}{2}}\} \text{ if } \rho \le \rho_0 \qquad (7.16a)$$

$$\frac{L}{d} = 1.2 \times \{11+1.5 \sqrt{30}\frac{0.55}{0.29}+3.2 \sqrt{30}(\frac{0.55}{0.29}-1)^{\frac{3}{2}}\} = 49.8$$

Allowable span/d ratio = 49.8
Actual span/d ratio = 2400/70 = 34.3

Hence the slab is satisfactory with respect to deflection.

(f) Cracking

For crack control the clear distance between bars is not to exceed 3 h = 300 mm.
The reinforcement spacing of 300 mm is satisfactory.

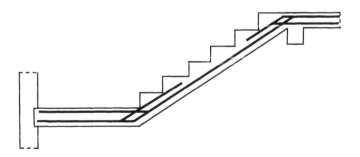

Fig. 8.117 Reinforcement details.

(g) Reinforcement
The reinforcement is shown in Fig. 8.117.

8.13 5 Analysis of Stair Slab as a Cranked Beam

Fig. 8.118 shows the stair slab as a cranked beam. It is assumed that the beam is fixed at the top and bottom to the landings.

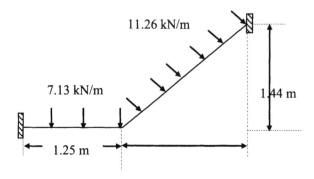

Fig. 8.118 Cranked beam.

(i) Landing slab

The overall thickness including the top and underside finish is 130 mm.
$$\text{Dead load} = 0.13 \times 1.25 \times 25 = 4.06 \text{ kN/m}$$
Taking the loaded length of the landing as 1100 mm and assuming that only 50 percent of the landing load goes to each flight,
$$\text{Imposed load} = 2 \times 1.1 \times (50\%) = 1.1 \text{ kN/m}$$
$$\text{Total design ultimate load} = 1.35 \times 4.06 + 1.5 \times 1.1 = 7.13 \text{ kN/m}$$

(ii) Stair slab
The slope length is $\sqrt{(1.775^2 + 1.44^2)} = 2.29$ m and the steps project 152 mm perpendicularly to the top surface of the waist. The average thickness including finishes is $100 + 152/2 + 30 = 206$ mm.
$$\text{Dead load} = 0.206 \times 2.29 \times 1.25 \times 25 = 14.74 \text{ kN}$$
$$\text{Imposed load} = 1.775 \times 1.1 \times 2.0 = 3.92 \text{ kN}$$
$$\text{Total design load} = 1.35 \times 14.74 + 1.5 \times 3.92 = 25.78 \text{ kN}$$
Spreading this load over a length of 2.29 m, loading is 11.25 kN/m.
The beam is analysed using a frame analysis program. Fig. 8.119 and Fig. 8.120 show respectively the bending moment and shear force diagrams.

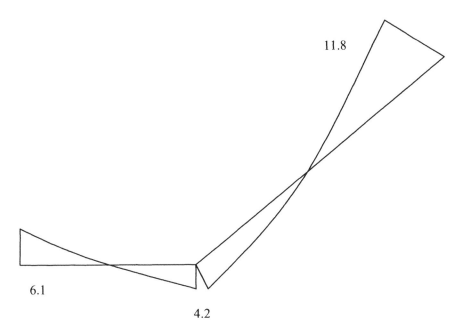

11.8

6.1

4.2

Fig. 8.119 Bending moment diagram.

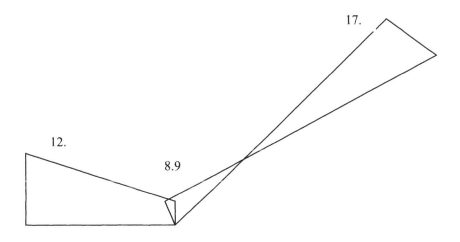

17.

12.

8.9

Fig. 8.120 Shear force diagram.

8.14 REFERENCES

Cope, R.J. and Clark, L.A. (1984). *Concrete Slabs: Analysis and Design*, Elsevier.

Goodchild, C.H. (1997). *Economic Concrete Frame Elements*. British Cement Association.

Hillerborg, A. (1996). *Strip Method Design Handbook*. E & FN Spon.

Hillerborg, A. (1975). *Strip Method of Design*. A Viewpoint Publication.

Jones, L.L. and Wood, R.H. (1967). *Yield Line Analysis of Slabs*. Thames and Hudson.

Park, R. and Gamble W.L. (1980). *Reinforced Concrete Slabs*. Wiley.

Timoshenko, S. and Woinowsky–Krieger, S. (1959). *Theory of Plates and Shells*. McGraw-Hill.

Great Britain (2004). *The Building Regulations 2000 Approved Document A: Structure*. Her Majesty's Stationery Office, 1985.

CHAPTER 9

COLUMNS

9. 1 TYPES, LOADS, CLASSIFICATION AND DESIGN CONSIDERATIONS

9.1.1 Types and Loads

Columns are structural members in buildings carrying roof and floor loads to the foundations. A column stack in a multi-storey building is shown in Fig. 9.1(a).

Columns primarily carry axial loads, but most columns are subjected to moment as well as axial load. Referring to the part floor plan in the figure, the internal column A is designed for predominantly axial load while edge columns B and corner column C are designed for axial load and appreciable moment.

Design of axially loaded columns is treated first. Then methods are given for design of sections subjected to axial load and moment. Most columns are termed short columns and fail when the material reaches its ultimate capacity under the applied loads and moments. Slender columns buckle and the additional moments caused by deflection must be taken into account in design.

The column section is generally

square or rectangular, but circular and polygonal columns are used in special cases. When the section carries mainly axial load it is symmetrically reinforced with four, six, eight or more bars held in a cage by links. It is not practical to cast vertically columns smaller than 200 mm square. Typical column reinforcement is shown in Fig. 9.1(b).

General requirements for design of columns are treated in section 5.8 of Eurocode 2. The provisions apply to columns where the greater cross sectional dimension does not exceed four times the smaller dimension.

The minimum size of a column must meet the fire resistance requirements given in *Eurocode 2: Design of Concrete Structures-Part 1-2: General Rules-Structural Fire Design.*

For example, from Table 2.10, Chapter 2, for a fire resistance period of 90 minutes, a fully exposed braced column must have a minimum dimension of 350 mm with the distance from the surface to centre of the steel of at least 53 mm.

9.1.2 Braced and Unbraced Columns

Lateral stability in braced reinforced concrete structures is provided by shear walls, lift shafts and stairwells. Fig. 9.2a shows a frame structure which is designed such that all horizontal load is resisted by a stiff lift shaft so that the column ends deflect very little. In a braced column the axial load and the bending moments at the ends of a column arise from the vertical loads acting on the beams. The horizontal loads do not affect the forces or deformation of the column. The columns **do not** contribute to the overall horizontal stability of the structure.

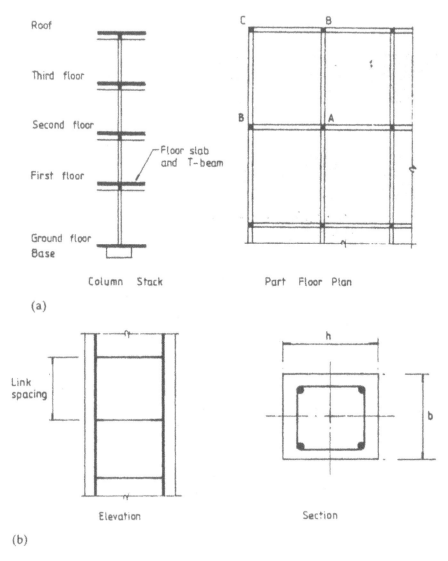

Fig. 9.1 (a) Building column; (b) column construction.

9.1.3 General Code Provisions

In unbraced structures, resistance to lateral forces is provided by bending in the columns and beams in that plane. The column ends can deflect laterally. Fig. 9.2b shows a typical unbraced structure. In a column in an unbraced structure, the axial force and moments in the column are caused not only by the vertical load on the beams but also by the lateral loads acting on the structure and additional moments due to the axial load being eccentric to the deflected column.

It is worth pointing out that most concrete buildings are designed as braced structures. Unbraced structures are rare and are used only if there is a need for uninterrupted floor space.

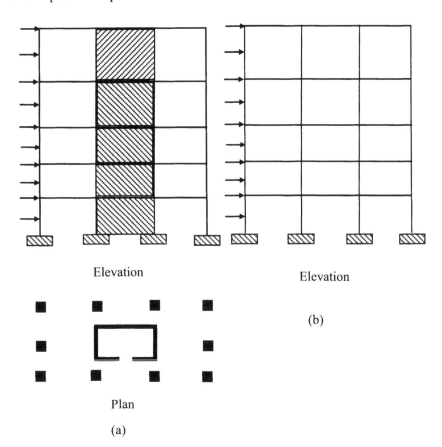

Elevation   Elevation

(b)

Plan

(a)

Fig. 9.2 (a) Braced structure. (b) Unbraced structure.

9.1.4 Practical Design Provisions

The following practical design considerations with regard to design of columns are extracted from section 9.5.2 of the Eurocode 2. The main points from the code are as follows.

(a) Minimum diameter of longitudinal bar
Clause 9.5.2(1) states that the longitudinal bar should have a diameter not less than 8 mm.

(b) Minimum area of reinforcement
Clause 9.5.2(2) states that the total amount of longitudinal reinforcement should not be less than $A_{s, min}$. $A_{s, min} = 0.10\ N_{Ed}\ /f_{yd}$ or $0.002\ A_c$ whichever is greater, where N_{Ed} = design axial force and A_c is the cross sectional area of concrete.

(c) Maximum area of reinforcement
Clause 9.5.2(3) states that the total area of longitudinal reinforcement should not exceed $A_{s, max}$:

$$A_{s, max} = 0.04\ A_c\ \text{outside laps}$$
$$A_{s, max} = 0.08\ A_c\ \text{at laps}$$

(d) Polygonal columns
Clause 9.5.2(4) states that there should be a longitudinal bar at each corner. In the case of circular columns, there should be a minimum of four bars.

(e) Requirements for links
It is necessary to provide transverse reinforcement like links, hoops, helical or spiral reinforcement to prevent the longitudinal reinforcement from buckling. The transverse reinforcement confines the concrete and therefore increases its compressive strength. Fig. 9.3 shows the use of links.
Clause 9.5.3 covers containment of compression reinforcement using transverse reinforcement:

1. The diameter of the transverse reinforcement should not be less than 6 mm or one-quarter of the diameter of the largest longitudinal bar whichever is greater.
2. The maximum spacing is to be $S_{cl, max}$. $S_{cl, max}$ = minimum of:
   - 20 times the diameter of the smallest longitudinal bar
   - The lesser dimension of the column
   - 400 mm
3. The maximum spacing in (2) above can be reduced by a factor of 0.6:
   - In sections within a distance equal to the larger dimension of the column cross section above or below a beam or slab.
   - Near lapped joints, if the maximum diameter of the longitudinal bar is greater than 14 mm. A minimum of three links should be evenly placed in the lap length.

4. Every longitudinal bar or bundle of bars placed in a corner should be held by a link. No bar within a compression zone should be further than 150 mm from a restrained bar.

5. If the direction of bars changes by less than equal to 1 in 12, any change in lateral forced can be ignored.

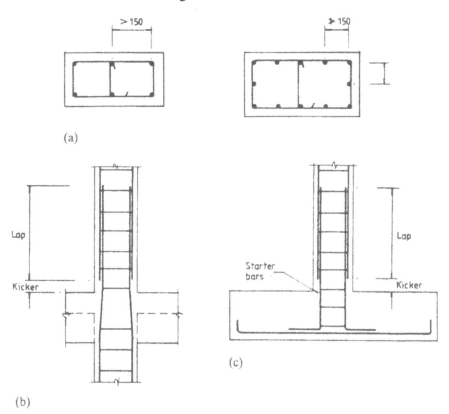

Fig. 9.3 (a) Arrangement of links; (b) column lap; (c) column base.

(f) Compression anchorage length, laps and butt joints
Section 5.3.5, Chapter 5 discusses the topic of laps. Table 5.6, Chapter 5 shows that in the case of compression, $\alpha_1 = \alpha_2 = \alpha_3 = 1$ and the design anchorage length

$$\ell_{bd} = \ell_{b, reqd} \geq \ell_{b, min}$$

Values of $\ell_{b, reqd}$ are shown in Table 5.5, Chapter 5.
The required anchorage length in compression is given by equation (8.7) of the code,

$$\ell_{b, min} > max \,(0.3 \,\ell_{b, reqd}; \, 10 \,\varphi; \, 100 \text{ mm}) \tag{8.7}$$

Using the value of $\ell_{b, reqd}$ from Table 5.5 of Chapter 5, Table 9.1 shows the values of $\ell_{b, min}$.

Table 9.1 Compression anchorage length $\ell_{b, min}$ values for f_{ck} = 25 and 30 MPa, f_{yk} = 500 MPa

φ	$\ell_{b, min}$ mm	
	f_{ck} = 25	f_{ck} = 30
6	100	100
8	100	100
10	120	108
12	144	130
16	192	173
20	240	216
25	300	270
32	384	346
40	528	468

Taking $\alpha_1 = \alpha_2 = \alpha_3 = 1$, the design lap length ℓ_0 from code equation (8.10) is given by

$$\ell_0 = \alpha_6 \, \ell_{b, reqd} \geq \ell_{0, min} \tag{8.10}$$

From equation (8.11) of the code,

$$\ell_{0, min} > \max (0.3 \, \alpha_6 \, \ell_{b, reqd}; \, 15 \, \varphi; \, 200 \text{ mm}) \tag{8.11}$$

$$1.5 \geq [\alpha_6 = \sqrt{(\rho_1/25)}] \geq 1.0$$

where ρ_1 is the percentage of reinforcement lapped with in 0.65 ℓ_0 from the centre of lap length (see Fig. 5.49, Chapter 5 or Fig. 8.8 of Eurocode 2).
Table 9.2 shows the required lap length ℓ_0.

Table 9.2 Lap length ℓ_0 values for f_{ck} = 25 and 30 MPa, f_{yk} = 500 MPa

Dia.	ℓ_0, mm				ℓ_0, mm			
	f_{ck}=25, MPa				f_{ck}=30, MPa			
	α_6				α_6			
	1	1.15	1.4	1.5	1	1.15	1.4	1.5
6	240	276	336	360	216	248	302	324
8	320	368	448	480	288	331	403	432
10	400	460	560	600	360	414	504	540
12	480	552	672	720	432	497	605	648
16	640	736	896	960	576	662	806	864
20	800	920	1120	1200	720	828	1008	1080
25	1000	1150	1400	1500	900	1035	1260	1350
32	1280	1472	1792	1920	1152	1325	1613	1728
40	1760	2024	2464	2640	1560	1794	2184	2340

See section 5.2.5.1, Chapter 5 and Fig. 5.52 for additional information about transverse reinforcement at a lap joint.
Clause 8.7.1(1)P of the code states that the load in compression bars may be transferred by welding, mechanical devices assuring load transfer in tension–compression or in compression only.

9.2 COLUMNS SUBJECTED TO AXIAL LOAD AND BENDING ABOUT ONE AXIS WITH SYMMETRICAL REINFORCEMENT

9.2.1 Code Provisions

The design of columns resisting moment and axial load is similar to the design in bending except that equilibrium should include bending moment and axial load. The bending moment is the sum of moment due to external loads and moment caused by the axial load acting eccentrically to the laterally deflected column.

9.2.2 Section Analysis: Concrete

A reinforced column section subjected to the ultimate axial load N and ultimate moment M is shown in Fig. 9.4. In most cases, columns are symmetrically reinforced because the direction of the moment in most cases is reversible. An additional reason is with unsymmetrical reinforcement there is always the danger of the smaller amount of steel being wrongly placed on the face requiring the larger reinforcement.

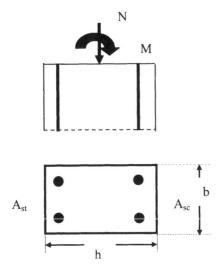

Fig. 9.4 Column subjected to axial force and moment.

Depending on the relative values of M and N, the following two main cases occur for analysis:

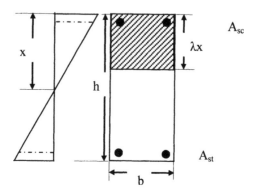

Fig. 9.5 Cross section partly in compression ($\lambda x < h$).

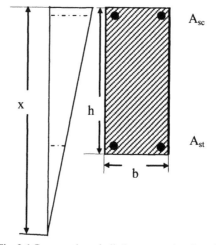

Fig. 9.6 Cross section wholly in compression ($\lambda x > h$).

Case 1: $\lambda x/h \leq 1$

In this case, compression on one side in the concrete and reinforcement and tension in the reinforcement on the other side occurs with the neutral axis lying between the rows of reinforcement. As shown in Fig. 9.5, this case occurs when $\lambda x \leq h$. In this case part of the column cross section is in compression and part in tension. Compressive force in concrete, C_c is give by

$$C_c = \alpha_{cc}\eta\, f_{cd}\, b\, \lambda x$$

$$\frac{C_c}{bh} = \alpha_{cc}\,\eta\, f_{cd}\,\lambda\frac{x}{h}$$

Taking moments about the mid-depth of the section, the moment of the compression force C_c is

$$M_c = C_c \, 0.5 \times (h - \lambda x) = 0.5 \alpha_{cc} \eta f_{cd} \, bh^2 \, \lambda(\tfrac{x}{h})(1 - \lambda(\tfrac{x}{h}))$$

$$\frac{M_c}{bh^2} = 0.5 \alpha_{cc} \eta f_{cd} \, bh^2 \, \lambda(\tfrac{x}{h})\{1 - \lambda(\tfrac{x}{h})\}$$

Case 2: λx/h > 1

In this case compression occurs over the whole section with the neutral axis lying at the edge or outside the section with both rows of steel bars in compression. As shown in Fig. 9.6, this case occurs when $\lambda x > h$.
Compressive force in concrete, C_c is given by

$$C_c = \alpha_{cc} \eta \, f_{cd} \, b \, h$$

$$\frac{C_c}{bh} = \alpha_{cc} \eta \, f_{cd}$$

Taking moments about the mid-depth of the section, the moment of the compression force is

$$\frac{M_c}{bh^2} = 0$$

9.2.3 Stresses and Strains in Steel

For all positions of the neutral axis, the strains in the compression and tension steels are

$$\varepsilon_s' = \varepsilon_{cu3} \frac{(x - d')}{x} = \varepsilon_{cu3}(1 - \frac{d'}{h}\frac{h}{x})$$

$$\varepsilon_s = \varepsilon_{cu3} \frac{(d - x)}{x} = \varepsilon_{cu3}(\frac{d}{h}\frac{h}{x} - 1)$$

The stresses in the compression and tension steels are

$$f_{sc} = E \, \varepsilon_{sc} \le f_{yd} \text{ and } f_{st} = E \, \varepsilon_{st} \le f_{yd}$$

where E = Young's modulus for steel.

Note that when $x > d$, then ε_{st} becomes negative, indicating that the stress in the 'tension' steel is actually compressive. The force C_s in compression and force T in tension steel are

$$C_s = A_{sc} \, f_{sc}, \; T = A_{st} \, f_{st}$$

9.2.4 Axial Force N and Moment M

The sum of the internal forces is

$$N = C_c + C_s - T$$

$$N = C_c + A_s' \, f_s' - A_s \, f_s$$

$$\frac{N}{bh} = \frac{C_c}{bh} + \frac{A_{sc}}{bh} f_{sc} - \frac{A_{st}}{bh} f_{st}$$

The sum of the moments of the internal forces about the centre line of the column is

$$M = M_c + C_s(0.5h - d') + T(d - 0.5h)$$

$$\frac{M}{bh^2} = \frac{M_c}{bh^2} + \frac{A_{sc}}{bh} f_{sc}(0.5 - \frac{d'}{h}) + \frac{A_{st}}{bh} f_{st}(\frac{d}{h} - 0.5)$$

Using the above equations, it is not possible to directly design a section to carry a given load and moment. It is necessary to assume a trial section and the required amount of steel can be determined using design charts constructed using the above equations.

9.2.5 Construction of Column Design Chart

A design curve can be drawn for a selected grade of concrete and reinforcing steel for a section with a given percentage of reinforcement, $100A_s/(bh)$, symmetrically placed at a given location d/h. The curve is formed by plotting values of $N/(bh)$ against $M/(bh^2)$ for various positions of the neutral axis x. Other curves can be constructed for percentages of steel ranging from a minimum value of 0.4% to a maximum value of 4% for vertically cast columns. The family of curves forms the design chart for that combination of materials and steel location. Separate charts are required for the same materials for different values of d/h which determines the location of the reinforcement in the section. Groups of charts are required for the various combinations of concrete and steel grades.

The process for construction of a design chart is demonstrated below.
1. Select materials: Concrete f_{ck} = 30 MPa, Reinforcement f_{yk} = 500 MPa.
2. Select a value of d/h = 0.95, d'/h = 0.05.
3. Select a total steel percentage $100A_s/(bh)$ = say, 4.
Let the steel be symmetrically placed and A_{sc}/bh = 0.02 and A_{st}/bh = 0.02.
The design chart is constructed by selecting different values of x/h and calculating the corresponding $N/(bh)$ and $M/(bh^2)$.

$$\frac{N}{bh} = \frac{C_c}{bh} + \frac{A_{sc}}{bh} f_{sc} - \frac{A_{st}}{bh} f_{st}$$

$$N/(bh) = C_c/(bh) + 0.02(f_{sc} - f_{st})$$

$$M/(bh^2) = \frac{M_c}{bh^2} + 0.02 \times 0.45 \times \{f_{sc} + f_{st}\}$$

For f_{ck} = 30 MPa, f_{cd} = 30/1.5 = 20 MPa, λ = 0.8, η = 1, α_{cc} = 1.0. ε_{cu3} = 0.0035.
Note that for U.K. practice α_{cc} = 0.85.

The equations for calculating $C_c/(bh)$ and $M_c/(bh^2)$ to be used depend upon the value of x/h. They can be summarised as follows.

$$\frac{C_c}{bh} = \alpha_{cc}\,\eta\,f_{cd}\,\lambda\frac{x}{h} = 16.0\frac{x}{h},\ \frac{x}{h} \le 1.25$$

$$\frac{C_c}{bh} = \alpha_{cc}\,\eta\,f_{cd} = 16.0,\ \frac{x}{h} > 1.25$$

$$\frac{M_c}{bh^2} = 0.5\alpha_{cc}\,\eta\,f_{cd}\,\lambda(\frac{x}{h})\{1 - \lambda(\frac{x}{h})\} = 8\frac{x}{h}(1 - 0.8\frac{x}{h}),\ \frac{x}{h} \le 1.25$$

$$\frac{M_c}{bh^2} = 0,\ \frac{x}{h} > 1.25$$

9.2.5.1 Typical Calculations for Rectangular Stress Block

(a) Choose material properties:

Concrete: f_{ck} = 30 MPa, f_{cd} = $f_{ck}/1.5$ = 20 MPa,

λ = 0.8, η = 1, α_{cc} = 1 (U.K. practice α_{cc} = 0.85)

ε_{cu3} = 0.0035

Steel: f_{yk} = 500 MPa, f_{yd} = $f_{yk}/1.15$ = 435 MPa

Young's modulus for steel, E_s = 200 × 10^3 MPa

(b) Choose a value for x/h:

(i) Let x/h = 0.4.

Using the equations for $x/h \le \lambda$

$$\frac{C_c}{bh} = 16.0\frac{x}{h}, = 6.4$$

$$\frac{M_c}{bh^2} = 8\frac{x}{h}(1 - 0.8\frac{x}{h}) = 2.176$$

$$\varepsilon_{sc} = \varepsilon_{cu3}\frac{(x - d')}{x} = 0.0035 \times (1 - 0.05\frac{h}{x}) = 0.00306$$

$$f_{sc} = 0.00306 \times 200 \times 10^3 = 613\,\text{MPa} > 435, f_{sc} = 435\,\text{MPa}$$

$$\varepsilon_{st} = \varepsilon_{cu3}\frac{(d - x)}{x} = 0.0035 \times (0.95\frac{h}{x} - 1) = 0.00481$$

$$f_{st} = 0.00481 \times 200 \times 10^3 = 963\,\text{MPa} > 435, f_{st} = 435\,\text{MPa}$$

$$N/(bh) = C_c/(bh) + 0.03\,(f_{sc} - f_{st}) = 4.86 + 0.0 = 6.40$$

$$\frac{M_c}{bh^2} = 8\frac{x}{h}(1 - 0.8\frac{x}{h}) = 2.176$$

$$M/(bh^2) = \frac{M_c}{bh^2} + 0.02 \times 0.45 \times \{f_{sc} + f_{st}\}$$

$$= 2.176 + 0.02 \times 0.45 \times \{435 + 435\} = 10.006$$

(ii) x/h = 1.4.

Using the equations for $x/h > \lambda$

$$\frac{C_c}{bh} = \alpha_{cc}\, \eta\, f_{cd} = 20.0$$

$$\frac{M_c}{bh} = 0$$

$$\varepsilon_{sc} = \varepsilon_{cu3}\frac{(x-d')}{x} = 0.0035 \times (1 - 0.05\frac{h}{x}) = 0.00333$$

$$f_{sc} = 0.00333 \times 200 \times 10^3 = 665\,\text{MPa} > 435, f_{sc} = 435\,\text{MPa}$$

$$\varepsilon_{st} = \varepsilon_{cu3}\frac{(d-x)}{x} = 0.0035 \times (0.95\frac{h}{x} - 1) = -0.00113$$

$$f_{st} = -0.00113 \times 200 \times 10^3 = -225\text{MPa}$$
$$N/(bh) = C_c/(bh) + 0.02\,(f_{sc} - f_{st})$$
$$N/(bh) = 20.0 + 0.02 \times \{435 - (-225)\} = 33.2$$
$$M/(bh^2) = 0 + 0.02 \times (0.45) \times \{435 - 225\} = 1.89$$

(iii) $x/h = 2.0$.
Using the equations for $x/h > \lambda$

$$\frac{C_c}{bh} = \alpha_{cc}\, \eta\, f_{cd} = 20.0$$

$$\frac{M_c}{bh} = 0$$

$$\varepsilon_{sc} = \varepsilon_{cu3}\frac{(x-d')}{x} = 0.0035 \times (1 - 0.05\frac{h}{x}) = 0.00331$$

$$f_{sc} = 0.00341 \times 200 \times 10^3 = 683\,\text{MPa} > 435, f_{sc} = 435\,\text{MPa}$$

$$\varepsilon_{st} = \varepsilon_{cu3}\frac{(d-x)}{x} = 0.0035 \times (0.95\frac{h}{x} - 1) = -0.00184$$

$$f_{st} = -0.00184 \times 200 \times 10^3 = -368\,\text{MPa}$$
$$N/(bh) = C_c/(bh) + 0.02\,(f_{sc} - f_{st})$$
$$N/(bh) = 20.0 + 0.02 \times \{435 - (-368)\} = 36.1$$
$$M/(bh^2) = 0 + 0.02 \times (0.45) \times \{435 - 368\} = 0.603$$

In a similar manner calculations can be carried out for other values of x/h. Calculations are most conveniently done using a spreadsheet. Using the results, a graph of N/ (bh) versus M/ (bh²) can be drawn as shown in Fig. 9.7.

As is to be expected, when x > d, the 'tension' reinforcement goes into compression. This naturally increases the value of N/ (bh) but drastically decreases the value of M/ (bh²). When the entire column section is under a compressive stress of $\alpha_{cc}\,\eta\,f_{cd}$ and the stress in both steels is f_{yd} compression, then the maximum value of N/ (bh) is attained and the corresponding value of M/ (bh²) is equal to zero.

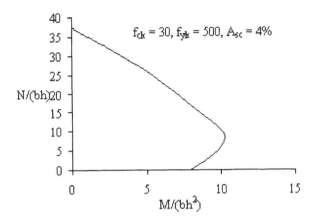

Fig. 9.7 Column design chart for 4% steel.

Curves for total steel percentages 1, 2, 3, and 4 can be plotted. The design chart is shown in Fig. 9.8. Other charts are required for different values of the ratio d/h to give a series of charts for a given concrete and steel strength. A separate series of charts is required for each combination of materials used.

It has to be noted that any combination of $\{N/(bh), M(bh^2)\}$ which lies on or inside the curve corresponding to a particular value of $A_{sc}/(bh)$ leads to a safe design.

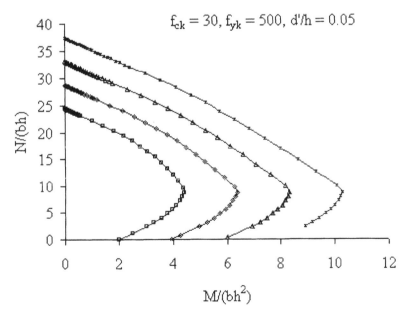

Fig. 9.8 Column design chart.

9.2.5.2 Column Design Using Design Chart

At a cross section in a column, design loads at ultimate are an axial load of
N = 1480 kN and a moment M = 54 kNm. The column section is 300 mm × 250
mm. Determine the area of steel required. The materials are f_{ck} = 30 MPa concrete
and f_{yk} = 500 MPa reinforcement.
Assume 25 mm diameter bars for the main reinforcement and 8 mm diameter links.
The cover on the links is 25 mm.
$$b = h = 250 \text{ mm}$$
$$d = 250 - 25 - 8 - 12.5 = 204 \text{ mm}$$
$$d/h = 204/250 = 0.82$$
Use the chart shown in Fig. 9.17 where d/h = 0.85.
$$N/ (bh) = 1480 \times 10^3/ (250 \times 300) = 16.4$$
$$M/ (bh^2) = 54 \times 10^6 (250 \times 300^2) = 2.4$$
For this combination of $\{N/ (bh), M/ (bh^2)\}$, the design chart for $100A_s/ (bh) = 2$
shown in Fig. 9.9 indicates a safe design.
$$A_{sc} = 2.0 \times 300 \times 250/ 100 = 1500 \text{ mm}^2$$
Provide four 25 mm diameter bars to give an area of 1963 mm².

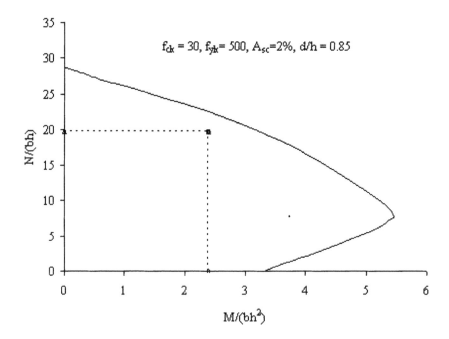

Fig. 9.9 Column design chart for example.

9.2.5.3 Three Layers of Steel Design Chart

The design chart shown in Fig. 9.7 strictly applies only to the case where the symmetrical reinforcement is placed on two opposite faces. Charts can be constructed for other arrangements of reinforcement. One such case is shown in Fig. 9.10 where eight bars are spaced evenly around the perimeter of the column. The total steel A_s is placed such that at the top and bottom rows steel is $0.375\,A_s$ (3 bars) and in the middle row it is $0.25\,A_s$ (2 bars).

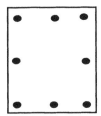

Fig. 9.10 Column with three rows of steel.

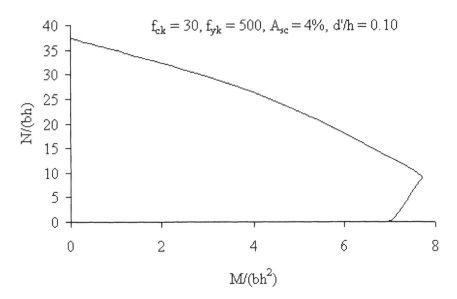

Fig. 9.11 Column design chart for three layers of reinforcement.

The contribution from concrete is calculated as in the previous section. There are however three strains to calculate.

$$\varepsilon_{sc} = \varepsilon_{cu3}\frac{(x - d')}{x} = \varepsilon_{cu3}\left(1 - \frac{d'}{h}\frac{h}{x}\right) \quad \text{(top layer)}$$

$$\varepsilon_{s1} = \varepsilon_{cu3}\frac{(0.5h - x)}{x} = \varepsilon_{cu3}(0.5\frac{h}{x} - 1) \text{ (middle layer)}$$

$$\varepsilon_{s2} = \varepsilon_{cu3}\frac{(d - x)}{x} = \varepsilon_{cu3}(\frac{d}{h}\frac{h}{x} - 1) \text{ (bottom layer)}$$

The stresses in the compression and tension steels are calculated from strains as before.

$$C_s = A_{sc}\,f_{sc}$$
$$T_1 = A_{st1}\,f_{st1}, \quad T_2 = A_{st2}\,f_{st2}$$
$$N = C_c + C_s - T_1 - T_2$$

$$\frac{N}{bh} = \frac{C_c}{bh} + \frac{A_{sc}}{bh}f_{sc} - \frac{A_{s1}}{bh}f_{st1} - \frac{A_{s2}}{bh}f_{st2}$$

The sum of the moments of the internal forces about the centre line of the column is

$$M = M_c + C_s(0.5h - d') + T_1(0.5h - 0.5h) + T_2(d - 0.5h)$$

$$= 0.5(h - \lambda x)\times C_c + C_s(0.5h - d') + T_2(d - 0.5h)$$

$$\frac{M}{bh^2} = \frac{M_c}{bh^2} + \frac{A_{sc}}{bh}f_{sc}(0.5 - \frac{d'}{h}) + \frac{A_{st2}}{bh}f_{st2}(\frac{d}{h} - 0.5)$$

Note that the middle layer steel has zero lever arm about the centre line and hence does not contribute to moment of resistance.
Fig. 9.11 shows the column design chart for this case.

9.3 COLUMNS SUBJECTED TO AXIAL LOAD AND BENDING ABOUT ONE AXIS: UNSYMMETRICAL REINFORCEMENT

An unsymmetrical arrangement of reinforcement provides the most economical solution for the design of a column subjected to a small axial load and a large moment about one axis. Such members occur in single storey reinforced concrete portals. Design charts for such cases can be constructed. If the total steel area is 4 percent, say, but is distributed such that the tension steel is 3 percent and compression steel is 1 percent, then the corresponding design chart is as shown in Fig. 9.12.

When the ratio $(x/h) \approx 1.18$, the stresses in both the steels are compressive and are respectively 195 MPa in 'tension' steel and 435 MPa in 'compression' steel. The forces in the two steels cause a moment equal to that caused by the compressive stress in concrete with $M/(bh^2) \approx 0$ and $N/(bh) = 29.1$. When the ratio $(x/h) = 2.25$, the stresses in the two steels are equal to $-f_{yd}$ and the entire column is almost in a state of uniform compression and the maximum value of $N/(bh) = 37.39$ is reached. However because of the fact that the compressive forces in the two steel are not equal, the force in the 'tension' steel gives rise to a negative value of $M/(bh^2) = -3.04$. However if the reinforcement is symmetrically distributed, then $N/(bh) = 37.39$ and $M/(bh^2)$ will be zero.

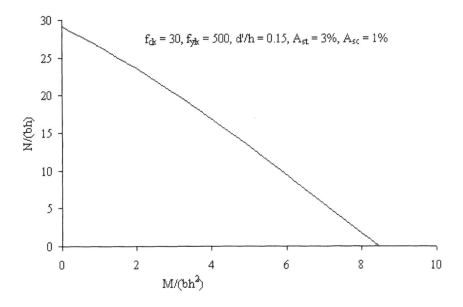

Fig. 9.12 Column design chart: unsymmetrical reinforcement.

9.3.1 Example of a Column Section Subjected to Axial Load and Moment: Unsymmetrical Reinforcement

At a cross section in a column, the ultimate design loads are an axial load $N = 230$ kN and moment $M = 244$ kNm. Design the reinforcement required using an unsymmetrical arrangement. The concrete is $f_{ck} = 30$ MPa and the reinforcement is $f_{yk} = 500$ MPa.

Assuming $d'/h = 0.15$ and $d/h = 0.85$ and because of the large moment, assume a rectangular section with $b = 300$ mm and $h = 350$ mm.

$$N/(bh) = 230 \times 10^3/(300 \times 350) = 2.19$$
$$M/(bh^2) = 244 \times 10^6/(300 \times 350^2) = 6.64$$

Assume $A_s/(bh) = 1\%$ and $A_s/(bh) = 2\%$ and draw the design chart as shown in Fig. 9.13. As shown in Fig. 9.12, the combination ($M/(bh^2) = 6.64$, $N/(bh) = 2.19$) is inside the interaction curve, indicating that it is a safe design.

Calculations show that at $(x/h) = 0.41$, $-f_{sc} = f_{st} = f_{yd}$, indicating that both steel yield. Approximately only a third of the column cross section is in compression.

$$N/(bh) = 2.20, M/(bh^2) = 6.77.$$

$A_s' = 0.01 \times 300 \times 400 = 1200$ mm^2, $A_s = 0.02 \times 300 \times 400 = 2400$ mm^2
Provide 3H25 on the compression face, $A_{sc} = 1473$ mm^2 and 5H25 on the tension face, $A_{st} = 2454$ mm^2.

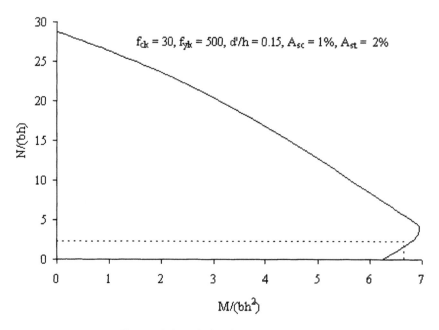

Fig. 9.13 Column design chart: $A_{st} = 2\%$, $A_{sc} = 1\%$.

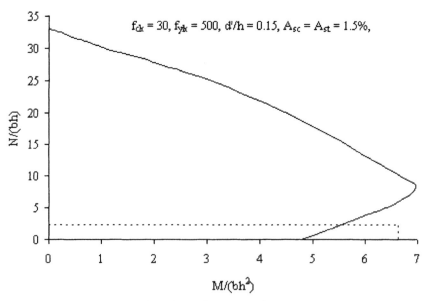

Fig. 9.14 Column design chart: $A_{st} = 1.5\%$, $A_{sc} = 1.5\%$.

If the column had been symmetrically reinforced, then for total of 3% steel, assuming $A_s'/(bh) = 1.5\%$ and $A_s/(bh) = 1.5\%$, the design chart is as shown in Fig. 9.14. As can be seen, it leads to an unsafe design.

9.4 COLUMN SECTIONS SUBJECTED TO AXIAL LOAD AND BIAXIAL BENDING

9.4.1 Outline of the Problem

When a column is subjected to an axial force and a bending moment about, say, the x-axis, the neutral axis is parallel to the x-axis. However when a column is subjected to an axial force and moments about the two axes, the neutral axis is inclined to the x-axis as shown in Fig. 9.15.

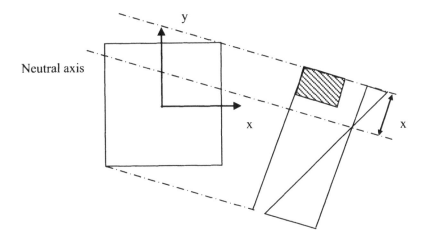

Fig. 9.15 Biaxial bending causes inclined neutral axis.

For a given location and direction of the neutral axis the strain diagram can be drawn with the maximum strain in the concrete of ε_{cu3}. The strains in the compression and tension steel can be found and the corresponding stresses determined from the stress–strain diagram for the reinforcement. The resultant forces C_s and T in the compression and tension steel and the force C_c in the concrete can be calculated and their locations determined. The net axial force is
$$N = C_c + C_s - T$$
Moments of the forces C_c, C_s and T are taken about the XX and YY axes to give M_x and M_y.

Thus a given section can be analysed for a given location and direction of the neutral axis and the axial force and biaxial moments that it can support can be determined. As in the case of axial load with moment about one axis only, a failure surface can be constructed. Calculations are naturally much more involved than in the case of axial load accompanied by moment about one axis.

9.4.1.1 Expressions for Contribution to Moment and Axial Force by Concrete

Fig. 9.16 shows a rectangular column b × h and reinforced with four bars.

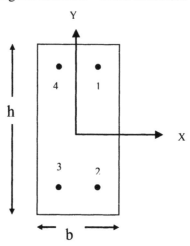

Fig. 9.16 Column subjected to axial load and biaxial moments.

Assuming the origin of coordinates at the centroid of the column cross section, the coordinates of the four bars can be calculated. The position of the neutral axis is governed by two parameters α and β as shown in Fig. 9.17. Assuming that the maximum compressive strain ε_{cu3} is at the top right hand corner of the column, the normal strain in the cross section is given by

$$\varepsilon = \varepsilon_{cu3} \{C_1 + C_2 (x/b) + C_3 (y/h)\}$$

The constants can be calculated from the boundary conditions as follows:

$$\varepsilon = \varepsilon_{cu3} \text{ at } (x/b = 0.5, y/h = 0.5),$$
$$\varepsilon = 0 \text{ at } (x/b = (0.5 - \beta), y/h = 0.5),$$
$$\varepsilon = 0 \text{ at } (x/b = 0.5, y/h = (0.5 - \alpha))$$

Solving for the constants:

$$C_1 = 1 - 1/ (2\beta) - 1/ (2\alpha), \quad C_2 = 1/\beta, \quad C_3 = 1/\alpha$$

$$\varepsilon = \varepsilon_{cu3} \{1 + \frac{(\frac{x}{b} - 0.5)}{\beta} + \frac{(\frac{y}{h} - 0.5)}{\alpha}\}$$

The strain in the bars can be calculated by substituting the appropriate coordinates of the bars. The stress σ in the bars is equal to σ = E ε but numerically not greater than f_{yd}.

Assuming a rectangular stress block with constant stress of $\alpha_{cc}\,\eta\,f_{cd}$ and a depth equal to λ times the depth of the neutral axis, depending on the position of the neutral axis, expressions for the compressive force and the corresponding moments about the x- and y-axes due to the compressive stress in the column can be derived as follows.

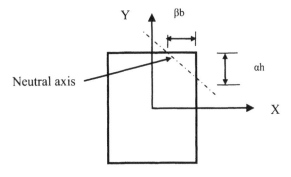

Fig. 9.17 Column with the neutral axis inclined to x-axis.

Case 1: $\lambda\beta \leq 1.0$ and $\lambda\alpha \leq 1.0$
From the triangular shape of the stress block shown in Fig. 9.18,
$$N_c = \alpha_{cc}\,\eta\,f_{cd}\,\{0.5 \times \lambda\,\alpha h \times \lambda\,\beta b\},$$
$$M_{xc} = N_c \times (0.5\,h - \lambda\,\alpha h/3),$$
$$M_{yc} = N_c \times (0.5\,b - \lambda\,\beta b/3)$$

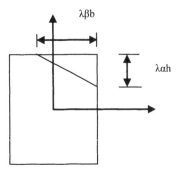

Fig. 9.18 Neutral axis position for Case 1.

Case 2: $\lambda\,\beta > 1.0$ and $\lambda\,\alpha \leq 1.0$
From the trapezoidal stress block shown in Fig. 9.19,
$$\alpha h_1 = \frac{\alpha h}{\beta b}(\lambda\beta b - b) = \alpha h(\lambda - \frac{1}{\beta})$$
$$N_c = \alpha_{cc}\,\eta\,f_{cd}\,\{0.5\,(\alpha h_1 + \lambda\alpha h)\,b\},$$
$$M_{xc} = N_c \times (0.5\,h - ybar),$$

$$M_{yc} = N_c \times (0.5\,b - xbar)$$

Position of centroid from right face of the trapezium:

$$xbar = \frac{b}{3}\frac{(2h_1 + \lambda h)}{(h_1 + \lambda h)}$$

Position of centroid from top face of the trapezium:

$$ybar = \frac{\alpha}{3}\frac{(h_1^2 + \lambda^2 h^2 + \lambda h_1 h)}{(h_1 + \lambda h)}$$

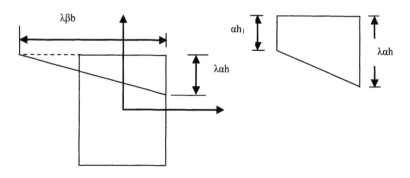

Fig. 9.19 Neutral axis position for case 2.

Case 3: $\lambda\beta \le 1.0$ and $\lambda\alpha > 1.0$

From the trapezoidal stress block shown in Fig. 9.20

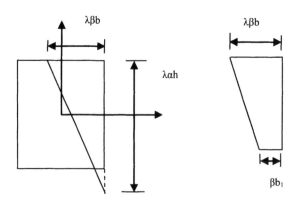

Fig. 9.20 Neutral axis position for case 3.

$$\beta b_1 = \frac{\beta b}{\alpha h}(\lambda\alpha h - h) = \beta b(\lambda - \frac{1}{\alpha})$$

$$N_c = \alpha_{cc}\,\eta\,f_{cd}\,\{0.5\,(\beta b_1 + \lambda\beta b)\,h\},$$

$$M_{xc} = N_c \times (0.5\,h - ybar),$$
$$M_{yc} = N_c \times (0.5\,b - xbar)$$

Position of centroid from right face of the trapezium:
$$xbar = \frac{\beta}{3} \frac{(b_1^2 + \lambda^2 b^2 + \lambda b_1 b)}{(b_1 + \lambda b)}$$

Position of centroid from top face of the trapezium:
$$ybar = \frac{h}{3} \frac{(2b_1 + \lambda b)}{(b_1 + \lambda b)}$$

Case 4: $\lambda\beta > 1.0$ and $\lambda\alpha > 1.0$
The five-sided stress block shown in Fig. 9.21 can be considered as compression over the entire column cross section with tension in the triangular area in the left hand bottom corner. Compression over the entire column does not give rise to any moment. Moment is caused purely by the tension in the triangular area.

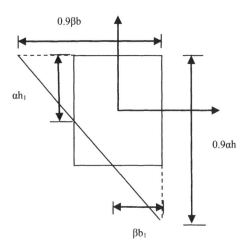

Fig. 9.21 Neutral axis position for case 4.

$$\alpha h_1 = \{\frac{\alpha h}{\beta b}(\lambda\beta b - b) = \alpha h(\lambda - \frac{1}{\beta})\} \le h$$
$$\beta b_1 = \{\frac{\beta b}{\alpha h}(\lambda\alpha h - h) = \beta b(\lambda - \frac{1}{\alpha})\} \le b$$
$$N_c = \alpha_{cc}\,\eta\,f_{cd}\,\{bh - 0.5\,(h - \alpha h_1)\,(b - \beta b_1)\}$$
$$M_{xc} = \alpha_{cc}\,\eta\,f_{cd}\,\frac{1}{2}(h - \alpha h_1)(b - \beta b_1)\{0.5h - \frac{1}{3}(h - \alpha h_1)\}$$

$$M_{yc} = \alpha_{cc} \, \eta f_{cd} \frac{1}{2} (h - \alpha h_1)(b - \beta b_1)\{0.5b - \frac{1}{3}(b - \beta b_1)\}$$

9.4.1.2 Example of Design Chart for Axial Force and Biaxial Moments

Consider a rectangular column b × h and reinforced with four bars as shown in Fig. 9.16. The total steel area A_s = 4% of bh, f_{ck} = 30 MPa, f_{yk} = 500 MPa, b/h = 0.5. The bars are located at 0.15h from top and bottom faces and at 0.35b from the sides. The coordinates (x/b, y/h) of the four bars are:
1: (0.2, 0.35), 2: (0.2, –0.35), 3: (–0.2, –0.35), 4: (–0.2, 0.35)
f_{ck} = 30 MPa, ε_{cu3} = 0.0035, λ = 0.8, η = 1, f_{cd} = 30/1.5 = 20 MPa,
f_{yk} = 500 MPa, f_{yd} = 500/1.15 = 435 MPa, E_s = 200 × 10³ MPa
Calculate N/ (bh), M_x/ (bh²) and M_y/ (b²h) for the following positions of the neutral axis.

(a) Assuming α =0.65, β = 0.95
Calculate the strains (positive is compressive) in steel from the equation

$$\varepsilon = 0.0035\{1 + \frac{(\frac{x}{b} - 0.5)}{\beta} + \frac{(\frac{y}{h} - 0.5)}{\alpha}\}$$

The strains in the four bars are respectively:
1.587 × 10⁻³, –2.182 × 10⁻³, –3.659 × 10⁻³, 0.113 × 10⁻³
Taking Young's modulus for steel E_s = 200 × 10³ MPa, the corresponding stresses are: 317, – 435, – 435 and 23 MPa
The contribution of the stresses in steel to:
N_s = (317 – 435 – 435 + 23) × (0.02 bh/4) = –2.652 bh (tensile)
M_{xs} = (317 + 435 + 435 + 23) × (0.02 bh/4) × 0.35h = 2.12 bh²
M_{ys} = (317 – 435 + 435 – 23) × (0.02 bh/4) × 0.2b = 0.294 b² h
Calculate the contribution of the compressive stress in concrete using:
α =0.65, β = 0.95, λ = 0.8, λα =0.52, λβ = 0.76
From the triangular stress block shown in Fig. 9.25, the contributions of the compressive stress in concrete to the forces are:
N_c = f_{cd} × 0.5 × (0.52 h × 0.76 b) = 3.952 bh
M_{xc} = N_c × (0.5 h – 0.52 h/3) = 1.291 bh²
M_{yc} = N_c × (0.5 b – 0.76 b/3) = 0.975 b²h
Adding the contribution of steel and concrete:
N/ (bh) = 3.952 – 2.652 = 1.3,
M_x/ (bh²) = 1.291 + 2.12 = 3.41,
M_y/ (b²h) = 0.975 + 0.294 = 1.269

(b) Assuming α =0.65, β = 1.35
The strains in the four bars are respectively:
1.915 × 10⁻³, –1.85 × 10⁻³, –2.89 × 10⁻³, 0.877 × 10⁻³
The corresponding stresses are: 383, –371, –435 and 175 MPa

The contribution of the stresses in steel to:

$$N_s = (383 - 371 - 435 + 175) \times (0.02\ bh/4) = -1.24\ bh\ \text{(tensile)}$$
$$M_{xs} = (383 + 371 + 435 + 175) \times (0.02\ bh/4) \times 0.35h = 2.39\ bh^2$$
$$M_{ys} = (383 - 371 + 435 - 175) \times (0.02\ bh/4) \times 0.2b = 0.273\ b^2 h$$

Calculate the contribution of the compressive stress in concrete using:

$$\alpha = 0.65,\ \beta = 1.35,\ \lambda = 0.8,\ \lambda\,\alpha = 0.52,\ \lambda\,\beta = 1.08$$

From the trapezoidal stress block shown in Fig. 9.26, the contributions of the compressive stress in concrete to the forces are:

$$\alpha h_1 = \frac{\alpha h}{\beta b}(\lambda \beta b - b) = \alpha h(\lambda - \frac{1}{\beta}) = 0.039\ h$$

$$h_1/h = 0.06$$

$$N_c = f_{cd} \times \{0.5\ (\alpha h_1 + \lambda \alpha h)\ b\} = 5.59\ bh$$

$$ybar = \frac{\alpha}{3} \frac{(h_1^2 + \lambda^2 h^2 + \lambda h_1 h)}{(h_1 + \lambda h)} = 0.174\ h$$

$$M_{xc} = N_c \times (0.5\ h - ybar) = 1.821\ bh^2$$

$$xbar = \frac{b}{3} \frac{(2h_1 + \lambda h)}{(h_1 + \lambda h)} = 0.357\ b$$

$$M_{yc} = N_c \times (0.5\ b - xbar) = 0.802\ b^2 h$$

Adding the contribution of steel and concrete:

$$N/ (bh) = 5.59 - 1.24 = 4.35,$$
$$M_x/ (bh^2) = 1.821 + 2.39 = 4.211,$$
$$M_y/ (b^2 h) = 0.802 + 0.273 = 1.075$$

(c) Assuming $\alpha = 1.2,\ \beta = 0.65$

The strains in the four bars are respectively:

$$1.447 \times 10^{-3}, -0.59 \times 10^{-3}, -2.75 \times 10^{-3}, -0.71 \times 10^{-3}$$

The corresponding stresses are: 289, -119, -435 and -141 MPa

The contribution of the stresses in steel to:

$$N_s = (289 - 119 - 435 - 141) \times (0.02\ bh/4) = -2.04\ bh\ \text{(tensile)}$$
$$M_{xs} = (289 + 119 + 435 - 141) \times (0.02\ bh/4) \times 0.35h = 1.232\ bh^2$$
$$M_{ys} = (289 - 119 + 435 + 141) \times (0.02\ bh/4) \times 0.2b = 0.748\ b^2 h$$

Calculate the contribution of the compressive stress in concrete using:

$$\alpha = 1.2,\ \beta = 0.65,\ \lambda = 0.8,\ \lambda\,\alpha = 0.96,\ \lambda\,\beta = 0.52$$

From the trapezoidal stress block shown in Fig. 9.27, the contributions of the compressive stress in concrete to the forces are:

$$\beta b_1 = \frac{\beta b}{\alpha h}(\lambda \alpha h - h) = \beta b(\lambda - \frac{1}{\alpha}) = -0.022\ b$$

$$b_1/b = -0.033$$

$$N_c = f_{cd} \times \{0.5\ (\beta b_1 + \lambda \beta b)\ h\} = 4.986\ bh$$

$$ybar = \frac{h}{3} \frac{(2b_1 + \lambda\ b)}{(b_1 + \lambda b)} = 0.319\ h$$

$$M_{xc} = N_c \times (0.5\ h - ybar) = 0.901\ bh^2$$

$$\text{xbar} = \frac{\beta}{3} \frac{(b_1^2 + \lambda^2 b^2 + \lambda b_1 b)}{(b_1 + \lambda b)} = 0.174\,b$$

$$M_{yc} = N_c \times (0.5\,b - \text{xbar}) = 1.625\,b^2 h$$

Adding the contribution of steel and concrete:

$$N/\,(bh) = 4.986 - 2.04 = 2.946$$
$$M_x/\,(bh^2) = 0.901 + 1.232 = 2.133$$
$$M_y/\,(b^2 h) = 1.625 + 0.748 = 2.373$$

(d) Assuming α =1.3, β = 1.5

Calculate the strains (positive is compressive) in steel.
The strains in the four bars are respectively:

$$2.396 \times 10^{-3},\ 0.512 \times 10^{-3},\ -0.42 \times 10^{-3},\ 1.463 \times 10^{-3}$$

The corresponding stresses are: 435, 102, –84 and 293 MPa
The contribution of the stresses in steel to:

$$N_s = (435 + 102 - 84 + 293) \times (0.02\,bh/4) = 3.74\,bh$$
$$M_{xs} = (435 - 102 + 84 + 293) \times (0.02\,bh/4) \times 0.35h = 1.25\,bh^2$$
$$M_{ys} = (435 + 102 + 84 - 293) \times (0.02\,bh/4) \times 0.2b = 0.331\,b^2 h$$

Calculate the contribution of the compressive stress in concrete using:

$$\alpha = 1.3,\ \beta = 1.5,\ \lambda = 0.8,\ \lambda\,\alpha = 1.04,\ \lambda\,\beta = 1.20,$$

From the trapezoidal stress block shown in Fig. 9.28, the contributions of the compressive stress in concrete to the forces are:

$$\alpha h_1 = \{\frac{\alpha h}{\beta b}(\lambda \beta b - b) = \alpha h(\lambda - \frac{1}{\beta})\} \le h\ ,\ \alpha h_1 = 0.1733\,h$$

$$\beta b_1 = \{\frac{\beta b}{\alpha h}(\lambda \alpha h - h) = \beta b(\lambda - \frac{1}{\alpha})\} \le b\ ,\ \beta b_1 = 0.046\,b$$

$$N_c = f_{cd}\{bh - 0.5\,(h - \alpha h_1)\,(b - \beta b_1)\} = 12.11\,bh$$

$$M_{xc} = f_{cd}\frac{1}{2}(h - \alpha h_1)(b - \beta b_1)\{0.5h - \frac{1}{3}(h - \alpha h_1)\} = 1.767\,bh^2$$

$$M_{yc} = f_{cd}\frac{1}{2}(h - \alpha h_1)(b - \beta b_1)\{0.5b - \frac{1}{3}(b - \beta b_1)\} = 1.435\,b^2 h$$

Adding the contribution of steel and concrete:

$$N/\,(bh) = 12.11 + 3.74 = 15.85$$
$$M_x/\,(bh^2) = 1.767 + 1.25 = 3.017$$
$$M_y/\,(b^2 h) = 1.435 + 0.331 = 1.766$$

9.4.1.3 Axial Force–Biaxial Moment Interaction Curve

Calculations similar to that in the previous section can be done. Assuming $f_{ck} = 30$ MPa, $f_{yk} = 500$ MPa, $A_s/\,(bh) = 4\%$, $x/b = 0.2$, $y/h = 0.35$. The results are shown in Table 9.3 for $M_y/\,(hb^2) = 2.0$. The interaction curve is shown in Fig. 9.22.

Table 9.3 Calculations for biaxial column design curve

α	β	$M_x/(bh^2)$	$N/(bh)$
2	1.5	2.7	27.4
1.9	1.5213	2.9	27.0
1.8	1.5381	3.1	26.5
1.7	1.5554	3.3	25.8
1.6	1.5726	3.6	25.0
1.5	1.5877	3.8	24.0
1.4	1.6	4.1	22.8
1.3	1.8377	4.1	22.6
1.25	1.814	4.4	21.5
1.2	1.8	4.6	20.5
1.25	1.5971	4.7	20.3
1.3	1.6014	4.5	21.3
1.0	1.7	5.5	15.5
0.8	1.6	6.4	9.5
0.7117	1.5	6.8	5.9
0.7	1.4	6.6	4.9
0.6852	1.3	6.4	3.6
0.678	1.25	6.3	2.9

9.4.2 Approximate Method Given in Eurocode 2

In the absence of interaction diagram as described in section 9.5.1, an approximate design method given in Eurocode 2, clause 5.8.9 can be used. The code gives two cases as follows.

Case 1: In this case if the code equations (5.38a) and (5.38b) are satisfied then biaxial bending can be ignored.

- As a first step, separate design in each principal direction, disregarding biaxial bending, is made. Imperfections need to be taken into account only in the direction where they will have the most unfavourable effect.

No further check is necessary if the following two conditions are satisfied.

- Code equation (5.38a) requiring that the ratio of slenderness

$$\lambda_y/\lambda_z \le 2.0 \ \text{and} \ \lambda_z/\lambda_y \le 2.0 \tag{5.38a}$$

- Code equation (5.38b) requiring that the relative eccentricities e_y/h and e_z/b satisfy the condition

$$(e_y/h_{eq})/(e_z/b_{eq}) \le 0.2 \ \text{or} \ (e_z/b_{eq})/(e_y/h_{eq}) \le 0.2 \tag{5.38b}$$

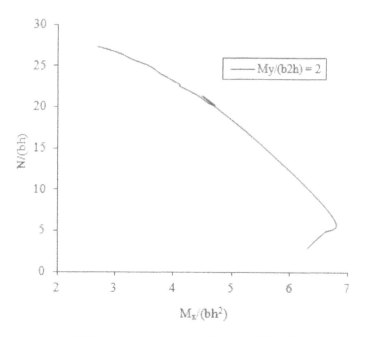

Fig. 9.22 Column design chart for axial force and biaxial bending.

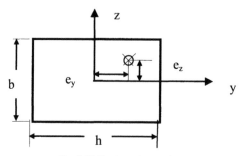

Fig. 9.23 Rectangular section.

In the above equations,

Rectangular sections: $b_{eq} = b$, $h_{eq} = h$

Circular sections, diameter d: $b_{eq} = h_{eq} = 0.866\ d$

General shape: $b_{eq} = \sqrt{(12\ I_{yy}/A)}$, $h_{eq} = \sqrt{(12\ I_{zz}/A)}$

where:

I_{yy} and I_{zz} are the second moments of area about the yy- and zz-axes respectively

A = cross sectional area

e_y and e_z are respectively the eccentricity along the y- and z-axes.

$e_z = M_{Edy}/N_{Ed}$, $e_y = M_{Edz}/N_{Ed}$

M_{Edy}, M_{Edz} and N_{Ed} are respectively the design values of bending moment about the y-axis, bending moment about the z-axis and axial load.

λ_y and λ_z are respectively slenderness ratios about the y- and z-axis.

Case 2: Biaxial bending needs to be considered.
If the conditions in code equations (5.38a) and (5.38b) are not fulfilled, then biaxial bending should be taken into account including second order effect in each direction. This can be done by using the code equation (5.39) given by

$$(\frac{M_{Edz}}{M_{Rdz}})^a + (\frac{M_{Edy}}{M_{Rdy}})^a \le 1.0$$

$$(5.39)$$

In the above equation, M_{Edy} and M_{Edz} are respectively the design values including second order effects of bending moment about the y- and z-axis. M_{Rdy} and M_{Rdz} are respectively the resistant values of bending moment about the y- and z-axis. For circular and elliptic cross sections, the exponent a = 2.0. For rectangular cross section, the exponent a is dependent on the ratio N_{Ed}/N_{Rd} as shown in Table 9.4. N_{Ed} = Design axial load. $N_{Rd} = A_c f_{cd} + A_s f_{yd}$. A_c = gross cross sectional area of concrete. A_s = Area of longitudinal reinforcement.

Table 9.4 Values of exponent a

N_{Ed}/N_{Rd}	0.1	0.7	1.0
a	1.0	1.5	2.0

9.4.2.1 Example of Design of Column Section Subjected to Axial Load and Biaxial Bending: Eurocode 2 Method

Design the reinforcement for the column section shown in Fig. 9.24 It is subjected to the following actions at ULS: N_{Ed} = 950 kN, M_{Edy} = 95 kN m, M_{Edz} = 65 kNm.
$N_{ED}/ (bh) = 7.92$, $MEdy/ ((hb^2) = 1.98$, $M_{Edz}/ ((h^2b) = 1.81$.
The material strengths are f_{ck} = 30 MPa for concrete and f_{yk} = 500 MPa reinforcement.
Assume the cover is 25 mm, links are 8 mm in diameter and main bars are H32 giving $A_s/ (bh) = 0.27\%$.

(a) As information about the height and end conditions of the column are not known, assume equation (5.38a) is satisfied.

(b) Check equation (5.38b).
$$e_z = M_{Edy}/N_{Ed} = 95/950 = 0.10 \text{ m} = 100 \text{ mm}$$
$$b_{eq} = b = 400 \text{ mm}, e_z/b_{eq} = 0.25$$
$$e_y = M_{Edz}/N_{Ed} = 65/950 = 0.068 \text{ m} = 68 \text{ mm}$$
$$h_{eq} = h = 300 \text{ mm}, e_y/h_{eq} = 0.23$$
$$(e_z/b_{eq})/ (e_y/h_{eq}) = 0.25/0.23 = 1.09 > 0.2$$
$$(e_y/h_{eq}) / (e_z/b_{eq}) = 0.23/0.25 = 0.92 > 0.2$$
Uniaxial design is inadmissible.

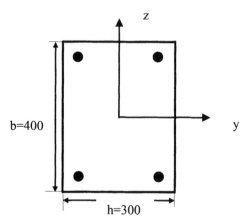

Fig. 9.24 Cross section of column subjected to axial load and biaxial moments.

Use equation (5.39) for biaxial design:
Effective depths:
$$b' = 400 - 25 - 8 - 32/2 = 351 \text{ mm}$$
$$h' = 300 - 25 - 8 - 32/2 = 251 \text{ mm}$$
$$N/ (bh) = 950 \times 10^3 / (400 \times 300) = 7.92$$
$$M_{Edy}/ (h\ b^2) = 95 \times 10^6/ (300 \times 400^2) = 1.98$$
$$M_{Edz}/ (b\ h^2) = 65 \times 10^6/ (400 \times 300^2) = 1.81$$
Assume 4H32 as steel.
$$A_c = 300 \times 400 = 12 \times 10^4 \text{ mm}^2$$
$$A_s = 4H32 = 3217 \text{ mm}^2$$
$$f_{cd} = 30/1.5 = 20 \text{ MPa, } f_{yd} = 500/1.15 = 435 \text{ MPa}$$
$$N_{Rd} = A_c\ f_{cd} + A_s\ f_{yd} = 3799.4 \text{ kN}$$
$$N_{Ed} = 950 \text{ kN}$$
$$N_{Ed}/N_{Rd} = 0.25$$

M_{Rdy}:
Ignore the effect of compression steel. Area of tension steel = 2H32 = 1609 mm^2.
Width (h) = 300 mm, effective depth (b') = 351 mm.
Determine the stress block depth, s:
$$300 \times s \times f_{cd} = A_s \times f_{yd}$$
$$300 \times s \times 20 = 1609 \times 435, s = 117 \text{ mm}$$
$$M_{Rdy} = A_s \times f_{yd} \times (351 - s/2) = 204.9 \text{ kNm}$$
$$M_{Edy}/M_{Rdy} = 95/204.9 = 0.46$$

M_{Rdz}:
Ignore the effect of compression steel. Area of tension steel = 2H32 = 1609 mm^2.
Width (b) = 400 mm, effective depth (h') = 251 mm.
Determine the stress block depth, s:
$$400 \times s \times f_{cd} = A_s \times f_{yd}$$
$$400 \times s \times 20 = 1609 \times 435, s = 88 \text{ mm}$$
$$M_{Rdz} = A_s \times f_{yd} \times (251 - s/2) = 145.1 \text{ kNm}$$

$$M_{Edz}/M_{Rdz} = 65/145.1 = 0.448$$

Calculate a by interpolation:

$$a = 1.0 + \frac{(1.5-1.0)}{(0.7-0.1)} \times (0.25-0.1) = 1.13$$

Check interaction relationship:

$$(\frac{M_{Edz}}{M_{Rdz}})^a + (\frac{M_{Edy}}{M_{Rdy}})^a \leq 1.0$$

$$(0.46)^{1.13} + (0.448)^{1.13} = 0.82 \leq 1.0$$

Design is safe.

Fig. 9.25 shows the uniaxial N–M chart for the designed section. The figure clearly shows the design is safe for both (N, M_y) and (N, M_z). However this does not necessarily indicate that the design is safe for (N, M_y, M_z) combination.

If it is decided to use the exact column design for biaxial moment, then a corresponding column design chart as shown in Fig. 9.26 needs to be constructed.

As a check, using α =1.3, β = 1.02 and the coordinates of the bars as (± 0.375h, ±0.333b)

Calculate the strains (positive is compressive) in steel.

The strains in the four bars are respectively:

$$2.592 \times 10^{-3}, \quad 0.573 \times 10^{-3}, \quad -1.71 \times 10^{-3}, \quad 0.306 \times 10^{-3}$$

The corresponding stresses are: 437, 115, – 437 and 61 MPa

The contributions of the stresses in steel to the column forces:

$$N_s = (437 + 115 - 343 + 61) \times (0.016\ bh/4) = 1.08\ bh$$
$$M_{xs} = (437 - 115 + 343 + 61) \times (0.016\ bh/4) \times 0.375h = 1.09\ bh^2$$
$$M_{ys} = (437 + 115 + 343 - 61) \times (0.016\ bh/4) \times 0.333b = 1.11\ b^2 h$$

The contributions of the compressive stress in concrete to the forces are calculated using λ = 0.8, α =1.3, β = 1.02, λα = 1.04 and λβ = 0.816, f_{cd} = 20.

From the trapezoidal stress block shown in Fig. 9.21,

$$\beta b_1 = \frac{\beta b}{\alpha h}(\lambda \alpha h - h) = \beta b(\lambda - \frac{1}{\alpha}) = 0.031\ b$$

$$N_c = f_{cd} \times \{0.5\ (\beta b_1 + \lambda\ \beta b)\ h\} = 8.47\ bh$$

$$ybar = \frac{h}{3}\frac{(2b_1 + \lambda b)}{(b_1 + \lambda b)} = 0.346\ h$$

$$M_{xc} = N_c \times (0.5\ h - ybar) = 1.308\ bh^2$$

$$xbar = \frac{\beta}{3}\frac{(b_1^2 + \lambda^2\ b^2 + \lambda\ b_1\ b)}{(b_1 + \lambda b)} = 0.272\ b$$

$$M_{yc} = N_c \times (0.5\ b - xbar) = 1.928\ b^2 h$$

Adding the contribution of steel and concrete:

$$N/\ (bh) = 8.47 + 1.08 = 9.55,$$
$$M_x/\ (bh^2) = 1.308 + 1.09 = 2.398,$$
$$M_y/\ (b^2 h) = 1.928 + 1.11 = 3.038$$

The required values are: N = 950 kN, M_x = 95 kNm, M_y = 65 kNm.

If b = 300mm and h = 400mm, the section is safe because

$$N/ (bh) = 7.92 < 9.55, \quad M_x/ (bh^2) = 1.98 < 2.398, \quad M_y/ (b^2h) = 1.81 < 3.038$$

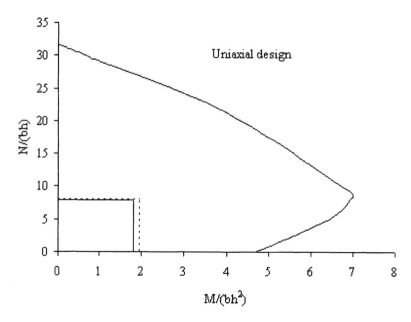

Fig. 9.25 Uniaxial N−M chart for section in Fig. 9.2.

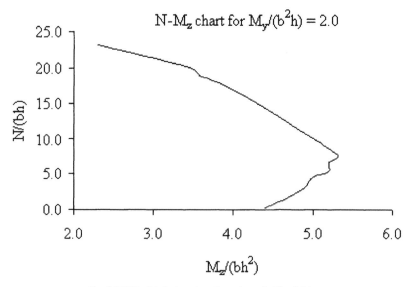

Fig. 9.26 Biaxial design chart for column in Fig. 9.24.

9.5 EFFECTIVE LENGTH OF COLUMNS

9.5.1 Effective Length

For a general column, effective length can be defined as the length of the column for which the elastic buckling load is the same as that of a pin-ended column. Euler buckling load P_E for a pin-ended column of length ℓ and flexural rigidity EI is $P_E = \pi^2$ EI $/\ell^2$. For a column fully restrained both in rotation and translation at the ends, the Euler buckling load is $P_E = 4 \pi^2$ EI $/\ell^2 = \pi^2$ EI $/ (0.5 \ell)^2$. The effective length ℓ_0 for a 'fixed' end column is therefore 0.5ℓ. Table 9.5 shows the effective lengths for some typical ideal end conditions.

Table 9.5 Effective lengths for various ideal end conditions

Restraints at End 1		Restraints at End 2		$\alpha = \ell_0/\ell$
Rotational	Translational	Rotational	Translational	
0	Infinite	0	Infinite	1.0
Infinite	Infinite	Infinite	Infinite	0.5
Infinite	Infinite	0	Infinite	0.7
Infinite	Infinite	Infinite	0	1.0
Finite	Infinite	Finite	Infinite	$0.5 < \alpha < 1.0$
Infinite	Infinite	0	0	2.0
Finite	Infinite	0	0	> 2.0
Finite	Finite	Finite	Finite	$0.5 < \alpha < \infty$

For a typical column in a braced structure as shown in Fig. 9.3, the effective length depends on the rotational restraint provided by the beams and columns joined to the column at each end. The effective length L_{eff} for a column in a braced structure lies in the region of 0.5 to 1.0 times the actual length.

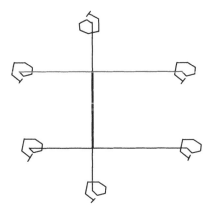

Fig. 9.27 Column in a braced structure.

Fig. 9.27 shows a typical column construction in a framed building. In section 5.8.3.2, the code gives the following equations for determining the effective length ℓ_0 of an isolated column.

(a) For braced members code equation (5.15) gives the equation for calculating the effective length.

$$\ell_0 = 0.5\ell\sqrt{\{1+\frac{k_1}{(0.45+k_1)}\}\{1+\frac{k_2}{(0.45+k_2)}\}}$$

(5.15)

k_1 and k_2 are the relative flexibilities of rotational restraints at ends 1 and 2 respectively, where:

$$k = \frac{\theta}{M}\frac{EI}{\ell}$$

EI = bending stiffness (flexural rigidity) of the column.
If there is a column above or below the column under consideration, then EI/ℓ should be replaced by EI/ℓ for column under consideration + EI/ℓ of the column above or below as appropriate.
θ = rotation of the restraining for a moment M.
Note: $k = 0$ represents fully restrained and $k = \infty$ represents pin-end. If $k_1 = k_2 = 0$, then $\ell_0 = 0.5\ \ell$ which corresponds to a fully fixed ended column.
Since fully rigid restraint is rare in practice, in clause 5.8.3.2(3), Eurocode 2 recommends a minimum value of 0.1 for k_1 and k_2.
If $k_1 = 0$, $k_2 = \infty$, then $\ell_0 = 0.5 \times \sqrt{2}\ell = 0.7\ell$ which corresponds to a column fully fixed at end 1 and pin-ended at end 2.
If $k_1 = \infty$, $k_2 = \infty$, then $\ell_0 = 0.5\times 2\ \ell = \ell$ which corresponds to a pin-ended column.
The above ideal end conditions, the equation gives the same values as in Table 9.1.
Fig. 9.28 shows the variation of ℓ_0/ℓ with $k_1 = k_2 = k$. As can be seen beyond a value of $k = 5.0$, any increase in the value of k has only a minor effect on the effective length.

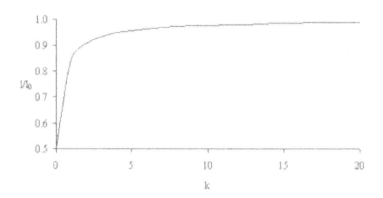

Fig. 9.28 Variation of ℓ_0/ℓ with $k_1 = k_2 = k$ for a braced column.

For a beam of span L, flexural rigidity EI and clamped at the far end as shown in Fig. 9.29(a), the relationship between M and θ is
$$M = 4 \ (EI/L) \ \theta \ \text{or} \ \theta/M = 0.25 \ L/ \ (EI)$$
Similarly, for a similar beam pinned at the far end as shown in Fig. 9.6(b), the relationship between M and θ is
$$M = 3 \ (EI/L) \ \theta \ \text{or} \ \theta/M = 0.33 \ L/ \ (EI)$$

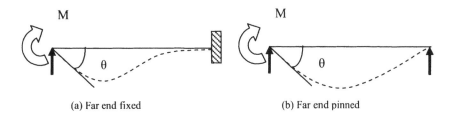

(a) Far end fixed   (b) Far end pinned

Fig. 9.29 Moment–rotation relationships.

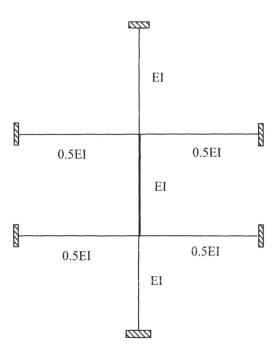

Fig. 9.30 A column sub-frame in a braced frame.

As an example for the column sub-frame shown in Fig. 9.30, let EI for beams be 50% of EI for columns. Span of beams equal to 1.5 times the height of columns.

For calculating k, at each end include two beams and one column (top or bottom). Therefore,

$$k_1 = k_2 = 2 \times \{0.33 \times \frac{1.5\ell}{0.5\,EI}\} \times [\frac{EI}{\ell} + \frac{EI}{\ell}] = 3.96$$

$$\ell_0 = 0.5\ell \sqrt{\{[1 + \frac{3.96}{(0.45+3.96)}][1 + \frac{3.96}{(0.45+3.96)}]\}} = 0.898\,\ell$$

$$1.0 \geq \ell_0/\ell \geq 0.5$$

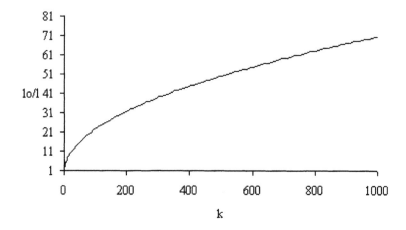

Fig. 9.31 Variation of ℓ_0/ℓ with $k_1 = k_2 = k$ for an unbraced column.

(b) For unbraced members code equation (5.16) gives the equation for calculating the effective length.

$$\ell_0 = \ell \times \max[\sqrt{\{1 + 10\frac{k_1\,k_2}{(k_1 + k_2)}\}}\;;\; \{1 + \frac{k_1}{(1 + k_1)}\} \times \{1 + \frac{k_2}{(1 + k_2)}\}] \tag{5.16}$$

Fig. 9.31 shows the variation of ℓ_0/ℓ with $k_1 = k_2 = k$.

For the unbraced frame shown in Fig. 9.32, using the same data as for the braced frame, except that the far ends of the beams are pinned and on rollers to allow sway,

$$k_1 = k_2 = 2 \times \{0.33 \times \frac{1.5\ell}{0.5\,EI}\} \times [\frac{EI}{\ell} + \frac{EI}{\ell}] = 3.96$$

$$\ell_0 = \ell \times \max[\sqrt{\{1+10\frac{3.96 \times 3.96}{(3.96 + 3.96)}\}} \; ; \; \{1+\frac{3.96}{(1 + 3.96)}\} \times \{1+\frac{3.96}{(1 + 3.96)}\}]$$

$$= \ell \times \max[4.56; \; 1.80]$$

$$= 4.56\,\ell$$

$$\infty \geq \ell_0 / \ell \geq 1.0$$

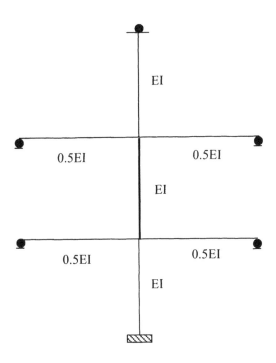

Fig. 9.32 A column sub-frame in an unbraced frame.

9.5.2 Long and Short Columns

Fig. 9.33 shows a column subjected to an axial load and end moments. When the column deforms, the axial load N_{Ed} is eccentric to the deformed column and causes a bending moment M_2, equal to $N_{Ed}\,\delta$, where δ is the lateral deflection of the column. M_2 is known as the second order moment and is clearly a function of the flexibility of the column. Axial forces and moments acting on the column from the loads acting on the structures including geometric imperfections are known as first order effects. The total bending moment is the sum of the bending moment M_{oEd} due to the external loads and the bending moment M_2 equal to $N\delta$. A column is considered short if only the first order effects need to be considered in its design.

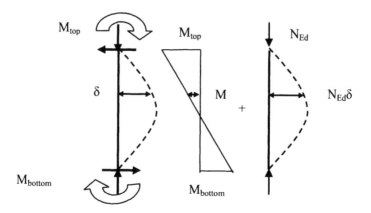

Fig. 9.33 Bending moment in a deformed column.

9.5.3 Slenderness Ratio

In general, the elastic buckling for a column is $P_E = \pi^2 \, EI \,/\ell_0^2$.

The axial stress in the column is $\sigma = \dfrac{P_E}{A} = \pi^2 \, E \dfrac{I}{A} \dfrac{1}{\ell_0^2}$.

The radius of gyration $i = \sqrt{(I/A)}$.

Therefore $\sigma = \pi^2 \, E \dfrac{1}{\left(\dfrac{\ell_0}{i}\right)^2} = \pi^2 \, E \dfrac{1}{\lambda^2}$.

$$\text{Slenderness ratio } \lambda = \ell_0/i \qquad (5.14)$$

In equation (5.13N), clause 5.8.3.1, the code classifies columns as short when the slenderness ratio λ about both axes are less than λ_{lim}

$$\lambda_{lim} = 20 \times A \times B \times C \frac{1}{\sqrt{n}}$$

$$(5.13N)$$

where

$A = \dfrac{1}{(1.0 + 0.2 \, \varphi_{ef})}$, (If the effective creep ratio φ_{ef} is not known take $A = 0.7$).

$B = \sqrt{(1 + 2\omega)}$, $\omega = (A_s \, f_{yd})/ (A_c \, f_{cd})$, (If the value of ω is not known take $B = 1.1$).

A_s = Total area of longitudinal reinforcement.

A_c = Total area of concrete in the cross section.

f_{yd} and f_{cd} respectively design strength of steel reinforcement and concrete.

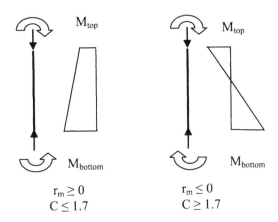

Fig. 9.34 First order moments.

$C = 1.7 - r_m$, r_m = moment ratio = M_{01}/M_{02}.

M_{01} and M_{02} are the first order moments.

M_{01} = min (M_{top}; M_{bottom}), M_{02} = max (M_{top}; M_{bottom}), $|M_{02}| \geq |M_{01}|$.

Note: If the end moments give tension on the same side, r_m is taken as positive (i.e., $C \leq 1.7$) otherwise negative (i.e., $C \geq 1.7$) as shown in Fig. 9.34.

(If the moment ratio r_m is not known take C = 0.7.)

Note that if r_m is positive, the column will bend in single curvature and is likely to buckle at a lower load than if r_m is negative when the column will try to buckle in double curvature whose buckling load is very much higher than that of a column buckling in single curvature.

n = relative normal force = $N_{Ed}/ (A_c\ f_{cd})$. N_{Ed} = Design axial force.

9.5.3.1 Example of Calculating the Effective Length of Columns

Specification: The lengths and proposed section dimensions for the columns and beams in a multi-storey building are shown in Fig. 9.35. Determine the effective lengths and slenderness ratios for the YY and ZZ axes for the lower column length AB, for the two cases where the structure is braced and unbraced. The connection to the base and the base itself are designed to resist the column moment.

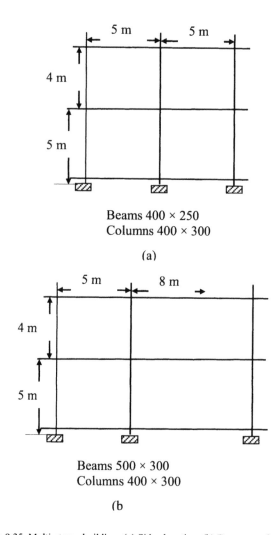

Beams 400 × 250
Columns 400 × 300

(a)

Beams 500 × 300
Columns 400 × 300

(b

Fig.9.35 Multi-storey building: (a) Side elevation. (b) Transverse frame.

(a) Effective length calculations in the longitudinal direction
Bending about z-z axis:

(i) Rotational stiffness of beams
L = 5m, section 250 × 400
$I = (250 \times 400^3/12) \times 10^{-12} = 1.33 \times 10^{-3}$ m^4
$4 \times I/L = 1.33 \times 10^{-3}$ m^3
Rotational stiffness of two beams $= 2 \times 1.33 \times 10^{-3} = 2.67 \times 10^{-3}$ m^3
(ii) Columns
Bottom column: L = 5m, section 300 × 400
$I = (300 \times 400^3/12) \times 10^{-12} = 1.60 \times 10^{-3}$ m^4

$I/L = 0.32 \times 10^{-3}$ m^3
Top column: L = 4m, section 300 × 300
$I = (300 \times 300^3/12) \times 10^{-12} = 0.675 \times 10^{-3}$ m^4
$I/L = 0.169 \times 10^{-3}$ m^3
Rotational stiffness of two columns = $(0.32 + 0.169) \times 10^{-3} = 0.489 \times 10^{-3}$ m^3
$k_{top} = 0.489/2.67 = 0.18$
$k_{bottom} = 0.1$ (fixed) (Recommended value. See clause 5.3.8.2(3))
$\ell = 5$ m, radius of gyration, $i_z = 400/\sqrt{12} = 116$ mm

(iii) Braced structure

$$\ell_0 = 0.5\ell\sqrt{\{1+\frac{k_1}{(0.45+k_1)}\}\{1+\frac{k_2}{(0.45+k_2)}\}}$$ (5.15)

$$\ell_0 = 3.08 \text{ m}$$
Slenderness ratio $\lambda = 3.08/0.116 = 26.6$

(iv) Unbraced structure

$$\ell_0 = \ell \times \max[\sqrt{\{1+10\frac{k_1 k_2}{(k_1 + k_2)}\}} \; ; \; \{1+\frac{k_1}{(1 + k_1)}\}\times\{1+\frac{k_2}{(1 + k_2)}\}]$$ (5.16)

$$\ell_0 = 6.41 \text{ m}$$
Slenderness ratio $\lambda = 6.41/0.116 = 55.3$

(b) Effective length calculations in the transverse direction
Bending about y-y axis:

(i) Rotational stiffness of beams
Beam on right: L = 8 m, section 300 × 500
$I = (300 \times 500^3/12) \times 10^{-12} = 3.125 \times 10^{-3}$ m^4
$4 \times I/L = 1.5625 \times 10^{-3}$ m^3
Beam on left: L = 5 m, section 300 × 500
$I = (300 \times 500^3/12) \times 10^{-12} = 3.125 \times 10^{-3}$ m^4
$4 \times I/L = 2.5 \times 10^{-3}$ m^3
Rotational stiffness of two beams = $(1.5625 + 2.5) \times 10^{-3} = 4.06 \times 10^{-3}$ m^3

(ii) Columns
Bottom column: L = 5m, section 300 × 400
$I = (400 \times 300^3/12) \times 10^{-12} = 0.90 \times 10^{-3}$ m^4
$I/L = 0.18 \times 10^{-3}$ m^3
Top column: L = 4m, section 300 × 300
$I = (300 \times 300^3/12) \times 10^{-12} = 0.675 \times 10^{-3}$ m^4
$I/L = 0.169 \times 10^{-3}$ m^3
Rotational stiffness of two columns = $(0.18 + 0.169) \times 10^{-3} = 0.349 \times 10^{-3}$ m^3
$k_{top} = 0.349/4.06 = 0.086$. Set it to 0.1, the minimum value.
$k_{bottom} = 0.1$ (fixed)
$\ell = 5$ m, i_y, radius of gyration = $300/\sqrt{12} = 87$ mm

(iii) Braced structure

$$\ell_0 = 0.5\ell \sqrt{\{1+\frac{k_1}{(0.45+k_1)}\}\{1+\frac{k_2}{(0.45+k_2)}\}}$$

(5.15)

$$\ell_0 = 2.95 \text{ m}$$

Slenderness ratio $\lambda = 2.95/0.087 = 33.96$.

(iv) Unbraced structure

$$\ell_0 = \ell \times \max[\sqrt{\{1+10\frac{k_1 k_2}{(k_1 + k_2)}\}} \; ; \; \{1+\frac{k_1}{(1 + k_1)}\} \times \{1+\frac{k_2}{(1 + k_2)}\}]$$

(5.16)

$$\ell_0 = 6.12 \text{ m}$$

Slenderness ratio, $\lambda = 6.12/0.087 = 70.39$.

9.5.4 Primary Moments and Axial Load on Column

Fig. 9.36 shows the characteristic loading on the column.

Design loading (ULS)
Case 1: Imposed load, primary. Wind load secondary
$N_{Ed} = 1.35 \times 765 + 1.5 \times 305 + 1.5 \times 0.6 \times 0 = 1490$ kN
Note M_{02} is numerically the larger and M_{01} is numerically the smaller of the moments at top and bottom of column.
$M_{02} = 1.35 \times 48 + 1.5 \times 28 + 1.5 \times 0.6 \times 37 = 140$ kNm
$M_{01} = -(1.35 \times 24 + 1.5 \times 14 + 1.5 \times 0.6 \times 37) = -87$ kNm

Case 2: Wind load, primary. Imposed load, secondary
$N_{Ed} = 1.35 \times 765 + 1.5 \times 0.7 \times 305 + 1.5 \times 0 = 1353$ kN
$M_{02} = 1.35 \times 48 + 1.5 \times 0.7 \times 28 + 1.5 \times 37 = 150$ kNm
$M_{01} = -(1.35 \times 24 + 1.5 \times 0.7 \times 14 + 1.5 \times 37) = -103$ kNm
Check whether column is short or long:
Using code equation (5.13N),

$$\lambda_{lim} = 20 \times A \times B \times C / \sqrt{n}$$

(5.13N)

Calculate A:
Age at loading, $t_0 = 3$ days (assumed)
Assume all four sides are exposed to atmosphere:
u, perimeter of the column $= 2 \times (300 + 400) = 1400$ mm
A_c, area of concrete in the cross section $= 400 \times 300 = 12 \times 10^4$ mm^2
$h_0 = 2 A_c/u = 171$ mm
Assuming class S cement, $f_{ck} = 30$ MPa, relative humidity RH = 50%, from Fig. 3.1 in the Eurocode 2, $\varphi (\infty, t_0) = 3.5$.
$\varphi_{ef} = \varphi (\infty, t_0) \times (M_{0Eqp}/M_{0Ed}) = 3.5 \times (98/119) = 2.88$
$A = 1/(1 + 0.2 \varphi_{ef}) = 0.63$.

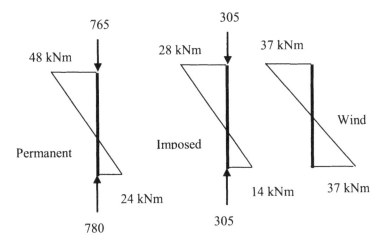

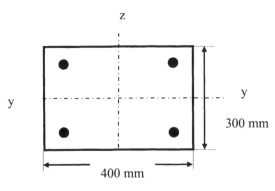

$A_s = 4H32$

Fig. 9.36 Loading on the column.

Calculate B:
$A_s = 4H32 = 3217$ mm^2, $f_{yd} = f_{yk}/1.15 = 500/1.15 = 435$ MPa
$A_s f_{yd} = 3217 \times 435 \times 10^{-3} = 1399.4$ kN
$A_c = 400 \times 300 = 12 \times 10^4$ mm^2, $f_{cd} = f_{ck}/1.5 = 30/1.3 = 20$ MPa
$A_c f_{cd} = 12 \times 10^4 \times 20 \times 10^{-3} = 2400$ kN
$\omega = A_s f_{yd}/A_c f_{cd} = 1399.4/2400 = 0.58$
$B = 1 + \omega = 1.58$

Calculate C:
Case 1: Imposed load, primary. Wind load secondary
$N_{Ed} = 1490$ kN
$n = 1490/2400 = 0.62$, $\sqrt{n} = 0.79$

$M_{02} = 140$ KNm, $M_{01} = -87$ kNm.
The moments cause tension on opposite sides. $r_m = M_{02}/M_{01} = -87/140 = -0.62$
$C = 1.7 - r_m = 1.7 - (-0.62) = 2.32$

Calculate λ_{lim}
Braced member: $\lambda_{lim} = 20 \times 0.63 \times 1.58 \times 2.32/ 0.79 = 58.5$
Actual $\lambda = 26.6$ and 33.96 in the longitudinal and transverse directions respectively
and both values are less than λ_{lim}. The column is short.

Unbraced member: In clause 5.8.3.1(1), Eurocode 2 suggests that for unbraced
members in general:
$C = 0.7$
$\lambda_{lim} = 20 \times 0.63 \times 1.58 \times 0.7/ 0.79 = 17.6$
Actual $\lambda = 55.3$ and 70.4 in the longitudinal and transverse directions respectively
and both values are greater than λ_{lim}. The column is long.

Case 2: Wind load, primary. Imposed load, secondary
$N_{Ed} = 1353$ kN
$n = 1353/2400 = 0.56$, $\sqrt{n} = 0.75$
$M_{02} = 150$ KNm, $M_{01} = -103$ kNm.
The moments cause tension on opposite sides. $r_m = -103/150 = -0.69$
$C = 1.7 - (-0.69) = 2.39$

Calculate λ_{lim}
Braced member: $\lambda_{lim} = 20 \times 0.63 \times 1.58 \times 2.39/ 0.75 = 63.4$
Actual $\lambda = 26.6$ and 33.96 in the longitudinal and transverse directions respectively
and both values are less than λ_{lim}. The column is short.

Unbraced member In clause 5.8.3.1(1), Eurocode2 suggests that for unbraced
members in general:
$C = 0.7$
$\lambda_{lim} = 20 \times 0.63 \times 1.58 \times 0.7/ 0.75 = 18.6$
Actual $\lambda = 55.3$ and 70.4 in the longitudinal and transverse directions respectively
and both values are greater than λ_{lim}. The column is long.

Short column design:
Geometrical imperfection:
In clause 5.2 (5), equation (5.1), the inclination θ_ℓ is given as
$$\theta_\ell = \theta_0 \, \alpha_h \, \alpha_m \qquad (5.1)$$
where $\theta_0 = 1/200$.
For an isolated member;
$\quad\quad \alpha_h = 2/\sqrt{\ell}$, $\ell =$ **actual length** (not effective length ℓ_0) of the column
$$2/3 \le \alpha_h \le 1$$
$$\alpha_m = 1$$
In clause 5.2(7), equation (5.2), the eccentricity e_i is given as
$$e_i = \theta_\ell \times \ell_0/2 \qquad (5.2)$$

where ℓ_0 = effective length.

Longitudinal

$$\text{Effective length } \ell_0 = 3.08 \text{ m, actual length } \ell = 5 \text{ m}$$
$$\alpha_h = 2/\sqrt{\ell} = 0.894, \; 2/3 \le \alpha_h \le 1$$
$$\theta_\ell = \theta_0 \, \alpha_h \, \alpha_m = (1/200) \times 0.894 \times 1 = 4.472 \times 10^{-3}$$
$$e_i = \theta_\ell \times \ell_0/2 = 4.472 \times 10^{-3} \times 3.08/2 = 6.887 \times 10^{-3} = 6.9 \text{ mm}$$

In clause 6.1(4) Eurocode 2 states that for cross sections with symmetrical reinforcement, the minimum eccentricity $e_0 = h/30 \ge 20$ mm, h = depth of the section.

$$h = 400 \text{ mm}$$
$$e_0 = h/30 = 400/30 = 13 \text{ mm} < 20 \text{ mm}$$

Design value of eccentricity = 20 mm.

Transverse

$$\text{Effective length } \ell_0 = 2.93 \text{ m, actual length } \ell = 5 \text{ m}$$
$$\alpha_h = 2/\sqrt{\ell} = 0.894, \; 2/3 \le \alpha_h \le 1$$
$$\theta_\ell = \theta_0 \, \alpha_h \, \alpha_m = (1/200) \times 0.894 \times 1 = 4.472 \times 10^{-3}$$
$$e_i = \theta_\ell \times \ell_0/2 = 4.472 \times 10^{-3} \times 2.95/2 = 6.596 \times 10^{-3} = 6.6 \text{ mm}$$

In clause 6.1(4) Eurocode 2 states that for cross sections with symmetrical reinforcement, the minimum eccentricity $e_0 = h/30 \ge 20$ mm, h = depth of the section.

$$h = 300 \text{ mm}$$
$$e_0 = h/30 = 300/30 = 10 \text{ mm} < 20 \text{ mm}$$

Design value of eccentricity = 20 mm.

Design loading (ULS)

Case 1: Imposed load, primary. Wind load secondary

$N_{Ed} = 1490$ kN, $M_{02} = M_{Top} = 140$ kNm, $M_{01} = M_{Bottom} = -87$ kNm
Geometrical imperfection: eccentricity e = 20 mm
Design for $N_{Ed} = 1490$ kN,
$$M_{Ed} = M_{02} + N_{Ed} \, e_i = 140 + 1490 \times 20 \times 10^{-3} = 169.8 \text{ kNm}$$
$$N_{Ed}/ (bh) = 12.4, \; M_{Ed}/ (bh^2) = 3.54$$

Case 2: Wind load, primary. Imposed load, secondary

$N_{Ed} = 1353$ kN, $M_{02} = M_{Top} = 150$ kNm, $M_{01} = M_{Bottom} = -103$ kNm
Geometrical imperfection: eccentricity $e_i = 20$ mm
Design for $N_{Ed} = 1353$ kN,
$$M_{Ed} = M_{02} + N_{Ed} \, e_i = 150 + 1353 \times 20 \times 10^{-3} = 177.0 \text{ kNm}$$
$$N_{Ed}/ (bh) = 11.3, \; M_{Ed}/ (bh^2) = 4.92$$

The plotted points in Fig. 9.37 fall inside the N–M curve indicating that the design is safe.

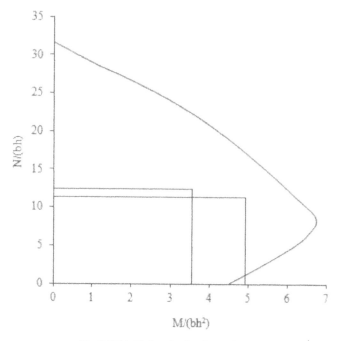

Fig. 9.37 N–M chart for the short column.

Check maximum and minimum steel areas:
$A_s = 4H32 = 3217 \text{ mm}^2$, $A_c = 300 \times 400 = 12 \times 10^4 \text{ mm}^2$
$A_s/A_c = 0.027$, $0.004 < A_s/A_c < 0.04$
Links: One quarter of diameter of main bar = 32/4 = 8 mm
Spacing:
20 times the diameter of main bar = 20 × 32 = 640 mm
Lesser dimension of column = 300 mm
Spacing = Min (640; 300; 400) = 300 mm
Provide H8 links at 300 mm c/c.

9.6 DESIGN OF SLENDER COLUMNS

9.6.1 Additional Moments Due to Deflection

In the primary analysis of the rigid frames, the secondary moments due to lateral
deflection are ignored. This effect is small for short columns but with slender
columns significant additional moments occur.
Eurocode in equation (5.32) allows for the differing first order moments M_{01} and
M_{02} to be replaced by an equivalent moment M_{0e}.

$$M_{0e} = 0.6\ M_{02} + 0.4\ M_{01} \geq 0.4\ M_{02} \qquad\qquad (5.32)$$

Numerical value of M_{02} > Numerical value of M_{01}.

Case 1: Imposed load, primary. Wind load secondary
$$M_{0e} = 0.6 \, M_{02} + 0.4 \, M_{01} \geq 0.4 \, M_{02}$$
$$= 0.6 \times 140 + 0.4 \times (-87) \geq 0.4 \times 140$$
$$= 49 < 56$$
$$M_{0e} = 56 \text{ kNm}$$

Case 2: Wind load, primary. Imposed load, secondary
$$M_{0e} = 0.6 \, M_{02} + 0.4 \, M_{01} \geq 0.4 \, M_{02}$$
$$= 0.6 \times 150 + 0.4 \times (-103) \geq 0.4 \times 150$$
$$= 49 < 60$$
$$M_{0E} = 60 \text{ kNm}$$

Using the method based on nominal curvature as given in section 5.8.8, the additional second order moment M_2 and the deflection e_2 are given by code equation (5.33) as

$$M_2 = N_{Ed} \, e_2, \quad e_2 = \frac{1}{r} \times \frac{\ell_0^2}{c}$$
(5.33)

c = a factor depending on the *total* curvature $1/r$ caused by sum of first and second order moments and c is normally taken as equal to π^2 on the assumption that the curvature distribution is sinusoidal.

For members with constant cross section and symmetrical reinforcement, code equation (5.34) gives the value of $1/r$ as

$$\frac{1}{r} = K_r \, K_\phi \, \frac{1}{r_0}$$
(5.34)

The correction factor K_r depending on the axial load is given by code equation (5.36) as

$$K_r = \frac{(n_u - n)}{(n_u - n_{bal})}$$
(5.36)

$$A_c \, f_{cd} = 300 \times 400 \times (30/1.5) \times 10^{-3} = 2400 \text{ kN}$$
$$A_s = 4H32 = 3217 \text{ mm}_2, \; A_s \, f_{yd} = 3217 \times (500/1.15) \times 10^{-3} = 1399 \text{ kN}$$

$$n = \frac{N_{Ed}}{A_c \, f_{cd}} = \frac{1490}{2400} = 0.62 \qquad \text{(case 1)}$$

$$n = \frac{N_{Ed}}{A_c \, f_{cd}} = \frac{1353}{2400} = 0.562 \qquad \text{(case 2)}$$

n_{bal} = value of n at maximum moment resistance. The code suggests that the value of 0.4 may be used.

$$n_u = 1 + \omega = 1 + A_s \, f_{yd}/ (A_c \, f_{cd}) = 1 + 1399/2400 = 1.58$$
$$K_r = \frac{(n_u - n)}{(n_u - n_{bal})} = \frac{(1.58 - 0.62)}{(1.58 - 0.4)} = 0.81$$
$$\text{(case 1)}$$

$$K_r = \frac{(n_u - n)}{(n_u - n_{bal})} = \frac{(1.58-0.56)}{(1.58-0.4)} = 0.86$$

(case 2)

The factor K_φ for taking account of creep is given by code equation (5.37) as

$$K_\varphi = 1 + \beta \varphi_{ef}$$

(5.37)

φ_{ef} = effective creep ratio = 2.88 (as previously calculated)

$$\beta = 0.35 + f_{ck}/200 - \lambda/150$$

$\beta = 0.35 + 30/200 - 55.3/150 = 0.131$ (longitudinal direction)

$\beta = 0.35 + 30/200 - 70.4/150 = 0.031$ (transverse direction)

$K_\varphi = 1 + 0.131 \times 2.88 = 1.38$ (longitudinal direction)

$K_\varphi = 1 + 0.031 \times 2.88 = 1.09$ (transverse direction)

$$\frac{1}{r_0} = \frac{f_{yd}}{E_s} \times \frac{1}{0.45d}$$

$f_{yd}/E_s = 435/(200 \times 10^3) = 2.175 \times 10^{-3}$

d = effective depth = 400 − 25 (cover) − 10(link) − 32/2 = 349 mm

$1/r_0 = = 1.39 \times 10^{-5}$

$$\frac{1}{r} = K_r\, K_\phi\, \frac{1}{r_0} = 0.81 \times 1.38 \times 1.39 \times 10^{-5} = 1.55 \times 10^{-5} \text{ (longitudinal directiion)}$$

$$\frac{1}{r} = K_r\, K_\phi\, \frac{1}{r_0} = 0.86 \times 1.09 \times 1.39 \times 10^{-5} = 1.30 \times 10^{-5} \text{ (transverse directiion)}$$

$$e_2 = \frac{1}{r} \times \frac{\ell_0^2}{c} = 1.55 \times 10^{-5} \times \frac{(6.41 \times 10^3)^2}{10} = 89\, mm \text{ (longitudinal direction)}$$

$$e_2 = \frac{1}{r} \times \frac{\ell_0^2}{c} = 1.30 \times 10^{-5} \times \frac{(6.12 \times 10^3)^2}{10} = 49\, mm \text{ (transverse direction)}$$

$M_2 = N_{Ed}\, e_2 = 1490 \times (89 \times 10^{-3}) = 133$ kNm (longitudinal direction, case 1)

$M_2 = N_{Ed}\, e_2 = 1353 \times (89 \times 10^{-3}) = 120$ kNm (longitudinal direction, case 2)

$M_2 = N_{Ed}\, e_2 = 1490 \times (49 \times 10^{-3}) = 73$ kNm (transverse direction, case 1)

$M_2 = N_{Ed}\, e_2 = 1353 \times (49 \times 10^{-3}) = 66$ kNm (transverse direction, case 2)

Total design forces in the Longitudinal direction:
Case 1: Imposed load, primary. Wind load secondary

$$M_2 = N_{Ed}\, e_2 = 133 \text{ kNm}$$
$$M_{0e} = 56 \text{ kNm}$$
$$M_{Ed} = 56 + 133 = 189 \text{ kNm}, \quad M_{Ed}/(bh^2) = 3.94,$$
$$N_{Ed} = 1490 \text{ kN}, \quad N_{Ed}/(bh) = 12.42$$

Case 2: Wind load, primary. Imposed load, secondary

$$M_2 = N_{Ed}\, e_2 = 120 \text{ kNm}$$
$$M_{0E} = 60 \text{ kNm}$$
$$M_{Ed} = 60 + 120 = 180 \text{ kNm}, \quad M_{Ed}/(bh^2) = 3.75,$$

$$N_{Ed} = 1353 \text{ kN}, N_{Ed}/ (bh) = 11.28$$

Fig. 9.37 shows the N–M chart for the column.

Note that from the chart N/ (bh) = 8.0 when M/ (bh^2) is a maximum value equal to 6.74. Therefore axial load at maximum moment is N_{bal} is given by

$$N_{Bal} = 8.0 \times 300 \times 400 \times 10^{-3} = 960 \text{ kN},$$
$$n_{bal} = N_{bal}/ (A_c f_{cd}) = 960/2400 = 0.4$$

This is the value assumed in calculating K_r. Therefore no revision of the design calculations for M_2 is needed.

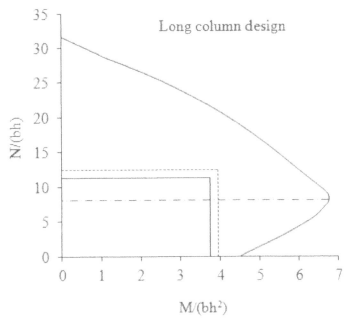

Fig. 9.38 N–M chart for the long column.

Check maximum and minimum steel areas:

$A_s = 4H32 = 3217 \text{ mm}^2$, $A_c = 300 \times 400 = 12 \times 10^4 \text{ mm}^2$

$A_s/A_c = 0.027$, $0.004 < A_s/A_c < 0.04$

Links: One quarter of diameter of main bar = 32/4 = 8 mm

20 times the diameter of main bar = 20 × 32 = 640 mm;

Lesser dimension of column = 300 mm

Spacing = Min (640; 300; 400) = 300 mm

Provide H8 links at 300 mm c/c.

CHAPTER 10

WALLS IN BUILDINGS

10.1 FUNCTIONS, TYPES AND LOADS ON WALLS

All buildings contain walls that function to carry loads, enclose and divide space, exclude weather and retain heat. Walls may be classified into the following types:

1. Internal non-load-bearing walls of block-work or light movable partitions that divide space only
2. External curtain walls that carry self-weight and lateral wind loads
3. External and internal infill walls in framed structures that may be designed to provide stability to the building but do not carry vertical building loads; the external walls would also carry lateral wind loads
4. Load-bearing walls designed to carry vertical building loads and horizontal lateral and in-plane wind loads and provide stability

The role of the wall is based on the type of building in which it is used. Building types and walls provided are as follows:

1. Framed buildings: wall types 1, 2 or 3
2. Load-bearing and shear wall building with no frame: wall types 1, 2 and 4
3. Combined frame and shear wall building: wall types 1, 2 and 4

Type (3) is the normal multi-storey building.

A wall is defined in Eurocode 2, clause 9.6.1, as a vertical load-bearing member whose length exceeds four times its thickness. This definition distinguishes a wall from a column. Loads are applied to walls in the following ways:

1. Vertical loads from roof and floor slabs or beams supported by the wall
2. Lateral loads on the vertical wall slab from wind, water or earth pressure
3. Horizontal in-plane loads from wind when the wall is used to provide lateral stability in a building as a shear wall

In the following sections, only Type 4 structural concrete walls are considered.

10.2 DESIGN OF REINFORCED CONCRETE WALLS

A reinforced concrete wall is a wall containing at least the minimum quantity of reinforcement required by clause 9.6.2. The reinforcement is taken into account in determining the strength of the wall. In the code EC2, there is very little guidance given for the design aspects specifically related to walls. The general assumption is that walls will be designed using the rules for the design of columns.

10.2.1 Wall Reinforcement

(a) Minimum and maximum area of vertical reinforcement
According to Eurocode 2, clause 9.6.2, the minimum and maximum amounts of reinforcement required for a reinforced concrete wall are $0.002A_c$ and $0.04A_c$ outside lap locations respectively. It is further stated that where minimum reinforcement controls design, half of this area should be located on each face. The distance between two adjacent vertical bars should not exceed three times the wall thickness or 400 mm, whichever is lesser.

(b) Area of horizontal reinforcement
According to Eurocode 2, clause 9.6.3, horizontal reinforcement should be provided at each face and should have a minimum area of 25% of the vertical reinforcement or $0.001 A_c$, whichever is greater. The spacing between two adjacent horizontal bars should not be greater than 400 mm.

(c) Provision of links
If the compression reinforcement in the wall exceeds $0.02A_c$, links must be provided through the wall thickness in accordance with the rules for columns in clause 9.5.3 which are:
1. The diameter of the transverse reinforcement should not be less than 6 mm or one-quarter of the diameter of the largest longitudinal bar whichever is greater.
2. The maximum spacing is to be $S_{cl, max}$.

$S_{cl, max}$ = minimum of
- 20 times the diameter of the smallest longitudinal bar
- The lesser dimension of the wall i.e. the thickness
- 400 mm

The maximum spacing should be reduced by a factor of 0.6 in the following cases.
- In sections within a distance equal to 4× thickness of wall above or below a beam or slab.
- Near lapped joints, if the diameter of the longitudinal bar is greater than 14 mm. A minimum of three bars evenly placed in the lap length is required.

Where the main reinforcement (i.e., vertical bars) is placed nearest to the wall faces, transverse reinforcement should be provided in the form of links with 4 per m² of the wall area.

10.2.2 General Code Provisions For Design

The design of reinforced concrete walls follows the rules for the design of columns. The general provisions are as follows.

(a) Axial loads
The axial load in a wall may be calculated assuming the beams and slabs transmitting the loads to it are simply supported.

(b) Effective height
Where the wall is constructed monolithically with adjacent elements, the effective height l_e should be assessed as though the wall were a column subjected to bending at right angles to the plane of the wall.

(c) Transverse moments
For continuous construction, transverse moments can be calculated using elastic analysis. The eccentricity is not to be less than h/30 or 20 mm where h is the wall thickness (Eurocode 2, clause 6.1(4)).

(d) In-plane moments
Moments in the plane of a single shear wall can be calculated from statics. When several walls resist forces, the proportion allocated to each wall should be in proportion to its stiffness.

Consider two shear walls connected by floor slabs and subjected to a uniform horizontal load, as shown in Fig. 10.1.

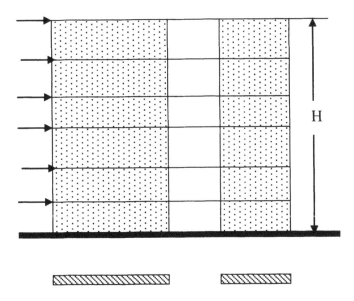

Fig. 10.1 Shear walls connected by floor slabs.

The walls acting as a cantilever of height H and loaded by a uniformly distributed load of p per unit length deflect by the same amount
$$\delta = pH^3/8EI$$

Thus the load is divided between the walls in proportion to their second moments of area:

$$\text{Wall 1: } p_1 = p\left[\frac{I_1}{I_1+I_2}\right], \text{ wall 2: } p_2 = p - p_1$$

A more accurate analysis for connected shear walls is given in Chapter 15.

As a more complex example, consider the plan of a shear wall assembly for a tall building shown in Fig. 10.2.

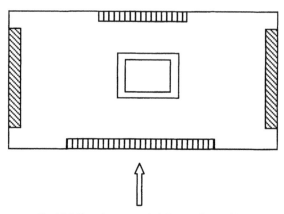

Fig. 10.2 Plan of a symmetrical shear wall assembly.

It consists of four shear walls and a central core. The arrangement is symmetrical about the vertical y-axis. The two walls parallel to the vertical axis are 8 m long and 200 mm thick. The two walls parallel to the horizontal x-axis axis are 12 m long and 200 mm thick. The central axis is core 4 m × 4 m and the walls are 150 mm thick.

For a total force F acting in the y-direction it is sensible to neglect any force resisted by the two walls parallel to the x-axis as their stiffness will be negligible compared with the stiffness of the core and the walls parallel to the y-axis.

The second moments of area about the x-axis of the two walls are:

$$I_{xxw} = 0.2 \times 8^3/12 = 8.53 \text{ m}^4$$

The second moment area about the x-axis of the core is

$$I_{xxc} = (4 \times 4^3/12 - 3.7 \times 3.7^3/12) = 5.72 \text{ m}^4$$
$$\Sigma I_{xx} = 2I_{xxw} + I_{xxc} = 22.78 \text{ m}^4$$
$$\text{Load taken by each wall} = (I_{xxw}/\Sigma I_{xx}) \times F = 0.38 \text{ F}$$
$$\text{Load taken by the core} = (I_{xxw}/\Sigma I_{xx}) \times F = 0.25 \text{ F}$$

If the arrangement of the walls is unsymmetrical, an arrangement not recommended, as shown in Fig. 10.3, then the building will experience a twisting moment. Assuming the dimensions of the walls and core as in the previous example, the second moments of are:

$$\text{Wall parallel to y-axis: } I_{xxwy} = 0.2 \times 8^3/12 = 8.53 \text{ m}^4$$
$$\text{Wall parallel to x-axis: } I_{xxwx} = 0.2 \times 12^3/12 = 28.8 \text{ m}^4$$

$$I_{xxc}=I_{yyc}=(4\times4^3/12-3.7\times3.7^3/12)=5.72\ m^4$$
$$\sum I_{xx}=I_{xxwy}+I_{xxc}=14.25\ m^4$$
$$\sum I_{yy}=I_{yywx}+I_{yyc}=34.52\ m^4$$

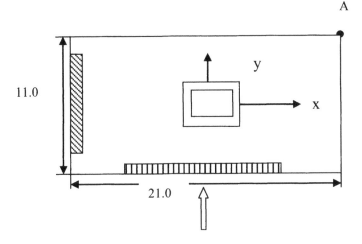

Fig. 10.3 Plan of an unsymmetrical shear wall assembly.

The centre of stiffness (also called as shear centre) is calculated as follows.
Taking the first moment of I_{xx} of all the elements about A,
$$I_{xxwy}\times(21.0-0.2/2)+I_{xxc}\times(21.0/2)=d_x\sum I_{xx}$$
$$8.53\times(21.0-0.2/2)+5.72\times(21.0/2)=14.25\times d_x$$
$$d_x=16.72\ m$$
Taking the first moment of I_{yy} of all the elements about A,
$$I_{yywx}\times(11.0-0.2/2)+I_{yyc}\times(11.0/2)=d_y\sum I_{yy}$$
$$28.8\times(11.0-0.2/2)+5.72\times(11.0/2)=34.52\times d_y$$
$$d_y=10.01\ m$$
The eccentricity e of the force F from the centre of stiffness is
$$e=16.72-21.0/2=6.22\ m$$
$$\text{Twisting moment, }T=Fe=6.22\ F$$
The force in any element is made of a 'direct' force and a force due to twisting moment.

Table 10.1 Forces in the walls

Wall	I_{xxi}	I_{yyi}	d_{xi}	d_{yi}	$I_{xxi}(d_x-d_{xi})^2+I_{yyi}(d_y-d_{yi})^2$
Wall y-axis	8.53	0	20.9	5.5	149.04
Wall x-axis	0	28.8	10.5	10.9	22.81
core	5.72	5.72	10.5	5.5	337.64
					$\sum 509.50$

Direct forces in the elements are:
$$\text{Wall in the y-direction}=(I_{xxwy}/\sum I_{xx})\times F=0.60\ F$$
$$\text{Wall in the x-direction}=0$$

$$\text{Core} = (I_{xxc}/\textstyle\sum I_{xx}) \times F = 0.40\ F$$

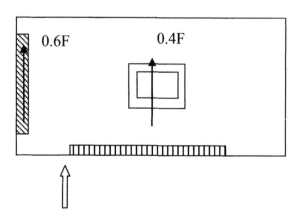

Fig. 10.4 Direct forces in the shear wall assembly.

Forces in the elements due to twisting moment are:

Taking $d_x = 16.72$ m and $d_y = 10.01$ m, table 10.1 can be completed.

Wall in the y-direction $= T \times (I_{xxwy} \times (d_x - d_{xi}))/\sum I_{xx} (d_x - d_{xi})^2 + I_{yy} (d_y - d_{yi})^2)$
$= -0.44F$

Wall in the x-direction $= T \times (I_{xxwx} \times (d_y - d_{yi}))/\sum I_{xx} (d_x - d_{xi})^2 + I_{yy} (d_y - d_{yi})^2)$
$= -0.31F$

Core in the y-direction $= T \times (I_{xxwy} \times (d_x - d_{xi}))/\sum I_{xx} (d_x - d_{xi})^2 + I_{yy} (d_y - d_{yi})^2) = 0.43F$

Core in the x-direction $= T \times (I_{xxwx} \times (d_y - d_{yi}))/\sum I_{xx} (d_x - d_{xi})^2 + I_{yy} (d_y - d_{yi})^2) = 0.32F$

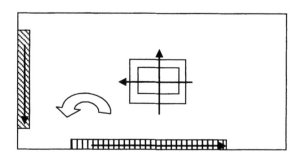

Fig. 10.5 Forces due to twisting moment in the shear wall assembly.

(iii) Total forces are:

Wall in the y-direction: $0.60F - 0.44F = 0.16F$
Wall in the x-direction: $0 - 0.31F = -0.12F$
Core in the y-direction: $0.40F + 0.43F = 0.83F$
Core in the x-direction: $0 + 0.32F = 0.32F$

(e) Reinforcement for walls in tension
If tension develops across the wall section, the reinforcement is to be arranged in two layers and the spacing of bars in each layer should comply with the bar spacing rules in section 9.6.2 and 9.6.3 of the code.

10.2.3 Design of Stocky Reinforced Concrete Walls

The design of stocky reinforced concrete walls is covered by the rules for the design of columns.

(a) Walls supporting transverse moment and uniform axial load
Where the wall supports a transverse moment and a uniform axial load, a unit length of wall can be designed as a column using column design charts discussed in Chapter 9.

(c) Walls supporting in-plane moments and axial load
The design for this case is set out in section 10.3.4 below.

(d) Walls supporting axial load and transverse and in-plane moments
The code states that the effects are to be assessed in three stages.

(i) In-plane Axial force and in-plane moments are applied. The distribution of force along the wall is calculated using elastic analysis assuming no tension in the concrete.

(ii) Transverse The transverse moments are calculated using the procedure set out in section 10.3.2(c).

(iii) Combined The effects of all actions are combined at various sections along the wall. The sections are checked using the general assumptions for beam design.

10.3 WALLS SUPPORTING IN-PLANE MOMENTS AND AXIAL LOADS

10.3.1 Wall Types and Design Methods

Some types of shear wall are shown in Fig. 10.6. The simplest type is the straight wall with uniform reinforcement as shown in Fig. 10.6(a). In practice the shear wall includes columns at the ends as shown in Fig. 10.6(e). Channel-shaped walls are also common as shown in Fig. 10.6(d), and other arrangements are used.
There are many procedures used to design a wall subjected to axial force and in-plane moment. Three design procedures are discussed.
1. Using an interaction chart
2. Assuming elastic stress distribution

3.   Assuming that only end zones resist moment

The methods are discussed briefly below. Examples illustrating their use are given.

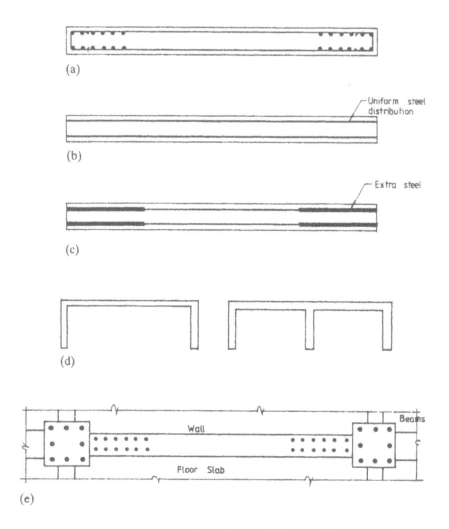

Fig. 10.6 (a) Wall reinforcement: (b) uniform strips of steel; (c) extra reinforcement in end zones; (d) channel-shaped shear walls; (e) shear wall between columns.

10.3.2 Interaction Chart

The chart construction is based on the assumptions for design of beams given in clause 6.1 of Eurocode 2. A straight wall with uniform reinforcement is considered.

For the purpose of analysis the vertical bars are replaced by uniform strips of steel running the full length of the wall as shown in Fig. 10.6 (b).

The chart shown in Fig. 10.8 is constructed using the following equations.

Assuming $f_{yk} = 500$ MPa and Young's modulus for steel is 200 GPa, then the strain ε_y when the stress is $f_{yd} = f_{yk}/1.15 = 435$ MPa is given by

$$\varepsilon_y = 435/(200 \times 10^3) = 2.174 \times 10^{-3}$$

If the maximum compressive strain in concrete is 0.0035 and the neutral axis depth is x, the strain in steel is equal to ε_y at a depth c from the neutral axis, where

$$c = (\varepsilon_y /0.0035)\, x = 0.6211\, x,$$
$$(x - c) = 0.3789\, x$$

Using the rectangular compressive stress block, $f_{cd} = f_{ck}/1.5 = 0.667\, f_{ck}$ and the thickness of the wall is b, the compressive force due to concrete $= 0.667 f_{ck} (0.8x)\, b$.

The steel in the wall is taken as A_{sc} mm^2/mm.

Three cases of neutral axis positions need to be considered.

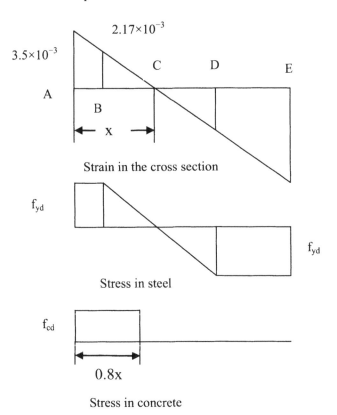

Fig. 10.7 (a) Strains and stresses in steel and concrete for Case 1.

Case 1: If $(x/h) \leq 0.6169$, as shown in Fig. 10.7(a)
$$AB = 0.3789\, x,\ BC = CD = 0.6211\, x,\ DE = h - 1.6211\, x$$

Note: DE = 0, if (h − 1.6211 x) = 0, x/h = 0.6169.
This is the maximum x/h for case 1.
Force in segments:
$$F_{AB} = 0.3789 \text{ x} \times (f_{yd} \times b \times A_{sc})$$
$$F_{BC} = 0.6211 \text{ x} \times (0.5 \times f_{yd} \times b \times A_{sc})$$
$$F_{CD} = -0.6211 \text{ x} \times (0.5 \times f_{yd} \times b \times A_{sc})$$
$$F_{DE} = (h - 1.6211 \text{ x}) \times (f_{yd} \times b \times A_{sc})$$
Summing up the forces in steel,
$$F_{steel} = (h - 1.432 \text{ x}) \times (f_{yd} \times b \times A_{sc})$$
Force due to concrete, $F_{concrete} = 0.8 \text{x} \times b \times f_{cd}$
Lever arms from the middle of the depth:
$$\ell_{AB} = 0.5 \text{ h} - 0.5 \times 0.3789 \text{ x} = 0.5 \text{h} - 0.1895 \text{ x}$$
$$\ell_{BC} = 0.5 \text{ h} - 0.3789 \text{ x} - (0.6211 \text{ x})/3 = = 0.5 \text{ h} - 0.5859 \text{ x}$$
$$\ell_{CD} = 0.5 \text{ h} - 0.3789 \text{ x} - 0.6211 \text{ x} - (2/3) \times (0.6211 \text{ x}) = 0.5 \text{ h} - 1.4141 \text{ x}$$
$$\ell_{DE} = 0.5 \text{ h} - 0.3789 \text{ x} - 2 \times 0.6211 \text{ x} - (h - 1.6211\text{x})/2 = -0.8106 \text{ x}$$
Lever arm for force in concrete, $\ell_{concrete} = 0.5 \text{h} - 0.4 \text{ x}$
N = Algebraic sum of all the axial forces.
M = Algebraic sum of the product of axial forces and corresponding lever arms.

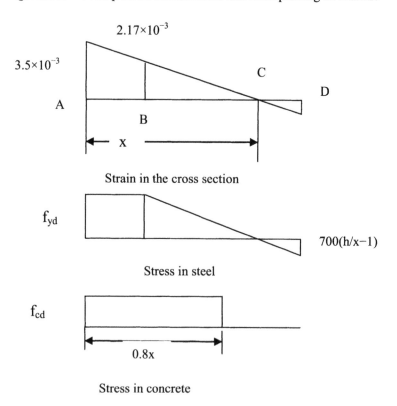

Fig. 10.7 (b) Strains and stresses in steel and concrete for Case 2.

Case 2: If $1.0 \geq (x/h) > 0.6169$, as shown in Fig. 10.7(b)
$$AB = 0.3789\ x,\ BC = 0.6211\ x,\ CD = h - x$$
Note: $CD = 0$ if $(h - x) = 0$, $x/h = 1.0$. This is the maximum x/h for case 2.
Force in segments:
$$F_{AB} = 0.3789\ x \times (f_{yd} \times b \times A_{sc})$$
$$F_{BC} = 0.6211\ x \times (0.5 \times f_{yd} \times b \times A_{sc})$$
$$F_{CD} = (h - x) \times (700 \times (h/x - 1) \times b \times A_{sc})$$
$$\text{Force due to concrete, } F_{concrete} = 0.8 x \times b \times f_{cd}$$
Lever arms from the middle of the depth:
$$\ell_{AB} = 0.5\ h - 0.5 \times 0.3789\ x = 0.5h - 0.1895\ x$$
$$\ell_{BC} = 0.5\ h - 0.3789\ x - (0.6211\ x)/3 = = 0.5\ h - 0.5859\ x$$
$$\ell_{CD} = 0.5\ h - 0.3789\ x - 0.6211\ x - (2/3) \times (h - x) = -0.1667\ h - 0.3333\ x$$
$$\text{Lever arm for force in concrete, } \ell_{concrete} = 0.5h - 0.4\ x$$
N = Algebraic sum of all the axial forces.
M = Algebraic sum of the product of axial forces and corresponding lever arms.

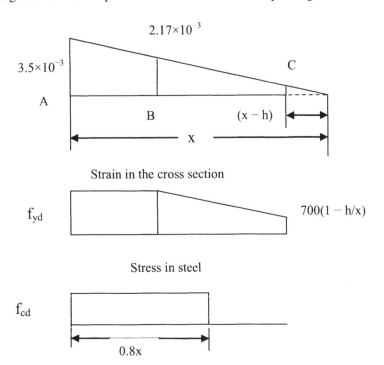

Strain in the cross section

Stress in steel

Stress in concrete

Fig. 10.7(c) Strains and stresses in steel and concrete for Case 3.

Case 3: If $2.6392 \geq (x/h) > 1.0$, as shown in Fig. 10.8(c)
$$AB = 0.3789\ x,\ BC = h - 0.3789\ x$$
Note: $BC = 0$, if $(h - 0.3789\ x) = 0$, $x/h = 2.6392$. This is the maximum x/h for Case 3.

Force in segments:
$$F_{AB} = 0.3789 \; x \times (f_{yd} \times b \times A_{sc})$$
$$F_{BC} = (h - 0.3789 \; x) \times [0.5 \times \{f_{yb} + \sigma\} \times b \times A_{sc}]$$
$$\sigma = 700 \times (1 - h/x)$$

Force due to concrete:
$$F_{concrete} = 0.8x \times b \times f_{cd}, \quad 0.8 \; x \le h$$
$$F_{concrete} = h \times b \times f_{cd}, \quad \quad 0.8 \; x > h$$

Lever arms from the middle of the depth:
$$\ell_{AB} = 0.5 \; h - 0.5 \times 0.3789 \; x = 0.5h - 0.1895 \; x$$
$$\ell_{BC} = 0.5 \; h - 0.3789 \; x - (h - 0.3789 \; x)/3 \times \{1 + \sigma/(\sigma + f_{yd})\}$$
$$= 0.1667 \; h - 0.2526 \; x - (h - 0.3789 \; x)/3 \times \{\sigma/(\sigma + f_{yd})\}$$

Lever arm for force in concrete, $\ell_{concrete} = 0.5h - 0.4 \; x, \; 0.8 \; x \le h$
$$\text{If } 0.8x > h, \; \ell_{concrete} = 0$$

N = Algebraic sum of all the axial forces.
M = Algebraic sum of the product of axial forces and corresponding lever arms.
The chart is shown in Fig. 10.8.

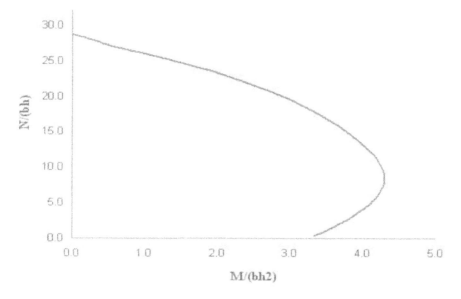

Fig. 10.8 Wall design chart: $f_{ck} = 30$ MPa, $f_{yk} = 500$ MPa, $A_{sc} = 2\%$.

A chart could also be constructed for the case where extra steel is placed in two zones at the ends of the walls as shown in Fig. 10.6(c). Charts could also be constructed for channel-shaped walls.

In design the wall is assumed to carry the axial load applied to it and the overturning moment from wind. The end columns, if existing, are designed for the loads and moments they carry. Very often the in-plane moment is treated as concentrated tension and compression forces in the end columns.

(a) Elastic stress distribution

A straight wall section, including columns if desired, or a channel-shaped wall is analyzed for axial load and moment using the properties of the gross concrete section in each case. The wall is divided into sections and each section is designed for the average direct load on it. Compressive forces are resisted by concrete and reinforcement. Tensile stresses are resisted by reinforcement only.

(b) Assuming that end zones resist moment

Reinforcement located in zones at each end of the wall is designed to resist the moment. The axial load is assumed to be distributed over the length of the wall.

10.3.3 Example of Design of a Wall Subjected to Axial Load and In-Plane Moment Using Design Chart

(a) Specification

The plan and elevation for a braced concrete structure are shown in Fig. 10.9.

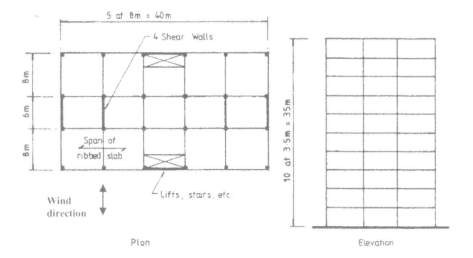

Fig. 10.9 Framing arrangement.

The total dead load of the roof and floors is 6 kN/m². The imposed load on roof is 1.5 kN/m² and that for each floor is 3.0 kN/m². The wind speed is 20 m/s and the building is located in a city centre. Design the transverse shear walls as straight walls without taking account of the columns at the ends. The load bearing part of the wall is 160 mm thick with 20 mm thick decorative tiles on both faces. The materials are f_{ck} = 30 MPa concrete and f_{yk} = 500 MPa for reinforcement.

Refer to *BS EN 1991-1-1: 2002 Eurocode 1: Actions on Structures Part -1-1: General actions-Densities, self-weight, imposed loads on buildings,* for more information.

(b) Type of wall: slenderness

The wall is 160 mm thick structurally and is braced. The slenderness is calculated as for columns as given in clause 5.8.3.2 of Eurocode 2. In this example it is assumed that the wall is 'stocky' and second order effects can be ignored.

(c) Dead and imposed loads on wall

Dead load: The dead load on the inner wall, given that the wall is 200 mm thick including finishes, is as follows.

Note that there are 10 floors including the roof and the plan area of each floor is 8×6 m. Total height of the building = 35 m.

$$\text{Roof and floor slabs: } 10 \times (6 \times 8) \times 6 \text{ kN/m}^2 = 2880 \text{ kN}$$
$$\text{Wall, 200 mm thick: } (0.2 \times 6 \times 35) \times 25 \text{ kN/m}^3 = 1050 \text{ kN}$$
$$\text{Total dead load at base: } 2880 + 1050 = 3930 \text{ kN}$$

Imposed load: The wall carries load from 10 floors. Therefore imposed load can be multiplied by a reduction factor α_n in accordance with BS EN 1991-1-1:2002 Part 1-1, equation (6.2)

$$\alpha_n = \frac{2 + (n - 2)\psi_0}{n} \tag{6.2}$$

where n = number of storeys (>2) above the structural element from the same category and $\Psi_0 = 0.7$ for office areas from Table A1.1 of the code. Substituting n = 10 and $\Psi_0 = 0.7$ in equation (6.2), $\alpha_n = 0.76$. The imposed load is

$$0.76 \times \{1.5 \text{ (Roof load)} + 3.0 \times 9 \text{ floors}\} (6 \times 8) = 1040 \text{ kN}$$

(d) Dead and imposed loads at each end of the wall from one transverse beam

The slabs span between the beams which are supported on the walls. On any transverse beam, acts, all the load acting on area 8×8 m. The beam reaction acts on the column.

$$\text{Dead load from roof and floor slab: } 10 \times \{(1/2) \times (8 \times 8)\} \times 6.0 = 1920 \text{ kN}$$
$$\text{Column (500 mm} \times \text{500 mm) at wall ends: } 35 \times 0.5 \times 0.5 \times 25 \text{ kN/m}^3 = 218.75 \text{ kN}$$
$$\text{Imposed load: } 0.76 \times \{(1/2) \times (8 \times 8)\} \times (1.5 + 3.0 \times 9) = 693.1 \text{ kN}$$

In summary, axial load from:

Dead load = 2880 (roof and floor) + 1050 (self weight of wall)
$$+ 2 \times (1920 + 218.75) \text{ (two transverse beams)} = 8207.5 \text{ kN}.$$
Imposed load: 1040 (roof and floors) + 2 × 693.12 (two transverse beams) =2426.2 kN.

(e) Wind load

Wind loads are specified in

BS EN 1991-1-4: 2005 + A1:2010 Eurocode 1: Actions on Structures. General actions. Wind actions

UK National Annex to Euro code 1: Actions on Structures. General actions. Wind actions
A useful reference on the code is Cook (2007).

The case for which wind load is calculated is wind acting normal to the 40 m width. The maximum height H = 35 m above the ground. The wind loads are *assumed* to be resisted equally by four shear walls. The calculations of wind loads are complex and reference must be made to the full code along with the National Annex.

(f) Wind load calculated using U.K. National Annex

The following is a very brief summary using the National Annex for the United Kingdom.

Step 1: The basic wind velocity v_b is calculated from the wind code equation (4.1) as
$$v_b = C_{dir} \times C_{season} \times v_{b,0} \qquad (4.1)$$
Conservatively, direction factor C_{dir} and season factor C_{season} can both be taken as 1.0
$$v_b = 1.0 \times 1.0 \times v_{b,0} = v_{b,0}$$

Step 2: The fundamental value of the basic wind velocity $v_{b,0}$ is given by the National Annex equation (NA.1) as
$$v_{b,0} = v_{b,map} \times C_{alt} \qquad (NA.1)$$
$Vb,_{map}$ = fundamental basic wind velocity in m/s given in the map of the country. The map for the United Kingdom is given in the National Annex in Fig. NA.1.
Assume $v_{b,map} \approx 20$ m/s.
C_{alt} = altitude correction factor.

Step 3: Conservatively, C_{alt} for any building height is given by National Annex equation (NA.2a) as
$$C_{alt} = 1 + 0.001 \times A \qquad (NA.2a)$$
A = Altitude of the site in meters above sea level.
If A = 100 m, C_{alt} = 1.1
$$Vb,_0 = v_{b,map} \times C_{alt} = 20 \times 1.1 = 22.0 \text{ m/s}$$
$$v_b = v_{b,0} = 22.0 \text{ m/s}$$

Step 4: The basic velocity pressure q_b is given by the wind equation (4.10) as
$$q_b = 0.5 \times \rho \times v_b^2 \text{ N/m}^2 \qquad (4.10)$$
ρ = density of air taken as 1.226 kg/m^3
$$q_b = 0.613 \times v_b^2 \times 10^{-3} = 0.613 \times 22.0^2 \times 10^{-3} = 0.30 \text{ kN/m}^2$$

Step 5: The peak wind pressure q_p (z) is given by National Annex equation (NA.3b) for sites in town terrain by

$$q_p(z) = c_e(z) \times c_{e,\,T} \times q_b \qquad\qquad\text{(NA.3b)}$$

Step 6: Exposure factor $c_e(z)$
Since the height h of the building is 35 m and the width of the building b is 40 m, h
< b. Therefore z = h = 35 m, h_{dis} = 0 for terrain category IV. From Fig. NA.7,
assuming 10 km from the shore line, $c_e(z) \approx 3.4$

Step 7: Exposure correction factor $c_{e,\,T}$
From Fig. NA.8, assuming 10 km from the shore line, $c_{e,\,T}(z) \approx 0.94$
$$q_p(z) = c_e(z) \times c_{e,\,T} \times q_b = 3.4 \times 0.94 \times 0.30 = 0.96 \text{ kN/m}^2$$

Step 8: The total pressure coefficient c_f is the sum of external pressure coefficient
c_{pe} and internal pressure coefficient c_{pi}. The values of the pressure coefficients are
calculated from Table 7.1 of the wind code. Values of $c_{pe,\,10}$ should be used for the
design of overall load bearing structure. In this case h = 35 m, b = 40 m,
h/b = 0.875. Interpolating from Table NA.4,
$$\text{Area D: } c_{pe,\,10} = 0.7 + (0.8 - 0.7) \times (0.875 - 0.25)/(1.0 - 0.25) = 0.78$$
$$\text{Area E: } c_{pe,\,10} = 0.3 + (0.5 - 0.3) \times (0.875 - 0.25)/(1.0 - 0.25) = 0.47$$
$$c_f = 0.78 + 0.47 = 1.25$$

Step 9: The total wind load wk = $q_p(z) \times c_f = 0.96 \times 1.25 = 1.20 \text{ kN/m}^2$.

(g) Wind load calculated using the Eurocode wind code
The following is a very brief summary using the BS EN 1991:2005 wind loading
code

Step 1: The mean wind velocity $v_m(z)$ at a height above the terrain is given by the
wind code equation (4.3) as
$$v_m(z) = c_r(z) \times c_0(z) \times v_b \qquad\qquad\text{(4.3)}$$
$c_r(z)$ = terrain roughness factor. This factor accounts for the variability of the
mean wind velocity at the site due to height above the ground level and ground
roughness of the terrain upwind of the structure.
$c_0(z)$ = Orography (terrain) factor taken as 1.0.

Step 2: $c_r(z)$ is defined by the wind code equation (4.4) as
$$c_r(z) = k_r \times \ell_n(z/z_0) \text{ for } z_{min} \le z \le z_{max} \qquad\qquad\text{(4.4)}$$
From Table 4.1 of the wind code, for terrain category IV (see figures in A.1 of the
code) which represents city environment, $z_0 = z_{min} = 1.0$, z = height of the building,
z_{max} = 200 m.

Step 3: k_r is defined by the wind code equation (4.5) for town areas as
$$k_r = 0.19 \times (z_0/z_{0,\,11})^{0.07} \qquad\qquad\text{(4.5)}$$
From Table 4.1 in the code, $z_0 = 1.0$, and $z_{0,\,11} = 0.05$

$$k_r = 0.19 \times (z_0/z_{0,\,\text{II}})^{0.07} = 0.19 \times (1.0/0.05)^{0.07} = 0.19 \times 1.23 = 0.23$$

Step 4: If $z = 35$ m,
$$c_r(z) = k_r \times \ell n\,(z/z_0) = 0.23 \times \ell n\,(35/1.0) = 0.23 \times 3.56 = 0.83$$

Step 5: The basic wind velocity v_b is calculated from the wind code equation (4.1) as
$$v_b = C_{\text{dir}} \times C_{\text{season}} \times v_{b,\,0} \qquad (4.1)$$
Conservatively, direction factor C_{dir} and season factor C_{season} can both be taken as 1.0
$$v_b = 1.0 \times 1.0 \times v_{b,\,0} = v_{b,\,0}$$

Step 6: The fundamental value of the basic wind velocity $v_{b,\,0}$ is given by the National Annex equation (NA.1) as
$$Vb,_0 = v_{b,\,\text{map}} \times C_{\text{alt}} \qquad (NA.1)$$
$Vb,_{\text{map}}$ = fundamental basic wind velocity in m/s given in the map of the country. The map for the United Kingdom is given in the National Annex in Fig. NA.1. Assume $v_{b,\,\text{map}} \approx 20$ m/s.
C_{alt} = altitude correction factor.

Step 7: Conservatively, C_{alt} for any building height is given by National Annex equation (NA.2a) as
$$C_{\text{alt}} = 1 + 0.001 \times A \qquad (NA.2a)$$
A = Altitude of the site in meters above sea level.
If $A = 100$ m, $C_{\text{alt}} = 1.1$
$$Vb,_0 = v_{b,\,\text{map}} \times C_{\text{alt}} = 20 \times 1.1 = 22.0 \text{ m/s}$$
$$v_b = v_{b,\,0} = 22.0 \text{ m/s}$$

Step 8: $v_m(z) = c_r(z) \times c_0(z) \times v_b = 0.83 \times 1.0 \times 22.0 = 18.3$.

Step 9: The basic velocity pressure q_b is given by the wind equation (4.10) as
$$q_b = 0.5 \times \rho \times v_b^2 \qquad (4.10)$$
ρ = density of air taken as 1.25 kg/m^3
$$q_b = 0.613 \times v_b^2 \times 10^{-3} = 0.613 \times 22.0^2 \times 10^{-3} = 0.30 \text{ kN/m}^2$$

Step 10: The peak velocity pressure $q_p(z)$ at height z is given by wind code equation (4.8)
$$q_p(z) = c_e(z) \times q_b \qquad (4.8)$$

Step 11: The value of $c_e(z)$ for different values of z and different terrain categories is given in Fig. 4.2 of the code. For $z = 35$ m and terrain category IV, $c_e(z) \approx 2.1$.

Step 12: $q_p(z) = c_e(z) \times q_b = 2.1 \times 0.30 = 0.63 \text{ kN/m}^2$

Step 13: The wind pressure acting on the external surface w_e is calculated from wind code equation (5.1)

$$w_e = q_p\,(z_e) \times c_{pe} \tag{5.1}$$

Step 14: The wind pressure acting on the internal surface w_i is calculated from wind code equation (5.1)

$$w_i = q_p\,(z_e) \times c_{pi} \tag{5.1}$$

Step 15: The wind force $F_{w,\,e}$ acting on the external surface is calculated from wind code equation (5.5)

$$W_e = c_s c_d \times \sum W_e \times A_{ref} \tag{5.5}$$

Step 16: The wind force $F_{w,\,i}$ acting on the internal surface is calculated from wind code equation (5.6)

$$F_{w,\,i} = \sum w_i \times A_{ref} \tag{5.6}$$

Step 17: From section 6.2(1)(c), for framed buildings with structural walls with height less than 100 m and less than four times the in-wind depth $c_s\,c_d = 1.0$. In this case h = 35 m, in-wind depth = 22 m. Therefore $c_s\,c_d = 1.0$.

Step 18: Frictional forces can be ignored when the total area of all surfaces parallel to the wind is equal to less than the total area of all external surfaces perpendicular to the wind.

Step 19: The total pressure coefficient c_f is the sum of external pressure coefficient c_{pe} and internal pressure coefficient c_{pi}. The values of the pressure coefficients are calculated from Table 7.1of the wind code. Values of $c_{pe,\,10}$ should be used for the design of overall load bearing structure. In this case h = 35 m, b = 40 m, h/b = 0.875. Interpolating from wind code Table 7.1,
 Zone D: External pressure c_{pe}
  $C_{pe,\,10} = 0.7 + (0.8 - 0.7) \times (0.875 - 0.25)/\,(1.0 - 0.25) = 0.78$
 Zone E: Internal pressure c_{pi}
  $c_{pe,\,10} = -0.3 + (-0.5 + 0.3) \times (0.875 - 0.25)/\,(1.0 - 0.25) = -0.47$

Step 20: The total wind load force $F_w = QP\,(z) \times (0.78 + 0.47) \times A_{ref}$
  $F_w = 0.63 \times (0.78 + 0.47) \times A_{ref}$
  F_w per wall = $0.79 \times (40 \times 35)\,/4 = 276.5$ kN

(h) Load combination: Using the design values of actions as shown in Table A1.2 (B) and recommended values of ψ factors for buildings given in Table A1.1 of *BS EN 1990:2002 Euro code - Basis of Structural Design*, the following design values of axial load N and in-plane moment M can be calculated. The wall is 160 mm thick by 6000 mm long. Taking the wind load as 1.2 kN/m^2 from the U.K. National Annex, horizontal wind load and the corresponding moment at the base are as follows:
  Total horizontal load per wall = $1.20 \times (40 \times 35)\,/4 = 420$ kN
  Moment = $420 \times 35/2 = 7350.0$ kNm

i. Load calculation using equation (6.10) of the code
Case 1:
Dead + Imposed as leading variable + Wind as accompanying variable
$\gamma_{Gj, sup} = 1.35, \gamma_{Q, 1} = 1.5, \gamma_{Q, i} = 1.5, \psi_{0, i} = 0.6$
$$N = 1.35 \times 8207.5 \text{ (dead)} + 1.5 \times 2426.2 \text{ (Imposed)} = 14719.4 \text{ kN}$$
$$M = 1.5 \times 0.6 \times 7350.0 = 6615.0 \text{ kNm}$$
$$b = 160 \text{ mm}, h = 6000 \text{mm}$$
$$N/ (bh) = 15.33, M/ (bh^2) = 1.15$$

Case 2:
Dead + Wind as leading variable + Imposed as accompanying variable
$\gamma_{Gj, sup} = 1.35, \gamma_{Q, 1} = 1.5, \gamma_{Q, i} = 1.5, \psi_{0, i} = 0.7$
$$N = 1.35 \times 8207.5 \text{ (dead)} + 1.5 \times 0.7 \times 2426.2 \text{ (Imposed)} = 13627.6 \text{ kN}$$
$$M = 1.5 \times 7350.0 = 11025.0 \text{ kNm}$$
$$b = 160 \text{ mm}, h = 6000 \text{mm}$$
$$N/ (bh) = 14.20, M/ (bh^2) = 1.91$$

ii. Load calculation using equation (6.10a) of the code
Case 1:
Dead + Imposed as leading variable + Wind as accompanying variable
$\gamma_{Gj, sup} = 1.35, \gamma_{Q, 1} = 1.5, \psi_{0, 1} = 0.7, \gamma_{Q, i} = 1.5, \psi_{0, i} = 0.6$
$$N = 1.35 \times 8207.5 \text{ (dead)} + 1.5 \times 0.7 \times 2426.2 \text{ (Imposed)} = 13627.6 \text{ kN}$$
$$M = 1.5 \times 0.6 \times 7350.0 = 6615.0 \text{ kNm}$$
$$b = 160 \text{ mm}, h = 6000 \text{mm}$$
$$N/ (bh) = 14.20, M/ (bh^2) = 1.15$$

Case 2:
Dead + Wind as leading variable + Imposed as accompanying variable
$\gamma_{Gj, sup} = 1.35, \gamma_{Q, 1} = 1.5, \psi_{0, 1} = 0.6, \gamma_{Q, i} = 1.5, \psi_{0, i} = 0.7$
$$N = 1.35 \times 8207.5 \text{ (dead)} + 1.5 \times 0.7 \times 2426.2 \text{ (Imposed)} = 13262.7 \text{ kN}$$
$$M = 1.5 \times 0.6 \times 7350.0 = 6615.0 \text{ kNm}$$
$$b = 160 \text{ mm}, h = 6000 \text{mm}$$
$$N/ (bh) = 14.20, M/ (bh^2) = 1.15$$

iii. Load calculation using equation (6.10b) of the code
Case 1:
Dead + Imposed as leading variable + Wind as accompanying variable
$\xi = 0.85, \gamma_{Gj, sup} = 1.35, \gamma_{Q, 1} = 1.5, \psi_{0, 1} = 0.7, \gamma_{Q, i} = 1.5, \psi_{0, i} = 0.6$
$$N = 0.85 \times 1.35 \times 8207.5 \text{ (dead)} + 1.5 \times 2426.2 \text{ (Imposed)} = 13057.4 \text{ kN}$$
$$M = 1.5 \times 0.6 \times 7350.0 = 6615.0 \text{ kNm}$$
$$b = 160 \text{ mm}, h = 6000 \text{mm}$$
$$N/ (bh) = 13.60, M/ (bh^2) = 1.15$$

Case 2:
Dead + Wind as leading variable + Imposed as accompanying variable
$\xi = 0.85, \gamma_{Gj, sup} = 1.35, \gamma_{Q, 1} = 1.5, \psi_{0, 1} = 0.6, \gamma_{Q, i} = 1.5, \psi_{0, i} = 0.7$
$$N = 0.85 \times 1.35 \times 8207.5 \text{ (dead)} + 1.5 \times 0.7 \times 2426.2 \text{ (Imposed)} = 11965.6 \text{ kN}$$
$$M = 1.5 \times 7350.0 = 11025 \text{ kNm}$$
$$b = 160 \text{ mm}, h = 6000 \text{mm}$$
$$N/ (bh) = 12.46, M/ (bh^2) = 1.91$$

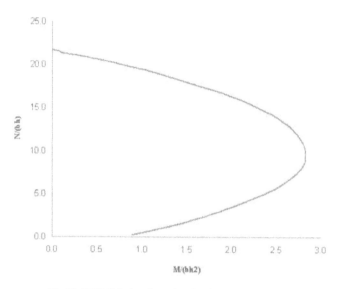

Fig. 10.10 Wall design chart: f_{ck} = 30, f_{yk} = 500, A_{sc} = 0.4%.

Wall design for load combinations in (f)

The wall is 160 mm thick by 6000 mm long. The design is made using the chart in Fig. 10.10 which is for a steel percentage of 0.4%. The most critical case is case 2 for equation (6.10b). The design using 0.4% of steel appears satisfactory.
The total steel area is given by
$$A_{sc} = (0.4/100) \times 160 \times 6000 = 3840 \text{ mm}^2$$
$$\text{Area per meter} = 3840/6.0 = 640 \text{ mm}^2/\text{m}$$
Provide one row of H10 bars at 100 mm centers. From Table 4.9, Chapter 4, area of steel provided is 785 mm^2/m.
The provided steel percentage is
$$[785/ (160 \times 1000)] \times 100 = 0.49\%$$

Table 10.2 Load combinations, wall design

Loads (N,M)	Eq. 6.10		Eq. 6.10a		Eq. 6.10b	
	Case 1	Case 2	Case 1	Case 2	Case 1	Case 2
N/ (bh)	15.33	14.20	14.20	14.20	13.60	*12.46*
M/ (bh^2)	1.15	1.91	1.15	1.15	1.15	*1.91*

(i) Check minimum and maximum steel percentages:
Minimum steel percentage required is 0.2% and maximum allowed is 4%.
The provided value of 0.49% satisfies the requirements.

(ii) Spacing of bars:
The distance between adjacent bars should not exceed Min (3 × wall thickness; 400 mm).
Wall thickness = 160 mm. Spacing should not exceed 400 mm. Actual spacing is 100 mm.

(iii) Horizontal reinforcement:
Steel area = max (25% of vertical steel; 0.1%).
25% of vertical steel = 0.12%.
Maximum spacing allowed = 400 mm.
0.12% = 0.12/100 × (thickness = 160) × (spacing = 400 mm) = 77 mm^2.
Provide H8 bars at 400 mm spacing as horizontal steel. Area of steel provided is 78.5 mm^2.

(iv) Transverse steel:
As the vertical steel is one layer only and the percentage does not exceed 2%, transverse steel is not required.

10.3.3.1 Example of Design of a Wall with Concentrated Steel in End Zones or Columns Subjected to Axial Load and In-Plane Moment

The plan of the wall is shown in Fig. 10.11.

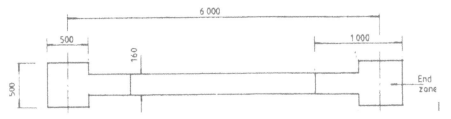

Fig. 10.11 Wall with end columns.

In the absence of a design chart to cover this case, the following approximate design procedure can be used. The wall simply carries the dead and imposed load acting on a plan area of 6 × 8 m. The end columns carry dead and imposed loads from the reactions from transverse beams. In addition they also resist the moment from the wind load.

(a) Wall Design:
Dead load: The wall section carries all the self weight and the dead and imposed loads which act directly on wall from the 10 floors including the roof over the plan area of each floor of 8 × 6 m.

Roof and floor slabs: 10 × (6 × 8) × 6 kN/m^2 = 2880 kN
Wall, 200 mm thick: (0.2 × 6 × 35) × 25 kN/m^3 = 1050 kN
Total dead load at base: 2880 + 1050 = 3930 kN

Imposed load: The wall carries load from 10 floors. Allowing for the reduction in total imposed load as calculated previously, the total imposed load is

0.76 × {1.5 (Roof load) + 3.0 × 9 floors} (6 × 8) = 1040 kN

Using Equation (6.10) for dead load and imposed combination as it gives the largest value of N,

$$\gamma G_{j, sup} = 1.35, \gamma Q,_1 = 1.5, \gamma Q,_i = 1.5$$
$$N = 1.35 \times 3930 \text{ (dead)} + 1.5 \times 1040 \text{ (Imposed)} = 6866 \text{ kN}$$

Clause 6.1(4) of Eurocode 2 recommends using a minimum eccentricity of 1/30 of the thickness or 20 mm, whichever is greater. $160/30 = 5.33 \text{ mm} < 20 \text{ mm}$

$$M = 1.5 \times 1040.0 \times 20 \times 10^{-3} = 31.2 \text{ kNm}$$
$$b = 160 \text{ mm}, h = 6000 \text{mm}$$
$$N/ (bh) = 7.15, M/ (bh^2) = 0.005$$

From the design chart shown in Fig. 10.10, minimum steel is all that is required. Provide H10 at 200 mm centers on both faces. From Table 4.9, Chapter 4, total steel area provided is 784 mm^2/m. Steel percentage is $784/ (1000 \times 200) \times 100 = 0.39\%$.

(i) Check minimum and maximum steel percentages:
Minimum steel percentage required is 0.2% and maximum allowed is 4%.
The provided value of 0.39% satisfies the requirements.

(ii) Spacing of bars:
The distance between adjacent bars should not exceed Min (3 × wall thickness; 400 mm). Wall thickness = 200 mm. Spacing should not exceed 400 mm. Actual spacing is 200 mm.

(iii) Horizontal reinforcement:
Steel area = max (25% of vertical steel; 0.1%).
25% of vertical steel = 0.1%.
Maximum spacing allowed = 400 mm.
$0.1\% = 0.1/100 \times (\text{thickness} = 200) \times (\text{spacing} = 400 \text{ mm}) = 80 \text{ mm}^2$.
On each face, steel area required is 40 mm^2.
On each face, provide H8 bars at 400 mm spacing as horizontal steel. Area of steel provided on each face is 50 mm^2.

(iv) Transverse steel:
As the vertical steel percentage does not exceed 2%, transverse steel is not required.

(b) Design of end columns:
The slabs span between the beams which are supported on the walls. The transverse beam supports all the load acting on area 8 × 8. The beam reaction acts on the column.

Dead load:
Dead load from roof and floor slab: $10 \times \{(1/2) \times (8 \times 8)\} \times 6.0 = 1920 \text{ kN}$
Self weight of Column (500 mm × 500 mm) at wall ends:
$35 \times 0.5 \times 0.5 \times 25 \text{ kN/m}^3 = 218.75 \text{ kN}$
Total dead load = $1920.0 + 218.75 = 2138.75 \text{ kN}$

Imposed load:
$0.76 \times \{(1/2) \times (8 \times 8)\} \times (1.5 + 3.0 \times 9) = 693.1 \text{ kN}$

Wind load:

Assume that the end columns resist the moment due to wind. The lever arm is 6.0 m. The equivalent axial force due to moment caused by wind is

$$\pm 7350.0 / 6.0 = \pm 1225 \text{ kN}$$

Take eccentricity = 20 mm, b = h = 500 mm

i. Load calculation using equation (6.10) of the code
Case 1:

Dead + Imposed as leading variable + wind as accompanying variable

$$\gamma G_{i, \text{sup}} = 1.35, \gamma Q, _1 = 1.5, \gamma Q, _i = 1.5, \psi_{0, i} = 0.6$$
$$N = 1.35 \times 2138.75 \text{ (dead)} + 1.5 \times 693.1 \text{ (Imposed)} \pm 1.5 \times 0.6 \times 1225 \text{ (wind)}$$
$$N = 5029.1 \text{ kN or } 2824.5 \text{ kN}$$
$$M = 100.6 \text{ kNm or } 56.55$$
$$N/(bh) = 20.12 \text{ or } 11.29, M/(bh^2) = 0.81 \text{ or } 0.45$$

Case 2:

Dead + wind as leading variable + imposed as accompanying variable

$$\gamma G_{i, \text{sup}} = 1.35, \gamma Q, _1 = 1.5, \gamma Q, _i = 1.5, \psi_{0, i} = 0.7$$
$$N = 1.35 \times 2138.75 \text{ (dead)} + 1.5 \times 0.7 \times 693.1 \text{ (Imposed)} \pm 1.5 \times 1225 \text{ (wind)}$$
$$N = 5452.6 \text{ kN or } 1777.6 \text{ kN}$$
$$M = 109.1 \text{ kNm or } 35.55 \text{ kNm}$$
$$N/(bh) = 21.81 \text{ or } 7.11, M/(bh^2) = 0.87 \text{ or } 0.28$$

ii. Load calculation using equation (6.10a) of the code
Case 1:

Dead + imposed as leading variable + wind as accompanying variable

$$\gamma G_{i, \text{sup}} = 1.35, \gamma Q, _1 = 1.5, \psi_{0, 1} = 0.7, \gamma Q, _i = 1.5, \psi_{0, i} = 0.6$$
$$N = 1.35 \times 2138.75 \text{ (dead)} + 1.5 \times 0.7 \times 693.1 \text{ (Imposed)} \pm 1.5 \times 0.6 \times 1225 \text{ (wind)}$$
$$N = 4717.6 \text{ kN or } 2512.6 \text{ kN}$$
$$M = 94.4 \text{ kNm or } 50.25 \text{ kNm}$$
$$N/(bh) = 18.87 \text{ or } 10.05, M/(bh^2) = 0.76 \text{ or } 0.40$$

Case 2:

Dead + wind as leading variable + imposed as accompanying variable

$$\gamma G_{i, \text{sup}} = 1.35, \gamma Q, _1 = 1.5, \psi_{0, 1} = 0.6, \gamma Q, _i = 1.5, \psi_{0, i} = 0.7$$
$$N = 1.35 \times 2138.75 \text{ (dead)} + 1.5 \times 0.7 \times 693.1 \text{ (Imposed)} \pm 1.5 \times 0.6 \times 1225 \text{ (wind)}$$
$$= 4717.6 \text{ kN or } 2512.6 \text{ kN}$$
$$M = 94.4 \text{ kNm or } 50.25 \text{ kNm}$$
$$N/(bh) = 18.87 \text{ or } 10.05, M/(bh^2) = 0.76 \text{ or } 0.40$$

iii. Load calculation using equation (6.10b) of the code
Case 1:

Dead + imposed as leading variable + wind as accompanying variable

$$\xi = 0.85, \gamma G_{i, \text{sup}} = 1.35, \gamma Q, _1 = 1.5, \psi_{0, 1} = 0.7, \gamma Q, _i = 1.5, \psi_{0, i} = 0.6$$
$$N = 0.85 \times 1.35 \times 2138.75 \text{ (dead)} + 1.5 \times 693.1 \text{ (Imposed)} \pm 1.5 \times 0.6 \times 1225 \text{ (wind)}$$
$$N = 4596.4 \text{ kN or } 2391.4 \text{ kN}$$

$$M = 91.9 \text{ kNm or } 47.8 \text{ kNm}$$
$$N/(bh) = 18.39 \text{ or } 9.57, M/(bh^2) = 0.74 \text{ or } 0.38$$

Case 2:

Dead + wind as leading variable + imposed as accompanying variable

$$\xi = 0.85, \gamma G_{j, \text{ sup}} = 1.35, \gamma Q_{,1} = 1.5, \psi_{0,1} = 0.6, \gamma Q_{,i} = 1.5, \psi_{0,i} = 0.7$$
$$N = 0.85 \times 1.35 \times 2138.75 \text{ (dead)} + 1.5 \times 0.7 \times 693.1 \text{ (Imposed)} \pm 1.5 \times 1225 \text{ (wind)}$$
$$N = 5019.5 \text{ kN or } 1344.5 \text{ kN}$$
$$M = 100.4 \text{ kNm or } 26.9 \text{ kNm}$$
$$N/(bh) = 20.1 \text{ or } 5.4, M/(bh^2) = 0.80 \text{ or } 0.22$$

Table 10.3 shows a summary of all load combinations. The two values refer to wind loading being additive or subtractive from the dead and imposed load combination.

Table 10.3 Load combinations, end column design

Load	Eq. 6.10		Eq. 6.10a		Eq. 6.10b	
	Case 1	*Case 2*	Case 1	Case 2	Case 1	Case 2
N/(bh)	20.1/11.	*21.81/7.1*	18.87/10.1	18.87/10.1	18.4/9.6	20.1/5.4
M/(bh²)	0.8/0.5	*0.9/0.3*	0.8/0.4	0.8/0.4	0.7/0.4	0.8/0.2

Fig. 10.12 shows the design chart for 2% steel. The load combination for case 2 is plotted, indicating that design is safe. Provide 8H32 bars as shown in Fig. 10.13.

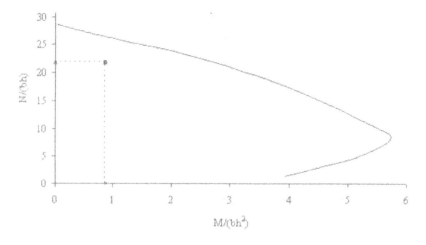

Fig. 10.12 N–M chart for end column design.

Design of links:

Column is 500 mm square with 8H32 bars as reinforcement.

Link diameter = H32/4 = 8 mm.

Spacing = min (20 × H32; (b = 500 mm); 400 mm) = 400 mm.

Provide links as shown in Fig. 10.13. All the bars are restrained by links.

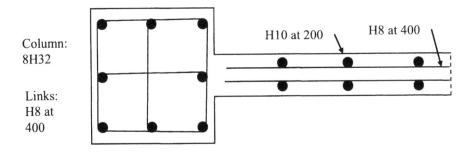

Column:
8H32

Links:
H8 at
400

H10 at 200   H8 at 400

Fig. 10.13 Reinforcement details in the wall and the end columns.

10.3.4 Design of a Wall Subjected to Axial Load and In-Plane Moment with Columns at the End

In the example in section 10.3.3.1, design was carried out assuming that the in-plane moment is resisted by the end columns only. A chart for the design of wall with end columns where the axial load and moment are resisted by the entire section can be constructed as follows.

Total depth of wall =h, thickness of wall = t

Reinforcement in the wall = $A_{sc, wall}$

End columns width = b, Area of end columns = A_{col}

Reinforcement in the end columns = $A_{sc, col}$

Assuming Young's modulus for steel is 200 GPA, f_{yk} = 500 MPa, the strain ε_{sy} in steel when the stress is $f_{yd} = f_{yk}/ (\gamma_s = 1.5) = 435$ MPa is given by
$$\varepsilon_{sy} = 435/ (200 \times 10^3) = 2.174 \times 10^{-3}$$
If the maximum compressive strain in concrete is $\varepsilon_{cu3} = 3.5 \times 10^{-3}$ and the neutral axis depth is x, the strain in steel is equal to ε_{sy} at a depth c from the neutral axis, where
$$c = (\varepsilon_{sy} / \varepsilon_{cu3}) x = 0.6211 x$$
$$(x - c) = 0.3789x$$

a. The entire left hand column steel will be at yield when (x − c) =0.3789 x = b the width of the column. Therefore x/b = 2.64.
$$X/h = (x/b) \times (b/h) = 2.64 \times b/h$$

b. The entire right hand column will be at yield in tension when the strain is equal to ε_{sy}, when (h − b − x) = 0.6211 x.
$$x/h = 0.6169 \times (1 - b/h)$$
$$\varepsilon_{cu3} \times E = (3.5 \times 10^{-3}) \times (200 \times 10^{-3}) = 700 \text{ MPa}$$

Three cases of neutral axis position are considered.

Note that in a trapezium as shown in Fig. 10.14, from the left hand side, the centroid is at
$$\bar{x} = \frac{a}{3} \times (1 + \frac{p_2}{p_1 + p_2})$$

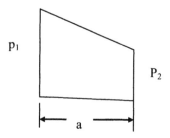

Fig. 10.14 Centroid of a trapezium.

Case 1: For simplicity, the first case is where the entire left hand column is in compression and the column steel is at yield.

Neutral axis depth lies between the limits such that and the entire left hand column steel is at yield (i.e., $x/h > 2.6415 \times b/h$) and the entire right hand column steel is at yield [i.e., $x \leq 0.6167 (h - b)$ or $x/h \leq 0.6167 (1 - b/h)$]. See Fig. 10.15.

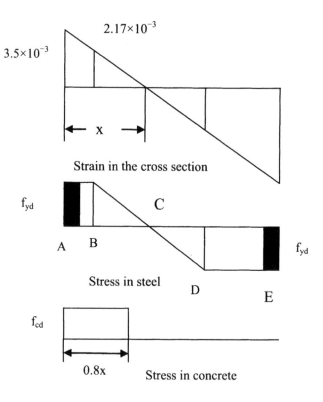

Fig. 10.15 Strains and stresses in steel and concrete for Case 1.

Force in segments:
AB = 0.3789 x, BC = CD = 0.6211 x, DE = h − 1.6211 x
Left hand column force (compressive): $f_{yd} \times A_{sc, col} + f_{cd} \times A_{col}$
Right hand column force (tensile): $- f_{yd} \times A_{sc, col}$
Concrete compressive force wall: $(0.8x - b) \times t \times f_{cd}$
Forces in the wall due to steel:

$$F_{AB} = (0.3789\ x - b) \times (f_{yd} \times A_{sc, wall})$$
$$F_{BC} = 0.6211\ x \times (0.5 \times f_{yd} \times A_{sc, wall})$$
$$F_{CD} = - F_{BC}$$
$$F_{DE} = - (h - 1.6211\ x - b) \times (f_{yd} \times A_{sc, wall})$$

Lever arms from the middle of the depth:

$$\ell_{\text{Left hand Column}} = 0.5h - 0.5\ b$$
$$\ell_{\text{Right hand Column}} = - (0.5h - 0.5\ b)$$
$$\ell_{AB} = 0.5\ h - b - 0.5 \times (0.3789\ x - b) - b$$
$$\ell_{BC} = 0.5\ h - 0.3789\ x - (0.6211\ x)/3$$
$$\ell_{CD} = 0.5\ h - 0.3789\ x - 0.6211\ x - (2/3) \times (0.6211\ x)$$
$$\ell_{DE} = 0.5\ h - 1.6211\ x - (h - 1.6211x - b)/2$$

Lever arm for force in concrete in wall, $\ell_{\text{concrete}} = 0.5h - 0.5 \times (0.8\ x - b)$.
N = Algebraic sum of all the axial forces.
M = Algebraic sum of the product of axial forces and corresponding lever arms.

Case 2: Two possible positions of the neutral axis need to be considered.

Case 2a: As shown in Fig. 10.16, the portion BC of the wall is in compression and portion CD of the wall is in tension. The right hand column is in tension.
Length CD = (h − b − x). CD = 0 when x/h = (1 − b/h).

Forces in segments:
Left hand column force (compressive): $f_{yd} \times A_{sc, col} + f_{cd} \times A_{col}$
Concrete compressive force wall: $(0.8x - b) \times t \times f_{cd}$
Forces in the wall due to steel:

$$F_{AB} = (0.3789\ x - b) \times (f_{yd} \times A_{sc, wall})$$
$$F_{BC} = 0.6211\ x \times (0.5 \times f_{yd} \times A_{sc, wall})$$

Strain ε at (h − x − b) = $-\varepsilon_{cu3} \times \dfrac{(h - x - b)}{x}$

Stress σ_1 at (h − x − b) = $-700 \times \left(\dfrac{(h - b)}{x} - 1 \right)$

$$F_{CD} = - (h - x - b) \times 0.5 \times \sigma_1 \times A_{sc, wall}$$

Stress σ_2 at (h − x) = $-700 \times \left(\dfrac{h}{x} - 1 \right)$ but numerically not greater than 435 MPa

Right hand column force (tensile): $= - 0.5 \times (\sigma_1 + \sigma_2) \times A_{sc, col}$

Lever arms from the middle of the depth:
$\ell_{\text{Left hand Column}} = 0.5h - 0.5\ b$

$\ell_{\text{Right hand Column}} = 0.5h - \left[h - b + \dfrac{b}{3} \times \left(1 + \dfrac{\sigma_2}{(\sigma_1 + \sigma_2)} \right) \right]$

$$\ell_{AB}= 0.5\,h - b - 0.5 \times (0.3789\,x - b)$$
$$\ell_{BC} = 0.5\,h - 0.3789\,x - (0.6211\,x)/3$$
$$\ell_{CD} = 0.5\,h - x - (2/3) \times (h - x - b)$$

Force in wall due to concrete, $F_{concrete} = (0.8x - b) \times t \times f_{cd}$.
Lever arm for force in concrete, $\ell_{concrete} = 0.5h - 0.4\,x - 0.5b$.
N = Algebraic sum of all the axial forces.
M = Algebraic sum of the product of axial forces and corresponding lever arms.

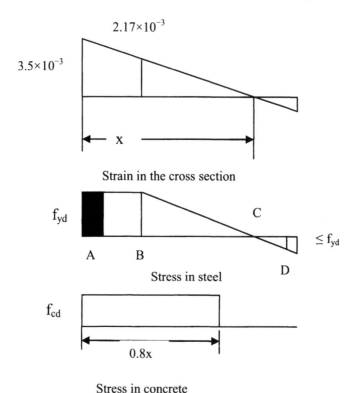

Fig. 10.16 Strains and stresses in steel and concrete for Case 2a.

Case 2b: As shown in Fig. 10.17, the portion BC of the wall is in compression and the right hand column is partly in tension and partly in compression.

Forces in segments:
Left hand column force (compressive): $f_{yd} \times A_{sc,\,col} + f_{cd} \times A_{col}$
Concrete compressive force wall: $(0.8x - b) \times t \times f_{cd}$
Forces in the wall due to steel:
$$F_{AB} = (0.3789\,x - b) \times (f_{yd} \times A_{sc,\,wall})$$
Strain ε_1 at right edge of the wall at $(x - h + b) = \varepsilon_{cu3} \times \dfrac{(x - h + b)}{x}$

Stress σ_1 at right edge of the wall $= 700 \times \left(1 - (1 - \frac{b}{h})\frac{h}{x}\right)$

$F_{BC} = (h - b - 0.3789\,x) \times 0.5 \times [f_{yd} + \sigma_1] \times A_{sc,\,wall})$

Strain ε_2 at $(h - x) = -\varepsilon_{cu3} \times \frac{(h-x)}{x}$

Stress σ_2 at $(h - x) = -700 \times \left(\frac{h}{x} - 1\right)$

Right hand column force (tensile): $= -700 \times \left((1 - 0.5\frac{b}{h})\frac{h}{x} - 1\right) \times A_{sc,\,col}$

Lever arms from the middle of the depth:

$\ell_{\text{Left hand Column}} = 0.5h - 0.5\,b$

$\ell_{\text{Right hand Column}} = 0.5h - [h - b + \frac{b}{3} \times \left(1 + \frac{\sigma_2}{(\sigma_1 + \sigma_2)}\right)]$

$\ell_{AB} = 0.5\,h - b - 0.5 \times (0.3789\,x - b)$

$\ell_{BC} = 0.5\,h - 0.3789\,x - (h - b - 0.3789\,x)/3 \times (1 + \sigma_1/(f_{yd} + \sigma_1))$

Force in wall due to concrete, $F_{concrete} = (0.8x - b) \times t \times f_{cd}$.

Lever arm for force in concrete, $\ell_{concrete} = 0.5h - (0.4\,x + 0.5b)$.

N = Algebraic sum of all the axial forces.

M = Algebraic sum of the product of axial forces and corresponding lever arms.

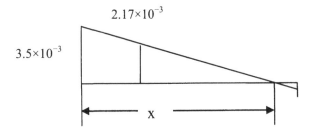

Strain in the cross section

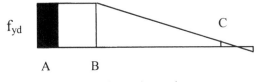

Stress in steel

Stress in concrete

Fig. 10.17 Strains and stresses in steel and concrete for Case 2b.

Case 3: $2.6389 \geq (x/h) > 1.0$, as shown in Fig. 10.18
$AB = 0.3789 \, x - b$, $BC = h - 0.3789 \, x - b$

Force in segments:
Left hand column force (compressive): $f_{yd} \times A_{sc, \, col} + f_{cd} \times A_{col}$
$F_{AB} = (0.3789 \, x - b) \times (f_{yd} \times A_{sc, \, wall})$
Stress σ at left edge of the right hand column: $\sigma_L = 700 \times (1 - (h - b)/x)$
Stress σ at right edge of the right hand column: $\sigma_R = 700 \times (1 - h/x)$
Average stress $= 700 \times [1 - (h/x) \times (1 + 0.5b/h)]$
Force in the right hand column $= 700 \times [1 - (h/x) \times (1 + 0.5b/h)] \times A_{sc, \, col}$
$F_{BC} = (h - 0.3789 \, x - b) \times [0.5 \times \{f_{yb} + \sigma_L\} \times A_{sc, \, wall}]$
Force due to concrete, $F_{concrete} = 0.8x \times t \times f_{cd}$, $0.8 \, x \leq h$
$\qquad\qquad\qquad F_{concrete} = h \times t \times f_{cd}$, $0.8 \, x > h$

Lever arms from the middle of the depth:
$\ell_{AB} = 0.5 \, h - 0.5 \times 0.3789 \, x = 0.5h - 0.1895 \, x$
$\ell_{BC} = 0.5 \, h - 0.3789 \, x - (h - 0.3789 \, x - b)/3 \times \{1 + \sigma_L/(\sigma L + f_{yd})\}$
Lever arm for force in concrete, $\ell_{concrete} = 0.5h - 0.4 \, x$, $0.8 \, x \leq h$.
If $0.8x > h$, $\ell_{concrete} = 0$.
N = Algebraic sum of all the axial forces.
M = Algebraic sum of the product of axial forces and corresponding lever arms.

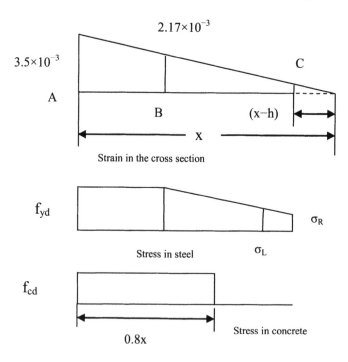

Fig. 10.18 Strains and stresses in steel and concrete for Case 3.

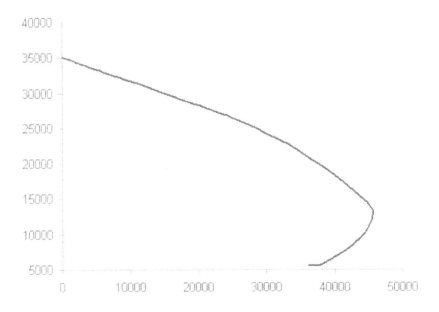

Fig. 10.19 Design chart for a wall with end columns.

Fig. 10.19 shows the design chart. It is constructed for the following parameters.
f_{ck} = 30 MPa, fcd = f_{ck}/1.5 = 20 MPa. f_{yk} = 500 MPa, f_{yd} = f_{yk}/1.15 = 435 MPa.
Total depth of wall, h = 6500 mm, thickness of wall, t = 160 mm.
Reinforcement in the wall, $A_{sc, wall}$ = 0.785 mm^2/mm.
Width of end columns, b = 500 mm, b/h = 500/6500 = 0.077.
Area of end columns, A_{col} = 2.55 × 10^6 mm^2.
Reinforcement in the column, $A_{sc, col}$ = 8H32 = 6434 mm^2.
The bar centre in the columns inset from the edges by 60 mm.

10.3.5 Design of a Wall Subjected to Axial Load, Out-of-Plane and In-Plane Moments

External shear walls are subjected not only to axial forces but also to in-plane moment due
to resisting wind forces. In addition there are also out-of-plane moments due to moments
caused by beams resting on the walls. In essence it is a case of axial force and biaxial
moments. The design procedure can be complex. The complexity is reduced by adopting a
simplified procedure as follows.
The wall is divided into a series of vertical strips. The axial force and in-plane moment is
substituted by a varying axial force in the vertical column strips. A vertical strip of the wall
is designed to resist the net axial force (due to the stress caused by axial force and in-plane
moment) and out-of-plane moment acting on that strip. The procedure is illustrated by an
example.

The section of a stocky reinforced concrete wall 150 mm thick and 4000 mm long is subjected to the following actions:

$$N = 4300 \text{ kN}$$
$$\text{In-plane moment, } M_y = 2100 \text{ kNm}$$
$$\text{Out-of-plane moment, } M_x = 244 \text{ kNm}$$

Design the reinforcement for the heaviest loaded end zone 500 mm long. The materials are $f_{ck} = 30$ MPa for concrete and $f_{yk} = 500$ MPa for reinforcement.

From an elastic analysis, the stress in the section due to axial force N and moment M_y are calculated as follows.

$$\sigma = \frac{N}{A} + \frac{M}{I} y$$

$$A = 150 \times 4000 = 6 \times 10^5 \text{ mm}^2$$
$$I = 150 \times 4000^3/12 = 8 \times 10^{11} \text{ mm}^4$$

$$\text{fibre stress} = \frac{4300 \times 10^3}{6 \times 10^5} + \frac{2100 \times 10^6 \times 2000}{8 \times 10^{11}}$$
$$= 7.17 + 5.25 = 12.42 \text{ MPa}$$

The stress at 250 mm from the end is

$$7.17 + 5.25 \times 1750/2000 = 11.76 \text{ MPa}$$

The axial load on the end zone is

$$11.76 \times 150 \times 500 \times 10^{-3} = 882 \text{ kN}$$

The bending moment on the end zone is

$$M = 224 \ (500/4000) = 28 \text{ kN m}$$

Design the end zone for an axial load moment combination using b = 150 mm, h = 500 mm

$$N/ (bh) = 882 \times 10^3/ (150 \times 500) = 11.76$$
$$M/ (bh^2) = 28 \times 10^6/ (500 \times 150^2) = 2.5$$

From a column design curve, $100 \ A_{sc}/ (bh) = 1.5$.
$A_{sc} = 1.5 \times 150 \times 500 \times 10^{-2} = 1125 \text{ mm}^2$.
Provide 4H20 to give an area of 1263 mm².

10.4 DESIGN OF PLAIN CONCRETE WALLS

10.4.1 Code Design Provisions

A plain wall contains either no reinforcement or less than 0.4% reinforcement. The reinforcement is not considered in strength calculations. The design procedure is given in section 12.0 of the code. The design procedure follows the same steps as for a reinforced concrete wall except for the following points.

(a) Material strength
The design value of compressive strength is defined in section 12.3.1 of Eurocode 2 as

$$f_{cd} = \alpha_{cc, pl} \times f_{ck}/\gamma_c$$

$$\alpha_{cc,\,pl} = 0.8$$

Note: $\alpha_{cc} = 1.0$ for reinforced concrete in the section 3.1.6(1) P of the code but the U.K. National Annex suggests a value of 0.85.

(b) Structural analysis

Since plain concrete has limited ductility, at ultimate limit state only elastic analysis with no redistribution is permitted. See section 12.5 of the code.

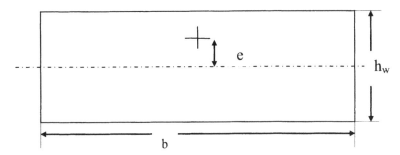

Fig. 10.20 Cross section of a wall.

The axial resistance N_{Rd} of a wall with uniaxial eccentricity e in the direction of the thickness h_w of the wall as shown in Fig. 10.20 is given by equation (12.2) of the code as

$$N_{Rd} = \eta\, f_{cd} \times b \times h_w \times (1 - 2e/h_w)$$

where
$\eta\, f_{cd}$ = design effective compressive strength.
b = overall width of the cross section.
h_w = overall depth of the cross section.
e = eccentricity of N_{Ed} in the direction of h_w.

As an example consider a wall b = 1 m, h_w = 200 mm, e = 30 mm, f_{ck} = 30 MPa.

$$\alpha_{cc,\,pl} = 0.8,\ \eta = 1.0,\ \gamma_c = 1.5$$
$$f_{cd} = \alpha_{cc,\,pl} \times f_{ck}/\gamma_c = 16\ \text{MPa}$$
$$N_{Rd} = \eta\, f_{cd} \times b \times h_w \times (1 - 2e/h_w) = 2240\ \text{kN/m}$$

10.5 REFERENCE

Cook, N.J. (2007). *Designers' Guide to EN 1991-1-4 Eurocode 1: Actions on structures, general actions part 1-4, Wind actions.* Thomas Telford.

CHAPTER 11

FOUNDATIONS

11.1 GENERAL CONSIDERATIONS

Foundations transfer loads from the building or individual columns to the earth.
Types of foundations commonly used are:
- Isolated bases for individual columns
- Combined bases for several columns
- Rafts for whole buildings which may incorporate basements

All the above types of foundations may bear directly on the ground or be supported
on piles. Only isolated and combined bases are considered in this chapter. The
type of foundation to be used depends on a number of factors such as
- Soil properties and conditions
- Type of structure and loading
- Permissible amount of differential settlement

The design of any foundation consists of two parts
- Geotechnical design to determine the safe bearing strength of the soil
- Structural design of the foundation using reinforced concrete

The Eurocode governing the geotechnical aspects of foundation design is
BS EN 1997-1:2004: Eurocode 7: Geotechnical Design —Part 1: General Rules.
Spread foundations are covered in section 6 and in Appendix D. Pile foundations
are covered in section 7 of the code.
A very useful reference on Eurocode 7 is Bond and Harris (2008).

The Eurocode governing the structural aspects of foundation design is *BS EN
1992-1-1:2004: Eurocode 2: Design of Concrete Structures Part 1: General Rules
and Rules for Buildings.*
For the vast majority of simple foundations, the two aspects can be treated
separately. However for some types of foundations, for example raft foundations,
the interaction between the structure and foundation might need to be taken into
account.

11.2 GEOTECHNICAL DESIGN

The following is a brief description of the important aspects of geotechnical design
as given in Eurocode 7.

11.2.1 Geotechnical Design Categories

In clause 2.1(14 to 21), Eurocode 7 gives three geotechnical categories for which geotechnical engineers might need to be involved in design.

a. **Geotechnical category 1** should only include small and relatively simple structures, for which it is possible to ensure that the fundamental requirements will be satisfied on the basis of experience and qualitative geotechnical investigations with negligible risk in terms of overall stability or ground movements and in ground conditions, which are known from comparable local experience to be sufficiently straight-forward. In these cases the procedures may consist of routine methods for foundation design and construction and structural engineers alone can take responsibility for geotechnical aspects of design.

b. **Geotechnical category 2** should include conventional types of structure and foundations like spread, raft and pile foundations with no exceptional risk or difficult or loading conditions. Routine procedures for field and laboratory testing and for design and execution may be used. In this case geotechnical design can be done by geotechnical or structural engineers.

c. **Geotechnical category 3** should include structures or parts of structures, which fall outside the limits of geotechnical categories 1 and 2. Geotechnical category 3 includes for example very large or unusual structures, structures involving abnormal risks, or unusual or exceptionally difficult ground or loading conditions, structures in highly seismic areas, structures in areas of probable site instability or persistent ground movements that require separate investigation or special measures. In this case responsibility for geotechnical design rests entirely with geotechnical engineers.

11.2.2 Geotechnical Design Approaches

In clause 2.4.7.3.4, Eurocode 7 gives three approaches to geotechnical design.
Design Approach 1: In this approach given in clause 2.2.7.3.4.2, partial factors are applied to *actions **and** to ground strength parameters*.
Design Approach 2: In this approach given in clause 2.2.7.3.4.3, partial factors are applied to *actions* **or** to the effects of actions **and** to *ground resistances*.
Design Approach 3: In this approach given in clause 2.2.7.3.4.4, partial factors are applied to actions **or** to the effects of actions from the structure **and** to ground *strength parameters*.
The three approaches can give very different results as there is no unanimity among geotechnical engineers as to which is the correct approach. The U.K. National Annex to Eurocode 7 permits only Design 1 approach.

11.2.3 Load Factors for Design 1 Approach

For design at the ultimate limit, the following limit states with their own combination of actions should normally be considered:

- EQU: Loss of *equilibrium* of the structure
- STR: Internal failure or excessive deformation of the *structure*
- GEO: Failure due to excessive deformation of the *ground*

Table 11.1 Partial factors on actions (γ_F) or the effects of actions (γ_E)

		Permanent Actions		Leading Variable Action	Accompanying Variable Action	
		Unfavourable	Favourable		Main (if any)	Others
		Combination 1				
STR/GEO	(6.10)	$1.35 \, G_k$	$1.0 \, G_k$	$1.5 \, Q_{k,1}$		$1.5 \, \psi_{0,i} \, Q_{k,i}$
	(6.10a)	$1.35 \, G_k$	$1.0 \, G_k$	.	$1.5 \, \psi_{01} \, Q_k$	$1.5 \, \psi_{0,i} \, Q_{k,i}$
	(6.10b)	$\xi \times 1.35 \, G_k$	$1.0 \, G_k$	$1.5 \, Q_{k,1}$		$1.5 \, \psi_{0,i} \, Q_{k,i}$
		Combination 2				
	(6.10)	$1.0 \, G_k$	$1.0 \, G_k$	$1.3 \, Q_{k,1}$	.	$1.3 \, \psi_{0,i} \, Q_{k,i}$
EQU	(6.10)	$1.1 \, G_k$	$0.90 \, G_k$	$1.5 \, Q_{k,1}$	.	$1.5 \, \psi_{0,i} \, Q_{k,i}$

Note: $\xi = 0.85$ but The U.K. National Annex gives $\xi = 0.925$.
Values for EQU are from Table A.1 of the code.
Values for STR/GEO are from Table A.3 of the code.

Table 11.2 Material property partial factors, γ_M

		Angle of Shearing Resistance, φ	Effective Cohesion, c	Undrained Shear Strength	Unconfined Strength	Bulk Density
		γ_φ	γ_c	γ_{cu}	γ_{qu}	γ_γ
STR/GEO	Combination 1	1.0	1.0	1.0	1.0	1.0
	Combination 2	1.25	1.25	1.4	1.4	1.0
EQU	EQU	1.25	1.25	1.4	1.4	1.0

Values for STR/GEO are from Table A.4 of the code.
Values for EQU are from Table A.2 of the code.

There are two sets of combinations used for the STR and GEO ultimate limit states. Table 11.1 shows the relevant partial factors on actions or effects of actions for persistent and transient design situations for STR/GEO and EQU situations. Exp. (6.10), Exp (6.10a) and Exp (6.10b) refer to expressions given in *BS EN 1990:2002, Eurocode-Basis of Structural Design*, Table A2.4 (B) and Table A2.4(C). Same values are also given in Tables A.1 and A.3 of the Eurocode 7.

Partial resistance factor γ_R for bearing for spread foundations is given in Table A.5 of the code. The γ_R is 1.4 for Design 2 approach and is 1.0 for Design 1 and Design 3 approaches.

Table 11.2 shows the material property factors γ_M.

In the following section examples of three design approaches for a rectangular footing on drained sandy soil are illustrated.

11.2.3.1 Example of Calculation of Bearing Capacity by Design 1 Approach

A rectangular footing 3.75 × 2.25 m and 700 mm thick supports the following column loads:
Permanent actions (Dead load): $G_k = 950$ kN
Leading variable action (Imposed load): $Q_k = 700$ kN
The footing rests on sandy soil with angle of shearing resistance $\varphi = 30°$.
Determine the allowable base pressure and compare it with the applied base pressure.

Solution:
Step 1: Calculate design loads.
Taking unit weight of concrete = 25 kN/m³, calculate the weight of the footing.
$$G_{k,\,footing} = 3.75 \times 2.25 \times 0.700 \times 25 = 147.7 \text{ kN}$$
$$G_k = 950 \text{ (applied)} + 147.7 \text{ (Footing)} = 1097.7 \text{ kN}$$
$$Q_k = 700 \text{ kN}$$
Use expression (6.10) as it gives the maximum design load.
$$\text{Design load for Combination 1: } \gamma_g = 1.35, \, \gamma_q = 1.5$$
$$V_{Ed} = 1.35 \times 1097.7 + 1.5 \times 700 = 2531.8 \text{ kN}$$
$$\text{Design load for Combination 2: } \gamma_g = 1.0, \, \gamma_q = 1.3$$
$$V_{Ed} = 1.0 \times 1097.7 + 1.3 \times 700 = 2007.7 \text{ kN}$$

Step 2: Calculate design base pressure:
$$\text{Area of footing, } A_{base} = 3.75 \times 2.25 = 8.44 \text{ m}^2$$
$$\text{Combination 1: Base pressure} = 2531.8 / 8.44 = 300 \text{ kN/m}^2 \text{ or kPa}$$
$$\text{Combination 2: Base pressure} = 2007.7 / 8.44 = 238 \text{ kN/m}^2 \text{ or kPa}$$

Step 3: Calculate design material properties.
As the footing rests on sandy soil, the only relevant material property is the angle of shearing resistance, φ.
Characteristic value $\varphi_k = 30°$. Note that the safety factor γ_φ is applied to tan φ_k **not** to φ_k.

Combination 1: $\gamma_\varphi = 1.0$, tan $\varphi_d =$ tan $\varphi_k / \gamma_\varphi =$ tan $30 = 0.577$, $\varphi_d = 30.0°$
Combination 2: $\gamma_\varphi = 1.25$, tan $\varphi_d =$ tan $\varphi_k / \gamma_\varphi =$ tan $30/1.25 = 0.462$, $\varphi_d = 24.8°$

Step 4: Calculate design bearing capacity factors. Use the equations in section D.4 of Annex D of Eurocode 7.

i.   Overburden factor N_q

$$N_q = e^{(\pi \times \tan\varphi_d)} \times \tan^2 (45 + \varphi_d/2)$$

              Combination 1: $\varphi_d = 30°$, $N_q = 18.40$

              Combination 2: $\varphi_d = 24.8°$, $N_q = 10.44$

ii.   Cohesion factor, N_c

$$N_c = (N_q - 1) \cot\varphi_d$$

              Combination 1: $\varphi_d = 30°$, $N_c = 30.14$

              Combination 2: $\varphi_d = 24.8°$, $N_c = 20.43$

    Note: If effective cohesion c = 0, there is no need to compute this factor.

iii.   Body weight factor N_γ

$$N_\gamma = 2 (N_q - 1) \tan\varphi_d$$

              Combination 1: $\varphi_d = 30°$, $N_\gamma = 20.09$

              Combination 2: $\varphi_d = 24.8°$, $N_\gamma = 8.72$

Step 5: Calculate design shape factors. Use the equations in Annex D of Eurocode 7.

Rectangular base: B = 2.25 m, L = 3.75, B/L = 0.6

i. $s_q = 1 + (B/L) \sin \varphi_d$

              Combination 1: $\varphi_d = 30°$, $s_q = 1.30$

              Combination 2: $\varphi_d = 24.8°$, $s_q = 1.25$

ii. $s_c = (s_q N_q - 1)/(N_q - 1)$

              Combination 1: $N_q = 18.40$, $s_q = 1.30$, $s_c = 1.317$

              Combination 2: $N_q = 10.44$, $s_q = 1.25$, $s_c = 1.276$

iii.   Note: If effective cohesion c = 0, there is no need to compute this factor.

    $s_\gamma = 1 - 0.3 (B/L)$

              Combination 1 and 2: B/L = 0.6, $s_\gamma = 0.82$

Step 6: Calculate the overburden pressure, q.

Taking the unit weight of soil as 18 kN/m³ and the safety factor $\gamma_\gamma = 1$

$$q = 18 \times \text{depth of footing} = \gamma_\gamma \times 18 \times 0.7 = 12.6 \text{ kN/m}^2 \text{ or kPa}$$

Step 7: Calculate the allowable q_{ult}:

$$q_{ult} = q \times N_q \times s_q + c \times N_c \times s_c + 0.5 \times \gamma \times B \times N_\gamma \times s_\gamma$$

Combination 1: qult = $12.6 \times 18.40 \times 1.30 + 0.0 \times 30.14 \times 1.317$

$+ 0.5 \times 18.0 \times 2.25 \times 20.09 \times 0.82 = 635$ kN/m² or kPa

Combination 2: qult = $12.6 \times 10.44 \times 1.25 + 0.0 \times 20.43 \times 1.276$

$+ 0.5 \times 18.0 \times 2.25 \times 8.72 \times 0.82 = 309$ kN/m² or kPa

Step 8: Compare applied to permissible base pressures:

      Combination 1: q_{ult} applied = 300 kPa, q_{ult} permissible = 635 kPa

           q_{ult} applied /q_{ult} permissible = 300/635 = 0.47

      Combination 2: q_{ult} applied = 238 kPa, q_{ult} permissible = 309 kPa

           q_{ult} applied /q_{ult} permissible = 238/309 = 0.77

Clearly combination 2, although it has a smaller value of applied base pressure, because of the factor of safety $\gamma_\varphi = 1.25$ applied to tan φ, it has a smaller value of the permissible base pressure. The footing size is adequate but can be reduced.

11.2.3.2 Example of Calculation of Bearing Capacity by Design 2 Approach

The Design 2 approach is similar to Design 1 approach except that only Combination 1 partial factors are used **both** for actions as well as for soil parameters. Finally an overall factor γ_R is used for bearing pressure.

Step 1: Calculate design loads:
$$G_k = 950 \text{ (applied)} + 147.7 \text{ (Footing)} = 1097.7 \text{ kN}$$
$$Q_k = 700 \text{ kN}$$
$$\gamma_g = 1.35, \ \gamma_q = 1.5$$
$$V_d = 1.35 \times 1097.7 + 1.5 \times 700 = 2531.8 \text{ kN}$$

Step 2: Calculate design base pressure:
$$\text{Area of footing, } A_{base} = 3.75 \times 2.25 = 8.44 \text{ m}^2$$
$$\text{Base pressure} = 2531.8 \ /8.44 = 300 \text{ kN/m}^2 \text{ or kPa}$$

Step 3: Calculate design material properties:
$$\text{Characteristic value } \varphi_k = 30^\circ.$$
$$\gamma_\varphi = 1.0, \ \tan \varphi_d = \tan \varphi_k / \gamma_\varphi = \tan 30 = 0.577, \ \varphi_d = 30.0^\circ$$

Step 4: Calculate design bearing capacity factors. Use the equations in Annex D of Eurocode 7.
i.   Overburden factor N_q:
$$N_q = e^{(\pi \times \tan\varphi_d)} \times \tan^2 (45 + \varphi_d/2)$$
$$\varphi_d = 30^\circ, \ N_q = 18.40$$
ii.   Cohesion factor, N_c
$$N_c = (N_q - 1) \cot\varphi_d$$
$$\varphi_d = 30^\circ, \ N_c = 30.14$$
Note: If effective cohesion c = 0, there is no need to compute this factor.
iii.   Body weight factor N_γ
$$N_\gamma = 2 (N_q - 1) \tan\varphi_d$$
$$\varphi_d = 30^\circ, \ N_\gamma = 20.09$$

Step 5: Calculate design shape factors. Use the equations in Annex D of Eurocode 7.
Rectangular base: B = 2.25 m, L = 3.75, B/L = 0.6
i. $sq = 1 + (B/L) \sin \varphi d$
$$\varphi_d = 30^\circ, \ s_q = 1.30$$
ii. $sc = (sq \ Nq - 1)/(Nq - 1)$
$$N_q = 18.40, \ s_q = 1.30, \ s_c = 1.317$$

Note: If effective cohesion c = 0, there is no need to compute this factor.

iii. $s\gamma = 1 - 0.3$ (B/L)
iv.   B/L = 0.6, $s\gamma = 0.82$

Step 6: Calculate the overburden pressure, q. Taking the unit weight of soil as 18 kN/m³ and the safety factor $\gamma_\gamma = 1$,
$$q = 18 \times \text{depth of footing} = \gamma_\gamma \times 18 \times 0.7 = 12.6 \text{ kN/m}^2 \text{ or kPa}$$

Step 7: Calculate q_{ult}:
$$q_{ult} = q \times N_q \times s_q + c \times N_c \times s_c + 0.5 \times \gamma \times B \times N_\gamma \times s_\gamma$$
$$q_{ult} = 12.6 \times 18.40 \times 1.30 + 0.0 \times 30.14 \times 1.317$$
$$+ 0.5 \times 18.0 \times 2.25 \times 20.09 \times 0.82 = 635 \text{ kN/m}^2 \text{ or kPa}$$

Step 8: Calculate $q_{allowable}$:
$$q_{ult} \text{ permissible} = q_{ult}/\gamma_R$$
$$\gamma_R = 1.4 \text{ for Design 2 from Table A.5 of the code.}$$
$$q_{ult} \text{ permissible} = 635/1.4 = 436 \text{ kPa}$$

Step 9: Compare applied to permissible base pressures:
$$q_{ult} \text{ applied} = 300 \text{ kPa, } q_{ult} \text{ permissible} = 436 \text{ kPa}$$
$$q_{ult} \text{ applied } /q_{ult} \text{ permissible} = 300/436 = 0.69$$
The footing size is adequate but can be reduced.

11.2.3.3 Example of Calculation of Bearing Capacity by Design 3 Approach

The Design 3 approach is similar to Design 1 approach except that Combination 1 partial factors are used only for actions but Combination 2 partial factors are used for soil parameters. Finally an overall factor $\gamma_R = 1$ is used for bearing pressure.

Step 1: Calculate design loads:
$$G_k = 950 \text{ (applied)} + 147.7 \text{ (Footing)} = 1097.7 \text{ kN}$$
$$Q_k = 700 \text{ kN}$$
$$\gamma_g = 1.35, \gamma_q = 1.5$$
$$V_d = 1.35 \times 1097.7 + 1.5 \times 700 = 2531.8 \text{ kN}$$

Step 2: Calculate design base pressure:
$$\text{Area of footing, } A_{base} = 3.75 \times 2.25 = 8.44 \text{ m}^2$$
$$\text{Base pressure} = 2531.8 /8.44 = 300 \text{ kN/m}^2 \text{ or kPa}$$

Step 3: Calculate design material properties.
As the footing rests on sandy soil, the only relevant material property is the angle of shearing resistance, φ.
Characteristic value $\varphi_k = 30°$. Note that the safety factor γ_φ is applied to $\tan \varphi_k$ not to φ_k.

Combination 2: $\gamma_\varphi = 1.25$, tan φ_d = tan $\varphi_k/ \gamma_\varphi$ = tan $30/1.25 = 0.462$, $\varphi_d = 24.8°$

Step 4: Calculate design bearing capacity factors. Use the equations in Annex D of Eurocode 7.

i.   Overburden factor N_q:

$$N_q = e^{(\pi \times tan\varphi_d)} \times tan^2 (45 + \varphi_d/2)$$

Combination 2: $\varphi_d = 24.8°$, $N_q = 10.44$

ii.   Cohesion factor, N_c

$$N_c = (N_q - 1) \, cot\varphi_d$$

Combination 2: $\varphi_d = 24.8°$, $N_c = 20.43$

Note: If effective cohesion $c = 0$, there is no need to compute this factor.

iii.   Body weight factor N_γ

$$N_\gamma = 2 \, (N_q - 1) \, tan\varphi_d$$

Combination 2: $\varphi_d = 24.8°$, $N_\gamma = 8.72$

Step 5: Calculate design shape factors. Use the equations in Annex D of Eurocode 7.

Rectangular base: B = 2.25 m, L = 3.75, B/L = 0.6

i. $s_q = 1 + (B/L) \, \sin \varphi_d$

ii.   Combination 2: $\varphi d = 24.8o$, sq = 1.25

iii.   sc = $(s_q \, N_q - 1)/(N_q - 1)$

Combination 2: $N_q = 10.44$, $s_q = 1.25$, $s_c = 1.276$

Note: If effective cohesion $c = 0$, there is no need to compute this factor.

iv. $s\gamma = 1 - 0.3$ (B/L)

Combination 2: B/L = 0.6, $s_\gamma = 0.82$

Step 6: Calculate the overburden pressure, q. Taking the unit weight of soil as 18 kN/m^3 and the safety factor $\gamma_\gamma = 1$,

$$q = 18 \times depth \ of \ footing = \gamma_\gamma \times 18 \times 0.7 = 12.6 \ kN/m^2 \ or \ kPa$$

Step 7: Calculate the allowable q_{ult}:

$$q_{ult} = q \times N_q \times s_q + c \times N_c \times s_c + 0.5 \times \gamma \times B \times N_\gamma \times s_\gamma$$

Combination 2: qult = $12.6 \times 10.44 \times 1.25 + 0.0 \times 20.43 \times 1.276$
$+ 0.5 \times 18.0 \times 2.25 \times 8.72 \times 0.82 = 309 \ kN/m^2 \ or \ kPa$

Step 8: Compare applied to permissible base pressures:

q_{ult} applied = 300 kPa, q_{ult} permissible = 309 kPa
q_{ult} applied /q_{ult} permissible = 300/309 = 0.97
The footing size is just adequate.

11.2.3.4 Comments on the Calculation of Bearing Capacity by Three Design Approaches

The three design approaches give very different permissible base pressures. The ratios of q_{ult} applied /q_{ult} permissible for Design approaches 1, 2 and 3 are respectively are 0.77, 0.69 and 0.97. The reason for Design approach 3 giving a low ratio is because it uses high partial factor for actions and soil parameters leading to high applied pressure and low permissible base pressure. On the other hand, the Design 1 approach for combination 2 uses low partial factor for actions and soil parameters leading to low applied pressure and low permissible base pressure. There is no guidance in Eurocode 7 on which is the most suitable design approach. For STR/GEO limit states, the U.K. National Annex allows only Design 1 approach.

11.3 SPREAD FOUNDATIONS

The geotechnical design of spread foundations like pad, strip and raft foundations is covered in section 6 of the Eurocode 7, part 1. The code gives three methods of design. They are

Table 11.3 Allowable bearing capacity (From BS 8004)

Category	Type of soil	Allowable bearing pressure kN/m^2 or kPa	Remarks
Non-cohesive soils	Dense gravel or dense gravel and sand	>600	Width of foundation not less than 1 m. Ground water level assumed to be below base of foundation.
	Medium gravel or medium gravel and sand	< 200–600	
	Loose gravel or loose gravel and sand	< 200	
	Compact sand	> 300	
	Medium dense sand		
	Loose sand	<100	
Cohesive soils	Very stiff boulder clay and hard clay	300–600	Susceptible to long term consolidation settlement.
	Stiff clay	150–300	
	Firm clay	75–150	
	Soft clay and silt	< 75	
	Very soft clay and silt	Not applicable	

(a) **Analytical method**: A commonly recognized analytical method including numerical method where relevant should be used.

(b) Semi-empirical method: A commonly recognized semi-empirical method such as bearing resistance estimation using pressure meter test should be used.

(c) Prescriptive method using presumed bearing resistance: A commonly recognized prescriptive method based on presumed bearing resistance should be used. When such a method is applied, the design result should be evaluated on the basis of comparable experience.

For a commonly used spread foundation, settlement will be the governing criterion. Traditionally, allowable bearing pressure has been used to control settlement. In general, site load tests and laboratory tests on soil samples should be carried out to determine the actual soil properties for foundation design. Where this is deemed unnecessary, values for relevant parameters for various soil types and conditions given in *BS 8004:1986: Code of Practice for Foundations* or similar publications can be used. Table 11.3 gives some typical values.

11.4 ISOLATED PAD BASES

11.4.1 General Comments

Isolated pad bases are square or rectangular slabs provided under individual columns. They spread the concentrated column load safely to the ground and may be axially or eccentrically loaded (Fig. 11.1 and Fig. 11.3). Mass concrete can be used for lighter foundations if the underside of the base lies inside a dispersal angle of 45°, as shown in Fig. 11.1(a). Otherwise a reinforced concrete pad is required (Fig. 11.1(b)).

There is little specific guidance given in Eurocode 2 for the design of footings. The following procedure is normally used.

1. When the base is axially loaded the load may be assumed to be uniformly distributed. The actual pressure distribution depends on the soil type; refer to soil mechanics textbooks.
2. When the base is eccentrically loaded, the reactions may be assumed to vary linearly across the base.

11.4.2 Axially Loaded Pad Bases

Refer to the axially loaded pad footing shown in Fig. 11(b) where the following symbols are used:

G_k = characteristic dead load from the column (kN)
Q_k = characteristic imposed load from the column (kN)
W = weight of the base (kN)
L, B = base length and breadth (m)
P_b = safe bearing pressure (kN/m^2 or kPa)

Note that the safe bearing pressure value is a serviceability value as it is used to control settlement of the foundation.

The required area is found from the characteristic loads including the weight of the base:

$$\text{Base area} = (G_k + Q_k + W)/P_b = L \times B \text{ m}^2$$

The design of the base is made for the ultimate load delivered to the base by the column shaft, i.e., the design load is $1.35\ G_k + 1.5\ Q_k$.

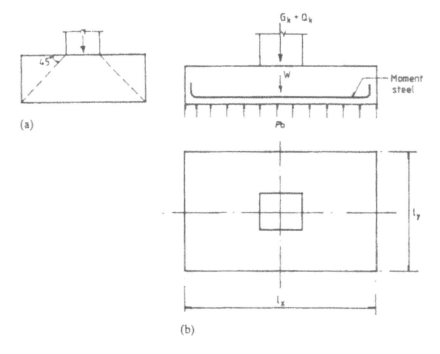

(a)

(b)

Fig. 11.1 (a) Mass concrete foundation; (b) reinforced concrete pad foundation.

(a) Bending

The critical section for bending is at the face of the column on a pad footing or the wall in a strip footing. The moment is taken on a section passing completely across a pad footing and is due to the ultimate loads on one side of the section. No redistribution of moments should be made. The critical sections are XX and YY in Fig. 11.2(a).

(b) Distribution of reinforcement

Because of the greater concentration of bending moment near the column than towards the edges, traditionally the practice has been to concentrate the reinforcement in a narrow width near the centre. The arbitrary rule is that if the distance from the centre line of the column to the edge of the pad exceeds

0.75 (c + 3d), two-thirds of the required reinforcement for the given direction should be concentrated within a zone from the centre line of the column to a distance 1.5 d from the face of the column. Here c is the column width and d is the effective depth of the base slab. Otherwise the reinforcement may be distributed uniformly over the entire width. The arrangement of reinforcement is shown in Fig. 11.2(b).

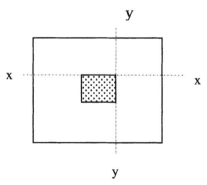

Fig. 11.2(a) Critical section for bending design.

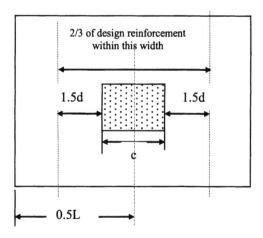

Fig. 11.2(b) Layout of flexural reinforcement.

(c) Shear on vertical section across full width of base
The vertical shear force is the sum of the loads acting outside the section considered. Shear stress is checked at a distance d from the face of the column as shown in Fig. 11.2(c).
The shear stress is

$$v = V/ (\ell \, d) \le v_{Rd, \, c}$$

where ℓ is the length L or width B of the base as appropriate.

It is normal practice to make the base sufficiently deep so that shear reinforcement is not required. The depth of the base is often controlled by the design for shear. Rules for members not requiring shear reinforcement are covered in clause 6.2.2 of the Eurocode. From equations (6.2a), (6.2b) and (6.3N) of the Eurocode 2,

$$v_{Rd, c} = C_{Rd, c} \times k \times (100 \times \rho_1 \times f_{ck})^{0.3333} \geq (v_{min} = 0.035 \times k^{1.5} \times \sqrt{f_{ck}})$$

$$C_{Rd, c} = 0.18/ (\gamma_c = 1.5) = 0.12, \, k = 1 + \sqrt{(200/d)} \leq 2.0, \, \rho_1 = A_{sl}/ (b_w \, d) \leq 0.02$$

It is normal practice to make the base sufficiently deep so that shear reinforcement is not required. The depth of the base is often controlled by the design for shear. If the shear stress calculation indicates the need for shear reinforcement, the solution is to increase the depth of the footing until no shear reinforcement is required.

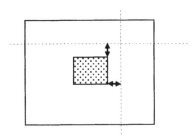

Fig. 11.2(c) Critical section for checking beam shear at d from the face of the column.

(d) Punching shear around the loaded area

Rules for checking for punching shear resistance are given in section 6.4 of the code. The punching shear force is the sum of the loads outside the periphery of the critical section shown in Fig. 11.2(d). The reader should refer Chapter 5, sections 5.1.10 to 5.1.13 dealing with the design of flat slabs for shear, where most of the concepts and equations are covered in detail.

The two checks on the shear stress are:

- First at the perimeter of the column.
$$[v_{Ed} = \text{Column load}/ (u_0 \, d)] < [v_{Rd, max} = 0.3 \times (1 - f_{ck}/250) \times f_{cd}]$$
u_0 = perimeter of the column = $2 \, (c_1 + c_2)$, c_1 and c_2 are side dimensions of the column.

If this requirement is not satisfied, then the thickness of the slab has to be increased till the requirement is satisfied.

- Next at perimeters from $r = d$ to $2d$ from the face of the column using the code equation (6.50).
$$v_{Ed} = V_{Ed. \, red}/ (u \times d) \leq (v_{Rd} = v_{Rd, c} \times 2d/a)$$

where

V_{Ed} = Total load outside the perimeter
$$V_{Ed} = \text{Column load} - p \times [2(c_1 + c_2) \times r + \pi \times r^2 + c_1 \times c_2]$$
$$u = 2 \times (\pi \times r + c_1 + c_2)$$
$$v_{Rd, c} = C_{Rd, c} \times k \times (100 \times \rho_1 \times f_{ck})^{0.3333} \geq (v_{min} = 0.035 \times k^{0.667} \times f_{ck})$$

$C_{Rd, c} = 0.18/ (\gamma_c = 1.5) = 0.12$, $k = 1 + \surd (200/d) \le 2.0$, $\rho_l = A_{sl}/ (b_w d) \le 0.02$
p = base pressure at ULS = column load / (area of base).
a = distance of the perimeter from the column face $d \le a \le 2\ d$.

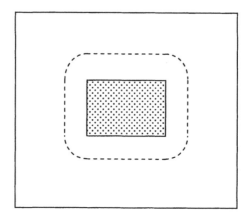

Fig. 11.2(d) Control perimeter for checking punching shear at 2d from column sides.

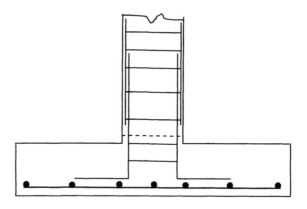

Fig. 11.2(e) Base reinforcement.

(e) Anchorage of column starter bars

Fig. 11.2(e) shows the arrangement. Apart from the reinforcement in the base, column bars extend at least an anchorage length to which column reinforcement is attached. It is common practice to cast along with the base a short length of the column. This is called as a kicker. This facilitates the positioning of the formwork for the column. However some people prefer not use a kicker (known as 'kickerless' construction) to speed up the construction process.

The required compression lap length $l_0 = l_{b, reqd} \ge$ max (15 × bar dia, 200 mm).
The reader should refer to section 5.2, Chapter 5 for further details.

(f) Minimum grade of concrete and nominal cover

The minimum grade of concrete to be used in foundations and the nominal cover to the reinforcement depend on many factors such as the presence of sulphates or chlorides in the soil, the type of cement used and so on. The reader should refer to sections 2.7 and 2.9, Chapter 2 for more details.

11.4.2.1 Example of Design of an Axially Loaded Base

(a) Specification

A column 400 mm × 400 mm carries a dead load of 800 kN and an imposed load of 300 kN. The safe bearing pressure is 200 kN/m². Design a square base to resist the loads. f_{ck} = 30 MPa and f_{yk} = 500 MPa. The exposure class is XC1/XC2. The minimum cover $c_{min, dur}$ = 25 mm. As the concrete is cast against a blinding layer, the cover is taken as 40 mm.

(b) Size of base

Assume the weight is 100 kN.
$$\text{Service load} = 800 + 300 + 100 = 1200 \text{ kN}$$
$$\text{Area of base} = 1200/200 = 6.0 \text{ m}^2. \text{ Make the base 2.5 m} \times 2.5 \text{ m.}$$

(c) Moment steel

$$\text{Ultimate load} = (1.35 \times 800) + (1.5 \times 300) = 1530 \text{ kN}$$
$$\text{Ultimate base pressure} = 1530/6.25 = 245 \text{ kN/m}^2$$

Note: The self weight of the footing is not included because the self weight and the corresponding base pressure will cancel themselves out when calculating the design forces for the base.

The critical section YY at the column face is shown in Fig. 11.2(a).
Length of the beyond the face of the column = (2.5 – 0.4)/2 = 1.05 m.
$$M_{yy} = 245 \times 1.05 \times 2.5 \times 1.05/2 = 337.6 \text{ kNm}$$

Try an overall depth of 650 mm with 16 mm bars both ways.
$$\text{The weight of the footing} = 2.5 \times 2.5 \times 0.65 \times 25 = 102 \text{ kN}$$
$$102 \text{ kN} \approx 100 \text{ kN assumed in design.}$$

The effective depth of the top layer of steel is
$$d = 650 - 40 - 16 - 16/2 = 586 \text{ mm}$$
$$f_{ck} = 30, \eta = 1, \lambda = 0.8$$
$$k = M/ (bd^2 f_{ck}) = 337.6 \times 10^6/ (2500 \times 586^2 \times 30) = 0.013 < 0.196$$

$$\frac{z}{d} = 0.5[1.0 + \sqrt{(1 - 3\frac{k}{\eta})}]$$

$$z/d = 0.99$$
$$f_{yk} = 500, f_{yd} = 500/1.15 = 435 \text{ MPa}$$
$$A_s = 337.6 \times 10^6/ (435 \times 0.99 \times 586) = 1338 \text{ mm}^2$$

Check minimum steel:

From equation (9.1N) of the code,
$$A_{s, min} = 0.26 \times (f_{ctm}/f_{yk}) \times bd \geq 0.0013 \; bd \qquad (9.1N)$$
$$f_{ctm} = 0.3 \times f_{ck}^{0.67} = 0.3 \times 30^{0.67} = 2.9 \; MPa, \; f_{yk} = 500 \; MPa,$$
$$b = 2500 \; mm, \; d = 586 \; mm$$
$$A_{s, min} = 0.26 \times (2.9/500) \times 2500 \times 586 \geq 0.0013 \times 2500 \times 586$$
$$A_{s, min} = 2209 \; mm^2$$
Area of H16 bar = 201 mm^2. Number of 16 mm bars = 2209/201 $\approx$ say, 11.
Provide 11H16 bars, A_s = 2212 mm^2.
The distribution of the reinforcement is determined to satisfy the rule.
$$3/4(c+3d) = 0.75 \; (400 + 3 \times 586) = 1619 \; mm$$
$$0.5 \; L = 2500/2 = 1250 \; mm < 1619 \; mm$$
The bars can be spaced equally at 240 mm centres.

The full anchorage length required past the face of the column. From Table 5.5, Chapter 5, for f_{ck} = 30 MPa, the anchorage length required is 36 bar diameters. Anchorage length = 36 × 16 = 576 mm. Adequate anchorage is available.

(d) Vertical shear
The critical section $Y_1 Y_1$ at d = 586 mm from the face of the column is shown in Fig. 11.2(c).
$$V_{Ed} = 245 \times 2.5 \times (1050 - 586) \times 10^{-3} = 284.2 \; kN$$
$$V_{Ed} = 284.2 \times 10^3 / (2500 \times 586) = 0.19 \; MPa$$
$$C_{Rd, c} = 0.18/ (\gamma_c = 1.5) = 0.12, \; k = 1 + \sqrt{(200/586)} = 1.58 \leq 2.0,$$
The bars extend 565 mm, i.e., more than d, beyond the critical section and so all the steel can be taken into account when calculating A_{sl}.
$$A_{sl} = 11H16 = 2212 \; mm^2, \; \rho_1 = A_{sl}/ (b_w \; d) = 2212/ (2500 \times 586) = 0.0015 \leq 0.02$$
$$C_{Rd, c} \times k \times (100 \times \rho_1 \times f_{ck})^{0.33} = 0.12 \times 1.58 \times (100 \times 0.0015 \times 30)^{0.33} = 0.31$$
$$V_{min} = 0.035 \times k^{1.5} \times \sqrt{f_{ck}} = 0.035 \times 1.58^{1.5} \times \sqrt{30} = 0.38 > 0.31$$
$$V_{Rd, c} = 0.38 \; MPa$$
$$(V_{Ed} = 0.19) < (V_{Rd, c} = 0.38)$$
The shear stress is satisfactory and no shear reinforcement is required.

(e) Punching shear
Check shear stress at column perimeter:
Column load = 1530 kN, u_0 = 2 × (400 + 400) = 1600 mm, d = 586 mm
Upward load = Base pressure × column area = 245 × 0.4 × 0.4 = 39.2 kN
$$V_{Ed} = (1530 - 39.2) \times 10^3 / (1600 \times 586) = 1.59 \; MPa$$
$$V_{Rd, max} = 0.3 \times (1 - f_{ck}/250) \times f_{cd} = 0.3 \times (1 - 30/250) \times 30/1.5 = 5.28 \; MPa$$
$$(V_{Ed} = 1.59) < (V_{Rd, max} = 5.28)$$
Slab depth is adequate.
Check punching shear on a perimeter at d to 2 d from the column face.
The critical perimeter is shown in Fig. 11.2(d).
Let a = distance of the perimeter from the column face. $d \leq a \leq 2 \; d$.
$u = 2 \times [(c_1 = 400) + (c_2 = 400)] + 2 \times \pi \times a$.
Let A = Area inside the perimeter.
$A = \pi a^2 + 2 \times [(c_1 = 400) + (c_2 = 400)] \times a + [(c_1 = 400) \times (c_2 = 400)] \; mm^2$.
p = base pressure at ULS = 245 kN/m^2 or kPa.

Column load at ULS = 1530 kN.

$V_{Ed, red}$ = Column load – A × p = 1530 – 245 × A × 10^{-6} kN.

$v_{Ed} = V_{Ed, red}/ (u \times d)$.

Table 11.4 shows the calculation of punching shear stress v_{Ed}. The maximum value at a = d is 0.32 MPa which is less than (see code equation (6.50)

$v_{Rd} = (v_{rd, c} \times 2d/a) = 0.76$ MPa. The slab does not require shear reinforcement.

Table 11.4 Calculation of punching shear stress v_{Ed}

a	A, m^2	u, mm	$V_{Ed,red}$, kN	v_{Ed}, MPa
586	2.18	5281.96	996.78	0.32
644.6	2.50	5650.15	918.30	0.28
703.2	2.84	6018.35	834.54	0.24
761.8	3.20	6386.54	745.49	0.20
820.4	3.59	6754.74	651.16	0.16
879	3.99	7122.93	551.54	0.13
937.6	4.42	7491.13	446.63	0.10
996.2	4.87	7859.32	336.44	0.07
1054.8	5.34	8227.52	220.96	0.05
1113.4	5.84	8595.71	100.19	0.02
1172	6.35	8963.91	−25.86	0.00

(f) Cracking

The required and provided areas of reinforcement are respectively 1338 mm^2 and 2209 mm^2. The loads at SLS and ULS are 1100 kN and 1300 kN respectively. The stress in steel at serviceability limit state is

$$f_s = \left[\frac{1338}{2209}\right] \times \left[\frac{1100}{1530}\right] \times 435 = 189 \text{ MPa}$$

From Table 7.2N of the code, the maximum bar diameter for 0.3 mm wide crack is 25 mm. From Table 7.3N of the code the maximum spacing of bars is 250 mm. Both the criteria are satisfied. No further checks are required.

(g) Reinforcement

The arrangement of reinforcement is shown in Fig. 11.3.

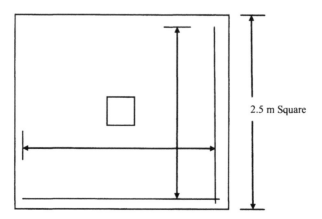

2.5 m Square

11H16 at 240 mm both ways

Fig. 11.3 Reinforcement arrangement.

11.5 ECCENTRICALLY LOADED PAD BASES

11.5.1 Vertical Soil Pressure at Base

Just as in the case of concentrically loaded bases, where the base pressure was assumed to be constant provided the footing was 'rigid' in comparison with the soil, in a similar manner the base pressure for eccentrically loaded pad bases may be assumed to vary linearly across the base for design purposes.

The characteristic loads on the base are the axial load N, moment M and horizontal load H arising from either shear on the column or horizontal forces developed at the base of portal frames as shown in Fig. 11.4. The base dimensions are length L, width B and depth h.

$$\text{Base area } A = B \times L$$
$$\text{Section modulus } Z = B \times L^2/6$$

The total vertical load is N + W and the moment at the underside of the base is (M + Hh). The maximum and minimum base pressures applied are

$$p_{max} = \frac{(N+W)}{A} + \frac{(M+Hh)}{Z}$$
$$p_{min} = \frac{(N+W)}{A} - \frac{(M+Hh)}{Z}$$

p_{max} should not exceed the safe bearing pressure.

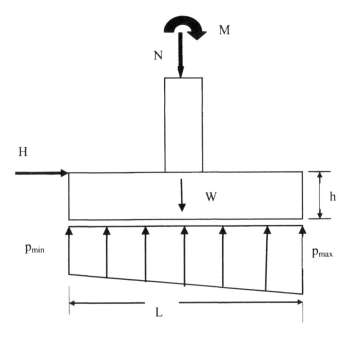

Fig. 11.4 Eccentrically loaded base.

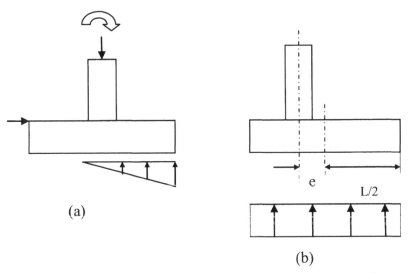

(a)

(b)

Fig. 11.5 Eccentrically loaded pads: (a) bearing on part of base; (b) base set eccentric to column.

The eccentricity e of the resultant reaction is

$$e = \frac{(M + Hh)}{(N + W)}$$

If $e \leq L/6$, there is pressure over the whole of the base, as shown in Fig. 11.4.
If $e > \ell/6$ only a part of the base bears on the ground, as shown in Fig. 11.5(a). This situation should be avoided as there is the danger of overloading the soil and also might lead to tipping over of the foundation. The column can be set eccentric to the column by, say, e_1 to offset the moments due to permanent loads and give uniform pressure, as shown in Fig. 11.5(b).

$$\text{Eccentricity } e_1 = (M + Hh)/N$$

11.5.2 Resistance to Horizontal Loads

Horizontal loads applied to bases are resisted by passive earth pressure against the end of the base, friction between the base and ground for cohesion-less soils such as sand, or adhesion for cohesive soils such as clay. In general, the load will be resisted by a combination of all actions. The ground floor slab can also be used to resist horizontal load. The forces are shown in Fig. 11.6.

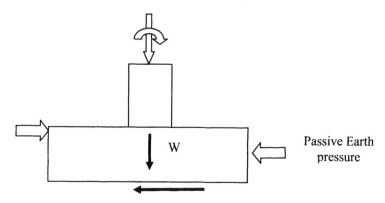

Fig. 11.6 Forces resisting horizontal force on the base.

Formulae from soil mechanics for calculating the resistance forces are given for the two cases of cohesionless and cohesive soils.

(a) Cohesionless soils
The passive earth pressure p at depth h is given

$$p = \gamma\, h\, k_p$$
$$k_p = (1 + \sin \varphi)/ (1 - \sin \varphi)$$

where φ is the angle of internal friction and γ is the soil density.

If p_1 and p_2 are passive earth pressures at the top and bottom of the base, then the passive resistance

$$R_{passive} = 0.5\, B\, h\, (p_1 + p_2)$$

If μ is the coefficient of friction between the base and the ground, generally taken as $\tan \varphi$, the frictional resistance is

$$R_{friction} = \mu\,(N + W)$$

(b) Cohesive soils

For cohesive soils $\varphi = 0$. Denote the cohesion at zero normal pressure c and the adhesion between the base and the load β. The resistance of the base to horizontal load is

$$R_{base} = 2cBh + 0.5B\,h\,(p_1 + p_2) + \beta\,L\,B$$

where the passive pressure p_1 at the top is equal to γh_1, the passive pressure p_2 at the bottom is equal to γh_2 and L is the length of the base. The resistance forces to horizontal loads derived above should exceed the factored horizontal loads applied to the foundation.

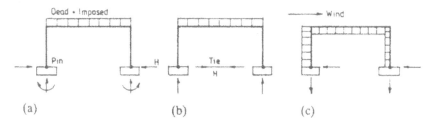

Fig. 11.7 (a) Portal base reactions; (b) force *H* taken by tie; (c) wind load and base reactions.

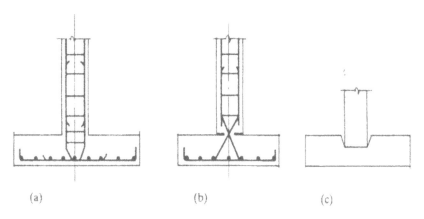

Fig. 11.8 (a) Fixed base; (b) pinned base; (c) pocket base.

In the case of portal frames it is often helpful to introduce a tie beam between bases to take up that part of the horizontal force due to portal action from dead and imposed loads as in the pinned base portal shown in Fig. 11.7(b). Wind load has to be resisted by passive earth pressure, friction or adhesion.

Pinned bases should be used where ground conditions are poor and it would be difficult to ensure fixity without piling. It is important to ensure that design assumptions are realized in practice.

11.5.3 Structural Design

The structural design of a base subjected to ultimate loads is carried out for the ultimate loads and moments delivered to the base by the column shaft. Pinned and fixed bases are shown in Fig. 11.8.

11.5.3.1 Example of Design of an Eccentrically Loaded Base

(a) Specification
The characteristic loads for an internal column footing in a building are given in Table 11.5. The proposed dimensions for the 450 mm square column and base (3600 × 2800 mm) are shown in Fig. 11.9. The base supports a ground floor slab 200 mm thick. The soil is firm well drained clay with the following properties:

$$\text{Unit weight} = 18 \text{ kN/m}^3,$$
$$\text{Safe bearing pressure} = 150 \text{ kN/m}^2,$$
$$\text{Cohesion} = 60 \text{ kN/m}^2$$

The materials to be used in the foundation are $f_{ck} = 30$ MPa and $f_{yk} = 500$ MPa.

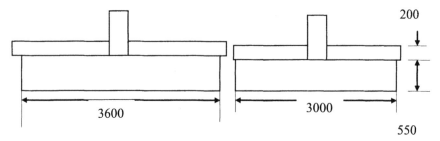

Fig. 11.9 Side and end elevations.

Table 11.5 Applied column loads and moments

	Vertical load, kN	Horizontal load, kN	Moment, kNm
Dead	770	35	78
Imposed	330	15	34

(b) Maximum base pressure on soil
The maximum base pressure is checked for the service loads.

$$\text{Weight of base + slab} = (550 + 200) \times 10^{-3} \times 3.6 \times 3.0 \times 25 = 202.5 \text{ kN}$$
$$\text{Total axial load} = 770 + 330 + 202.5 = 1302.5 \text{ kN}$$
$$\text{Total moment} = 78 + 34 + 0.550 \times (35 + 15) = 139.5 \text{ kN m}$$
$$\text{Base area } A = 3.0 \times 3.6 = 10.8 \text{ m}^2$$
$$\text{Section modulus } Z = 3.0 \times 3.6^2/6 = 6.48 \text{ m}^3$$
$$\text{Maximum base pressure} = 1302.5/10.8 + 139.5/6.48 = 120.6 + 21.5 = 142.1 \text{ kN/m}^2$$
$$\text{Maximum base pressure} < (\text{safe bearing pressure} = 150 \text{ kN/m}^2)$$

(c) Resistance to horizontal load

Check the passive earth resistance assuming no ground slab.

No adhesion, $\beta = 0$, ($h_1 = 0$, $p_1 = 0$), ($h_2 = 0.550$, $p_2 = 18 \times 0.550 = 9.9$)

The passive resistance is

$$= 2c\ B\ h + 0.5\ B\ h\ (p_1 + p_2) + \beta\ L\ B$$
$$= \{2 \times 60 \times 3.0 \times 0.550\} + \{0.5 \times 3.0 \times 0.5 \times (0 + 9.9)\} + 0$$
$$= 198 + 7.4 = 205.4\ kN$$

Factored horizontal load $= (1.35 \times 35) + (1.5 \times 15) = 69.75\ kN$

Passive resistance $> 69.75\ kN$

The resistance to horizontal load is satisfactory.

The reduction in moment on the underside of the base due to the horizontal reaction from the passive earth pressure has been neglected.

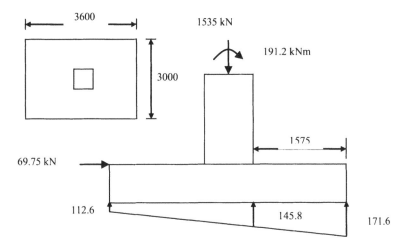

Fig. 11.10 (a) Base pressures and plan of footing.

(d) Design of the moment reinforcement

The design is carried out for the ultimate loads from the column.

(i) Long-span moment steel

Axial load $N = (1.35 \times 770) + (1.5 \times 330) = 1535\ kN$

Horizontal load $H = (1.35 \times 35) + (1.5 \times 15) = 69.75\ kN$

Moment $M = (1.35 \times 78) + (1.5 \times 34) + (0.5 \times 69.75) = 191.2\ kNm$

Maximum pressure $= 1535/10.8 + 191.2/6.48 = 171.6\ kN/m^2$

Minimum pressure $= 1535/10.8 - 191.2/6.48 = 112.6\ kN/m^2$

The pressure distribution is shown in Fig. 11.10.

At the face of the column pressure is

Pressure $= 112.6 + (171.6 - 112.6) \times (3.6 - 1.575)/3.6 = 145.8\ kN/m^2$

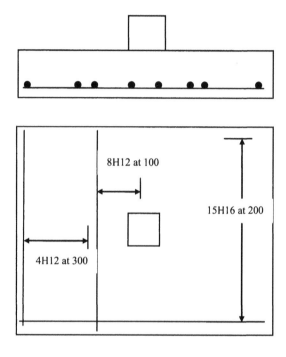

Fig. 11.11 Reinforcement in the base. (Only a symmetrical half of the short direction steel is shown.

Moment at the face of the column is

$M_y = 145.8 \times 3.0 \times 1.575^2/2 + 0.5(171.6 - 145.8) \times 3.0 \times 1.575 \times (2/3) \times 1.575$
$= 606.5$ kNm

If the cover is 40 mm and 16 mm diameter bars are used, the effective depth for the bottom layer is

$$d = 550 - 40 - 16/2 = 502 \text{ mm}$$
$$k = M/ (bd^2 \, f_{ck}) = 606.5 \times 10^6/ (3000 \times 502^2 \times 30) = 0.027 < 0.196$$

$$\frac{z}{d} = 0.5[1.0 + \sqrt{(1 - 3\frac{k}{\eta})}]$$

$$z/d = 0.98$$
$$f_{yk} = 500, \, f_{yd} = 500/1.15 = 435 \text{ MPa}$$
$$A_s = 606.5 \times 10^6/ (435 \times 0.98 \times 502) = 2834 \text{ mm}^2$$

Check minimum steel:

From equation (9.1N) of the code,

$$A_{s, \, min} = 0.26 \times (f_{ctm}/f_{yk}) \times bd \geq 0.0013 \, bd$$
$$f_{ctm} = 0.3 \times f_{ck}^{0.67} = 0.3 \times 30^{0.67} = 2.9 \text{ MPa}, \, f_{yk} = 500 \text{ MPa},$$
$$b = 3000 \text{ mm}, \, d = 502 \text{ mm}$$
$$A_{s, \, min} = 0.26 \times (2.9/500) \times 3000 \times 502 \geq 0.0013 \times 3000 \times 502$$
$$A_{s, \, min} = 2271 \text{ mm}^2 < 2834$$
Provide 15H16. $A_s = 2834 \text{ mm}^2$.
$$0.75 \, (c + 3d) = 0.75 \, (450 + 3 \times 502) = 1467 \text{ mm},$$

$$L/2 = 3000/2 = 1500 \text{ mm}$$
$$0.75 (c + 3d) < \ell_x$$

The difference between 1467 mm and 1500 mm is small enough to be ignored and steel can be distributed uniformly. Provide 15 bars at 200 mm centres to give a total steel area of 2834 mm².

(ii) Short-span moment steel

Average pressure $= 0.5 \times (171.6 + 112.6) = 142.1 \text{ kN/m}^2$
Moment $M_x = 142.1 \times 3.6 \times 1.275^2/2 = 415.8 \text{ kNm}$
Using H12 bars,
Effective depth $d = 550 - 40 - 16 - 12/2 = 488 \text{ mm}$
$k = M/ (bd^2 \ f_{ck}) = 415.8 \times 10^6/ (3600 \times 488^2 \times 30) = 0.016 < 0.196$

$$\frac{z}{d} = 0.5[1.0 + \sqrt{(1 - 3\frac{k}{\eta})}]$$

$z/d = 0.99$
$f_{yk} = 500, f_{yd} = 500/1.15 = 435 \text{ MPa}$
$A_s = 415.8 \times 10^6/ (435 \times 0.99 \times 488) = 1979 \text{ mm}^2$

Check minimum steel:
From equation (9.1N) of the code,

$$A_{s, \, min} = 0.26 \times (f_{ctm}/f_{yk}) \times bd \geq 0.0013 \, bd$$
$f_{ctm} = 0.3 \times f_{ck}^{0.67} = 0.3 \times 30^{0.67} = 2.9 \text{ MPa}, \, f_{yk} = 500 \text{ MPa},$
$b = 3600 \text{ mm}, d = 488 \text{ mm}$
$A_{s, \, min} = 0.26 \times (2.9/500) \times 3600 \times 488 \geq 0.0013 \times 3600 \times 488$
$A_{s, \, min} = 2649 \text{ mm}^2 > 1979 \text{ mm}^2$
Provide 24H12. A_s provided $= 2488 \text{ mm}^2$
$0.75(c + 3d) = 1436 < (\ell_x = 1800 \text{ mm})$

Place two-thirds of the bars (16 bars) in the central zone 1450 mm wide. Provide 16H12 at 100 mm over a width of 1500 mm. In the outer strips 870 mm wide provide 4H12 at 300 mm centres.

The arrangement of bars is shown in Fig. 11.11. *Note that for clarity, only a symmetrical half of steel in the short direction is shown.*

(e) Vertical shear

Long span: The vertical shear stress is checked at $d = 502 \text{ mm}$ from the face of the column.
Pressure $= 112.6 + (171.6 - 112.6) \times (3.6 - 1.575 + 0.502)/3.6 = 154.0 \text{ kN/m}^2$
Shear at a distance d from the face of the column is
$V_{Ed} = 0.5(154.0 + 171.6) \times 3.0 \times (1.575 - 0.502) = 524.1 \text{ kN}$
$v_{Ed} = 524.1 \times 10^3/ (3000 \times 502) = 0.35 \text{ MPa}$
$C_{Rd, \, c} = 0.18/ (\gamma_c = 1.5) = 0.12, k = 1 + \sqrt{(200/502)} = 1.63 \leq 2.0$
The bars extend 502 mm, i.e., more than d, beyond the critical section and so all the steel can be taken into account when calculating A_{sl}.
$A_{sl} = 15H16 = 3016 \text{ mm}^2, \rho_1 = A_{sl}/ (b_w \, d) = 3016/ (3000 \times 502) = 0.002 \leq 0.02$

$$C_{Rd, c} \times k \times (100 \times \rho_1 \times f_{ck})^{0.33} = 0.12 \times 1.63 \times (100 \times 0.002 \times 30)^{0.33} = 0.36$$
$$v_{min} = 0.035 \times k^{1.5} \times \sqrt{f_{ck}} = 0.035 \times 1.63^{1.5} \times \sqrt{30} = 0.40 > 0.36$$
$$v_{Rd, c} = 0.40 \text{ MPa}$$
$$(v_{Ed} = 0.35) < (v_{Rd, c} = 0.40)$$

No shear reinforcement is required.

Short span:

Average pressure $= 0.5(171.6 + 112.6) = 142.1 \text{ kN/m}^2$

The average pressure acts over an area of dimensions

$$\{(3000 - 450)/2 - 488 = 787 \text{ mm}\} \times 3600 \text{ mm}$$

Shear at a distance d from the face of the column is

$$V_{Ed} = 142.1 \times 3.6 \times 0.787 = 416.0 \text{ kN}$$
$$v_{Ed} = 416.0 \times 10^3 / (3600 \times 488) = 0.24 \text{ MPa}$$
$$C_{Rd, c} = 0.18/ (\gamma_c = 1.5) = 0.12, k = 1 + \sqrt{(200/488)} = 1.64 \leq 2.0$$

The bars extend 488 mm, i.e. more than d, beyond the critical section and so all the steel can be taken into account when calculating A_{sl}.

$$A_{sl} = 24H12 = 2714 \text{ mm}^2, \rho_1 = A_{sl}/ (b_w d) = 2714/ (3600 \times 488) = 0.0016 \leq 0.02$$
$$C_{Rd, c} \times k \times (100 \times \rho_1 \times f_{ck})^{0.33} = 0.12 \times 1.64 \times (100 \times 0.0016 \times 30)^{0.33} = 0.33$$
$$v_{min} = 0.035 \times k^{1.5} \times \sqrt{f_{ck}} = 0.035 \times 1.64^{1.5} \times \sqrt{30} = 0.40 > 0.33$$
$$v_{Rd, c} = 0.40 \text{ MPa}$$
$$(v_{Ed} = 0.24) < (v_{Rd, c} = 0.40)$$

No shear reinforcement is required.

(f) Punching shear and maximum shear

Check punching shear around column perimeter:

$$\text{Column perimeter, } u_0 = 2(c_1 + c_2) = 1800 \text{ mm}$$
$$d = 495 \text{ mm}$$
$$\text{Column axial force} = 1535 \text{ kN}$$
$$v_{Rd, max} = 0.3 \times (1 - f_{ck}/250) \times f_{cd} = 0.3 \times (1 - 30/250) \times (30/1.5) = 5.28 \text{ MPa}$$
$$\text{Shear stress around column perimeter} = 1535 \times 10^3 / (1800 \times 495)$$
$$= 1.72 \text{ MPa} < (v_{Rd, max} = 5.28 \text{MPa})$$

Thickness of the slab is acceptable.

Check punching shear on perimeters at Nd from the face of the column, where $1 \leq N \leq 2$.

Fig. 11.12 shows the punching perimeter considered. c_1 and c_2 are respectively the dimensions of the column *parallel* and *perpendicular* to the eccentricity of the load.

Using the data from (e) above,

$$\text{Average } d = 0.5 (502 + 488) = 495 \text{ mm}$$
$$\text{Average pressure} = (171.6 + 112.6)/2 = 141.8 \text{ kN/m}^2$$
$$\text{Area inside the perimeter } A = c_1 \times c_2 + 2 \times (c_1 + c_2) \times Nd + \pi \times (Nd)^2$$

where $c_1 = c_2 = 450$ mm.

$$\text{Upward thrust from base pressure} = 141.8 \times A \text{ kN}$$
$$\text{Pressure at column face is } p_2 = 29.8 \times (c_1/L) = 3.688 \text{ kN/m}^2$$

where $c_1 = 450$ mm, L = 3600 mm.

$$\text{Pressure at punching shear perimeter, } p_1 = 29.8 \times (c_1 + 2 Nd)/L \text{ kN/m}^2$$

Perimeter length, $u = 2(c_1 + c_2 + \pi \times Nd)$

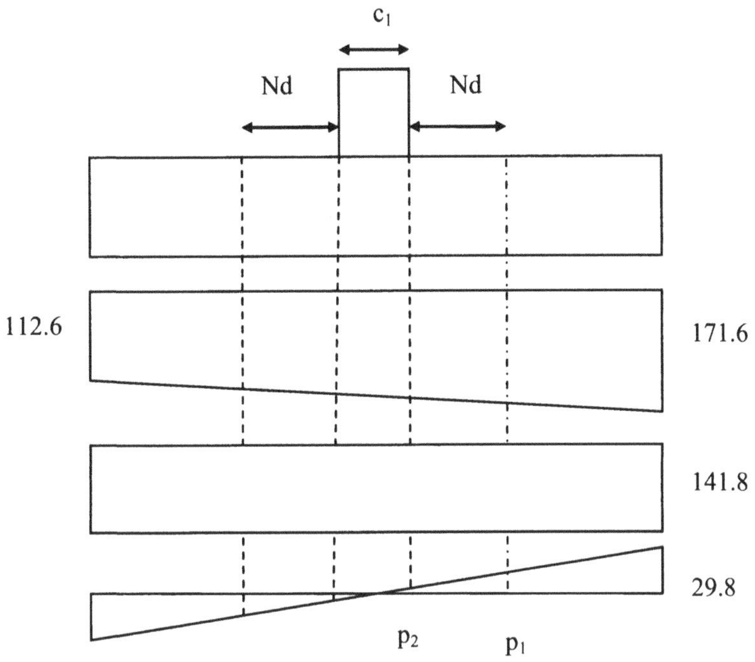

c_1

Nd   Nd

112.6   171.6

141.8

29.8

p_2 p_1

Fig. 11.12 Pressure distribution at the base at ULS.

Moment caused by the linear pressure distribution in the three areas as shown in Fig. 11.13(a) are:

$$\text{Area A: } M_a = \frac{2}{3} \times p_1 \times (0.5c_1 + Nd)^2 \times c_2$$

$$\text{Area B: } M_b = \frac{1}{3} \times p_2 \times c_1^2 \times Nd$$

Area C: The necessary equations for calculating M_c can be derived as follows. Total force F acting on the area shown in Fig. 11.13(b) due to linear pressure distribution variation in the x-direction is given by

$$F = \int_{x=0}^{x=a} \{p_2 + (p_1 - p_2) \times \frac{x}{a}\} dx \int_{y=0}^{y=\sqrt{(a^2 - x^2)}} dy$$

$$F = \frac{a^2}{12} \times [(3\pi - 4)p_2 + 4p_1]$$

$$M_y = \int_{x=0}^{x=a} \{p_2 + (p_1 - p_2) \times \frac{x}{a}\} x \, dx \int_{y=0}^{y=\sqrt{(a^2 - x^2)}} dy$$

The moment M_y about the vertical axis is given by

$$M_y = \frac{a^3}{48} \times [(16 - 3\pi)p_2 + 3\pi\, p_1]$$
$$M_c = 4M_y + 2 \times F \times c_1$$

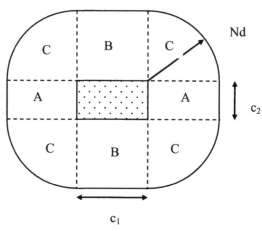

Fig. 11.13(a) Punching shear perimeter.

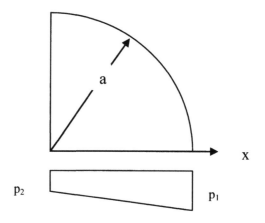

Fig. 11.13(b) Pressure distribution on a quarter circular area.

$V_{Ed,\,red}$ = Column axial force – upward thrust from base pressure
$M_{Ed,\,red}$ = Moment on the column – $(M_a + M_b + M_c)$
$$v_{Ed} = \beta \times (V_{Ed} / (u_1 \times d)$$
$$u_1 = 2(c_1 + c_2 + \pi \times Nd)$$
β is given by code equation (6.51).

$$\beta = 1 + k \times \left\{\frac{M_{Ed}}{V_{Ed,red}}\right\} \times \{\tfrac{u}{w}\} \qquad\qquad (6.51)$$

From Table 6.1 of the code, for $c_1/c_2 = 1$, $k = 0.6$.

Rewriting code equation (6.41) for a perimeter at a distance of a instead of $2d$ from the column face,

$$W = c_1 c_2 + 2 c_2 a + 0.5 c_1^2 + 4a^2 + \pi c_1 a$$

Calculations can be done using a spreadsheet. The results are shown in Table 11.6.

Table 11.6 Punching shear stress calculations

N_d	p_1	A	N-Soil	$V_{Ed,red}$	M_a	M_b	M_c	M-soil	$M_{Ed,red}$	W_1	u_1	V_{Ed}
495	11.8	1.9	265	1270	1.8	0.1	2.6	4.6	187	2.4	4.9	0.62
545	12.6	2.1	300	1235	2.2	0.1	3.5	5.9	185	2.7	5.2	0.56
594	13.4	2.4	338	1197	2.7	0.1	4.6	7.4	184	3.1	5.5	0.51
644	14.2	2.7	378	1157	3.2	0.2	5.9	9.3	182	3.4	5.8	0.46
693	15.0	3.0	420	1115	3.8	0.2	7.5	11.5	180	3.8	6.2	0.42
743	15.9	3.3	465	1070	4.5	0.2	9.4	14.0	177	4.2	6.5	0.39
792	16.7	3.6	511	1024	5.2	0.2	11.6	17.0	174	4.6	6.8	0.35
842	17.5	3.9	560	975	6.0	0.2	14.2	20.3	171	5.1	7.1	0.32
891	18.3	4.3	611	924	6.8	0.2	17.1	24.2	167	5.5	7.4	0.29
941	19.1	4.7	664	871	7.8	0.2	20.6	28.6	163	6.0	7.7	0.26
990	19.9	5.1	720	816	8.8	0.2	24.5	33.5	158	6.5	8.0	0.23

Calculations have been done using, $p_2 = 3.688$ kN/m^2 and $d = 495$ mm.

$$C_{Rd,c} = 0.18/(\gamma_c = 1.5) = 0.12, \quad k = 1 + \sqrt{(200/495)} = 1.64 \le 2.0,$$

$$\text{Average } 100A_s/(bd) = \sqrt{(0.20 \times 0.16)} = 0.18$$

$$C_{Rd,c} \times k \times (100 \times p_1 \times f_{ck})^{0.33} = 0.12 \times 1.64 \times (0.18 \times 30)^{0.33} = 0.35$$

$$v_{min} = 0.035 \times k^{1.5} \times \sqrt{f_{ck}} = 0.035 \times 1.64^{1.5} \times \sqrt{30} = 0.40 > 0.35$$

$$v_{Rd,c} = 0.40 \text{ MPa}$$

From code equation (6.50),

$$v_{Rd} = v_{Rd,c} \times (2d/a) = 0.40 \times (2d/a)$$

At 'a' = d, $v_{Rd} = 0.80$ MPa which is greater than 0.62 MPa

At a = 2d, $v_{Rd} = 0.40$ MPa which is greater than 0.23 MPa

Note that in the above calculations, an allowance has been made for the reduction of the net moment by the moment from the soil reaction. This is reasonable because the column force has been reduced by the upward pressure from the soil reaction. The differences are unlikely to be significant. In this example for perimeter at d from the column, $M_{Ed} = 191.2$ kNm, $M_{Ed,red} = 186.6$ kNm. The code equation (6.51) uses only M_{Ed} rather than $M_{Ed,red}$.

(g) Sketch of reinforcement

The reinforcement is shown in Fig. 11.11.

11.5.3.2 Example of Design of a Footing for a Pinned Base Steel Portal

(a) Specification
The column base reactions for a pinned base rigid steel portal for various load cases are shown in Fig. 11.14. Determine the size of foundation for the two cases of independent bases and tied bases. The soil is firm clay with the following properties:

$$\text{Unit weight} = 18 \text{ kN/m}^3,$$
$$\text{Safe bearing pressure} = 150 \text{ kN/m}^2,$$
$$\text{Cohesion and adhesion} = 50 \text{ kN/m}^2$$

(b) Independent base
The base is first designed for dead + imposed load. The proposed arrangement of the base is shown in Fig. 11.14(a). The base is 2 m long by 1.2 m wide by 0.5 m deep. The finished thickness of the floor slab is 180 mm. The unfactored loads on the soil are:

$$\text{Weight of base} = (0.5 + 0.18) \times 2 \times 1.2 \times 25 = 40.8 \text{ kN}$$
$$\text{Vertical load} = 103 + 84 + 40.8 = 227.8 \text{ kN}$$
$$\text{Horizontal load} = 32.4 + 40.3 = 72.7 \text{ kN}$$
$$\text{Moment} = 72.7 \times 0.5 = 36.4 \text{ kN m}$$
$$\text{Area} = 2 \times 1.2 = 2.4\text{m}^2$$
$$\text{Section modulus} = 1.2 \times 2^2/6 = 0.8 \text{ m}^3$$

The maximum vertical pressure is
$$227.8/2.4 + 36.4/0.8 = 140.4 \text{ kN/m}^2$$

The resistance to horizontal load is
$$= 2 \text{ c B h} + 0.5 \text{ B h } (p_1 + p_2) + \beta \text{ L B}$$
$$= (2 \times 50 \times 1.2 \times 0.5) + 0.5 \times 1.2 \times 0.5 \times (0 + 18 \times 0.5) + 50 \times 2 \times 1.2$$
$$= 60 + 2.7 + 120 = 182.7 \text{ kN}$$

The maximum factored horizontal load is
$$(1.35 \times 32.4) + (1.5 \times 40.3) = 104.2 \text{ kN} < 182.7 \text{ kN}$$

The base is satisfactory with respect to resistance to sliding.

Check the dead + imposed + wind load internal suction on the right hand side base.
$$\text{Vertical load} = 103 + 84 + 40.8 - 29.4 = 198.4 \text{ kN}$$
$$\text{Horizontal load} = 32.4 + 40.3 + 2.4 = 75.1 \text{ kN}$$
$$\text{Moment} = 75.1 \times 0.5 = 37.6 \text{ kNm}$$
$$\text{Maximum pressure} = 198.4/2.4 + 37.6/0.8 = 129.7 \text{ kN/m}^2$$

The reinforcement for the base can be designed and the shear stress checked as in the previous example.

(c) Tied base
The proposed base is shown in Fig. 11.15(b). The trial size for the base is 1.2 m × 1.2 m × 0.5 m deep and tie rods are provided in the ground slab. The horizontal tie resists the reaction from the dead and imposed loads. For this case,

$$\text{Weight of the base} = (0.5 + 0.18) \times 1.2 \times 1.2 \times 25 = 24.5 \text{ kN}$$

$$\text{Vertical load} = 103 + 84 + 24.5 = 211.5 \text{ kN}$$
$$\text{Maximum pressure} = 211.5/1.2^2 = 146.9 \text{ kN/m}^2$$

The main action of the wind load is to cause uplift and the slab has to resist a small compression from the net horizontal load when the dead load and wind load internal pressure are applied at left hand base.

(d) Design of tie

To find the steel area for the tie using $f_{yk} = 500$ MPa reinforcement,

$$\text{Ultimate load} = (1.35 \times 32.4) + (1.5 \times 40.3) = 104.2 \text{ kN}$$
$$f_{yd} = f_{yk}/1.15 = 435 \text{ MPa}$$
$$A_s = 104.2 \times 10^3/435 = 239.5 \text{ mm}^2$$

Provide two 16 mm diameter bars to give an area of 402 mm².

If the steel column base bearing plate is 400 mm × 400 mm, the underside of the base lies within the 45° load dispersal lines. Theoretically no reinforcement is required but 8H12 each way would provide 0.15% reinforcement which should be more than minimum requirement.

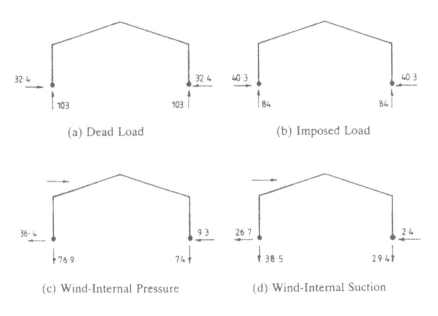

Fig. 11.14 Pinned portal frame reactions (characteristic reactions): (a) dead load; (b) imposed load; (c) wind, internal pressure; (d) wind, internal suction.

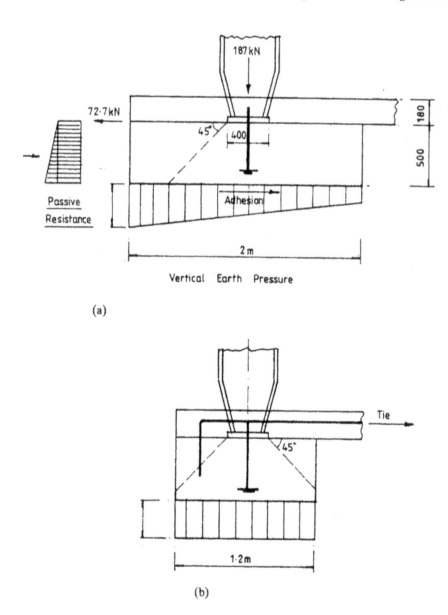

Fig. 11.15 Unfactored loads and base pressures: (a) independent base; (b) tied base.

11.6 WALL, STRIP AND COMBINED FOUNDATIONS

11.6.1 Wall Footings

Typical wall footings are shown in Fig. 11.16(a) and Fig. 11.16(b).   In Fig. 11.16(a) the wall is cast integral with the footing. The critical section for moment is at Y_1Y_1, the face of the wall, and the critical section for shear is at Y_2Y_2, d from the face of the wall. A 1 m length of wall is considered and the design is made on similar lines to that for a pad footing.

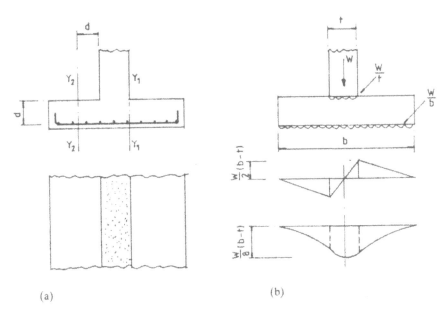

(a)   (b)

Fig. 11.16 (a) Wall and footing integral; (b) wall and footing separate.

If the wall is separate from the footing, e.g., a brick wall, the base is designed for the maximum moment at the centre and maximum shear at the edge, as shown in Fig. 11.16(b). The wall distributes the load W/t per unit length to the base and the base distributes the load W/b per unit length to the ground, where W is the load per unit length of wall, t is the wall thickness and b is the base width. The maximum shear at the edge of the wall is

$$W (b - t)/ (2b)$$

The maximum moment at the centre of the wall is

$$\frac{W}{b}\frac{b}{2}\frac{b}{4} - \frac{W}{t}\frac{t}{2}\frac{t}{4} = \frac{W}{8}(b-t)$$

11.6.2 Shear Wall Footings

If the wall and footing resist an in-plane horizontal load, e.g., when the wall is used as a shear wall to stabilize a building, the maximum pressure at one end of the wall is found assuming a linear distribution of base pressure (Fig. 11.17). The footing is designed for the average base pressure on, say, 0.5 m length at the end subjected to maximum base pressure. Define the following variables:

$$W = \text{total load on the base}$$
$$H = \text{horizontal load at the top of the wall}$$
$$h = \text{height of the wall}$$
$$b = \text{width of the base}$$
$$\ell = \text{length of the wall and base}$$
$$\text{Base area } A = b\ell$$
$$\text{Section modulus } Z = b\ell^2/6$$
$$\text{Maximum pressure} = W/A + H\,h/Z$$

If the footing is on firm ground and is sufficiently deep so that the underside of the base lies within 45° dispersal lines from the face of the wall, reinforcement need not be provided. However, it would be very advisable to provide at least minimum reinforcement at the top and bottom of the footing to control cracking in case some settlement should occur.

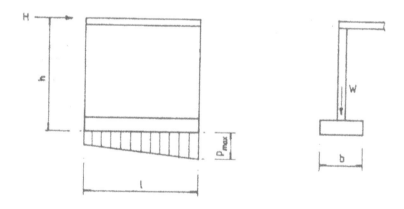

Fig. 11.17 Shear wall footing.

11.6.3 Strip Footings

A continuous strip footing is used under closely spaced rows of columns as shown in Fig. 11.18 where individual footings would be close together or overlap.

If the footing is concentrically loaded, the pressure is uniform. If the column loads are not equal or not uniformly spaced and the base is assumed to be rigid, moments of the loads can be taken about the centre of the base and the pressure distribution can be determined assuming that the pressure varies uniformly. These cases are shown in Fig. 11.18.

In the longitudinal direction, the footing may be analysed for moments and shears by the following methods.

1. Assume a rigid foundation. Then the shear at any section is the algebraic sum of the column forces acting down and the base pressure acting up on one side of the section, and the moment at the section is the corresponding sum of the moments of the forces on one side of the section.
2. A more accurate analysis may be made if the flexibility of the footing and the assumed elastic response of the soil are taken into account. The footing is analysed as a so-called beam on an elastic foundation.

In the transverse direction the base may be designed along lines similar to that for a pad footing.

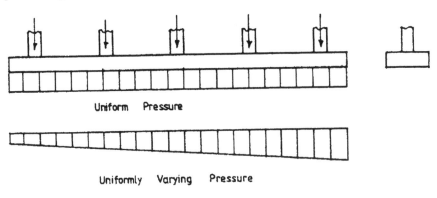

Uniform   Pressure

Uniformly   Varying   Pressure

Fig. 11.18 Continuous strip footing.

11.6.4 Combined Bases

Where two columns are close together and separate footings would overlap, a combined base can be used as shown in Fig. 11.19(a). Again, if one column is close to an existing building or sewer it may not be possible to design a single pad footing, but if it is combined with that of an adjacent footing a satisfactory base can engineered. This is shown in Fig. 11.19(b).

If possible, the base is arranged so that its centre line coincides with the centre of gravity of the loads because this will give a uniform pressure on the soil. In a general case with an eccentric arrangement of loads, moments of forces are taken about the centre of the base and the maximum soil pressure is determined from the total vertical load and moment at the underside of the base. The pressure is assumed to vary uniformly along the length of the base.

In the longitudinal direction the actions for design may be found from statics. In the transverse direction, the critical moment and shear are determined in the same way as for a pad footing. Punching shears at the column face and at d to 2 d from the column face must also be checked.

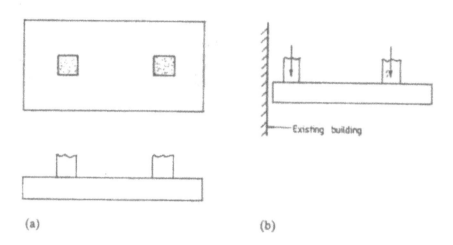

(a)   (b)

Fig. 11.19 (a) Combined base; (b) column close to existing building.

11.6.4.1 Example of Design of a Combined Base

(a) Specification
Design a rectangular base to support two columns carrying the following loads:

Column 1: dead load = 310 kN, imposed load = 160 kN
Column 2: dead load = 430 kN, imposed load = 220 kN

The columns are each 350 mm square and are spaced at 2.5 m centres. The width of the base is not to exceed 2.0 m. The safe bearing pressure on the ground is 160 kN/m². Take f_{ck} = 30 MPa concrete and f_{yk} = 500 MPa.

(b) Base arrangement and soil pressure
Assume the weight of the base is 130 kN. Various load conditions are examined. It is assumed here that the imposed loads on the columns are independent loads and therefore carry different load factors. *If this is not the case, then a single load factor should be applied for both the loads.*

(i) Case 1: Dead + imposed load on both columns. Use SLS values

Axial load = (310 + 160) + (430 + 220) + 130 = 1250 kN
Area of base = 1250/160 = 7.81 m²
Length of base = 7.81/2.0 = 3.91 m

Choose 4.5 m × 2.0 m × 0.6 m deep base.
The weight of the base is (4.5 × 2.0 × 0.6 × 25) = 135.0 kN ≈ 130 kN.

$$\text{Area} = 4.5 \times 2.0 = 9.0 \text{ m}^2$$
$$\text{Section modulus} = 2.0 \times 4.5^2/6 = 6.75 \text{ m}^3$$

The base is arranged so that the centre of gravity of the loads coincides with the centre line of the base, in which case the base pressure will be uniform. This arrangement will be made for the maximum ultimate loads.

The ultimate loads are

$$\text{Column 1: load} = 1.35 \times 310 + 1.5 \times 160 = 658.5 \text{ kN}$$
$$\text{Column 2: load} = 1.35 \times 430 + 1.5 \times 220 = 910.5 \text{ kN}$$

The distance of the centre of gravity from column 1 is

$$x = (910.5 \times 2.5)/ (658.5 + 910.5) = 1.45 \text{ m}$$

The base arrangement is shown in Fig. 11.20.

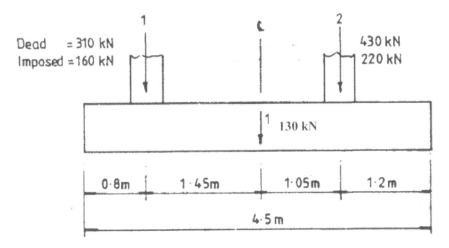

Fig. 11.20 Combined base dimensions and column loads.

The soil pressure is checked for service loads for case 1:

$$\text{Direct vertical load} = 310 + 160 + 430 + 220 + 130 = 1250 \text{ kN}$$

Since the centroid of the loads does not exactly coincide with the centroid of the base, check for maximum pressure which is non-uniform. The moment about the centreline of the base is

$$M = (430 + 220) \times 1.05 - (310 + 160) \times 1.45 = 1.0 \text{ kNm}$$

The moment is very small and can be ignored. The base pressure is practically constant.

$$\text{Base pressure} = 1250/9.0 = 138.9 \text{ kN/m}^2 < 160.0$$

(ii) Case 2: Column 1, dead + imposed load; column 2, dead load only. Use SLS values

$$\text{Axial load} = (310+160) + (430 + 0) + 130.0 = 1030 \text{ kN}$$
$$\text{Moment} = M = (430+0) \times 1.05 - (310 + 160) \times 1.45 = -230 \text{ kN m}$$
$$\text{Maximum pressure} = 1030/9.0 + 230.0/6.75 = 148.5 \text{ kN/m}^2 < 160.0 \text{ kN/m}^2$$

Maximum base pressure occurs toward the column 1 side.

(iii) Case 3: Column 1: dead load only; column 2: dead + imposed load. Use SLS values

Axial load = (310+0) + (430 + 220) + 130 = 1090 kN

Moment = M = (430+220) × 1.05 – (310 + 0) × 1.45 = 233 kN m

Maximum pressure = 1090/9.0 + 233.0/6.75 = 155.6 kN/m² < 160.0 kN/m²

Maximum base pressure occurs toward the column 2 side.

The base is satisfactory with respect to soil pressure.

(c) Analysis for actions in longitudinal direction at ULS

The cover is 40 mm, and the bars, say, 20 mm in diameter, giving an effective depth d of 550 mm. Using the 'Macaulay bracket notation', the shear force V and moment M in the longitudinal direction due to ultimate loads are calculated by statics.

As shown in Fig. 11.21, p_1 and p_2 are the base pressure at left and right hand ends respectively. W_1 and W_2 are the column loads and p is the base pressure at a distance x from left hand end.

$$p = p_1 + (p_2 - p_1)\frac{x}{4.5}$$

$$V = 2\{\frac{(p_1 + p)}{2}x\} - W_1\langle x - 0.8\rangle^0 - W_2\langle x - 3.3\rangle^0$$

$$M = 2\{p_1\frac{x^2}{2} + \frac{(p - p_1)}{2}\frac{x^2}{3}\} - W_1\langle x - 0.8\rangle - W_2\langle x - 3.3\rangle$$

The maximum design moments are at the column face and between the columns, and maximum shears are at d from the column face. Calculations are best done using spreadsheets. The load cases are as follows. The weight of the base is ignored as the corresponding base pressure will cancel the pressure due to the weight of the base.

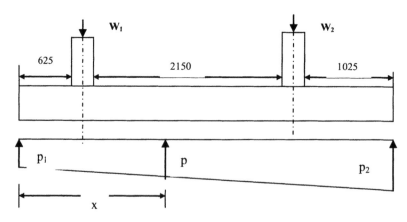

Fig. 11.21 Combined footing.

In the following load factors from Equation (6.10), Table A2.4 (B) from *BS EN 1990:2002, Eurocode-Basis of Structural Design* are used.

$\gamma_{Gj, sup} = 1.35$, $\gamma Gj_{, inf} = 1.0$, $\gamma_{Q, 1} = 1.5$, $\psi_{0, i} = 0.7$

Six loading cases are discussed in detail.

Case 1A: Maximum load on both columns with column 1 carrying leading variable load.

Treat G_k as unfavourable on both columns, Q_k on column 1 as leading variable action and Q_k on column 2 as accompanying variable action. Loads are as shown in Fig. 11.20.

$$W_1: 1.35 \times 310 + 1.5 \times 160 = 658.5 \text{ kN}$$
$$W_2: 1.35 \times 430 + 1.5 \times 0.7 \times 220 = 811.5 \text{ kN}$$
$$W_1 + W_2 = 658.5 + 811.5 = 1470.0 \text{ kN},$$
$$\text{Moment } M = 811.5 \times 1.05 - 658.5 \times 1.45 = -102.75 \text{ kN m}$$
$$p_1 = 1470.0/9.0 + 102.75/6.75 = 178.6 \text{ kN/m}^2$$
$$p_2 = 1470.0/9.0 - 102.75/6.75 = 148.1 \text{ kN/m}^2$$

Table 11.7 Shear and moment calculation for case 1A.

x	V	M	Remarks
0.075	26.8	1.0	d from left face of column 1
0.625	220.6	69.2	Left face of column 1
0.975	−316.7	52.5	Right face of column 1
1.525	−129.5	−70.1	d from right face of column 1
1.89	0	**−95.1**	Maximum negative moment
2.575	**216.3**	−23.2	d from left face of column 2
3.125	391.6	144.2	Left face of column 2
3.475	−310.6	**158.4**	Right face of column 2
4.025	−142.1	34.1	d from right face of column 2

The results are shown in Table 11.7. Fig. 11.22 and Fig. 11.23 show respectively the shear force and bending moment diagrams.

Design values: shear force = 216.3 kN, moment = 158.4 kNm and −95.1 kNm.

Case 1B: Maximum load on both columns with column 2 carrying leading variable load.

Treat G_k as unfavourable on both columns, Q_k on column 2 as leading variable action and Q_k on column 1 as accompanying variable action. Loads are as shown in Fig. 11.20.

$$W_1: 1.35 \times 310 + 1.5 \times 0.7 \times 160 = 586.5 \text{ kN}$$
$$W_2: 1.35 \times 430 + 1.5 \times 220 = 910.5 \text{ kN}$$
$$W_1 + W_2 = 586.5 + 910.5 = 1497.0 \text{ kN},$$
$$\text{Moment } M = 910.5 \times 1.05 - 586.5 \times 1.45 = 105.6 \text{ kN m}$$
$$p_1 = 1497.0/9.0 - 105.6/6.75 = 150.7 \text{ kN/m}^2$$
$$p_2 = 1497.0/9.0 + 105.6/6.75 = 181.96 \text{ kN/m}^2$$

The results are shown in Table 11.8. Fig. 11.24 and Fig. 11.25 show respectively shear force and bending moment diagrams. Design values: shear force = 235.7 kN, moment = 188.8 kNm and –85.4 kNm.

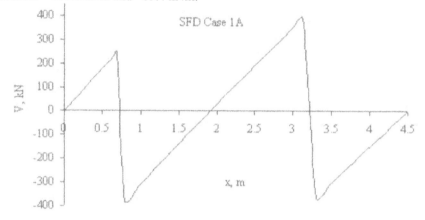

Fig. 11.22 Shear force diagram for case 1A.

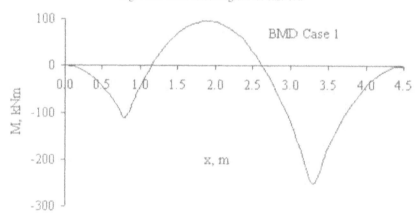

Fig. 11.23 Bending moment diagram for case 1A.

Table 11.8 Shear and moment calculation for case 1B

x	V	M	Remarks
0.075	22.6	0.8	d from left face of column 1
0.625	191.1	59.4	Left face of column 1
0.975	–286.0	42.8	Right face of column 1
1.525	–110.7	–66.5	d from right face of column 1
1.89	0	**–85.4**	Maximum negative moment
2.575	**235.7**	–2.3	d from left face of column 2
3.125	423.3	178.8	Left face of column 2
3.475	–365.7	**188.8**	Right face of column 2
4.025	–171.3	40.9	d from right face of column 2

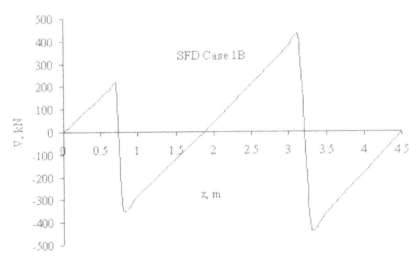

Fig. 11.24 Shear force diagram for case 1B.

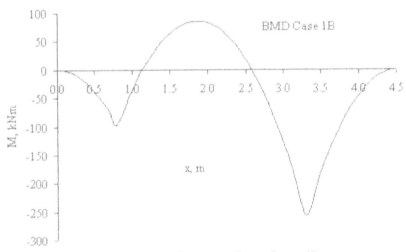

Fig. 11.25 Bending moment diagram for case 1B.

Case 2A: Maximum load on column 1 and minimum load on column 2 with column 1 carrying leading variable load.

Treat G_k as unfavourable on column 1 and as favourable on column 2, Q_k on column 1 as leading variable action and Q_k on column 2 as accompanying variable action. Loads are as shown in Fig. 11.20.

$$W_1: 1.35 \times 310 + 1.5 \times 160 = 658.5 \text{ kN}$$
$$W_2: 1.0 \times 430 + 1.5 \times 0.7 \times 220 = 661.0 \text{ kN}$$
$$W_1 + W_2 = 658.5 + 661.0 = 1319.5 \text{ kN,}$$
$$\text{Moment } M = 661.0 \times 1.05 - 658.5 \times 1.45 = -260.8 \text{ kN m}$$

$$p_1 = 1319.5\,/9.0 + 260.8\,/6.75 = 185.2 \text{ kN/m}^2$$
$$p_2 = 1319.5\,/9.0 - 260.8\,/6.75 = 108.0 \text{ kN/m}^2$$

The results are shown in Table 11.9. Fig. 11.26 and Fig. 11.27 show respectively shear force and bending moment diagrams. Design values: shear force = 181.7 kN, moment = 119.7 kNm and –95.3 kNm.

Table 11.9 Shear and moment calculation for case 2A

x	V	M	Remarks
0.075	27.7	1.0	d from left face of column 1
0.625	224.9	71.0	Left face of column 1
0.975	–313.6	55.6	Right face of column 1
1.525	–133.4	–66.9	d from right face of column 1
2.04	0	**–95.3**	Maximum negative moment
2.575	**181.7**	–38.2	d from left face of column 2
3.125	331.7	103.4	Left face of column 2
3.475	–239.3	**119.7**	Right face of column 2
4.025	–106.3	25.2	d from right face of column 2

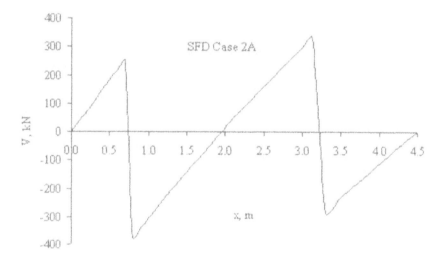

Fig. 11.26 Shear force diagram for case 2A.

Case 2B: Maximum load on column 1 and minimum load on column 2 with column 2 carrying leading variable load.

Treat G_k as unfavourable on column 1 and as favourable on column 2, Q_k on column 2 as leading variable action and Q_k on column 1 as accompanying variable action. Loads are as shown in Fig. 11.20.

$$W_1: 1.35 \times 310 + 1.5 \times 0.7 \times 160 = 586.5 \text{ kN}$$
$$W_2: 1.0 \times 430 + 1.5 \times 220 = 760.0 \text{ kN}$$
$$W_1 + W_2 = 586.5 + 760.0 = 1346.5 \text{ kN},$$

Moment $M = 760.0 \times 1.05 - 586.5 \times 1.45 = -52.4$ kN m
$$p_1 = 1346.5 / 9.0 + 52.4 / 6.75 = 157.4 \text{ kN/m}^2$$
$$p_2 = 1346.5 / 9.0 - 52.4 / 6.75 = 141.9 \text{ kN/m}^2$$

The results are shown in Table 11.10. Fig. 11.28 and Fig. 11.29 show respectively shear force and bending moment diagrams. Design values: shear force = 201.1 kN, moment = 150.2 kNm and −84.9 kNm.

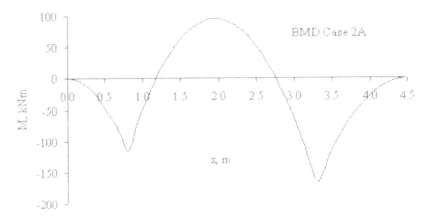

Fig. 11.27 Bending moment diagram for case 2A.

Table 11.10 Shear and moment calculation for case 2B

x	V	M	Remarks
0.075	23.6	1.0	d from left face of column 1
0.625	195.4	61.2	Left face of column 1
0.975	−282.9	45.9	Right face of column 1
1.525	−114.5	−63.3	d from right face of column 1
2.04	0	**−84.9**	Maximum negative moment
2.575	**201.1**	−17.2	d from left face of column 2
3.125	363.4	138.1	Left face of column 2
3.475	−294.4	**150.2**	Right face of column 2
4.025	−135.5	32.1	d from right face of column 2

Case 3A: Minimum load on column 1 and maximum load on column 2 with column 1 carrying leading variable load.

Treat G_k as unfavourable on column 2 and as favourable on column 1, Q_k on column 1 as leading variable action and Q_k on column 2 as accompanying variable action. Loads are as shown in Fig. 11.20.

$$W_1: 1.0 \times 310 + 1.5 \times 160 = 550.0 \text{ kN}$$
$$W_2: 1.35 \times 430 + 1.5 \times 0.7 \times 220 = 811.5 \text{ kN}$$
$$W_1 + W_2 = 550.0 + 811.5 = 1361.5 \text{ kN},$$
Moment $M = 811.5 \times 1.05 - 550.0 \times 1.45 = 54.6$ kN m

$$p_1 = 1361.5 /9.0 - 54.6 /6.75 = 143.2 \text{ kN/m}^2$$
$$p_2 = 1361.5 /9.0 + 54.6 /6.75 = 159.4 \text{ kN/m}^2$$

The results are shown in Table 11.11. Fig. 11.30 and Fig. 11.31 show respectively shear force and bending moment diagrams. Design values: shear force = 211.4 kN, moment = 166.3 kNm and −79.9 kNm.

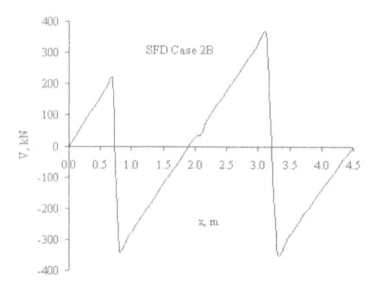

Fig. 11.28 Shear force diagram for case 2B.

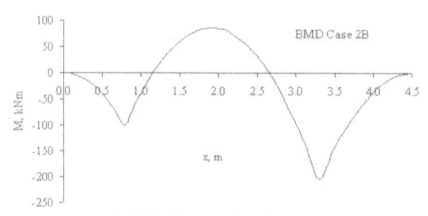

Fig. 11.29 Bending moment diagram for case 2B.

Case 3B: Minimum load on column 1 and maximum load on column 2 with column 2 carrying leading variable load.

Treat G_k as unfavourable on column 2 and as favourable on column 1, Q_k on column 2 as leading variable action and Q_k on column 1 as accompanying variable action. Loads are as shown in Fig. 11.20.

$$W_1: 1.0 \times 310 + 1.5 \times 0.7 \times 160 = 478.0 \text{ kN}$$
$$W_2: 1.35 \times 430 + 1.5 \times 220 = 910.5 \text{ kN}$$
$$W_1 + W_2 = 478.0 + 910.5 = 1388.9 \text{ kN},$$
$$\text{Moment } M = 910.5 \times 1.05 - 478.0 \times 1.45 = 262.9 \text{ kN m}$$
$$p_1 = 1388.9 / 9.0 - 262.9 / 6.75 = 115.3 \text{ kN/m}^2$$
$$p_2 = 1388.9 / 9.0 + 262.9 / 6.75 = 193.2 \text{ kN/m}^2$$

The results are shown in Table 11.12. Fig. 11.32 and Fig. 11.33 show respectively shear force and bending moment diagrams. Design values: shear force = 230.6 kN, moment = 196.6 kNm and −70.8 kNm.

Table 11.11 Shear and moment calculation for case 3A

x	V	M	Remarks
0.075	21.5	1.0	d from left face of column 1
0.625	180.4	56.2	Left face of column 1
0.975	−267.3	41.0	Right face of column 1
1.525	−104.9	−61.5	d from right face of column 1
2.04	0	**−79.9**	Maximum negative moment
2.575	**211.4**	−6.3	d from left face of column 2
3.125	380.2	156.3	Left face of column 2
3.475	−322.8	**166.3**	Right face of column 2
4.025	−150.4	36.1	d from right face of column 2

Design values: shear force = 211.4 kN, moment = 166.3 kNm and −79.9 kNm.

Table 11.12 Shear and moment calculation for case 3B

x	V	M	Remarks
0.075	17.4	1.0	d from left face of column 1
0.625	150.9	46.4	Left face of column 1
0.975	−236.7	31.3	Right face of column 1
1.525	−86.1	−57.9	d from right face of column 1
2.04	0	**−70.8**	Maximum negative moment
2.575	**230.6**	−14.6	d from left face of column 2
3.125	411.7	190.8	Left face of column 2
3.475	−378.0	**196.6**	Right face of column 2
4.025	−179.8	42.7	d from right face of column 2

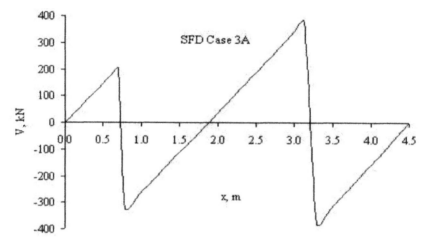

Fig. 11.30 Shear force diagram for case 3A.

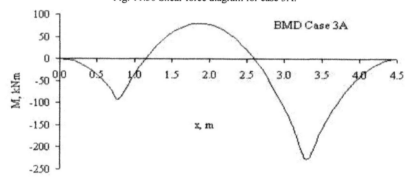

Fig. 11.31 Bending moment diagram for case 3A.

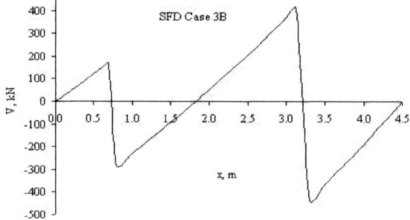

Fig. 11.32 Shear force diagram for case 3B.

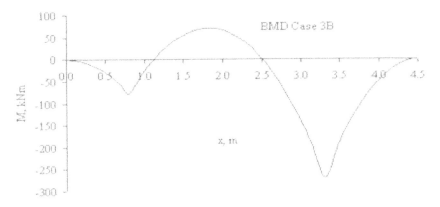

Fig. 11.33 Bending moment diagram for case 3B.

(d) Design of longitudinal reinforcement

(i) Bottom steel
The maximum moment from case 3B is (Fig. 11.33):
$$M = 196.6 \text{ kNm}$$
$$k = M/ (bd^2 f_{ck}) = 196.6 \times 10^6/ (2000 \times 550^2 \times 30) = 0.011 < 0.196$$
$$\frac{z}{d} = 0.5[1.0 + \sqrt{(1 - 3\frac{k}{\eta})}]$$
$$z/d = 0.99$$
$$f_{yk} = 500, f_{yd} = 500/1.15 = 435 \text{ MPa}$$
$$A_s = 196.6 \times 10^6/ (435 \times 0.99 \times 550) = 830 \text{ mm}^2$$
Check minimum steel:
From equation (9.1N) of the code,
$$A_{s, min} = 0.26 \times (f_{ctm}/f_{yk}) \times bd \geq 0.0013 \text{ bd}$$
$$f_{ctm} = 0.3 \times f_{ck}^{0.67} = 0.3 \times 30^{0.67} = 2.9 \text{ MPa}, f_{yk} = 500 \text{ MPa}$$
$$b = 2000 \text{ mm}, d = 550 \text{ mm}$$
$$A_{s, min} = 0.26 \times (2.9/500) \times 2000 \times 550 \geq 0.0013 \times 2000 \times 550$$
$$A_{s, min} = 1659 \text{ mm}^2 > 830 \text{ mm}^2$$
Provide minimum reinforcement.
Provide 9H16 at 240 mm centres to give a total area of 1809 mm².
$$(\ell_c = 1000\text{mm}) < \{0.75(c + 3d) = 0.75(350 + 3 \times 550) = 1500 \text{ mm}\}$$
Reinforcement should be spread uniformly across the width.

(ii) Top steel
The maximum moment from case 2A is (Fig. 11.27),
$$M = 95.3 \text{ kNm}$$
Provide minimum reinforcement as above.

(e) Transverse reinforcement
At ULS, the base pressure distribution in kN/m² is shown in Fig. 11.34.

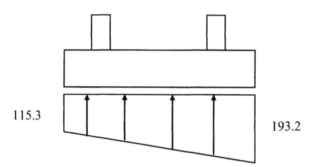

Fig. 11.34 Base pressure distribution in case 3B.

The maximum pressure under the base is for case 3B. The bending moment along the length of 4.5 m is variable. In order to calculate a moment which is reasonable, the average pressure over a width of 0.5 m of the footing length is calculated.
The pressure at 0.5 m from the end is
$$115.3 + (193.2 - 115.3) \times 4.0/4.5 = 184.5 \text{ kN/m}^2$$
The average pressure on a 0.5 m length at the heavier end is
$$(193.2 + 184.5) /2 = 188.9 \text{ kN/m}^2$$
$$(2000 - 350)/2 = 825 \text{ mm, d} = 550 - 16 = 534 \text{ mm}$$
The moment at the face of the columns on a 0.5 m length strip at the heaviest loaded end is
$$M = \{188.9 \times (0.5 \times 0.825) \times 0.825/2 = 32.14 \text{ kNm}$$
$$k = M/ (bd^2 f_{ck}) = 32.14 \times 10^6/ (500 \times 534^2 \times 30) = 0.008 < 0.196$$
$$\tfrac{z}{d} = 0.5[1.0 + \sqrt{(1 - 3\tfrac{k}{\eta})}]$$
$$z/d = 0.99$$
$$f_{yk} = 500, f_{yd} = 500/1.15 = 435 \text{ MPa}$$

$A_s = 32.14 \times 10^6/ (435 \times 0.99 \times 534) = 140 \text{ mm}^2$.
Check minimum steel:
From equation (9.1N) of the code,
$$A_{s, min} = 0.26 \times (f_{ctm}/f_{yk}) \times bd \geq 0.0013 \text{ bd}$$
$$f_{ctm} = 0.3 \times f_{ck}^{0.67} = 0.3 \times 30^{0.67} = 2.9 \text{ MPa, } f_{yk} = 500 \text{ MPa,}$$
$$b = 500 \text{ mm, d} = 534 \text{ mm}$$
$$A_{s, min} = 0.26 \times (2.9/500) \times 500 \times 534 \geq 0.0013 \times 500 \times 534$$
$$A_{s, min} = 402 \text{ mm}^2 > 140 \text{ mm}^2$$
Provide minimum reinforcement. Total steel for a width of 4500 mm is
$402 \times 4500/500 = 3624 \text{ mm}^2$.
Provide 19H16 at 245 mm centres to give a total area of 3820 mm².
Reinforcement should be spread uniformly across the length of the base.
Fig. 11.35 shows the reinforcement arrangement. Note that same reinforcement is provided at top and bottom faces.

(f) Vertical shear

The maximum vertical shear from case 1 is
$$V_{Ed} = 235.7 \text{ kN}$$
$$v_{Ed} = 235.7 \times 10^3/ (2000 \times 550) = 0.21 \text{ MPa}$$
$$A_{sl} = 9H16 = 1810 \text{ mm}^2$$
$$100 \times \rho_1 = 100 \times 1810/ (2000 \times 550) = 0.165 < 2.0$$
$$C_{Rd, c} = 0.18/ (\gamma_c = 1.5) = 0.12, k = 1 + \sqrt{(200/550)} = 1.60 \le 2.0,$$
$$C_{Rd, c} \times k \times (100 \times \rho_1 \times f_{ck})^{0.33} = 0.12 \times 1.60 \times (0.165 \times 30)^{0.33} = 0.33$$
$$v_{min} = 0.035 \times k^{1.5} \times \sqrt{f_{ck}} = 0.035 \times 1.60^{1.5} \times \sqrt{30} = 0.39 > 0.33$$
$$v_{Rd, c} = 0.39 \text{ MPa}$$
$$(v_{Ed} = 0.21) < (v_{Rd, c} = 0.39)$$

No shear reinforcement is required.

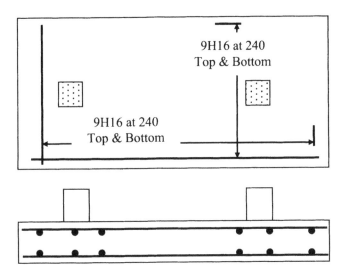

Fig. 11.35 Reinforcement in the base slab.

(g) Punching shear

Check punching shear at column perimeter. Table 11.13 shows the column loads, base pressure at column centre line and
$$V_{Ed} = \text{column load} - \text{base pressure} \times \text{column area}.$$
The maximum value is 889.4 kN for column 2 from case 3B.
$$u_0 = 2 \times (350 + 350) = 1400 \text{ mm}, d = 550 \text{ mm}$$
$$v_{Ed} = 889.4 \times 10^3/ (1400 \times 550) = 1.16 \text{ MPa}$$
$$v_{Rd, max} = 0.3 \times (1 - f_{ck}/250) \times f_{cd} = 0.3 \times (1 - 30/250) \times (30/1.5) = 5.28 \text{ MPa}$$
$$v_{Ed} < v_{Rd, max}.$$

The thickness of slab is adequate.

Table 11.13 Punching loads around column perimeter

Case	Column 1			Column 2		
	Load	Pressure	V_{Ed}	Load	Pressure	V_{Ed}
Case 1A	658.5	173.1	637.3	811.5	156.2	792.4
Case 1B	586.5	156.3	567.4	910.5	173.6	889.2
Case 2A	658.5	171.5	637.4	661,0	128.6	645.3
Case 2B	586.5	154.6	567.6	760.0	145.9	742.1
Case 3A	550.0	146.1	532.1	811.5	155.1	792.5
Case 3B	478.0	129.2	462.2	910.5	172.5	**889.4**

Check punching shear is checked at perimeters at d to 2 d from the face of a column.

At d = 550 mm from the face of the column,

$$u = \text{perimeter} = 2 \times (350 + 350) + 2 \times \pi \times 550 = 4856 \text{ mm}$$

$$A = \text{Area under perimeter} = [4 \times 350 \times (350/2 + 550) + \pi \times 550^2] \times 10^{-6} = 1.965 \text{ m}^2$$

Column load = 910.5 kN, base pressure at centre line of column= 172.5 kN/m^2

$$V_{Ed, red} = 910.5 - 172.5 \times 1.965 = 571.5 \text{ kN}$$
$$v_{Ed} = 571.5 \times 10^3/ (4856 \times 550) = 0.21 \text{ MPa}$$
$$A_{sl} \text{ in x-direction} = 9H16 = 1810 \text{ mm}^2$$
$$100 \times \rho_x = 100 \times 1810/ (2000 \times 550) = 0.165$$
$$A_{sl} \text{ in y-direction} = 19H16 = 3992 \text{ mm}^2$$
$$100 \times \rho_y = 100 \times 3992/ (4500 \times 550) = 0.16$$
$$100\rho_1 = \sqrt{(0.165 \times 0.16)} = 0.162 < 2.0$$
$$C_{Rd, c} = 0.18/ (\gamma_c = 1.5) = 0.12, k = 1 + \sqrt{(200/550)} = 1.60 \leq 2.0,$$
$$C_{Rd, c} \times k \times (100 \times \rho_1 \times f_{ck})^{0.33} = 0.12 \times 1.60 \times (0.162 \times 30)^{0.33} = 0.33$$
$$v_{min} = 0.035 \times k^{1.5} \times \sqrt{f_{ck}} = 0.035 \times 1.60^{1.5} \times \sqrt{30} = 0.39 > 0.33$$
$$v_{Rd, c} = 0.39 \text{ MPa}$$
$$(v_{Ed} = 0.21) < (v_{Rd} = v_{Rd, c} \times \{2d/ (a = d)\} = 0.78)$$

At 1.5 d from the face of the column, the perimeter touches the edge of the slab on the width side. The punching shear is less critical than the vertical shear in this case. The slab is safe against punching shear failure.

(h) Sketch of reinforcement
The reinforcement is shown in Fig. 11.35. A complete mat has been provided at the top and bottom. Some U-spacers are required to fix the top reinforcement in position.

11.7 PILED FOUNDATIONS

11.7.1 General Considerations

When a solid bearing stratum such as rock is deeper than about 3 m below the base level of the structure, a foundation supported on end-bearing piles will provide an economical solution. Where the bedrock is too deep to obtain end bearing, foundations can also be supported on friction piles by skin friction between the pile sides and the soil. The main types of piles are as follows:

1. Precast reinforced or prestressed concrete piles driven into the required position.
2. Cast-in-situ reinforced concrete piles placed in holes formed either by
   (a) Driving a steel tube with a plug of dry concrete or packed aggregate at the end into the soil
   (b) Boring a hole and lowering a steel tube to follow the boring tool as a temporary liner. A reinforcement cage is inserted and the tube is withdrawn after the concrete is placed.

In practice, many other methods are also used. See Bowles (1995).

Short bored plain concrete piles are used for light loads such as carrying ground beams to support walls. Deep cylinder piles are used to carry large loads and can be provided under basement and raft foundations. A small number of cylinder piles can give a more economical solution than a large number of ordinary piles.

The safe load that a pile can carry can be determined by test loading a pile or using a pile formula that gives the resistance calculated from the energy of the driving force and the final set or penetration of the pile per blow. In both cases the ultimate load is divided by a factor of safety of from 2 to 3 to give the safe load. Safe loads depend on the size and depth and whether the pile is of the end-bearing or friction type. The pile can be designed as a short column if lateral support from the ground is adequate. However, if ground conditions are unsatisfactory, it is better to use test load results. The group action of piles should be taken into account because the group capacity can be considerably less than the summed capacities of the individual piles.

The manufacture and driving of piles are carried out by specialist firms that guarantee to provide piles with a given bearing capacity on the site. Safe loads for precast and cast-in-situ piles vary from 100 kN to 1500 kN. Piles are also used to resist tension forces and the safe load in tension is often taken as one-third of the safe load in bearing. Piles in groups are generally spaced at 0.8 to 1.5 m apart. Sometimes piles are driven at an inclination to resist horizontal loads in poor ground conditions. Rakes of 1 in 5 to 1 in 10 are commonly used in building foundations. In an isolated foundation, the pile cap transfers the load from the column shaft to the piles in the group. The cap is cast around the tops of the piles and the piles are anchored into it by projecting bars. Some arrangements of pile caps are shown in Fig. 11.36(a) and Fig. 11.36(b).

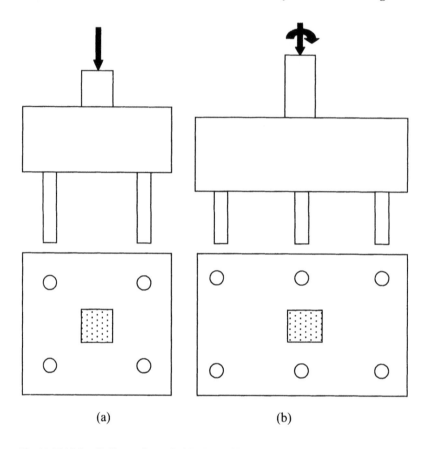

Fig. 11.36 (a) Small pile cap for vertical load; (b) pile group resisting axial load and moment.

11.7.2 Loads in Pile Groups

In general pile groups are subjected to axial load, moment and horizontal loads. The pile loads are as follows.

(a) Axial load
When the load is applied at the centroid of the group, it is assumed to be distributed uniformly to all piles by the pile cap, which is taken to be rigid. This gives the load per pile as

$$F_a = (P + W)/N$$

where P is the axial load from the column, W is the weight of the pile cap and N is the number of piles.

(b) Moment on a group of vertical piles

The pile cap is assumed to rotate about the centroid of the pile group and the pile loads resisting moment vary uniformly from zero at the centroidal axis to a maximum or minimum for the piles farthest away. Referring to Fig. 11.36(b), the second moment of area about the YY axis is

$$I_y = 2(x^2 + x^2) = 4x^2$$

where x is the pile spacing in x direction. The maximum load due to moment on piles A in tension and C in compression is

$$F_m = \pm \frac{M x}{I_y} = \pm \frac{M}{4x}$$

For a symmetrical group of piles spaced at $\pm x_1, \pm x_2 \dots \pm x_n$ *perpendicular* to the centroidal axis YY, the second moment of area of the piles about the YY axis is

$$I_y = 2(x_1^2 + x_2^2 + \dots + x_n^2)$$

The maximum pile load is

$$F_m = \pm \frac{M x_n}{I_y}$$

If the pile group is subjected to bending about both the XX and YY axes moments of inertia are calculated for each axis. The pile loads from bending are calculated for each axis as above and summed algebraically to give the resultant pile loads. The loads due to moment are combined with those due to vertical load.

(c) Horizontal load

In building foundations where the piles and pile cap are buried in the soil, horizontal loads can be resisted by friction, adhesion and passive resistance of the soil. Ground slabs that tie foundations together can be used to resist horizontal reactions due to rigid frame action and wind loads by friction and adhesion with the soil and so can relieve the pile group of horizontal load. However, in the case of isolated foundations in poor soil conditions where the soil may shrink away from the cap in dry weather or in wharves and jetties where the piles stand freely between the deck and the sea bed, the piles must be designed to resist horizontal load. Pile groups resist horizontal loads by

1. Bending in the piles
2. Using the horizontal component of the axial force in inclined piles

These cases are discussed below.

(d) Pile in bending

A group of vertical piles subjected to a horizontal force H applied at the top of the piles is shown in Fig. 11.37. The piles are assumed to be fixed at the top and bottom. The deflection of the pile cap is shown in Fig. 11.37(b).

$$\text{Shear per pile } V = H/N$$

$$\text{Moment } M_1 \text{ in each pile} = H h_1 / (2N)$$

where N is the number of piles and h_1 is the length of pile between fixed ends.

The horizontal force is applied at the top of the pile cap of depth h_2 and this causes a moment $H h_2$ at the pile tops. When vertical load and moment are also applied, the resultant pile loads are a combination of those caused by the three actions. The total vertical load P + W is distributed equally to the piles. The total moment $M + H\,h_2$ is resisted by vertical loads in the piles and the analysis is carried out as set out in section 11.5.2(b) above. The pile is designed as a reinforced concrete column subjected to axial load and moment. If the pile is clear between the cap and ground, additional moment due to slenderness may have to be taken into account. If the pile is in soil, complete or partial lateral support may be assumed.

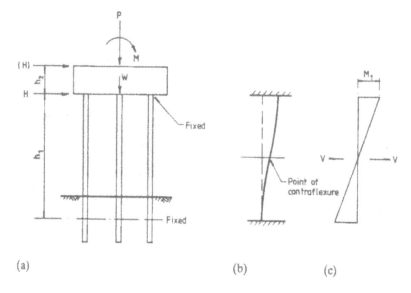

(a)   (b)   (c)

Fig. 11.37(a) Pile group; (b) deflection; (c) moment diagram.

(e) Resistance to horizontal load by inclined piles
An approximate method used to determine the loads in piles in a group subjected to axial load, moment and horizontal load where the horizontal load is resisted by inclined piles is set out. In Fig. 11.38 the foundation carries a vertical load *P*, moment *M* and horizontal load *H*. The weight of the pile cap is *W*.
  The loads *F* in the piles are calculated as follows.

(i) Vertical loads, pile loads F_v
The sum of vertical loads is P+W.

$$F_{v2} = F_{v3} = \frac{(P+W)}{8}, \quad F_{v1} = F_{v4} = \frac{(P+W)}{8}\frac{\sqrt{(R^2+1)}}{R}$$

(ii) Horizontal loads, pile loads F_H
The horizontal load is assumed to be resisted by pairs of inclined piles as shown in Fig. 11.38(b). The sum of the horizontal loads is *H*.

$$F_{H2} = F_{H3} = 0, \; -F_{H1} = F_{H4} = \frac{H}{4}\sqrt{(R^2+1)}$$

(iii) Moments, pile loads F_M

The second moment of area is

$$I_y = 2[(0.5S)^2 + (1.5S)^2]$$

The sum of the moments is

$$M' = M + Hh$$

$$-F_{M2} = F_{M3} = \frac{0.55S\,M'}{I_y}, \quad -F_{M1} = F_{M4} = \frac{1.55S\,M'}{I_y}\frac{\sqrt{(R^2+1)}}{R}$$

The maximum pile load is $F_{v4} + F_{H4} + F_{M4}$

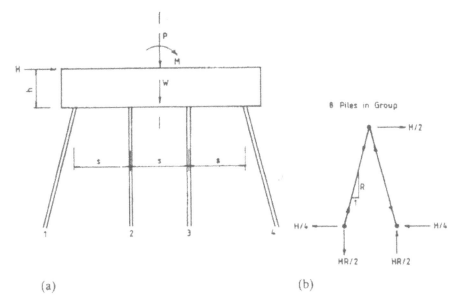

Fig. 11.38(a) Pile group; (b) resistance to horizontal load.

11.7.2.1 Example of Loads in Pile Group

The analysis using the approximate method set out above is given for a pile group to carry the loads and moment from a 6 m long shear wall similar to the one designed in Chapter 10, section 10.3.4.1.

The design actions for service loads are assumed to be as follows:

Axial load = 9592 kN
Moment = 5657 kNm
Horizontal load = 281 kN

The proposed pile group consisting of 18 piles inclined at 1 in 6 is shown in Fig. 11.39.

The weight of the base is 610 kN. For the vertical loads F_{v1} to F_{v6}

$$\frac{(9592 + 610)}{16} \frac{\sqrt{(6^2 + 1)}}{6} = 574.6 \, kN$$

For the horizontal loads:

$-F_{H1} = -F_{H2} = -F_{H3} = F_{H4} = F_{H5} = F_{H6} = 281 \times \sqrt{(6^2 + 1)}/18 = 94.9 \, kN$

The second moment of area is $2 \times (0.6^2 + 1.8^2 + 3.0^2) = 76.6$.

The moment at the pile top is $5657 + (281 \times 1.2) = 5994 \, kNm$.

$$-F_{M1} = F_{M6} = \frac{5994 \times 3}{76.6} \frac{\sqrt{(6^2 + 1)}}{6} = 238 \, kN$$

$$-F_{M2} = F_{M5} = \frac{5994 \times 1.8}{76.6} \frac{\sqrt{(6^2 + 1)}}{6} = 142.8 \, kN$$

$$-F_{M3} = F_{M4} = \frac{5994 \times 0.6}{76.6} \frac{\sqrt{(6^2 + 1)}}{6} = 47.6 \, kN$$

The maximum pile load is

$$F_6 = 574.6 + 94.9 + 238 = 907.5 \, kN$$

The pile group and pile cap shown in Fig. 11.35 can be analysed using a plane frame computer program. The large size of the cap in comparison with the piles ensures that it acts as a rigid member. The pile may be assumed to be pinned or fixed at the ends.

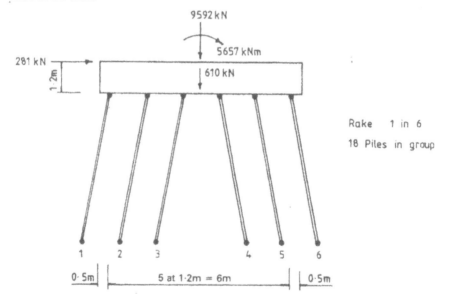

Fig. 11.39 Pile group.

11.7.3 Design of Pile Caps

Pile caps are 'deep beams'. In a deep beam, the ratio of span to depth is nearer unity. The bending stress distribution is complex and departs significantly from the linear stress distribution. The usual checks for bending and shear strengths are no longer valid. Pile caps are therefore designed using the Strut–tie method which is covered in section 6.5 of the Eurocode 2. In this method the forces are assumed to be resisted by a series of concrete struts and steel ties. The truss might be 2 D or 3 D triangulated form. Fig. 11.40 shows a simple strut–tie model for a small pile cap such as that shown in Fig. 11.30(a). In the diagram the struts are shown by heavy lines and ties by 'broken' lines. More details about the strut–tie method are given in Chapter 18.

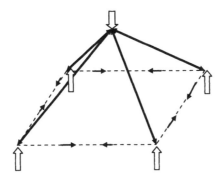

Fig. 11.40 Strut–tie model for a pile cap.

11.6 REFERENCES

Bond, Andrew and Harris, Andrew. (2008). *Decoding Eurocode 7*. Taylor & Francis.

Bowles, Joseph. E. (1995). *Foundation Analysis and Design*, 5th ed. McGraw-Hill.

Calavera, Jose. (2012). *Manual for Detailing Reinforced Concrete Structures to EC2*. Spon Press.

CHAPTER 12

RETAINING WALLS

12.1 WALL TYPES AND EARTH PRESSURE

12.1.1 Types of Retaining Walls

Retaining walls are structures used to retain mainly earth but also other materials which would not be able to stand vertically unsupported. The wall is subjected to overturning due to pressure of the retained material.

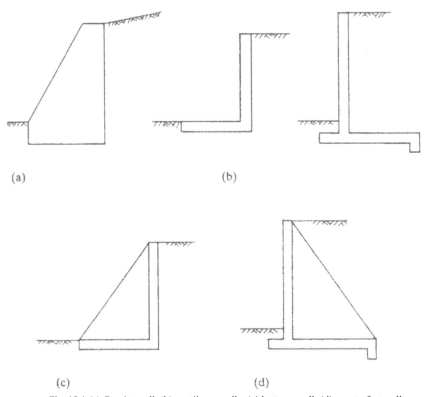

Fig. 12.1 (a) Gravity wall; (b) cantilever walls; (c) buttress wall; (d) counterfort wall.

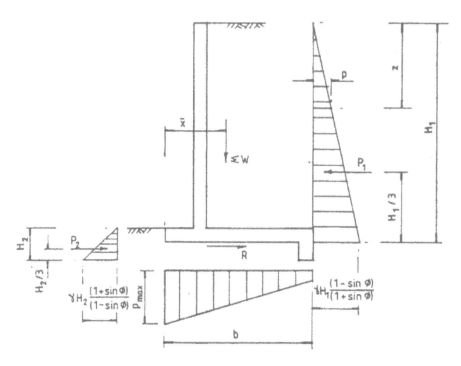

Fig. 12.2(a) Earth pressure: cohesion less soil (c = 0).

The types of retaining walls are as follows:

1.   In a **gravity wall** stability is provided by the weight of concrete in the wall.
2.   In a **cantilever wall** the wall slab acts as a vertical cantilever. Stability is provided by the weight of structure and earth on an inner base or the weight of the structure only when the base is constructed externally.
3.   In **counterfort and buttress walls** the vertical slab is supported on three sides by the base and counterforts or buttresses. Stability is provided by the weight of the structure in the case of the buttress wall and by the weight of the structure and earth on the base in the counterfort wall.

Examples of retaining walls are shown in Fig. 12.1. Detailed designs for cantilever and counterfort retaining walls are given.

12.1.2 Earth Pressure on Retaining Walls

The Eurocode governing the geotechnical aspects of retaining wall design is section 9 and Annex C of *BS EN 1997-1:2004: Eurocode 7: Geotechnical Design —Part 1: General Rules.*
A very useful reference on Eurocode 7 is Bond and Harri (2008).

(a) Active soil pressure

Active soil pressures are given for the two extreme cases of soil such as a cohesionless soil like sand and a cohesive soil like clay (Fig. 12.2). General formulae are available for intermediate cases. The formulae given apply to drained soils and reference should be made to textbooks on soil mechanics for pressure where the water table rises behind the wall. The soil pressures given are those due to a level backfill. If there is a surcharge of q kN/m² on the soil behind the wall, this is equivalent to an additional soil depth of z = q /γ where γ is the unit weight in kN/m³. The textbooks give solutions for cases where there is sloping backfill.

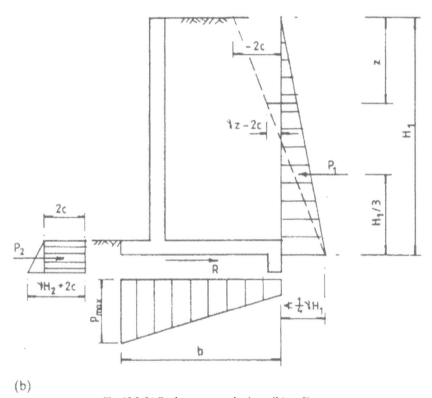

(b)

Fig. 12.2 (b) Earth pressure: cohesive soil ($\varphi = 0$).

(i) Cohesionless soil, c = 0 (Fig. 12.2(a)): The horizontal pressure at any depth z is given by

$$p = K_a\,(\gamma z + q)$$
$$K_a = \frac{(1-\sin\phi)}{(1+\sin\phi)}$$

where γ is the unit weight of soil in kN/m³, q = uniformly distributed surcharge in kN/m², φ is the angle of internal friction and K_a is the coefficient of active earth pressure.

The horizontal force P_1 on the wall of height H_1 is

$$P_1 = \frac{1}{2}K_a\gamma H_1^2 + K_aqH_1$$

Note Eurocode 7, Annex C, Fig. C.1.1 gives active earth pressure coefficient for horizontal surface with various values of friction between soil and wall.

(ii) Cohesive soil, $\varphi= 0$ (Fig. 12.2(b)): The pressure at any depth z is given theoretically by

$$p = \gamma z + q - 2c$$

where c is the cohesion at zero normal pressure. This expression gives negative values near the top of the wall. In practice there are cracks at the top of the soil normally filled with water.

(b) Wall stability against overturning
Referring to Fig. 12.2 the vertical loads are made up of the weight of the wall stem and base and the weight of backfill on the base. Front fill on the outer base has been neglected. Surcharge would need to be included if present. If the centre of gravity of these loads is x from the *toe* of the wall, the stabilizing moment with respect to overturning about the toe is ΣWx with a beneficial partial safety factory $\gamma_f= 1.0$. The overturning moment due to the active earth pressure is $\gamma_f\,P_1H_1/3$ with an adverse partial safety factor $\gamma_f = 1.4$. The stabilizing moment from passive earth pressure has been neglected. For the wall to satisfy the requirement of stability

$$\Sigma Wx \geq \gamma_f P_1 H_1/3$$

(c) Vertical pressure under the base
The vertical pressure under the base is calculated for service loads. For a 1 m length of cantilever wall with base width b transmitting forces to the foundation

Area $A = b$ m^2, section modulus $Z = b^2/6$ m^3.

If ΣM is the sum of the moments of all vertical forces ΣW about the centre of the base and of the active pressure on the wall, then

$$\Sigma M = \Sigma W(x - b/2) - P_1 H_1/3$$

where x = the centre of gravity of vertical loads *from* the *toe* of the wall.

The passive pressure in front of the base has again been neglected. The maximum pressure is

$$P_{max} = \frac{\Sigma W}{A} + \frac{\Sigma M}{Z}$$

This should not exceed the safe bearing pressure on the soil.

(d) Resistance to sliding (Fig. 12.2)
The resistance of the wall to sliding is as follows.

(i) Cohesionless soil: The friction R between the base and the soil is $\mu \Sigma W$, where μ is the coefficient of friction between the base and the soil ($\mu = \tan \varphi$). The

passive earth pressure force P_2 against the front of the wall from a depth H_2 of soil
is

$$P_2 = \frac{1}{2}\gamma H_2^2 \frac{(1+\sin\phi)}{(1-\sin\phi)}$$

Note Eurocode 7, Annex C, Fig. C.2.1 gives passive earth pressure coefficient for
horizontal surface with various values of friction between soil and wall.

(ii) Cohesive soils: The adhesion R between the base and the soil is βb where β is
the adhesion in kN/m^2. The passive earth pressure P_2 is

$$P_2 = 0.5\gamma H_2^2 + 2cH_2$$

A downstand nib can be added, as shown in Fig. 12.2 to increase the resistance
to sliding through passive earth pressure.
For the wall to be safe against sliding

$$1.5\,P_1 < P_2 + R$$

where P_1 is the horizontal active earth pressure on the wall.

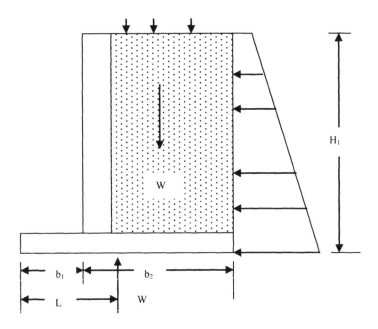

Fig. 12.3 Model for initial sizing.

12.2 DESIGN OF CANTILEVER WALLS

12.2.1 Initial Sizing of the Wall

Before a retaining wall can be designed, it is necessary to assume initial dimensions whose accuracy can be checked by detailed calculations. Often designers assume initial dimensions based on previous experience. However, initial dimensions of the wall can be determined from the following approximate equations.

Ignoring the difference in unit weight between soil and concrete and the weight of the toe slab of width b_1, for a unit length of wall the total gravity load W is approximately given by

$$W = \gamma\, b_2\, H_1 + q\, b_2$$

The total horizontal force P_1 is given by

$$P_1 = 0.5\, K_a \gamma\, H_1^2 + K_a\, q\, H_1$$

where q = surcharge in kN/m^2 and K_a = coefficient of active earth pressure.

(i) Resistance to sliding

Ignoring any contribution from passive earth pressure and using load factors of $\gamma_Q = 1.5$ on P_1 as it is an adverse load and $\gamma_{G,\,inf} = 1.0$ on W as it is a beneficial load, for resistance against sliding,

$$\mu\, W \ge 1.5\, P_1$$

Substituting for W and P_1,

$$\{\mu\, \gamma\, b_2\, H_1 + q\, b_2\} \ge (\gamma_Q = 1.5) \times \{0.5\, K_a\, \gamma\, H_1^2 + Ka\, q \times H_1\}$$

Simplifying,

$$\frac{b_2}{H_1}\{\mu + \frac{q}{\gamma H_1}\} \ge 1.5\, K_a\, \{0.5 + \frac{q}{\gamma H_1}\}$$

(ii) Zero tension in the base pressure

Taking moments about the toe of the wall,

$$W\,(b_1 + b_2/2) - 0.5\, K_a \gamma\, H_1^2\,(H_1/3) - K_a\, q\, H_1\,(H_1/2) = W\, L$$
$$L = b_1 + 0.5\, b_2 - b_2\, K_a\,(H_1/b_2)^2\{1/6 + q/\,(2\,\gamma H_1)\}/\,[1 + q/\,(\gamma H_1)]$$

Eccentricity e of W with respect to the centre of the base is

$$e = 0.5(b_1 + b_2) - L$$
$$e = (1/6)\, b_2\, K_a\,(H_1/b_2)^2\{1 + 3q/\,(\gamma H_1)\}/\,[1 + q/\,(\gamma H_1)] - 0.5\, b_1$$

For no tension to develop at the heel, W must lie in the middle third of the base. Therefore

$$e \le (b_1 + b_2)/6$$

$$\frac{b_1}{b_2} \ge 0.25\, K_a\, (\frac{H_1}{b_2})^2 [\frac{1 + 3\dfrac{q}{\gamma H_1}}{1 + \dfrac{q}{\gamma H_1}}] - 0.25$$

12.2.2 Design Procedure for a Cantilever Retaining Wall

For a given height of earth to be retained, the steps in the design of a cantilever retaining wall are as follows.

1. Assume a breadth for the base. This can be calculated from the equations developed in section 12.2.1. A nib is often required to increase resistance to sliding.
2. Calculate the horizontal earth pressure on the wall. Considering all forces, check stability against overturning and the vertical pressure under the base of the wall. Calculate the resistance to sliding and check that this is satisfactory. A load factor $\gamma_O = 1.5$ is applied to the horizontal loads for the overturning and sliding check. The maximum vertical pressure is calculated using service loads and this should not exceed the safe bearing pressure.
3. Reinforced concrete design for the wall is made for ultimate loads using appropriate load factors. Surcharge if present may be classed as either dead or imposed load depending on its nature.
Referring to Fig. 12.4, the structural design consists of the following.

(a) Cantilever wall: Calculate shear forces and moments caused by the horizontal earth pressure. Design the vertical moment steel for the inner (earth side) face and check the shear stresses. Minimum secondary steel is provided in the horizontal direction for the inner face and both vertically and horizontally for the outer face.

(b) Inner footing (heel slab): The net moment due to vertical loads on the top and earth pressure on the bottom face causes tension in the top and reinforcement is designed for this position.

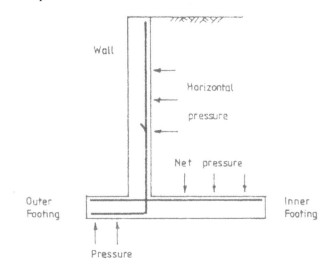

Fig. 12.4 Three parts of the cantilever retaining wall.

*(c) **Outer footing (toe slab):*** The moment due to earth pressure at the bottom face causes tension in the bottom face.

The moment reinforcement for the three parts is shown in Fig. 12.4.

12.2.3 Example of Design of a Cantilever Retaining Wall

(a) Specification
Design a cantilever retaining wall to support a bank of earth 3.5 m high. The top surface is horizontal behind the wall but it is subjected to a dead load surcharge of 15 kN/m^2. The soil behind the wall is well-drained sand with the following properties:

$$\text{Unit weight } \gamma = 18.0 \text{ kN/m}^3$$
$$\text{Angle of internal friction } \varphi = 30°$$

The material under the wall has a safe bearing pressure of 100 kN/m^2. The coefficient of friction μ between the base and the soil is 0.5. Design the wall using $f_{ck} = 30$ MPa concrete and $f_{yk} = 500$ MPa reinforcement.

Active earth pressure coefficient:
$$K_a = (1 - \sin \varphi)/ (1 + \sin \varphi) = (1 - 0.5)/ (1 + 0.5) = 0.3333$$

(b) Check preliminary sizing

(i) Check minimum stem thickness
For 1 m length of the wall, bending moment M at the base of the cantilever is
$$M = 0.5 \, K_a\gamma \, H^2 \, (H/3) + K_a \, q \, H \, (H/2)$$
Substituting $K_a = 0.333$, $\gamma = 18.0$ kN/m^3, $H = 3.5$ m, $q = 15$ kN/m^2,
$$M = 0.5 \times 0.3333 \times 18.0 \times 3.5^2 \times 3.5/3 + 0.3333 \times 15 \times 3.5 \times 3.5/2$$
$$M = 42.88 + 30.62 = 73.50 \text{ kN m/m}$$
In order for there to be no need for compression steel,
$$M < 0.196 \, bd^2 \, f_{ck}$$
Taking b = 1000 mm, $f_{cu} = 30$ N/mm^2,

$$d > \sqrt{\frac{M}{(0.196 \times b \times f_{ck})}} = \sqrt{\frac{73.50 \times 10^6}{0.196 \times 1000 \times 30}} = 112$$

Take a value of d much larger than this to reduce the amount of steel required. However it should not be so large that minimum steel requirement is greater than the calculated steel area. 0Assume total stem thickness of 250 mm. Same thickness is assumed for the base slab as well.

(ii) Check resistance to sliding
$$H_1 = 3.5 + 0.25 = 3.75 \text{ m,}$$
$$q/ (\gamma \, H_1) = 15.0/ (17.6 \times 3.75) = 0.227$$

$$K_a = 0.333$$
$$\mu = 0.5$$

Use equation from section 12.2.1.1, to calculate width b_2.

$$\frac{b_2}{H_1}\{\mu + \frac{q}{\gamma H_1}\} \geq 1.5 K_a \{0.5 + \frac{q}{\gamma H_1}\}$$

$$(b_2/H_1)\{0.5+0.227\} \geq 1.5 \times 0.333 \times (0.5+0.227)$$
$$b_2/H_1 \geq 0.50$$
$$b_2 \geq 1.875 \text{ m,}$$
$$\text{Take } b_2 = 2.05 \text{ m,}$$
$$b_2 / H_1 = 0.55$$

(iii) Check eccentricity

$$b_2 = 2.05,$$
$$H_1/b_2 = 1.829$$
$$q/ (\gamma H_1) = 0.227$$
$$K_a = 0.333$$

Use equation from section 12.2.1.1, to calculate width b_1.

$$\frac{b_1}{b_2} \geq 0.25 K_a (\frac{H_1}{b_2})^2 [\frac{1+3 \dfrac{q}{\gamma H_1}}{1+ \dfrac{q}{\gamma H_1}}] - 0.25$$

$$(b_1/b_2) \geq 0.25 \times 0.333 \times (1.829)^2 [\{1+ 3 \times 0.227\}/ \{1 + 0.227\}] - 0.25$$
$$(b_1/b_2) \geq 0.132,$$
$$b_1 \geq 0.27 \text{ m, Take } b_1 = 0.8 \text{ m.}$$

The proposed arrangement of the wall is shown in Fig. 12.5. Wall and base thicknesses are assumed to be 250 mm. A 0.6 m nib has been added under the wall to assist in the prevention of sliding.

(c) Wall stability
Consider 1 m length of wall. The horizontal pressure at depth z from the top is
$$p = K_a (\gamma z + q) = 0.333(18.0 z + 15)$$
The horizontal pressure at the base $(z = 3.75 \text{ m}) = 27.5 \text{ kN/m}^2$
The horizontal pressure at the top $(z = 0) = 5 \text{ kN/m}^2$.
The weight of wall, base and earth and the corresponding moments about the toe of the wall for stability calculations are given in Table 12.1. Clockwise moments are taken as positive.

(i) Maximum soil pressure
$$\text{Width of base } b = 2.85 \text{ m}$$
For 1 m length of wall, area
$$A = 2.85 \text{ m}^2$$
$$\text{Section modulus } Z = 2.85^2/6 = 1.35 \text{ m}^3$$

Taking moments of all forces about the toe A, the centroid of the base pressure from A is at a distance L.

$$L \times 185.40 = 324.31 - 87.89 = 236.42$$
$$L = 236.42/185.40 = 1.275 \text{ m}$$
$$\text{Eccentricity, } e = B/2 - L = 2.85/2 - 1.275 = 0.15 < 2.85/6$$

Hence no tension is developed at C.

The base is acted on by

$$\text{Vertical load} = 185.40 \text{ kN}$$
$$\text{Moment M} = 185.40 \times e = 27.81 \text{ kNm.}$$

The maximum soil pressure at end of the toe slab calculated for *service load* is

$$185.40/ (A = 2.85) + 27.81/ (Z = 1.35) = 8565 \text{ kN/m}^2 < 100 \text{ kN/m}^2$$

This is satisfactory, as the maximum pressure is less than the safe bearing capacity of soil.

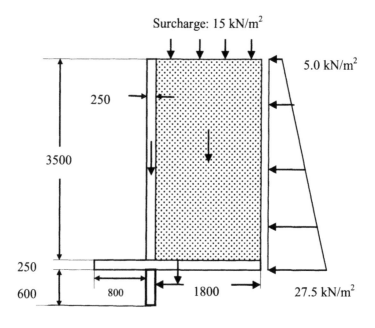

Surcharge: 15 kN/m²

5.0 kN/m²

250

3500

250

600   800   1800   27.5 kN/m²

Fig. 12.5 Forces acting on the retaining wall.

(ii) Stability against overturning
The stabilizing (beneficial) moment due to gravity loads about the toe A of the wall has a partial safety factor $\gamma_{G, \text{inf}} = 1.0$ and the disturbing (adverse) moment due to horizontal loads has a partial safety factor $\gamma_Q = 1.5$. The net stabilizing moment is

$$(324.31 \times 1.0 - 87.89 \times 1.5) = 192.48 > 0$$

The wall is considered safe against overturning.

(iii) Resistance to sliding

The forces resisting sliding are the friction under the base and the passive resistance for a depth of earth of 850 mm to the top of the base. The gravity loads are beneficial loads but the horizontal load is an adverse load. Ignoring the passive pressure, for the wall to be safe against sliding

$$(\mu = 0.5) \times \{(\gamma_{G, \, inf} = 1.0) \times 185.40\} > \{(\gamma_Q = 1.5) \times 60.0\},$$
$$\text{i.e. } 92.70 > 90.0$$

The resistance to sliding is satisfactory. There was no need for the nib but is included for additional protection. No reliance is placed on passive earth pressure.

Table 12.1 Stability calculations (cantilever wall)

Type of Load	Load (kN)	Distance to centroid from A, m	Moment about A (kNm)
HORIZONTAL (Active earth pressure)			
Surcharge	$5 \times 3.75 = 18.75$	$3.75/2 = 1.875$	-35.16
Triangular	$0.5 \times 3.75 \times (27.5 - 5) = 42.19$	$3.75/3 = 1.25$	-52.73
Σ	$18.75 + 42.19 = 60.94$		$-35.16 - 52.73 = 87.89$
VERTICAL (Gravity)			
Wall + Nib	$(3.75 + 0.6) \times 0.25 \times 25 = 27.19$	$0.8 + 0.25/2 = 0.925$	25.15
Base	$2.85 \times 0.25 \times 25 = 17.81$	$2.85/2 = 1.425$	25.38
Back fill	$1.8 \times 3.5 \times 18.0 = 113.40$	$0.8 + 0.25 + 1.8/2 = 1.95$	221.13
Surcharge	$15 \times 1.8 = 27$	$0.8 + 0.25 + 1.8/2 = 1.95$	52.65
Σ	185.40		324.31

(iv) Overall comment:
The wall section is satisfactory. The maximum soil pressure under the base controls the design.

(d) Structural design of wall, heel and toe slabs

(1) Cantilever wall slab

(a) Bending design:
At serviceability limit state, the horizontal pressure at the base ($z = 3.5$ m) is
$$p = K_a (\gamma z + q) = 0.333(18.0 \times 3.5 + 15) = 26.0 \text{ kN/m}^2$$
At the top ($z = 0$) is 5 kN/m^2.
$$\text{Average pressure} = 0.5 \times (26.0 + 5.0) = 15.50 \text{ kN.m}^2$$

At ultimate limit state using $\gamma_Q = 1.5$, at the base of the cantilever, shear force V and moment M are

$$V = 15.50 \times 3.5 \times (\gamma_f = 1.5) = 81.38 \text{ kN}$$
$$M = \{(26.0 - 5.0) \times 0.5 \times 3.5 \times 3.5/3 + 5.0 \times 3.5 \times 3.5/2\} \times (\gamma_Q = 1.5)$$
$$M = 110.25 \text{ kNm/m}$$

Assume that the cover is 40 mm and the diameter of the bars is 16mm. Effective depth d is

$$d = 250 - 40 - 8 = 202 \text{ mm}$$
$$k = M/(bd^2 f_{ck}) = 110.25 \times 10^6/(1000 \times 202^2 \times 30) = 0.09 < 0.196$$
$$\frac{z}{d} = 0.5[1.0 + \sqrt{(1 - 3\frac{k}{\eta})}]$$
$$k = 0.09, \eta = 1.0, z/d = 0.927$$
$$A_s = 110.25 \times 10^6/(0.927 \times 202 \times 0.87 \times 500) = 1353 \text{ mm}^2/\text{m}$$

Provide 12 mm diameter bars at 80 mm centres to give a steel area of $(\pi/4 \times 12^2) \times (1000/80) = 1414 \text{ mm}^2/\text{m}$.

Check minimum steel. From equation (9.1N) of the code,

$$A_{s, \min} = 0.26 \times (f_{ctm}/f_{yk}) \times b \times d \geq 0.0013\, b \times d$$
$$f_{ctm} = 0.3 \times f_{ck}^{0.67} = 0.3 \times 30^{0.67} = 2.9 \text{ MPa}, f_{yk} = 500 \text{ MPa},$$
$$b = 1000 \text{ mm}, d = 202 \text{ mm}$$
$$A_{s, \min} = 0.26 \times (2.9/500) \times 1000 \times 202 \geq 0.0013 \times 1000 \times 202$$
$$A_{s, \min} = 305 \text{ mm}^2$$

provided steel is greater than the minimum percentage of steel.

$$\text{Moment at SLS} = \text{Moment at ULS}/(\gamma_Q = 1.5) = 110.25/1.5 = 73.5 \text{ kNm/m}$$
$$\text{Stress in steel at SLS} = (M_{SLS}/M_{ULS}) \times (A_{s, \text{reqd}}/A_{s, \text{Provided}}) \times f_{yd}$$
$$= (73.5/110.25) \times (1353/1414) \times (500/1.15) = 277 \text{ MPa}$$

Check maximum bar diameter and spacing of steel permitted: For steel stress at SLS of 280 MPa and for a maximum crack width of 0.3 mm, maximum spacing allowed from code Table 7.3N is 150 mm and from code Table 7.2N maximum bar size is 12mm. The provided steel area satisfies both criteria.

(b) Curtailment of flexural steel

Determine the depth z from the top where the spacing of 12 mm bars can be doubled to 160 mm. Steel area at 160 mm c/c is equal to $(\pi/4 \times 12^2) \times (1000/160) = 707 \text{ mm}^2/\text{m}$.

The corresponding moment of resistance is approximately

$$M = 0.5 \times 110.25 = 55.13 \text{ kNm}$$

This moment occurs at a depth z from top given by

$$55.13 = 1.5 \times K_a (\gamma z^3/6 + 15 \times z^2/2)$$
$$55.13 = 1.5 \times 0.333 \times (18.0 \times z^3/6 + 15 \times z^2/2)$$

Solving by trial and error, z = 2.67 m,

$$d = 250 - 40 - 6 = 204 \text{ mm}$$
$$M = 55.13$$
$$k = 55.13 \times 10^6/(1000 \times 204^2 \times 30) = 0.044 < 0.196$$
$$\frac{z}{d} = 0.5[1.0 + \sqrt{(1 - 3\frac{k}{\eta})}]$$
$$k = 0.044, \eta = 1.0, z/d = 0.966$$

$$A_s = 55.13 \times 10^6 / (0.966 \times 202 \times 0.87 \times 500) = 650 \text{ mm}^2/\text{m}$$
$$A_s = 650 \text{ mm}^2/\text{m} > (A_{s, min} = 305 \text{ mm}^2/\text{m})$$

From Table 5.5, Chapter 5, the required anchorage length for $f_{ck} = 30$ MPa is 36 bar diameters which is equal to $36 \times 12 = 432$ mm. For anchorage requirements, bars are to extend an anchorage length beyond the theoretical cut off point. Therefore alternate bars need to continue up to a distance from top of

$$= 2670 - 432 = 2238 \text{ mm}$$

Stop bars off bars at a distance from base equal to

$$= 3500 - 2238 = 1262 \text{ mm, say } 1300 \text{ mm.}$$

(c) Shear check

At serviceability limit state, the horizontal pressure p at d from the base is

$$p = K_a (\gamma z + q) = 0.333[18.0 \times (3.5 - 0.202) + 15] = 24.80 \text{ kN/m}^2$$
$$\text{At the top } (z = 0). \quad p = 0.333[18.0 \times 0 + 15] = 5 \text{ kN/m}^2$$
$$\text{Average pressure} = 0.5 \times (24.80 + 5.0) = 14.90 \text{ kN.m}^2$$

At ultimate limit state using $\gamma_Q = 1.5$, at d from the base of the cantilever, shear force V_{Ed} is

$$V_{Ed} = 14.90 \times (3.5 - 0.202) \times (\gamma_Q = 1.5) = 73.71 \text{ kN/m}$$
$$V_{Ed} = 73.71 \times 10^3 / (1000 \times 202) = 0.37 \text{ MPa}$$
$$\rho_1 = A_{sl} / (b_w \, d) \le 0.02$$
$$100 \, \rho_1 = 100 \times 1414 / (1000 \times 202) = 0.70 < 2.0$$
$$C_{Rd, c} = 0.18 / (\gamma_c = 1.5) = 0.12, \, k = 1 + \sqrt{(200/202)} = 1.995 \le 2.0,$$
$$V_{Rd, c} = C_{Rd, c} \times k \times (100 \times \rho_1 \times f_{ck})^{0.3333} \ge (v_{min} = 0.035 \times k^{1.5} \times \sqrt{f_{ck}})$$
$$= 0.12 \times 1.995 \times (0.70 \times 30)^{0.3333} \ge (v_{min} = 0.035 \times 1.995^{1.5} \times \sqrt{30})$$
$$= 0.66 \ge 0.54$$
$$V_{Rd, c} = 0.66 \text{ MPa} > (v_{Ed} = 0.37 \text{ MPa})$$

The shear stress is satisfactory.

(d) Distribution steel

Clause 9.3.1(2) of the code recommends that in slabs, secondary reinforcement of not less than 20% of the principal reinforcement and at a spacing less than or equal to 3.5 h or 400 mm whichever is lesser.
In this case main steel is 1413 mm²/m. 20% of this value is 283 mm²/m. 10 mm bars at 275 mm provide a steel area of 286 mm²/m. h = 250 mm, 3.5 h = 875 mm. For crack control on the outer face, provide 10 mm diameter bars at 275 mm centres each way. For ease of construction it is better to provide steel fabric. From Table 8.3, Chapter 8, a square mesh A393 10 mm bars at 200 mm centre both ways provides a steel area of 393 mm²/m.

(2) Inner footing (heel slab)

In order to determine the appropriate load factors to be used, it is necessary to consider the effect of gravity loads and earth pressure loads on the bending moment caused in the heel slab.

From Table 12.1, gravity loads provide:

Vertical load = 184.40 kN,
Moment about the Toe A = 324.31 kNm (Clockwise)

The centroid of the base pressure due to *gravity loads only* from A is at a distance L.

$$L \times 184.40 = 324.31, \text{ giving } L = 1.76 \text{ m},$$
$$\text{Eccentricity, } e = 2.85/2 - L = -0.334$$

The base is acted on by a

$$\text{Vertical load} = 184.40 \text{ kN}$$
$$\text{Moment } M = 184.40 \times e = 61.60 \text{ kNm (clockwise)}$$
$$184.40 / (A = 2.85) = 64.70 \text{ kN/m}^2$$
$$M / (Z = 1.35) = 45.62 \text{ kN/m}^2$$

On the top of the heel slab there is surcharge of 15 kN/m^2 and a height of soil equal to 3.5 m and self weight of slab of 250 mm. The total downward load is

$$\{15 + 3.5 \times (\gamma = 18.0) + 0.25 \times 25\} = 84.25 \text{ kN/m}^2$$

The bending moment on the base due to horizontal earth pressure is

$$M = 87.89 \text{ kNm/m (anticlockwise)}$$
$$M / (Z = 1.35) = 65.10 \text{ kN/m}^2$$

The pressures due to gravity loads and horizontal earth pressure are shown in Fig. 12.6.

The net effect of gravity loads is to produce tension on the bottom of the slab while the base pressure due to horizontal loads produces tension on the top of the slab. Therefore gravity loads are beneficial loads with a load factor of $\gamma_{G, \text{ inf}} = 1.0$ while earth pressure loads are adverse with a load factor of $\gamma_Q = 1.5$ to be applied. Using these load factors, the base pressure at right and left ends of the base slab are

$$\text{Left end} = 64.70 - 45.62 + 65.10 \times 1.5 = 116.73 \text{ kN/m}^2$$
$$\text{Right end} = 64.70 + 45.62 - 65.10 \times 1.5 = 12.67 \text{ kN/m}^2$$

The base pressure at the junction of the heel slab and cantilever is

$$= 12.67 + (116.73 - 12.67) \times (1.8/2.85) = 78.20 \text{ kN/m}^2$$

Fig. 12.7 shows the forces acting on the heel slab.

(a) Bending design

Referring to Fig. 12.7, moment M at the face of the wall is

$$M = 0.5 \times (84.25 - 12.67) \times 1.8^2 - 0.5 \times (78.20 - 12.67) \times 1.8 \times 1.8/3$$
$$= 80.57 \text{ kN m/m}$$
$$k = M / (bd^2 f_{ck}) = 80.57 \times 10^6 / (1000 \times 202^2 \times 30) = 0.066 < 0.196$$
$$\frac{z}{d} = 0.5[1.0 + \sqrt{(1 - 3\frac{k}{\eta})}]$$
$$k = 0.044, \eta = 1.0, z/d = 0.948$$
$$A_s = 80.57 \times 10^6 / (0.948 \times 202 \times 0.87 \times 500) = 967 \text{ mm}^2/\text{m}$$

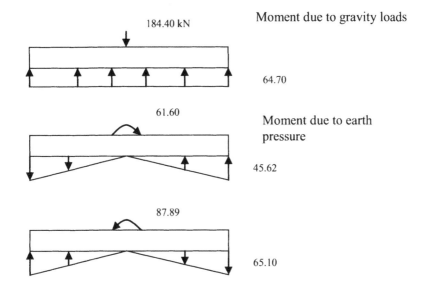

184.40 kN

Moment due to gravity loads

64.70

61.60

Moment due to earth pressure

45.62

87.89

65.10

Fig. 12.6 Forces on the base slab due to gravity and earth pressure forces.

Check for minimum steel from equation (9.1N) of the code.

$$A_{s, min} = 0.26 \times (f_{ctm}/f_{yk}) \times b \times d \geq 0.0013\ b \times d$$
$$f_{ctm} = 0.3 \times f_{ck}^{0.67} = 0.3 \times 30^{0.67} = 2.9 \text{ MPa}, f_{yk} = 500 \text{ MPa},$$
$$b = 1000 \text{ mm}, d = 202 \text{ mm}$$
$$A_{s, min} = 0.26 \times (2.9/500) \times 1000 \times 202 \geq 0.0013 \times 1000 \times 202$$
$$A_{s, min} = 305 \text{ mm}^2$$

provided steel is greater than the minimum percentage of steel.
Provide 12 mm bars at 100 mm centre. $A_s = 1130 \text{ mm}^2/\text{m}$.

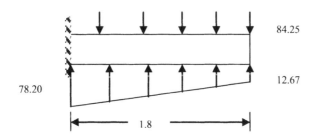

84.25

12.67

78.20

1.8

Fig. 12.7 Pressures in kN/m² acting on heel slab at ULS.

Moment at SLS: Using unit load factors for all loads,
Left end = 64.70 − 45.62 + 65.10 = 84.18 kN/m²
Right end = 64.70 + 45.62 − 65.10 = 45.22 kN/m²
The base pressure at the junction of the heel slab and cantilever is
45.22 + (84.18 − 45.22) × (1.8/2.85) = 69.83 kN/m²

Fig. 12.8 shows the forces acting on the heel slab.

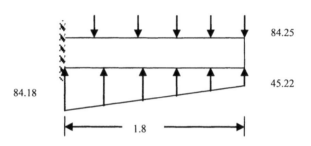

Fig. 12.8 Pressures in kN/m² acting on heel slab at SLS.

Referring to Fig. 12.8, moment M at the face of the wall is
$$M = 0.5 \times (84.25 - 45.22) \times 1.8^2 - 0.5 \times (84.18 - 45.22) \times 1.8 \times 1.8/3$$
$$= 42.54 \text{ kN m/m}$$
Stress in steel at SLS $= (M_{SLS}/M_{ULS}) \times (A_{s, \text{reqd}}/A_{s, \text{Provided}}) \times f_{yd}$
$$= (42.54/80.57) \times (967/1130) \times (500/1.15) = 197 \text{ MPa}$$
Check maximum bar diameter and spacing of steel permitted. For steel stress at SLS of 200 MPa and for a maximum crack width of 0.3 mm, maximum spacing allowed from code Table 7.3N is 250 mm and from code Table 7.2N maximum bar size is 25 mm. The provided steel area satisfies both criteria.

(b) Shear Check
The base pressure at d from the junction of the heel slab and cantilever is
$$= 12.67 + (116.73 - 12.67) \times \{(1.8 - 0.202)/2.85\} = 71.02 \text{ kN/m}^2$$
Referring to Fig. 12.7, the shear force V_{Ed} at d from the base slab-wall junction is
$$V_{Ed} = \{(84.25 - 12.67) - 0.5 \times (71.02 - 12.67)\} \times (1.8 - 0.202)$$
$$= 67.76 \text{ kN /m}$$
$$v_{Ed} = 67.76 \times 10^3/ (1000 \times 202) = 0.32 \text{ MPa}$$
$$\rho_1 = A_{sl} / (b_w d) \leq 0.02$$
$$100 \, \rho_1 = 100 \times 1130/ (1000 \times 202) = 0.56 < 2.0$$
$$C_{Rd, c} = 0.18/ (\gamma_c = 1.5) = 0.12, \ k = 1 + \sqrt{(200/202)} = 1.995 \leq 2.0,$$
$$v_{Rd, c} = C_{Rd, c} \times k \times (100 \times \rho_1 \times f_{ck})^{0.3333} \geq (v_{min} = 0.035 \times k^{1.5} \times \sqrt{f_{ck}})$$
$$= 0.12 \times 1.995 \times (0.56 \times 30)^{0.3333} \geq (v_{min} = 0.035 \times 1.995^{1.5} \times \sqrt{30})$$
$$= 0.61 \geq 0.54$$
$$v_{Rd, c} = 0.61 \text{ MPa} > (v_{Ed} = 0.32 \text{ MPa})$$
The shear stress is satisfactory.

(c) Distribution steel
Clause 9.3.1(2) of the code recommends that in slabs, secondary reinforcement of not less than 20% of the principal reinforcement at a spacing less than or equal to 3.5 h or 400 mm whichever is lesser.
In this case main steel is 1130 mm²/m. 20% of this value is 226 mm²/m. H10 bars at 325 mm provide a steel area of 242 mm²/m. h = 250 mm, 3.5 h = 875 mm.

For crack control on the outer face provide H10 bars at 325 mm centres each way. For ease of construction it is better to provide steel fabric. From Table 8.3, Chapter 8, a square mesh A393 H10 bars at 200 mm centre both ways provides a steel area of 393 mm²/m.

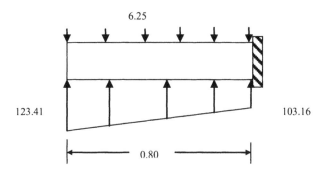

6.25

123.41   103.16

0.80

Fig. 12.9 Pressures in kN/m² acting on toe slab.

(3) Outer Footing (toe slab)

As shown in Fig. 12.6, both gravity and horizontal loads acting on the base slab produce tension on the bottom of the slab. Therefore both loads are adverse and take a load factor of $\gamma_{G, sup} = 1.35$ and $\gamma_Q = 1.50$. The only beneficial load is due to self weight. Using these load factors, the base pressure at right and left ends of the base slab are

Left end $= (64.70 - 45.62) \times 1.35 + 65.10 \times 1.5 = 123.41$ kN/m²
Right end $= (64.70 + 45.62) \times 1.35 - 65.10 \times 1.5 = 51.28$ kN/m²

The base pressure at the junction of the toe slab and cantilever

$= 51.25 + (123.41 - 51.25) \times (2.85 - 0.80)/2.85 = 103.16$ kN/m²
Self weight load $= 0.25 \times 25 = 6.25$ kN/mm²

Fig. 12.9 shows the forces acting on the toe slab.

The moment at the face of the wall is:

$M = 0.5 \times (103.16 - 6.25) \times 0.8^2 + 0.5 \times (123.41 - 103.16) \times 0.8 \times (2/3) \times 0.8$
$= 35.33$ kN m/m

Reinforcement from the wall which is designed for a moment of 110.25 kNm/m will be anchored in the toe slab and will provide the moment steel here. From Table 5.5, Chapter 5, the required anchorage length for $f_{ck} = 30$ MPa is 36 bar diameters which is equal to $36 \times 12 = 432$ mm. This will be provided by the bend and a straight length of bar along the toe slab.

Shear stress:

The base pressure p at d from the junction of the toe slab and cantilever s

$p = 51.25 + (123.41 - 51.25) \times (2.85 - 0.80 + 0.202)/2.85 = 108.27$ kN/m²
$V_{Ed} = \{(108.27 - 6.25) + (123.41 - 108.27) \times 0.5\} \times (0.8 - 0.202) = 65.35$ kN

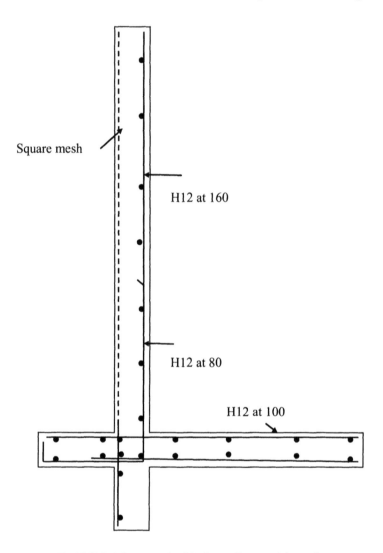

Fig. 12.10 Reinforcement detail in the cantilever retaining wall.

The flexural steel and the dimensions of the toe slab are same as for the stem which is safe for shear force of 73.71 kN. This is satisfactory. The distribution steel is 10 mm diameter bars at 240 mm centres.

(4) Nib
The passive earth pressure coefficient $K_p = 1/K_a = 3.0$.
The earth pressures at the top and bottom of the nib are
$$\text{Top: } K_p \, \gamma \, z = 3 \times 18.0 \times 0.25 = 13.5 \text{ kN/m}^2$$
$$\text{Bottom: } K_p \, \gamma \, z = 3 \times 18.0 \times 0.85 = 45.90 \text{ kN/m}^2$$
Referring to Fig. 12.5 the shear and moment in the nib using a load factor of

$\gamma_Q = 1.5$ are as follows:
$$V = 1.5 \times (13.5 + 45.90) \times 0.6/2 = 26.73\text{kN}$$
$$M = 1.5 \times \{13.5 \times 0.6^2/2 + (45.90 - 13.5) \times 0.5 \times 0.6 \times (2/3) \times 0.6\}$$
$$= 9.48 \text{ kNm/m}$$

The values are quite small. The minimum reinforcement is 305 mm²/m. Provide 10 mm diameter bars at 250 mm centres ($A_s = 314$ mm²/m) to lap onto the main wall steel. The distribution steel is 10 mm diameter bars at 250 mm centres.

Sketch of the wall reinforcement
A sketch of the wall with the reinforcement designed above is shown in Fig. 12.10. Note that the reinforcement is organized to produce a 3-D cage which can be easily fabricated.

12.3 COUNTERFORT RETAINING WALLS

12.3.1 Stability Check and Design Procedure

A counterfort retaining wall is shown in Fig. 12.10. The spacing of the counterforts is usually made equal to the height of the wall. The following comments are made regarding the design.

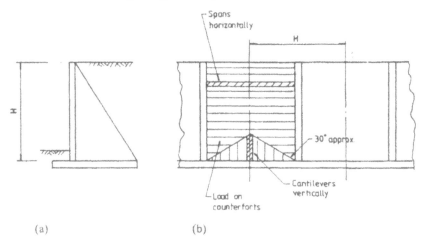

Fig. 12.10 (a) Section: (b) back of wall.

(a) Stability
Consider as one unit a centre-to-centre length of panels taking into account the weight of the counterfort. The horizontal earth acting on this unit together with the gravity loads must provide satisfactory resistance to overturning and sliding. The calculations are made in a similar way to those for a cantilever wall.

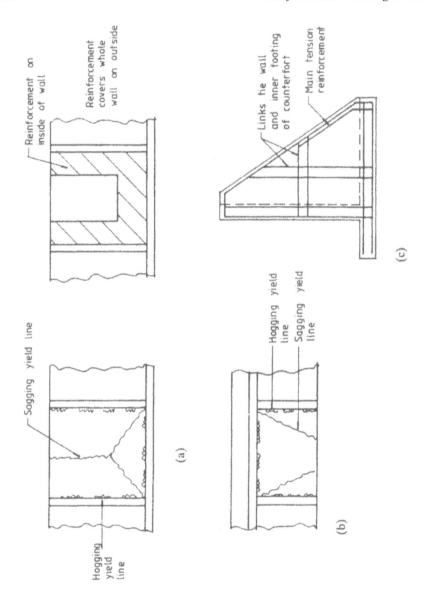

Fig. 12.11 Counterfort wall: (a) Yield line pattern and reinforcement in wall; (b) yield line pattern in base slab; (c) reinforcement in counterfort.

(b) Wall slab

The slab is thinner than that required for a cantilever wall. It is built in on three edges and free at the top. It is subjected to a triangular load due to the active earth pressure. The lower part of the wall cantilevers vertically from the base and the upper part spans horizontally between the counterforts. A load distribution commonly adopted between vertically and horizontally spanning elements is shown in Fig. 12.10. The finite element method could also be used to analyse the wall to determine the moments for design. Yield line analysis and Hillerborg's strip methods are used in the example that follows.

(c) Base slab

Like the wall slab, the base slab behind the vertical wall is built-in on three sides and free on the fourth. The loading is trapezoidal in distribution across the base due to the net effect of the weight of earth down and earth pressure under the base acting upwards. As in the case of the wall slab, near the junction with the wall, the forces are resisted by cantilever action while away from this junction, the load is resisted by beam action with the strips spanning between the counterforts. Like the wall slab, the moments in the base slab can be determined using yield line analysis or Hillerborg's strip methods.

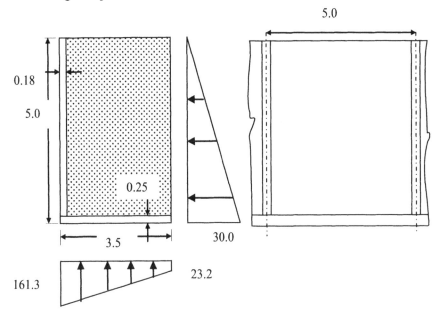

Fig. 12.12 Forces acting on the structure at SLS.

(d) Outer footing (toe slab)

If provided, it is designed as a cantilever in a manner similar to cantilever retaining wall.

(e) Counterforts

Counterforts support the wall and base slabs and are designed as vertical cantilevers of varying T-beam sections. The load on the counterforts is from the wall slab spanning between the counterforts. A design is made for the section at the base and one or more sections up the height of the counterfort. Link reinforcement must be provided between the wall slab and inner base slab and the counterfort to transfer the loading. Reinforcement for the counterfort is shown in Fig. 12.11(c).

12.3.2 Example of Design of a Counterfort Retaining Wall

(a) Specification
A counterfort retaining wall has a height from the top to the underside of the base of 5 m and a spacing of counterforts of 5 m. The backfill is level with the top of the wall. The earth in the backfill is granular with the following properties:

Unit weight $\gamma = 18.0$ kN/m^3

Angle of internal friction $\varphi = 30°$

Coefficient of active earth pressure $K_a = 0.333$

Coefficient of friction between the soil and concrete $\mu = 0.5$

Safe bearing pressure of the soil under the base $= 170$ kN/m^2

The construction materials are $f_{ck} = 30$ MPa concrete and $f_{yk} = 500$ MPa reinforcement.

(b) Trial section
The proposed section for the counterfort retaining wall is shown in Fig. 12.12. Wall slab is made 180 mm thick and the counterfort and base slab are both 250 mm thick.

(c) Stability
Consider a 5 m length of wall centre to centre of counterforts. The horizontal earth pressure at depth z is

$K_a \, \gamma \, z = 0.333 \times 18.0 \times z = 6.0 \, z$ kN/m^2

The pressure at $z = 5$ m is 30.0 kN/m^2

The loads are shown in Fig. 12.12. The stability calculations are given in Table 12.2. Clockwise moments are considered as positive.

(i) Maximum soil pressure
The properties of the base are as follows:

Area $A = 3.5 \times 5 = 17.5$ m^2

Section modulus $Z = 5 \times 3.5^2/6 = 10.21$ m^3

Taking moments of all forces about the toe A, the centroid of the base pressure from A is at a distance L.

$L \times 1613.88 = 2745.54 - 626.25$, L = 1.31 m

Eccentricity, e $= 3.5/2 - L = 0.44 < 3.5/6$

No tension is developed at C.

The base is acted on by

Vertical load $= 1613.88$ kN, moment $= 1613.88 \times$ e $= 705.0$ kNm.

The maximum soil pressure at A calculated for *service load* is

$1613.88 / (A = 17.5) + 705.0 / (Z = 10.21) = 161.27$ kN/m$^2 < 170$ kN/m^2

The minimum soil pressure *calculated* for *service load* is

$1613.88 / (A = 17.5) - 705.0 / (Z = 10.21) = 23.17$ kN/m^2

Width of base is sufficient to prevent bearing capacity failure.

(ii) Stability against overturning
The stabilizing moment due to gravity loads about the toe A of the wall has a partial safety factor $\gamma_{G,\,inf} = 1.0$ and the disturbing moment due to horizontal loads has a partial safety factor $\gamma_Q = 1.5$.
$$(2745.54 \times 1.0 - 626.25 \times 1.5) = 1806 > 0$$
The wall is very safe against overturning.

(iii) Resistance to sliding
The forces resisting sliding are the friction under the base. For the wall to be safe against sliding
$$(\mu = 0.5)\ \{1613.88 \times (\gamma_{G,\,inf} = 1.0)\} > (\gamma_Q = 1.5) \times 375.0$$
$$806.94 > 562.5$$
The resistance to sliding is satisfactory.

Table 12.2 Stability calculations for counterfort wall (all loads characteristic)

Type of Load	Load, (kN)	Distance to centroid from A, (m)	Moment about A, (kNm)
HORIZONTAL (Active earth pressure)			
Triangular	$0.5 \times 5 \times 5 \times 30.0 =$ 375.00	$5/3 = 1.67$	-626.25
VERTICAL (Gravity)			
Wall	$5 \times 0.18 \times (5.0 - 0.25) \times$ $25 = 106.88$	$0.18/2 = 0.09$	9.62
Base	$5 \times 0.25 \times 3.5 \times 25 =$ 109.38	$3.5/2 = 1.75$	191.41
Back fill	$(5.0 - 0.25) \times (3.5 - 0.18)$ $\times (5.0 - 0.25) \times 18.0 =$ 1348.34	$0.18 + (3.5 - 0.18)/2$ $= 1.84$	2480.94
Counterfort	$0.5 \times (3.5 - 0.18) \times$ $(5.0 - 0.25) \times 0.25 \times 25 =$ 49.28	$0.18 + (3.5 - 0.18)/3 = 1.29$	63.57
Σ	1613.88		2745.54

(iv) Overall comment
The wall section is satisfactory. The maximum soil pressure under the base controls the design.

12.3.3 Design of Wall Slab Using Yield Line Method

Mode 1 yield line mechanism
The yield line solution is given for a square wall with a triangular load. The yield line pattern for Mode 1 is shown in Fig. 12.13. Parameter α, locating point F, controls the collapse pattern. Deflection at F is Δ. It is assumed that the slab will be isotropically reinforced with moment of resistance for both positive (tension on the outer face) and negative (tension on the earth face) being equal to m.

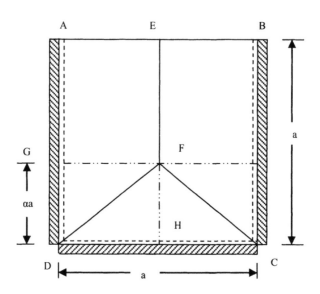

Fig. 12.13 Mode 1 yield line pattern for wall slab.

(i) Energy dissipated in the yield lines

1. Rigid regions AEFD and BEFC
Both rotate about y-axis only.
$\quad$ (i) Negative yield lines: $\ell_y = a$, $m_y = m$, $\theta_y = \Delta / (0.5a)$
$\quad$ (ii) Positive yield lines: $\ell_y = a$, $m_y = m$, $\theta_y = \Delta / (0.5a)$
Energy dissipated
$$E_1 = 2\{m \times a \times \Delta / (0.5a) + m \times a \times \Delta / (0.5a)\} = 8\ m\ \Delta$$

2. Rigid region DFC
Rotates about x-axis only.
$\quad$ (i) Negative yield lines: $\ell_x = a$, $m_x = m$, $\theta_x = \Delta / (\alpha a)$
$\quad$ (ii) Positive yield lines: $\ell_x = a$, $m_x = m$, $\theta_x = \Delta / (\alpha a)$
Energy dissipated
$$E_2 = \{m \times a \times \Delta / (\alpha\ a) + m \times a \times \Delta / (\alpha\ a)\} = (2/\alpha)\ m\ \Delta$$

3. Total energy dissipated

$$E = E_1 + E_2 = (8 + 2/\alpha)\ m\ \Delta$$

(ii) External work done

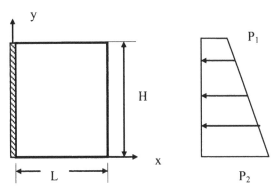

Fig. 12.14 Rectangular region.

Case 1: Rectangular region with rotation about the y-axis, Fig. 12.14. Deflection is Δ at the edge L from the y-axis.

Pressure distribution $p = p_2 - (p_2 - p_1)\ (y/H)$, rotation $\theta_y = \Delta/L$

Force F on an element $dx \times dy = p \times dx \times dy$

Displacement of the force $F = \theta_y \times x$

Total work done $W = \iint \theta_y\ x\ p\ dx\ dy$

Limits for x = 0 and L, limits for y = 0 and H

$$W = \frac{\Delta}{L} \int_0^L x\ dx \int_0^H \left\{ p_2 - (p_2 - p_1)\ \frac{y}{H} \right\} dy$$

Carrying out the integration: $W = \Delta \times HL \times \frac{(p_1 + p_2)}{4}$

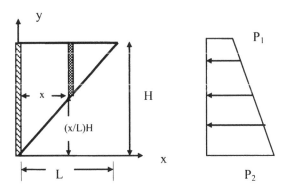

Fig. 12.15 Triangular region.

Case 2: Triangular region with rotation about the y-axis, Fig. 12.15. Deflection is Δ at the edge L from the y-axis.

Pressure distribution $p = p_2 - (p_2 - p_1)(y/H)$, rotation $\theta_y = \Delta/L$

Force F on an element $dx \times dy = p \times dx \times dy$

Displacement of the force $F = \theta_y \times x$

Total work done $W = \iint \theta_y \, x \, p \, dx \, dy$

Limits for $x = 0$ and L, limits for $y = (x/L)H$ and H

$$W = \frac{\Delta}{L} \int_0^L x \, dx \int_{\frac{x}{L}H}^{H} \{p_2 - (p_2 - p_1)\frac{y}{H}\} \, dy$$

$$W = \Delta \frac{H}{L} \int_0^L x \, [\, \{p_2 \left(1 - \frac{x}{L}\right) - \frac{(p_2 - p_1)}{2}\left(1 - \frac{x^2}{L^2}\right)\}] \, dx$$

Carrying out the integration, $W = \Delta \times H L \times \frac{(3p_1 + p_2)}{24}$

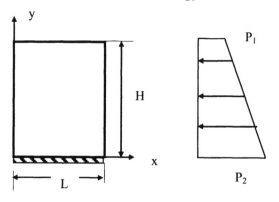

y

H

L

x

P_1

P_2

Fig. 12.16 Rectangular region.

Case 3: Rectangular region with rotation about the x-axis, Fig. 12.16. Deflection is Δ at the edge H from the x-axis.

Pressure distribution $p = p_2 - (p_2 - p_1)(y/H)$, rotation $\theta_x = \Delta/H$

Force F on an element $dx \times dy = p \times dx \times dy$

Displacement of the force $F = \theta_x \times y$

Total work done $W = \iint \theta_x \, y \, p \, dx \, dy$

Limits for $x = 0$ and L, limits for $y = 0$ and H

$$W = \frac{\Delta}{H} \int_0^L dx \int_0^H \{p_2 - (p_2 - p_1)\frac{y}{H}\} y \, dy$$

Carrying out the integration, $W = \Delta \times H L \times \frac{(2p_1 + p_2)}{6}$

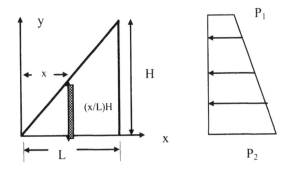

Fig. 12.17 Triangular region.

Case 4: Triangular region with rotation about the x-axis, Fig. 12.17. Deflection is Δ at the edge H from the x-axis.

Pressure distribution $p = p_2 - (p_2 - p_1)(y/H)$, rotation $\theta_x = \Delta/H$

Force F on an element $dx \times dy = p \times dx \times dy$

Displacement of the force $F = \theta_x \times y$

Total work done $W = \iint \theta_x \, y \, p \, dx \, dy$

Limits for $x = 0$ and L, limits for $y = 0$ and $(x/L)H$

$$W = \frac{\Delta}{H} \int_0^L dx \int_0^{\frac{x}{L}H} \{p_2 - (p_2 - p_1)\frac{y}{H}\} y \, dy$$

Carrying out the integration, $W = \Delta \times H L \times \dfrac{(p_1 + p_2)}{12}$

Using the above integrals, the work done in various regions can be calculated.

(i) Rigid region AEGF: From Fig. 12.13,

Substituting $L = a/2$, $H = a(1 - \alpha)$. $p_1 = 0$ and $p_2 = H\gamma$

$$W_1 = \Delta \times H L \times \frac{(p_1 + p_2)}{4} = \Delta \gamma a^3 \frac{(1-\alpha)^2}{8}$$

(ii) Rigid region GFD: From Fig. 12.14,

Substituting $L = a/2$, $H = a\alpha$, $p_1 = (1 - \alpha)a\gamma$ and $p_2 = a\gamma$

$$W_2 = \Delta \times H L \times \frac{(3p_1 + p_2)}{24} = \Delta \gamma a^3 \frac{(4\alpha + 3\alpha^2)}{48}$$

(iii) Rigid region FDH: From Fig. 12.16,

Substituting $L = a/2$, $H = a\alpha$, $p_1 = (1 - \alpha)a\gamma$ and $p_2 = a\gamma$

$$W_3 = \Delta \times H L \times \frac{(p_1 + p_2)}{12} = \Delta \gamma a^3 \frac{(2\alpha - \alpha^2)}{24}$$

4. Total work done W

$$W = 2(W_1 + W_2 + W_3)$$

$$W = \frac{\gamma \Delta}{24} a^3 (6 - 4\alpha + \alpha^2)$$

5. Moment m

Equating the work done by external loads to the energy dissipated at the yield lines,

$$m = \frac{\gamma a^3}{48} \frac{(6\alpha - 4\alpha^2 + \alpha^3)}{(4\alpha + 1)}$$

For a maximum m, dm/dα = 0

$$(4\alpha + 1)(6 - 8\alpha + 3\alpha^2) - 4(6\alpha - 4\alpha^2 + \alpha^3) = 0$$

Simplifying, $6 - 8\alpha - 13\alpha^2 + 8\alpha^3 = 0$, α = 0.483312

$$m = 0.014762 \ \gamma a^3,$$

Substituting a = 5.0 m, γ = 18.0,

$$m = 33.22 \ \text{kNm/m}$$

Using a load factor γ_Q on the earth pressure of 1.5 and also increasing the calculated moment by 10% to account for the formation of corner levers,

$$m = (33.22 \times 1.5) \times 1.1 = 54.80 \ \text{kNm/m}$$

Mode 2 yield line mechanism

The yield line pattern for Mode 1 is shown in Fig. 12.18. Parameter α, locating point F, controls the collapse pattern. Deflections at F and E are Δ. It is assumed that the slab will be isotropically reinforced with moment of resistance for both positive and negative being equal to m.

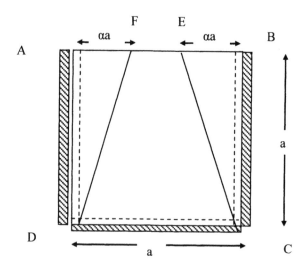

Fig. 12.18 Mode 2 yield line pattern for wall slab.

(i) Energy dissipated in the yield lines

1. Rigid region AFD and BEC

Both rotate about y-axis only.

(i) Negative yield lines: $\ell_y = a$, $m_y = m$, $\theta_y = \Delta/(\alpha\, a)$
(ii) Positive yield lines: $\ell_y = a$, $m_y = m$, $\theta_y = \Delta/(\alpha\, a)$

Energy dissipated E_1 is

$$E_1 = 2\{m \times a \times \Delta/(\alpha\, a) + m \times a \times \Delta/(\alpha\, a)\} = 4\,(m/\alpha)\,\Delta$$

2. Rigid region DFEC

Rotates about x-axis only.

(i) Negative yield lines: $\ell_x = a$, $m_x = m$, $\theta_x = \Delta/a$
(ii) Positive yield lines: $\ell_x = 2\alpha\, a$, $m_x = m$, $\theta_x = \Delta/a$

Energy dissipated E_2 is

$$E_2 = \{m \times a \times \Delta/a + m \times 2\,\alpha\, a \times \Delta/a\} = (2\alpha + 1)\, m\, \Delta$$

3. Total energy dissipated

$$E = E_1 + E_2 = \{(4/\alpha + (2\,\alpha+1)\}\, m\, \Delta$$

4. Work done

Rigid region AFD: From Fig. 12.14,
Substituting $L = a\,\alpha$, $H = a$, $p_1 = 0$ and $p_2 = a\,\gamma$

$$W_1 = \Delta \times H\,L \times \frac{(3\,p_1 + p_2)}{24} = \Delta\,\gamma\,a^3\,\frac{\alpha}{24}$$

Region FDCE: Divide it into two triangles and a rectangle:
Each triangle, from Fig. 12.17,
Substituting $L = \alpha\, a$, $H = a$, $p_1 = 0$ and $p_2 = a\,\gamma$

$$W_2 = \Delta \times H\,L \times \frac{(p_1 + p_2)}{12} = \Delta\,\gamma\,a^3\,\frac{\alpha}{12}$$

Rectangle, from Fig. 12.16,
Substituting $L = a\,(1 - 2\,\alpha)$, $H = a$, $p_1 = 0$ and $p_2 = a\,\gamma$

$$W_3 = \Delta \times H\,L \times \frac{(2p_1 + p_2)}{6} = \Delta\,\gamma\,a^3\,\frac{(1-2\alpha)}{6}$$

Total work done

$$W = 2(W_1 + W_2) + W_3 = W_2 = \Delta\,\gamma\,a^3\,\frac{(2-\alpha)}{12}$$

4. Moment m

Equating the work done by external loads to the energy dissipated at the yield lines,

$$m = \frac{\gamma a^3}{12}\,\frac{(2\alpha - \alpha^2)}{(4 + \alpha + 2\alpha^2)}$$

For a maximum m, $dm/d\alpha = 0$

$$(2 - 2\alpha)(4 + \alpha + 2\alpha^2) - (2\alpha - \alpha^2)(1 + 4\alpha) = 0$$

Simplifying,

$$6 - 8\alpha - 5\alpha^2 = 0,\ \alpha = 0.697 > 0.5$$

Take $\alpha = 0.5$

$$m = 0.013\,\gamma a^3,$$

Substituting $a = 5.0$ m, $\gamma = 18.0$,

$$m = 28.13\ \text{kNm/m}$$

Using a load factor γ_Q on the earth pressure of 1.5 and also increasing the calculated moment by 10% to account for the formation of corner levers, m = (28.13 × 1.5) × 1.1 = 46.41 kNm/m which is smaller than the value of m for mode 1. Mode 1 controls design.

5. Reinforcement
Use 12 mm diameter bars and 40mm cover.
$$d = 180 - 40 - 12/2 = 134 \text{ mm}$$
According to clause 5.6.2 of Eurocode 2, plastic analysis of slabs without explicit check on rotation capacity can be made provided that $x_u/d \le 0.25$ for $f_{ck} \le 50$ MPa and reinforcing steel is either Class B or Class C. If $x_u/d = 0.25$,
$$M_u = b \times 0.8 \, x_u \times f_{cd} \times (d - 0.4 \, x_u) = 0.12 \, bd^2 \, f_{ck}$$
$$k = M/(bd^2 f_{ck}) = 54.80 \times 10^6/(1000 \times 134^2 \times 30) = 0.102 < 0.12$$
$$\frac{z}{d} = 0.5[1.0 + \sqrt{(1 - 3\frac{k}{\eta})}]$$
$$k = 0.102, \eta = 1.0, z/d = 0.917$$
$$A_s = 54.80 \times 10^6/(0.917 \times 134 \times 0.87 \times 500) = 1026 \text{ mm}^2/\text{m}$$
Check the minimum reinforcement required. From equation (9.1N) of the code,
$$A_{s, min} = 0.26 \times (f_{ctm}/f_{yk}) \times b \times d \ge 0.0013 \, b \times d$$
$$f_{ctm} = 0.3 \times f_{ck}^{0.67} = 0.3 \times 30^{0.67} = 2.9 \text{ MPa}, f_{yk} = 500 \text{ MPa},$$
$$b = 1000 \text{ mm}, d = 134 \text{ mm}$$
$$A_{s, min} = 0.26 \times (2.9/500) \times 1000 \times 134 \ge 0.0013 \times 1000 \times 132$$
$$A_{s, min} = 202 \text{ mm}^2$$
provided steel is greater than the minimum percentage of steel.
Provide 12 mm bars at 100 mm centre to give a steel area of $A_s = 1130$ mm²/m. The same steel is provided in each direction on the outside and inside of the wall. The steel on the outside of the wall covers the whole area. The points of cut-off of the bars on the inside of the wall may be determined by finding the size of a slab simply supported on three sides and one edge free that has the same ultimate moment of resistance m = 54.80 kNm/m as the whole wall. This slab has the same yield line pattern as the wall slab.
Taking the moment of resistance of negative yield lines as zero,
$$\text{Total energy dissipated, } E = (4 + 1/\alpha) \, m \, \Delta$$
$$\text{Total work done, } W = \frac{\gamma \Delta}{24} a^3 (6 - 4\alpha + \alpha^2)$$
Equating E = W and substituting α = 0.483312 and m = 54.80 kNm/m, a = 4.69 m. Therefore a slab of similar shape to the clamped edge slab considered but clamped edges being replaced by simply supported edges and side length equal to 4.69 m instead of 5.0 m can carry the required load.
In the horizontal direction, top face bars can be stopped from the sides at 0.5 (5.0 – 4.69) + anchorage length of 36 bar diameter = 0.732 m i.e,. 750 mm say.
In the vertical direction bars can be stopped from the base slab at (5.0 – 4.69) + anchorage length of 36 bar diameter = 0.886 m i.e., 900 mm say.

Maximum Spacing: Taking the moment ratio of M at ULS/M at SLS = γ_Q =1.5, the steel stress f_s at SLS is given by

$$f_s = \frac{M \text{ at } SLS}{M \text{ at } ULS} \times \frac{A_{s,required}}{A_{s,provided}} \times f_{yd} = \frac{1}{1.5} \times \frac{1026}{1130} \times \frac{500}{1.15} = 263 \text{ MPa}$$

From Table 7.2N and Table 7.3N of the code, for a maximum crack width of 0.3 mm, maximum bar size is 16 mm and maximum spacing is 175 mm. These requirements are satisfied by the steel provided.

In the above only one mode of collapse has been investigated. As the yield line method is an upper bound method other possible yield line patterns need to be investigated before finalizing the reinforcement.

12.3.4 Design of Base Slab Using Yield Line Method

(i) Base pressure calculation at the ultimate
The properties of the base are:
$$\text{Area } A = 17.5 \text{ m}^2$$
$$\text{Section modulus } Z = 10.21 \text{ m}^3$$
The forces at SLS per meter length are shown in Table 12.2.

Total gravity load = 1613.88 kN, corresponding moment = 2745.54 kNm
Earth pressure load = 375 kN (Horizontal), corresponding moment = −626.25 kNm
Taking moments of all forces about the toe A, the centroid of the base pressure from A is at a distance L.

Case (a): Load factor is $\gamma_Q = 1.5$ for earth pressure as an unfavourable load and $\gamma_{G, \text{inf}} = 1.0$ for gravity load as a favourable load,
$$L \times 1613.88 \times 1.0 = 2745.54 \times 1.0 - 626.25 \times 1.5, L = 1.12 \text{ m}$$
$$\text{Eccentricity, } e = 3.5/2 - L = 0.63 > 3.5/6.$$
Tension is developed at C.
The base is acted on by a
$$\text{Vertical load} = 1613.88 \times 1.0 = 1613.88 \text{ kN}$$
$$\text{Moment} = (1613.88 \times 1.0) \times e = 1016.74 \text{ kNm}$$
The maximum soil pressure at A and minimum soil pressure at C are
$$1613.88 / 17.5 \pm 1016.74 / 10.21 = 191.8 \text{ and } -7.40 \text{ kN/m}^2$$
The negative pressure is very small and can be neglected.

Case (b): Load factor is $\gamma_Q = 1.0$ for earth pressure as a favourable load and $\gamma_{G, \text{inf}} = 1.35$ for gravity load as an unfavourable load,
$$L \times 1613.88 \times 1.35 = 2745.54 \times 1.35 - 626.25 \times 1.0, L = 1.41 \text{ m}$$
$$\text{Eccentricity, } e = 3.5/2 - L = 0.34 < 3.5/6.$$
No tension develops at C.
The base is acted on by
$$\text{Vertical load} = 1613.88 \times 1.35 = 2178.74 \text{ kN}$$
$$\text{Moment} = (1613.88 \times 1.35) \times e = 740.77 \text{ kNm}$$
The maximum soil pressure at A and minimum soil pressure at C are
$$2178.74 / 17.5 \pm 740.77 / 10.21 = 197.05 \text{ and } 51.95 \text{ kN/m}^2$$

Case (c): Load factor is $\gamma_Q = 1.5$ for earth pressure and $\gamma_{G, \, inf} = 1.35$ for gravity load as both are unfavourable loads,

$$L \times 1613.88 \times 1.35 = 2745.54 \times 1.35 - 626.25 \times 1.5, \, L = 1.27 \text{ m}$$
$$\text{Eccentricity, e} = 3.5/2 - L = 0.48 < 3.5/6$$

No tension develops at C.
The base is acted on by

$$\text{Vertical load} = 1613.88 \times 1.35 = 2178.74 \text{ kN}$$
$$\text{Moment} = (1613.88 \times 1.35) \times e = 1045.79 \text{ kNm}$$

The maximum soil pressure at A and minimum soil pressure at C are

$$2178.74 / 17.5 \pm 1045.79 / 10.21 = 226.93 \text{ and } 22.07 \text{ kN/m}^2$$

Using Case (b) as it gives comparatively small base pressure so that the moment causing tension above will be large, the yield line solution is given for a rectangular base slab with a trapezoidal load due to base pressure on the bottom face and a uniform load due to self weight of the slab and soil on the slab at the top face. The uniformly distributed load at the top is

$$\text{Base slab} = 0.25 \times 25 \times 1.35 = 8.4 \text{ kN/m}^2$$
$$\text{Weight of soil} = 4.75 \times 18.0 \times 1.35 = 115.43 \text{ kN/m}^2$$
$$\text{Total} = 123.83 \text{ kN/m}^2$$

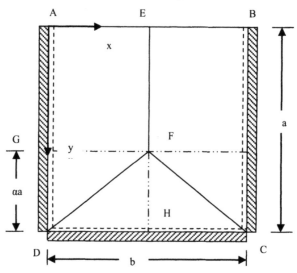

Fig. 12.19 Mode 1 collapse of base slab.

Two modes need to be investigated.

Mode 1: The yield line pattern shown in Fig. 12.19.
Parameter α, locating the position of point F, controls the pattern. Deflection at F is Δ. It is assumed that the slab will be isotropically reinforced with moment of

resistance for both positive (tension at bottom) and negative (tension at top) being equal to m.

The pressure at a point distant y from the free edge is

$$p = 123.83 - \{51.95 + (197.05 - 51.95)y/3.5\} = 71.88 - 41.46 \, y$$

Energy dissipated in the yield lines:

1. Rigid region AEFD and BEFC

Both rotate about y-axis only.

(i) Negative yield lines: $\ell_y = a$, $m_y = m$, $\theta_y = \Delta/ (0.5b)$
(ii) Positive yield lines: $\ell_y = a$, $m_y = m$, $\theta_y = \Delta/ (0.5b)$

Energy dissipated E_1 is

$$E_1 = 2\{m \times a \times \Delta/ (0.5b) + m \times a \times \Delta/ (0.5b)\} = 8 \, m \, (a/b) \, \Delta$$

2. Rigid region DFC

Rotates about x-axis only.

(i) Negative yield lines: $\ell_x = b$, $m_x = m$, $\theta_x = \Delta/ (\alpha \, a)$
(ii) Positive yield lines: $\ell_x = a$, $m_x = m$, $\theta_x = \Delta/ (\alpha \, a)$

Energy dissipated E_2 is

$$E_2 = \{m \times b \times \Delta/ (\alpha \, a) + m \times b \times \Delta/ (\alpha \, a)\} = (2b/\alpha \, a) \, m \, \Delta$$

3. Total energy dissipated E

$$E = E_1 + E_2 = \{8a/b + 2b/ (\alpha \, a)\} \, m \, \Delta$$

Substituting a = 3.5, b = 5.0, total energy dissipated $E = (5.6 + 2.8571/\alpha) \, m \, \Delta$

External work done:

Set up the coordinate axes (x, y) with origin at A.

Using the above integrals, the work done in various regions can be calculated.

Rigid region AEGF: From Fig. 12.13,

Substituting L = 5.0/2, H = 3.5 (1 − α).

Pressure at any point = 71.88 − 41.46 y

$y = 0$, $p_1 = 71.88$, $y = 3.5 (1 - \alpha)$, $p_2 = -73.23 + 145.11 \, \alpha$

$$W_1 = \Delta \times H L \times \frac{(p_1 + p_2)}{4} = \Delta(-2.953 + 320.38 \, \alpha - 317.43 \, \alpha^2)$$

Rigid region GFD: From Fig. 12.14,

Substituting L = 5.0/2, H = 3.5 α,

Pressure at any point = 71.88 − 41.46 y

$y = 3.5 (1 - \alpha)$, $p_1 = -73.23 + 145.11 \, \alpha$

$y = 3.5$, $p_2 = -73.23$

$$W_2 = \Delta \times H L \times \frac{(2 p_1 + p_2)}{24} = \Delta (-106.79 \, \alpha + 158.71 \, \alpha^2)$$

Rigid region FDH: From Fig. 12.16,

Substituting L = 5.0/2, H = 3.5 α,

Pressure at any point = 71.88 – 41.46 y

$$y = 3.5 (1 - \alpha), \quad p_1 = -73.23 + 145.11 \, \alpha$$
$$y = 3.5, \quad p_2 = -73.23$$
$$W_3 = \Delta \times H \, L \times \frac{(p_1 + p_2)}{12} = \Delta \,(-106.79\alpha + 105.81 \, \alpha^2)$$

Total work done:

$$W = 2(W_1 + W_2 + W_3)$$
$$W = \Delta(-5.90 + 213.60\,\alpha - 105.82\,\alpha^2)$$

Moment m

Equating the work done by external loads to the energy dissipated at the yield lines,

$$m = \frac{(-5.90\,\alpha + 213.60\,\alpha^2 - 105.82\alpha^3)}{(5.6\alpha + 2.8571)}$$

For a maximum m, $\alpha = 1.0$, m = 12.05 kNm/m

Mode 2: The yield line pattern shown in Fig. 12.20. Making the same assumptions as for Mode 1, the pressure p at any point y from the free edge is
p = 71.88 – 41.46 y

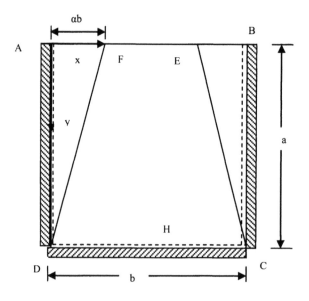

Fig. 12.20 Mode 2 collapse of base slab.

Energy dissipated in the yield lines:

1. Rigid region AFD and BEC

Both rotate about y-axis only.

$\quad\quad$ (i) Negative yield lines: $\ell_y = a$, $m_y = m$, $\theta_y = \Delta/(\alpha b)$

$\quad\quad$ (ii) Positive yield lines: $\ell_y = a$, $m_y = m$, $\theta_y = \Delta/(\alpha b)$

Energy dissipated E_1 is

$$E_1 = 2\{m \times a \times \Delta/(\alpha\,b) + m \times a \times \Delta/(\alpha\,b)\} = 4m\,a/(\alpha\,b)\,\Delta$$

2. Rigid region DFEC

Rotates about x-axis only.

$\quad\quad$ (i) Negative yield lines: $\ell_x = b$, $m_x = m$, $\theta_x = \Delta/a$

$\quad\quad$ (ii) Positive yield lines: $\ell_x = 2\alpha b$, $m_x = m$, $\theta_x = \Delta/a$

Energy dissipated E_2 is

$$E_2 = \{m \times b \times \Delta/a + m \times 2\,\alpha b \times \Delta/a\} = (2\alpha + 1)\,(b/a)\,m\,\Delta$$

3. Total energy dissipated

$$E = E_1 + E_2 = \{(4a/\,\alpha b + (2\,\alpha+1)\,(b/a)\}\,m\,\Delta$$

Substituting $a = 3.5$, $b = 5.0$, Energy dissipated is

$$E = (2.8/\,\alpha +1.4286 + 2.8571\,\alpha)\,m\,\Delta$$

External work done:

Set up the coordinate axes (x, y) with origin at A.

Rigid region AFD

From Fig. 12.14,

Substituting $L = 5\alpha$, $H = 3.5$,

$\quad\quad$ Pressure at any point $= 71.88 - 41.46\,y$

$\quad\quad$ $y = 0$, $p_1 = 71.88$, $y = 3.5$, $p_2 = -73.23$

$$W_1 = \Delta \times H\,L \times \frac{(3\,p_1 + p_2)}{24} = \Delta(103.84\ \alpha)$$

Region FDCE

Divide it into two triangles and a rectangle:

Each Triangle: From Fig. 12.17,

Substituting $L = 5\alpha$, $H = 3.5$,

$\quad\quad$ Pressure at any point $= 71.88 - 41.46\,y$

$\quad\quad$ $y = 0$, $p_1 = 71.88$, $y = 3.5$, $p_2 = -73.23$

$$W_2 = \Delta \times H\,L \times \frac{(p_1 + p_2)}{12} = \Delta(-1.969\ a)\frac{a}{12}$$

Rectangle: From Fig. 12.16,

$\quad\quad$ Substituting $L = 5(1 - 2\,\alpha)$, $H = 3.5$,

$\quad\quad$ $y = 0$, $p_1 = 71.88$, $y = 3.5$, $p_2 = -73.23$

$$W_3 = \Delta \times H\,L \times \frac{(2\,p_1 + p_2)}{6} = \Delta\,(205.71 - 411.43\alpha)$$

Total work done:

$$W = 2(W_1 + W_2) + W_3 = \Delta(205.71 - 207.68\alpha)$$

4. Moment m

Equating the work done by external loads to the energy dissipated at the yield lines,

$$m = \frac{(205.71\alpha - 207.68\alpha^2)}{(2.8 + 1.4286\alpha + 2.8571\alpha^2)}$$

For a maximum m, $\alpha = 0.40$, m = 12.821 kNm/m
Mode 2 gives marginally higher value of m = 12.81 kNm/m. Increasing the calculated moment by 10% to account for the formation of corner levers,
$$m = 12.81 \times 1.1 = 14.09 \text{ kNm/m}.$$

5. Reinforcement

Use 10 mm diameter bars and 40 mm cover.
$$d = 250 - 40 - 10/2 = 205 \text{ mm}$$

According to clause 5.6.2 of Eurocode 2, plastic analysis of slabs without explicit check on rotation capacity can be made provided that $x_u/d \leq 0.25$ for $f_{ck} \leq 50$ MPa and reinforcing steel is either Class B or Class C. If $x_u/d = 0.25$,

$$M_u = b \times 0.8 \ x_u \times f_{cd} \times (d - 0.4 \ x_u) = 0.12 \ bd^2 \ f_{ck}$$
$$k = M/ (bd^2 f_{cu}) = 14.09 \times 10^6/ (1000 \times 205^2 \times 30) = 0.011 < 0.12$$
$$\frac{z}{d} = 0.5[1.0 + \sqrt{(1 - 3\frac{k}{\eta})}]$$
$$k = 0.011, \ \eta = 1.0, \ z/d = 0.99$$
$$A_s = 14.09 \times 10^6/ (0.99 \times 205 \times 0.87 \times 500) = 160 \text{ mm}^2/\text{m}$$

Check the minimum steel area required. From equation (9.1N) of the code,
$$A_{s, \, min} = 0.26 \times (f_{ctm}/f_{yk}) \times b \times d \geq 0.0013 \ b \times d$$
$$f_{ctm} = 0.3 \times f_{ck}^{\, 0.67} = 0.3 \times 30^{\, 0.67} = 2.9 \text{ MPa}, \ f_{yk} = 500 \text{ MPa},$$
$$b = 1000 \text{ mm}, \ d = 134 \text{ mm}$$
$$A_{s, \, min} = 0.26 \times (2.9/500) \times 1000 \times 205 \geq 0.0013 \times 1000 \times 205$$
$$A_{s, \, min} = 309 \text{ mm}^2$$

provided steel is less than the minimum percentage of steel.
As most of the moment is coming from gravity loads with a load factor of $\gamma_{G, \, sup} = 1.35$, the ratio of moment at SLS/ moment at ULS $\approx 1/1.35 = 0.74$.
Stress in steel at SLS = $0.74 \times (160/309) \times (500/1.15) = 225$ MPa.
From Tables 7.2N and 7.3N of the code, in order to limit crack widths to 0.3 mm, maximum spacing is 225 mm and maximum bar size is 16 mm. Provide 12 mm bars at 225 mm centre to give a steel area of $A_s = 503$ mm^2/m. The same steel is provided in each direction on the outside and inside of the wall. The steel on the outside of the wall covers the whole area.

12.3.5 Base Slab Design Using Hillerborg's Strip Method

Although yield line method was used in design in the previous sections, it is not the ideal method. Hillerborg's strip method offers a better alternative. At the junction of the base slab with the wall slab, load is resisted by cantilever action but at a distance away from the base, load is resisted by clamped beam action with the slab spanning between the counterforts. As shown in Fig. 12.21, the 3.5 m × 5.0 m base slab is divided into a set of 14 strips, each 250 mm wide. The pressure at any level y from the free edge is given by the equation,

$$p = 71.88 - 41.46 \, y$$

It is assumed that load lying in a triangle with the sides at an inclination of approximately 30° to the horizontal (see Fig. 12.21) is resisted by cantilever action. The rest of the load is resisted by horizontal clamped beam action.

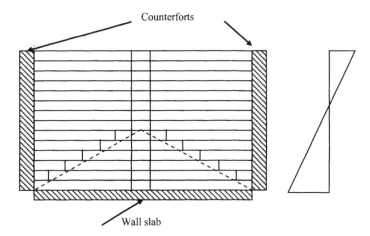

Fig. 12.21 Division of base slab into `horizontal` strips.

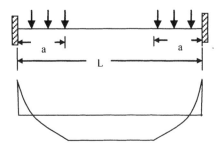

Fig. 12.22 Load on `horizontal` strips.

12.3.5.1 `Horizontal` Strips in Base Slab

The `horizontal` strips span between the counterforts. The strips towards the base are loaded as shown in Fig. 12.22. The bending moment at the support and mid-span of each strip is calculated using the equation

$$\text{Moment at support} = \frac{qL^2}{12} C_1 \, , \, C_1 = \alpha^2(6 - 4\alpha),$$

$$\text{Moment at mid-span} = \frac{qL^2}{24}C_2, \; C_2 = 8\alpha^3$$

where q = uniformly distributed load in kN/m, L = span = 5 m, α = a/L, a = loaded length.

Table 12.3: Bending moments (kNm/m) in horizontal strips in base slab

Strip	y	p	a	alpha	C_1	C_2	M-Supp	M-Span	A_S-Supp	A_S-Span
1	0.125	66.7	2.5	0.5	1	1	139.0	69.5	429	204
2	0.375	56.3	2.5	0.5	1	1	117.4	58.7	356	171
3	0.625	46.0	2.5	0.5	1	1	95.8	47.9	286	138
4	0.875	35.6	2.5	0.5	1	1	74.2	37.1	218	106
5	1.125	25.2	2.5	0.5	1	1	52.6	26.3	152	77
6	1.375	14.9	2.5	0.5	1	1	31.0	15.5	89	77
7	1.625	4.5	2.5	0.5	1	1	9.4	4.7	77	77
8	1.875	−5.9	2.5	0.5	1	1	−12.2	−6.1	77	77
9	2.125	−16.2	2.165	0.43	0.80	0.65	−27.0	−11.0	77	77
10	2.375	−26.6	1.732	0.35	0.55	0.33	−30.7	−9.2	88	77
11	2.625	−37.0	1.299	0.26	0.33	0.14	−25.8	−5.4	77	77
12	2.875	−47.3	0.866	0.17	0.16	0.04	−15.7	−2.0	77	77
13	3.125	−57.7	0.433	0.09	0.04	0.01	−5.1	−0.3	77	77
14	3.375	−68.1	0	0.00	0.00	0.00	0.0	0.0	77	77

Detailed calculations are shown in Table 12.3. Fig. 12.23 shows the bending moment distribution in the `horizontal` strips.

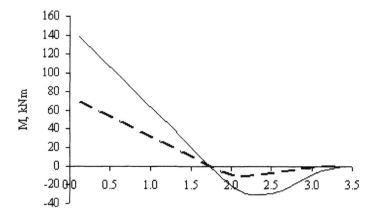

Fig. 12.23 Bending moment (kNm/m) in `horizontal` strips in base slab.
Full line = support moment, broken line = span moment.

At the *support*, the maximum bending moment causing tension on the *bottom* of the slab is 139.0 kNm/m and the maximum bending moment at *mid-span* causing tension on the *top* of the slab is 69.5 kNm/m in strip 1. Similarly, at the support the maximum bending moment causing tension on the *top* of the slab is 30.7 kNm/m in strip 10 and the maximum bending moment causing tension on the *bottom* of the slab is 11.0 kNm/m in strip 9.

$$d = 250 - 40 - 10/2 = 205 \text{ mm}$$
$$M = 139.0 \text{ kNm/m}$$
$$k = M/ (bd^2 f_{ck}) = 139.0 \times 10^6/ (1000 \times 205^2 \times 30) = 0.11 < 0.12$$
$$\frac{z}{d} = 0.5[1.0 + \sqrt{(1 - 3\frac{k}{\eta})}]$$
$$\eta = 1.0, \text{ z/d} = 0.91$$
$$A_s = 139.0 \times 10^6/ (0.91 \times 205 \times 0.87 \times 500) = 1715 \text{ mm}^2/\text{m}$$

Over a width of 250 mm, $A_s = 1715 \times 0.25 = 429 \text{ mm}^2$
Provide 3H16, $A_s = 603 \text{ mm}^2$.

Check for minimum steel from equation (9.1N) of the code,
$$A_{s, \text{ min}} = 0.26 \times (f_{ctm}/f_{yk}) \times b \times d \geq 0.0013 \ b \times d$$
$$f_{ctm} = 0.3 \times f_{ck}^{0.67} = 0.3 \times 30^{0.67} = 2.9 \text{ MPa}, \ f_{yk} = 500 \text{ MPa},$$
$$b = 1000 \text{ mm}, d = 202 \text{ mm}$$
$$A_{s, \text{ min}} = 0.26 \times (2.9/500) \times 1000 \times 205 \geq 0.0013 \times 1000 \times 205$$
$$A_{s, \text{ min}} = 309 \text{ mm}^2/\text{m}$$

Over a width of 250 mm, $A_s = 309 \times 0.25 = 77 \text{ mm}^2$
As most of the moment is from gravity loads,
$$\text{Moment at SLS} = \text{Moment at ULS}/ (\gamma_{G, \text{ sup}} = 1.35)$$
$$\text{Stress in steel at SLS} = (M_{SLS}/M_{ULS}) \times (A_{s, \text{ reqd}}/A_{s, \text{ Provided}}) \times f_{yd}$$
$$= (1/1.35) \times (429/603) \times (500/1.15) = 229 \text{ MPa}$$

Check maximum bar diameter and spacing of steel permitted: For steel stress at SLS of 240 MPa and for a maximum crack width of 0.3 mm, maximum spacing allowed from code Table 7.3N is 200 mm and from code Table 7.2N maximum bar size is 16 mm. The provided steel area satisfies both criteria.
Similar calculations can be done for the required steel in other strips.

12.3.5.2 Cantilever Moment in Base Slab

The cantilever moment is determined by taking a series of vertical strips. The strips cantilever from the wall slab. The cantilever moment is greatest in the middle vertical strip. Pressures occur only in strips 9 to 14. The bending moment M at the base of the cantilever is given by the product of the pressures on the 250 mm wide strips and the distance from the base to the centre of the strips. Pressures at the centre of strips are given in Table 12.3.

$$\text{Horizontal width of strip} = 830 \text{ mm}$$
$$M = 0.250 \times \{16.22 \times 1.375 + 26.59 \times 1.125 + 36.95 \times 0.875$$
$$+ 47.32 \times 0.625 + 57.68 \times 0.375 + 68.05 \times 0.125\} = 35.22 \text{ kNm/m}$$
$$k = M/ (bd^2 f_{ck}) = 35.22 \times 10^6/ (1000 \times 205^2 \times 30) = 0.028 < 0.196$$

$$\frac{z}{d} = 0.5[1.0 + \sqrt{(1 - 3\frac{k}{\eta})}]$$

$$k = 0.028,\ \eta = 1.0,\ z/d = 0.98$$

$$A_s = 29.93 \times 10^6 / (0.98 \times 205 \times 0.87 \times 500) = 343\ mm^2/m$$

H10 at 225 mm gives $A_s = 349\ mm^2/m$.

Chech the minimum steel needed. From equation (9.1N) of the code,

$$A_{s,\ min} = 0.26 \times (f_{ctm}/f_{yk}) \times b \times d \geq 0.0013\ b \times d$$

$$f_{ctm} = 0.3 \times f_{ck}^{\ 0.67} = 0.3 \times 30^{\ 0.67} = 2.9\ MPa,\ f_{yk} = 500\ MPa,$$

$$b = 1000\ mm,\ d = 202\ mm$$

$$A_{s,\ min} = 0.26 \times (2.9/500) \times 1000 \times 205 \geq 0.0013 \times 1000 \times 205$$

$$A_{s,\ min} = 309\ mm^2/m$$

provided steel is greater than the minimum percentage of steel.

As most of the moment is from gravity loads,

$$\text{Moment at SLS} = \text{Moment at ULS} / (\gamma_{G,\ sup} = 1.35)$$

$$\text{Stress in steel at SLS} = (M_{SLS}/M_{ULS}) \times (A_{s,\ reqd}/A_{s,\ Provided}) \times f_{yd}$$

$$= (1/1.35) \times (343/349) \times (500/1.15) = 317\ MPa$$

Check maximum bar diameter and spacing of steel permitted:

For steel stress at SLS of 317 MPa and for a maximum crack width of 0.3 mm, maximum spacing allowed from code Table 7.3N is 100 mm and from code Table 7.2N maximum bar size is 10 mm. Provide H10 at 100 c/c.

Similar calculations can be done for the required steel in other strips.

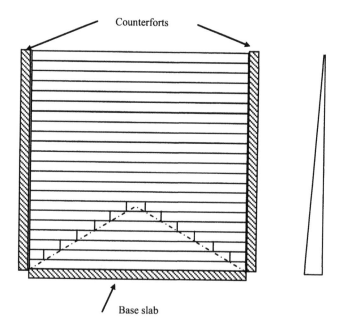

Fig. 12.24 Horizontal strips in vertical wall slab.

12.3.6 Wall Slab Design Using Hillerborg's Strip Method

Wall design is done similar to the base design. As the height of the wall is 5.0 m, it is divided into 20 strips each 250 mm wide as shown in Fig. 12.24.
The pressure at any level y from the top is equal to $1.5 \, \gamma K_a y$, where $\gamma = 18.0$ kN/m^3, K_a = coefficient of earth pressure = 0.33, load factor for earth pressure $\gamma_Q = 1.5$. Therefore p = 9.0 y.
It is assumed that load lying in a triangle with the sides at an inclination of approximately 30° to the horizontal is resisted by vertical cantilever action. Calculations are shown in Table 12.4 and Fig. 12.25 shows the bending moment distribution in the horizontal strips. The calculation of steel in the strips is done as for the base.

Table 12.4 Bending moment (kNm/m) in horizontal strips of vertical wall slab

Strip	y	p	a	alpha	C_1	C_2	Mspan	M-Supp	A_s-Span	A_s-supp
1	0.125	1.1	2.5	0.5	1	1	1.2	2.3	309	309
2	0.375	3.4	2.5	0.5	1	1	3.5	7.0	309	309
3	0.625	5.6	2.5	0.5	1	1	5.9	11.7	309	309
4	0.875	7.9	2.5	0.5	1	1	8.2	16.4	309	309
5	1.125	10.1	2.5	0.5	1	1	10.5	21.1	309	309
6	1.375	12.4	2.5	0.5	1	1	12.9	25.8	309	309
7	1.625	14.6	2.5	0.5	1	1	15.2	30.4	309	348
8	1.875	16.9	2.5	0.5	1	1	17.6	35.1	309	403
9	2.125	19.1	2.5	0.50	1	1	19.9	39.8	309	458
10	2.375	21.4	2.5	0.50	1	1	22.2	44.5	309	513
11	2.625	23.6	2.5	0.50	1	1	24.6	49.2	309	569
12	2.875	25.9	2.5	0.50	1	1	26.9	53.9	309	625
13	3.125	28.1	2.5	0.50	1	1	29.3	58.5	334	681
14	3.375	30.3	2.5	0.50	1	1	31.6	63.2	362	738
15	3.625	32.6	2.2	0.43	0.8	0.7	22.0	54.3	309	631
16	3.875	34.8	1.7	0.35	0.6	0.3	12.1	40.2	309	462
17	4.125	37.1	1.3	0.26	0.3	0.1	5.4	25.9	309	309
18	4.375	39.3	0.9	0.17	0.2	0.0	1.7	13.0	309	309
19	4.625	41.6	0.4	0.09	0.0	0.0	0.2	3.7	309	309
20	4.875	43.8	0	0.00	0.0	0.0	0.0	0.0	309	309

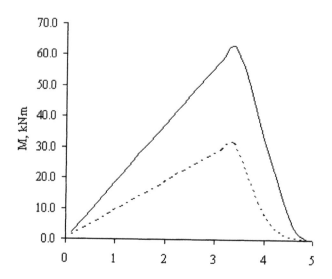

Fig. 12.25 Bending moment (kNm/m) in the horizontal strips in the vertical wall slab.
Full line = support moment, broken line = span moment.

12.3.6.1 Cantilever Moment in Vertical Wall Slab

The cantilever moment is greatest in the central vertical strip. Pressures occur only in strips 15 to 20. The bending moment M at the base of the cantilever is given by the product of the pressures at the 250 mm wide strips and the distance from the base to the centre of the strips.

$$M = 0.250 \times (32.59 \times 1.375 + 34.84 \times 1.125 + 37.09 \times 0.875 + 39.34 \times 0.625$$
$$+ 41.58 \times 0.375 + 43.83 \times 0.125) = 40.53 \text{ kNm/m}$$

Steel required can be calculated in the same way as was done for base slab.

12.3.7 Counterfort Design Using Hillerborg's Strip Method

The reactions from the horizontal strips of the wall slab act as horizontal forces on the counterfort. At any level, the force R on the counterfort from the 250 mm wide strips on either side of the counterfort is (Fig. 12.17)

$$R = 2 \times p \times a \times 0.250$$

This is calculated at the centre of each strip. From the calculated value of R, shear force and bending moment at different levels in the counterfort can be calculated. The detailed calculations are shown in Table 12.5. The distribution of shear force and bending moment are shown in Fig. 12.26 and Fig. 12.27.

Table 12.5 Shear force and bending moment in counterfort

SF (kN)	M (kNm)	h (mm)	d (mm)	k	z/d	A_s (mm²)
1.40	0.00	263	215	0	1	99
5.62	0.35	429	381	0.0003	1.00	175
12.64	1.76	595	547	0.0008	1.00	252
22.48	4.92	761	713	0.0013	1.00	328
35.12	10.54	927	879	0.0018	1.00	405
50.57	19.32	1093	1045	0.0024	1.00	481
68.84	31.96	1259	1211	0.0029	1.00	557
89.91	49.17	1425	1377	0.0035	1.00	634
113.79	71.65	1591	1543	0.0040	1.00	710
140.48	100.10	1757	1709	0.0046	1.00	786
169.99	135.22	1923	1875	0.0051	1.00	863
202.30	177.71	2089	2041	0.0057	1.00	939
237.42	228.29	2255	2207	0.0062	1.00	1016
275.35	287.64	2421	2373	0.0068	0.99	1092
310.63	356.48	2587	2539	0.0074	0.99	1168
340.80	434.14	2753	2705	0.0079	0.99	1245
364.89	519.34	2919	2871	0.0084	0.99	1321
381.92	610.56	3085	3037	0.0088	0.99	1398
390.93	706.04	3251	3203	0.0092	0.99	1474
390.93	803.77	3417	3369	0.0094	0.99	1550

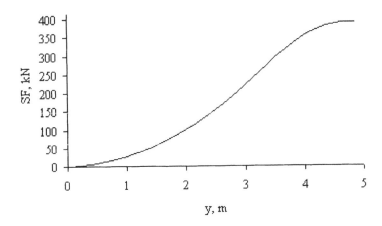

Fig. 12.26 Shear force in counterfort.

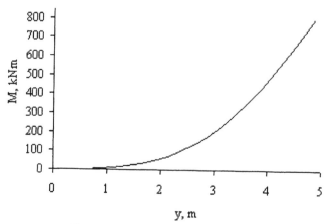

Fig. 12.27 Bending moment in counterfort.

The depth of the counterfort is 180 mm at top and increasing to 3500 mm at the bottom. Assuming 40 mm cover and 16 mm bars, the effective depth at different levels is calculated. The width of the counterfort is 250 mm. The back of the counterfort is inclined at an angle θ to the horizontal where from Fig. 12.11,
$$\theta = \tan^{-1}(4750/33200) = 55°.$$
At any level, the area of steel required is given by
$$A_s = M/ (z \times 0.87 \times f_y \times \sin 55)$$
Note that because of the fact that the tension steel is placed parallel to the back of the counterfort as shown in Fig. 12.11, only the vertical component the force in the steel is taken into account. The required area of steel is very small because of the very large effective depth of the counterfort.
The minimum steel is calculated as
$$A_{s,\,min} = 0.26\times (f_{ctm}/f_{yk}) \times b \times d \geq 0.0013\, b \times d$$
$$f_{ctm} = 0.3 \times f_{ck}^{\,0.67} = 0.3 \times 30^{\,0.67} = 2.9 \text{ MPa},\ f_{yk} = 500 \text{ MPa},$$
$$b = 250 \text{ mm},$$
$$A_{s,\,min} = 0.26\times (2.9/500) \times 250 \times d /\sin 55 \geq 0.0013 \times 250 \times d/\sin 55$$
Minimum steel governs in all cases. Table 12.5 shows the steel required at different levels. 4H25 will give a steel area of 1964 mm². As shown in Fig. 12.11, it is essential to tie the counterfort and the wall slab together by horizontal links to resist the force R in tension. Similarly, the counterforts must be anchored to the base slab by vertical links as shown in Fig. 12.11.

12.4 REFERENCE

Bond, Andrew and Harris, Andrew. (2008). *Decoding Eurocode 7*. Taylor & Francis.

CHAPTER 13

DESIGN OF STATICALLY
INDETERMINATE STRUCTURES

13.1 INTRODUCTION

Design of structures in structural concrete involves satisfying
- The serviceability limit state (SLS) criteria
- The ultimate limit state (ULS) criteria

Design for ULS is concerned with safety and this means ensuring that the ultimate load of the structure is at least equal to the design ultimate load. The theoretical principles used in design at ULS are based on classical theory of plasticity which was developed for the design of steel structures with unlimited ductility. Fig. 13.1 shows the moment–curvature relationship for a steel section. As can be seen, once the ultimate or plastic moment capacity is reached, for further changes in curvature and hence increasing deformation, the moment capacity is maintained provided that the compression flanges do not buckle.

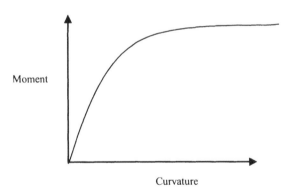

Fig. 13.1 Moment–curvature relationship for a steel section.

Assuming that unlimited ductility can be relied upon, then according to the theory of plasticity, at ultimate load the state of stress has to satisfy the following three conditions.

(a) Equilibrium condition: The state of stress must be in equilibrium with the ultimate load. One convenient way of obtaining a set of stresses in equilibrium with external loads is to do an elastic analysis of a structure under a load equal to the ultimate load. It does not in any way imply that the designed structure behaves elastically under the applied ultimate load. Theoretically it is permissible to use elastic analysis or any variation of it as long as the stresses are in equilibrium with the external loads. The implication of this statement for the design of reinforced concrete structures will be discussed later.

The Eurocode 2 permits three methods of structural analysis for the calculation of a set of 'stresses' or stress resultants like bending and twisting moments, axial and shear forces in equilibrium with the applied load for design at the ULS. They are:

(i) Linear elastic analysis: In clause 5.4 of the Eurocode 2 it is stated that linear elastic analysis may be used for the determination of 'stresses both for ULS as well SLS designs. It may be carried out assuming uncracked sections with linear stress–strain relationship using mean value of modulus of elasticity.

(ii) Linear elastic analysis with limited redistribution: In clause 5.5 of the Eurocode 2 this method is limited to design at ULS only. This will be discussed in detail in the rest of this chapter.

(iii) Plastic analysis: In clause 5.6 of the Eurocode 2 this method is limited to design at ULS only provided the ductility at critical sections is sufficient for the envisaged collapse mechanism to form.

(b) Yield Condition: The state of stress must not violate the yield condition for the material. This means for example, that for any combinations of bending moment and axial force, the capacity of the column should not exceeded the limits as defined by column design chart (section 9.3, Chapter 9). In members in framed structures primarily subjected to bending moment and shear forces, adequate reinforcement is provided such that the moment and shear capacity of the section is at least equal to the design forces at that section.

(c) Mechanism Condition: Sufficient yielded zones must be present to convert the structure in to a mechanism, indicating that there is no reserve load capacity left. In the case of framed structures this means that there must be sufficient plastic hinges and in the case of plate structures sufficient 'yield lines' (Chapter 8) to convert the structure in to a mechanism.

When using the methods based on the classical theory of plasticity to design structures in structural concrete, it is important to recognize the fact that unlike steel, reinforced concrete is a material of very limited ductility. Fig. 13.2 shows by the discontinuous line the moment–curvature relationship for a reinforced concrete section. After the maximum moment capacity is reached, the capacity is maintained for a limited increase in curvature beyond the curvature at maximum capacity. For curvature beyond this value, the moment capacity *decreases*. It is therefore necessary to ensure at no section is the curvature so large that the moment capacity decreases significantly before the structure collapses.

The need to pay attention to ductility and its effect on ultimate strength as well as serviceability behaviour is explained by two examples.

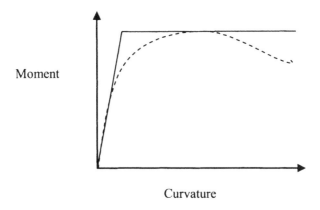

Fig. 13.2 Idealized and actual moment–curvature relationship.

13.2 DESIGN OF A PROPPED CANTILEVER

Consider the design of a propped cantilever of 6 m span as shown in Fig. 13.3. It is required to support at mid-span an ultimate load W equal to 100 kN. The design can be carried out in several ways as follows.

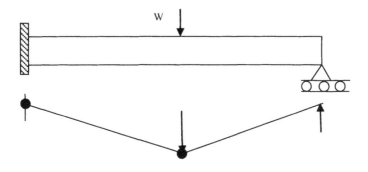

Fig. 13.3 Propped cantilever.

(a) Design 1 based on elastic bending moment distribution

In a propped cantilever supporting a midspan load W over a span L, from elastic analysis, the moments at support and mid-span are respectively 3WL/16 and 5WL/32. If W = 100 kN and L = 6 m, the corresponding moments are 112.5 kNm and 93.75 kNm respectively. If the beam is designed for these moments, then assuming for simplicity that moment–curvature is elastic-perfectly plastic as shown by full line in Fig. 13.2, plastic hinges will form *simultaneously* at the support and mid-span sections and the beam will collapse. Up to the collapse load, there is no rotation at the built-in support and the beam behaves as an elastic structure right up to collapse.

Of course this is a very simplified picture as to what really happens when a beam is tested, because cracking and other non-linear behaviours start almost from the beginning and the moment–curvature is more like that shown by dotted line in Fig. 13.2. However the grossly simplified elastic-perfectly assumption for moment–curvature relationship is sufficient for the present discussion.

(b) Design 2 based on modified elastic bending moment distribution
Instead of designing the beam using the elastic moment distribution, let the beam be designed for a support moment equal to 80% of elastic value of 112.5 kNm, i.e., 90 kNm. The moment at mid-span for equilibrium at the ultimate load is given by (WL/4 – support moment /2) = 100 × 6/4 – 90/2 = 105 kNm which is 112% of the corresponding moment of resistance at mid-span in Design 1.

In the elastic state, the maximum bending moment is at the support. Since the design moment at the support is 90 kNm, which is only 80% of the corresponding elastic moment at a load of 100 kN, the first plastic hinge will form at the support at a load of 80 kN. Up to the stage when the first plastic hinge forms, the beam behaves as an elastic propped cantilever and the rotation at the built in support is zero. The moment at the mid-span is 5/32 × (80 × 6) = 75 kNm which is less than the moment capacity of the section which is 105 kNm.

For a load greater than 80 kN since a plastic hinge has formed at the support, the moment there cannot increase any further but moment at mid-span can increase until a second plastic hinge forms at mid-span and the beam collapses. Therefore for the load stage from 80 kN to 100 kN, the beam behaves as if loaded by a concentrated load at mid-span and a support moment equal to 90 kNm. The *additional behaviour* of the beam beyond a load of 80 kN can be computed by treating the beam as a simply supported beam. During this stage, the support section continues to rotate. The elastic rotation θ at the support in a simply supported beam of flexural rigidity EI and loaded at mid-span by a load P is given by

$$\theta = P \, L^2 / (16 \; EI)$$

Substituting

$$P = (100 - 80) = 20 \text{ kN and } L = 6, \text{ EI } \theta = 45$$

At this stage the moment at mid-span is equal to the moment capacity of 105 kNm and the beam collapses by the formation of plastic hinges at the support and at mid-span.

Comparing the two designs, both beams collapse by the formation of plastic hinges at support and at mid-span. However in Design 1, the two plastic hinges form simultaneously and there was no rotation at the built in support right up to collapse. However in Design 2 with the support moment capacity of only 80% of the elastic value as used in Design 1, the support section has to rotate from the load equal to 80 kN at which the first plastic hinge forms right up to collapse load of 100 kN with the moment at the support remaining at the value of 90 kNm. The support section had to undergo substantial rotation while continuing to maintain a moment of 90 kNm. In other words, the section needs to have sufficient ductility between 80 kN to ultimate load of 100 kN to ensure that there is no decrease in moment capacity.

(c) Design 3 based on greater modification to elastic bending moment distribution than Design 2

In this case the beam is designed for a support moment of 67.5 kNm (60% of elastic value of 112.5 kNm). The moment at mid-span for equilibrium at the ultimate load is equal to $(100 \times 6/4 - 67.5/2) = 116.25$ kNm. Carrying out the calculations as was done for Design 2, the first plastic hinge forms at the support at a load of 60 kN. Up to the stage when the first plastic hinge forms, the beam behaves as propped cantilever and the rotation at the support is zero. The moment at mid-span is $5 \times 60 \times 6/32 = 56.25$ kNm which is less than the capacity of the section which is 116.25 kNm. Since a plastic hinge has formed at the support at 60 kN, the moment there cannot increase any further. However, since the ultimate load to be supported is 100 kN, for the load stage from 60 kN to 100 kN, the beam behaves as if loaded by the concentrated load at mid-span and a support moment equal to 67.5 kNm. Substituting $P = (100 - 60) = 40$ kN and $L = 6$, EI $\theta = 90$. At this stage the moment at the mid-span also reaches a value equal to the moment capacity of 116.25 kNm and the beam collapses by the formation of plastic hinges at the support and mid-span.

Comparing the three designs, all three beams collapse by the formation of plastic hinges at support and at mid-span. However at the stage when the load on the beam is at its ultimate value, the rotation θ at the built in support for the three designs considered are EI $\theta = 0$, 45 and 90 respectively. Thus the smaller the designed support moment capacity is compared with the elastic value, the larger is the rotation at the support. This is shown in Fig. 13.4.

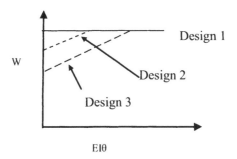

Fig. 13.4 Load–support rotation relationship.

During the stage when the support is rotating from load at which first plastic hinge forms to ultimate load, the moment at the support has to remain constant at the designed value. The larger the load range, the larger the resulting rotation and greater is the demand placed on the ductility of the section. Sections that yield earlier in the loading history are the ones where there is the possibility of moment capacity reducing due to increasing curvature. The greater the difference between the load at which the first plastic hinge forms and the ultimate load, the greater will be the required plastic hinge rotation. It is important therefore that the difference between the ultimate load and the load at which the first section yields is made as small as possible.

What the above example has demonstrated is that it is perhaps possible to design a structure using a bending moment distribution different from the elastic moment distribution, provided sufficient ductility could be assured. Otherwise the assumption that the moment will remain constant during the rotation of the plastic hinge becomes invalid leading to unsafe design. *It is therefore desirable during designing that the stress distribution used in design departs from elastic stress distribution as little as possible.*

13.3 DESIGN OF A CLAMPED BEAM

The idea that although a design might satisfy the ULS criteria, it might be unacceptable from an SLS point of view is demonstrated by the following example. Consider the design of a beam spanning a distance L between two walls and subjected to a uniformly distributed load q. Fig. 13.5 shows three bending moment diagrams, all of which are in equilibrium with a load of q. From an ultimate limit state (ULS) point of view, one can design the beam using any one of the three bending moment distributions.

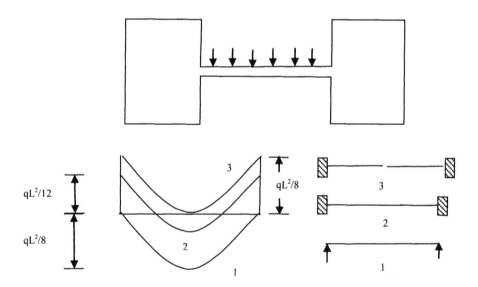

Fig. 13.5 Alternative designs for a clamped beam.

Design 1: Assume that the beam behaves as a simply supported beam. The bending moment at mid-span is $qL^2/8$. In this case clearly only steel at the bottom face is required. The moment of resistance at the support is zero and the first plastic hinges at the supports form at essentially zero load while the plastic hinge at mid-span forms at the ultimate load. The support hinge starts rotating right from the start leading to large cracks there. While this design is satisfactory from a ULS point of view, it is clearly an unsatisfactory design from a serviceability limit state (SLS) point of view.

Design 2: Assume that the beam behaves as a clamped beam. From elastic analysis, bending moment at the junction with the wall is $qL^2/12$ and at mid-span is $qL^2/24$. The plastic hinges at support and at mid-span form simultaneously. This design is satisfactory from both the ULS and SLS points of view, because the design corresponds to the behaviour of the beam taking into account the proper boundary conditions.

Design 3: Assume that the beam behaves as a pair of cantilevers. Bending moment at the junction with the wall is $qL^2/8$. In this case clearly only steel at the top face is required. The moment of resistance at the mid-span is zero and the first plastic hinge at mid-span forms at essentially zero load while the plastic hinges at supports form at the ultimate load. The mid-span hinge starts rotating right from the start leading to large cracks there. While this design is satisfactory from a ULS point of view, it is clearly an unsatisfactory design from a serviceability limit state (SLS) point of view.

As shown in Fig. 13.5, the bending moment distribution in Design 2 is the elastic distribution and requires both top and bottom reinforcement. The bending moment distributions used in Design 1 requires only bottom reinforcement and Design 3 requires only top reinforcement. They are extreme variations of the elastic moment distribution. This example demonstrates the need to pay particular attention to both ULS and SLS aspects, keeping in mind the rather limited ductility available in the case of reinforced concrete sections. Once again the example demonstrates that using elastic distribution of moments is likely to lead to satisfactory designs from both ULS and SLS points of view.

13.4 WHY USE ANYTHING OTHER THAN ELASTIC VALUES IN DESIGN?

One question that naturally arises is why not simply use the elastic values of moments and avoid all problems of ductility? The reason for using values of moments other than the elastic values is purely a matter of convenience. Generally at support sections in frame structures, flat slabs and such structures there is considerable congestion of steel due to the fact that flexural steel in two directions at top and bottom of the beam or slab are required. In addition, steel in the column and shear links need to be accommodated in the same congested area. Therefore any reduction of steel in this zone is an advantage from the point of view of detailing. Using moments at supports smaller than the elastic values helps in mitigating the problem. Elastic stress fields often contain zones of stress concentration and it is useful to modify these stress distributions in the interests of a smoothed out stress distribution which leads to a more convenient detailing of reinforcement.

13.5 DESIGN USING REDISTRIBUTED ELASTIC MOMENT IN EUROCODE 2

Considerable experimental evidence shows that a satisfactory design can be obtained on the basis of reasonably small adjustments to the elastic bending moment distribution. To take account of the factors that influence design using redistribution of elastic moments, in clause 5.5, the Eurocode 2 sets out the procedure for adjusting the elastic moment distribution for design. This process is called moment redistribution. This section states that a redistribution of moments obtained by a rigorous elastic analysis may be carried out provided that the following conditions hold:

(a) Equilibrium between internal and external forces is maintained under all appropriate combinations of design ultimate load. This generally means that any reduction in support moment should be accompanied by an increase in span moment. This amounts to an increase in span moment of one half of the reduction in support moment.

(b) In continuous beams and slabs which
  1. Are subjected predominantly to flexure.
  2. Have the range of lengths of adjacent spans in the range of 0.5 to 2.
  3. Redistribution should not be carried out in circumstances where the rotation capacity cannot be defined with confidence. This situation occurs at corners in prestressed concrete frames.
  4. Redistribution of bending moments may be carried out without explicit check on the rotation capacity, provided that code equations (5.10a) and (5.10b) are satisfied.

$$\delta \geq 0.44 + 1.25(0.6 + \frac{0.0014}{\varepsilon_{cu2}})\,(x_u/d) \quad \text{for } f_{ck} \leq 50 \text{ MPa} \quad (5.10a)$$

$$\delta \geq 0.54 + 1.25(0.6 + \frac{0.0014}{\varepsilon_{cu2}})(x_u/d) \quad \text{for } f_{ck} > 50 \text{ MPa} \quad (5.10b)$$

$$\delta \geq 0.7 \text{ where Class B and Class C reinforcement is used}$$
$$\delta \geq 0.8 \text{ where Class A reinforcement is used}$$

$$\varepsilon_{cu2} = 0.0035 \text{ for } f_{ck} \leq 50 \text{ MPa}$$
$$\varepsilon_{cu2} = \left[2.6 + 35 \times \left\{\frac{(90 - f_{ck})}{100}\right\}^4\right] \times 10^{-3} \text{ for } f_{ck} > 50 \text{ MPa}$$

where
δ = the ratio of redistributed moment/Elastic bending moment.
x_u = the maximum depth of the neutral axis at ULS after redistribution.

Table 4.5, Table 4.6 and Table 4.8 of Chapter 4 show respectively the relationship between δ and x_u/d, z/d and k = M/ (bd^2 f$_{ck}$) for different values of f$_{ck}$.

In general in framed structures if Class B or C reinforcements which have high ductility are used, then reductions *of moments up to 30 percent of the elastic moments can be tolerated* without making excessive demands on the ductility of the structure. On the other hand if Class A steel is used then reductions are limited to 20% only. It is worth pointing out that as demonstrated in section 13.2, *ductility demand is increased by the use of moment values smaller than the elastic values*. However as the ductility demand is unaffected if moments are increased above elastic values, *there is no limit to the use of moment values larger than the elastic values*. In the case of flexural members, one way of ensuring that sufficient ductility is available is to limit the maximum depth of neutral axis as is reflected in equations (5.10a) and (5.10b) of Eurocode 2. *Larger reduction in moments from the elastic values will require smaller maximum depth of neutral axis so that steel yields well before concrete reaches maximum strain*.

13.6 DESIGN USING PLASTIC ANALYSIS IN EUROCODE 2

According to clause 5.6.2 of the Eurocode 2, design of beams, frames can be done at ULS without explicit check on the rotation capacity provided the following conditions are all satisfied.

(a) The area of tensile reinforcement is limited such that at any section
$$x_u/d \leq 0.25 \text{ for } f_{ck} \leq 50 \text{ MPa}$$
$$x_u/d \leq 0.15 \text{ for } f_{ck} \geq 55 \text{ MPa}$$
(b) Reinforcing steel is either Class B or C
(c) The ratio of moments at intermediate supports to the moments in the span is between 0.5 and 2.

13.7 SERVICEABILTY CONSIDERATIONS WHEN USING REDISTRIBUTED ELASTIC MOMENTS

Elastic bending moment at ULS: Fig. 13.6 shows the elastic bending moment distribution in a uniformly loaded clamped beam. If the ultimate design load is q, then from elastic analysis the bending moments at the support and mid-span are respectively $qL^2/12$ and $qL^2/24$. The points of contraflexure are 0.211 L from the fixed ends.

Elastic bending moment at SLS: At ULS the load factor for dead and live loads are respectively 1.35 and 1.5 or an average value of approximately 1.4. The load at SLS is $q/1.4 \approx 0.7$ q. Since at SLS, the beam is more likely to behave elastically, the bending moments at SLS is 0.7 times the bending moment values calculated by elastic analysis at ULS.

Redistributed bending moment at ULS: If the bending moments at the supports are redistributed by 30% when using Class B or C steel, then at ULS the bending moments at the support and mid-span after redistribution are respectively

0.7qL²/12 and (qL²/8 − 0.7qL²/12) = 1.6 qL²/24. For this distribution of bending moments the points of contraflexure are 0.135 L from the fixed end.

Fig. 13.6 shows the SLS and redistributed bending moments at ULS. Because of the shift in the position of contraflexure points, at certain sections in the beam the bending moment at SLS is *larger* than the redistributed bending moments at ULS. During design, it is necessary to ensure that the moment of resistance is equal to larger of the SLS and redistributed ULS moment at the section. Although Eurocode 2 does not make any recommendation about design at SLS when using redistributed moments, it is sensible that the resistance moment at any section should be at least 70 percent of the moment at that section obtained from an elastic maximum moments diagram covering all appropriate combinations design ultimate load

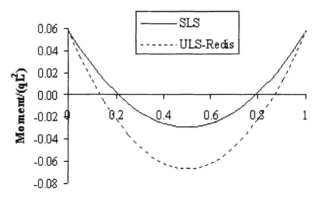

Fig. 13.6 Bending moment distribution at ULS and SLS.

13.8 CONTINUOUS BEAMS

13.8.1 Continuous Beams in *Cast-in-Situ* Concrete Floors

Continuous beams are a common element in *cast-in-situ* construction. A reinforced concrete floor in a multi-storey building is shown in Fig. 13.7. The floor action to support the loads is as follows:

(a) The one-way slab is supported on the edge frame, intermediate T-beams and centre frame. It spans transversely across the building.

(b) Intermediate T-beams on line AA span between the transverse end and interior frames to support the floor slab.

(c) Transverse end frames DD and interior frames EE span across the building and carry loads from intermediate T-beams and longitudinal frames.

(d) Longitudinal edge frames CC and interior frame BB support the floor slab.

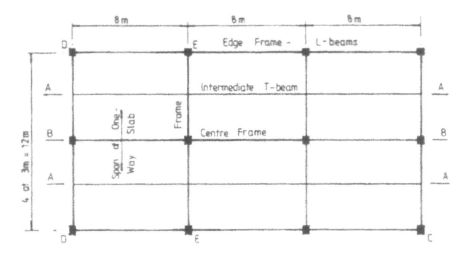

Fig. 13.7 Floor in a multi-storey building.

In the days when computer programs were not readily available, several simplified methods for the analysis of rigid-jointed frames were developed. It is no longer necessary to resort to these methods. The steps in design of continuous beams are the same as those set out in Chapter 4, section 4.4.3 for simple beams except for the limit on the depth of the neutral axis depending on the amount of redistribution done.

13.8.2 Loading on Continuous Beams

13.8.2.1 Arrangement of Loads to Give Maximum Moments

The loading to be applied to the continuous beam to give the most adverse conditions at any section along the beam can be found using qualitative influence lines obtained using Muller-Breslau's principle. It shows that in any continuous beam, the following two basic loading patterns need to be investigated.

(a) Maximum moment in a span of a beam occurs when that span and every alternate span are loaded by $(\gamma_G \, G_k + \gamma_Q \, Q_k)$ and the rest of the spans by $1.0 \, G_k$.

(b) Maximum moment at a support in a beam occurs when spans on either side of the support and every alternate span are loaded by $(\gamma_G\ G_k + \gamma_Q\ Q_k)$ and the rest of the spans by $1.0\ G_k$.

13.8.2.2 Eurocode 2 Arrangement of Loads to Give Maximum Moments

In order to reduce the number of load cases to be analysed, Eurocode 2, clause 5.1.3 prescribes the following simplified load arrangements for buildings.

(a) Alternate spans are loaded with $(\gamma_G\ G_k + \gamma_Q\ Q_k)$ and the rest of the spans carrying $\gamma_G\ G_k$.

(b) Any two adjacent spans are loaded with are loaded with $(\gamma_G\ G_k + \gamma_Q\ Q_k)$ and the rest of the spans carrying $\gamma_G\ G_k$.

13.8.2.3 The U.K. National Annex Arrangement of Loads to Give Maximum Moments

The U.K. National Annex allows the following load combinations.

(a) Alternate spans are loaded with $(\gamma_G\ G_k + \gamma_Q\ Q_k)$ and the rest of the spans carrying $\gamma_G\ G_k$.

(b) All spans are loaded with $(\gamma_G\ G_k + \gamma_Q\ Q_k)$.

13.8.2.4 Example of Critical Loading Arrangements

The total dead load on the floor in Fig. 13.7 including an allowance for the ribs of the T-beams, screed, finishes, partitions, ceiling and services is 6.6 kN/m^2 and the imposed load is 3 kN/m^2. Calculate the design load and set out the load arrangements for the continuous T-beam on lines AA and BB.

$$G_k = 3 \times 6.6 = 19.8\ \text{kN/m}$$
$$Q_k = 3 \times 3 = 9\ \text{kN/m}$$
$$1.35\ G_k + 1.5\ Q_k = (1.35 \times 19.8) + (1.5 \times 9) = 40.23\ \text{kN/m}$$
$$1.35\ G_k = (1.35 \times 19.8) = 26.73\ \text{kN/m}$$

The loading arrangements are shown in Fig. 13.8 for Eurocode 2 loading and in Fig. 13.9 for the U.K. National Annex loading.

13.8.2.5 Loading from One-Way Slabs

Continuous beams supporting slabs designed as spanning one-way can be considered to be uniformly loaded. Fig. 13.10 shows a set of single span main beams spanning between the columns. Another set of long continuous beams span between short beams. The slab rests on the long beams and span in the direction of short beams. Because of the fact that slabs span in one direction only, the loads on the long beams are simply reactions from the slabs and can be considered to be uniformly distributed. Note that some two-way action occurs at the corners of one-way slabs.

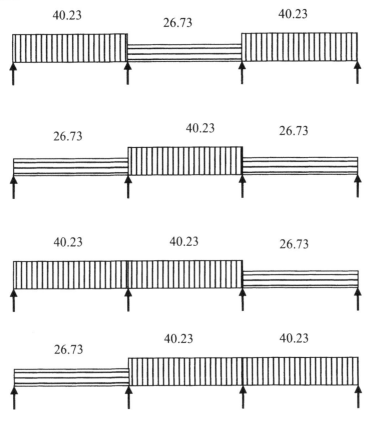

Fig. 13.8 Loading arrangements: Eurocode 2.

13.8.2.6 Loading from Two-Way Slabs

If the slab is designed as spanning two ways, the four edge beams assist in carrying the loading. The load distribution normally assumed for analyses of the edge beams is shown in Fig. 13.11 where lines at 45° are drawn from the corners of the slab. This distribution gives triangular and trapezoidal loads on the edge beams as shown in the Fig. 13.11.

The fixed end moments for the two load cases shown in Fig. 13.11(b) and Fig. 13.11(c) are as follows.

(a) Trapezoidal load: The fixed end moments are determined by splitting the total load into a uniform central portion and two triangular end portions each, where W_1 is the total load on one span of the beam, ℓ_x is the short span of the slab and ℓ_y is the long span of the slab.

The fixed end moments for the two spans in the beam on AA are:

$$M_1 = \frac{W_1}{(\ell_y - 0.5\ell_x)} \frac{1}{96\ell_y} [8\ell_y^3 - 4\ell_y\ell_x^2 + \ell_x^3]$$

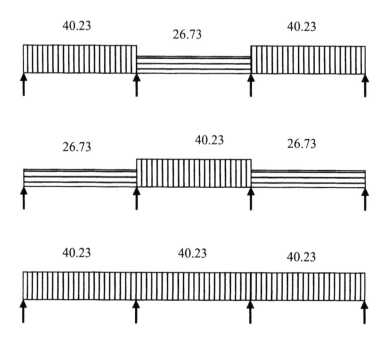

Fig. 13.9 Loading arrangements: The U.K. National Annex.

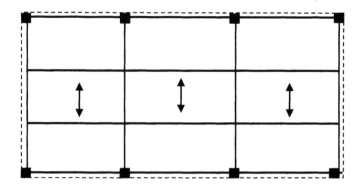

Fig. 13.10 Floor plan of one-way spanning slabs.

(b) Triangular load: The fixed end moments for the two spans in the beam on line BB in Fig. 13.11(a) is

$$M_2 = \frac{5W_2 \ell_x^2}{48}$$

where W_2 is the total load on one span of the beam.

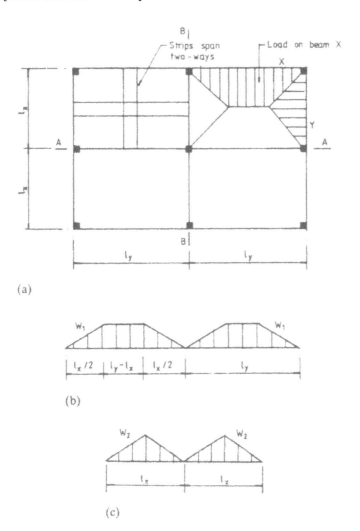

(a)

(b)

(c)

Fig. 13.11 (a) Floor plan; (b) beam on AA; (c) beam on BB.

13.8.2.7 Analysis for Shear and Moment Envelopes

Analyses using the matrix stiffness method can be used to find the shear forces and bending moments for design.

13.9 EXAMPLE OF ELASTIC ANALYSIS OF CONTINUOUS BEAM

(a) Specification

Analyse the continuous beam for the three load cases shown in Fig. 13.8 and Fig. 13.9 and draw the separate shear force and bending moment diagrams. Construct the maximum shear force and bending moment envelopes.

(b) Analysis by Stiffness method

(i) Fixed end moments for the Eurocode 2 loading are:

Case 1: Spans AB and CD: $M = 40.23 \times 8^2/12 = 214.56$ kNm
Span BC: $M = 26.73 \times 8^2/12 = 142.56$ kNm
Case 2: Spans AB and CD: $M = 26.73 \times 8^2/12 = 142.56$ kNm
Span BC: $M = 40.23 \times 8^2/12 = 214.56$ kNm
Case 3: Spans AB and BC: $M = 40.23 \times 8^2/12 = 214.56$ kNm
Span CD: $M = 26.73 \times 8^2/12 = 142.56$ kNm
Case 4: Span AB: $M = 26.73 \times 8^2/12 = 142.56$ kNm
Spans BC and CD: $M = 40.23 \times 8^2/12 = 214.56$ kNm

Assuming uniform flexural rigidity EI and equal spans of 8 m, the stiffness matrix K and the load vectors F for the four load cases are:

$$K = \frac{EI}{8}\begin{bmatrix} 4 & 2 & 0 & 0 \\ 2 & 8 & 2 & 0 \\ 0 & 2 & 8 & 2 \\ 0 & 0 & 2 & 4 \end{bmatrix}\begin{bmatrix} \theta_A \\ \theta_B \\ \theta_C \\ \theta_D \end{bmatrix}, \; F = \begin{bmatrix} 214.56 & 142.56 & 214.56 & 142.56 \\ -72.0 & 72.0 & 0 & 72.0 \\ 72.0 & -72.0 & -72.0 & 0 \\ -214.56 & -142.56 & -142.56 & -214.56 \end{bmatrix}$$

(ii) The fixed end moments for the U.K. National Annex loading are:

Case 1: Spans AB and CD: $M = 40.23 \times 82/12 = 214.56$ kNm
Span BC: $M = 26.73 \times 8^2/12 = 142.56$ kNm
Case 2: Spans AB and CD: $M = 26.73 \times 8^2/12 = 142.56$ kNm
Span BC: $M = 40.23 \times 8^2/12 = 214.56$ kNm
Case 3: Spans AB, BC and CD: $M = 40.23 \times 8^2/12 = 214.56$ kNm

Assuming uniform flexural rigidity EI and equal spans of 8 m, the stiffness matrix K and the load vectors F for the three load cases are:

$$K = \frac{EI}{8} \begin{bmatrix} 4 & 2 & 0 & 0 \\ 2 & 8 & 2 & 0 \\ 0 & 2 & 8 & 2 \\ 0 & 0 & 2 & 4 \end{bmatrix} \begin{bmatrix} \theta_A \\ \theta_B \\ \theta_C \\ \theta_D \end{bmatrix}, F = \begin{bmatrix} 214.56 & 142.56 & 214.56 \\ -72.0 & 72.0 & 0 \\ 72.0 & -72.0 & 0 \\ -214.56 & -142.56 & -214.56 \end{bmatrix}$$

Using clockwise moment as positive, the bending moments at the ends of a span are obtained from the following equations

$$M_{left} = \text{Fixed end moment} + (EI/L)\ \{4\ \theta_{Left} + 2\ \theta_{Right}\}$$
$$M_{Right} = \text{Fixed end moment} + (EI/L)\ \{2\ \theta_{Left} + 4\ \theta_{Right}\}$$

The reactions R are given by

$$R_{left} = 0.5\ q\ L + (M_{left} - M_{right})/L$$
$$R_{Right} = 0.5\ q\ L - (M_{left} - M_{right})/L$$

The shear force V and bending moment M at a section x from the left hand support are given by

$$V = R_{left} - q\ x$$
$$M = M_{left} - R_{left}\ x + 0.5\ q\ x^2$$

where L = span (8 m) and q = uniformly distributed loading.

The results for Eurocode 2 loading are summarised in Table 13.1. Detailed calculations for beam AB and beam BC are shown in Tables 13.2 and 13.3 respectively. The bending moment diagrams, bending moment envelope and shear force envelope for the four load cases are shown respectively in Fig. 13.12(a), Fig. 13.12(b) and Fig. 13.12(c).

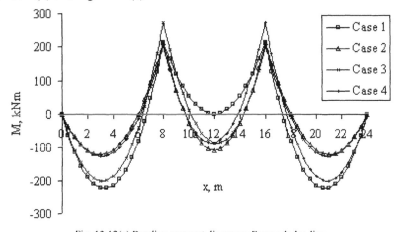

Fig. 13.12(a) Bending moment diagrams: Eurocode loading.

Table 13.1 Summary of elastic analysis: Eurocode 2 loading

	Beam AB Cases				Beam BC Cases			
	1	2	3	4	1	2	3	4
R_{Left}	134	80	127	82	107	161	170	152
R_{Right}	188	134	195	132	107	161	152	170
M_{Left}	0	0	0	0	−214	−214	−272	−200
M_{Right}	214	214	272	200	214	214	200	272
M_{Max} in Span	−224	−120	−200	−126	0	139	−87	−87
M_{max} at x	3.33	3.0	3.16	3.07	4	4	4.22	3.78

Table 13.2 Elastic moment and shear calculations for beam AB: Eurocode 2 loading

x	Case 1	Case 2	Case 3	Case 4	Moment		Shear force	
					Max.	Min.	Max.	Min.
0	0	0	0	0	0	0	134	80
1	−114	−67	−107	−69	−67	−114	94	53
2	−188	−107	−173	−110	−107	−188	54	27
3	−221	−120	−200	−126	−120	−221	14	0
4	−215	−107	−186	−114	−107	−215	−25	−34
5	−168	−67	−132	−76	−67	−168	−52	−74
6	−81	0	−38	−11	0	−81	−78	−114
7	47	94	97	81	97	47	−105	−155
8	214	214	272	200	272	200	−132	−195

Table 13.3 Elastic moment and shear calculations for beam BC: Eurocode 2 loading

x	Case 1	Case 2	Case 3	Case 4	Moment		Shear force	
					Max.	Min.	Max.	Min.
0	214	214	272	200	272	200	170	107
1	121	74	122	68	122	68	130	80
2	54	−27	13	−24	54	−27	90	54
3	14	−88	−57	−75	14	−88	49	27
4	0	−108	−86	−86	0	−108	9	−9
5	14	−88	−75	−57	14	−88	−27	−49
6	54	−27	−23	13	54	−27	−54	−90
7	121	74	68	122	122	68	−80	−130
8	214	214	200	272	272	200	−107	−170

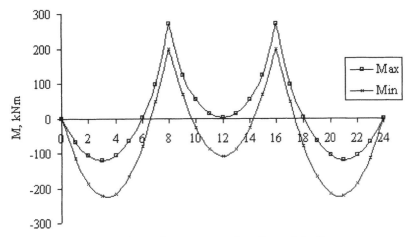

Fig. 13.12(b) Bending moment envelope: Eurocode loading.

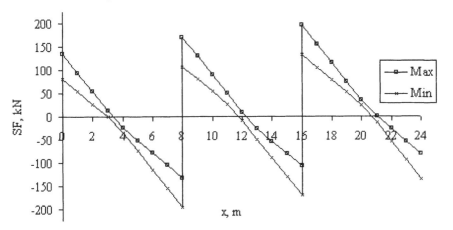

Fig. 13.12(c) Shear force envelope: Eurocode loading.

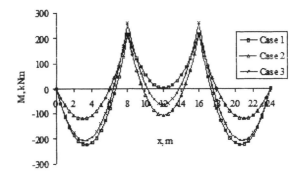

Fig. 13.13(a) Bending moment diagrams. U.K. National Annex loading.

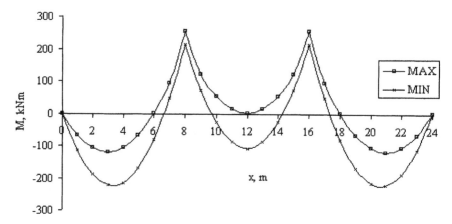

Fig. 13.13(b) Bending moment envelope: The U.K. National Annex loading.

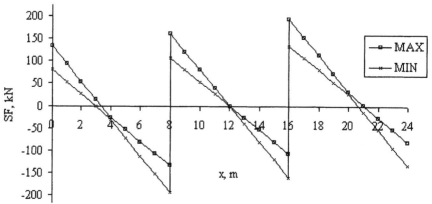

Fig. 13.13(c) Shear force envelope: The U.K. National Annex loading.

Table 13.4 Summary of elastic analysis: U.K. National Annex loading

	Beam AB			Beam BC		
	Case 1	Case 2	Case 3	Case 1	Case 2	Case 3
R_{Left}	134	80	129	107	161	161
R_{Right}	188	134	193	107	161	161
M_{Left}	0	0	0	−214	−214	−258
M_{Right}	214	214	258	214	214	258
M_{Max} in Span	216	75	206	0	139	64
M_{max} at x	3.33	3.0	3.20	4.0	4.0	4.0

Table 13.5 Elastic moment and shear calculations for beam AB: U.K. National Annex loading

x	Case 1	Case 2	Case 3	Moment		Shear force	
				Max.	Min.	Max.	Min.
0	0	0	0	0	0	134	80
1	−114	−67	−109	−67	−114	94	53
2	−188	−107	−177	−107	−188	54	27
3	−221	−120	−205	−120	−221	14	0
4	−215	−107	−193	−107	−215	−27	−32
5	−168	−67	−141	−67	−168	−54	−72
6	−81	0	−48	0	−81	−80	−113
7	47	94	85	94	47	−107	−153
8	214	214	257	257	214	−134	−193

Table 13.6 Elastic moment and shear calculations for beam BC: U.K. National Annex loading

x	Case 1	Case 2	Case 3	Moment		Shear force	
				Max.	Min.	Max.	Min.
0	214	214	257	257	214	161	107
1	121	74	117	308	74	121	80
2	54	−27	16	375	−27	81	54
3	14	−88	−44	415	−88	40	27
4	0	−108	−64	428	−108	0	0
5	14	−88	−44	415	−88	−27	−40
6	54	−27	16	375	−27	−54	−81
7	121	74	117	308	74	−80	−121
8	214	214	258	258	214	−107	−161

The results for U.K. National Annex loading are summarised in Table 13.4. Detailed calculations for beam AB and beam BC are shown in Tables 13.5 and 13.6 respectively. The bending moment diagrams, bending moment envelope and shear force envelope for the four load cases are shown respectively in Fig. 13.13(a), Fig. 13.13(b) and Fig. 13.13(c).

13.10 EXAMPLE OF MOMENT REDISTRIBUTION FOR CONTINUOUS BEAM

As explained in section 13.4, redistribution gives a more even arrangement for the reinforcement, relieving congestion at supports. It might also lead to a saving in the amount of reinforcement required.

(a) Specification
Referring to the three-span continuous beam analysed in section 13.7 above, redistribute the moments after making a 30% reduction in the maximum hogging moment at the interior support. Draw the envelopes for maximum shear force and bending moment.

(b) Moment redistribution: Eurocode 2 loading
In the case of loading according to Eurocode 2, the maximum elastic hogging moment over the support B in case 3 (Table 13.4) or over the support C in case 4 is 271.84 kNm. If this is *reduced* by 30%, the hogging moment over the support is

$0.7 \times 271.84 = 190.29$. The support section is therefore designed for a moment of 190.29 kNm. Reducing the hogging moment increases the corresponding span moments.

The results are summarised in Tables 13.7 and detailed calculations for beams AB and BC are shown in Tables 13.8 and Table 13.9 respectively. Fig. 13.14(a), Fig. 13.14(b) and Fig. 13.13(c) show respectively the redistributed bending moment diagrams, moment envelope and shear force diagrams.

When Fig. 13.12(a) and Fig. 13.14(a) are compared, it is noted that the maximum hogging moment from the elastic bending moment envelope has been reduced by the moment redistribution. However the maximum sagging moment from the elastic bending moment envelope has been increased by the moment redistribution. This is because any reduction in hogging moment leads to an increase in the sagging moment at mid-span by half that amount. As stated before, the object of moment redistribution is to reduce congestion of reinforcement at the supports.

Table 13.7 Summary of redistributed analysis: Eurocode 2 loading

	Beam AB				Beam BC			
	Case 1	Case 2	Case 3	Case 4	Case 1	Case 2	Case 3	Case 4
R_{Left}	137	83	137	83	107	161	161	161
R_{Right}	185	131	185	131	107	161	161	161
M_{Left}	0	0	0	0	−190	−190	−190	−190
M_{Right}	190	190	190	190	190	190	190	190
M_{Max} in Span	*−234*	−129	−234	−129	24	132	132	132
M_{max} at x	3.41	3.11	3.41	3.11	4.0	4.0	4.0	4.0

Table 13.8 Redistributed moment and shear calculations for beam AB: Eurocode 2 loading

x	Case 1	Case 2	Case 3	Case 4	Moment		Shear force	
					Max.	Min.	Max.	Min.
0	0	0	0	0	0	0	137	83
1	−117	−70	−117	−70	−70	−117	97	56
2	−194	−113	−194	−113	−113	−194	57	30
3	−230	−129	−230	−129	−129	−230	16	3
4	−227	−119	−227	−119	−119	−227	−24	−24
5	−183	−82	−183	−82	−82	−183	−51	−64
6	−99	−18	−99	−18	−18	−99	−77	−104
7	26	73	26	73	73	26	−104	−145
8	190	190	190	190	190	190	−131	−185

Table 13.9 Redistributed moment and shear calculations for beam BC: Eurocode 2 loading

x	Case 1	Case 2	Case 3	Case 4	Moment		Shear force	
					Max.	Min.	Max.	Min.
0	191	190	190	190	191	190	161	107
1	97	50	50	50	97	50	121	80
2	30	−51	−51	−51	30	−51	81	54
3	−10	−111	−111	−111	−10	−111	40	27
4	−24	−132	−132	−132	−24	−132	0	0
5	−10	−111	−111	−111	−10	−111	−27	−40
6	30	−51	−51	−51	30	−51	−54	−81
7	97	50	50	50	97	50	−80.	−121
8	190	190	190	190	190	190	−107	−161

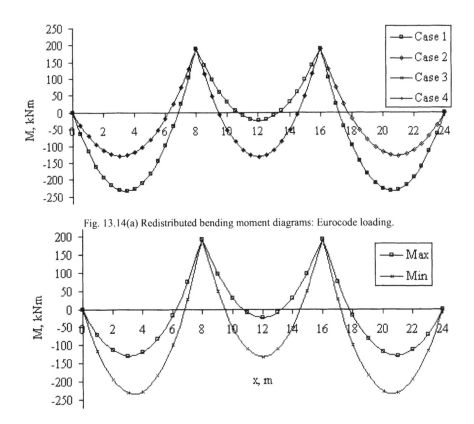

Fig. 13.14(a) Redistributed bending moment diagrams: Eurocode loading.

Fig. 13.14(b) Redistributed bending moment envelope: Eurocode loading.

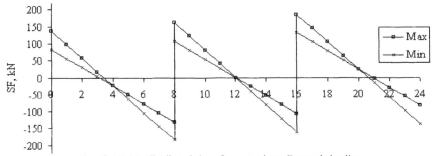

Fig. 13.14(c) Redistributed shear force envelope: Eurocode loading.

(c) Moment redistribution: The U.K. National Annex loading

In the case of loading according to the U.K. National Annex, the maximum elastic hogging moment over supports B and C in case 3 (Table 13.6) is 257.47 kNm. If this is *reduced* by 30 percent, the hogging moment over the support is $0.7 \times 257.47 = 180.23$. The support section is therefore designed for a moment of

180.23 kNm. Reducing the hogging moment increases the corresponding span moments.

Table 13.10 Summary of redistributed elastic analysis: U.K. National Annex loading

	Beam AB			Beam BC		
	Case 1	Case 2	Case 3	Case 1	Case 2	Case 3
R_{Left}	138	84	138	107	161	161
R_{Right}	184	130	184	107	161	161
M_{Left}	0	0	0	−180	−180	−180
M_{Right}	180	180	180	180	180	180
M_{Max} in Span	238	133	238	34	142	142
M_{max} at x	3.44	3.16	3.44	4.0	4.0	4.0

Table 13.11 Redistributed elastic moment and shear calculations for beam AB: The U.K. National Annex loading

x	Case 1	Case 2	Case 3	Moment		Shear force	
				Max.	Min.	Max.	Min.
0	0	0	0	0	0	138	84
1	−118	−71	−118	−71	−118	98	58
2	−196	−115	−196	−115	−196	58	31
3	−234	−133	−234	−133	−234	18	4
4	−232	−124	−232	−124	−232	−23	−23
5	−189	−88	−189	−88	−189	−49	−63
6	−106	−25	−106	−25	−106	−76	−103
7	17	64	17	64	17	−103	−143
8	180	180	180	180	180	−130	−184

Table 13.12 Redistributed elastic moment and shear calculations for beam BC: The U.K. National Annex loading

x	Case 1	Case 2	Case 3	Moment		Shear force	
				Max.	Min.	Max.	Min.
0	180	180	180	180	180	161	107
1	87	40	40	87	40	121	80
2	20	−61	−61	20	−61	81	54
3	−20	−121	−121	−20	−121	40	27
4	−34	−142	−142	−34	−142	0	0
5	−20	−121	−121	−20	−121	−27	−40
6	20	−61	−61	20	−61	−54	−81
7	87	40	40	87	40	−80	−121
8	180	180	180	180	180	−107	−161

The results are summarised in Tables 13.10 and detailed calculations for beams AB and BC are shown in Tables 13.11 and Table 13.12 respectively. Fig. 13.15(a), Fig. 13.15(b) and Fig. 13.15(c) show respectively the redistributed bending moment diagrams, moment envelope and shear force diagrams.

(d) Shear force envelope for design
Redistribution of moments alters the shear force distribution. In order to guard against the possibility of redistribution **not** occurring, for design at any section one

need to take the larger of the elastic and redistributed shear force. Table 13.13 and Table 13.14 show the design shear envelope values for beams AB and BC respectively for Eurocode loading. Tables 13.15 and Table 13.16 show the design shear envelope values for beams AB and BC respectively for the U.K. National Annex loading. The final values for design are highlighted. Fig. 13.14(c) and Fig. 13.15(c) show the corresponding shear envelope values.

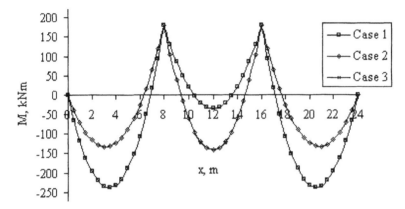

Fig. 13.15(a) Redistributed bending moment diagrams: The U.K. National Annex loading.

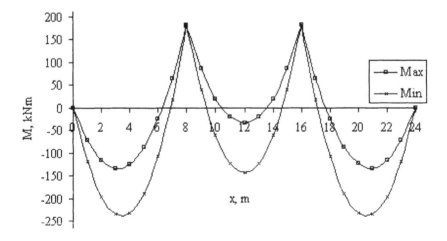

Fig. 13.15(b) Redistributed bending moment envelope: The U.K. National Annex loading.

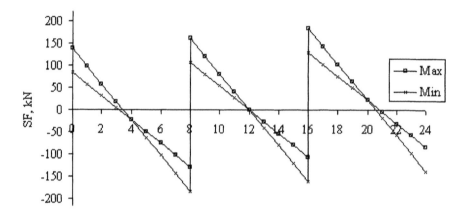

Fig. 13.15(c) Redistributed shear force envelope: U.K. National Annex loading.

Table 13.13 Design shear envelope for beam AB: Eurocode loading

x	Redistributed		Elastic		Design	
	Max.	Min.	Max.	Min.	Max.	Min.
0	137	83	134	80	**137**	83
1	97	56	94	53	**97**	56
2	57	30	54	27	**57**	30
3	16	3	14	0	**16**	3
4	−24	−24	−25	−34	−25	**−34**
5	−51	−64	−52	−74	−52	**−74**
6	−77	−104	−78	−114	−78	**−114**
7	−104	−145	−105	−155	−105	**−155**
8	−131	−185	−132	−195	−132	**−195**

13.14 Design shear force envelopes for beam BC: Eurocode loading

x	Redistributed		Elastic		Design	
	Maximum	Minimum	Maximum	Minimum	Maximum	Minimum
0	161	107	170	107	**170**	107
1	121	80	130	80	**130**	80
2	81	54	90	54	**90**	54
3	40	27	49	27	**49**	27
4	0	0	9	−9	**9**	−9
5	−27	−40	−27	−49	−27	**−49**
6	−54	−81	−54	−90	−54	**−90**
7	−80	−121	−80	−130	−80	**−130**
8	−107	−161	−107	−170	−107	**−170**

Table 13.15 Design shear envelope for beam AB: The U.K. National Annex loading

	Redistributed		Elastic		Design	
x	Maximum	Minimum	Maximum	Minimum	Maximum	Minimum
0	138	84	134	80	**138**	84
1	98	58	94	53	**98**	58
2	58	31	54	27	**58**	31
3	18	4	14	0	**18**	4
4	−23	−23	−27	−32	−27	**−32**
5	−49	−63	−54	−72	−54	**−72**
6	−76	−103	−80	−113	−80	**−113**
7	−103	−143	−107	−153	−107	**−153**
8	−130	−184	−134	−193	−134	**−193**

Table 13.16 Design shear force envelope for beam BC: The U.K. National Annex loading

x	Redistributed		Elastic		Design	
	Maximum	Minimum	Maximum	Minimum	Maximum	Minimum
0	161	107	161	107	**161**	107
1	121	80	121	80	**121**	80
2	81	54	81	54	**81**	54
3	40	27	40	27	**40**	27
4	0	0	0	0	**0**	0
5	−27	−40	−27	−40	−27	**−40**
6	−54	−81	−54	−81	−54	**−81**
7	−80	−121	−80	−121	−80	**−121**
8	−107	−161	−107	−161	−107	**−161**

13.11 CURTAILMENT OF BARS

The curtailment of bars may be carried out in accordance with the detailed provisions set out in Eurocode 2, clauses 9.2.1.3. The anchorage of tension bars at the simply supported ends is dealt with in clauses 9.2.1.4 of the code. The anchorage of bottom reinforcement at intermediate supports is dealt with in clauses 9.2.1.5 of the code. More details are given in sections 5.2 and 5.3 of Chapter 5.

13.12 EXAMPLE OF DESIGN FOR THE END SPAN OF A CONTINUOUS BEAM

(a) Specification
Design the end span of the continuous beam analysed in section 13.8. The design is to be made for the shear forces and moments obtained after 30% redistribution from the elastic analysis has been made for the Eurocode 2 loading. The shear force and moment envelopes are shown in Fig. 13.14. The materials are $f_{ck} = 30$ MPa and $f_{yk} = 500$ MPa.

(b) Design of moment steel
The assumed beam sections for mid-span and over the interior support are shown in Fig. 13.16(a) and Fig. 13.16(b) respectively. The cover for exposure class

XC2/XC3 from code Table 4.4N is 25 mm. The axis distance for a fire resistance period of R120 from the extract from Table 2.8 shown below is about 40 mm for continuous beams.

R 120	b = 200	300	450	500	130
	a = 45	35	35	30	

(i) Section near the centre of the span: From Table 13.9, design moment
$$M = 233.7 \text{ kNm.}$$
Using equations 5.7 of the Eurocode 2, calculate the effective width.
$$\text{Spacing of beams, b} = 3\text{m (See Fig. 13.7)}$$
$$\text{Clear span } \ell_1 = 8 \text{ m}$$
$$\text{Effective span } \ell_0 = 0.85 \times \ell_1 = 6.8 \text{ m (See Fig. 4.10)}$$
$$b_1 = b_2 = 0.5 \times (3000 - 250) = 1375 \text{ mm (See Fig. 4.11)}$$
$$b_{eff, 1} = b_{eff, 2} = 0.2 \times 1375 + 0.1 \times \ell_0 = 955 \text{ mm} < 0.2 \times \ell_0$$
$$b_{eff} = 2 \times 955 + 250 = 2160 \text{ mm}$$
$$d \approx 450 - 25 \text{ (cover)} - 10 \text{ (links)} - 25/2 = 400 \text{ mm}$$
$$\text{Axis distance} = 25 + 10 + 25/2 = 48 \text{ mm (Satisfactory)}$$
The moment of resistance of the section when the entire flange is in compression is
$$M_{Flange} = f_{cd} \, b_{eff} \, h_f \, (d - h_f/2)$$
$$M_{Flange} = (30/1.5) \times 2160 \times 125 \times (400 - 0.5 \times 125) \times 10^{-6} = 1822.5 \text{ kNm}$$
$$M < M_{flange}$$
The neutral axis lies in the flange. The beam can be designed as a rectangular beam. Although 30% redistribution has been done at supports, the moment in the span has increased. Therefore there are no problems of rotation capacity. Therefore maximum value of k = 0.196.
$$k = 233.7 \times 10^6 / (30 \times 2160 \times 400^2) = 0.023 < 0.196$$

$$\frac{z}{d} = 0.5[1.0 + \sqrt{(1 - 3\frac{k}{\eta})}]$$

Substituting $\eta = 1$, k = 0.023, z/d = 0.98
$$A_s = 233.7 \times 10^6 / (0.98 \times 400 \times 0.87 \times 500) = 1371 \text{ mm}^2$$
Provide 3H25, $A_s = 1473 \text{ mm}^2$.
Check for minimum steel:
$$A_{s, min} = 0.26 \times (f_{cm}/f_{yk}) \times b_t \, d \qquad\qquad (9.1\text{N})$$

$b_t = 250$, $f_{yk} = 500$ MPa, $f_{cm} = 0.30 \times f_{ck}^{0.667} = 2.9$MPa, $A_{s, min} = 151 \text{ mm}^2$

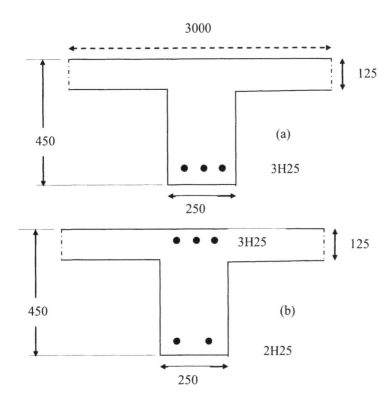

Fig. 13.16 (a) T-beam at mid-span; (b) rectangular beam over support.

The moment of resistance after stopping off one H25 bar is calculated where
$$d = 400 \text{ mm}, z = 0.98d \text{ and } A_s = 981 \text{ mm}^2.$$
The moment of resistance is
$$M_R = (0.87 \times 500) \times (0.98 \times 400) \times 981 \times 10^{-6} = 167 \text{ kNm}$$
The design moment envelope using the data in Table 13.9 for beam AB is
$$M = 137.1 \text{ x} - 40.23 \times x^2/2$$
If M = 167.0 kNm, x =1.59 m and 5.22 m from the simply supported end.
The shift a_1 in moment diagram to allow for tensile stress caused by shear is
$$a_1 = 0.5 \text{ z cot } \theta \tag{9.2}$$
Taking $z \approx 0.98$ d = 400mm, cot θ = 2.5, a_1 = 500 mm.
From Table 5.5, ℓ_{bd} = 36 φ = 36 × 25 = 900 mm.
x = 1.59 – 0.5 – 0.9 = 0.19 m and x = 8.0 – 5.22 – 0.5 – 0.9= 1.38 m from the ends.
2H25 bars are carried to the end.   According to Eurocode 2 clause 6.2.3(7),
equation (6.18), the additional tensile force in the longitudinal reinforcement due to
inclined cracks caused by shear is
$$\Delta F_{Ed} = 0.5 \text{ V}_{Ed} \text{ cot } \theta \tag{6.18}$$
Taking $V_{Ed} \approx$ reaction – load over 1.59 m,
   At the left hand end: V_{Ed} = 137.1 kN – 40.23 × 1.59 = 73 kN, cot θ ≈ 2.5

At the right hand end: $V_{Ed} = 137.1$ kN $- 40.23 \times 5.22 = -73$ kN, cot $\theta \approx 2.5$

$$\Delta F_{Ed} = 0.5 \, V_{Ed} \cot \theta = 91 \text{ kN}$$

Using the data from Table 5.5, bond strength $f_{bd} = 3.0$ MPa, anchorage length for 2H25 bars is

$$91 \times 10^3 = 2 \times \pi \phi \times l_{bd,reqd} \times f_{bd}$$

ϕ = bar diameter = 25 mm, $\ell_{bd, \, reqd} = 193$ mm.

This can be provided by a hook at the end of bar at the left and carrying the bars to the next span at the right.

Check minimum steel using Eurocode 2 equation (9.1N):

$f_{cm} = 0.3 \times f_{ck}^{\,0.667} = 2.9$ MPa, $f_{yk} = 500$ MPa, $b_t = 250$ mm, d = 400 mm

$$A_{s, \, min} = 0.26 \times (f_{cm}/f_{yk}) \times b_t \, d = 151 \text{ mm}^2 \qquad (9.1N)$$

(ii) Section at the interior support: The beam acts as a rectangular beam at the support. As 30% redistribution has been to reduce the moment at the support, from Table 4.8, Chapter 4, for $\delta = 0.7$, maximum value of k = 0.102.

The design moment from Table 13.10 is 190.32 kN m.

$$k = 190.32 \times 10^6 / (30 \times 250 \times 400^2) = 0.159 > 0.102$$

Compression reinforcement is required.

Moment resisted without compression steel:

$$M_{sr} = 0.102 \times 250 \times 400^2 \times 30 \times 10^{-6} = 122.4 \text{ kNm}$$
$$M_{compression \, steel} = 190.32 - 122.4 = 67.92 \text{ kNm}$$

Check whether compression steel yields:

$$\text{From Table 4.5, } x_u/d = 0.208, \, x_u = 83 \text{ mm}$$
$$d' = 30 \text{ (cover)} + 10 \text{ (link)} + 25/2 = 52.5 \text{ mm}$$

Calculate the strain in compression steel.

$$\varepsilon_{sc} = \varepsilon_{cu3} \frac{(x - d')}{x} = 3.5 \times 10^{-3} \times \frac{(83 - 52.5)}{83} = 1.286 \times 10^{-3}$$

$$\text{Stress in steel, } f'_s = \varepsilon_{sc} \times (E_s = 200 \times 10^3) = 257 \text{ MPa}$$
$$A_s' = (M - M_{sr})/\{f_s \, (d - d')\}$$
$$A_s' = (190.32 - 122.4) \times 10^6 / \{257 \times (400 - 52.5)\} = 761 \text{ mm}^2$$
$$\text{Provide 2H25 giving } A'_s = 982 \text{ mm}^2.$$

Compression in concrete due to singly reinforced moment = $C_{sr} = k_c$ bd f_{ck}

From Table 4.7, $k_c = 0.1110$

$$C_{sr} = k_c \text{ bd } f_{ck} = 0.1110 \times 250 \times 400 \times 30 \times 10^{-3} = 333 \text{ kN}$$
$$C_{sc} = A'_s \, f'_s = 761 \times 257 \times 10^{-3} = 195.6 \text{ kN}$$
$$T = A_s \, 0.87 \, f_{yk} = C_{sr} + C_{sc}$$
$$A_s \times (500/1.15) \times 10^{-3} = 333.0 + 195.6$$
$$A_s = 1216 \text{ mm}^2$$
$$\text{Provide 3H25 giving } A_s = 1473 \text{ mm}^2.$$

The compression reinforcement will be provided by carrying 2H25 mid-span bars through to the support. For tension reinforcement, provide 3H25.

The theoretical and actual cut-off points for one of the three top bars are determined. The moment of resistance of the section with 2H25 bars and an effective depth $d = 400$ mm is calculated. Assuming that steel yields, equate the total tensile force T and the total compressive force C.

$$T = 0.87 \times 500 \times 981 \times 10^{-3} = 426.74 \text{ kN}$$
$$C = (30/1.5) \times (0.8x) \times 250 \times 10^{-3} = 4x \text{ kN}$$
$$\text{Equating T} = C, x = 107 \text{ mm}$$
$$x/d = 107/400 = 0.27 > (0.208 \text{ for } \delta = 0.7)$$

Assuming that steel does not yield and taking $x_u/d = 0.208$,

$$C = f_{cd} \times 0.8 \ x_u \times b = (30/1.5) \times 0.8 \times (0.208 \times 400) \times 250 \times 10^{-3} = 332.8 \text{ kN}$$
$$T = f_s \times 981 \times 10^{-3} = 0.981 \ f_s \text{ kN}$$
$$\text{Equating T} = C, f_s = 339 \text{ MPa}$$
$$\text{Lever arm: } z/d = 1 - 0.4 \ x/d = 0.92$$

The moment of resistance is

$$M_R = C \ z = 332.8 \times (0.92 \times 400) \times 10^{-3} = 122.5 \text{ kNm}$$

From the data in Table 13.9, for case 1 loading, the equation for the moment is given by

$$M = -190.3 + 184.7 \ x - 40.23 \times x^2/2$$

If M = 122.5 kNm, x = 0.76 m. Shifting the moment diagram by $a_1 = 0.5$ m, and $\ell_{bd} = 36\varphi = 36 \times 25 = 900$ mm, carry three bars to $0.76 + 0.5 + 0.9 = 2.16$ m from the support.

$$V_{Ed} = 184.7 - 40.23 \times 1.26 = 134 \text{ kN}$$
$$\Delta F_{Ed} = 0.5 \ V_{Ed} \cot \theta = 0.5 \times 134 \times 2.5 = 168 \text{ kN}$$

Using the data from Table 5.5, bond strength $f_{bd} = 3.0$ MPa, anchorage length for 2H25 bars is

$$168 \times 10^3 = 2 \times \pi \varphi \times l_{bd,reqd} \times f_{bd}$$

φ = bar diameter = 25 mm, $\ell_{bd, \ reqd} = 357$ mm.

As the bars are carried right to the ends, there is enough anchorage length. The bar cut-offs are shown in Fig. 13.17.

(c) Design of shear reinforcement

(i) Simply supported end: From the design shear force envelope:
Check for maximum shear stress: From the data in Table 13.10

$$V_{Ed} = 137 \text{ kN}$$

Using code equations (6.9), (6.6N) and (6.11bN),

$$V_{Rd, max} = \alpha_{cw} \times b_w \times z \times v_1 \times f_{cd}/ (\cot\theta + \tan \theta) \qquad (6.9)$$
$$f_{cd} = 30/1.5 = 20 \text{ MPa}, b_w = 250 \text{ mm}, z \approx 0.9 \ d = 360 \text{ mm}$$
$$v_1 = v = 0.6 \ (1 - f_{ck}/250) = 0.528, \alpha_{cw} = 1.0, \qquad (6.6N) \text{ and } (6.11.bN)$$
$$V_{Rd, max} = 950.4/ (\cot\theta + \tan \theta) \text{ kN}$$
$$\text{Equating } V_{Ed} \text{ to } V_{Rd, max}$$
$$(\cot\theta + \tan \theta) = 6.929$$

$\cot\theta = 0.3$ and 6.8 both of which are outside the range. Taking $\cot\theta = 2.5$ to give the smallest value of $V_{Rd, max} = 327.7$ kN $> V_{Ed}$.
Section is satisfactory and shear links can be designed.

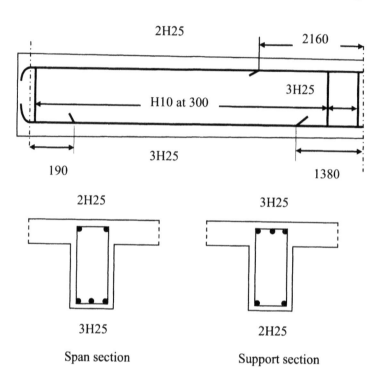

Fig. 13.17 Reinforcement details for the continuous beam.

Shear V at d from support
$$V = 143.71 - 40.23 \times (d = 400) \times 10^{-3} = 127.62 \text{ kN}$$
Check if shear reinforcement is needed. Use equation (6.2a), (6.2b) and (6.6N) of the code.
$$V_{Rd, c} = [C_{Rd, c} \times k \times (100 \, \rho_1 \times f_{ck})^{0.33}] \, b_w \, d \geq v_{min} \times b_w \, d \quad \text{(6.2.a) and (6.2.b)}$$
$$C_{Rd, c} = 0.18/1.5 = 0.12, \, k = 1 + \sqrt{(200/d)} = 1 + \sqrt{(200/400)} = 1.71 < 2.0$$
$$A_s = 2H25 = 982 \text{ mm}^2, \, 100 \, \rho_1 = 100 \times 982/ (250 \times 400) = 0.98 < 2.0$$
$$v_{min} = 0.035 \times k^{1.5} \times f_{ck}^{0.5} = 0.43 \text{ MPa} \quad \text{(6.3N)}$$
$$V_{Rd, c} = (63.38 \text{ kN} > 42.86) < V_{Ed}$$
Shear reinforcement is needed.
$$V_{Rd, s} = (A_{sw}/s) \times z \times f_{ywd} \times \cot\theta \quad \text{(6.8)}$$
Using H10 for links, A_{sw} = area of two legs = 157 mm²
$$z = 0.9d = 360 \text{ mm}, \, f_{ywd} = 0.8 \, f_{ywk} = 0.8 \times 500 = 400 \text{ MPa}, \, \cot\theta = 2.5$$
Equating $V_{Rd, s} = V_{Ed}$, s = 354 mm
Maximum spacing s = 0.75d = 300 mm
Using s = 300 mm, $V_{Rd, s}$ = 150.7 kN
$$100 \, \rho_w = 100 \times A_{sw}/ (s \times b_w) = 0.21 \quad \text{(9.4)}$$
$$100 \, \rho_{w, min} = (8 \times \sqrt{f_{ck}})/f_{yk} = 0.09 < 0.21 \quad \text{(9.5N)}$$

(ii) *Near the internal support*: From the shear force envelope, the maximum shear is $V_{Ed} = 184.71$ kN $< (V_{Rd, max} = 327.7$ kN)

The shear at $d = 400$ mm from the support is

$$V = 184.71 - 40.23 \times 400 \times 10^{-3} = 168.62 \text{ kN}$$

Carrying out the calculations as for end support, s = 268 mm.

Shear capacity with links at 300mm is $V_{Rd, s} = 150.7$ kN.

This shear force occurs at $(184.71 - 150.7)/40.23 = 0.85$ m from the support. Over this length link spacing can be reduced to 250 mm. Over the rest of the beam link spacing is 300 mm.

According to clause 9.5.3(3), on the bottom face where the reinforcement is in compression the link spacing must not exceed the smaller of 20×25 (bar diameter) = 500 mm or 400 mm. Link spacing of 250 mm satisfies the requirement.

(d) Deflection

The allowable value for the span/effective depth ratio can be calculated using the code equations (7.16a) and (7.16b) for normal cases.

The equations have been derived on the basis of the following assumptions:

- The maximum stress in steel σ_s at SLS is 310 MPa for $f_{yk} = 500$ MPa. The L/d from equation should be multiplied by $310/\sigma_s$ where

$$\frac{310}{\sigma_s} = \frac{A_{s,prov}}{A_{s,reqd}}$$

- For flanged sections, b/b_w exceeds 3, then L/d from equation should be multiplied by 0.8.
- For beams and slabs other than flat slabs, where the effective span L_{eff} exceeds 7 m and the beam supports partitions liable to be damaged due to excessive deflections, L/d values from equation should be multiplied by $7/L_{eff}$.

$$\frac{L}{d} = K[11 + 1.5 \sqrt{f_{ck}} \frac{\rho_0}{\rho - \rho'} + \frac{1}{12} \sqrt{f_{ck}} \sqrt{\frac{\rho'}{\rho}}] \text{ if } \rho > \rho_0 \qquad (7.16b)$$

where

L/d = limit of span/effective depth ratio

K = a factor to account for different structural systems

$$\rho_0 = 10^{-3} \times \sqrt{f_{ck}}$$

ρ = tension reinforcement ratio to resist the maximum moment due to design loads

$$K = 1.3 \text{ for end span of continuous beam.}$$
$$\rho_0 = 10^{-3} \times \sqrt{f_{ck}} = 5.48 \times 10^{-3}$$
$$\rho = 3H25/ (b_w \times d) = 1473/ (250 \times 400) = 14.73 \times 10^{-3} > \rho_0$$
$$\rho_0/\rho = 0.37$$
$$\rho' = 2H25/ (b_w \times d) = 982/ (250 \times 400) = 9.82 \times 10^{-3}$$

Note: Although the steel at top is not taken into consideration in calculating moment of resistance, its effect can be included when calculating deflection.

$$\sqrt{(\rho'/\rho)} = 0.82$$

$$A_{s\ Provided} = 3H25 = 1473\ mm^2,\ A_{s\ Required} = 1371\ mm^2$$
$$A_{s\ Provided}\ /A_{s\ Required} = 1.07$$
$$b/b_w = 2160/250 = 8.64 > 3.0.\ \text{Multiply L/d calculated by 0.8}$$
$$L/d = 1.3 \times [11 + 1.5 \times 5.48 \times 5.48/\ (14.73 - 9.82) + 5.48 \times 0.82/12] \times 1.07 \times 0.8$$
$$L/d = 22.9$$
$$\text{Allowable span/d ratio} = 22.9$$
$$\text{Actual span/d ratio } 8000\ /\ 400 = 20.0$$

The beam is satisfactory with respect to deflection.

(e) Cracking

$$1.35\ G_k + 1.5\ Q_k = 40.23\ kN/m$$
$$G_k + Q_k = 28.8\ kN/m$$
$$\sigma_s \approx (28.8/40.23) \times 0.87 \times 500/1.42 = 311\ MPa \approx 310\ MPa\ \text{(assumed)}$$

From Table 7.3N of the code, maximum bar spacing is 100 mm for a maximum crack width of 0.3 mm.

The clear distance between bars on the tension faces at mid-span and over the support is $[250 - 2 \times \{25\ (cover) + 10(link) + 25/2\}]/2 = 78\ mm$. This does not exceed the 100 mm permitted.

(f) Sketch of the beam

A sketch of the beam with the moment and shear reinforcement and curtailment of bars is shown in Fig. 13.17.

13.13 EXAMPLE OF DESIGN OF A NON-SWAY FRAME

(a) Specification

Fig. 13.18 shows a typical frame supporting a loading bay. The frames are spaced at 4 m centres. The floor consists of 250 mm thick precast slabs simply supported on top of beams. The imposed load is 10 kN/m². The beams are 300 × 600 mm and columns are 300 mm × 300 mm. The material strengths are $f_{ck} = 30$ MPa and $f_{yk} = 500$ MPa.

(b) Loads

Permanent load, G_k:

$$\text{Beam self weight} = 0.3 \times 0.6 \times 24 = 4.32\ kN/m$$
$$\text{Precast planks: } 0.125 \times 4.0 \times 24 = 24.00\ kN/m$$
$$G_k = 4.32 + 24.0 = 28.32\ kN/m$$

Imposed load, Q_k: $Q_k = 10 \times 4 = 40\ kN/m$
$$(1.35G_k + 1.5\ Q_k) = 1.35 \times 28.32 + 1.5 \times 40 = 98.23\ kN/m$$
$$1.0\ G_k = 28.32\ kN/m$$

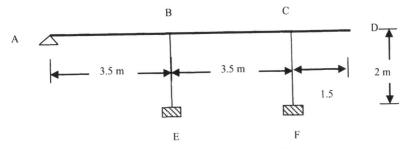

Fig. 13.18 Non-sway rigid-jointed frame.

(c) Elastic analysis

$$I_{\text{beams}} = 0.3 \times 0.6^3/12 = 5.4 \times 10^{-3} \text{ m}^4$$
$$I_{\text{columns}} = 0.3 \times 0.3^3/12 = 0.675 \times 10^{-3} \text{ m}^4$$
$$\text{Beams, I/L:} = 5.4 \times 10^{-3}/3.5 = 1.5429 \times 10^{-3} \text{ m}^3$$
$$\text{Columns, I/L:} = 0.675 \times 10^{-3}/2.0 = 0.3375 \times 10^{-3} \text{ m}^3$$

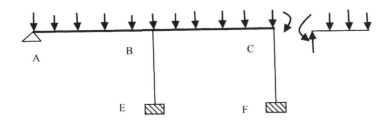

Fig. 13.19 Simplified frame used in the analysis.

In order to simplify the computation, the structure analysed is as shown in Fig. 13.19. The cantilever CD is not included in the stiffness matrix but the moment induced by the cantilever on the rest of the frame in taken into account when computing the load vector and rotations at the joints A, B and C.
The simplified stiffness matrix K is given by

$$K = E \times 10^{-3} \begin{bmatrix} 6.1716 & 3.0858 & 0 \\ 3.0858 & 13.6932 & 3.0858 \\ 0 & 3.0858 & 7.5216 \end{bmatrix} \begin{bmatrix} \theta_A \\ \theta_B \\ \theta_C \end{bmatrix}$$

The structure is analysed for the following four load cases:

Case 1: $(1.35G_k + 1.5 Q_k)$ on AB and BC, $1.0 G_k$ on CD
    Fixed end moments AB and BC $= 98.23 \times 3.5^2/12 = 100.28$ kNm
        Fixed end moment in CD $= 28.32 \times 1.5^2/2 = 31.86$ kNm

Case 2: $1.0 G_k$ on AB, $(1.35G_k + 1.5 Q_k)$ on BC and CD

Fixed end moments AB = 28.32 × 3.5²/12 = 28.91 kNm
Fixed end moments BC = 98.23 × 3.5²/12 = 100.28 kNm
Fixed end moment in CD = 98.23 × 1.5²/2 = 110.51 kNm

Case 3: $(1.35G_k + 1.5\ Q_k)$ on AB and CD, $1.0G_k$ on BC
Fixed end moments AB = 98.23 × 3.5²/12 = 100.28 kNm
Fixed end moments BC = 28.32 × 3.5²/12 = 28.91 kNm
Fixed end moment in CD = 98.23 × 1.5²/2 = 110.51 kNm

Case 4: $1.0\ G_k$ on AB and CD, $(1.4G_k + 1.6\ Q_k)$ BC
Fixed end moments AB = 28.32 × 3.5²/12 = 28.91 kNm
Fixed end moments BC = 98.23 × 3.5²/12 = 100.28 kNm
Fixed end moment in CD = 28.32 × 1.5²/2 = 31.86

The load vectors for the four cases are

$$F = \begin{bmatrix} 100.28 & 28.91 & 100.28 & 28.91 \\ 0 & 71.37 & -71.37 & 71.37 \\ -68.42 & 10.23 & 81.60 & -68.42 \end{bmatrix}$$

The displacement vectors for the four cases are

$$E \times 10^{-3} \begin{bmatrix} \theta_A \\ \theta_B \\ \theta_C \end{bmatrix} = \begin{bmatrix} 17.264 & 2.263 & 23.389 & 0.780 \\ -2.032 & 4.844 & -14.282 & 7.808 \\ -8.250 & -0.627 & 16.708 & -12.30 \end{bmatrix}$$

Table 13.17 Summary of elastic analysis

	Case 1	Case 2	Case 3	Case 4
M_{BA}	141	66	84	80
M_{BC}	-138	-72	-66	-90
M_{CB}	43	111	88	49
M_{CD}	-32	-111	-111	-32
M_{BE}	-3	7	-19	11
M_{CF}	-11	-1	23	-17
Axial: BE	411	229	239	256
Axial: CF	187	330	203	203

The results are summarised in Table 13.17 to Table 13.20. The elastic bending
moment diagrams, the moment envelope and shear force envelope for beam AB
are shown in Fig. 13.20(a), Fig. 13.20(b) and Fig. 13.20(c) respectively. The
elastic bending moment diagrams, the moment envelope and shear force envelope
for beam BC are shown in Fig. 13.21(a), Fig. 13.21(b) and Fig. 13.21(c)
respectively.

Table 13.18 Summary of elastic analysis: End moments and reactions

	Beam AB				Beam BC			
	Case 1	Case 2	Case 3	Case 4	Case 1	Case 2	Case 3	Case 4
R_{Left}	132	31	148	27	138	72	66	90
R_{Right}	212	68	196	72	43	111	88	49
M_{Left}	0	0	0	0	199	161	43	184
M_{Right}	141	66	84	80	145	183	56	160
M_{Max} in Span	−88	−17	−111	−13	−64	−59	33	−82
M_{max} at x	1.34	1.09	1.50	1	2.03	1.64	1.52	1.87

Table 13.19 Elastic moment and shear calculations for beam AB

x	Case 1	Case 2	Case 3	Case 4	Moment		Shear force	
					Max.	Min.	Max.	Min.
0.0	0.	0	0	0	0	0.	148	27
0.5	−54	−1	−62	−10	−10	−62	99	13
1.0	−83	−17	−99	−13	−13	−99	50	−2
1.5	−87	−14	−111	−8	−8	−111	1	−16
2.0	−67	−5	−99	3	3	−99	−26	−65
2.5	−22	12	−63	21	21	−63	−40	−114
3.0	47	35	−1	47	47	−1	−54	−163
3.5	141	66	84	80	141	66	−68	−212

Table 13.20 Elastic moment and shear calculations for beam BC

x	Case 1	Case 2	Case 3	Case 4	Moment		Shear force	
					Max.	Min.	Max.	Min.
0.0	138	72	66	90	138	66	199	43
0.5	51	4	48	10	51	4	150	29
1.0	−12	−39	37	−45	37	−45	101	15
1.5	−50	−58	33	−75	33	−75	52	1
2.0	−64	−53	36	−81	36	−81	3	−36
2.5	−53	−23	46	−62	46	−62	−28	−85
3.0	−17	32	64	−19	64	−19	−42	−134
3.5	43	111	88	49	111	43	−56	−183

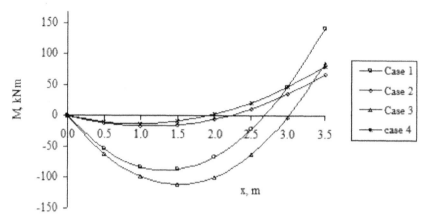

Fig. 13.20(a) Elastic bending moment diagrams: Beam AB.

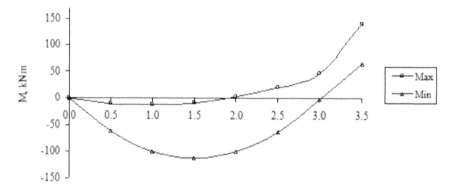

Fig. 13.20(b) Elastic bending moment envelope: Beam AB.

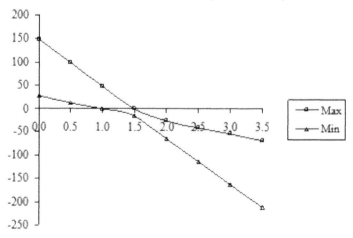

Fig. 13.20(c) Elastic shear force envelope: Beam AB.

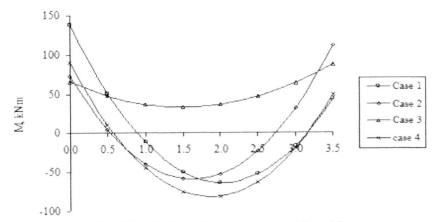

Fig. 13.21(a) Elastic bending moment diagrams: Beam BC.

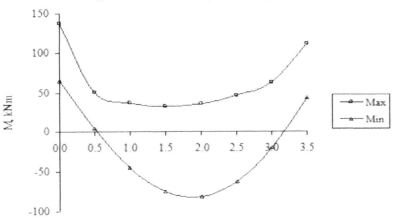

Fig. 13.21(b) Elastic bending moment envelope: Beam BC.

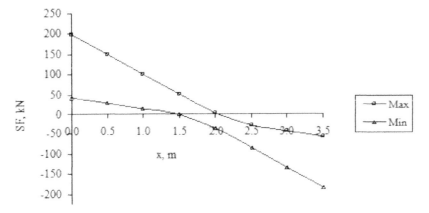

Fig. 13.21(c) Elastic shear force envelope: Beam BC.

(d) Redistribution

(i) Beam CD
The maximum moment M_{CD} at the root of the cantilever is 110.51. Since CD is a cantilever, this moment cannot be reduced.

(ii) Beam AB
The support moment $M_{BA} = 141.02$ from case 1 is the largest value of hogging moment considering all load cases and can be reduced to 110.51, the moment at the root of the cantilever. The percentage reduction is
$$(1 - 110.51/141.02) = 21.6\%$$
The redistributed support moments are as follows.
> **Case 1**: $M_{BA} = 110.51$ (changed from elastic value of 141.02)
> **Case 2**: $M_{BA} = 65.78$ (Unchanged from elastic value)
> **Case 3**: $M_{BA} = 110.51$ (changed from elastic value of 84.31)
> **Case 4**: $M_{BA} = 110.51$ (changed from elastic value of 79.50)

Note that in case 3 and case 4, the support moment has been increased. This will result in a *reduction* in span moment.

(iii) Beam BC
Hogging moment $M_{BC} = 138.27$ from case 1 can also be reduced to 110.51 so that the same top steel over the column BE serves for both moments M_{BA} and M_{BC}.
Elastic moment M_{CB} from case 2 is 117.62 and this can be reduced also to 110.51 so that the same top steel over the column CF for both moments M_{CB} and M_{CD}.
The redistributed moments are as follows.
> **Case 1**: $M_{BC} = 110.51$ (changed from elastic values of 138.27), $M_{CB} = 43.10$
> **Case 2**: $M_{BC} = 110.51$, $M_{CB} = 110.51$ (changed from elastic values of 72.32 and 111.36 respectively)
> **Case 3**: $M_{BC} = 110.51$, $M_{CB} = 110.51$ (changed from elastic values of 65.49 and 87.95 respectively)
> **Case 4**: $M_{BC} = 110.51$(changed from elastic value of 90.05), $M_{CB} = 48.47$

Note that at support C, in cases 1 and 4, no redistribution has been done. The reason for this is that the moment in the cantilever is small. If the support moment M_{CB} is raised to 110.51, then it will result in a very large moment in column CF.

The results of redistribution are shown in Table 13.21 to Table 13.23. The redistributed bending moment diagrams, the moment envelope and shear force envelope for beam AB are shown in Fig. 13.22(a), Fig. 13.22(b) and Fig. 13.22(c) respectively. The redistributed bending moment diagrams, the moment envelope and shear force envelope for beam BC are shown in Fig. 13.23(a), Fig. 13.23(b) and Fig. 13.23(c) respectively.

Table 13.21 Summary of redistributed elastic analysis: End moments and reactions

	Beam AB				Beam BC			
	Case 1	Case 2	Case 3	Case 4	Case 1	Case 2	Case 3	Case 4
R_{Left}	140	31	140	18	191	172	50	190
R_{Right}	204	68	204	81	153	172	50	154
M_{Left}	0	0.	0	0	111	111	111	111
M_{Right}	111	66	111	111	43	111	111	49
M_{Max} in span	−100	−17	−100	−6	−76	−40	67	−73
M_{max} at x	1.43	1.09	1.43	0.64	1.95	1.75	1.75	1.93

Table 13.22 Redistributed elastic moment and shear calculations for beam AB

x	Case 1	Case 2	Case 3	Case 4	Moment		Shear force	
					Max.	Min.	Max.	Min.
0.0	0	0	0	0	0	0	140	18
0.5	−58	−12	−58	−6	−6	−58	91	4
1.0	−91	−17	−91	−4	−4	−91	42	−10
1.5	−100	−14	−100	5	5	−100	−7	−25
2.0	−84	−5	−84	21	21	−84	−26	−56
2.5	−44	12	−44	44	44	−44	−40	−105
3.0	21	35	21	74	74	21	−54	−154
3.5	111	66	111	111	111	66	−68	−204

Table 13.23 Redistributed elastic moment and shear calculations for beam BC

x	Case 1	Case 2	Case 3	Case 4	Moment		Shear force	
					Max.	Min.	Max.	Min.
0.0	111	111	111	111	111	111	191	50
0.5	27	37	89	28	89	27	142	35
1.0	−32	−12	75	−30	75	−32	93	21
1.5	−66	−37	68	−63	68	−66	44	7
2.0	−75	−37	68	−72	68	−75	−5	−25
2.5	−60	−12	75	−57	75	−60	−21	−74
3.0	−21	37	89	−16	89	−21	−35	−123
3.5	43	111	111	49	111	43	−50	−172

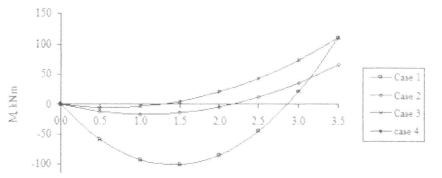

Fig. 13.22(a) Redistributed bending moment diagrams for beam AB.

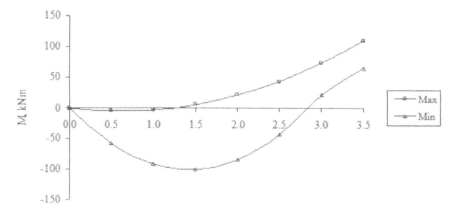

Fig. 13.22(b) Redistributed bending moment envelope for beam AB.

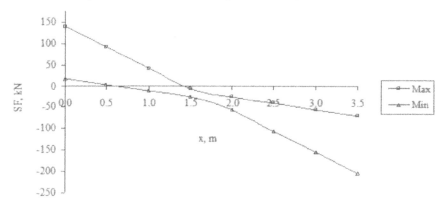

Fig. 13.22(c) Redistributed shear force envelope for beam AB.

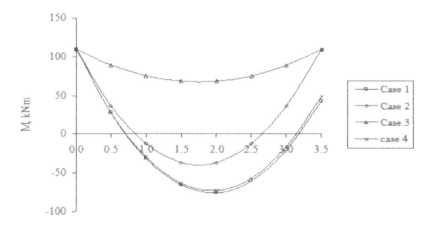

Fig. 13.23(a) Redistributed bending moment diagrams for beam BC.

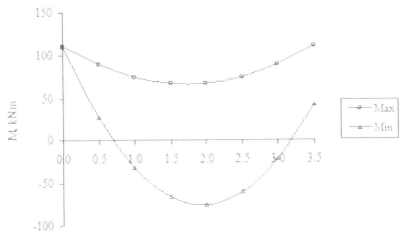

Fig. 13.23(b) Redistributed bending moment envelope for beam BC.

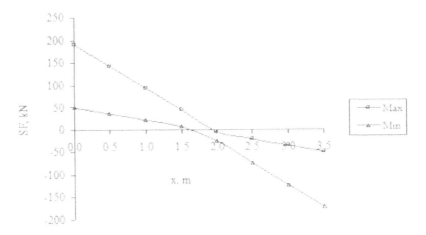

Fig. 13.23(c) Redistributed shear force envelope for beam BC.

(e) Design moment envelopes

The design maximum (or minimum) value of bending moment at a section is obtained as follows.

The maximum (or minimum) moment value at a section is obtained by considering all load cases. For a particular load case the values to be considered are:

- If there is no redistribution, then the elastic moment value
- If there is redistribution then the larger of $0.7 \times$ elastic value or the corresponding redistributed moment value.

The design moment envelopes for beams AB and BC are shown in Fig. 13.24 and Fig. 13.25 respectively. The results are summarised in Table 13.24 and Table 13.25 for beam AB and beam BC respectively.

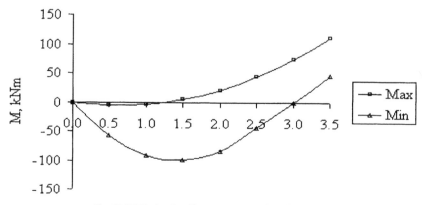

Fig. 13.24 Design bending moment envelope for beam AB.

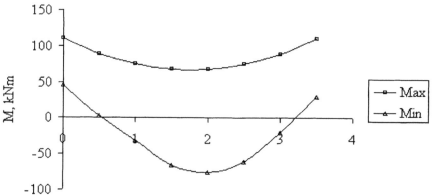

Fig.13.25 Design bending moment envelope for beam BC.

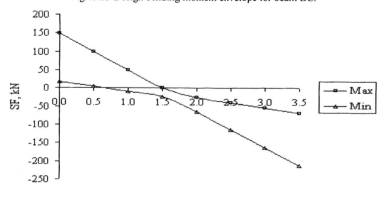

Fig. 13.26 Design shear force envelope for beam AB.

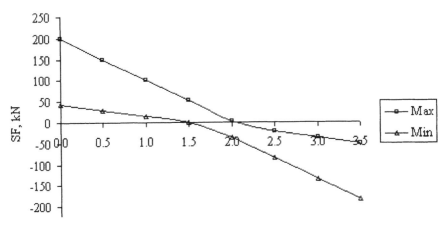

Fig. 13.27 Design shear force envelope for beam BC.

Table 13.24 Design bending moment and shear force: Beam AB

	Bending Moment		Shear force	
x	Max	Min	Max	Min
0	0	0	**148**	18
0.5	-6	-58	99	4
1.0	-4	-91	50	-10
1.5	5	-100	1	-25
2.0	21	-84	-26	-65
2.5	44	-44	-40	-114
3.0	74	-1	-54	-163
3.5	111	46	**-68**	**-212**

Table 13.25 Design bending moment and shear force: Beam BC

	Bending Moment		Shear force	
x	Max	Min	Max	Min
0	111	46	**199**	43
0.5	89	3	150	29
1.0	75	-32	101	15
1.5	68	-66	52	1
2.0	68	-75	3	-36
2.5	75	-60	-21	-85
3.0	89	-21	-35	-134
3.5	111	30	-50	**-183**

(f) Design of moment steel

(i) Section near the centre of span AB
Assuming 20 mm bars and a cover of 30 mm and 8 mm diameter shear links
$$d = 600 - 20/2 - 8 - 30 = 552 \text{ mm}$$
Maximum redistributed moment from the design moment envelope is 100.23 kNm at approximately 1.5 m from left hand support from Case 3.
   The corresponding moment before redistribution is 111.21 kNm.
$$\delta = 100.23/111.21 = 0.90$$
As there is a decrease in the redistributed moment from the corresponding elastic value, it is necessary to check the maximum depth of neutral axis depth x_u permitted. From code equation (5.10a),
$$\delta \geq 0.44 + 1.25\ x_u/d,\ x_u/d \leq 0.37$$
Substituting $x_u/d = 0.37$, as a singly reinforced section, the maximum moment allowable is
$$M_{SR} = f_{cd} \times b \times 0.8\ x_u \times (d - 0.4\ x_u) = 0.168\ bd^2\ f_{ck}$$
$$k = 100.51 \times 10^6/ (300 \times 552^2 \times 30) = 0.037 < 0.168$$
No compression steel is required.

$$\frac{z}{d} = 0.5[1.0 + \sqrt{(1 - 3\frac{k}{\alpha_{cc}\eta})}]$$

Substituting $\alpha_{cc} = 1$, $\eta = 1$, $k = 0.037$, $z/d = 0.97$
$$A_s = 110.51 \times 10^6 / (0.97 \times 552 \times 0.87 \times 500) = 475 \text{ mm}^2$$
Provide 3H16, $A_s = 603 \text{ mm}^2$.

(ii) Section over support B
Maximum redistributed moment from the design moment envelope is 100.51 kNm from Case 1. The corresponding moment before redistribution is 141.02 kNm.
$$\delta = 100.51/141.02 = 0.71$$
As there is a decrease in the redistributed moment from the corresponding elastic value, it is necessary to check the maximum depth of neutral axis depth x_u permitted. From code equation (5.10a),
$$\delta \geq 0.44 + 1.25\ x_u/d,\ x_u/d \leq 0.22$$
Substituting $x_u/d = 0.37$, as a singly reinforced section, the maximum moment allowable is
$$M_{SR} = f_{cd} \times b \times 0.8\ x_u \times (d - 0.4\ x_u) = 0.107\ bd^2\ f_{ck}$$
$$k = 100.51 \times 10^6/ (300 \times 552^2 \times 30) = 0.037 < 0.107$$
No compression steel is required.

$$\frac{z}{d} = 0.5[1.0 + \sqrt{(1 - 3k)}]$$

Substituting $k = 0.037$, $z/d = 0.97$
$$A_s = 110.51 \times 10^6 / (0.97 \times 552 \times 0.87 \times 500) = 475 \text{ mm}^2$$
Provide 3H16, $A_s = 603 \text{ mm}^2$.

(iii) Section near the centre of span BC

Maximum redistributed moment from the design moment envelope from Case 1 is 75.50 kNm at approximately 2.0 m from left hand support. The corresponding moment before redistribution is 81.88 kNm from Case 4. The moment after redistribution has decreased. Following the steps for Beam AB,

$$\delta = 75.50/81.88 = 0.92$$

As there is a decrease in the redistributed moment from the corresponding elastic value, it is necessary to check the maximum depth of neutral axis depth x_u permitted. From code equation (5.10a),

$$\delta \geq 0.44 + 1.25 \, x_u/d, \, x_u/d \leq 0.38$$

Substituting xu/d = 0.37, as a singly reinforced section, the maximum moment allowable is

$$M_{SR} = f_{cd} \times b \times 0.8 \, x_u \times (d - 0.4 \, x_u) = 0.172 \, bd^2 \, f_{ck}$$
$$k = 75.50 \times 10^6/ (300 \times 552^2 \times 30) = 0.028 < 0.172$$

No compression steel is required.

$$\frac{z}{d} = 0.5[1.0 + \sqrt{(1-3k)}]$$

Substituting k = 0.037, z/d = 0.98

$$A_s = 75.50 \times 10^6 / (0.98 \times 552 \times 0.87 \times 500) = 321 \text{ mm}^2$$

Provide 2H16, $A_s = 402 \text{ mm}^2$.

(iv) Section over support C

Maximum redistributed moment from the design moment envelope is 110.51 kNm for Case 2. The corresponding moment before redistribution is 111.36 kNm. Following the steps for Beam AB,

$$\delta = 110.51/111.36 = 0.99$$

As there is a decrease in the redistributed moment from the corresponding elastic value, it is necessary to check the maximum depth of neutral axis depth x_u permitted. From code equation (5.10a),

$$\delta \geq 0.44 + 1.25 \, x_u/d, \, x_u/d \leq 0.44$$

Substituting xu/d = 0.37, as a singly reinforced section, the maximum moment allowable is

$$M_{SR} = f_{cd} \times b \times 0.8 \, x_u \times (d - 0.4 \, x_u) = 0.193 \, bd^2 \, f_{ck}$$
$$k = 100.51 \times 10^6/ (300 \times 552^2 \times 30) = 0.037 < 0.193$$

No compression steel is required.

$$\frac{z}{d} = 0.5[1.0 + \sqrt{(1 - 3\frac{k}{\alpha_{cc}\eta})}]$$

Substituting α_{cc} = 1, η = 1, k = 0.037, z/d = 0.97

$$A_s = 100.51 \times 10^6 / (0.97 \times 552 \times 0.87 \times 500) = 432 \text{ mm}^2$$

Provide 3H16, $A_s = 603 \text{ mm}^2$.

By rationalizing the steel area calculations, for simplicity, provide 3H16 at both top and bottom for the beams including the cantilever.

Check minimum and maximum steel areas:
Maximum steel provided is 3H16 = 603 mm^2.
Minimum. steel provided is 2H16 = 402 mm^2

$\qquad$ f_{ck} = 30 MPa, f_{ctm} = 2.9 MPa, f_{yk} = 500 MPa, b_t = 300 mm, d = 532 mm

$\qquad$ $A_{s,\,min}$ = 0.26 × (f_{ctm} /f_{yk}) × b_t d = 241 mm^2 $\qquad\qquad$ (9.1N)

$\qquad$ $A_{s,\,max}$ = 0.04 A_c = 0.04 × 300 × 600 = 7200 mm^2

Steel provided satisfies the required limitations.

(g) Design shear envelopes
Design shear force is the larger of the elastic and redistributed values. The design shear force envelopes for beams AB and BC are shown in Fig. 13.26 and Fig. 13.27 respectively. The results are also shown in Table 13.26 and Table 13.27 for beams AB and BC respectively.

(h) Design of shear reinforcement
Using the data from Table 13.24 and Table 13.25, calculate shear forces at d = 552 mm from the face of the column/support,
Beam AB:
At end A: Elastic value = 147.81 – 98.23 × 0.552 = **93.59** (Case 3)
$\qquad$ Redistributed value = 140.23 – 98.23 × 0.552 = 86.01 (Cases 1 and 3)
At end B: Elastic value = 212.19 – 98.23 × 0.552 = **157.97** (Case 1)
$\qquad$ Redistributed value = 203.48 – 98.23 × 0.552 = 149.26 (Cases 1 and 3)

Beam BC:
At end B: Elastic value = 199.10 – 98.23 × 0.552 = **144.88** (Case 1)
$\qquad$ Redistributed value = 191.16 – 98.23 × 0.552 = 136.94 (Case 2)
At end C: Elastic value = 183.05 – 98.23 × 0.552 = **128.83** (Case 1)
$\qquad$ Redistributed value = 171.90 – 98.23 × 0.552 = 117.68 (Case 2)

Check for maximum shear stress: From the data in Table 13.10
$\qquad\qquad$ Maximum V_{Ed} = 157.97 kN at end B for beam AB.
Using code equations (6.9), (6.6N) and (6.11.bN)

$\qquad\qquad$ $V_{Rd,\,max}$ = α_{cw} × b_w × z × v_1 × f_{cd}/ (cotθ + tan θ) $\qquad\qquad$ (6.9)

f_{cd} = 30/1.5 = 20 MPa, b_w = 300 mm, z ≈ 0.9 d = 497 mm, α_{cw} = 1.0

$\qquad\qquad$ v_1 = v = 0.6 (1 – f_{ck}/250) = 0.528, $\qquad\qquad\qquad$ (6.6N)

$\qquad\qquad$ $V_{Rd,\,max}$ = 1574.9/ (cotθ + tan θ) kN

$\qquad\qquad\qquad$ Equating V_{Ed} to $V_{Rd,\,max}$

$\qquad\qquad\qquad$ (cotθ + tan θ) = 9.98

cotθ = 0.10 and 9.88 both of which are outside the range. Taking cotθ =2.5 to give the smallest value of $V_{Rd,\,max}$ = 543.1 kN > V_{Ed}.
Section is satisfactory and shear links can be designed.

Check whether shear reinforcement is needed:
Use equation (6.2a), (6.2b) and (6.6N) of the code.
As the tension and compression steel is 3H16 over the entire span, a common value of $V_{Rd,\,c}$ can be calculated which is applicable over the entire span.

$$V_{Rd,c} = [C_{Rd,c} \times k \times (100 \, \rho_1 \times f_{ck})^{0.33}] \, b_w \, d \geq v_{min} \times b_w \, d \qquad (6.2a)$$

$$C_{Rd,c} = 0.18/1.5 = 0.12, \, k = 1 + \sqrt{(200/d)} = 1 + \sqrt{(200/552)} = 1.60 < 2.0$$

$$A_s = 3H216 = 603 \text{ mm}^2, \, 100 \, \rho_1 = 100 \times 603/ (300 \times 552) = 0.36 < 2.0$$

$$v_{min} = 0.035 \times k^{1.5} \times f_{ck}^{0.5} = 0.43 \text{ MPa} \qquad (6.3N)$$

$$V_{Rd,c} = (70.28 \text{ kN} > 64.25) < V_{Ed}$$

Shear reinforcement is needed.

$$V_{Rd,s} = (A_{sw}/s) \times z \times f_{ywd} \times \cot\theta \qquad (6.8)$$

Using H8 for links, A_{sw} = area of two legs = 101 mm^2.

$$z = 0.9d = 497 \text{ mm}, \, f_{ywd} = 0.8 \, f_{ywk} = 0.8 \times 500 = 400 \text{ MPa}, \, \cot\theta = 2.5$$

$$V_{Rd,s} = 49964/s$$

Equating $V_{Rd,s} = V_{Ed}$, calculate, s.

Maximum spacing s = 0.75 d = 414 mm.

Table 13.26 Spacing of links

	Beam AB		Beam BC	
	End A	End B	End B	End C
V_{Ed}	93.59	157.97	144.88	128.83
s, mm	534	316	345	388

By rationalizing the link spacings calculated, for simplicity, provide 8 mm diameter two leg links at 300 mm c/c throughout the beams including the cantilever.

$$\rho_w = A_{sw}/(s \times b_w) = 101/(300 \times 300) = 1.12 \times 10^{-3} \qquad (9.4)$$

$$\rho_{w,min} = 0.08 \, \sqrt{f_{ck}}/f_{yk} = 0.876 \times 10^{-3} \qquad (9.5N)$$

(i) Deflection

Over the entire span, tension and compression steel is provided by 3H16.
$A_s = A_s' = 603 \text{ mm}^2$.

$$\rho = \rho' = 100 \times 603/ (300 \times 552) = 0.36\%$$

The allowable value for the span-to-effective depth ratio can be calculated using the code equations (7.16a) and (7.16b) for normal cases.

$$\frac{L}{d} = K[11 + 1.5 \sqrt{f_{ck}} \frac{\rho_0}{\rho} + 3.2 \sqrt{f_{ck}} (\frac{\rho_0}{\rho} - 1)^{\frac{3}{2}}] \text{ if } \rho \leq \rho_0 \qquad (7.16a)$$

Substituting

K = 1.3 for end span of continuous beam.

$$\rho_0 = 10^{-3} \times \sqrt{f_{ck}} = 0.548\%$$

$$\rho_0/\rho = 1.522$$

$$A_{s \, Provided} = 3H25 = 1473 \text{ mm}^2, \, A_{s \, Required} = 1371 \text{ mm}^2$$

$$A_{s \, Provided}/A_{s \, Required} = 1.07$$

$$L/d = 1.3 \times [11 + 1.5 \times 5.48 \times 1.522 + 3.2 \times 5.48 \times (1.522 - 1)^{1.5}]$$

$$L/d = 39.2$$

(i) Beam AB

$$A_{s \, required} = 475 \text{ mm}^2, \, A_{s \, provided} = 603 \text{ mm}^2$$

Allowable L/d = 39.2 × 603/475 = 49.8

Actual L/d = 3500 / 522 = 6.7

The beam is satisfactory with respect to deflection.

(ii) Beam BC

$$A_{s\ required} = 321\ mm^2,\ A_{s\ provided} = 603\ mm^2$$
$$\text{Allowable } L/d = 39.2 \times 603/321 = 73.6$$
$$\text{Actual } L/d = 3500 / 522 = 6.7$$

The beam is satisfactory with respect to deflection.

(iii) Cantilever CD

$K = 0.3$ for a cantilever

$$L/d = 0.3 \times [11 + 1.5 \times 5.48 \times 1.522 + 3.2 \times 5.48 \times (1.522 - 1)^{1.5}]$$
$$L/d = 9.05$$
$$A_{s\ required} = 475\ mm^2,\ A_{s\ provided} = 603\ mm^2,$$
$$\text{Allowable } L/d = 9.05 \times 603/475 = 11.5$$
$$\text{Actual } L/d = 1500 / 522 = 2.9$$

The cantilever is satisfactory with respect to deflection.

(j) Cracking

Taking the stress in bars at SLS as approximately f_{yd} divided by an average load factor equal to $(1.35 + 1.5)/2 = 1.42$, the stress is the bars is equal to $0.87 \times 500/1.42 = 306$ MPa. From Table 7.3N of the code, maximum bar spacing is 100 mm for a maximum crack width of 0.3 mm.

The clear distance between bars on the tension faces at mid-span and over the support is

$$(300 - 2 \times 30 \text{ (cover)} - 2 \times 8 \text{ (link)} - 16)/2 - 16 = 88 \text{ mm}$$

The beam is satisfactory with respect to crack control.

(k) Column design

Table 13.27 shows the values of axial load N and moment M at the top of the columns for the four cases considered. The values are shown both for elastic analysis as well as for the redistributed analysis. The figures in brackets are $M/ (bh^2)$ and $N/ (bh)$ using $b = h = 300$ mm.

Table 13.27 Axial force and moments in columns

		Case 1		Case 2		Case 3		Case 4	
		Elas.	Redis	Elas.	Redis	Elas.	Redis	Elas.	Redis
BE	N	**411**	395	229	240	239	253	256	271
		(4.6)	(4.4)	(2.5)	(2.7)	(2.7)	(2.8)	(2.8)	(3.0)
	M	2.8	0	6.5	36	18.8	0	10.6	0
		(0.1)	(0)	(0.2)	(1.3)	(0.7)	(0)	(0.4)	(0)
CF	N	187	195	330	319	203	197	203	197
		(2.1)	(2.2)	(3.7)	(3.5)	(2.3)	(2.2)	(2.3)	(2.3)
	M	0.6	11.2	36.0	0	59.4	0	6.0	16.6
		(0)	(0.4)	(1.3)	(0)	(2.2)	(0)	(0.2)	(0.6)

Check whether the column is short:

Column BE: 300 mm square, $\ell = 2000$ mm, $N_{Ed} = 411$ kN,
$$n = 411 \times 10^3 / (300^2 \times 20) = 0.228$$
$k_1 = 0$ at E as E is clamped.
$$\theta/M = 0.25 \times (L/I)_{BC} + 0.333 \times (L/I)_{BA} = 0.25/1.5429 + 0.33/1.5429 = 0.378$$
$$k_2 = 0.378 \times (I/L)_{BC} = 0.378 \times 0.3375 = 0.128$$
$$\ell_0 = 0.5 \, \ell \times \sqrt{[1 + 0.128/(0.45 + 0.128)]} = 0.552 \, \ell = 1104 \text{ mm}$$
$$i = \sqrt{(I/A)} = \sqrt{\{300^4/[12 \times 300^2]\}} = 87 \text{ mm}$$
$$\lambda = \ell_0/i = 12.7$$
$$\lambda_{min} = 20 \times (A = 0.7) \times (B = 1.1) \times (C = 0.7)/\sqrt{0.228} = 22.6$$
Column is short.

Column CF: 300 mm square, $\ell = 2000$ mm, $N_{Ed} = 330$ kN,
$$n = 411 \times 10^3 / (300^2 \times 20) = 0.183$$
$$k_1 = 0 \text{ at F as F is clamped}$$
$$\theta/M = 0.25 \times (L/I)_{CB} = 0.25/1.5429 = 0.162$$
$$k_2 = 0.162 \times (I/L)_{CF} = 0.162 \times 0.3375 = 0.055$$
$$\ell_0 = 0.5 \, \ell \times \sqrt{[1 + 0.055/(0.45 + 0.055)]} = 0.526 \, \ell = 1053 \text{ mm}$$
$$i = \sqrt{(I/A)} = \sqrt{\{300^4/[12 \times 300^2]\}} = 87 \text{ mm}$$
$$\lambda = \ell_0/i = 12.1$$
$$\lambda_{min} = 20 \times (A = 0.7) \times (B = 1.1) \times (C = 0.7)/\sqrt{0.228} = 22.6$$
Column is short.

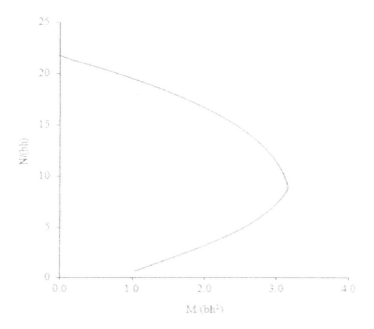

Fig. 13.28 Column design chart.

The column design chart shown in Fig. 13.28 for $f_{ck} = 30$ MPa, $f_{yk} = 500$ MPa and
$d/h = 0.95$ shows that only minimum steel equal to $A_{sc}/$ (bh) = 0.4% is required.
$A_{sc} = (0.4/100) \times 300^2 = 360$ mm^2. Provide one H12 bar in each corner.
$A_{sc} = 452$ mm^2. Provide 6 mm diameter links spaced at $20 \times H12 = 240$ mm c/c
(see clause 9.5.3 of the Eurocode 2).

13.14 APPROXIMATE METHODS OF ANALYSIS

In the examples of continuous beam and non-sway frame analysed in the previous
sections, the relative flexural rigidity EI was assumed in order to carry out the
elastic analysis. In the case of statically indeterminate structures, information
about the relative stiffness of members is required before analysis can be carried
out. In many cases experience can be used to guess at the relative size of members.
However it is convenient to have approximate methods of analysis which allow a
designer to estimate the relative stiffness of members. Approximate methods of
analysis convert a statically indeterminate structure into a statically determinate
structure by assuming the position of points of contraflexure. This enables the
determination of inevitably approximate values of bending moment and shear
forces in the structure without the need to know the relative stiffness of members.

13.14.1 Analysis for Gravity Loads

Analysis for gravity loads is done by assuming the points of contraflexure in the
individual beams.

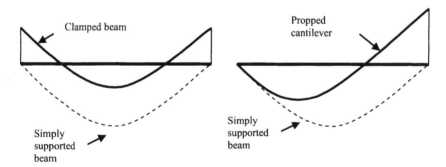

Fig. 13.29 Bending moment diagrams: (a) clamped beam; (b) propped cantilever.

If a beam is continuous at both ends, its behaviour will be between the
behaviour of a clamped beam at one extreme and that of a simply supported beam
at the other extreme. As shown in Fig. 13.29(a), in a clamped beam of span L
subjected to a uniformly distributed load q, the contraflexure points are at 0.21 L
from the ends. The support and mid-span moments are respectively, $qL^2/12$ and
$qL^2/24$. In the corresponding simply supported beam, the points of zero moment

are at the support and the moment at mid-span is $qL^2/8$. Assuming the position of contraflexure at approximately at 0.1 L from the ends,

Mid-span moment = $q(0.8L)^2/8 = 0.64(qL^2/8)$

Support moment = $q(0.1L)^2/2 + 0.4qL \times 0.1L = 0.36(qL^2/8)$

If a beam is continuous at one end only, its behaviour will be between the behaviour of a propped cantilever at one extreme and that of a simply supported beam at the other extreme. As shown in Fig. 13.29(b), in a propped cantilever of span L clamped at the right hand end and subjected to a uniformly distributed load q, the contraflexure point is at 0.25 L from the clamped end. The support and mid-span moments are respectively, $qL^2/8$ and $qL^2/16$. In the corresponding simply supported beam, the points of zero moment are at the support and the moment at mid-span is $qL^2/8$. Assuming the position of contraflexure at approximately 0.15 L from the ends,

Span moment = $q(0.85L)^2/8 = 0.723(qL^2/8)$

Support moment = $q(0.15L)^2/2 + 0.5 \times 0.85$ all $\times 0.15 L = 0.60(qL^2/8)$

The following two examples show how these values can be used to analyse beams subjected to uniformly distributed loading.

13.14.2 Analysis of a Continuous Beam for Gravity Loads

Fig. 13.30 shows a three-span continuous beam. The beam spans are 8 m each. Assuming the position of contraflexure points, the beams are analysed for the four cases of loading as shown in Fig. 13.8.
It is given that $1.35G_k + 1.5Q_k = 40.23$ kN/m and $1.35 G_k = 26.73$ kN/m.

Case 1: End beams carry 40.23 kN/m and central span carries 26.73 kN/m.
End spans:

Support moment = $0.60 \times 40.23 \times 8^2/8 = $ **193** kNm

Span moment = $0.723 \times 40.23 \times 8^2/8 = $ **223** kNm.

Central span:

Support moment = $0.36 \times 26.73 \times 8^2/8 = 77$ kNm

Span moment = $0.64 \times 26.73 \times 8^2/8 = 137$ kNm

Case 2: End spans carry 26.73 kN/m and central span carries 40.23 kN/m.
End spans:

Support moment = $0.60 \times 26.73 \times 8^2/8 = 128$ kNm

Span moment = $0.723 \times 26.73 \times 8^2/8 = 155$ kNm

Central span:

Support moment = $0.36 \times 40.23 \times 8^2/8 = 116$ kNm

Span moment = $0.64 \times 40.23 \times 8^2/8 = 206$ kNm

Case 3: First and second beams carry 40.23 kN/m and end span carries 26.73 kN/m.
First span:

Support moment = $0.60 \times 40.23 \times 8^2/8 = $ **193** kNm

Span moment = 0.723 × 40.23 × 8²/8 = **223** kNm.

Central span:

Support moment = 0.36 × 40.23 × 8²/8 = 116 kNm
Span moment = 0.64 × 40.23 × 8²/8 = 206 kNm

Last span:

Support moment = 0.60 × 40.23 × 8²/8 = **193** kNm
Span moment = 0.723 × 40.23 × 8²/8 = **223** kNm.

Case 4: First span carries 26.73 kN/m and the second and last spans carry
    40.23 kN/m.

First span:

Support moment = 0.60 × 26.73 × 8²/8 = **128** kNm
Span moment = 0.723 × 26.73 × 8²/8 = 155 kNm

Central span:

Support moment = 0.36 × 40.23 × 8²/8 = 116 kNm
Span moment = 0.64 × 40.23 × 8²/8 = 206 kNm

Last span:

Support moment = 0.60 × 40.23 × 8²/8 = **193** kNm
Span moment = 0.723 × 40.23 × 8²/8 = **223** kNm.

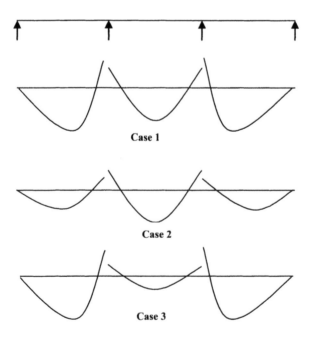

Fig. 13.30 Approximate bending moment distribution.

From the above analyses, the support needs to be designed for approximately
193 kNm and the mid-span for 223 kNm. Exact elastic analysis assuming uniform

flexural rigidity shows that the maximum support moment is 272 kNm and span moment in end spans is 220 kNm.

13.14.3 Analysis of a Rectangular Portal Frame for Gravity Loads

Fig. 13.31 shows a single bay portal frame subjected to gravity loads on the beam. The bending moment distribution can be obtained by assuming the points of contraflexure in the beam. The moment at the tops of the columns will be the same as in the beam. If the columns are fixed at the base, then the moment at the base of columns is half that of the moment at the top of columns.

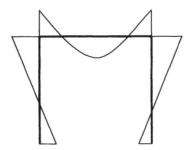

Fig. 13.31 Rectangular portal frame.

13.14.4 Analysis for Wind Loads by Portal Method

Analysis of portal frames for the wind load is made for the whole frame assuming points of contraflexure at the mid-height of columns and at mid-span of beams. In the portal method the horizontal shear force in each storey is assumed to be divided between the bays in proportion to their spans. The shear force in each bay is then divided equally between the columns. The column end moments are the column shear force multiplied by one-half the column height. Beam moments balance the column moments. The method is considered to be applicable to building frames of regular geometry up to 25 storeys high with a height-to-width ratio of less than five. Variations in beam spans and column heights should be small. The application of the method is shown by the analyses of the frame for wind loads shown in Fig. 13.32. The dimensions of the frame are:

Widths of bays: Bay 1 = L, Bay 2 = 2 L.
Column heights: Top storey = 2h, middle storey = 1.5 h and bottom storey = h.

Shear force Q in top storey = 1.5 W
Shear force Q in bottom storey = 1.5 W + 2.5 W = 4.0 W
Shear force Q in bottom storey = 1.5 W + 2.5 W + 3.5 W = 7.5 W

Total shear force Q in any storey is shared by the two bays in proportion to their widths. Bay 1 = Q/3, Bay 2: = 2Q/3.

The shear force in any bay is shared equally between the two columns of the bay.
$$\text{Shear force in left column} = 0.5\ (Q/3) = Q/6$$
$$\text{Shear force in middle column} = 0.5(Q/3 + 2Q/3) = 0.5Q$$
$$\text{Shear force in right column} = 0.5(2Q/3) = Q/3$$

The bending moment M at the top and bottom of columns in any storey is given by
$$M = \text{shear force in the column} \times (\text{height of column }/2)$$
Top storey:
$$\text{Left column: } M = (Q/6) \times h/2 = Qh/12$$
$$\text{Middle column: } M = (Q/2) \times h/2 = Qh/4$$
$$\text{Right column: } M = (Q/3) \times h/2 = Qh/6$$
Middle storey:
$$\text{Left column: } M = (Q/6) \times 1.5\ h/2 = Qh/8$$
$$\text{Middle column: } M = (Q/2) \times 1.5\ h/2 = 3Qh/8$$
$$\text{Right column: } M = (Q/3) \times 1.5\ h/2 = Qh/4$$
Bottom storey:
$$\text{Left column: } M = (Q/6) \times 2.0\ h/2 = Qh/6$$
$$\text{Middle column: } M = (Q/2) \times 2.0\ h/2 = Qh/2$$
$$\text{Right column: } M = (Q/3) \times 2.0\ h/2 = Qh/3$$

Table 13.28 shows the shear force and bending moments in the columns.

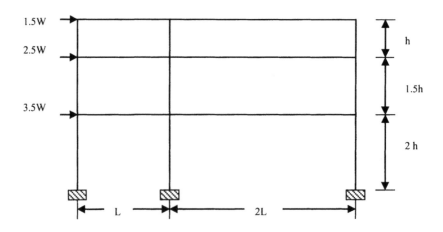

Fig. 13.32 Rigid-jointed frame subjected to lateral loads.

(b) Bending moments in beams
As shown in Fig. 13.33, the bending moments at the ends of the left beam are equal to the sum of the bending moments at the ends of the columns on the left of the connecting beam. Similarly the bending moments at the ends of the right beam are

equal to the sum of the bending moments at the ends of the columns on the right of the connecting beam. Table 13.29 shows the bending moments in the beams. Fig. 13.34 shows the bending moment distribution in the frame.

Table 13.28 Shear forces and bending moment in columns

	Shear in storey Q/W	Shear in columns/W			Moment in columns/(Wh)		
		Left	Middle	Right	Left	Middle	Right
Top	1.5	0.25	0.75	0.50	0.125	0.375	0.25
Middle	1.5 + 2.5 = 4.0	0.67	2.0	1.33	0.50	1.50	1.0
Bottom	4.0 + 3.5 = 7.5	1.25	3.75	2.5	1.25	3.75	2.5

Table 13.29 Moments in beams

Location	Left beam	Right beam
Top storey	0.125 Wh	0.25 Wh
Middle storey	0.625 Wh	1.25 Wh
Bottom storey	1.75 Wh	3.50 Wh

Table 13.30 Axial forces in columns

Level	Beam moments/(Wh)		Reactions × (L/Wh)		Axial force in column × (L/Wh)
	Beam-Left	Beam-Right	Beam-Left	Beam-Right	Left and right columns
Top	0.125	0.25	0.25	0.25	0.25
Middle	0.625	1.25	1.25	1.25	0.25+1.25 = 1.50
Bottom	1.75	3.50	3.50	3.50	0.25+1.25+3.50 =5.0

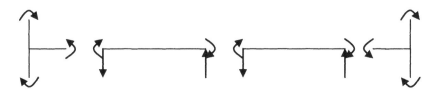

Fig. 13.33 Moments at joints and in beams.

(c) Axial forces in columns

As shown in Fig. 13.34, from the bending moments in the beams, reactions R at the ends of the beam is given by

$$R = 2 \times \text{bending moments/span.}$$

From the reactions in the beam, axial forces in the columns can be determined.

Table 13.30 shows the axial forces in columns. Note that the axial force in the middle column is zero and the axial force in the left column is tensile while in the right column it is compressive.

(d) Axial forces in beams

As shown in Fig. 13.35, by considering at joints equilibrium in the horizontal direction of shear forces in the columns and the axial forces in the beams, axial forces in beams can be determined. Table 13.31 shows the axial forces in beams.

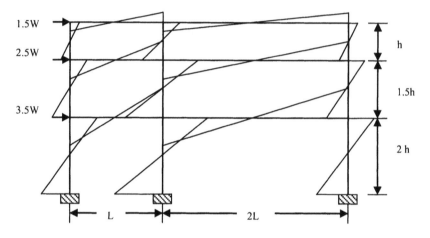

Fig. 13.34 Bending moment distribution in the frame.

Fig. 13.35 Axial force in beams.

Table 13.31 Axial forces in beams

	Load at joint/W	Shear columns/W		Axial force in beam/W	
		Left	Right	Beam-Left	Beam-right
Top	1.5	0.25	0.5	1.5 – 0.25 = 1.25	0.5
Middle	2.5	0.67	1.33	2.5 + 0.25 – 0.67 = 2.08	1.33 – 0.5 = 0.83
Bottom	3.5	1.25	2.5	3.5 + 0.67 – 1.25 = 2.92	2.5 – 1.33 = 1.17

CHAPTER 14

REINFORCED CONCRETE FRAMED BUILDINGS

14.1 TYPES AND STRUCTURAL ACTION

Commonly used single-storey and medium-rise reinforced concrete framed structures are shown in Fig. 14.1. Tall multi-storey buildings are discussed in Chapter 15. Only *cast-in-situ* rigid-jointed frames are dealt with, but the structures shown in the figure could also be precast.

The loads are transmitted by roof and floor slabs and walls to beams and to rigid frames and through the columns to the foundations. In *cast-in-situ* buildings with monolithic floor slabs, the frame consists of flanged beams and rectangular columns. However, it is common practice to base the analysis on the rectangular beam section, but in the design for sagging moments the flanged section is used. If precast slabs are used the beam sections are rectangular.

Depending on the floor system and framing arrangement adopted, the structure may be idealized into a series of plane frames in each direction for analysis and design. Such a system where two-way floor slabs are used is shown in Fig. 14.2; the frames in each direction carry part of the load. In the complete three-dimensional frame, torsion occurs in the beams and biaxial bending in the columns. These effects are small and it is usually only necessary to design for the maximum moment about the critical axis. In rectangular buildings with a one-way floor system, the transverse rigid frame across the shorter plan dimension carries the load. Such a frame is shown in the design example in section 14.5.

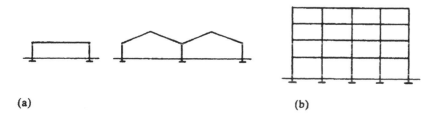

(a)   (b)

Fig. 14.1 (a) Single storey; (b) multi-storey.

Resistance to horizontal wind loads is provided by:

1.   In braced structures by shear walls, lift shaft and stairs

2.   In unbraced structures by bending of rigid-jointed frames

The analysis for combined shear wall, rigid frame systems is discussed in Chapter 15.

In multi-storey buildings, the most stable arrangement is obtained by bracing with shear walls in two directions. Stairwells, lift shafts, permanent partition walls as well as specially designed external shear walls can be used to resist the horizontal loading. Shear walls should be placed symmetrically with respect to the building axes. If this is not done the shear walls must also be designed to resist the resulting torque. The concrete floor slabs act as large horizontal diaphragms to transfer loads at floor levels to the shear walls. For economic reasons, the overall stability of a multi-storey building should not depend on unbraced frames alone. Shear walls in a multi-storey building are shown in Fig. 14.2.

Foundations for multi-storey buildings may be separate pad or of strip type. However, rafts or composite raft and basement foundations are more usual. For raft type foundations the column base may be taken as fixed for frame analysis. The stability of the whole building must be considered and the stabilizing moment from dead loads should prevent the structure from overturning.

Separate pad type foundations should only be used for multi-storey buildings if foundation conditions are good and differential settlement will not occur. For single-storey buildings, separate foundations are usually provided and, in poor soil conditions, pinned bases can be more economical than fixed bases. The designer must be satisfied that the restraint conditions assumed for analysis can be achieved in practice. If a fixed base settles or rotates, a redistribution of moments occurs in the frame.

14.2 BUILDING LOADS

The load on buildings is due to dead, imposed, wind, dynamic, seismic and accidental loads. Normally multi-storey buildings for office or residential purposes are designed for dead, imposed and wind loads. Earthquake loads are considered in earthquake-prone areas. The design is checked and adjusted to allow for the effects of accidental loads. The types of loads are discussed briefly.

14.2.1 Dead Load

Dead load is due to the weight of roofs, floors, beams, walls, columns, floor finishes, partitions, ceilings, services etc. The load is estimated from assumed section sizes and allowances are made for further dead loads that are additional to the structural concrete.

14.2.2 Imposed Load

Imposed load depends on the occupancy or use of the building and includes distributed loads, concentrated loads, impact, inertia and snow. Loads for all types of buildings are given in *BS EN 1991-1-4: 2002 Eurocode 1: Actions on Structures. General actions. Densities, self weight, imposed loads for buildings.*

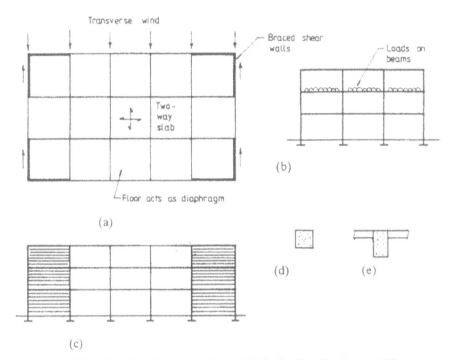

Fig. 14.2 (a) Plan; (b) rigid transverse frame; (c) side elevation; (d) column; (e) T-beam.

14.2.3 Wind Load

Wind load on buildings is estimated in accordance with *BS EN 1991-1-4: 2005 + A1:2010 Eurocode 1: Actions on Structures. General actions. Wind actions* and *UK National Annex to Eurocode 1: Actions on Structures. General actions. Wind actions.*

For full details and a worked example see section 10.3.4 of Chapter 10. The following is a summary of the basic factors to be included in wind load calculation. Wind loads should be calculated for lateral and longitudinal directions to obtain loads on frames or shear walls to check the stability in each direction. In asymmetrical buildings it may be necessary to investigate wind from all directions.

14.2.3 .1 Wind Load Calculated Using the U.K. National Annex

The following is a very brief summary of calculations steps using the U.K. National Annex.

Step 1: The basic wind velocity v_b is calculated from the wind code equation (4.1) as

$$v_b = C_{dir} \times C_{season} \times v_{b,0} \qquad (4.1)$$

Conservatively, direction factor C_{dir} and season factor C_{season} can both be taken as 1.0

$$v_b = 1.0 \times 1.0 \times v_{b,0} = v_{b,0}$$

Step 2: The fundamental value of the basic wind velocity $v_{b,0}$ is given by the National Annex equation(NA.1) as

$$v_{b,0} = v_{b,map} \times C_{alt} \qquad (NA.1)$$

$v_{b,map}$ = fundamental basic wind velocity in m/s given in the map for the United Kingdom given in the National Annex in Fig. NA.1.
C_{alt} = altitude correction factor.

Step 3: Conservatively, C_{alt} for any building height is given by National Annex equation (NA.2a) as

$$C_{alt} = 1 + 0.001 \times A \qquad (NA.2a)$$

A = Altitude of the site in meters above sea level.

Step 4: The basic velocity pressure q_b is given by the wind equation (4.10) as

$$q_b = 0.5 \times \rho \times v_b^2 \ N/m^2 \qquad (4.10)$$

ρ = density of air taken as 1.226 kg/m^3.

Step 5: The peak wind pressure $q_p(z)$ is given by National Annex equation (NA.3b) for sites in town terrain by

$$q_p(z) = c_e(z) \times c_{e,T} \times q_b \qquad (NA.3b)$$

Step 6: Exposure factor $c_e(z)$ is calculated from Fig. NA.7. Exposure correction factor $c_{e,T}$ is calculated from Fig. NA.8.

Step 7: The total pressure coefficient c_f is the sum of external pressure coefficient c_{pe} and internal pressure coefficient c_{pi}. The values of the pressure coefficients are calculated from Table 7.1 of the wind code. The total wind load is given by

$$w_k = q_p(z) \times c_f$$

14.2.3 .2 Wind Load Calculated Using the Eurocode

The following is a very brief summary of calculations steps using the Eurocode.

Step 1: The mean wind velocity $v_m(z)$ at a height above the terrain is given by the wind code equation (4.3) as
$$v_m(z) = c_r(z) \times c_0(z) \times v_b \qquad (4.3)$$
$c_r(z)$ = terrain roughness factor. This factor accounts for the variability of the mean wind velocity at the site due to height above the ground level and ground roughness of the terrain upwind of the structure.
$c_0(z)$ = orography (terrain) factor taken as 1.0.

Step 2: $c_r(z)$ is defined by the wind code equation (4.4) as
$$c_r(z) = k_r \times \ell_n(z/z_0) \text{ for } z_{min} \le z \le z_{max} \qquad (4.4)$$

Step 3: k_r is defined by the wind code equation (4.5) for town areas as
$$k_r = 0.19 \times (z_0/z_{0,11})^{0.07}$$

Step 4: The basic wind velocity v_b is calculated from the wind code equation (4.1) as
$$v_b = C_{dir} \times C_{season} \times v_{b,0} \qquad (4.1)$$
Conservatively, direction factor C_{dir} and season factor C_{season} can both be taken as 1.0
$$v_b = 1.0 \times 1.0 \times v_{b,0} = v_{b,0}$$

Step 5: The fundamental value of the basic wind velocity $v_{b,0}$ is given by the National Annex equation (NA.1) as
$$v_{b,0} = v_{b,map} \times C_{alt} \qquad (NA.1)$$
$v_{b,map}$ = fundamental basic wind velocity in m/s given in the map of the country. The map for the United Kingdom is given in the National Annex in Fig. NA.1.
C_{alt} = altitude correction factor.

Step 6: Conservatively, C_{alt} for any building height is given by National Annex equation (NA.2a) as
$$C_{alt} = 1 + 0.001 \times A \qquad (NA.2a)$$
A = altitude of the site in meters above sea level.

Step 7: The basic velocity pressure q_b is given by the wind equation (4.10) as
$$q_b = 0.5 \times \rho \times v_b^2 \text{ N/m}^2 \qquad (4.10)$$
ρ = density of air taken as 1.25 kg/m^3
$$q_b = 0.613 \times v_b^2 \times 10^{-3} = 0.613 \times 22.0^2 \times 10^{-3} = 0.30 \text{ kN/m}^2$$

Step 8: The peak velocity pressure $q_p(z)$ at height z is given by wind code equation (4.8)
$$q_p(z) = c_e(z) \times q_b \qquad (4.8)$$

Step 9: The value of $c_e(z)$ for different values of z and different terrain categories is given in Fig. 4.2 of the code.

Step 10: The wind pressure acting on the external surface w_e is calculated from wind code equation (5.1)

$$w_e = q_p(z_e) \times c_{pe} \qquad (5.1)$$

Step 11: The wind pressure acting on the internal surface w_i is calculated from wind code equation (5.1)

$$W_i = q_p(z_e) \times c_{pi} \qquad (5.1)$$

Step 12: The wind force $F_{w,e}$ acting on the external surface is calculated from wind code equation (5.5)

$$F_{w,e} = c_s c_d \times \sum w_e \times A_{ref} \qquad (5.5)$$

Step 13: The wind force $F_{w,i}$ acting on the internal surface is calculated from wind code equation (5.6)

$$F_{w,i} = \sum w_i \times A_{ref} \qquad (5.6)$$

Step 14: From section 6.2(1)(c), for framed buildings with structural walls with height less than 100 m and less than four times the in-wind depth, $c_s\,c_d = 1.0$.

Step 15: Frictional forces can be ignored when the total area of all surfaces parallel to the wind is equal to less than the total area of all external surfaces perpendicular to the wind.

Step 16: The total pressure coefficient c_f is the sum of external pressure coefficient c_{pe} and internal pressure coefficient c_{pi}. The values of the pressure coefficients are calculated from Table 7.1 of the wind code. Values of $c_{pe,\,10}$ should be used for the design of overall load bearing structure.

14.2.4 Use of Influence Lines to Determine Positioning of Gravity Loads to Cause Maximum Design Moments

Muller-Breslau's principle can be used to obtain qualitative influence lines as deflection curves. The influence lines help to place the loads in correct positions in the structure in order to produce maximum moments in members. The method is explained by three examples as follows.

(a) Fig. 14.3(a) shows the positioning of gravity loads to give maximum bending moment in the span of beam AB. This is known as chequer board load pattern. The deflection curve is obtained by creating a hinge at the mid-point in the beam AB and applying equal and opposite couples as shown in Fig. 14.3(b).

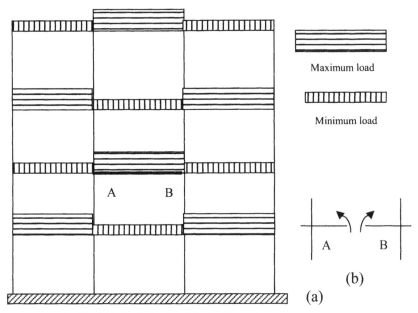

Fig. 14.3 Gravity load positioning for maximum span moment.

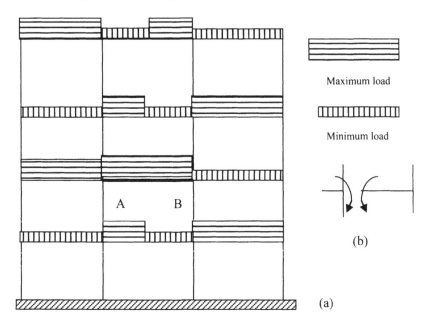

Fig. 14.4 Gravity load positioning for maximum support moment at A in beam AB.

(b) Fig. 14.4(a) shows the positioning of gravity loads to give maximum bending moment at the support A of beam AB. The deflection curve is obtained by creating a hinge at the end A of beam AB and applying equal and opposite couples as shown in Fig. 14.4(b). Note that apart from beam AB, the other beams in the middle column of the frame are loaded with part of the beam with the maximum load and the other part with minimum load.

(c) Fig. 14.5(a) shows the positioning of gravity loads to give maximum bending moment in column AB. The deflection curve is obtained by creating a hinge at the end A of column AB and applying equal and opposite couples as shown in Fig. 14.5(b). Note that maximum moment in column is obtained when the beam on one side of the column carries maximum load and the beam on the other side carries minimum load.

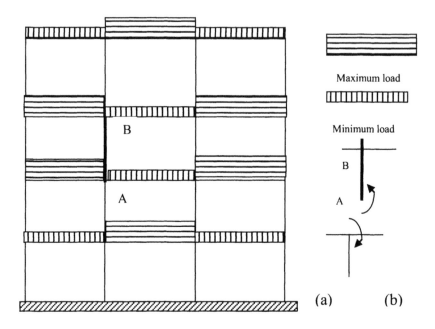

Fig. 14.5 Gravity load positioning for maximum moment in column AB.

14.2.5 Use of Sub-Frames to Determine Moments in Members

The loading patterns shown in Fig. 14.4 to Fig. 14.6 show clearly that in a large rigid-jointed frame structure, the number of load patterns to be considered become too numerous for practical design. Fortunately, the moments and shears under gravity loads can be determined using only part of the structure. The reason for this is that generally the loads at any chosen floor level affect mainly the moments and shear forces in members at that level. The loads on floor levels other than the chosen level have minimum effect. This can be shown by a simple example.
Fig. 14.6 shows a typical 2-D rigid-jointed building frame. Fig. 14.7 shows a sub-frame isolated from a much larger structure shown in Fig. 14.6. For simplicity,

assume that (EI/L) is same for all members of the sub-frame and that the far ends of the members are clamped. Under a gravity load of q per unit length on the beam of span L at the centre, the moment distribution is as shown in Fig. 14.8.

The moments at the supports at each end in the loaded beam are approximately $0.86 \, qL^2/12$ and the moment at the mid-span is $0.64 \, qL^2/12$. The moment at the fixed end of unloaded beams and column is only $0.14 \, qL^2/12$. In reality the moment value will be even smaller than this because the far ends are not fully clamped but attached to the members of the larger frame of which the sub-frame is only a part. The moment distribution in the rest of the structure of which the sub-frame is a part, is caused by the moment of $0.14 \, qL^2/12$ acting as the applied load at the ends of the members where the sub-frame is attached to the overall frame. This shows that the influence of the load q on the rest of the structure is quite small and can be ignored in the interests of simplicity.

When using structural analysis computer programs for the determination of moments in beams under gravity loading, it is convenient to include all the beams at any one level with the columns above and below as shown in Fig. 14.9. The ends of the columns can be taken as clamped unless pinned condition reflects reality better.

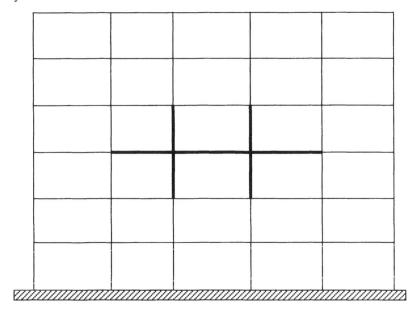

Fig. 14.6 A typical 2-D rigid-jointed building frame.

Clause 5.1.3 states that under gravity and imposed load combination, the following load arrangements should be investigated:
1. Alternate spans carrying $\gamma_G \, G_k + \gamma_Q \, Q_k$ and other spans carrying $\gamma_G \, G_k$

2.   Any two adjacent spans carrying $\gamma_G\, G_k + \gamma_Q\, Q_k$ and other spans carrying $\gamma_G\, G_k$ only

The first load arrangement is to ensure maximum span moments are calculated and the second load arrangement is to ensure that the maximum support moments are calculated.

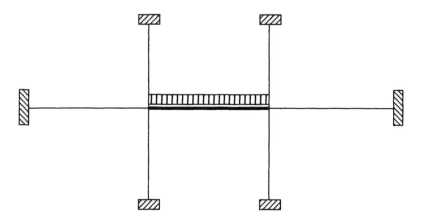

Fig. 14.7 A typical sub-frame for analysis of moments in beams under gravity loading.

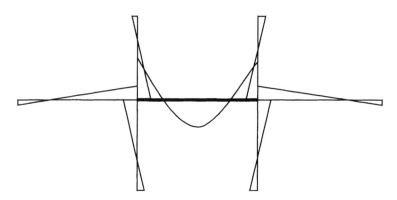

Fig. 14.8 Bending moment distribution in the sub-frame under gravity loading.

The sub-frame to be used for calculating column moments is shown in Fig. 14.10. Note that the beams on one side carries maximum load ($\gamma_G\, G_k + \gamma_Q\, Q_k$) and the other side carries minimum load $\gamma_G\, G_k$.

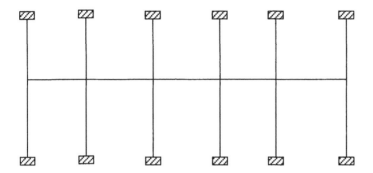

Fig. 14.9 Sub-frame for determining the bending moment distribution in beams under gravity loading.

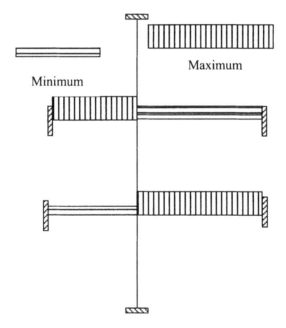

Fig. 14.10 A typical sub-frame for analysis of moments in columns under gravity loading.

14.2.6 Load Combinations

Separate loads must be applied to the structure in appropriate directions and various types of loading combined with partial safety factors selected to cause the most severe design condition for the member under consideration. In general the following load combinations should be investigated.

(a) Dead load G_k + imposed load Q_k

1. Alternate spans carrying maximum load of 1.35 G_k + 1.5 Q_k and other spans carrying maximum load of 1.35 G_k
2. Any two adjacent spans carrying 1.35 G_k + 1.5 Q_k and other spans carrying 1.35 G_k only

(b) Dead load G_k + wind load W_k

If dead load and wind load effects are additive the load combination is 1.35 G_k + 1.5 W_k. However, if the effects are in opposite directions the critical load combination is 1.0G_k − 1.5W_k.

The distributed wind pressure is applied as a concentrated load at floor levels as shown in Fig. 14.11.

(c) Dead load G_k + imposed load Q_k + wind load W_k

(See *BS EN 1990:2002, Eurocode– Basis of Structural Design.* The recommended values of ψ factors for buildings are shown in Table A1.2(B) and Table A1.1.)

(i) If imposed load is the main variable and wind action is treated as accompanying action, then the load combination is

$$1.35\ G_k + 1.5\ Q_k + 1.5 \times 0.6 \times W_k$$

(ii) If wind load is the main variable and imposed load is treated as accompanying action, then the load combination is

$$1.35\ G_k + 1.5 \times 0.7 \times Q_k + 1.5 \times W_k$$

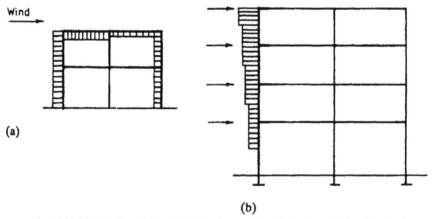

(a)

(b)

Fig. 14.11 Wind loading (a) Loads distributed on surfaces; (b) loads applied at floor levels.

14.2.6.1 Example of Load Combinations

Fig. 14.12 shows a building supported on pinned base columns spaced 10 m in both directions. Calculate the maximum bending moment and axial force (compression and tension) in the columns. It is given that

$$G_k = 15.0 \text{ kN/m}^2, \ Q_k = 12.5 \text{ kN/m}^2, \ W_k = 1.0 \text{ kN/m}^2$$

Wind acts on the face AB or GD and can act from left to right or vice versa.

In considering the load combinations to be considered for a particular stress resultant (bending moment, axial force, shear force), it is necessary to be aware of what effect a particular load on a particular part of the structure has on the stress resultant calculated in order that appropriate load factors can be applied.
In this example the following effects can be noted.

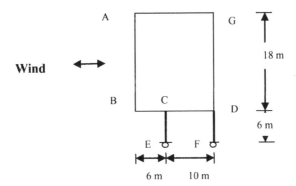

Fig. 14.12 Building supported by pinned base columns.

(a) Left hand column:
- Vertical load acting on the entire plan area BCD will cause compression in the column
- Wind blowing from right to left causes compression in the column
- Wind blowing from left to right causes tension in the column

(b) Right hand column:
- Vertical load acting on a plan area CD will cause compression
- Vertical load on plan area BC will cause tension
- Wind blowing from right to left causes tension in the column
- Wind blowing from left to right causes compression in the column

Load combinations:

(i) Dead load G_k + imposed load Q_k
There are four cases to be considered. They are

a. $(1.35 \, G_k + 1.5 \, Q_k)$ on the plan area BCD:

Total vertical load = $(1.35 \times 15 + 1.5 \times 12.5) \times 16 \times 10 = 6240$ kN
Taking moments about F,
Axial force in CE = $\{6240 \times 16/2\}/10 = 4992$ kN
Axial force in DF = $6240 - 4992 = 1248$ kN

b. $(1.35 \ G_k + 1.5 \ Q_k)$ on the plan area CD and $1.0 \ G_k$ on plan area BC:
Total vertical load on plan area CD
$= (1.35 \times 15 + 1.5 \times 12.5) \times 10 \times 10 = 3900$ kN
Total vertical load on plan area BC
$= (1.0 \times 15) \times 6 \times 10 = 900$ kN
Taking moments about F,
Axial force in CE = $\{3900 \times 10/2 + 900 \times (10 + 6/2)\}/10 = 3120$ kN
Axial force in DF = $3900 + 900 - 3120 = 1680$ kN

c. $1.0 \ G_k$ on the plan area CD and $(1.35 \ G_k + 1.5 \ Q_k)$ on plan area BC:
Total vertical load on plan area CD
$= (1.0 \times 15) \times 10 \times 10 = 1500$ kN
Total vertical load on plan area BC
$= (1.35 \times 15 + 1.5 \times 12.5) \times 6 \times 10 = 2340$ kN
Taking moments about F,
Axial force in CE = $\{1500 \times 10/2 + 2340 \times (10 + 6/2)\}/10 = 3792$ kN
Axial force in DF = $1500 + 2340 - 3792 = 48$ kN

d. $1.0 \ G_k$ on the plan area BCD:
Total vertical load = $(1.0 \times 15) \times 16 \times 10 = 2400$ kN
Taking moments about F,
Axial force in CE = $\{2400 \times 16/2\}/10 = 1920$ kN
Axial force in DF = $2400 - 1920 = 480$ kN

Column forces: From the above four combinations for dead and imposed loads, the maximum and minimum forces in the columns are:
Column CE
Maximum compressive force = 4992 kN (from a)
Minimum compressive force = 1920 kN (from d)
Column DF
Maximum compressive force = 1680 kN (from b)
Minimum compressive force = 48 kN (from c)

(ii) Dead load and wind load: $[(1.35 \ or \ 1.0) \ G_k \pm 1.5 \ W_k]$
For convenience, the calculations are done by considering wind and dead loads separately.

a. Wind load $1.5 \ W_k$ only
Wind load = $1.5 \times 1.0 \times 18 \times 10 = 270$ kN
Axial force in columns = $\pm 270 \times (6 + 18/2)/10 = \pm 405$ kN
Shear force in columns = $\pm 270/2 = \pm 135$ kN

Moment in columns $= \pm 135 \times 6 = \pm 810$ kNm

b. $1.35\ G_k$ on the plan area BCD:

Total vertical load $= (1.35 \times 15) \times 16 \times 10 = 3240$ kN

Taking moments about F,

Axial force in CE $= \{3240 \times 16/2\}/10 = 2592$ kN

Axial force in DF $= 3240 - 2592 = 648$ kN

c. $1.35\ G_k$ on the plan area CD and $1.0\ G_k$ on plan area BC:

Total vertical load on plan area CD $= (1.35 \times 15) \times 10 \times 10 = 2025$ kN

Total vertical load on plan area BC $= (1.0 \times 15) \times 6 \times 10 = 900$ kN

Taking moments about F,

Axial force in CE $= \{2025 \times 10/2 + 900 \times (10 + 6/2)\}/10 = 2183$ kN

Axial force in DF $= 2025 + 900 - 2183 = 743$ kN

d. $1.0\ G_k$ on the plan area CD and $1.35\ G_k$ on plan area BC:

Total vertical load on plan area CD $= (1.0 \times 15) \times 10 \times 10 = 1500$ kN

Total vertical load on plan area BC $= (1.35 \times 15) \times 6 \times 10 = 1215$ kN

Taking moments about F,

Axial force in CE $= \{1500 \times 10/2 + 1215 \times (10+6/2)\}/10 = 2330$ kN

Axial force in DF $= 1500 + 1215 - 2330 = 385$ kN

e. $1.0\ G_k$ on the plan area BCD:

Total vertical load $= (1.0 \times 15) \times 16 \times 10 = 2400$ kN

Taking moments about F,

Axial force in CE $= \{2400 \times 16/2\}/10 = 1920$ kN

Axial force in DF $= 2400 - 1920 = 480$ kN

Column forces: From the above five cases for dead and wind load combination, the critical axial force and moment combinations are:

Column CE

Maximum compressive force $= 2592$ (from b) $+ 405$ (from a) $= 2997$ kN

Moment in column $= \pm 810$ kNm

Minimum compressive force $= 1920$ (from e) $- 405$ (from a) $= 1515$ kN

Moment in column $= \pm 810$ kNm

Column DF

Maximum compressive force $= 743$ (from c) $+ 405$ (from a) $= 1148$ kN

Moment in column $= \pm 810$ kNm

Minimum compressive force $= 385$ (from d) $- 405$ (from a) $= -20$ kN

Moment in column $= \pm 810$

(iii) [Dead + imposed + wind] load: Imposed load main variable

$$1.35G_k + 1.5\ Q_k + 1.5 \times 0.6 \times W_k$$

For convenience, the calculations are done by considering wind, dead + imposed loads separately.

(a) Wind load

$$\text{Wind load} = 1.5 \times 0.6 \times 1.0 \times 18 \times 10 = 162 \text{ kN}$$
$$\text{Axial force in columns} = \pm 216 \times (6 + 18/2)/10 = \pm 324 \text{ kN}$$
$$\text{Shear force in columns} = \pm 162/2 = \pm 81 \text{ kN}$$
$$\text{Moment in columns} = \pm 81 \times 6 = \pm 486 \text{ kNm}$$

(b) Dead + imposed load on the plan area BCD

$$\text{Total vertical load} = (1.35 \times 15 + 1.5 \times 12.5) \times 16 \times 10 = 6240 \text{ kN}$$
Taking moments about F,
$$\text{Axial force in CE} = \{6240 \times 16/2\}/10 = 4992 \text{ kN}$$
$$\text{Axial force in DF} = 6240 - 4992 = 1248 \text{ kN}$$

(iv) [Dead + imposed + wind] load: Wind load main variable

$$1.35 G_k + 1.5 \times 0.7 \times Q_k + 1.5 W_k$$

For convenience, the calculations are done by considering wind, dead and imposed loads separately.

(a) Wind load

$$\text{Wind load} = 1.5 \times 1.0 \times 18 \times 10 = 270 \text{ kN}$$
$$\text{Axial force in columns} = \pm 270 \times (6 + 18/2)/10 = \pm 405 \text{ kN}$$
$$\text{Shear force in columns} = \pm 270/2 = \pm 135 \text{ kN}$$
$$\text{Moment in columns} = \pm 135 \times 6 = \pm 810 \text{ kNm}$$

(b) Dead + imposed load on the plan area BCD:

$$\text{Total vertical load} = (1.35 \times 15 + 1.5 \times 0.7 \times 12.5) \times 16 \times 10 = 5340 \text{ kN}$$
Taking moments about F,
$$\text{Axial force in CE} = \{5340 \times 16/2\}/10 = 4272 \text{ kN}$$
$$\text{Axial force in DF} = 5340 - 4272 = 1068 \text{ kN}$$

Column forces: From the above four cases for dead + imposed + wind combination, the critical axial force and moment combinations are:

Column CE

Maximum compressive force = 4992 (from b) + 405 (from c) = 5397 kN
Moment in column = ± 810 kNm
Minimum compressive force = 4272 (from d) − 405 (from c) = 3867 kN
Moment in column = ± 810 kNm

Column DF

Maximum compressive force = 1248 (from b) + 405 (from c) = 1653 kN
Moment in column = ± 810 kNm
Minimum compressive force = 1068 (from d) − 405 (from c) = 663 kN
Moment in column = ± 810 kNm

Design values: From a design point of view, the critical combinations *probably* are
Column CE: (N = 1515 kN, M = 810 kNm) or (N = 5397 kN, M = 810 kNm)

Column DF: (N = – 20 kN, M = 810 kNm) or (N = 1653 kN, M = 810 kNm)

This example shows that even in a simple structure, the number of load cases to be considered may become quite large. In large scale structures, a good understanding of the behaviour of the structure under consideration is necessary in order to limit the number of load cases to be considered, by eliminating load cases clearly not critical.

14.3 ROBUSTNESS AND DESIGN OF TIES

Clause 9.10.1 of the Eurocode 2 states that situations should be avoided where damage to a small area or failure of a single element could lead to collapse of major parts of the structure. Provision of effective ties is one of the precautions necessary to prevent progressive collapse. The layout also must be such as to give a stable and robust structure.

14.3.1 Types of Ties

The types of tie are
1. Peripheral ties
2. Internal ties
3. Horizontal ties to columns and walls
4. Vertical ties particularly in panel buildings

The types and location of ties are shown in Fig. 14.13.

14.3.2 Design of Ties

Steel reinforcement provided for a tie can be designed to act at its characteristic strength. Reinforcement provided for other purposes may form the whole or part of the ties.

14.3.3 Internal Ties

Internal ties are to be provided at the roof and all floors in two directions at right angles. They are to be continuous throughout their length and anchored to peripheral ties. The ties may be spread evenly in slabs or be grouped in beams or walls. Ties in walls are to be within 0.5 m of the top or bottom of the floor slab.
The ties should be capable of resisting a design tensile force of 10 kN/m in each direction.

14.3.4 Peripheral Ties

A continuous peripheral tie is to be provided at each floor and at the roof. This tie is to be located within 1.2 m of the edge of the building or within the perimeter wall. The tie is to resist a force:
$$F_{tie, per} = \ell \times q_1 \leq q_2$$
where $q_1 = 10$ kN/m, $q_2 = 70$ kN, ℓ = length of end span.

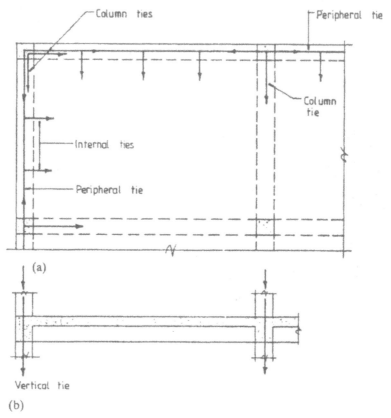

(a)

(b)

Fig. 14.13 Building ties: (a) Plan; (b) section.

14.3.5 Horizontal Ties to Columns and Walls

Edge columns and walls should be tied horizontally to the structure at each floor and roof level. The ties should be capable of resisting a tensile force $f_{tie, fac} = 20$ kN per meter of the facade. For columns the tie force need not exceed $F_{tie, col} = 150$ kN.

Corner columns are to be tied in two directions. Steel provided for peripheral ties may be used as the horizontal tie.

Where the peripheral tie is located within the walls the internal ties are to be anchored to it.

14.3.6 Vertical Ties

Vertical ties are required in buildings of five or more storeys. Each column and load bearing wall is to be tied continuously from foundation to roof. The tie is to be capable of carrying a tensile force equal to the ultimate dead and imposed load carried by the column or wall from one floor.

14.4 FRAME ANALYSIS

14.4.1 Methods of Analysis

Modern frame analysis is achieved by plane frame computer programs based on the matrix stiffness method (see Bhatt (1981) and Bhatt (1986)). Especially for vertical loading analysis, sub-frames are used to simplify the number of load cases to be analysed.

Analysis assumes elastic behaviour. Clause 5.4 of Eurocode 2 states that linear elastic analysis may be carried out assuming

- Uncracked cross sections
- Linear stress–strain relationships
- Mean value of modulus of elasticity

In beam–slab floor construction it is normal practice to base the beam stiffness on a uniform rectangular section consisting of the beam depth by the beam rib width. The flanged beam section is taken into account in the beam design for sagging moments near the centre of spans.

As explained in section 13.5, Chapter 13, clause 5.5 of Eurocode 2 states that limited redistribution while maintaining equilibrium may be applied to the results of elastic analysis. In continuous beams and slabs which have the ratio of lengths of adjacent spans in the range 0.5 to 2.0, there is no need to check explicitly for rotation capacity provided that

$$\delta \geq 0.44 + 1.25(0.6 + \frac{0.0014}{\varepsilon_{cu2}}) \text{ for } f_{ck} \leq 50 \text{ MPa} \tag{5.10a}$$

$$\delta \geq 0.54 + 1.25(0.6 + \frac{0.0014}{\varepsilon_{cu2}}) \text{ for } f_{ck} > 50 \text{ MPa} \tag{5.10b}$$

$$\geq 0.7 \text{ where Class B and Class C reinforcement is used}$$

$$\geq 0.8 \text{ where Class A reinforcement is used}$$

$$\varepsilon_{cu2} = 0.0035 \text{ for } f_{ck} \leq 50 \text{ MPa}$$

$$\varepsilon_{cu2} = \left[2.6 + 35 \times \left\{\frac{(90 - f_{ck})}{100}\right\}^4\right] \times 10^{-3} \text{ for } f_{ck} > 50 \text{ MPa}$$

where δ = ratio of redistributed moment/Elastic bending moment and x_u = the maximum depth of the neutral axis at ULS after redistribution.   Table 4.5, Table 4.6 and Table 4.8 of Chapter 4 show respectively the relationship between δ and x_u/d, z/d and $k = M/ (bd^2 f_{ck})$ for different values of f_{ck}.

It is further stated in clause 5.5(6) of Eurocode 2 that for design of columns, elastic moments from frame analysis should be used ***without any redistribution***.

In rigid-frame analysis, the sizes for members must be chosen from experience; or established by an approximate design before the analysis can be carried out. Ratios of stiffness of the final member sections should be checked against those estimated and the frame should be reanalysed if it is found necessary to change the sizes of members significantly.

14.4.2 Example of Simplified Analysis of Concrete Framed Building Under Vertical Load

(a) Specification
The cross section of a reinforced concrete building is shown in Fig. 14.14(a). The frames are at 4.5 m centres, the length of the building is 36 m and the column bases are fixed.   Preliminary sections for the beams and columns are shown in Fig. 14.14(b). The floor and roof slabs are designed to span one way between the frames.   Longitudinal beams are provided between external columns of the roof and floor levels only.

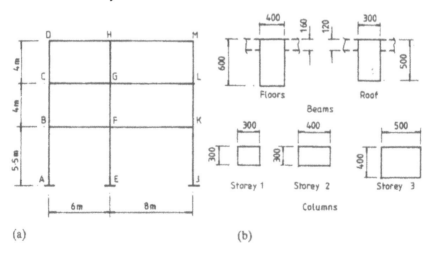

Fig. 14.14 (a) Cross section; (b) assumed member sections.

The dead and imposed loadings are as follows:
Roof:
$$\text{Total dead load} = 4.3 \text{ kN/m}^2$$
$$\text{Imposed load} = 1.5 \text{ kN/m}^2$$

Floor:

$$\text{Total dead load} = 6.2 \text{ kN/m}^2$$
$$\text{Imposed load} = 3.0 \text{ kN/m}^2$$

Wind load: Wind load on buildings is estimated in accordance with *BS EN 1991-1-4: 2005 + A1:2010 Eurocode 1: Actions on Structures. General actions. Wind actions* and *UK National Annex to Eurocode 1: Actions on Structures. General actions. Wind actions.*
The location is on the outskirts of a city in the northeast of the U.K. The materials characteristic strengths are f_{ck} = 30 MPa for concrete and f_{yk} = 500 MPa for reinforcement. Determine the design actions for the beam BFK and column length FE for an internal frame for the two cases where the frame is braced and unbraced.

(b) Loading
Using a frame spacing of 4.5 m c/c, the characteristic loads are as follows.

Dead load:

$$\text{Roof } 4.3 \times 4.5 = 19.4 \text{ kN/m}$$
$$\text{Floors } 6.2 \times 4.5 = 27.9 \text{ kN/m}$$

Imposed load:

$$\text{Roof } 1.5 \times 4.5 = 6.8 \text{ kN/m}$$
$$\text{Floors } 3.0 \times 4.5 = 14.5 \text{ kN/m}$$

(c) Section properties
The beam and column properties are given in Table 14.1.

Table 14.1 Section properties

Member	b × d	L (mm)	I (mm^4)	I/L
Columns AB, EF, JK	400 × 500	5500	4.17×10^9	7.58×10^5
Columns BC, FG, KL	300 × 400	4000	1.60×10^9	4.0×10^5
Columns CD, GH,LM	300 × 300	4000	0.675×10^9	1.6875×10^5
Beam, DH	300 × 500	6000	3.125×10^9	5.208×10^5
Beam, HM	300 × 500	8000	3.125×10^9	3.906×10^5
Beams GL, FK	400 × 600	8000	7.20×10^9	9.0×10^5
Beams CG, BF	400 × 600	6000	7.20×10^9	12.0×10^5

(d) Sub-frame analysis for braced frame
The sub-frame used in this example is shown in Fig. 14.15. The frame is analysed for the three load cases using the matrix stiffness method of analysis. From characteristic dead and imposed loads,

$$(1.35G_k + 1.5\ Q_k) = (1.35 \times 27.9 + 1.5 \times 14.5) = 59.42\ \text{kN/m},$$
$$1.35\ G_k = 37.67\ \text{kN/m}$$

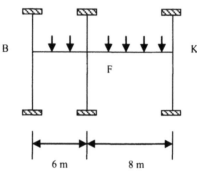

Fig. 14.15 Simplified sub-frame for beam BFK.

(e) Braced frame
The following load cases are required for beam BFK for the braced frame.
Case 1: $(1.35G_k + 1.5\ Q_k)$ on BF and $1.35G_k$ on FK
Case 2: $(1.35G_k + 1.5\ Q_k)$ on FK and $1.35G_k$ on BF
Case 3: $(1.35G_k + 1.5\ Q_k)$ on BF and FK
The fixed end moments are:
Case 1:

$$\text{Span BF} = 59.42 \times 6^2/12 = 178.26\ \text{kNm},$$
$$\text{Span FK} = 37.67 \times 8^2/12 = 200.91\ \text{kN m}$$

Case 2:

$$\text{Span BF} = 37.67 \times 6^2/12 = 113.01\ \text{kNm}$$
$$\text{Span FK} = 59.42 \times 8^2/12 = 316.91\ \text{kN m}$$

Case 3:

$$\text{Span BF} = 59.42 \times 6^2/12 = 178.26\ \text{kNm}$$
$$\text{Span FK} = 59.42 \times 8^2/12 = 316.91\ \text{kN m}$$

The stiffness matrix K and load vector F for the three load cases are

$$K = E \times 10^5 \begin{bmatrix} 94.32 & 24.0 & 0 \\ 24.0 & 130.32 & 18.0 \\ 0 & 18.0 & 82.32 \end{bmatrix} \begin{bmatrix} \theta_B \\ \theta_F \\ \theta_K \end{bmatrix}, F = \begin{bmatrix} 178.26 & 113.91 & 178.26 \\ 22.65 & 203.90 & 138.65 \\ -200.91 & -316.91 & -316.91 \end{bmatrix}$$

The displacement vectors for the three load cases are

$$E \times 10^5 \begin{bmatrix} \theta_B \\ \theta_F \\ \theta_K \end{bmatrix} = \begin{bmatrix} 1.8328 & 0.6865 & 1.5357 \\ 0.1788 & 2.0313 & 1.3537 \\ -2.4797 & -4.2939 & -4.1457 \end{bmatrix}$$

where E = Young's modulus.

The shear force and bending moment diagrams for beam BFK are shown respectively in Fig. 14.16 and Fig. 14.17 and the results of analysis are shown in Table 14.2.

Table 14.2 Bending moments for different loadings

M	Case 1	Case 2	Case 3
M_{BF}	−86.00	−31.31	−72.06
M_{FB}	−230.83	−226.99	−280.10
Span-BF	199.88	85.79	173.49
M_{FK}	−239.11	−384.47	−342.80
M_{KF}	−114.86	−198.89	−192.03
Span-FK	127.58	188.21	210.93

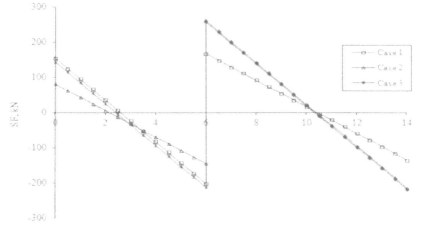

Fig. 14.16 Shear force diagrams for beam BFK in sub-frame.

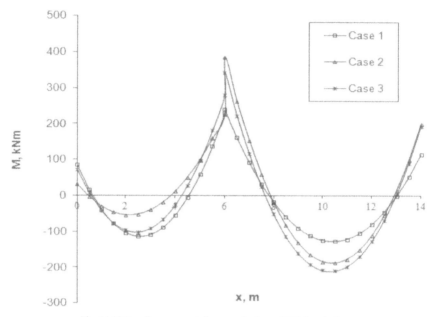

Fig. 14.17 Bending moment diagrams for beam BFK in sub-frame.

(f) Asymmetrically loaded columns

The moments for a column are calculated assuming that column and beam ends remote from the junction considered are fixed. Fig. 14.18 shows the sub-frames for determining column moments. In the case of end columns, maximum load of 1.35 G_k + 1.5 Q_k is applied on the connecting beam. In the case of the middle column, maximum load of 1.35 G_k + 1.5 Q_k is applied on the longer connecting beam and only 1.35 G_k is applied on the shorter connecting beam. Fig. 14.19 shows the bending moment diagrams in columns and connecting beams.

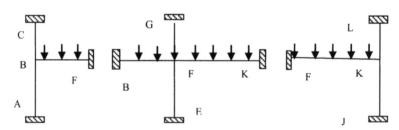

Fig. 14.18 Loading on column sub-frames.

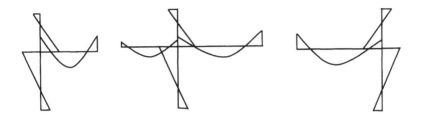

Fig. 14.19 Bending moment diagrams for column sub-frame.

The stiffness matrix and the load vectors and the corresponding joint rotations for the three cases are:

Asymmetrical Column ABC:
$$\text{Span BF} = 59.42 \times 6^2/12 = 178.26 \text{ kNm}$$
$$E \times 10^5 \times 94.32 \times \theta_B = 178.26, \ E \times 10^5 \times \theta_B = 1.89$$
$$\text{Moments: } M_{BA} = 57, \ M_{BA} = 29, \ M_{BC} = 30, \ M_{CB} = 15$$

Asymmetrical Column JKL:
$$\text{Span KF} = 59.42 \times 8^2/12 = 316.91 \text{ Kn m}$$
$$E \times 10^5 \times 82.32 \times \theta_K = -316.91, \ E \times 10^5 \times \theta_K = -3.85$$
$$\text{Moments: } M_{KJ} = -117, \ M_{JK} = -58, \ M_{KL} = -62, \ M_{LK} = -31$$

Asymmetrical Column EFG:
$$\text{Span FK} = 59.42 \times 8^2/12 = 316.91 \text{ kN m}$$
$$\text{Span FB} = 37.67 \times 6^2/12 = 113.01 \text{ kN m}$$
$$E \times 10^5 \times 130.32 \times \theta_F = (316.91 - 113.01), \ E \times 10^5 \times \theta_F = 1.565$$
$$\text{Moments: } M_{FE} = 48, \ M_{EF} = 24, \ M_{FG} = 25, \ M_{GF} = 13$$

If a slightly more complicated sub-frame is adopted as shown in Fig. 14.20, then the results are as follows.

Column ABCD:
$$\text{Span BF and CG} = 59.42 \times 6^2/12 = -178.26 \text{ kNm},$$

$$10^5 \times E \begin{bmatrix} 70.75 & 8.0 \\ 8.0 & 94.32 \end{bmatrix} \begin{bmatrix} \theta_C \\ \theta_B \end{bmatrix} = \begin{bmatrix} 178.26 \\ 178.26 \end{bmatrix}$$

$$10^5 \times E \begin{bmatrix} \theta_C \\ \theta_B \end{bmatrix} = \begin{bmatrix} 2.328 \\ 1.692 \end{bmatrix}$$

$M_{BA} = 51 (57), \ M_{AB} = 26(29), \ M_{BC} = 46 (30), \ M_{CB} = 51(15), \ M_{CD} = 16, \ M_{DC} = 8$.
Figures in brackets correspond to simplified sub-frames with one beam only.

Column EFGH:
$$\text{Span FK} = 59.42 \times 8^2/12 = 316.91 \text{ kNm}$$
$$\text{Span GL} = 37.67 \times 8^2/12 = 200.91 \text{ kNm}$$

$$\text{Span GC} = 59.42 \times 6^2/12 = 178.26 \text{ kNm}$$
$$\text{Span BF} = 37.67 \times 6^2/12 = 113.01 \text{ kNm}$$

$$10^5 \times E \begin{bmatrix} 106.75 & 8.0 \\ 8.0 & 130.32 \end{bmatrix} \begin{bmatrix} \theta_G \\ \theta_F \end{bmatrix} = \begin{bmatrix} 22.65 \\ 203.90 \end{bmatrix}$$

$$10^5 \times E \begin{bmatrix} \theta_G \\ \theta_F \end{bmatrix} = \begin{bmatrix} 0.095 \\ 1.559 \end{bmatrix}$$

$M_{FE} = 47$ (70), $M_{EF} = 23(24)$, $M_{FG} = 26$ (25), $M_{GF} = 14$ (13), $M_{GH} = 1$, $M_{HG} = 0.5$.
Figures in brackets correspond to simplified sub-frames with one beam only.

Column JKLM:
$$\text{Span KF} = 59.42 \times 8^2/12 = 316.91 \text{ kNm}$$
$$\text{Span LG} = 59.42 \times 8^2/12 = 316.91 \text{ kNm}$$

$$10^5 \times E \begin{bmatrix} 58.75 & 8.0 \\ 8.0 & 82.32 \end{bmatrix} \begin{bmatrix} \theta_L \\ \theta_K \end{bmatrix} = \begin{bmatrix} -316.91 \\ -316.91 \end{bmatrix}$$

$$10^5 \times E \begin{bmatrix} \theta_L \\ \theta_K \end{bmatrix} = \begin{bmatrix} -4.935 \\ -3.370 \end{bmatrix}$$

$M_{KJ} = -102$ (−117), $M_{JK} = -51(-59)$, $M_{KL} = -93$ (−62), $M_{LK} = -106(-31)$,
$M_{LM} = -33$, $M_{ML} = -17$
Figures in brackets correspond to simplified sub-frames with one beam only.

It can be concluded from the above results that simplified sub-frames shown in Fig. 14.18 give sufficiently accurate results for the column for which he simplified sub-frame is used.

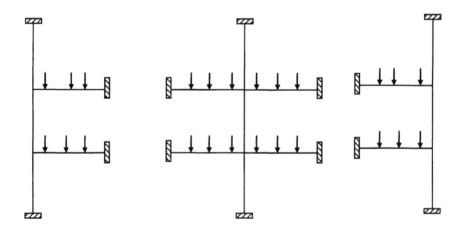

Fig. 14.20 Modified column sub-frames.

(g) Unbraced frame-analysis for vertical loads

The analysis for vertical loads can be made in the same way as for the braced frame assuming that horizontal displacement of joints can be ignored.

Case 1: Taking the wind load as the leading variable, taking ψ_0 for imposed load as 0.7, the load in this case is $1.35\, G_k + 1.5 \times 0.7 \times Q_k$.

$$1.35\, G_k + 1.5 \times 0.7 \times Q_k = (1.35 \times 27.9 + 1.5 \times 0.7 \times 14.5) = 52.89 \text{ kN/m,}$$

If all the spans are loaded by the maximum load

$$\text{Span BF} = 52.89 \times 6^2/12 = 158.67 \text{ kNm}$$
$$\text{Span FK} = 52.89 \times 8^2/12 = 282.08 \text{ kN m}$$

Case 2: Taking the imposed load as the leading variable, taking ψ_0 for the wind load as 0.6, the load in this case is $1.35\, G_k + 1.5 \times Q_k$.

$$1.35\, G_k + 1.5 \times 0.7 \times Q_k = (1.35 \times 27.9 + 1.5 \times 14.5) = 59.42 \text{ kN/m,}$$

If all the spans are loaded by the maximum load

$$\text{Span BF} = 52.89 \times 6^2/12 = 178.25 \text{ kNm}$$
$$\text{Span FK} = 59.42 \times 8^2/12 = 316.88 \text{ kN m}$$

The stiffness matrix K, load vector F and displacement vector are

$$K = E \times 10^5 \begin{bmatrix} 94.32 & 24.0 & 0 \\ 24.0 & 130.32 & 18.0 \\ 0 & 18.0 & 82.32 \end{bmatrix} \begin{bmatrix} \theta_B \\ \theta_F \\ \theta_K \end{bmatrix}, F = \begin{bmatrix} 158.67 & 178.25 \\ 123.41 & 138.63 \\ -282.08 & -316.88 \end{bmatrix}$$

$$E \times 10^5 \begin{bmatrix} \theta_B \\ \theta_F \\ \theta_K \end{bmatrix} = \begin{bmatrix} 1.3761 & 1.5461 \\ 1.2033 & 1.3519 \\ -3.6898 & -1.5414 \end{bmatrix}$$

The resulting moments are:

Case 1:

Beams: $M_{BF} = -63.92$, $M_{FB} = 250.14$, $M_{FK} = -304.84$, $M_{KF} = 170.10$,
Columns: $M_{BG} = 22.20$, $M_{BA} = 41.72$, $M_{FJ} = 18.75$, $M_{FE} = 35.95$,
$M_{KJ} = -112.38$, $M_{KM} = -57.72$

Case 2:

Beams: $M_{BF} = -71.81$, $M_{FB} = 281.02$, $M_{FK} = -342.48$, $M_{KF} = 191.10$,
Columns: $M_{BG} = 24.94$, $M_{BA} = 46.87$, $M_{FJ} = 21.07$, $M_{FE} = 40.39$,
$M_{KJ} = -126.26$, $M_{KM} = -64.85$

14.4.3 Example of Simplified Analysis of Concrete Framed Building for Wind Load by Portal Frame Method

(a) Wind loads

The wind loads are calculated using *UK National Annex to Eurocode 1: Actions on Structures. General actions. Wind actions.*

The case for which wind load is calculated is wind acting normal to the 40 m width. The total pressure p on the building is 0.86 kN/m². Fig. 14.21 shows the wind load acting as a uniformly distributed load over the height of the building.

Total horizontal load per frame = 0.86 × (4.5 × 13.5) = 52.25 KN

The total load is distributed at the storey levels in proportion as follows:
Roof level = 52.25 × (0.5 × 4)/13.5 = 7.74 KN
Second floor level = 52.25 × 4.0 /13.5 = 15.48 KN
First floor level = 52.25 × {0.5(4.0 + 5.5)}/13.5 = 18.38 kN
Fig. 14.22 shows the concentrated loads acting at roof and storey levels.

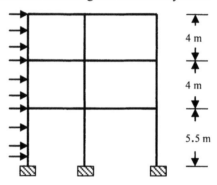

Fig. 14.21 Rigid frame subjected to wind load.

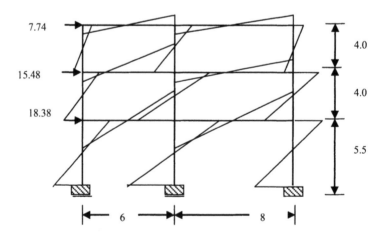

Fig. 14.22 Loads at joints and bending moment diagram.

(b) Analysis by portal method
The moments and shear forces due to wind load are calculated using the portal method explained in Chapter 13, section 13.12.4.

(i) Shear in bays
Total shear Q in each storey is divided between the bays in proportion to their spans.

Bay 1: Span = 6 m, Bay 2: Span = 8m
Shear in Bay 1 = Q × 6/ (6 + 8) = 0.4286 Q
Shear in Bay 2 = Q × 8/ (6 + 8) = 0.5714 Q

(ii) Shear in columns
The shear in each bay is divided equally between the columns. The shear in an interior column is the sum of the contributions from adjacent bays.

Shear in left column = 0.5 × 0.4286 Q = 0.2143 Q
Shear in middle column = 0.5 × (0.4286 + 0.5714) Q = 0.5Q
Shear in right column = 0.5 × 0.5714 Q = 0.2857 Q

(iii) Bending moment in columns
Bending moment at the top and bottom of a column is equal to product of shear in the column and storey height/2.

Table 14.3a summarises the storey shear forces and shear forces in columns. Table 14.3b shows bending moment in columns.

Table 14.3a Shear forces in kN in columns

Location	Q (kN)	Storey height	Shear in columns (kN)		
			Left	Middle	Right
Top	7.74	4.0	1.66	3.87	2.21
Middle	7.74+15.48 = 23.22	4.0	4.97	11.61	6.63
Bottom	23.22 + 18.38 = 41.60	5.5	8.92	20.80	11.89

Table 14.3b Bending moments in kNm in columns

Location	Storey height	Shear in columns			Moment in columns		
		Left	Middle	Right	Left	Middle	Right
Top	4.0	1.66	3.87	2.21	3.32	7.74	4.42
Middle	4.0	4.97	11.61	6.63	9.94	23.22	14.26
Bottom	5.5	8.92	20.80	11.89	24.53	57.2	32.70

Table 14.4 Moments in beams

Location	Left beam	Right beam
Top	3.32	4.42
Middle	14.26	17.68
Bottom	34.47	45.96

(iv) Bending moment in beams
As shown in Fig. 13.28, the bending moments at the ends of the left beam are equal to sum of the bending moments at the ends of the columns on the left connecting the beam. Similarly the bending moments at the ends of the right beam are equal to sum of the bending moments at the ends of the columns on the right connecting

the beam. Table 14.5 shows the bending moments in the beams. Fig. 14.12 shows the bending moment distribution in the frame.

(v) Axial force in columns

Since there is no distributed load on the beams and the contraflexure point is at mid-span, reaction R at the ends of the beam is given by

$$R = 2 \times \text{bending moments at the ends/span}$$

From the reactions in the beam, axial forces in the columns can be determined. Table 14.5 shows the axial forces in columns. Forces reverse in sign if the wind blows from right to left.

(vi) Axial forces in beams

As shown in Fig. 13.33, Chapter 13, from the shear forces in the columns and by considering the horizontal force equilibrium at the joints, axial forces in beams can be determined. Table 14.6 shows the axial forces in beams.

Table 14.5 Axial forces in columns

Level	Beam moments		Reactions in Beams	Axial Force in Columns
	Left Beam	Right Beam		
Top	3.32	4.42	1.11	1.11
Middle	14.26	17.68	4.42	1.11 + 4.42 = 5.53
Bottom	34.47	45.96	11.49	5.53 + 11.49 = 17.02

Table 14.6 Axial forces in beams

	Load at Joint	Shear in columns		Axial Force in Beam	
		Left	Right	Left beam	Right beam
Top	7.74	1.66	2.21	7.74 − 1.66 = 6.08	1.66
Middle	15.48	4.97	6.63	15.48 + 1.66 − 4.97 = 12.17	6.63 − 2.21 = 4.42
Bottom	18.38	8.92	11.89	18.38 + 4.97 − 8.92 = 14.43	11.89 − 6.63 = 5.26

14.5 BUILDING DESIGN EXAMPLE

14.5.1 Example of Design of Multi-Storey Reinforced Concrete Framed Buildings

(a) Specification

The framing plans for a multi-storey building are shown in Fig. 14.23. The main dimensions, structural features, loads, materials etc. are set out below.

(b) Overall dimensions
The overall dimensions are 36 m × 22 m in plan × 36 m high

Length: six bays at 6 m each; total 36 m
Breadth: three bays, 8 m, 6 m, 8 m; total 22 m
Height: ten storeys, nine at 3.5 m + one at 4.5 m

(c) Roof and floors
The floors and roof are constructed in one-way ribbed slabs spanning along the length of the building. Slabs are made solid for 300 mm on either side of the beam supports.

(d) Stability
Stability is provided by shear walls at the lift shafts and staircases in the end bays.

(e) Fire resistance
All elements are to have a fire resistance period of 2 hours.

(f) Loading condition

Roof imposed load: 1.5 kN/m^2
Floors imposed load: 3.0 kN/m^2
Finishes, roof: 1.5 kN/m^2
Finishes, floors, partitions, ceilings, services: 3.0 kN/m^2
Parapet: 2.0 kN/m
External walls at each floor: 6.0 kN/m

The load due to self-weight is estimated from preliminary sizing of members. The imposed load contributing to axial load in the columns is reduced by 50% for a building with ten floors including the roof as permitted by Table 2 of BS 6399-1, 1996.

(g) Exposure conditions
External moderate
Internal mild

(h) Materials
$f_{ck} = 30$ MPa concrete and $f_{yk} = 500$ MPa reinforcement.

(i) Foundations
Pile foundations are provided under each column and under the shear walls.

(j) Scope of work
The work carried out covers analysis and design for
1. Transverse frame members at floor 2 outer span only
2. An internal column between floors 1 and 2

The design is to meet requirements for robustness. In this design, the frame is taken as completely braced by the shear walls in both directions. A link-frame analysis can be carried out to determine the share of wind load carried by the rigid frames (Chapter 15). The design for dead and imposed loads will be the critical design load case.

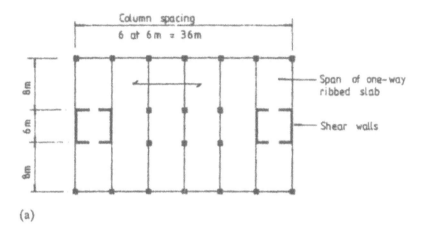

(a)

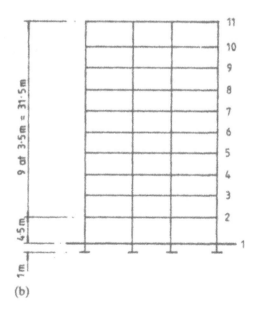

(b)

Fig. 14.23 (a) Floor plan and roof plan; (b) end elevation.

(k) Preliminary sizes and self-weight of members

(1) Floor and roof slab

The one-way ribbed slab is designed first. The size is shown in Fig. 14.24. The weight of the ribbed slab 0.5 m wide × 1.0 m is

$$25 \times [(0.5 \times 0.275) - (0.375 \times 0.215)] = 1.422 \text{ kN/m}$$

Load per unit area = $1.422/ (0.5) = 2.84 \text{ kN/m}^2$

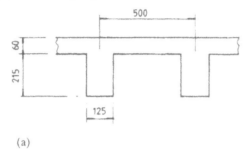

(a)

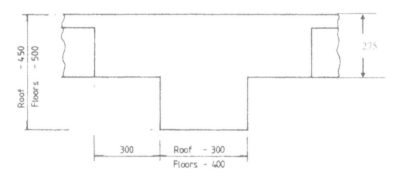

(b)

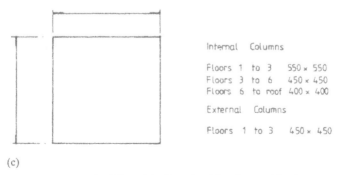

(c)

Fig. 14.24 (a) Roof and floor slab; (b) roof and floor beams; (c) columns.

(2) Beam sizes
Beam sizes are specified from experience:

$$\text{Depth} = \text{span}/15 = 500 \text{ mm,}$$
$$\text{Width} = 0.6 \times \text{depth} = 400 \text{ mm, say}$$

Preliminary beam sizes for roof and floors are shown in Fig. 14.14. The weights of the beams including the solid part of the slab are:

$$\text{Roof beams, } 25 \times [(0.3 \times 0.45) + (0.6 \times 0.275)] = 7.5 \text{ kN/m}$$
$$\text{Floor beams, } 25 \times [(0.4 \times 0.5) + (0.6 \times 0.275)] = 9.13 \text{ kN/m.}$$

(3) Column sizes
Preliminary sizes are shown in Fig. 14.14. Self-weights are as follows.

$$\text{Floors 1 to 3: } 0.55^2 \times 25 = 7.56 \text{ kN/m}$$
$$\text{Floors 3 to 7: } 0.45^2 \times 25 = 5.06 \text{ kN/m}$$
$$\text{Floor 7 to roof: } 0.4^2 \times 25 = 4.0 \text{ kN/m}$$

(l)Vertical loads
(1) Roof beam
The floor slab extends to only $(6000 - 2 \times 300 - 300) = 5100 \text{ mm}$

$$\text{Dead load (slab, beam, finishes), } G_k = (2.84 \times 5.1) \text{ floor slab}$$
$$+ 7.5 \text{ beam self weight} + (1.5 \times 6) \text{ finishes} = 31.0 \text{ kN/m}$$
$$\text{Imposed load, } Q_k = 6 \times 1.5 = 9 \text{ kN/m}$$

(2) Floor beams
The floor slab extends to only $(6000 - 2 \times 300 - 400) = 5000 \text{ mm}$

$$\text{Dead load, } G_k = (2.84 \times 5) \text{ floor slab} + 9.13 \text{ beam self-weight}$$
$$+ (3 \times 6) \text{ finishes} = 41.33 \text{ kN/m}$$
$$\text{Imposed load, } Q_k = 6 \times 3 = 18 \text{ kN/m}$$

(3) Internal column below floor 2
An estimation of the load on the column can be calculated as follows.
The entire load over a width $(8+6)/2 = 7 \text{ m}$ is carried by the internal column.
The dead load is:

$$\text{Beam: } 7 \times [31 \text{ roof} + (41.33 \times 9 \text{ floors})] = 2820.8 \text{ kN}$$
$$\text{Column self weight: } 3.5[7.56 + 4(5.06 + 4.0)] = 153.30 \text{ kN}$$
$$\text{Total: } 2974.10 \text{ kN}$$
$$\text{Imposed load} = 7 \times [9 \text{ roof} + (18 \times 9 \text{ floors})] = 1197 \text{ kN}$$

According to Equation (6.2) in *1991-1-1:2002(E) Eurocode 1: Actions on Structures*, for columns and walls the total imposed loads from several storeys may be multiplied by the reduction factor α_n where

$$\alpha_n = \frac{2 + (n - 2)\psi_0}{n}$$

n = number of storeys ($n > 2$) above the loaded structural elements from the same category.
$\psi_0 = 0.7$ for imposed loads (see *Table A1.1, BS EN 1990:2002 Eurocode. Basis of Structural Design*).

Substituting $n = 10$, $\alpha_n = 0.76$

Reduced imposed load $= 0.76 \times 1197 = 909.72$ kN

(m) Sub-frame analysis

(1) Sub-frame

The sub-frame consisting of the beams and columns above and below the floor level 1 is shown in Fig. 14.25. The properties of members are shown in Table 14.7.

(2) Loads and load combinations

$$(1.35 \, G_k + 1.5 \, Q_k) = (1.35 \times 41.33) + (1.5 \times 18) = 82.8 \text{ kN/m}$$
$$1.35 \, G_k = 55.8 \text{ kN/m}$$

Table 14.7 Section properties

Member	L (mm)	I (mm^4)	I/L(mm^3)
BC, MN	4500	3.417×10^9	7.5938×10^5
EF, HK	4500	7.626×10^9	16.9456×10^5
AB, LM	3500	3.417×10^9	9.7634×10^5
DE, GH	3500	7.626×10^9	21.787×10^5
BE, HM	8000	4.167×10^9	5.208×10^5
EH	6000	4.167×10^9	6.945×10^5

Note: For columns BC, EF and MN, the length is taken as 4.5 m, although it is shown as 5.5 m in Fig. 14.23 and Fig. 14.25. It is assumed that they are deformable only over 4.5 m.

Case 1: Spans BE and HM with maximum load and span EH with minimum load. The fixed end moments are:

Spans BE and HM: $82.8 \times 8^2/12 = 441.6$ kNm
Span EH: $55.8 \times 6^2/12 = 167.4$ kNm

Case 2: Spans BE and HM with minimum load and span EH with maximum load. The fixed end moments are:

Spans BE and HM: $55.8 \times 8^2/12 = 297.60$ kNm
Span EH: $82.8 \times 6^2/12 = 248.4$ kNm

Case 3: Spans BE and EH with maximum load and span HM with minimum load. The fixed end moments are:

Span BE: $82.8 \times 8^2/12 = 441.6$ kNm
Span EH: $82.8 \times 6^2/12 = 248.4$ kNm
Span HM: $55.80 \times 8^2/12 = 297.60$ kNm

Case 4: Spans EH and HM with maximum load and span BE with minimum load. The fixed end moments are:

Span BE: $55.80 \times 8^2/12 = 297.60$ kNm

$$\text{Span EH: } 82.8 \times 6^2/12 = 248.4 \text{ kNm}$$
$$\text{Span HM: } 82.8 \times 8^2/12 = 441.6 \text{ kNm}$$

(3) Analysis

The stiffness matrix relationship is given by

$$E \times 10^6 \begin{bmatrix} 9.026 & 1.0416 & 0 & 0 \\ 1.0416 & 20.3540 & 1.3888 & 0 \\ 0 & 1.3888 & 20.3540 & 1.0416 \\ 0 & 0 & 1.0416 & 9.026 \end{bmatrix} \begin{bmatrix} \theta_B \\ \theta_E \\ \theta_{II} \\ \theta_M \end{bmatrix} = \begin{bmatrix} M_B \\ M_E \\ M_{II} \\ M_M \end{bmatrix}$$

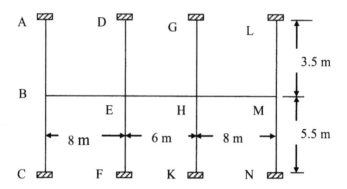

Fig. 14.25 Sub-frame for beam moment analysis.

The load vectors for the four cases are

$$F = \begin{bmatrix} 441.60 & 297.60 & 441.60 & 297.60 \\ -274.20 & -49.20 & -193.2 & -49.2 \\ 274.20 & 49.20 & 49.20 & 193.20 \\ -441.60 & -297.60 & -297.60 & -441.60 \end{bmatrix}$$

Solving, the rotation values are shown in Table 14.8.

Table 14.8 Joint rotations

Rotation	Case 1	Case 2	Case 3	Case 4
$\theta_B \times E \times 10^6$	50.9165	33.4830	50.3573	33.5462
$\theta_E \times E \times 10^6$	-17.2545	-4.4332	-12.4089	-4.9807
$\theta_H \times E \times 10^6$	17.2555	4.4338	4.9813	12.4099
$\theta_M \times E \times 10^6$	-50.9174	-33.4833	-33.5465	-50.3580

Table 14.9 shows the values of the bending moments in the symmetrical half of the sub-frame. The shear force diagrams and bending moment diagrams are shown in Fig. 14.26 and Fig. 14.27 respectively. The corresponding envelopes are shown in Fig. 14.28 and Fig. 14.29.

Table 14.9 Bending moments in kNm in the sub-frame

Moment	Case 1	Case 2	Case 3	Case 4
M_{BE}	−353.50	−232.47	−349.62	−232.90
M_{EB}	458.69	323.24	468.20	322.17
Span BE	257.35	169.70	254.82	169.98
M_{EH}	−191.37	−254.56	−275.95	−245.00
M_{HE}	191.37	254.56	245.00	275.96
Span EH	59.73	118.04	112.29	112.29
M_{BA}	198.85	130.76	196.66	131.01
M_{BC}	154.66	101.71	152.96	101.90
M_{ED}	−150.37	−38.63	−108.14	−43.41
M_{EF}	−116.96	−30.05	−84.11	−33.76

(n) Design of outer span of beam BEH

(1) Design of moment reinforcement
Section at mid-span: According to Table 4.1 of the code, for concrete inside buildings with moderate or high humidity, the exposure category is XC3. According to Table 4.4N of the code, the minimum cover is 25 mm. According to Table 5.6 of *BS EN 1992-1-2:2004 Eurocode 2: Design of concrete structures Part 1-2: General rules-Structural fire design*, for a fire resistance 2 h, the required axis distance is 35 mm.
The beam is a T-beam as shown in Fig. 14.24(b) with a rib width, b_w = 400 mm, total flange width, b_f = 1000 mm, total depth, h = 500 mm, flange thickness, h_f = 275 mm.
Assume H25 bars and H10 diameter links:
$$\text{Axis distance} = 25 + 10 + 25/2 = 47.5 \text{ mm} > 35 \text{ mm}$$
$$d = 500 - 25 - 10 - 25/2 = 452.5 \text{mm}$$
Maximum span moment occurs for case 1 loading. The support moments from Table 14.10 are
$$M_{BE} = 353.50, M_{EB} = 458.69 \text{ kNm, load } q = 82.8 \text{ kN/m, Span} = 8 \text{ m}$$
Reaction at left support
$$V_{left} = 0.5 \times 82.8 \times 8 + (353.50 - 458.69)/8 = 318.05 \text{ kN}$$
Shear force is zero when
$$318.05 - 82.8 \, a = 0, a = 3.84 \text{ m}$$
Maximum span moment
$$M_{span} = -353.50 + 318.05 \times 3.84 - 82.8 \times 3.84^2/2 = 257.34 \text{ kNm}$$
Check the moment capacity if the entire flange is in compression.
$$M_{flange} = b_f \times h_f \times f_{cd} \times (d - 0.5 \, h_f) = 1754.4 \text{ kNm} > 257.34 \text{ kNm}$$

The stress block is inside the flange. Design the beam as a rectangular beam of width 1000 mm.

$$k = M/ (f_{ck} b d^2) = 257.34 \times 10^6 / (30 \times 1000 \times 452.5^2) = 0.042 < 0.196$$

$$\frac{z}{d} = 0.5[1.0 + \sqrt{(1-3k)}]$$

Substituting k = 0.042, z/d = 0.967

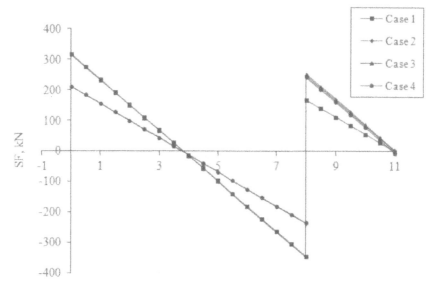

Fig. 14.26 Shear force diagrams for a symmetrical half of beam.

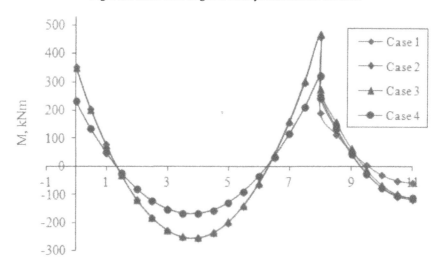

Fig. 14.27 Bending moment diagrams for a symmetrical half of beam.

$$A_s = 257.34 \times 10^6 / (0.967 \times 452.5 \times 0.87 \times 500) = 1352 \text{ mm}^2$$

Provide 3H25 to give an area of 1473 mm² (Fig. 14.18).

Check minimum steel: $f_{cm} = 0.3 \times f_{ck}^{\ 0.667} = 2.9$ MP, $f_{yk} = 500$ MPa, $b_t = 400$ mm, $d = 453$ mm.

$$A_{s, min} = 0.26 \times (f_{cm}/f_{yk}) \times b_t d = 273 \text{ mm}^2 \qquad (9.1N)$$

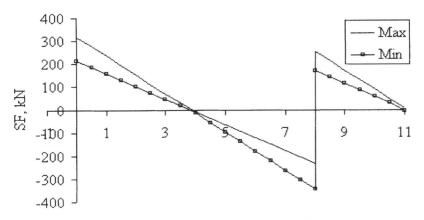

Fig. 14.28 Shear force envelope for a symmetrical half of beam.

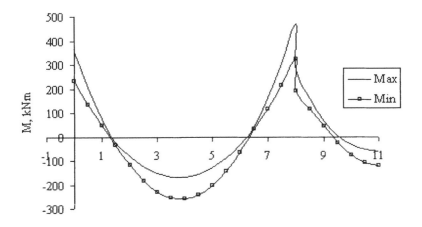

Fig. 14.29 Bending moment envelope for a symmetrical half of beam.

Section at outer support: Maximum support moment occurs for case 1 loading. From Table 14.10,

$$M_{BE} = -353.50 \text{ kN m.}$$

At support tension is at the top. The flange will be in tension. The beam section is therefore rectangular.

b = 400 mm. Provide for H25 bars; $d = 452.5$ mm.

$$k = M / (f_{ck} b d^2) = 353.50 \times 10^6 / (30 \times 400 \times 452.5^2) = 0.144 < 0.196$$

$$\frac{z}{d} = 0.5[1.0 + \sqrt{(1 - 3k)}] = 0.877$$

$A_s = 353.50 \times 10^6 / (0.877 \times 452.5 \times 0.87 \times 500) = 2048 \text{ mm}^2$
Provide 3H32 = 2413 mm^2.

Section at inner support: Maximum support moment occurs for case 3 loading.
From Table 14.10, $M_{EB} = 468.20$ kN m.
Assuming H32 mm bars,
$$d = 500 - 25 - 10 - 32/2 = 449 \text{ mm}$$
$$k = M / (f_{ck} b \, d^2) = 468.20 \times 10^6 / (400 \times 449^2 \times 30) = 0.194 < 0.196$$

$$\frac{z}{d} = 0.5[1.0 + \sqrt{(1 - 3\frac{k}{\eta})}] = 0.823$$

$A_s = 468.20 \times 10^6 / (0.823 \times 452.5 \times 0.87 \times 500) = 2890 \text{ mm}^2$
Provide 4H32 = 3217 mm^2.
Fig. 14.30 shows the flexural reinforcement at different sections of the beam.
Note that in clause 9.2.1.2(2), the Eurocode allows for the reinforcement in the flange to be spread over the effective width.

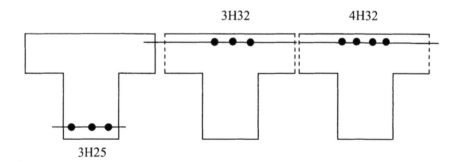

3H32   4H32

3H25

Fig. 14.30 Beam BE: (a) Mid-span; (b) outer column; (c) inner column.
Note: Links not shown for clarity.

(2) Curtailment and anchorage
As moments have been calculated by detailed analysis, the cut-off points will be calculated accurately.

Top steel-outer support: Refer to Fig. 14.30. The section has 3H32 bars at the top. Determine the positions along the beam where the one bar can be cut off.
The moment of resistance of the section with 2H32 bars ($A_s = 1609$ mm^2) at top in tension is calculated.
Assume that tension steel yields and the stress in the steel is 0.87 f_{yk}.
        Yield strain in steel, $\varepsilon_{sy} = 0.87 \times 500/E = 2.175 \times 10^{-3}$
Equate the forces in the section:
        $0.87 \times 500 \times 1609 = (f_{cd} = 20) \times 400 \times (0.8x)$

$$x = 109 \text{ mm}$$

Check strain in tension steel

$$\varepsilon_s = 0.0035 \times (452.5 - x)/x = 10.98 \times 10^{-3} > \text{yield strain in steel, } \varepsilon_{sy}$$

Therefore tension steel yields.

Taking moment about the tension steel, moment of resistance is given by

$$M_R = f_{cd} \times b \times (0.8 \text{ x}) \times (d - 0.8x/2)$$
$$= [20 \times 400 \times (0.8 \times 109) \times (452.5 - 0.4 \times 109)] \times 10^{-6}$$
$$= 285.25 \text{ kNm}$$

Using case 1 loads, the theoretical cut-off point for one bar is given by the solution of the equation

$$-282.25 = -353.50 + 318.05 \times x - 82.8 \times x^2/2, \text{ giving } x = 0.23 \text{ m}$$

The position of point of contraflexure is given by

$$0 = -353.50 + 318.05 \times x - 82.8 \times x^2/2, \text{ giving } x = 1.35 \text{ m}$$

Because of the effect of shear, the bending diagram will be shifted away from the position of maximum moment in the direction of the decreasing moment by $a_1 = 0.5 \text{ z cot } \theta$ (see clause 9.2.1.3, equation (9.2)). Cotθ can be taken as approximately 2.5 and z as approximately equal to the effective depth d. The shift is therefore approximately 1.25 d. Taking d = 452.5 mm, a_1 = 1.25 d = 567 mm From Table 5.5, Chapter 5, the anchorage length is 32 bar diameters = 1024 mm. One out of three bars will be cut at 230 + 1024 + 567 = 1825 mm from the support. Two bars will be stopped at 1350 + 1024 + 567 = 2945 mm from the support.

Top steel-inner support: The section has 4T32 bars at the top. Determine where 2H32 can be stopped.

Calculate the moment of resistance of the section with 2H32 bars (A_s = 1609 mm^2) at top in tension is calculated. As in (i) above, M_R= 285.25 kNm.

Using case 3 loads,

$$M_{BE} = 349.62, M_{EB} = 468.20 \text{ kNm, load q} = 82.8 \text{ kN/m, Span} = 8 \text{ m}$$

Reaction at left support

$$V_{left} = 0.5 \times 82.8 \times 8 + (349.62 - 468.20)/8 = 315.18 \text{ kN}$$

The theoretical cut-off point for two bars is given by the solution of the equation

$$-282.25 = -349.62 + 315.18 \times x - 82.8 \times x^2/2, \text{ giving } x = 7.39 \text{ m}$$

The position of point of contraflexure is given by

$$0 = -349.62 + 315.18 \times x - 82.8 \times x^2/2, \text{ giving } x = 6.27 \text{ m}$$

Two out of four bars will be cut at (8000 – 7390) + 1024 + 567 = 2205 mm from the support.

Two more bars will be stopped at (8000 – 6270) + 1024 + 567 = 3325 mm from the support.

(o) Summary top steel arrangement

Outer support: From the outer support the outer 2H32 will be stopped at 1825 mm from support and the inner 1H32 will be stopped at 2945 mm from support.

Inner support: From the inner support the inner 2H32 will be stopped at 2205 mm from support and the outer 2H32 will be stopped at 3375 mm from support.

Link bars: 2H12 approximately 3000 mm will connect between the 2H32 from the outer support to 2H32 from the inner support. This will to provide 'hanger bars' to the shear links.

Bottom steel in span: The section has 3H25 bars at the bottom. Determine the positions along the beam where the one bar can be cut off.

The moment of resistance of the section with 2H25 bars ($A_s = 982$ mm^2) at the bottom face in tension is calculated.

Assume that tension steel yields and the stress in the steel is 0.87 f_{yk}.

Equate the forces in the section:
$$0.87 \times 500 \times 982 = (f_{cd} = 20) \times 400 \times (0.8x)$$
$$x = 67 \text{ mm}$$

Check strain in tension steel
$$\varepsilon_s = 0.0035 \times (452.5 - x)/x = 20.14 \text{ x } 10^{-3} > \text{ yield strain in steel, } \varepsilon_{sy}.$$
Therefore tension steel yields.

Taking moment about the tension steel, moment of resistance is given by
$$M_R = f_{cd} \times b \times (0.8 \text{ x}) \times (d - 0.8x/2)$$
$$= [20 \times 400 \times (0.8 \times 67) \times (452.5 - 0.4 \times 67)] \times 10^{-6}$$
$$= 182.54 \text{ kNm}$$

Using case 1 loads, the theoretical cut-off point for one bar is given by the solution of the equation
$$182.54 = -353.50 + 318.05 \times x - 82.8 \times x^2/2, \text{ giving } x = 2.50 \text{ and } 5.19 \text{ m}$$

The position of point of contraflexure is given by
$$0 = -353.50 + 318.05 \times x - 82.8 \times x^2/2, \text{ giving } x = 1.35 \text{ m and } 6.33 \text{ m}$$

Because of the effect of shear, the bending diagram will be shifted away from the position of maximum moment in the direction of the decreasing moment by $a_l = 567$ mm

From Table 5.5, Chapter 5, the anchorage length is 36 bar diameters $36 \times 25 = 900$ mm.

One out of three bars can be stopped at $2500 - 900 - 567 = 1033$ mm from the outer support and at $(8000 - 6330) - 900 - 567 = 203$ mm from the inner support. However the curtailed length are quite short. In the interests of convenience and also of crack control, all the three bars will be carried onto the supports.

(3) Design of Shear Reinforcement

Inner support: From case 3 loading, shear force at support
$$V_{Ed} = 346.02 \text{ kN}$$

Check whether section strength is adequate at support, $V_{Ed} < V_{Rd, max}$

From equation (6.9) of the code
$$V_{Rd,max} = \alpha_{cw} b_w \, z \, v_1 \, f_{cd} \, \frac{1}{(cot\theta + tan\theta)} \tag{6.9}$$

From equation (6.11aN), $\alpha_{cw} = 1$, $b_w = 400$ mm, $z \approx 0.9d = 404$ mm,
$$v_1 = 0.6(1 - f_{ck}/250) = 0.6 \, (1 - 30/25) = 0.528 \tag{6.6N}$$
$f_{cd} = 30/1.5 = 20$ MPa. Equate $V_{Ed} = V_{Rd, max}$ and determine θ
$$V_{Rd, max} = 1706.5/ (cot \, \theta + tan \, \theta)$$

$$\cot \theta = 2.5, \ V_{Rd, \ max} = 588.5 \ kN$$
$$\cot \theta = 1.0, \ V_{Rd, \ max} = 853.2 \ kN$$

Choosing $\cot \theta = 2.5$ for minimum shear reinforcement,
$$V_{Rd, \ max} = 588.5 \ kN > 346.02 \ kN$$

Section size is adequate.

ii. Check whether shear reinforcement is required at d from the face of support i.e., $V_{Ed} > V_{Rd, \ c}$

Distance from centre line of support (column = 550 mm wide) to a distance (d = 449 mm) from the support
$$= 550/2 + 449 = 724 \ mm$$

The shear at d from the face of the support is
$$V_{Ed} = 346.02 - 82.8 \times (0.724) = 286.1 \ kN.$$
$$b_w = 400 \ mm, \ d = 449 \ mm$$

From 4H32, $A_{sl} = 4 \times 804 = 3217 \ mm^2$

Using equations (6.2a) and (6.2b) of the code,

$$C_{Rd,c} = \frac{0.18}{(\gamma_c = 1.5)} = 0.12$$

$$k = 1 + \sqrt{\frac{200}{449}} = 1.67 \le 2.0$$

$$100\rho_1 = 100 \times \frac{3217}{400 \times 449} = 1.79 < 2.0$$

$$v_{min} = 0.035 \times 1.79^{1.5} \times \sqrt{30} = 0.46 \ MPa \qquad (6.3N)$$

$$V_{Rd,c} = [0.12 \times 1.67 \times \{1.79 \times 30\}^{1/3}] \times 400 \times 449 \times 10^{-3}$$

$$\ge [0.756] \times 400 \times 449 \times 10^{-3}$$

$$V_{Rd,c} = 135.7 \ge 82.6 \ kN \qquad (6.2a)$$

$V_{Rd, \ c} = 135.7 \ kN < 286.1 \ kN$. Therefore shear reinforcement is required.

Design of shear reinforcement:

Ensuring that $V_{Rd, \ s} \ge V_{Ed}$, and choosing 2-leg H10 links,
$A_{sw} = 157 \ mm^2$, $\cot \theta = 2.5$, $z = 0.9d$, $f_{ywd} = 500/1.15 = 435 \ MPa$, $V_{Ed} = 286.1 \ kN$
Using code equation (6.8),

$$V_{Rd,s} = \frac{z}{s} A_{sw} f_{ywd} \cot \theta \qquad (6.8)$$

$$V_{Rd,s} = \frac{(0.9 \times 449)}{s} \times 157 \times 435 \times 2.5 \times 10^{-3}$$

$$V_{Rd,s} = \frac{68961}{s} \ kN \ge (V_{Ed} = 286.1 kN)$$

$$s \le 241 \ mm$$

Maximum spacing $s \le (0.75 \ d = 0.75 \times 449 = 337 \ mm)$.

Using a spacing of 225 mm,

$$V_{Rd,s} = \frac{68961}{225}\,kN = 306.5\,kN$$

Check minimum steel requirement

$$\frac{A_{sw}}{s\,b_w} \ge \frac{0.08\,\sqrt{f_{ck}}}{f_{yk}}$$

(9.4) and (9.5N)

$$\frac{157}{s \times 400} \ge \frac{0.08 \times \sqrt{30}}{500}\,,\ s < 448mm$$

Shear capacity with minimum shear reinforcement

If s = 337 mm, the maximum spacing allowed

$$V_{Rd,s} = \frac{68961}{337} = 205kN$$

This value of shear force occurs at 1.35 m from the outer support and from the inner support at $(8.0 - 6.30) = 1.70$ m.

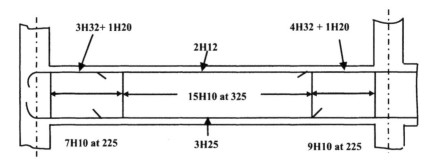

Fig. 14.31 Reinforcement detail.

Rationalization of link spacings

The following rationalization of link spacing will be adopted:

1. From face of outer support to a distance of 1350 mm, provide 7H10 at 225 mm c/c.
2. From face of inner support to a distance of 1800 mm, provide 9H10 at 225 mm c/c.
3. Centre portion over a distance of $(8000 - 1350 - 1800) = 4850$ mm provide 15H10 at 325 mm c/c.

The link spacing is shown in Fig. 14.31.

(4) Deflection

$$L/d = \text{limit of span/effective depth ratio}$$
$$K = 1.3 \text{ for end span of continuous beam}$$
$$\rho_0 = 10^{-3} \times \sqrt{f_{ck}} = 5.477 \times 10^{-3}$$

$$\text{Tension steel, } \rho = 3H25/(400 \times 452.5) = 8.136 \times 10^{-3}$$
$$\text{Compression steel, } \rho' = 2H12/(400 \times 452.5) = 1.25 \times 10^{-3}$$
$$b/b_w = 1000/400 = 2.5 < 3.0$$
$$L_{eff} = 8 \text{ m, } 7/L_{eff} = 0.875$$
$$A_{s, \text{ prov}} = 3H25 = 1472 \text{ mm}^2$$
$$A_{s, \text{ reqd}} = 1352 \text{ mm}^2$$

$$\text{If } f_{yk} = 500 \text{ MPa, } \frac{310}{\sigma_s} = \frac{A_{s,prov}}{A_{s,reqd}} = \frac{1472}{1352} = 1.089$$

$$\frac{L}{d} = K[11 + 1.5 \sqrt{f_{ck}} \frac{\rho_0}{\rho - \rho'} + \frac{1}{12} \sqrt{f_{ck}} \sqrt{\frac{\rho'}{\rho}}] \text{ if } \rho > \rho_0$$

(7.16b)

$$\frac{L}{d} = 1.3 \times [11 + 1.5 \sqrt{30} \frac{5.477}{8.136 - 1.25} + \frac{1}{12} \sqrt{30} \sqrt{\frac{1.25}{8.136}}] \times 0.875 \times 1.089 = 21.94$$

Allowable span/d ratio = 21.9
Actual span/d ratio = 8000/452.5 = 17.7

The beam is satisfactory with respect to deflection.

(5) Cracking

Steel stress at SLS:
At ULS, loading, $q = (1.35 \, G_k + 1.5 \, Q_k) = (1.35 \times 41.33) + (1.5 \times 18) = 82.8$ kN/m
At SLS, loading, $q = (G_k + Q_k) = 41.33 + 18 = 59.33$ kN/m

$$\sigma_s = f_{yd} \times (59.33/82.8) \times (A_{s, \text{ reqd}}/A_{s, \text{ prov}}) = 286 \text{ MPa}$$

Referring to Table 7.3N in the code, the clear spacing between bars in the tension zone should not exceed 150 mm.

Outer support -top steel

3H32 bars with 25 mm cover and 10 mm links. Spacing between bars is
$$= [400 - 2 \times (25 + 10) - 32]/2 = 149 \text{ mm} < 150 \text{ mm}$$
If the inner 3H32 bar is curtailed, spacing between bars is
$$= [400 - 2 \times (25 + 10) - 32] = 298 \text{ mm} > 150 \text{ mm}$$
In the interests of crack control, add an additional 25 mm bar to link with middle 25 bar from outer support steel.

Inner support -top steel

4T32 bars with 25 mm cover and 10 mm links. Spacing between bars is
$$= [400 - 2 \times (25 + 10) - 32]/3 = 99 \text{ mm} < 150 \text{ mm}$$
If the inner 2H32 bars are curtailed, spacing between bars is
$$= [400 - 2 \times (25 + 10) - 32] = 298 \text{ mm} > 150 \text{ mm}$$
In the interests of crack control, add an additional H20 mm bar to link with middle H32 bar from outer support steel.

Bottom steel: Spacing between bars is
$$= [400 - 2 \times (25 + 10) - 25]/2 = 153 \text{ mm} \approx 150 \text{ mm}$$

(6) Arrangement of reinforcement

The final arrangement of the reinforcement at the top and bottom is shown in Fig. 14.32.

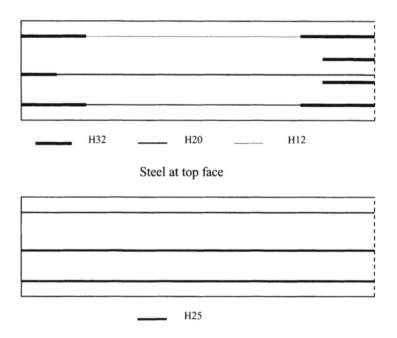

Steel at top face

Steel at bottom face

Fig. 14.32 Arrangement of flexural reinforcement.

(p) Design of the lower length of the centre column

(1) Design loads and moments

Roof beams:
$$G_k = 31.0 \text{ kN/m}, \ Q_k = 9.0 \text{ kN/m}$$
$$1.35 \ G_k = 41.85 \text{ kN}, \ 1.5 \ Q_k = 13.5 \text{ kN/m}$$

Floor beams:
$$G_k = 41.33 \text{ kN/m}, \ Q_k = 18.0 \text{ kN/m}$$
$$1.35 \ G_k = 55.80 \text{ kN}, \ 1.5 \ Q_k = 27.0 \text{ kN/m}$$

Three load cases will be analysed.

Case A: All spans carry $1.35 \ G_k$.

Case B: Only spans BE and HM carry $1.5 \ Q_k$.

Case C: Only spans BE and EH carry $1.5 \ Q_k$.

The results of the analysis are shown in Tables 14.10 (Rotations) and Table 4.11 (Moments).

Table 14.10 Joint rotations

Rotation	Case A 1.35 G_k on all spans	Case B 1.5Q_k on spans BE and HM	Case C 1.5Q_k on spans BE and EH
$\theta_B \times E \times 10^6$	33.9790	16.9375	16.3783
$\theta_E \times E \times 10^6$	−8.7314	−8.5231	−3.6774
$\theta_H \times E \times 10^6$	8.7321	8.5234	−3.7508
$\theta_M \times E \times 10^6$	−33.9795	−16.9379	0.4330

Table 14.11 Bending moments in kNm in the sub-frame

Moment	Case A	Case B	Case C
M_{BE}	−235.91	−117.59	−113.71
M_{EB}	314.80	143.89	153.40
M_{EH}	−179.53	−11.84	−96.43
M_{HE}	179.53	11.84	65.47
M_{BA}	132.70	66.15	63.96
M_{BC}	103.21	51.45	49.75
M_{ED}	−76.09	−74.28	−32.05
M_{EF}	−59.18	−57.77	−24.93

Analysis Case 1 = Case A + Case B:

Permanent load analysis: From the frame analysis for the case A when all the floor beams are loaded with $1.35G_k = 55.80$ kN/m, the reaction from the floor beams are:
Floor beam BE: $R_E = 0.5 \times 55.80 \times 8 + (314.80 − 235.91)/8 = 233.06$ kN.
Floor beam EH: $R_E = 0.5 \times 55.80 \times 6 − (179.53 − 179.53)/6 = 167.40$ kN.
The compression force on the column = 233.06 + 167.40 = 400.46 kN.
The column moment $M_{EF} = 59.18$ kNm.
On the roof beam, 1.35 $G_k = 41.85$ kN/m.
The compression force on the column = $400.46 \times (41.85/55.80) = 300.35$ kN.
Total load on column from roof + 9 floors = $300.35 + 9 \times 400.46 = 3904.49$ kN.

Imposed load analysis: From the frame analysis for the case B when the floor beams BE and HM alone are loaded with $1.5Q_k = 27.0$ kN/m, the reaction from the floor beams are:
Floor beam BE: $R_E = 0.5 \times 27 \times 8 + (143.89 − 117.59)/8 = 111.29$ kN.
Floor beam EH: $R_E = 0$ kN.

The compression force on the column = 111.29 + 0 = 111.29 kN.
The column moment M_{EF} = 57.77 kNm.
On the roof beam, 1.5 Q_k = 13.5.0 kN/m.
The compression force on the column = 111.29 × (13.5/27) = 55.65 kN.
Total load on column from roof + 9 floors = 55.65 + 9 × 111.29 = 1057.26 kN.
Apply the reduction factor α_n = 0.76.
Total load from imposed load on the column = 1057.26 × (α_n = 0.76) = 803.51 kN.

Column self weight: (γ_g = 1.35) × 3.5 × [7.56 + 4(5.06 + 4.0)] = 206.96 kN.
Total column axial load, N = 3904.49 + 803.51 + 206.96 = 4914.96 kN.
Column moment, M_{EF} = 59.18 + 57.77 = 116.95 kNm.
Design the column for N = 4915 kN and M = 117 kNm.

Analysis Case 2 = Case A + Case C:

Permanent load analysis: The results will be as for case 1.

Imposed load analysis: From the frame analysis for the case C when the floor beams BE and EH alone are loaded with 1.5Q_k = 27.0 kN/m, the reaction from the floor beams are:
Floor beam BE: R_E = 0.5 × 27 × 8 + (153.40 – 113.71)/8 = 112.96 kN.
Floor beam EH: R_E = 0.5 × 27 × 6 + (96.43 – 65.47)/6 = 86.16 kN.
The compression force on the column = 112.96 + 86.16 = 199.12 kN.
The column moment M_{EF} = 24.93 kNm.
On the roof beam, 1.5 Q_k = 13.5 kN/m.
The compression force on the column = 199.12 × (13.5/27) = 99.56 kN.
Total load on column from roof + 9 floors = 99.56 + 9 × 199.12 = 1891.64 kN.
Apply the reduction factor α_n = 0.76.
Total load from imposed load on the column = 1891.64 × α_n = 1437.65 kN.
Column self weight: As under case 1 = 206.96 kN.
Total column axial load, N = 3904.49 + 1437.65 + 206.96 = 5549.09 kN.
Column moment, M_{EF} = 59.18 + 24.93 = 84.11 kNm.
Design the column for N = 5550 kN and M = 84 kNm.

(2) Effective length and slenderness
Considering the braced column sub-frame shown in Fig. 14.33, use code equation (5.15) to calculate the effective length. k_1 and k_2 are the relative flexibilities of rotational restraints at ends 1 and 2 respectively.
where:

$$k = \frac{\theta}{M} \frac{EI}{\ell}$$

EI = bending stiffness (flexural rigidity) of the column.
If there is a column above or below the column under consideration, then EI/ℓ should be replaced by EI/ℓ for column under consideration + EI/ℓ of the column above or below as appropriate.

θ = rotation of the restraining for a moment M.

For a beam of span L, flexural rigidity EI and clamped at the far end the relationship between M and θ is M = 4 (EI/L) θ or θ/M = 0.25 L/ (EI).

For the column sub-frame shown in Fig. 14.32, I/L for beams EB and EH are respectively 5.208 × 10^5 and 6.945 × 10^5. I/L for columns ED and EF are respectively 21.787 × 10^5 and 16.9456 × 10^5.

k_1 = 0.1 minimum value for clamped at the base as recommended by the code.

k_2 = 0.25 × (1/5.208 + 1/6.945) × (21.787 + 16.9456) = 3.25.

$$\ell_0 = 0.5\ell\sqrt{\{1+\frac{k_1}{(0.45+k_1)}\}\{1+\frac{k_2}{(0.45+k_2)}\}} = 0.745\ell$$

(5.15)

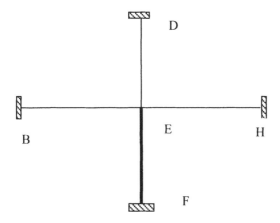

Fig. 14.33 A column sub-frame in a braced frame.

(q) Slenderness ratio

In equation (5.13N), clause 5.8.3.1, the code classifies columns as short when the slenderness ratio λ about both axes are less than λ_{lim}

$$\lambda_{lim} = 20 \times A \times B \times C \frac{1}{\sqrt{n}}$$

(5.13N)

where

$A = \dfrac{1}{(1.0+0.2\varphi_{ef})}$, (If the effective creep ratio φ_{ef} is not known, use A = 0.7)

$$B = \sqrt{(1+2\omega)} \ , \ \omega = (A_s \ f_{yd})/ (A_c \ f_{cd})$$

Assume the steel value in the column as A_s = 8H25 = 3927 mm²

$$f_{yd} = 500/1.15 = 435 \text{ MPa}$$
$$A_c = 550 \ 2 = 30.25 \times 10^4 \text{ mm}^2$$
$$f_{cd} = 30/1.5 = 20 \text{ MPa}$$
$$\omega = 0.282$$
$$B = 1.25$$

$$C = 1.7 - r_m, \quad r_m = \text{moment ratio} = M_{01}/M_{02}$$

Take $r_m = 0.5$ as M_{01} is one half of M_{02} and of same sign producing tension on opposite faces. r_m is negative. $C = 1.7 - (-0.5) = 2.2$.

$n = $ relative normal force $= N_{Ed}/(A_c f_{cd})$.

$N_{Ed} = $ Design axial force $= 4916$ kN, $A_c = 550 \times 550$, $f_{cd} = 20$ MPa, $n = 0.81$.

$$\lambda_{lim} = 20 \times 0.7 \times 1.25 \times 2.2 \times \frac{1}{\sqrt{0.81}} = 42.8$$

$$\ell_0 = 0.745 \times (\ell = 4500) = 3352 \text{ mm}$$
$$i = \sqrt{(I/A)} = \sqrt{(550^4/12)/(550^2)} = 159 \text{ mm}$$
$$\lambda = 3352/159 = 21.1 < \lambda_{min}$$

The column is short.

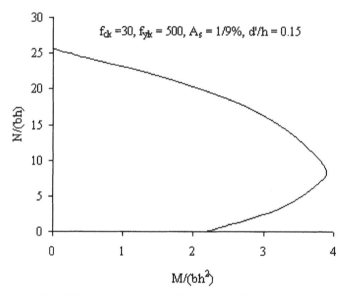

$f_{ck} = 30$, $f_{yk} = 500$, $A_s = 1/9\%$, $d'/h = 0.15$

Fig. 18.34 Column design chart ($f_{ck} = 30$, $f_{yk} = 500$, $d/h = 0.85$).

(1) Column reinforcement

Fig. 18.34 shows the design chart for a column with three layers of steel as shown in Fig. 18.35. The chart is constructed for $f_{ck} = 30$, $f_{yk} = 500$, $d/h = 0.85$. It can be seen that the design is safe for the following two load cases.

Case 1:
$$N/(bh) = 4915 \times 10^3/550^2 = 16.25$$
$$M/(bh^2) = 117 \times 10^6/550^3 = 0.71$$

Case 2:
$$N/(bh) = 5550 \times 10^3/550^2 = 18.35$$
$$M/(bh^2) = 84 \times 10^6/550^3 = 0.51$$

Provide 8H25 to give an area of 3928 mm^2.
$$100A_{sc}/bh = 1.3$$
Check minimum and maximum areas of steel:
Minimum steel: Max $(0.10 \times N_{ED}/f_{yd}; 0.002\ A_c)$
$$= \text{Max } (0.10 \times 5545 \times 10^3/435; 0.002 \times 550^2) = 1274\ mm^2$$
Maximum steel: $0.04\ A_c = 0.04 \times 550^2 = 12100\ mm^2$.
Both are satisfied.
Link diameter = max (6 mm, diameter of largest bar = 25/4). Use H8 links.
The maximum spacing is to be the minimum of
- 20 times the diameter of the smallest longitudinal bar = 20 × 25 = 500 mm
- Lesser dimension of the column = 550 mm
- 400 mm

Use 400 mm spacing for links.
Every longitudinal bar placed in a corner should be held by a link. No bar within a compression zone should be further than 150 mm from a restrained bar. Note that the bars at the centre of sides are held by separate horizontal and vertical links to satisfy this criterion.

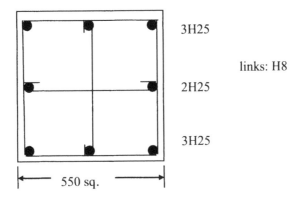

3H25

links: H8

2H25

3H25

550 sq.

Fig. 14.35 Column reinforcement.

(r) Robustness: design of ties
The design must comply with the requirements of sections 9.10.2 of the code regarding robustness and the design of ties. These requirements are examined.

(1) Internal ties
According to clause 9.10.2.3 of the code, the ties must be able to resist a tensile force in 20 kN/m width and can be spread evenly in the slabs or in beams. These must be provided at each floor and roof level in two directions approximately at right angles.
$$\text{Steel area} = 20 \times 10^3 / (0.87 \times 500) = 46\ mm^2/m$$
Provide one H12 bar every 2250 mm in the topping of the ribbed slab. Steel provided is equal to 50 mm^2/m.

(2) Peripheral ties
According to clause 9.10.2.2 of the code, the peripheral ties must resist a maximum force $F_{tie, per}$ of 70 kN.
$$\text{Steel area} = 70 \times 10^3 / (0.87 \times 500) = 161 \text{ mm}^2$$
This will be provided by an extra H16 bar in the edge L-beams running around the building. Area provided = 201 mm^2.

(3) External column tie
According to clause 9.10.2.4 of the code, edge columns should be tied horizontally to the structure at each floor and roof level. The force to be resisted is $F_{tie, fac}$ equal to 20 kN/m. For the column the maximum force F_{col} = 150 kN. The required steel area is
$$\text{Steel area} = 150 \times 10^3 / (0.87 \times 500) = 345 \text{ mm}^2$$
Provide 2H16 = 402 mm^2 per column.
The corner columns must be anchored in two directions at right angles.

14.6 REFERENCES

Bhatt, P. (1981). *Problems in Structural Analysis by Matrix Methods.* Construction Press.

Bhatt, P. (1986). *Programming the Matrix Analysis of Skeletal Structures.* Ellis Horwood.

CHAPTER 15

TALL BUILDINGS

15.1 INTRODUCTION

For the structural engineer the major difference between low and tall buildings is the influence of the horizontal loads due to wind and earthquake on the design of the structure. Lateral deflection of a tall concrete building is generally limited to H/1000 to H/200 of the total height H of the building. In the case of tall buildings, in addition to limiting this so called lateral drift, attention has to be focused on the comfort of the occupants because vibratory motion may induce mild discomfort to acute nausea.

Another aspect that needs to be addressed in tall buildings is the vertical movement due to creep and shrinkage in addition to that due to elastic shortening. These movements can cause distress in non-structural elements and must be allowed for in detailing.

This chapter is mainly concerned with the elastic static analysis of tall structures subject to lateral loads. An attempt is made to explain the complex behaviour of such structures and to suggest simplified methods of analysis of those types of structures which do not require full 3-D analysis. The behaviour of individual planar bents and the interaction between shear walls and rigid-jointed frames will be examined in detail as it highlights the complexity involved in the analysis of three-dimensional structures subjected to horizontal forces.

15.2 ASSUMPTIONS FOR ANALYSIS

The structural form of a building is inherently three-dimensional. The development of efficient methods of analysis for tall structures is possible only if the usual complex combination of many different types of structural members can be reduced or simplified whilst still representing accurately the overall behaviour of the structure. A necessary first step is therefore the selection of an idealized structure that includes only the significant structural elements with their dominant modes of behaviour. It is often possible to ignore the asymmetry in a structural floor plan of a building, thereby making a three-dimensional analysis unnecessary. One common justifiable assumption is that floor slabs are fully rigid in their own planes but flexible out-of-plane. Consequently, all vertical members at any level

This chapter is a modified version of the chapter on Tall Buildings by Hoenderkamp from *Reinforced Concrete: Design, Theory and Examples* by T.J MacGinley, 1990, Spon Press.

are subject to the same components of translation and rotation in the horizontal plane. This does not hold true for very long narrow buildings and for slabs which have their widths drastically reduced at one or more locations.   Similarly contributions from the out-of-plane stiffness of floor slabs and structural bents can be neglected because of their low stiffness compared with in-plane stiffness.

15.3 PLANAR LATERAL LOAD RESISTING ELEMENTS

15.3.1 Rigid-Jointed Frames

The most common type of planar bent used for medium height structures is the rigid-jointed frame.   These frames are economic for buildings up to about 25 storeys.   Beyond that height, control of drift becomes problematic and requires uneconomically large members.

15.3.2 Braced Frames

The lateral stiffness of a rigid frame can be improved significantly by providing diagonal members.   In fact such structures could be economic in case of very tall structures.   Bracing can be either in storey height–bay width module or they could extend over many bays and storeys.   Fig. 15.1 shows rigid frame and braced frames.

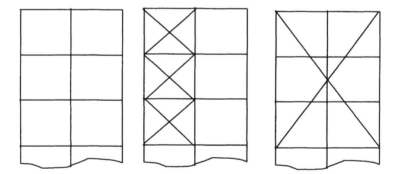

Fig. 15.1 (a) Rigid–jointed frame; (b) braced frame; (c) braced frame with large diagonal bracing.

15.3.3 Shear Walls

The simplest form of bracing against horizontal loading is the plane cantilevered shear wall.   The main difficulty with shear walls is their solid form which tends to restrict planning where wide open internal spaces are required. They are particularly suitable for hotel and residential buildings requiring repetitive floor plans. This allows the walls to be vertically continuous and they can serve both as room dividers and also provide sound and fire insulation.

15.3.4 Coupled Shear Walls

As shown in Fig. 15.2, a coupled shear wall structure is a shear wall with openings. The two halves of the wall could be connected by beams or slabs at each floor level. For analysis purposes, coupled shear walls are treated as rigid frames. However compared to the width of a column in a rigid frame, the width of the wall is very large. To allow for the large width of the walls, the beams connecting the walls are assumed to be rigid over half the width of the walls as shown in Fig. 15.3.

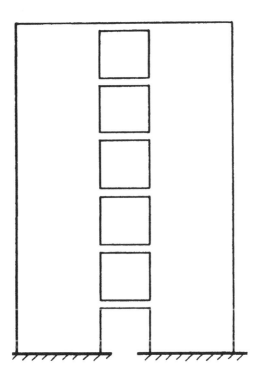

Fig. 15.2 Coupled shear wall.

15.3.5 Wall-Frame Structures

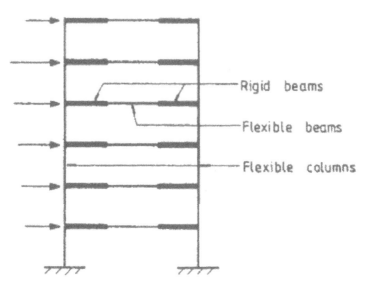

Fig. 15.3 Rigid-jointed frame model for a coupled shear wall.

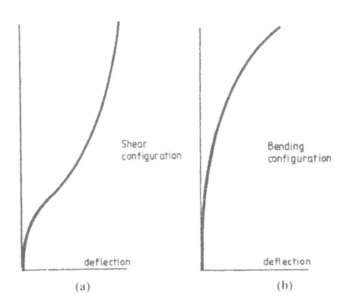

Fig. 15.4 Shear mode and bending mode of deflection of rigid-jointed frames and walls.

When rigid-jointed structures which deflect in a shear mode as shown in Fig. 15.4(a) are combined with shear walls which deflect in a flexural mode as shown in Fig. 15.4(b), they are constrained to adopt a common deflected shape because of the horizontally stiff girders and slabs. As a consequence, the two horizontal load resisting structural forms interact, especially at the top to produce a stiffer and stronger structure than a simple addition of the stiffnesses of the two elements would indicate. This combined form has been found to be suitable for structures in the 40 to 60 storey range.

15.3.6 Framed Tube Structures

In this type of structure, the lateral load resisting system consists of moment resisting rigid-jointed frames in two orthogonal directions which form a closed tube around the perimeter of the building plan as shown in Fig. 15.5. The frame consists of closely spaced columns at around 2 to 4 m centres joined by deep girders. The lateral load is carried by the perimeter frames but gravity load is shared between internal columns and perimeter frames. Perimeter frames aligned in the direction of the lateral load act as the webs and the frames normal to the direction of loading act as the flanges of the massive box cantilever. Inevitably, with a wide flange, shear lag effect as shown in Fig. 15.5 makes the flanges much less efficient.

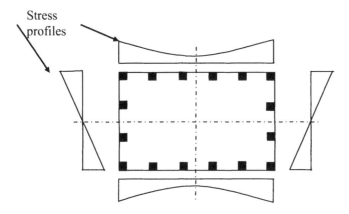

Fig. 15.5 Framed tube structure.

15.3.7 Tube-in-Tube Structures

This is similar to framed tube structures except that apart from the perimeter tube there is an internal tube formed of a service core and lift cores as shown in Fig. 15.6. Both tubes participate in resisting lateral loads.

15.3.8 Outrigger-Braced Structures

Fig. 15.7 shows an outrigger structure. This consists of a central core which could be shear walls which form part of the elevator and service cores or a braced frame. The core is connected to the perimeter columns by horizontal cantilevers or outriggers. Under horizontal loads the core bends and rotation of the core is restrained by the outrigger trusses through tension and compression in the perimeter columns. In effect the outriggers considerably increase the effective depth of the building and provide a very stiff structure. The number of outriggers up the height is generally limited to a maximum of about four. This type of structure has been found to be efficient in the design of buildings in the 40 to 70 storey range.

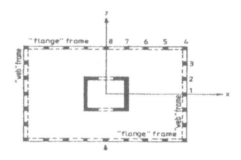

Fig. 15.6 Tube-in-tube structure.

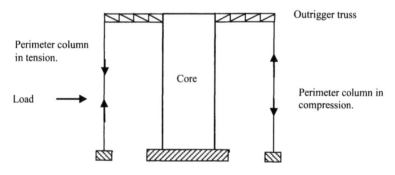

Fig. 15.7 Outrigger braced structure.

15.4 INTERACTION BETWEEN BENTS

The analysis of a tall building structure subject to horizontal and vertical loads is a three-dimensional problem. In many cases, however, it is possible to simplify and reduce this to a 2-D problem by splitting the structure into several smaller two-dimensional components which then allow a less complicated planar analysis to be carried out.

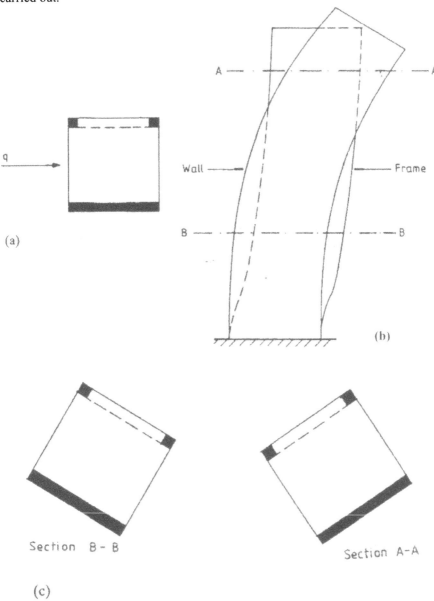

Fig. 15.8 (a) Structural floor plan, (b) deflected profiles and (c) floor rotations.

The procedure for subdividing a three-dimensional structure requires some knowledge of the sway behaviour of individual bents subjected to lateral loads. As stated in section 15.3.5, rigid frames subject to lateral load will mainly deflect in a shear configuration and shear walls will adopt a flexural configuration under identical loading conditions. These types of behaviour describe extreme cases of deflected shapes along the height of the structures. Other bents such as coupled walls will show a combination of the two deflection curves. In general they behave as flexural bents in the lower region of the structure and show some degree of shear behaviour in the upper storeys. Combining several bents with characteristically different types of behaviour in a single three-dimensional structure will inevitably complicate the lateral load analysis. It would be incorrect to isolate one of the bents and subject it to a percentage of the horizontal loading.

Fig. 15.8(a) shows the structural floor plan of a multi-storey building that consists of a single one bay frame combined with a shear wall. The symmetrically applied lateral load will cause the structure to rotate owing to the distinctly different characteristics of the two bents. A side view of the deflections of both cantilevers is shown in Fig. 15.8(b). Fig. 15.8(c) shows rotation of sections taken at different levels. It shows that it cannot be assumed not only that the rotation of the floor plans continuously increases along the height in one direction but also that the structure has a single centre of rotation. To deal with these complications, a more sophisticated three-dimensional analysis will be necessary.

15.5 THREE-DIMENSIONAL STRUCTURES

15.5.1 Classification of Structures for Computer Modelling

In many cases it is possible to simplify the analysis of a three-dimensional tall building structure subject to lateral load by considering only small parts which can be analysed as two-dimensional structures. This type of reduction in the size of the problem can be applied to many different kinds of buildings. The degree of reduction that can be achieved depends mainly on the layout of the structural floor plan and the location, in plan, of the horizontal load resultant. The analysis of tall structures as presented here is divided into three main categories on the basis of the characteristics of the structural floor plan.

15.5.1.1 Category I: Symmetric Floor Plan with Identical Parallel Bents Subject to a Symmetrically Applied Lateral Load Q

The structure shown in Fig. 15.9(a) comprises six rigid–jointed frames, four in the y-direction and two in the x-direction. Because of symmetry about the y-axis, all beams and columns at a particular floor level will have identical translations in the y-direction when subjected to load q. There will be no deflections in the x-direction. For the analysis of this model consisting of four identical rigid frames parallel to the applied load, it will be sufficient to analyse only one frame subjected to a quarter of the total load.

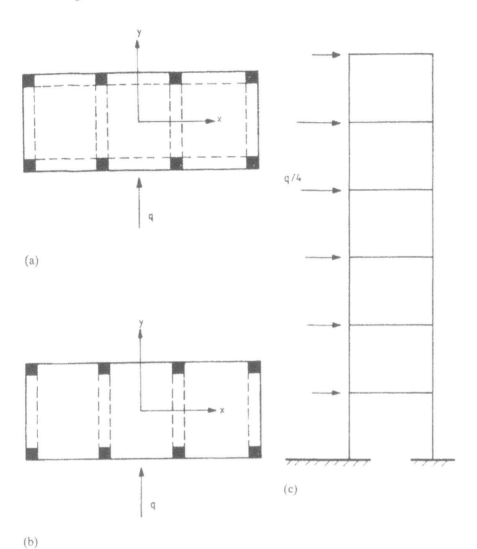

Fig. 15.9 (a) Structural floor plan of tall rigid frame building; (b) simplified floor plan; (c) one–bay rigid frame computer model.

15.5.1.2 Category II: Symmetric Structural Floor Plan with Non-identical Bents Subject to a Symmetric Horizontal Load Q

The lateral load–resisting component of the structure shown in Fig. 15.10(a) comprises two rigid frames and two shear walls orientated parallel to the direction of the horizontal load q. The behaviour of the structure is similar to Category I structures except that for the analysis a symmetrical half of the structure needs to be analysed. In addition the shear wall and the rigid-jointed frame need to be connected in line such that the horizontal deflections of the two elements at any level are identical. The two structures can be linked by members of high axial stiffness to achieve the required compatibility of deflections.

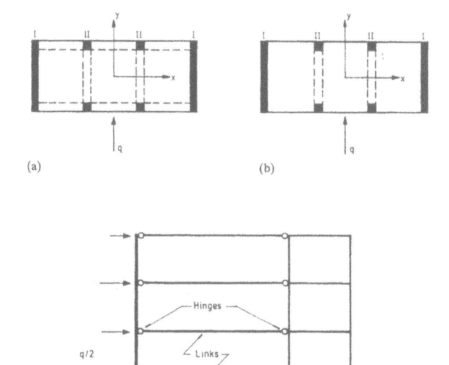

Fig. 15.10(a) Structural floor plan of frame–wall building; (b) simplified floor plan; (c) computer model of linked bents in a single plane.

Note that as long as symmetry about the y-axis is maintained, it is possible to cope with any variation in geometry with height of different frames and walls. A setback in the upper storeys for all exterior bays in the floor plan shown in Fig. 15.11(a) will still allow a plane frame analysis for the linked bents shown in Fig. 15.11(b). If the setback causes a loss of symmetry about the y-axis; however, the structure will rotate in the horizontal plane and a full three-dimensional analysis will be necessary.

15.5.1.3 Category III: Non-symmetric Structural Floor Plan with Identical or Non-identical Bents Subject to a Lateral Load Q

A category III structure, of which an example floor plan is shown in Fig. 15.12, will rotate in the horizontal plane regardless of the location of the lateral load. It cannot be reduced to a plane frame problem and a complete three-dimensional analysis is required.

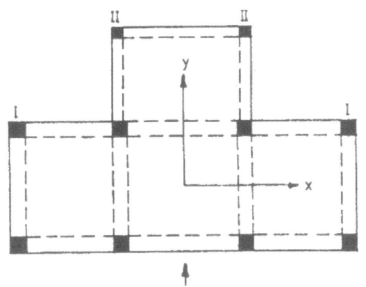

Fig. 15.11(a) Structural floor plan of rigid frame building.

15.6 ANALYSIS OF FRAMED TUBE STRUCTURES

Framed tube structures shown in Fig. 15.5 can be analysed as a pair of cantilevers lying in the same plane. However it is necessary to allow for the shear lag effect in columns in the 'flange' frame. This can be allowed for by treating the 'web' frame and the 'flange' frame as two cantilevers as shown in Fig. 15.13. The lateral load is applied to the web frame. At the junction between the two frames, at each storey level the web frame is connected to the flange frame through a set of `rigid` vertical springs. This ensures that at the junction between the two frames, both frames move in the vertical z-direction by the same amount. However the displacement in the y-direction of web frame is different from the deflection of the flange frame in the x-direction, although in the analysis they lie in the same plane. The compatibility of deflections is valid only in the vertical direction.

15.7 ANALYSIS OF TUBE-IN-TUBE STRUCTURES

The distribution of horizontal load between the inner core and the perimeter tube of a tube–in–tube structure (Fig. 15.14) depends on the characteristics of the floor system connecting the vertical elements.   Two assumptions about these connections can be made, resulting in different computer models.

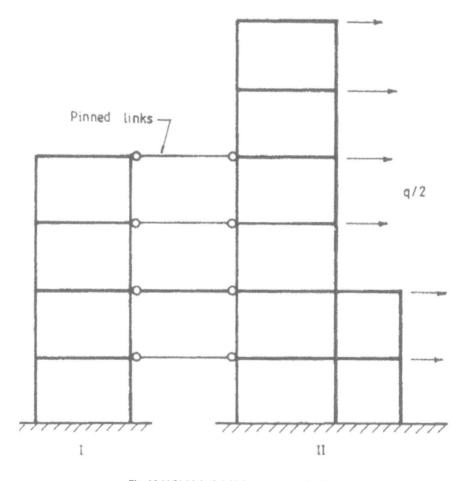

Fig. 15.11(b) Linked rigid frames in a single plane.

(a) The interior beams and/or floors are effectively pin-connected to the cores and columns: If the structural floor plan is symmetric about the y-axis as shown in Fig. 15.14(a), the structure can be classified under category II and a plane frame analysis is possible. Only half the structure needs to be considered. As shown in Fig. 15.14(b), half the core is bent B, i.e., one channel-shaped cantilever wall is bundled with its exterior columns of the two exterior frames perpendicular to the direction of the load. In calculating the second moment of area of the channel section, allowance has to be made for shear lag effect by assuming a reduced effective width for flange width. Together they can be modelled as a single flexural cantilever with a combined bending stiffness represented by the wall and columns 1 and 2. One rigid frame parallel to the direction of the load, bent A is then connected to it in a single plane by means of rigid links at each floor level. The two-dimensional model is to be subjected to half the lateral load.

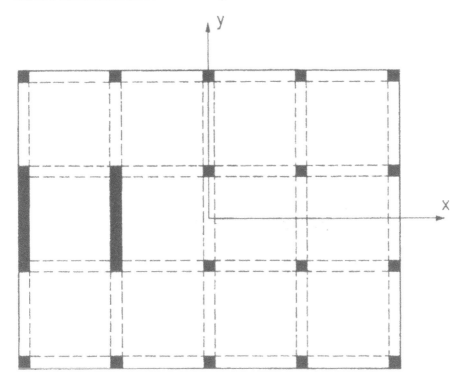

Fig. 15.12 Non-symmetric structural floor plan.

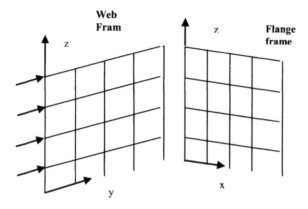

Fig. 15.13 Web and flange frames.

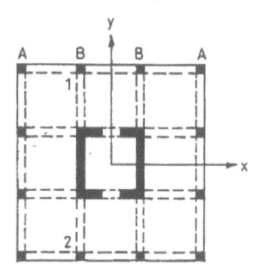

Fig. 15.14(a) Structural floor plan of a tube-in-tube building.

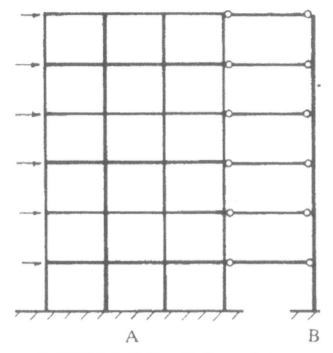

Fig. 15.14(b) Rigid-jointed frame linked to core and columns.

(b) Beams spanning from the exterior columns to the cores can be considered rigidly connected: The channel-shaped shear wall which is parallel to the load and rigidly connected to floor beams will behave as a wide column and must be modelled as such. Flexural column elements are located on the neutral axis of the wall but in the plane of the bent. The moment of inertia of these members should represent the full section of the channel-shaped wall. Rigid arms are then attached in two directions at each floor level. Floor beams are rigidly connected to these arms and the columns of the perpendicular frames. The plane frame model of half the structure subjected to half the horizontal loading is shown in Fig. 15.14(c). The short deep beams connecting the shear walls at each floor level will not influence the deflection behaviour of the structure in the y-direction since both walls adopt exactly the same deflection profile when subjected to lateral load.

When the core is turned through 90°, without loss of symmetry, a wide arm column model is still possible. The flexible column elements are to be placed on the neutral axis of the channel-shaped section but in the plane of the bent to be analysed. The second moment of area of this element should represent only one-half of one channel-shaped section. The two unequal rigid arms at each floor level add up to the width of the 'flange' of the channel-shaped cantilever. Beams connecting the wide column to other walls or columns can then be rigidly jointed to the arms.

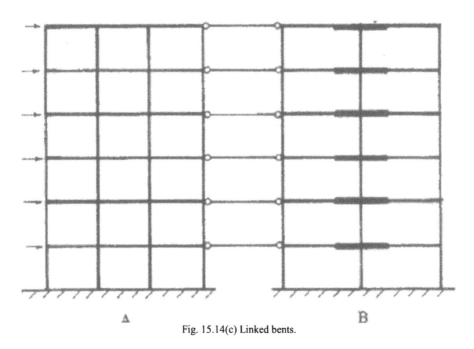

Fig. 15.14(c) Linked bents.

15.8 REFERENCES

Schueller, Wolfgang. (1977). *High-Rise Building Structures*. Wiley.

Stafford Smith, Bryan and Coull, Alex. (1991). *Tall Building Structures: Analysis and Design*. Wiley.

Taranath, Bungale S. (2009). *Reinforced Concrete Design of Tall buildings*. CRC Press.

CHAPTER 16

PRESTRESSED CONCRETE

16.1 INTRODUCTION

Prestressed concrete structures can be defined as concrete structures where external compressive forces are applied to overcome tensile stresses caused by unavoidable loads due to gravity, wind, etc. In other words, it is precompressed concrete meaning that compressive stresses are introduced into areas where tensile stresses might develop under working load and this precompression is introduced even before the structure begins its working life.

This chapter gives a brief introduction to the basic aspects of prestressed concrete design. For a more extensive treatment, the reader is referred to Bhatt (2011).

One of the disadvantages of reinforced concrete is that tensile cracks due to bending occur even under working loads. This has four major disadvantages.

- Cracks encourage corrosion of steel.
- A cracked concrete beam is much more flexible than an uncracked beam. This means that when using a reinforced concrete beam, one could have serviceability problems due to deflection or even due to cracking if too slender a beam is used.
- Cracked concrete is not, on the whole, contributing to strength but rather it is simply adding to dead weight.
- The width of the cracks is to a large extent governed by the strain in reinforcing steel. If high tensile steel is used as reinforcement, then the resulting width of the cracks at working loads would be unacceptable. Ordinary reinforced concrete precludes the utilization of high strength steel and the resulting possible economies.

Clearly the above problems can be overcome if we can apply external compressive forces to the beam to prevent it from cracking or even better if the external compressive forces can be applied so as to neutralize the stresses created by applied loads under serviceability conditions, a very efficient structure can be designed.

Consider the simply supported beam supporting loads as shown in Fig. 16.1. Bending moment at a section XX produces tensile and compressive stresses at bottom and top fibres respectively. If a compressive force is applied at the centroidal axis, it sets up uniform compression throughout the beam cross section. It does neutralize the tensile stresses at the bottom portion of the beam caused by bending but it has the disadvantage of increasing the total compressive stresses at the top face. If however the compressive force is applied towards the bottom face at an eccentricity of **e** from the centroidal axis, then in addition to an axial force of P, a bending moment equal to Pe of a nature *opposite* to that caused by external loads is created. The total stresses due to the bending moment M and the prestress P at an eccentricity e are

$$f_{top} = -\frac{P}{A} + \frac{Pe}{z_t} - \frac{M}{z_t}, \quad f_{bottom} = -\frac{P}{A} - \frac{Pe}{z_b} + \frac{M}{z_b}$$

where z_t and z_b are the section moduli of the cross section for top and bottom fibres respectively and A is the cross sectional area of the beam.

As shown in Fig. 16.1, by proper manipulation of the values of P and e, it is possible to ensure that at working loads, the entire cross section is in compression. It is often stated that one tonne of prestressing steel can result in up to 15 times the amount of building that is made possible by one tonne of structural steel.

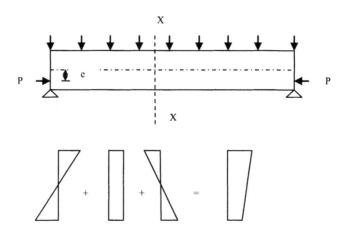

Fig. 16.1 Stresses due to prestress and external loads.

16.2 APPLYING PRESTRESS

There are two main methods of prestressing. They are called pretensioning and posttensioning.

16.2.1 Pretensioning

This is used for producing precast prestressed concrete products such as bridge beams, double T-beams for floors, floor slabs, railway sleepers, etc. In this method, as shown in Fig. 16.2, the process consists of the following steps.

- Any reinforcing steel such as links etc. are threaded onto the high tensile steel cables. The cables are tensioned or jacked to the desired force between abutments. The cable is anchored using a simple barrel and wedge device. Because of the fact that the cables are tensioned before concrete is cast, the name pretensioning is used for this process.
- The formwork is assembled around the steel cables.
- Concrete is placed in the moulds around it and is allowed to cure to gain desired level of strength. This is often speeded up using steam curing. This also enables the prestressing bed to be reused quickly for another job.
- Steel is released from the abutments, transferring the compressive force to the concrete through the bond between steel and concrete.

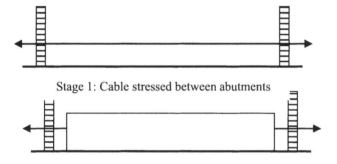

Stage 1: Cable stressed between abutments

Stage 2: Concrete cast and allowed to harden

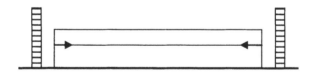

Stage 3: Cable released from the abutments

Fig. 16.2 Basic stages of pretensioning.

In practice a large number of identical units are cast at the same time using what is known as longline production method.

It is worth noting that when the force in the cable is transferred to concrete, it contracts. Because of the full bond between concrete and steel, steel also suffers the same contraction leading to a certain loss of stress from the stress at the time of jacking. This is known as loss of prestress at transfer and is generally of the order of 10%. Thus

$$P_{Transfer} \approx 0.9\, P_{jack}$$

where

P $_{Transfer}$ = Total force in the cable after initial loss of stress due to compression of concrete.

P$_{jack}$ = Total force used at the time of initial jacking the cable between the abutments.

16.2.1.1 Debonding

One of the disadvantages of having the same prestressing force P and eccentricity e at all sections is that while normally the external loads produce large bending stresses at the mid-span of the simply supported beam but small stresses towards the supports, the stresses due to prestressing remains constant at all sections. This clearly defeats the idea of tailoring stresses due to prestressing to match the stresses due to external loads. This disadvantage can be overcome to a great extent by two methods as follows.

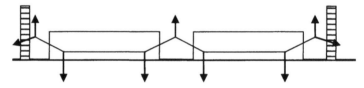

Fig. 16.3 Deflected tendons.

(a) Deflected Tendons: As shown in Fig. 16.3, the cable is deflected along its length by pulling the cable up or down as necessary. However this is generally not preferred because of extra cost.

(b) Debonding: In this method by preventing bond from developing between concrete and steel by sheathing some of the cables in plastic tubing as shown in Fig. 16.4, both the prestressing force and eccentricity can be varied in a stepwise fashion along the span. This is generally the preferred option due to low cost.

The transfer of force between concrete and steel takes place gradually. The force transfer takes place due to two basic actions.

- Bond between concrete and steel plays an important part. It is therefore essential to ensure that the cable is clean and free from loose rust and that concrete is well compacted.
- The cable is stretched and therefore has a very slightly reduced diameter due to the Poisson effect. However when the cable is released from the abutments, the wire regains its original diameter at the ends. This creates a certain amount of wedging action and in addition frictional forces also come into play.

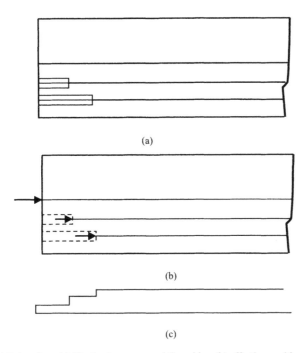

(a)

(b)

(c)

Fig. 16.4 Debonding: (a) Plastic sleeves around the cables; (b) effective position of where prestress starts; (c) variation of prestress and eccentricity along span.

16.2.1.2 Transmission Length

As shown in Fig. 16.2, once the cable is released from the abutments, the force in the cable becomes zero at the ends of the cable. However away from the ends, the bond between cable and concrete prevents the cable from regaining its original length. As shown in Fig. 16.5, the force in the cable gradually builds up to its full value over a certain length. This is known as transmission length. This varies depending on the surface characteristics of the cables and the strength of concrete. It is generally of the order of about 50 diameters for 7-wire strands.

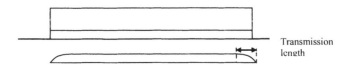

Transmission length

Fig. 16.5 Gradual build-up of force in the cable.

16.2.2 Posttensioning

One of the limitations of pretensioning is that normally the cables need to remain straight because the cable is pretensioned before concrete hardens. This limitation can be overcome if as shown in Fig. 16.6, the cable is laid to any desired profile inside a metal ducting fixed to the required profile to the reinforcement cage with the permanent anchorages also positioned at the ends of the duct. Afterwards, concrete is cast and once it has hardened the cable is tensioned and anchored to the concrete using permanent external anchors rather than relying on bond between the cable and the concrete as in the case of pretensioning. This is the basic idea of posttensioning. Because of the fact that the cables are tensioned after the concrete has hardened, the system is known as posttensioning. Finally, the duct is filled with a colloidal grout under pressure in order to establish bond between concrete and steel and also as protection against corrosion.

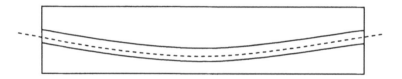

Stage 1: Cable inside the duct but not tensioned and concrete is cast

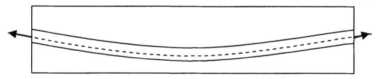

Stage 2: Concrete has hardened and cable is tensioned but not anchored

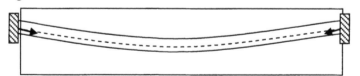

Stage 3: Cable is tensioned and anchored at the ends

Fig. 16.6 Different stages in posttensioning.

There are various types of anchors used in practice but they are generally of two types.
*   The bar is threaded at the ends and anchoring is by a nut bearing on concrete. The threads are rolled rather than cut to reduce stress concentration. The main advantage of this system is that prestress can be applied in stages to suit design considerations or losses can be compensated at any time prior to grouting. The anchorage is completely positive and there is no loss of prestress at the transfer stages.

- Anchoring is done using a system of cones and wedges. In this case, there is loss of prestress at transfer stage because of the slip between the cable and the wedge before the wedges 'bite in'.

16.2.3 External Prestressing

One of the disadvantages of traditional posttensioning is that there is no guarantee that the ducts are properly filled with grout to prevent corrosion and if the steel corrodes, cables cannot be replaced. In order to overcome these problems, external prestressing as shown in Fig. 16.7 is used. The cables are on the *outside* of the beam and the eccentricity is varied using saddles at appropriate places to obtain the required profile. This is similar to the use of deflected tendons in a pretensioned system. This system allows replacement of cables as required and also allows the use of additional cables at a later stage in order to strengthen the structure. Although the term external prestressing is used, it is not necessary for the cables to be outside the structure. For example in the case of box girders, cables can be placed inside the void of the box girder.

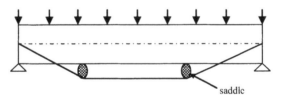

Fig. 16.7 External prestressing.

16.2.4 Unbonded Construction

Because of the relative unreliability of grouting to prevent corrosion of the cables and also because of the fact grouting is a time-consuming job and sufficient time has to elapse for the bond between the cables (also called tendons or strands) and concrete to become effective, a common form of construction used in practice is to dispense altogether with the bond between the concrete and steel. Cables used in this form of construction are coated by grease and encased in a plastic sleeve. The plastic sleeve acts as the duct and the construction process is similar to normal posttensioning. The main advantage of this unbonded system is speed of construction as no grouting is done. However this is not a particularly structurally efficient system because the ultimate bending capacity tends to be only about 70% of a corresponding beam using bonded construction. Nevertheless, unbonded posttensioned slabs are a very common form of construction.

16.2.5 Statically Indeterminate Structures

Because the bending moment due to external loading in a simply supported beam is parabolic, the cable profile is also parabolic. One of the advantages of posttensioning is that the cable profile can be varied so that the bending moment due to external loading can be neutralized by varying the eccentricity to match the shape of the bending moment diagram. A posttensioning system is essential when constructing prestressed statically indeterminate structures. Fig. 16.8 shows the cable profile in a two span continuous beam.

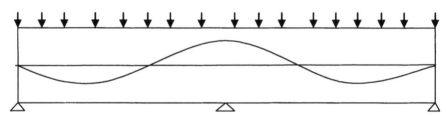

Fig. 16.8 Cable profile in a two span continuous beam.

16.2.6 End Block

One important aspect of posttensioning that needs special attention is the area where cables are anchored. Because of the fact that many cables are anchored to the same anchorage block of a relatively small size, large compressive forces are transferred at the anchorage block. Depending on the number of cables anchored and their diameter, the force at an anchor block can vary from 100 to 12000 kN. This large transfer of force has the same effect as driving a wedge into a block of wood and has the tendency to burst the concrete transversely near the anchorage. The bursting forces have to be resisted using a large number of links near the anchors. This area of beam is known as an end block.

16.3 MATERIALS

16.3.1 Concrete

Concrete used for prestressing work is generally of much higher quality than that used for reinforced concrete work. Concrete of grade C50 or over is common. Certain deformational properties of concrete affect the design of prestressed concrete structures and it is necessary to understand them. One of the important properties of concrete is creep. Creep is defined as the increase of strain with time when the stress is held constant. Under the action of compressive stress due to prestress, concrete continues to deform. Because of the bond between steel and concrete, steel also experiences compressive strain which reduces the tension in the cables. In addition to creep, shrinkage of concrete also contributes to the loss of prestress. This long-term loss can be as high as 25% of the initial stress. Thus

$$P_{\text{Service}} \approx 0.75 \ P_{\text{jack}}$$

where

P_{Service} = total prestress remaining in the long-term under working load conditions, after all the losses have taken place.

P_{jack} = Total load in the cables at the time of jacking.

It should be noted that these long-term losses of prestress are common to both pre and posttensioning systems.

One important effect of creep is increased deflections. Because in a prestressed concrete member a greater part of the cross section is in compression compared to the corresponding reinforced concrete section, long-term creep movements are increased.

16.3.2 Steel

Compared to normal high yield steel bars used in reinforced concrete which has a characteristic tensile stress f_{yk} of about 500 MPa, prestressing steel is usually cold drawn high tensile steel wires or alloy steel bars with a characteristic tensile stress f_{pk} tensile stress of about 1800 MPa. Apart from the fact that steel used in prestressing work is of higher strength, it is also much less ductile compared with reinforcing bars. Steel used in prestressed concrete is available in the form of:

- Wires from 7 mm to 3 mm diameter. In order to improve bond, wires are often indented. This is called crimping.
- Tendons used today are almost always 7-wire strand made from six wires spun round a straight central wire. The overall nominal diameter varies from 12.5 mm to 18 mm. Two basic shapes of cables are available. In standard cables the individual wires maintain their circular cross section. In order to reduce the overall diameter, the standard strand can be passed through a die to compress the cable and reduce the voids. The final shape of the individual wires is trapezoidal rather than circular. This type of cable is called drawn and has, for the same nominal diameter, a higher amount of steel in the cross section.
- Bars: 25 mm to 50 mm diameter. Two types are common. In one type, the bar has ribs along its entire length. The ribs are rolled rather than cut to reduce stress concentration problems. The ribs act as threads for coupling purposes. In the other type the bar is smooth but has threads rolled only at the ends for coupling or for anchorage purposes.

16.3.2 .1 Relaxation of Steel

Just as concrete exhibits time-dependent deformation due to creep, steel exhibits a property called relaxation. If the strain in steel is maintained constant, then the stress required to maintain that strain reduces with time. This property is known as relaxation. Relaxation is thus loss of stress under constant strain. Generally tests are conducted for duration of 1000 hours (about 42 days) at a temperature of 20°C and at an initial stress of 70% of the actual tensile strength of the steel to determine Relaxation properties. Final relaxation loss in the long term normally taken as about 57 years is expressed as a multiple of the 1000 hour loss. Relaxation also contributes to the loss of prestress in the long-term.

Heat treatment is used to improve the elastic and yield properties of strands. In clause 3.3.2(4), two classes of relaxation for strands are defined. They are:

Class 1: Wire or strand, ordinary relaxation. In order to remove residual stresses due to cold drawing, the strand is heated to about 350° C and allowed to cool slowly.
Class 2: Wire or strand, low relaxation. The strand is heated to about 350° C *while the strand is under tension* and allowed to cool slowly. This process is known as strain tempering.
Class 3: Hot rolled and processed bars are classed as relaxation class 3.

16.4 DESIGN OF PRESTRESSED CONCRETE STRUCTURES

Although design of prestressed concrete structures has to satisfy both serviceability and ultimate limit state criteria, there is a fundamental difference in the approach to the design of reinforced and prestressed concrete structures. The normal design procedure for a reinforced concrete structure is to design the structure for ultimate limit state and then check that the structure behaves satisfactorily at serviceability limit state. On the other hand, the normal design procedure for a prestressed concrete structure is to design the structure for the serviceability limit state and then check that the structure behaves satisfactorily at ultimate limit state. The reason for this difference is that generally speaking in prestressed concrete structures, serviceability limit state conditions are much more critical than the conditions at ultimate limit state. Thus generally structures designed for serviceability limit state also satisfy the ultimate limit state criteria, but not the other way around.

16.5 LIMITS ON PERMISSIBLE STRESSES IN CONCRETE

Since prestressed concrete structures are primarily designed to satisfy the serviceability limit state, it is necessary to limit the stresses in concrete and steel. The structure is assumed to behave elastically and the two critical conditions to be considered are

- Stress state at transfer of prestress: At this stage the loads acting are the self weight of the structure and prestress with only elastic shortening during transfer having taken place.
- Stress state at serviceability limit state when the loads acting are the dead and live loads along with the prestress with all the long-term losses assumed to have taken place.

16.5.1 Permissible Compressive and Tensile Stress in Concrete at Transfer

Let $f_{ck}(t)$ be the compressive strength at the time of stressing in the case of posttensioned members or at force transfer in the case of pretensioned members. Let the **numerical value** of the permissible stress in compression at the extreme fibre be f_{tc}. Clause 5.10.2.2(5) of the code states that $f_{tc} \leq 0.6\ f_{ck}(t)$.

For pretensioned elements the stress at the time of transfer of prestress may be increased to $0.7\ f_{ck}(t)$.

Similarly, the permissible stress in tension, f_{tt} at transfer is limited to $f_{ctm}(t)$.

$$f_{ctm}(t) = f_{ck}(t)^{0.667}, \quad f_{ck}(t) \leq 50\ \text{MPa}$$
$$f_{ctm}(t) = 2.12\ \ell n\ [1.8 + 0.1\ f_{ck}(t)],\ f_{ck}(t) > 50\ \text{MPa}$$

16.5.2 Permissible Compressive and Tensile Stress in Concrete at Serviceability Limit State

Expressions for permissible stresses in compression f_{sc} and in tension f_{st} at serviceability limit state are similar to the expressions given above for transfer state except that $f_{ck}(t)$ is equal to f_{ck} the compressive strength at 28 days. However in order to ensure that creep deformation is linear, compressive stress under quasi-permanent loads should be limited to $0.45\ f_{ck}$.

16.6 LIMITS ON PERMISSIBLE STRESSES IN STEEL

As described in section 16.3.2, many different types of prestressing steel are available in the form of wires, bars, tendons, etc. The characteristic tensile stress fpk varies from about 1030 MPa for hot rolled bars to 1860 MPa for strands. The value of Young's modulus $E = (195 \pm 10)\ \text{kN/mm}^2$.

16.6.1 Maximum Stress at Jacking and at Transfer

The permissible stresses are given in clause 5.10.2.1.

Stress at jacking $\sigma_{p,\ max}$ is limited to
$$\sigma_{p,\ max} = \min\ (0.8\ f_{pk};\ 0.9\ f_{p0.1k})$$

where:

f_{pk} = characteristic tensile strength of prestressing steel.

$f_{p0.1k}$ = characteristic tensile strength of prestressing steel at 0.1% strain
        $\approx 0.85\ f_{pk}$.

Substituting in the expression for $\sigma_{p,\ max}$ = min $_{(0.8\ f_{pk};\ 0.9\ f_{p0.1k})}$ = 0.77 f_{pk}.

Overstressing is permitted if the force in the jack can be measured to an accuracy of ±5% of the final value of the prestressing force. In such cases the stress in steel can be increased to $\sigma_{p,\ max}$ = 0.95 $f_{p0.1k} \approx 0.808\ f_{pk}$.

16.7 EQUATIONS FOR STRESS CALCULATION

In a statically determinate structure, the stresses at top and bottom fibres are given by the following equations. The sign convention used is as follows.

Eccentricity e is *positive below* the neutral axis.

Tensile stresses are positive and compressive stresses are negative.

Bending moment causing sagging is considered positive.

16.7.1 Transfer State

At transfer, the external load acting is normally only the self weight. With a large prestress force applied below the neutral axis, the beam will hog up. The critical stress conditions are

- Tension is critical at the top of the beam
- Compression is critical at the bottom of the beam.

The expressions for the stress at top and bottom fibres are given by

$$f_{top} = [-\frac{P_t}{A} + \frac{P_t e}{z_t} - \frac{M_{self\,weight.}}{z_t}] \leq f_{tt} \qquad (C16.1)$$

$$f_{bottom} = [-\frac{P_t}{A} - \frac{P_t e}{z_b} + \frac{M_{self\,weight.}}{z_b}] \geq f_{tc} \qquad (C16.2)$$

where f_{tt} and f_{tc} are the permissible stresses at transfer in tension and compression respectively. The first subscript, t stands for transfer and the second subscript, (t or c) stands for tension or compression respectively. Note that f_{tc} is a compressive stress, and is a negative number. P_t = prestress at transfer. $P_t = \alpha\ P_{jack}$, where $\alpha \approx 0.90$ because at transfer, there is generally loss of prestress of about 10%. z_t and z_b are the section moduli of the beam for top and bottom fibres respectively.

16.7.2 Serviceability Limit State

At working loads, external loads on the beam increase due to live loads and other dead loads. In addition due to long-term loss the prestress in the cables also decreases. These effects cause the beam to sag. The critical stress conditions are:

- Compression is critical at the top of the beam.
- Tension is critical at the bottom of the beam.

The expressions for the stress at top and bottom fibres are given by

$$f_{top} = [-\frac{P_s}{A} + \frac{P_s e}{z_t} - \frac{M_{service}}{z_t}] \geq f_{sc} \tag{C16.3}$$

$$f_{bottom} = [-\frac{P_s}{A} - \frac{P_s e}{z_b} + \frac{M_{service}}{z_b}] \leq f_{st} \tag{C16.4}$$

where

f_{st} and f_{sc} are the permissible stresses at service in tension and compression respectively. The first subscript, s stands for service and the second subscript (t or c) stands for tension or compression respectively. $M_{service}$ = moment at serviceability limit state. It includes the effect of self-weight, live loads, etc. P_s = prestress at service = βP_{jack} where $\beta \approx 0.75$ because of about 25% prestress is lost due to elastic shortening, creep, shrinkage and relaxation.

16.7.3 Example of Stress Calculation

Fig. 16.9 shows a pretensioned symmetric double T-beam. It is used as a simply supported beam to support a total characteristic load (excluding self weight) of 45 kN/m over a span of 10 m. It is prestressed by a total force of P_{jack} = 1450 kN. The constant eccentricity e is 390 mm. Calculate the stresses at mid-span and support at transfer and serviceability limit state. $f_{ck}(t)$ = 30 MPa and f_{ck} = 40 MPa. Assume loss at transfer and serviceability limit state as respectively 10% and 25% of the jacking force.

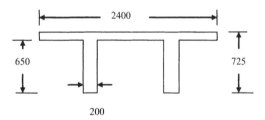

Fig. 16.9 A double T-beam.

(i) Section Properties
Area of cross section
$$A = 2400 \times 75 + 2 \times (200 \times 650) = 44 \times 10^4 \text{ mm}^2$$
Position of the centroid from the bottom: Taking moment about the bottom of the webs,
$$A y_b = 2400 \times 75 \times (725 - 75/2) + 2 \times (200 \times 650 \times 650/2) = 2.083 \times 10^8 \text{ mm}^3$$
Distances from the centroidal axis to bottom and top fibres are:
$$y_b = 473 \text{ mm}, y_t = 725 - y_b = 252 \text{ mm}$$

Second moment of area, I:
$$I = [2400 \times 75^3/12 + 2400 \times 75 \times (y_t - 75/2)^2] + 2 \times [200 \times 650^3/12$$
$$+ 200 \times 650 \times (650/2 - y_b)^2] = 2.322 \times 10^{10} \text{ mm}^4$$
$$z_t = I/y_t = 92.12 \times 10^6 \text{ mm}^3$$
$$z_b = I/y_b = 49.08 \times 10^6 \text{ mm}^3$$

(ii) Calculation of moments
Unit weight of concrete = 25 kN/m^3
Self weight = $(44 \times 10^4) \times 10^{-6} \times 25 = 11.0$ kN/m
$M_{\text{self weight}} = 11.0 \times 10^2/8 = 137.5$ kNm at mid-span
Total load on the beam (including self weight)
= 11.0 + 45.0 = 56.0 kN/m
$M_{\text{service}} = 56 \times 10^2/8 = 700$ kNm at mid-span

(iii) Prestress
$$P_t = 0.90 \, P_{\text{jack}} = 0.9 \times 1450 = 1305 \text{ kN}$$
$$P_s = 0.75 \, P_{\text{jack}} = 0.9 \times 1450 = 1087.5 \text{ kN}$$

(iv) Permissible stresses

(a) Transfer
$$f_{tt} = 0.30 \times 30^{0.67} = 2.9 \text{ MPa}$$
$$f_{tc} = -0.6 \, f_{ck}(t) = -0.6 \times 30 = -18.0 \text{ MPa}$$

(b) Service
$$F_{st} = 0.30 \times 50^{0.67} = 4.1 \text{ MPa}$$
$$f_{sc} = -0.6 \, f_{ck} = -0.6 \times 40 = -24.0 \text{ MPa}$$

(v) Stress calculation at transfer
$$P_t = 1305 \text{ kN}, e = 390 \text{ mm}$$
Expressions for stresses at top and bottom fibres are given by equations (C16.1) and (C16.2).

(a) Support: At support self weight moment is zero because of the simply supported condition.
$$f_{\text{top}} = -1305 \times 10^3/(44 \times 10^4) + 1305 \times 10^3 \times 390/(92.12 \times 10^6)$$
$$= -2.97 + 5.53 = 2.56 < 2.70 \text{ MPa}$$

$$f_{\text{bottom}} = -1305 \times 10^3/(44 \times 10^4) - 1305 \times 10^3 \times 390/(49.08 \times 10^6)$$
$$= -2.97 - 10.37 = -13.33 > -17.50 \text{ MPa}$$

(b) Mid-span:
Self weight moment = 137.5 kNm
$$f_{\text{top}} = -1305 \times 10^3/(44 \times 10^4) + 1305 \times 10^3 \times 370/(92.12 \times 10^6)$$
$$- 137.5 \times 10^6/(92.12 \times 10^6)$$
$$= -2.97 + 5.53 - 1.49 = 1.07 < 2.90 \text{ MPa}$$

$$f_{bottom} = -1305 \times 10^3 / (44 \times 10^4) - 1305 \times 10^3 \times 370 / (49.08 \times 10^6)$$
$$+ 137.5 \times 10^6 / (49.08 \times 10^6)$$
$$f_{bottom} = -2.97 - 10.37 + 2.80 = -10.54 > -18.50 \text{ MPa}$$

Note: Stresses at the supports are *larger* than at mid-span.

(vi) Stress calculation at serviceability limit state
$$P_s = 1087.5 \text{ kN, e} = 390 \text{ mm}$$

Expressions for stresses at top and bottom fibres are given by equations (C16.3) and (C16.4).

(a) Support: At support moment is zero because of the simply supported condition.
$$f_{top} = -1087.5 \times 10^3 / (44 \times 10^4) + 1087.5 \times 10^3 \times 390 / (92.12 \times 10^6)$$
$$= -2.47 + 4.60 = 2.13 < 4.1 \text{ MPa}$$

$$f_{bottom} = -1087.5 \times 10^3 / (44 \times 10^4) - 1087.5 \times 10^3 \times 390 / (49.08 \times 10^6)$$
$$= -2.47 - 8.64 = -11.11 > -24.0 \text{ MPa}$$

(b) Mid-span:
$$\text{Serviceability limit state moment} = 700.0 \text{ kNm}$$
$$f_{top} = -1087.5 \times 10^3 / (44 \times 10^4) + 1087.5 \times 10^3 \times 390 / (92.12 \times 10^6)$$
$$- 700.0 \times 10^6 / (92.12 \times 10^6)$$
$$= -2.47 + 4.60 - 7.60 = -5.47 > -24.0 \text{ MPa}$$

$$f_{bottom} = -1087.5 \times 10^3 / (44 \times 10^4) - 1087.5 \times 10^3 \times 370 / (49.08 \times 10^6)$$
$$+ 700.0 \times 10^6 / (49.08 \times 10^6)$$
$$= -2.47 - 8.64 + 14.26 = 3.15 < 4.1 \text{ MPa}$$

Note: Stresses at the supports are *smaller* than at transfer condition. The stresses at mid-span are larger than at transfer condition. In addition the state of stress has reversed from tension at top to tension at bottom and vice versa.

16.8 DESIGN FOR SERVICEABILITY LIMIT STATE

For a given structural configuration and loads, design in prestressed concrete for serviceability limit state requirements involves two items:
- A suitable section
- Choice of prestress and corresponding eccentricity

16.8.1 Initial Sizing of Section

Consider the four equations (C16.1) to (C16.4) associated with the calculation of stresses at top and bottom fibres at a cross section under transfer and serviceability conditions. In these equations

$$P_{transfer} = \alpha\, P_{Jack}, \; \alpha \approx 0.90$$
$$P_{Service} = \beta\, P_{Jack}, \; \beta \approx 0.75$$
$$P_{transfer} = \eta\, P_{Service},$$
$$\eta = \beta/\alpha \approx 0.83$$

Expressing P_t in terms of P_s, equations (C16.1) and (C16.2) can be expressed as

$$[-\frac{P_s}{A} + \frac{P_s e}{z_t} - \eta\frac{M_{self\,weight.}}{z_t}] \le \eta\, f_{tt} \tag{C16.5}$$

$$[-\frac{P_s}{A} - \frac{P_s e}{z_b} + \eta\frac{M_{self\,weight.}}{z_b}] \ge \eta\, f_{tc} \tag{C16.6}$$

Eliminating P_s and e from equations (C16.5) and (C16.3),

$$[\frac{M_{service}}{z_t} - \frac{\eta\, M_{self\,weight.}}{z_t}] \le \{\eta\, f_{tt} - f_{sc}\}$$

$$z_t \ge \frac{M_{service} - \eta M_{self\,weight}}{\eta\, f_{tt} - f_{sc}}$$

$$z_t \ge \frac{(M_{service} - M_{self\,weight}) + (1-\eta)M_{self\,weight}}{\eta\, f_{tt} - f_{sc}} \tag{C16.7}$$

Similarly, eliminating P_s and e from equations (C16.6) and (C16.4),

$$\frac{M_{service}}{z_b} - \frac{\eta\, M_{self\,weight.}}{z_b} \le f_{st} - \eta f_{tc}$$

$$z_b \ge \frac{M_{service} - \eta M_{self\,weight}}{f_{st} - \eta\, f_{tc}}$$

$$z_b \ge \frac{(M_{Service} - M_{self\,weight}) + (1-\eta)M_{self\,weight}}{f_{st} - \eta\, f_{tc}} \tag{C16.8}$$

Initially the self weight moment is not known. However, $(M_{service} - M_{self\,weight})$ represents the moment due to external loads which are known. As $\eta \approx 0.83$, the effect of including $M_{self\,weight}$ has a small impact on the required section moduli. Therefore for an initial estimate, it is reasonable to take $M_{self\,weight}$ as zero. After an initial section has been decided upon, if necessary, the required value of section modulus can be recalculated.

16.8.1.1 Example of Initial Sizing

Calculate the section moduli required for a simply supported beam to support a characteristic load of 45 kN/m (excluding self weight) over a span of 10 m. It is given that the allowable stresses are:

$$f_{tt} = 2.9 \text{ MPa}, \; f_{tc} = -18.0 \text{ MPa}$$
$$f_{st} = 4.1 \text{ MPa}, \; f_{sc} = -24.0 \text{ MPa}$$

The loss of prestress at transfer and service can be taken as 10% and 25% of the force at jacking.

$$M_{service} - M_{self\ weight} = 45 \times 10^2/8 = 562.5 \text{ kNm at mid-span.}$$
$$\eta = 0.9/0.75 = 0.83$$
$$\eta\ f_{tt} - f_{sc} = 0.83 \times 2.9 - (-24.0) = 26.4 \text{ MPa}$$
$$f_{st} - \eta\ f_{tc} = 4.1 - 0.83 \times (-18.0) = 19.0 \text{ MPa}$$

Ignoring $M_{self\ weight}$ for an initial estimate of moduli, from equations (C16.7) and (C16.8),

$$z_t \geq 562.5 \times 10^6/26.4 = 21.31 \times 10^6 \text{ mm}^3$$
$$z_b \geq 562.5 \times 10^6/19.0 = 29.61 \times 10^6 \text{ mm}^3$$

If it is decided to choose a T-section shown in Fig. 16.10, the section properties can be expressed as functions of the two parameters (h_f/h) and (b_w/b) as follows. Table 16.1 gives the section properties.

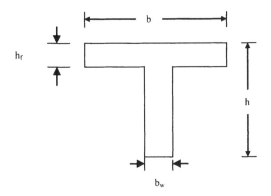

Fig. 16.10 T-section.

$$\alpha = \frac{h_f}{h}, \beta = \frac{b_w}{b}$$
$$A = bh[\alpha + \beta(1-\alpha)]$$

$$y_b = h\frac{[0.5\beta(1-\alpha)^2 + \alpha(1-0.5\alpha)]}{[\alpha + \beta(1-\alpha)]}$$

$$y_t = h\frac{0.5[\beta(1-\alpha)^2 + 0.5\alpha^2]}{[\alpha + \beta(1-\alpha)]}$$

$$I = \frac{bh^3}{12}[\alpha^3 + 12\alpha(\frac{y_t}{h} - 0.5\alpha)^2 + \beta(1-\alpha)^3 + 12\beta(1-\alpha)(0.5 - 0.5\alpha - \frac{y_b}{h})^2]$$

$$z_t = \frac{I}{y_t}, z_b = \frac{I}{y_b}$$

Choosing

$$h_f/h = 0.1, b_w/b = 0.2,$$

a variety of section sizes is possible, for example:

(i) $b = 2000$ mm, $h = 595$ say 600 mm, $b_w = 400$ mm, $h_f = 60$ mm
$$z_t = 0.076 \times 2000 \times 600^2 = 54.72 \times 10^6 \text{ mm}^3$$
$$z_b = 0.045 \times 2000 \times 600^2 = 32.40 \times 10^6 \text{ mm}^3$$

(ii) $b = 1500$ mm, $h = 686$ say 700 mm, $b_w = 300$ mm, $h_f = 70$ mm
$$z_t = 0.076 \times 1500 \times 700^2 = 55.86 \times 10^6 \text{ mm}^3$$
$$z_b = 0.045 \times 1500 \times 700^2 = 33.08 \times 10^6 \text{ mm}^3$$

Table 16.1 Section properties of T-beams

h_f/h	b_w/b	$A/(bh)$	y_b/h	y_t/h	$I/(bh^3)$	$z_t/(bh^2)$	$z_b/(bh^2)$
0.1	0.1	0.19	0.713	0.287	0.018	0.063	0.025
0.2	0.1	0.28	0.757	0.243	0.019	0.079	0.025
0.3	0.1	0.37	0.755	0.245	0.019	0.079	0.026
0.1	0.2	0.28	0.629	0.371	0.028	0.076	0.045
0.2	0.2	0.36	0.678	0.322	0.031	0.098	0.046
0.3	0.2	0.44	0.691	0.309	0.032	0.103	0.046
0.1	0.3	0.37	0.585	0.415	0.037	0.088	0.062
0.2	0.3	0.44	0.627	0.373	0.041	0.109	0.065
0.3	0.3	0.51	0.644	0.356	0.042	0.117	0.065
0.1	0.4	0.46	0.559	0.441	0.044	0.100	0.079
0.2	0.4	0.52	0.592	0.408	0.049	0.119	0.082
0.3	0.4	0.58	0.609	0.391	0.050	0.127	0.082

If a double T-section (Fig. 16.9) is desired, the web width b_w can be shared between two webs with the width of each web equal to $0.5b_w$.

Having chosen a section, its self weight can be calculated. For example for the section:

$$b = 1500 \text{ mm}, h = 700 \text{ mm}, b_w = 300 \text{ mm}, h_f = 70 \text{ mm}$$
$$A = 294.0 \times 10^3 \text{ mm}^2$$
$$\text{Self weight} = 7.056 \text{ kN/m}$$
$$M_{\text{self weight}} = 88.2 \text{ kNm}$$
$$(1 - \eta) M_{\text{self weight}} = 15.00 \text{ kNm}$$

Using the self weight moment, the revised required section moduli become
$$z_t \geq (562.5 + 15.00) \times 10^6/26.4 = 21.88 \times 10^6 \text{ mm}^3$$
$$z_b \geq (562.5 + 15.00) \times 10^6/19.0 = 30.40 \times 10^6 \text{ mm}^3$$

The section modulus z_b of the chosen section is $33.08 \times 10^6 \text{ mm}^3$ which is only slightly greater than the required value of $30.40 \times 10^6 \text{ mm}^3$. The chosen section is adequate.

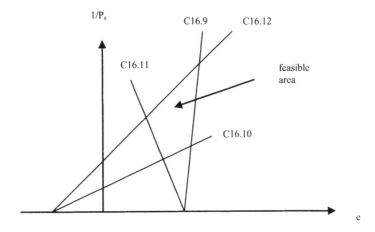

Fig. 16.11 Magnel diagram (the numbers against the lines correspond to equation numbers).

16.8.2 Choice of Prestress and Eccentricity

Having chosen a section, the next step is to choose the required value of prestress and eccentricity such that none of the stress criteria are violated. By dividing throughout by $1/P_s$, equations (C16.3) to (C16.6) can be rewritten as follows

$$-\frac{1}{A}+\frac{e}{z_t} \le \eta \langle \frac{M_{self\,weight.}}{z_t}+f_{tt} \rangle \frac{1}{P_s} \tag{C16.9}$$

$$-\frac{1}{A}-\frac{e}{z_b} \ge \eta \langle -\frac{M_{self\,weight.}}{z_b}+f_{tc} \rangle \frac{1}{P_s} \tag{C16.10}$$

$$-\frac{1}{A}+\frac{e}{z_t} \ge \langle \frac{M_{service}}{z_t}+f_{sc} \rangle \frac{1}{P_s} \tag{C16.11}$$

$$-\frac{1}{A}-\frac{e}{z_b} \le \langle -\frac{M_{service}}{z_b}+f_{st} \rangle \frac{1}{P_s} \tag{C16.12}$$

If the inequality signs are replaced by an equality sign, a plot of e versus $1/P_s$ of each equation is a straight line and the plot of all the four equations encloses a quadrilateral as shown in Fig. 16.11. Any choice of e and P_s inside the quadrilateral satisfies all the four stress criteria. This plot is known as a Magnel diagram.

16.8.2.1 Example of Construction of Magnel Diagram

Fig. 16.9 shows a pretensioned symmetric double T-beam. It is used as a simply supported beam to support a total characteristic load (excluding self weight) of

45 kN/m over a span of 10 m. Construct the Magnel diagram for the mid-span section using the following data. Assume loss at transfer and serviceability limit state as respectively 10% and 25% of the jacking force.

(i) Section properties:

$$A= 44 \times 10^4 \text{ mm}^2$$
$$z_t = 92.12 \times 10^6 \text{ mm}^3$$
$$z_b = 49.08 \times 10^6 \text{ mm}^3$$
$$1/A = 227.27 \times 10^{-8},$$
$$1/z_b = 2.035 \times 10^{-8},$$
$$1/z_t = 1.086 \times 10^{-8}$$

(ii) Moments at mid-span:

$$M_{\text{self weight}} = 137.5 \text{ kNm}$$
$$M_{\text{self weight}}/z_t = 137.5 \times 10^6/ (92.12 \times 10^6) = 1.49 \text{ MPa}$$
$$M_{\text{self weight}}/z_b = 137.5 \times 10^6/ (49.08 \times 10^6) = 2.80 \text{ MPa}$$

$$M_{\text{service}} = 694.5 \text{ kNm}$$
$$M_{\text{service}}/z_t = 700.0 \times 10^6/ (92.12 \times 10^6) = 7.60 \text{ MPa}$$
$$M_{\text{service}}/z_b = 700.0 \times 10^6/ (49.08 \times 10^6) = 14.26 \text{ MPa}$$

(iii) Permissible stresses:

$$f_{tt} = 2.9 \text{ MPa}, \ f_{tc} = -18.0 \text{ MPa}, \ f_{st} = 4.1 \text{ MPa}, \ f_{sc} = -24.0 \text{ MPa}$$

(iv) Prestress losses:

$$\eta = 0.75/0.9 = 0.83$$

(v) Solution:
Substituting the above values in equations (C16.9) to (C16.11), the following four linear equations are obtained.

$$-227.27 + 1.086 \ e = 0.83(1.49 + 2.9) \ 10^8/P_s = 3.65 \ (10^8/P_s)$$
$$-227.27 - 2.035 \ e = 0.83(-2.80 - 18.0) \ 10^8/P_s = -17.26 \ (10^8/P_s)$$
$$-227.27 + 1.086 \ e = (7.60 - 24.0) \ 10^8/P_s = -16.40 \ (10^8/P_s)$$
$$-227.27 - 2.035 \ e = (-14.26 + 4.1) \ 10^8/P_s = -10.16 \ (10^8/P_s)$$

Fig. 16.12 shows the Magnel diagram.

16.8.2.2 Example of Choice of Prestress and Eccentricity

Fig. 16.12 shows the feasible region. Any combination of P_s and e within the feasible region will satisfy all the four stress criteria. Unfortunately practical limitations of cover, etc., reduce the extent of the feasible area. In the above example $y_b = 473$ mm. Assuming that cables require a cover of approximately 50 mm, the maximum eccentricity e_{max} allowable is only ($y_b - 50$) = 423 mm. This limitation is shown in Fig. 16.12 by the vertical line.

In choosing a value of P_s and e two important points to keep in mind are:

- Choose as small a value of P_s (i.e., as large a value of $10^8/P_s$) and as large a value of e as possible. This will keep the costs down. In this example this is approximately $10^8/P_s \approx 70$ and e ≈ 420 mm.
- It is not advisable to work right at the edge of the feasible region as it does not allow for any flexibility in the arrangement of cables in the cross section.

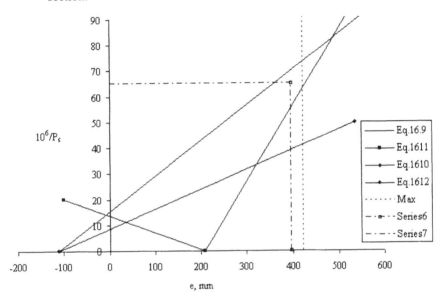

Fig. 16.12 Magnel diagram for mid-span.

(i) Determination of number of cables (strands, tendons) required

Note: Especially in bridge construction it is common to group many 7-wire strands into a single cable in a duct with an appropriate anchorage. A value of P_s can be chosen which is near the top right hand part of the feasible region. If $10^8/P_s \approx 65$, $P_s \approx 1539$ kN and $P_{jack} = P_s/0.75 \approx 2051$kN.

Table 16.2 Properties of 7-wire strands

Properties	Strand type			
	N	S	N	S
Nominal diameter, mm	13	13	15	15
Nominal area, mm²	93	100	140	150
Tensile strength, MPa	1860	1860	1860	1860
Minimum breaking load, kN	173	186	260	279
Minimum relaxation	2.5	2.5	2.5	2.5

Table 16.3 Properties of VSL cables with 7-wire strands

No. of strands	Anchor unit	Corrugated steel strip sheath: Range of Internal dia./ External dia.	Range of Shift mm
1	6−1	25/30	5
2	6−2	40/45	9
3	6−3	40/45	6
4	6−4	45/50	7
5 to 7	6−7	50/57 to 55/62	9 to 7
8 to 12	6−12	65/72 to 75−82	9 to 11
13 to 15	6−15	80/87	13 to 10
16 to 19	6−19	85/92 to 90/97	25 to 18
20 to 22	6−22	100/107	14 to 22
23 to 27	6−27	100/107 to 110/117	13 to 16
28 to 31	6−31	110/117 to 120/127	15 to 19
32 to 37	6−37	120/127 to 130/137	16 to 22
38 to 43	6−43	140/147	21 to 25
44 to 55	6−55	150/157 to 160/167	23 to 29

In practice for posttensioned construction of bridges, strands are grouped into cables with suitable anchorages. The variety of cables and the associated anchorages depend on the manufacturer. For example VSL anchorages are available for cables containing 1, 2, 3, 4, 7, 12, 15, 19, 22, 27, 31, 37, 43 and 55 strands. The CCL anchorages allow for cables of 4,7,12, 19 and 22 strands of 13 mm diameter or 4, 7, 12, 19, 27 and 37 strands of 15 mm diameter. The cables are placed inside ducts before concreting and the ducts are grouted afterwards to prevent corrosion of the strands. During tensioning the strands in a cable tend to bunch together onto the tension side of the duct. This alters the position of the centroid of the tensile force with respect to the centroid of the duct as shown in Fig. 5.3. The shift depends primarily on the diameter of the duct and the number of strands in the cable. Manufacturers provide data on this. Table 16.3 shows the data for VSL cables.

Having calculated the value of P_{jack}, the next stage is to choose the type and number of cables required. Table 16.2 gives the strengths of various types of 7-wire strands. If 7-wire drawn strand of 15.0 mm nominal diameter is chosen, f_{pk} = 1860 MPa, the net cross sectional area = 150 mm^2 and breaking load is 279 kN. Jacking stress should not normally exceed 77% of the breaking load of the tendon but may be increased to 80.8% provided additional consideration is given to safety (see section 16.6.1).

Force per cable at jacking $= 0.77 \times 279 = 214.8$ kN
Number of cables required $= P_{jack}/214.8 = 2051/214.8 \approx 9.6$

For symmetry, choose 10 cables. Force per cable $= 2051/10 = 205.1$ kN.
Stress in cable $= 205.1 \times 10^3/150 = 1367$ MPa, $= 0.735\ f_{pk}$.

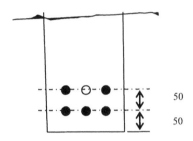

Fig. 16.13 Arrangement of cables in the web.

(ii) Determination of eccentricity

Assuming that cables can be placed in the webs in horizontal layers at 50 mm intervals vertically with three cables per layer, 10 cables can be accommodated with five cables in each web as shown in Fig. 16.13.

The resultant eccentricity

$$e = y_b - (50 \times 3 + 100 \times 2)/5 = 473 - 70 = 403 \text{ mm}$$

The point corresponding to

$$10^8/P_s = 65,\ e = 403 \text{ mm}$$

is inside the feasible region. Therefore the arrangement and force in the cables are satisfactory.

16.8.2.3 *Example of Debonding*

If it is decided to debond some cables towards the support, then a Magnel diagram has to be drawn for the support section as well. At a support section, the critical condition is at transfer. Conditions at service are not critical because of the long - term losses in the prestress. Since there are no moments acting at supports, the two critical equations are:

$$-227.27 + 1.086\ e = 0.83(0 + 2.9)\ 10^8/P_s = 2.41\ (10^8/P_s)$$
$$-227.27 - 2.035\ e = 0.83(0 - 18.0)\ 10^8/P_s = -14.94\ (10^8/P_s)$$

Fig. 16.14 shows the Magnel diagram at the support. The feasible area is *not* a closed polygon.

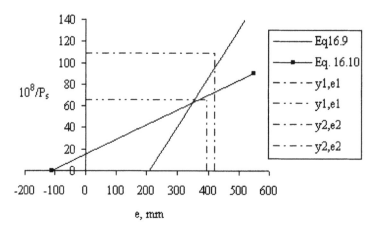

Fig. 16.14 Magnel diagram at support.

As shown in Fig. 16.14, the point corresponding to $10^8/P_s = 65$, $e = 403$ mm is outside the feasible region and cannot be accepted. In order to bring the point inside the feasible region, preserving symmetry of cable position, remove two cables from each web. The number of strands is reduced from ten to six. The corresponding $10^8/P_s$ is given by

$$10^8/P_s = 65 \times (10/6) = 108.3$$
$$e = 473 - 50 = 423 \text{ mm}$$

As shown in Fig. 16.14, the point (423, 108) lies inside the feasible region and can be accepted. Thus two cables in each web need to be debonded towards the support.

16.9 COMPOSITE BEAMS

In a very common form of bridge construction, precast pretensioned beams are erected first and the in-situ concrete is cast on top of them using formwork which is supported on the precast beams. The formwork is just left in place. This type of formwork is called sacrificial formwork. Beams are placed at approximately 1 m apart. Once the in-situ concrete has hardened, the beam and the deck slab act as a composite structure. Fig. 16.15 shows a typical bridge superstructure using inverted T-beams.

In this type of beam, the weight of the slab and associated permanent formwork is carried wholly by the precast beam. However once the slab hardens, then all subsequent loads acting on the slab will be resisted by the pretensioned beam acting in conjunction with the cast-in-situ slab. The cast-in-situ slab acts as the compression flange of the composite I-beam.

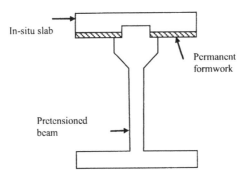

Fig. 16.15 Composite beam.

Since the object is to calculate the value of P_s and e so that the stresses in the precast section are within permissible limits, the stresses are calculated in the precast section only.

(i) Transfer stage: Prestress acts on the precast beam. The only external moment is due to the self weight of the beam. The governing equations are (C16.1) and (C16.2) (repeated here for convenience).

$$f_{top} = [-\frac{P_t}{A} + \frac{P_t e}{z_t} - \frac{M_{self\,weight.}}{z_t}] \le f_{tt} \qquad (C16.1)$$

$$f_{bottom} = [-\frac{P_t}{A} - \frac{P_t e}{z_b} + \frac{M_{self\,weight.}}{z_b}] \ge f_{tc} \qquad (C16.2)$$

(ii) Serviceability limit: The weights of slab and precast beam are supported by the precast beam. The live loads are supported by composite beam. In addition to the stresses caused by the loads, one needs to include the stresses caused by the shrinkage of the in-situ slab. Stresses due to shrinkage occur because as the cast-in-situ slab dries, it shrinks and forces the precast beam to bend.

$$f_{top} = [-\frac{P_s}{A} + \frac{P_s e}{z_t} - \frac{M_{Dead}}{z_t} - \frac{M_{live}}{Comp.z_{top\,of\,precast}} - Shrink\;stress] \ge f_{sc} \;(C16.13)$$

$$f_{bottom} = [-\frac{P_s}{A} - \frac{P_s e}{z_b} + \frac{M_{Dead}}{z_b} + \frac{M_{live}}{Comp.z_b} + Shrink\;stress] \le f_{st} \qquad (C16.14)$$

The stresses due to shrinkage of the slab can be calculated as follows as shown in Fig. 16.16.

Step 1: Assume that the slab is disconnected from the precast beam.

Step 2: Let the free shrinkage of the slab be ε_{sh}. If the area of cross section of the slab is A_{slab} and the Young's modulus of the concrete in the slab allowing for creep is E_c, the force **tensile** F required to restrain this shrinkage is given by

$$F = \varepsilon_{sh} \cdot \times E_c \times A_{slab}$$

This force acts only on the slab.

Step 3: As this force does not exist in the final composite beam, the stress induced in the composite beam is given by a **compressive** force F is applied at the centre of the slab.

$$\sigma_{top} = -\frac{F}{A_{comp}} - \frac{Fa}{Comp.z_{top\ of\ precast}}$$

$$\sigma_{bottom} = -\frac{F}{A_{comp}} + \frac{Fa}{Comp.z_b}$$

where a is the eccentricity of the force F with respect the centroid of the composite beam.

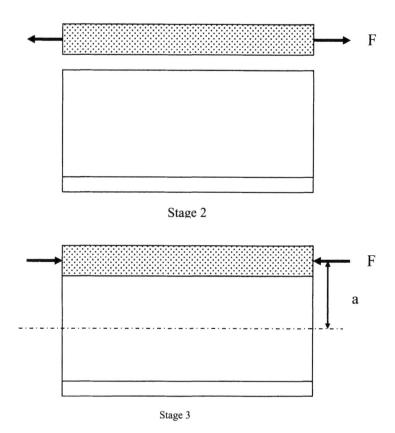

Stage 2

Stage 3

Fig. 16.16 Shrinkage stresses in composite beam.

16.9.1 Magnel Equations for a Composite Beam

Fig. 16.17 shows a composite beam. The precast pretensioned inverted T-beam is made composite with a cast-in-situ slab acting as the top flange of the composite beam. It is used as a simply supported beam over a span of 24 m.

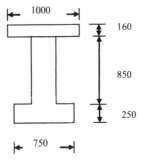

Fig. 16.17 Composite beam section.

The section properties on precast and composite beam are as follows.

(a) Precast beam:
$$\text{Area} = 4.425 \times 10^5 \text{ mm}^2$$
$$y_b = 442 \text{ mm}$$
$$y_t = 658 \text{ mm}$$
$$I = 4.90 \times 10^{10} \text{ mm}^4$$
$$z_b = 111.0 \times 10^6 \text{ mm}^3$$
$$z_t = 74.5 \times 10^6 \text{ mm}^3$$
$$\text{self weight} = 11.06 \text{ kN/m}$$

(b) Composite beam:
$$A_{composite} = 6.025 \times 10^5 \text{ mm}^2$$
$$y_{b \text{ Composite}} = 638 \text{ mm},$$
$$y_t \text{ to top of precast} = 1100 - 638 = 462 \text{ mm}$$
$$I_{Composite} = 11.33 \times 10^{10} \text{ mm}^4$$
$$z_{b \text{ Composite}} = 177.6 \times 10^6 \text{ mm}^3$$
$$\text{Composite } z_{t \text{ to top of precast}} = 245.2 \times 10^6 \text{ mm}^3$$
$$\text{self weight} = 15.06 \text{ kN/m}$$

(c) Loads:
$$\text{Self weight of precast} = 11.06 \text{ kN/m}$$
$$q_{dead} = \text{Weight of (Composite beam + permanent formwork)}$$
$$q_{dead} = 15.06 + \text{say } 1.2 = 16.26 \text{ kN/m}$$
$$q_{live} = 18.2 \text{ kN/m}$$

(d) Moments and stresses at mid-span:

Self weight:

$$q_{\text{Self weight}} = 11.06 \text{ kN/m}$$
$$M_{\text{self weight}} = 11.06 \times 24^2/8 = 796.32 \text{ kNm}$$
$$\frac{M_{\text{Self weight}}}{z_b} = \frac{796.32 \times 10^6}{111.0 \times 10^6} = 7.2 \text{ MPa}$$
$$\frac{M_{\text{Self weight}}}{z_t} = \frac{796.32 \times 10^6}{74.5 \times 10^6} = 10.7 \text{ MPa}$$

Total dead load:

$$q_{\text{dead}} = 16.26 \text{ kN/m}$$
$$M_{\text{Dead}} = 16.26 \times 24^2/8 = 1170.72 \text{ kNm}$$
$$\frac{M_{\text{Dead}}}{z_b} = \frac{1170.72 \times 10^6}{111.0 \times 10^6} = 10.6 \text{ MPa}$$
$$\frac{M_{\text{Dead}}}{z_t} = \frac{1170.76 \times 10^6}{74.5 \times 10^6} = 15.7 \text{ MPa}$$

Live load:

$$q_{\text{Live}} = 18.2 \text{ kN/m}$$
$$M_{\text{Live}} = 18.2 \times 24^2/8 = 1310.4 \text{ kNm}$$
$$\frac{M_{\text{Live}}}{\text{Comp} z_b} = \frac{1310.4 \times 10^6}{177.6 \times 10^6} = 7.4 \text{ MPa}$$
$$\frac{M_{\text{Live}}}{\text{Comp} z_{t \text{ to top of precast}}} = \frac{1310.4 \times 10^6}{245.0 \times 10^6} = 5.3 \text{ MPa}$$

Shrinkage stresses: Assume:
$$\text{Top} = -1.7 \text{ MPa}$$
$$\text{Bottom} = 0.6 \text{ MPa}$$

(e) Magnel Equations: Magnel equations consider the stresses in precast section only. Using precast beam properties,

$$1/A = 1/ (4.425 \times 10^5) = 226.0 \times 10^{-8}$$
$$1/z_b = 1/ (111.0 \times 10^6) = 0.90 \times 10^{-8},$$
$$1/z_t = 1/ (74.5 \times 10^6) = 1.34 \times 10^{-8}$$

(f) Losses
Take 10% loss at transfer and 25% long-term.
$$\eta = (1 - 0.1)/ (1 - 0.25) = 0.83$$

(g) Permissible stresses: Assume

$$f_{tt} = 2.9 \text{ MPa}, \; f_{tc} = -18.0 \text{ MPa}, \; f_{st} = 4.1 \text{ MPa}, \; f_{sc} = -24.0 \text{ MPa}$$

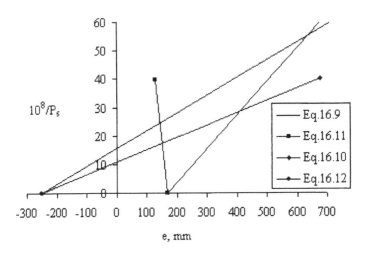

Fig. 16.18 Magnel diagram for the composite beam section.

(h) Stress conditions and Magnel equations:

At transfer:

Top: (Equation C16.1)

$$(-226.0 \times 10^{-8} + 1.34e \times 10^{-8})P_s - 0.83 \times 10.7 \leq 0.83 \times 3.0$$
$$-226.0 + 1.34 \, e \leq 11.37 \times (10^8/P_s)$$

Bottom: (Equation C16.2)

$$(-226.0 \times 10^{-8} - 0.90e \times 10^{-8})P_s + 0.83 \times 7.2 \geq 0.83 \times (-18.0)$$
$$-226.0 - 0.90 \, e \geq -20.92 \times (10^8/P_s)$$

At Service:

Top: (Equation C16.13)

$$(-226.0 \times 10^{-8} + 1.34e \times 10^{-8})P_s - 15.7 - 5.3 - 1.7 \geq -24.0$$
$$-226.0 + 1.34 \, e \geq 1.3 \times (10^8/P_{sc})$$

Bottom: (Equation C16.14)

$$(-226.0 \times 10^{-8} - 0.90e \times 10^{-8})P_s + 10.6 + 7.4 + 0.6 \leq 4.1$$
$$-226.0 - 0.90 \, e \leq -14.5 \times (10^8/P_s)$$

The Magnel diagram for the above set of four equations is shown in Fig. 16.18.

16.10 POSTTENSIONED BEAMS: CABLE ZONE

In pretensioned beams, the strands are straight (except when cables are deflected) and due to debonding, prestress and eccentricity vary along the span in a stepwise manner. In posttensioned beams, cables take a curved profile. Thus the *eccentricity can vary* along the span but the *prestress remains constant* (if losses in prestress along the span can be ignored). The permissible eccentricity at any section can be calculated by rearranging equations (C16.9) to (C16.12) as follows.

$$e \le \frac{z_t}{A} + \eta(M_{self\,weight.} + z_t f_{tt})\frac{1}{P_s} \tag{C16.15}$$

$$e \le -\frac{z_b}{A} + \eta(M_{self\,weight.} - z_b f_{tc})\frac{1}{P_s} \tag{C16.16}$$

$$e \ge \frac{z_t}{A} + (M_{service} + z_t f_{sc})\frac{1}{P_s} \tag{C16.17}$$

$$e \ge -\frac{z_b}{A} + (M_{service} - z_b f_{st})\frac{1}{P_s} \tag{C16.18}$$

16.10.1 Example of a Posttensioned Beam

Fig. 16.9 shows a posttensioned symmetric double T-beam. It is used as a simply supported beam to support a total load (excluding self weight) of 45 kN/m over a span of 10 m. From the Magnel diagram at mid-span, the value of $10^8/P_s = 65$, giving $P_s = 1538.5$ kN. Assume loss at transfer and serviceability limit state as respectively 10% and 25% of the jacking force.

(i) Section properties:
$$A = 44 \times 10^4 \text{ mm}^2$$
$$z_t = 92.12 \times 10^6 \text{ mm}$$
$$z_b = 49.08 \times 10^6 \text{ mm}^3$$
$$z_t/A = 209.36 \text{ mm}$$
$$z_b/A = 111.56 \text{ mm}$$

(ii) Moments (at position x from left hand support):
$$S_{self\,weight} = 11.0 \text{ kN/m}$$
$$M_{self\,weight} = 55.0\,x - 5.5\,x^2$$
$$S_{service} = 56.0 \text{ kN/m}$$
$$M_{service} = 280.0\,x - 28.0\,x^2$$

(iii) Permissible stresses:
$$f_{tt} = 2.9 \text{ MPa},\ f_{tc} = -18.0 \text{ MPa},\ f_{st} = 4.1 \text{ MPa},\ f_{sc} = -24.0 \text{ MPa}$$

(iv) Loss of prestress:

$$\eta = 0.75/0.9 = 0.83$$

(v) Prestress:

$$P_s = 1538.5 \text{ kN}$$

(vi) Limits on eccentricity:

Substituting the above values in equations (C16.15) to (C16.18)

$$e \leq 209.36 + 0.83 \times [(55.0x - 5.5x^2) \times 10^6 + 92.12 \times 10^6 \times 2.9]/(1538.5 \times 10^3)$$

$$e \leq -111.56 + 0.83 \times [(55.0x - 5.5x^2) \times 10^6 - 49.08 \times 10^6 \times (-18.0)]/(1538.5 \times 10^3)$$

$$e \geq 209.36 + [(280.0x - 28.0x^2) \times 10^6 + 92.12 \times 10^6 \times (-24.0)]/(1538.5 \times 10^3)$$

$$e \geq -111.56 + [(280.0x - 28.0x^2) \times 10^6 - 49.08 \times 10^6 \times 4.1]/(1538.5 \times 10^3)$$

Simplifying

$$e \leq \{353.48 + 29.67 x - 2.967 x^2\}$$
$$e \leq \{365.04 + 29.67 x - 2.967 x^2\}$$
$$e \geq \{-1227.68 + 182.0 x - 18.2 x^2\}$$
$$e \geq \{-242.36 + 182.0 x - 18.2 x^2\}$$

Noting that eccentricity is positive below the neutral axis, lower limits for e (i.e., eccentricity $\geq$) are governed by equations (C16.15) and (C16.16). Clearly equation (C16.15) gives a lower value than equation (C16.16). Similarly the upper limits for e (i.e. eccentricity $\leq$) are governed by equations (C16.17) and (C16.18). Clearly equation (C16.18) gives a larger value than equation (C16.17). Thus the feasible cable zone lies between the curves corresponding to equations (C16.15) and (C16.18). Cables placed inside the feasible zone thus satisfy all the stress criteria. Assuming a minimum cover to the cables of 50 mm, the maximum and minimum values of e attainable are equal to

$$y_b - 50 = 423 \text{ mm}$$
$$y_t - 50 = 202 \text{ mm.}$$

Fig. 16.19 shows the feasible region. The range of e at the support and mid-span are:

Ends: $-242(-202) \leq e \leq 353$, gap = 555mm
Mid-span: $212 \leq e \leq 428(423)$, gap = 211 mm

Figures in brackets show the practical values.

As can be seen, the feasible region has a narrower width at mid-span (211 mm) than at the supports (555 mm). The cables can be raised towards the supports providing an upward force which helps in counteracting the applied shear force.

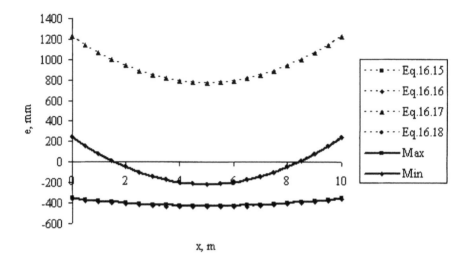

Fig. 16.19 Feasible cable zone.

16.11 ULTIMATE MOMENT CAPACITY

One aspect of design of prestressed sections which is different from the procedure used in the case of a reinforced concrete section is that the designs are carried out for SLS and the designed section is ***checked*** to ensure that ULS conditions are also satisfied. The calculations for determining the ultimate moment capacity are similar to the ultimate moment capacity calculation in the case of reinforced concrete section as explained in Chapter 4, section 4.6.2. As in the case of reinforced concrete sections, the compressive stress distribution in concrete is assumed to be that given by rectangular stress block assumption with the maximum stress equal to f_{cd} and the maximum strain equal to 0.0035. The main difference from the calculations for a reinforced concrete section is in calculating the strains in steel. In the case of reinforced concrete sections, the strain in steel is due to bending. However, in the case of prestressed concrete sections, because the cables are pretensioned before the application of load, the total strain in the cable is the sum of the prestrain due to prestress $P_{service}$ and the strain due to applied bending.

16.11.1 Example of Ultimate Moment Capacity Calculation

Fig. 16.20 shows the cross section of a precast pretensioned inverted T-beam made composite with a cast-in-situ slab. The beam is used to carry a total *factored* uniformly distributed dead load of 20 kN/m and 30 kN/m live load over a simply supported span of 24 m. Calculate the ultimate moment capacity of the section.

(a) Specification

The properties of the beam are as follows. Total prestressing force P_s at service is 3712 kN applied at an eccentricity of 283 mm. The prestress is applied by 32 number 15 mm diameter 7-wire standard strands with an ultimate breaking load of 279 kN.

The 32 strands are positioned as follows:

- 10 cables at 60 mm from the soffit
- 14 cables at 110 mm from the soffit
- 6 cables at 160 mm from the soffit
- 2 cables at 1000 mm from the soffit

The cross sectional area A_{ps} of cable

$$A_{ps} = 150.0 \text{ mm}^2$$
$$f_{ck} \text{ for precast beam} = 40 \text{ MPa}$$
$$f_{ck} \text{ for in-situ slab} = 30 \text{ MPa}$$

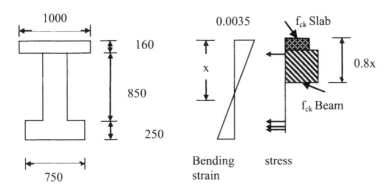

Fig. 16.20 Composite beam.

(b) Stress–strain relationship

The stress–strain relationship for prestressing steel is given in clause 3.3.6, Fig. 3.10 of the code and is shown in Fig. 16.21.

Two options are given.

1. A bilinear curve with an initial straight line with a modulus of elasticity E_p equal to 195 GPa and inclined branch with a strain limit of $\varepsilon_{ud} = 0.02$.

2. A bilinear curve with an initial straight line with a modulus of elasticity E_p equal to 195 GPa and a horizontal branch with no strain limit. $f_{p\,0.1\,k} = 0.9\,f_{pk}$, ε_{uk} = characteristic strain of prestressing steel at maximum load, $\gamma_s = 1.15$.

It is simple and convenient to use the option (2) for design.

$f_{pk} = 1860$ MPa, $f_{p0.1\,k} = 0.9 \times 1860 = 1674$ MPa, $f_{pd} = 1674/1.15 = 1456$ MPa

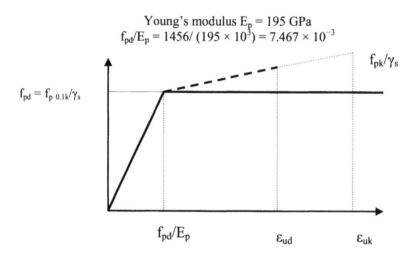

Fig. 16.21 Stress–strain relationship for prestressing steel.

(c) Prestrain calculation

f_{pe} = prestress in the cables = P_s/Total area of cables
$$= 3712 \times 10^3 / (32 \times 150) = 773 \text{ MPa}$$
ε_{pe} = prestrain in the cables = $f_{pe} / E = 773 / (195 \times 10^3) = 3.96 \times 10^{-3}$

Table 16.4 Data for cables

Layer	c = depth from soffit, mm	No. in layer	a =1260 – c, mm
1	60	10	1200
2	110	14	1150
3	160	6	1100
4	1000	2	260

(d) Stress and strain in cables

For a given depth x of neutral axis, at a depth a from the compression face, strain ε_b due to bending
$$\varepsilon_b = 0.0035 \times (a - x)/x = 3.5 \times 10^{-3} \, (a/x - 1.0)$$
Total depth of composite beam = 1100 + 160 = 1260 mm
$$a = 1260 - \text{distance to the layer from soffit}$$
Table 16.4 summarises the data for all the cables.

Total strain ε at a depth a from the compression face
$$\varepsilon = \varepsilon_{pe} + \varepsilon_b = \{3.96 + 3.5(a/x - 1)\} \times 10^{-3}$$
From Fig. 16.19, for a given strain ε, the corresponding stress σ is given by the following equations.
$$0 < \varepsilon \le 7.467 \times 10^{-3}, \sigma = \varepsilon \times E_p$$

$$\varepsilon > 7.467 \times 10^{-3}, \sigma = 1456 \text{ MPa}$$

(e) Compressive stress in concrete
Using the rectangular stress block, the depth of the stress block is 0.8x. The compressive stress in concrete is f_{cd}.

(f) Determination of neutral axis depth x
The determination of the neutral axis depth is a trial and error process. The steps involved are as follows.

(i) Assume a value for neutral axis depth, x

(ii) Calculate at different levels a, the bending strain in ε_b in the cables
$$\varepsilon_b = 3.5 \times 10^{-3} (a/x - 1.0)$$

(iii) Calculate the total strain $\varepsilon = \varepsilon_{pe} + \varepsilon_b$, $\varepsilon_{pe} = 4.28 \times 10^{-3}$

(iv) Calculate the stress σ in the cables

(v) Calculate the total tensile force F in each layer
$$F = \sigma \times (\text{Area of cable} = 150 \text{ mm}^2) \times \text{No. of cables in the layer}$$

(vi) Total tensile force $T = \Sigma F$

(vi) Calculate the total compressive force C:

   (a) If $x \leq$ depth of slab (= 160 mm), $C = f_{cd \text{ Slab}} \times 1000 \times (0.8 \, x)$

   (b) If $x >$ (depth of slab = 160 mm), $C_{Slab} = f_{cd \text{ Slab}} \times (1000 \times 160)$,
   $C_{Beam} = f_{cd \text{ beam}} \times (0.8x - 160) \times 300$, $C = C_{Slab} + C_{Beam}$
$f_{cd, \text{ slab}} = 30/1.5 = 20 \text{ MPa}$, $f_{cd, \text{ beam}} = 40/1.5 = 26.7 \text{ MPa}$

(vii) Check whether $T = C$. If not go back to step (i) and repeat. If $T > C$ then choose a larger value of x and vice versa. Larger value of x increases the compression area and also reduces the bending strain in steel.

(g) Trial 1
Assume $x = 600$ mm.
Table 16.5 summarizes the calculation of forces in layers.
$F = \sigma \times$ No. of cables in the layer $\times 150 \times 10^{-3}$ kN
$$C_{slab} = 20 \times 1000 \times 160 \times 10^{-3} = 3200 \text{ kN}$$
$$C_{beam} = 26.7 \times (0.8x - 160) \times 300 \times 10^{-3} = 2560 \text{ kN}$$
$$C = C_{slab} + C_{beam} = 3200 + 2560 = 5760 \text{ kN}$$
$$T - C = 52 \text{ kN}$$
Since $T > C$, increase the value of x and repeat.

Table 16.5 Force calculation in the cables for Trial 1

Layer	$\varepsilon_b \times 10^3$	$(\varepsilon = \varepsilon_{pe} + \varepsilon_b) \times 10^3$	σ, MPa	F, kN
1	3.50	7.46	1266	1899
2	3.21	7.17	1266	2658
3	2.92	6.88	1266	1139
4	−1.98	1.98	385	116
				$T = \Sigma F = 5812$

(h) Trial 2

Assume $x = 700$ mm.

Table 16.6 summarises the calculation of forces in layers.

Table 16.6 Force calculation in the cables for Trial 2

Layer	$\varepsilon_b \times 10^3$	$(\varepsilon = \varepsilon_{pe} + \varepsilon_b) \times 10^3$	σ, MPa	F, kN
1	2.5	6.46	1260	1890
2	2.25	6.21	1211	2543
3	2.0	5.96	1162	1046
4	−2.2	1.76	343	103
				$T = \Sigma F = 5582$

$$C_{slab} = 20 \times 1000 \times 160 \times 10^{-3} = 3200 \text{ kN}$$
$$C_{beam} = 26.7 \times (0.8x - 160) \times 300 \times 10^{-3} = 3200 \text{ kN}$$
$$C = C_{slab} + C_{beam} = 3200 + 3200 = 6400 \text{ kN}$$
$$T - C = -819 \text{ kN}$$

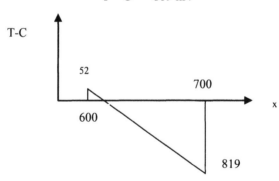

Fig. 16.22 Linear interpolation.

(i) Linear interpolation

Since there are two values of neutral axis depth for which values of $(T - C)$ are known, linear interpolation between $x = 600$ and $x = 700$ can be done to determine the value of x for which $T - C = 0$.

From Fig. 16.22,

$$x = 600 + \frac{(700 - 600)}{\{52 - (-819)\}} 52 = 606$$

(j) Calculation of tensile and compressive forces at $x = 606$ mm

Table 16.7 shows the force calculation in the cables.

Table 16.7 Force calculation in cables for interpolated value of x

Layer	$\varepsilon_b \times 10^3$	$(\varepsilon = \varepsilon_{pe} + \varepsilon_b) \times 10^3$	σ MPa	F, kN
1	3.43	7.39	1266	1899
2	3.14	7.10	1266	2658
3	2.85	6.81	1266	1139
4	-2.00	1.96	383	115
				$T = \Sigma F = 5811$

$$C_{slab} = 20 \times 1000 \times 160 \times 10^{-3} = 3200 \text{ kN}$$
$$C_{beam} = 26.7 \times (0.8x - 160) \times 300 \times 10^{-3} = 2598 \text{ kN}$$
$$C = C_{slab} + C_{beam} = 3200 + 2598 = 5798 \text{ kN}$$
$$T - C = 12 \text{ kN which is small enough to be ignored.}$$

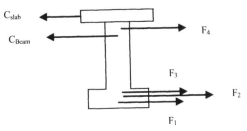

Fig. 16.23 Forces in the cross section.

(k) Calculation of ultimate moment M_u

Since the total T and C form a couple, the ultimate moment is calculated by taking moments about any convenient point of the tensile and compressive forces. Taking moments about the top of the cross section of the forces shown in Fig. 16.23, the ultimate moment capacity is equal to 4897.53 kNm. Detailed calculations are shown in Table 16.8.

Table 16.8 Calculation of ultimate moment capacity, M_u

	Force, kN	Lever arm = Distance from top, m	Moment, kNm
C_{slab}	-3200	$(160/2 = 80) \times 10^{-3}$	-256
C_{Beam}	-2598	$\{160+(0.8x - 160)/2 = 322.4\} \times 10^{-3}$	-838
F_1	1899	1200×10^{-3}	2279
F_2	2658	1150×10^{-3}	3057
F_3	1139	1100×10^{-3}	1252
F_4	115	260×10^{-3}	30
			Σ 5524

The applied bending moment at ultimate load is

$$M \text{ at ULS} = 20 \times 24^2/8 \text{ due to dead load} + 30 \times 24^2/8 \text{ due to live load}$$
$$= 3600 \text{ kNm} < 5524 \text{ kNm}$$

The applied moment is less than the ultimate moment capacity M_u. The beam has sufficient capacity to resist the applied bending moment at ULS.
Calculations such as this are best done using spreadsheets.

16.12 SHEAR CAPACITY OF A SECTION WITHOUT SHEAR REINFORCEMENT AND UNCRACKED IN FLEXURE

Sections are said to be uncracked in flexure if the flexural tensile stress is smaller than $f_{ctk0.05}/\gamma_c$. In sections which are uncracked in flexure, it is necessary to limit the maximum principal tensile stress in the web to a value f_{ctd}.

$$f_{ctd} = f_{ctm}/\gamma_c, \gamma_c = 1.5$$
$$f_{ctm} = 0.30 \times f_{ck}^{0.667}, f_{ck} \le 50 \text{ MPa}$$
$$= 2.12 \times \ell n \, [1.8 + 0.1 f_{ck}], f_{ck} > 50 \text{ MPa}$$
$$f_{ctk0.05} = 0.7 \, f_{ctm}$$

If f_{cp} is the compressive stress due to prestress at the neutral axis, then the state of stress at the neutral axis is as shown in Fig. 16.24. The maximum principal tensile stress equal to f_{ctd} for the biaxial state of stress shown in Fig. 16.24 is given by

$$f_{ctd} = -\frac{\sigma_{cp}}{2} + \sqrt{\{(-\frac{\sigma_{cp}}{2})^2 + \tau^2\}}$$

From which the permissible value of τ can be calculated as follows.

$$f_{ctd} + \frac{\sigma_{cp}}{2} = \sqrt{\{(-\frac{\sigma_{cp}}{2})^2 + \tau^2\}}$$

$$(f_{ctd} + \frac{\sigma_{cp}}{2})^2 = (-\frac{\sigma_{cp}}{2})^2 + \tau^2$$

$$f_{ctd}^2 + f_{ctd} \times \sigma_{cp} = \tau^2$$

$$\tau = \sqrt{\{f_{ctd}^2 + f_{ctd}\sigma_{cp}\}}$$

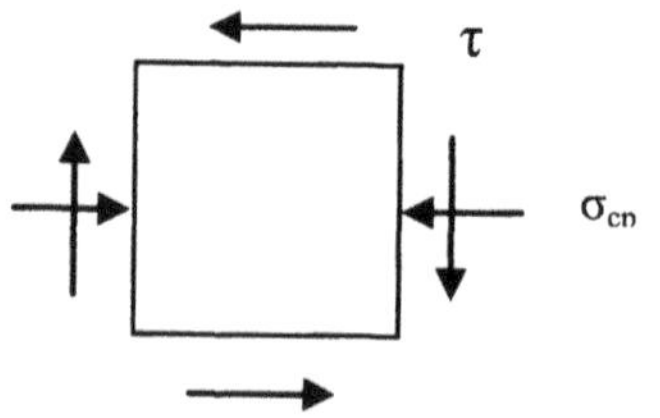

Fig. 16.24 Normal and shear stresses at the neutral axis.

The elastic shear stress τ is given by $\tau = \dfrac{S}{I \times b_w} V_{Rd,c}$

where
I = second moment of area of the cross section.
S = first moment of area about the centroidal axis of the area of section above the section where the shear stress is calculated (see hatched area in Fig. 16.25).
b_w = width of the section where the shear stress is calculated.

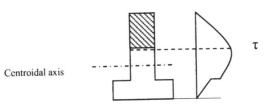

Centroidal axis

Fig. 16.25 Shear stresses at a level in the cross section.

$$V_{Rd,c} = \frac{I \times b_w}{S} \sqrt{\{f_{ctd}^2 + f_{ctd}\sigma_{cp}\}}$$

Eurocode 2 in equation (6.4) gives the following formula for calculating $V_{Rd,\,c}$, the shear capacity due to concrete alone in sections uncracked in flexure,

$$V_{Rd,c} = \frac{I \times b_w}{S} \sqrt{\{f_{ctd}^2 + \alpha_\ell\, f_{ctd}\sigma_{cp}\}} \qquad (6.4)$$

α_ℓ = is a factor to allow for the reduction in the value of σ_{cp} due to the transmission length effect.
$\alpha_\ell = \ell_x/\ell_{pt2} \le 1.0$ for pretensioned tendons.
ℓ_x = distance from the start of the transmission length.
ℓ_{pt2} = upper bound value of the transmission length.
The bond stress f_{pbt} at the time of release of tendons is give for 7-wire tendons by code equation (8.15) as
$f_{bpt} = (\eta_{p1} = 3.2) \times (\eta_1 = 1.0$ for good bond otherwise 0.7$) \times 0.7\, f_{ctm}$ (t)/γ_c.
The basic value for transmission length ℓ_{pt} is given by code equation (8.16) as
$$\ell_{pt} = \alpha_1 \times \alpha_2 \times \phi \times (\sigma_{pmo}/f_{pbt}) \qquad (8.16)$$
α_1 = 1.0 for gradual release or 1.25 for sudden release.
α_2 = 0.19 for wire strands.
ϕ = nominal diameter of tendon.
σ_{pmo} = tendon stress just after release.
$$\ell_{pt2} = 1.2\, \ell_{pt} \qquad (8.18)$$

As an example if f_{ck} = 40 MPa,
$$f_{ctm} = 0.3 \times f_{ck}^{0.667} = 3.5 \text{ MPa}$$
Assume tendons released at t = 3 days.
$$f_{ctm} (t) = \beta_{cc} \times f_{ctm} \qquad (3.1)$$

$$\beta_{cc} = \exp\left\{s\left[1 - \sqrt{\frac{28}{t}}\right]\right\} \qquad (3.2)$$

Assume s = 0.2 for the class R type of cement used.
t = 3 days, when the tendon is released.

$$\beta_{cc} = 0.66$$
$$f_{ctm}(t) = (\beta_{cc} = 0.66) \times (f_{ctm} = 3.5) = 2.3 \text{ MPa}$$
$$f_{ctd}(t) = 0.7 \times [f_{ctm}(t) = 2.3]/(\gamma_c = 1.5) = 1.1 \text{ MPa}$$
$$f_{bpt} = (\eta_{pl} = 3.2) \times (\eta_l = 1.0 \text{ or good bond}) \times [f_{ctd}(t) = 1.1] = 3.4 \text{ MPa} \quad (8.15)$$

For 7-wire 15 mm strand, take f_{yp} = 1860 MPa.
Assume that the strand is stressed to 0.75 f_{yk} and at release there is a 10% loss.
σ_{pmo} = tendon stress just after release ≈ 0.9 × 0.75 × 1860 = 1256 MPa.
$\ell_{pt} = (\alpha_1 = 1.0) \times (\alpha_2\ 0.19) \times (\varphi\ 15) \times [\sigma_{pmo} = 1256)/(f_{pbt} = 3.4)]$
= 1051 mm (8.16)

$$\ell_{pt2} = 1.2\,\ell_{pt} = 1.2 \times 1051 = 1261 \text{ mm} \quad (8.18)$$
$$\ell_{pt2} = 1261/(\varphi = 15 \text{ mm}) \approx 84 \text{ tendon diameters}$$

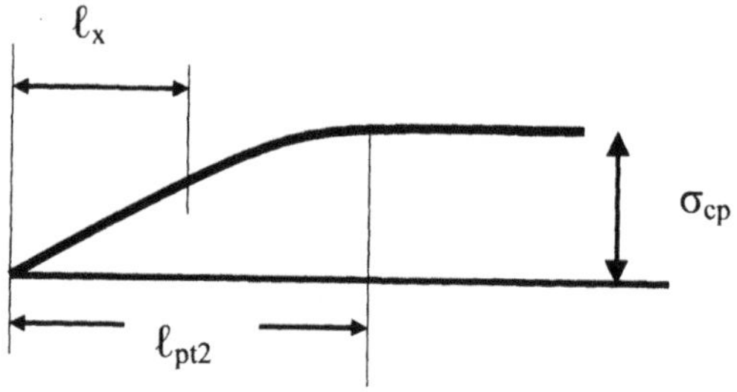

Fig. 16.26 Transmission length.

16.12.1 Example of Calculating Ultimate Shear Capacity $V_{rd,\,c}$

Calculate $V_{Rd,\,c}$ at the support section of the beam shown in Fig. 16.27. The beam prestressed with a prestress at service of 1856 kN. The beam is made from concrete with f_{ck} = 40 MPa.

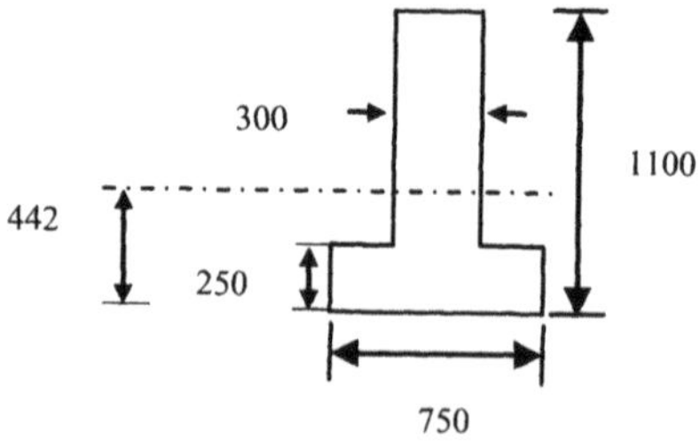

Fig. 16.27 Precast prestressed beam.

Area of cross section, $A = (750 - 300) \times 250 + 1100 \times 300 = 4.425 \times 10^5 \text{ mm}^2$
First moment of area, $A\bar{y}$ about the soffit = $(750 - 300) \times 250^2/2 + 300 \times 1100^2/2$

$$A\bar{y} = 1.956 \times 10^8 \text{ mm}^3$$

Centriodal axis from the soffit $= A\bar{y}/A = 442$ mm

Second moment of area, I about the centroidal axis:

$$I = (750 - 300) \times 250^3/12 + (750 - 300) \times 250 \times (442 - 250/2)^2$$
$$+ 300 \times 1100^3/12 + 300 \times 1100 \times (1100/2 - 442)^2 = 4.882 \times 10^{10} \text{ mm}^4$$

σ_{cp} = compressive stress at centroidal axis of the beam due to prestress of 1856 kN:

$$\sigma_{cp} = 1856 \times 10^3/4.424 \times 10^5 = 4.2 \text{ MPa}$$
$$f_{ck} = 40 \text{ MPa}, \, f_{ctm} = 0.30 \times f_{ck}^{0.667} = 3.5 \text{ MPa}$$
$$\gamma_c = 1.5, \, f_{ctd} = f_{ctm}/\gamma_c = 2.3 \text{ MPa}$$

Assuming $\alpha_\ell = 0.75$,

$$\sqrt{\{f_{ctd}^2 + \alpha_\ell \, f_{ctd}\sigma_{cp}\}} = 4.98\text{MPa}$$

S = First moment of area about the centroidal axis of the hatched area above the centroidal axis shown in Fig. 16.28.

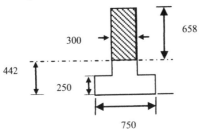

Fig. 16.28 Area above the centroidal axis.

$$S = 300 \times 658^2/2 = 6.50 \times 10^7 \text{ mm}^3, \text{ web width, } b_w = 300 \text{ mm}$$

$$V_{Rd,c} = \frac{I \times b_w}{S} \sqrt{\{f_{ctd}^2 + \alpha_\ell \, f_{ctd}\sigma_{cp}\}} = 1122\text{kN} \qquad (6.4)$$

16.12.2 Example of Calculating Ultimate Shear Capacity $V_{rd, c}$ for a Composite Beam

In the example in section 16.12.1, the cross section involved was that of a precast prestressed section only. In the case of a composite beam, prestress acts on the precast cross section. The dead load is carried by the precast section but live load is carried by the composite section. In order to calculate VRd, c it is necessary to proceed from the first principals. Fig. 16.29 shows a composite beam cross section. The precast beam shown in Fig. 16.27 is made composite with a cast-in-situ slab 1000 mm wide × 160 mm deep. fck of slab concrete is 30 MPa and fck of precast beam is 40 MPa. The precast beam is stressed with a prestress of 1856 kN at an eccentricity of 202 mm.

Solution: Two sets of calculations will be carried out. The first one based on the maximum principal tensile stress at the centroidal axis of the precast beam and the second one at the centroidal axis of the composite beam. The smaller of the two values will govern design.

Cross sectional properties:

Precast beam: (From section 16.12.1)

Depth of beam = 1100 mm, $A_{precast}$ = 4.425 ×105 mm^2, ybar = 442 mm, $I_{precast}$= 4.882 × 1010 mm^4.

Composite section: In order to allow for the difference in the value of $f_{ck, \, slab}$ = 30 MPa and $f_{ck, \, beam}$ = 40 MPa, the width of the slab is taken as 1000 × 30/40= 750 mm.

$$\text{Depth of beam} = 1100 + 160 = 1260 \text{ mm,}$$

$$A_{composite} = A_{Precast} + A_{\, slab} = 4.425 \times 10^5 + 750 \times 160 = 5.625 \times 10^5 \text{ mm}^2$$

ybar = [4.425 ×10^5× 442 + 750 ×160 × (1100 + 160/2)]/ (5.625×10^5) = 599 mm
$I_{composite}$ = (750 – 300) × 250^3/12 + (750 – 300) × 250 × (599 – 250/2)^2
+ 300 ×1100^3/12 + 300 ×1100 × (1100/2 – 599)^2
+ 750 × 160^3/12 + 750 × 160 × (1100 + 160/2 – 599)= 10.07 ×10^{10} mm^4

(1) $V_{Rd, \, c}$ based on the maximum principal tensile stress at the centroidal axis of the precast beam

At the centroidal axis of precast beam, σ_{cp}= 1856 ×10^3/ (4.425 ×10^5) = 4.2 MPa.
f_{ck} = 30 MPa, f_{ctd} = 2.3 MPa and α_ℓ = 075.
Maximum shear stress τ permitted = $\sqrt{}$ (2.3^2 + 0.75 × 2.3 × 4.2) = 3.5 MPa.

(a) For determining shear stress due to dead load in the precast beam at beam centroidal axis, take moment about the centroidal axis of the beam of the area above beam centroidal axis (see Fig. 16.28):
S = 300 × (1100 – 442)^2/2 = 6.50 ×10^7 mm^3, b$_w$ = 300 mm

$$\tau_{dead} = \frac{V_{dead} \times S}{I_{precast} \times b_w} = \frac{V_{dead} \times 10^3 \times 6.60 \times 10^7}{4.882 \times 10^{10} \times 300} = 4.506 \times V_{dead} \times 10^{-3}$$

(b) For determining shear stress due to live load in the composite beam at the precast beam centroidal axis level, take moment about the centroidal axis of composite beam of the area above precast beam centroidal axis (see Fig. 16.29): Web depth above the precast beam centroidal axis = 1100 – 442 = 658 mm

$$\text{ybar}_{Composite \, beam} - \text{ybar}_{precast \, beam} = 599 - 442 = 157 \text{ mm}$$

S = 750 × 160 × (1260 – 599 – 160/2) for slab + 300 × 658 × (658/2 – 157) web
S = 10.37 × 10^7

$$\tau_{live} = \frac{V_{live} \times S}{I_{composite} \times b_w} = \frac{V_{live} \times 10^3 \times 10.37 \times 10^7}{10.07 \times 10^{10} \times 300} = 3.43 \times V_{live} \times 10^{-3}$$

$$\tau = 3.5 = (4.506 \times V_{Dead} + 3.43 \times V_{live}) \times 10^{-3}$$

If at ULT, $V_{Dead} = 240$ kN, then $V_{Live} = 705$ kN

$$V_{Rd, c} = V_{Dead} + V_{Live} = 945 \text{ kN}$$

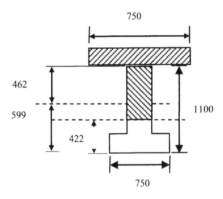

Fig. 16.29 Area above the centroidal axis of precast beam.

(2) $V_{Rd, c}$ based on the maximum principal tensile stress at the centroidal axis of the composite beam:

Stress due to prestress at the centroidal axis of the component beam:

$$\sigma_{cp} = -[-\frac{P_s}{A_{Precastbeam}} + \frac{P_s \times e}{I_{Precastbeam}}(y_{b \, Composite \, beam} - y_{b \, Precastbeam})]$$

$$\sigma_{cp} = [\frac{1856 \times 10^3}{4.425 \times 10^5} - \frac{1856 \times 10^3 \times 202}{4.88 \times 10^{10}}(599 - 442)] = 3.0 \text{MPa}$$

$f_{ck} = 30$ MPa, $f_{ctd} = 2.3$ MPa and $\alpha_\ell = 075$.
Maximum shear stress τ permitted $= \sqrt{(2.3^2 + 0.75 \times 2.3 \times 3.0)} = 3.2$ MPa.

(c) For determining shear stress in the precast beam at composite beam centroidal axis, take moment about the centroidal axis of the beam of the area above composite beam centroidal axis (see Fig. 16.30):

ybar-composite beam – ybar-precast beam = 599 – 442 =157 mm
Depth above composite beam centroidal axis = 1100 – 599 = 501 mm
$S = 300 \times 501 \times (501/2 + 157) = 6.125 \times 10^7$ mm^3

$$\tau_{dead} = \frac{V_{dead} \times S}{I_{precast} \times b_w} = \frac{V_{dead} \times 10^3 \times 6.125 \times 10^7}{4.882 \times 10^{10} \times 300} = 4.182 \times V_{dead} \times 10^{-3}$$

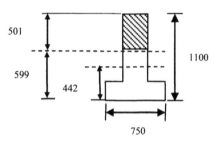

Fig. 16.30 Area above the centroidal axis of composite beam.

(d) For determining shear stress in the composite beam at the composite beam centroidal axis, take moment about the centroidal axis of composite beam of the area above composite beam centroidal axis (see Fig. 16.31):

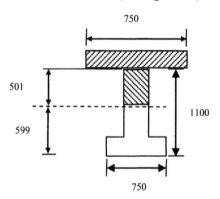

Fig. 16.31 Area above the centroidal axis of composite beam.

Web depth above the composite beam centroidal axis = 1100 − 599 = 501 mm
S = 750 × 160 × (1260 − 599 − 160/2) for slab + 300 × 501²/2 for web
S = 10.74 ×10⁷

$$\tau_{live} = \frac{V_{live} \times S}{I_{composite} \times b_w} = \frac{V_{live} \times 10^3 \times 10.74 \times 10^7}{10.07 \times 10^{10} \times 300}$$
$$= 3.56 \times V_{live} \times 10^{-3}$$

$$\tau = 3.2 = (4.182 \times V_{Dead} + 3.56 \times V_{live}) \times 10^{-3}$$

If at ULT, V_{Dead} = 240 kN, then V_{Live} = 617 kN

$$V_{Rd,\,c} = V_{Dead} + V_{Live} = 857 \text{ kN}$$
Taking the smaller of two values, $V_{Rd,\,c}$ = 857 kN.

If as an approximation, it is assumed that all loads are carried by the composite section, then

$$V_{Rd,c} = \frac{10.07 \times 10^{10} \times 300}{10.74 \times 10^{7}} \times 3.2 \times 10^{-3} = 900 kN$$

900 kN is only 5% larger than the more accurate value of 857 kN.

16.13 SHEAR CAPACITY OF SECTIONS WITHOUT SHEAR REINFORCEMENT AND CRACKED IN FLEXURE

In the case of sections where the flexural tensile stress is greater than f_{ctd} the shear capacity of the section is given by the following semi-empirical formula given by code equations (6.2a), (6.2b) and (6.3N).

$$V_{Rd,c} = [C_{Rd,c} k\{100 \rho_1 f_{ck}\}^{1/3} + k_1 \sigma_{cp}]b_w d \geq [v_{min} + k_1 \sigma_{cp}]b_w d \qquad (6.2a)$$

$$C_{Rd,c} = \frac{0.18}{(\gamma_c = 1.5)} = 0.12 , k = 1 + \sqrt{\frac{200}{d}} \leq 2.0 , \rho_1 = \frac{A_{sl}}{b_w d} \leq 0.02$$

$k_1 = 0.15$, A_{sl} = area of tensile reinforcement which extends a length of (design anchorage length l_{bd} + effective depth) beyond the section where the shear capacity is being calculated.

$$v_{min} = 0.035 k^{1.5} \sqrt{f_{ck}} \qquad (6.3N)$$

$$\sigma_{cp} = \text{Axial force, } P_s/A_c \leq 0.2 f_{cd}$$

16.13.1 Example of Calculating Ultimate Shear Capacity $V_{rd, c}$

Example: For the double T-beam shown in Fig. 16.32, calculate the shear capacity at a section which is cracked in bending. The tension steel consists in each web of three 7-wire 15 mm diameter tendons with an effective cross sectional area of 150 mm². The prestress in each web is 462 kN. The cables are placed at 50 mm from the soffit. The beam is made from concrete f_{ck} = 40 MPa.

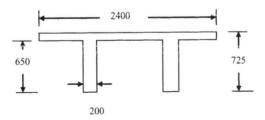

Fig. 16.32 Double T-beam.

Solution:
$$d = 725 - 50 = 875 \text{ mm}, k = 1 + \sqrt{(200/875)} = 1.48 < 2.0$$
$$A_{sl} = 3 \times 150 = 450 \text{ mm}^2 \text{ per web}$$
$$100\rho_1 = 100 \times 450/ (200 \times 875) = 0.257 < 2.0$$
$$f_{cd} = f_{ck}/\gamma_c = 40/1.5 = 26.7 \text{ MPa},$$
$$0.2 f_{cd} = 5.34 \text{ MPa}$$
$$v_{min} = 0.035 k^{1.5} \sqrt{f_{ck}} = 0.40 \text{ MPa}$$
$$\sigma_{cp} = 462 \times 10^3/ (200 \times 725) = 3.2 \text{ MPa} < 0.2 f_{cd}$$
$$V_{Rd,c} = [0.12 \times 1.48 \times \{0.257 \times 40\}^{1/3} + 0.15 \times 3.2] \times 200 \times 875 \times 10^{-3}$$
$$\geq [0.40 + 0.15 \times 3.2] \times 200 \times 875 \times 10^{-3}$$
$$V_{Rd, c} = [0.386 + 0.48] \times 175 \geq [0.40 + 0.48] \times 175$$
$$V_{Rd, c} = 154 \text{ kN per web}$$
Total shear capacity $= 2 \times 154 = 308$ kN.

16.14 SHEAR CAPACITY WITH SHEAR REINFORCEMENT

For members with shear reinforcement, the shear resistance is the smaller value of
$$V_{Rd,s} = \frac{A_{sw}}{s} \times z \times f_{ywd} \cot\theta \tag{6.8}$$
and
$$V_{Rd,max} = \alpha_{cw} \times b_w \times z \times v_1 \times f_{cd} \times \frac{1}{(\cot\theta + \tan\theta)} \tag{6.9}$$
A_{sw} = cross sectional area of shear reinforcement.
s = spacing of links.
f_{ywd} = design yield strength of shear reinforcement.
v_1 = strength reduction factor for concrete cracked in shear.
$$v_1 = 0.6(1 - f_{ck}/250) \tag{6.6N}$$
α_{cw} is defined as follows:
$$\alpha_{cw} = (1 + \sigma_{cp}/f_{cd}), \ 0 < (\sigma_{cp}/f_{cd}) \leq 0.25 \tag{6.11.aN}$$
$$\alpha_{cw} = 1.25, \ 0.25 < (\sigma_{cp}/f_{cd}) \leq 0.5 \tag{6.11.bN}$$
$$\alpha_{cw} = 0.25(1 - \sigma_{cp}/f_{cd}), \ 0.5 < (\sigma_{cp}/f_{cd}) \leq 1.0 \tag{6.11.cN}$$
θ = inclination of the concrete strut to the beam axis. $1 \leq \cot\theta \leq 2.5$.

16.14.1 Example of Calculating Shear Capacity with Shear Reinforcement

Calculate the shear reinforcement for the beam in section 16.13.1 if the shear force
to be resisted is $V_{Ed} = 580$ kN per web.
$$f_{ck} = 40 \text{ MPa}, f_{cd} = f_{ck}/\gamma_c = 40/1.5 = 26.7 \text{ MPa}, v_1 = 0.6(1 - f_{ck}/250) = 0.504$$
$$f_{ywd} = 500/ (\gamma_s = 1.15) = 435 \text{ MPa}$$
Use 2-leg H10 links. $A_{sw} = 157 \text{ mm}^2$
$$\sigma_{cp} = 462 \times 10^3/ (200 \times 725) = 3.2 \text{ MPa}, \sigma_{cp}/f_{cd} = 0.12$$
$$\alpha_{cw} = (1 + \sigma_{cp}/f_{cd}) = 1.12$$
Equate V_{Ed} to $V_{Rd, max}$ and determine the value of $\cot\theta$.

Take z = 0.9d = 0.9 × 875 = 788 mm.
b_w = 200 mm

$$V_{Rd,max} = \alpha_{cw} \times b_w \times z \times v_1 \times f_{cd} \times \frac{1}{(cot\theta + tan\theta)}$$ (6.9)

$V_{Rd, max}$ = 1.12 × 200 ×788 × 0.504 × 26.7× 10^{-3} ×1/ (cotθ + tanθ)
$V_{Rd, max}$ = 2356.5×1/ (cotθ + tanθ) kN = (V_{Ed} = 580 kN)
(cot θ + tan θ) = 4.06
$cot^2θ$ – 4.06 cot θ + 1 = 0, cot θ = 0.26 or 3.80
Both values are outsize the limits for cot θ. Limiting cot θ = 2.5 for maximum shear capacity with shear reinforcement, determine the spacing of reinforcement.
If cot θ = 2.5, $V_{Rd, max}$ = 3534.7/ (2.5 + 0.4) = 1219 kN.
Equating $V_{Rd, s}$ to V_{Ed}, determine the spacing of links.

$$V_{Rd\,s} = \frac{A_{sw}}{s} \times z \times f_{ywd}\, cot\theta = \frac{157}{s} \times 788 \times 435 \times 2.5 \times 10^{-3} = \frac{134541}{s}\, kN$$

$V_{Rd, s}$ = (V_{Ed} = 580 kN)
s = 232 mm
Maximum s = 0.75d = 0.75 × 875 = 656 mm (code equation 9.6N).
Check minimum shear link requirement:

$$A_{sw}/(s \times b_w) > 0.08 \times \sqrt{f_{ck}}/f_{yk}$$ (9.4 and 9.5N)

$$157/ (232 \times 200) > 0.08 \times \sqrt{40}/500$$

$$3.38 \times 10^{-3} > 1.01 \times 10^{-3}$$

Minimum requirement is satisfied.

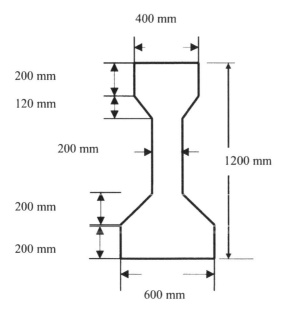

Fig. 16.33 A bridge I-beam.

16.14.2 Example of Design for Shear for a Bridge Beam

Design the shear reinforcement for the bridge I-beam shown in Fig. 16.33. The beam is simply supported over a span of 20 m. The relevant section properties are

$A = 412 \times 10^3$ mm^2, $y_b = 530$ mm, $y_t = 670$ mm, $I = 6.17 \times 10^{10}$ mm^4

$Z_b = 116.6 \times 10^6$ mm^3, $Z_t = 92.1 \times 10^6$ mm^3

At ultimate limit state the uniformly distributed load on the beam is q = 53.0 kNm. The beam is prestressed with twenty 7-wire 13 mm strands spaced as follows: 6 at 50 mm, 6 at 100 mm, 6 at 200 mm and 2 at 1100 mm from the soffit giving $P_S = 2037$ kN, e = 315 mm

Ignoring the cable at 1000 mm from the soffit, effective depth,
d = 1200 − (6 × 50 + 6 × 100 + 6 × 200) / 18 = 1083 mm.
The beam is made from concrete $f_{ck} = 40$ MPa.
$f_{ctm} = 0.3 \times 40^{0.667} = 3.5$ MPa, $f_{ctd} = f_{ctm}/ (\gamma_c = 1.5) = 2.3$ MPa,
$f_{ck, 0.05} = 0.7 \times f_{ctm} = 2.45$ MPa, $f_{ck, 0.05} / (\gamma_c = 1.5) = 1.7$ MPa

Step 1: Calculate the shear capacity of concrete alone at an uncracked section
At the centroidal axis, $\sigma_{cp} = N_{Ed}/A = P_s/A$

$$\sigma_{cp} = 2037.0 \times 10^3/ (412.0 \times 10^3) = 4.94 \text{ MPa}$$

$$f_{ctd} = 2.3 \text{ MPa. Assume } \alpha_1 = 0.5$$

$$\tau = \sqrt{\{f_{ctd}^2 + \alpha_\ell\, f_{ctd}\sigma_{cp}\}} = 3.2 \text{MPa}$$

S = 400 × 200 × (1200 − ybar − 200/2) for top flange
+ 0.5× (400 − 200) × 120 × (1200 − ybar − 200 − 120/3) for top triangle
+ 200 × (1200 − ybar − 200)²/2 for the web
$S = (45.6 + 5.16 + 22.09) \times 10^6 = 72.85 \times 10^6$ mm³
$b_W = 200$ mm

$$V_{Rd,c} = \frac{I \times b_w}{S} \sqrt{\{f_{ctd}^2 + \alpha_\ell\, f_{ctd}\sigma_{cp}\}}$$

$$= \frac{6.17 \times 10^{10} \times 200}{72.85 \times 10^6} \times 3.2 \times 10^{-3} = 542 \text{kN}$$

(6.4)

Step 2: Check the start of the cracked section

If the flexural stress is greater than $f_{ck, 0.05} / (\gamma_c =1.5) = 1.7$ MPa, the section should be considered as cracked.

$$1.7 = -\frac{P_s}{A} - \frac{P_s \times e}{Z_b} + \frac{q \times 0.5 \times (Lx - x^2)}{Z_b}$$

$$1.7 = -\frac{2037 \times 10^4}{412.0 \times 10^3} - \frac{2037 \times 10^4 \times 315}{116.6 \times 10^6} + \frac{53.05 \times 0.5 \times (20x - x^2) \times 10^6}{116.6 \times 10^6}$$

$$1.7 = -4.94 - 5.50 + 0.227 \times (20x - x^2)$$

Simplifying, $x^2 - 20x + 53.48 = 0$. Solving the quadratic equation, $x = 3.2$ m and 16.8 m. The beam should be considered as cracked beyond 3.2 m from the supports. At distances less than 3.2 m from supports, shear capacity of the section is the uncracked value of $V_{Rd, c} = 542$ kN.

Step 3: Calculate the shear capacity of concrete at a cracked section without shear reinforcement

$$C_{Rd, c} = 0.12$$

At the neutral axis, $\sigma_{cp} = N_{Ed}/A = P_s/A = 2037.0 \times 10^3/ (412.0 \times 10^3) = 4.94$ MPa.
Average stress in stress block, $f_{cd} = 22.7$ MPa. $0.2 f_{cd} = 4.5$ MPa.

$$\sigma_{cp} = 4.94 > (0.2 \ f_{cd} = 4.5)$$

Take $\sigma_{cp} = 4.5$ MPa in further calculations.

$$k_1 = 0.15$$

Note: Only the steel areas in the 'tension zone' should be included. The two cables at 1100 mm from the soffit are therefore not included. $d = 1083$ mm.

$$k = 1 + \sqrt{(200/d)} = 1 + \sqrt{(200/1083)} = 1.43 < 2.0$$
$$v_{min} = 0.035 \ k^{1.5} \ \sqrt{f_{ck}} = 0.035 \times 1.43^{1.5} \times \sqrt{40} = 0.38 \text{ MPa} \qquad \text{(6.3N)}$$
$$A_{sl} = 18 \text{ number 7-wire 13 mm strands each 100 mm}^2$$
$$= 18 \times 100 = 1800 \text{ mm}^2$$
$$b_w = 200 \text{ mm, } d = 1083 \text{ mm}$$
$$100\rho_t = (100 \times 1800)/ (200 \times 1083) = 0.82 < 2.0$$

Substituting the values into (9.6),

$$V_{Rd,c} = [C_{Rd,c} \ k\{100\rho_1 \ f_{ck}\}^{1/3} + k_1 \ \sigma_{cp}]b_w d \geq [v_{min} + k_1 \ \sigma_{cp}]b_w d \qquad \text{(6.2a)}$$
$$V_{Rd, c} = [0.12 \times 1.43 \times (0.82 \times 40)0.33 + 0.15 \times 4.94] \times 200 \times 1083 \times 10^{-3}$$
$$V_{Rd, c} \geq [0.38 + 0.15 \times 4.94] \times 200 \times 1083 \times 10^{-3}$$
$$V_{Rd, c} = 279.5 \geq 242.8 \text{ kN}$$
$$V_{Rd, c} = 279.5 \text{ kN}$$

Step 4: Design necessary shear reinforcement

The shear force V_{Ed} at the support is $53.0 \times 20/2 = 530.0$ kN.
The shear capacity of a section cracked in flexure = 279.5 kN. The shear capacity of a section uncracked in flexure = 542 kN. The beam is cracked beyond 3.2 m from supports. The applied shear force is equal to shear force capacity of the cracked section without shear reinforcement at

$$279.5 = 530 - 53 \times x, x = 4.73 \text{ m}$$

No shear reinforcement (except nominal reinforcement) is required beyond 4.9 m from the supports. Shear reinforcement is required only from 3.2 m to 4.9 m from the supports.
Shear V at d from support = $530.0 - q d = 530 - 53 \times 1083 \times 10^{-3} = 472.6$ kN.

This value is much less than the shear capacity of 542 kN for a section uncracked in flexure. Shear at 3.1 m from the support $= 530 - 53 \times 3.1 = 365.7$ kN. Therefore design shear reinforcement for $V_{Ed} = 365.7$ kN.

$f_{ck} = 40$ MPa, $f_{cd} = f_{ck}/\gamma_c = 40/1.5 = 26.7$ MPa, $v_1 = 0.6(1 - f_{ck}/250) = 0.504$.

$$f_{ywd} = 500/ (\gamma_s = 1.15) = 435 \text{ MPa}$$

Use 2-leg H10 links. $A_{sw} = 157$ mm^2.

At the neutral axis, $\sigma_{cp} = N_{Ed}/A = P_s/A = 2037.0 \times 10^3/ (412.0 \times 10^3) = 4.94$ MPa.

$$\sigma_{cp}/f_{cd} = 4.94/26.7 = 0.185 < 0.25$$
$$\alpha_{cw} = (1 + \sigma_{cp}/f_{cd}) = 1.185 \tag{6.11.aN}$$
$$z \approx 0.9d = 0.9 \times 1083 = 975 \text{ mm}, b_w = 200 \text{ mm}$$

$$V_{Rd.max} = \alpha_{cw} \times b_w \times z \times v_1 \times f_{cd} \times \frac{1}{(\cot\theta + \tan\theta)} \tag{6.9}$$

$$V_{Rd, max} = 1.185 \times 200 \times 975 \times 0.504 \times 26.7 \times 10^{-3}/(\cot\theta + \tan\theta)$$
$$V_{Rd, max} = = 3109.5/ (\cot\theta + \tan\theta) \text{ kN}$$
$$\cot\theta = 2.5, V_{Rd, max} = 1072.2 \text{ kN}$$
$$\cot\theta = 1.0, V_{Rd, max} = 1557.8 \text{ kN}$$

Equating $V_{Rd, s}$ from (6.8) to V_{Ed}, determine the spacing of links.

$$V_{Rd.s} = \frac{A_{sw}}{s} \times z \times f_{ywd} \cot\theta = \frac{157}{s} \times 975 \times 435 \times 2.5 \times 10^{-3} = \frac{166469}{s} \text{ kN}$$

$$V_{Rd, s} = (V_{Ed} = 365.7 \text{ kN})$$
$$s = 455 \text{ mm}$$

Maximum $s = 0.75d = 0.75 \times 975 = 731$ mm.

Provide 2-leg H10 links at 450 mm c/c.

Check minimum shear link requirement:

$$A_{sw}/(s \times b_w) > 0.08 \times (\sqrt{f_{ck}})/f_{yk} \tag{9.4 and 9.5N}$$
$$157/ (450 \times 200) > 0.08 \times (\sqrt{40})/500$$
$$1.74 \times 10^{-3} > 1.01 \times 10^{-3}$$

Minimum requirement is satisfied.

16.14.3 Example of Design for Shear for a Composite Beam

Design shear reinforcement for composite beam shown in Fig. 16.34.

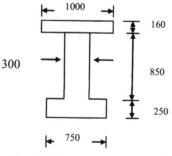

Fig. 16.34 Composite beam section.

The beam is simply supported over a span of 24 m.   The relevant   section properties are as follows.

Precast beam: (From section 16.12.1)

Depth of beam = 1100 mm, $A_{precast}$ = 4.425 ×10^5 mm^2, ybar = 442 mm,

$I_{precast}$ = 4.882 × 10^{10} mm^4, z_b = 101.40 × 10^6 mm^3, z_t = 68.11 × 10^6 mm^3

$f_{ck, beam}$ = 40 MPa

Cast-in-situ slab: $f_{ck, slab}$ = 30 MPa. Width = 1000 mm, thickness = 160 mm.
Composite section: For calculating the composite section properties, in order to allow for the difference in the strengths of concrete in the precast beam and the cast-in-situ slab, the width of the slab is taken as:

1000 × ($f_{ck, slab}/f_{ck, beam}$) = 1000 × 30/40= 750 mm.

Depth of beam = 1260 mm, $A_{composite}$ = 5.625 ×10^5 mm^2, ybar = 599 mm
$I_{composite}$ = 10.07 ×10^{10} mm^4, z_b = 168.11 ×10^6 mm^3, zt = 152.34 ×10^6 mm^3

At ultimate limit state the uniformly distributed load on the beam is
Dead load = 20 kN/m and live load is 30 kNm.
Total prestressing force P_s at service is 3712 kN applied at an eccentricity of 283 mm. The prestress is applied by 32 number 15.0 mm diameter 7-wire standard strands.
The 32 strands are positioned as follows:

10 cables at 60 mm from the soffit
14 cables at 110 mm from the soffit
6 cables at 160 mm from the soffit
2 cables at 1000 mm from the soffit.

The cross sectional area A_{ps} of cable

A_{ps} = 150.0 mm^2

f_{ck} for precast beam = 40 MPa, f_{ctm} = 0.3 ×40 $^{0.667}$ = 3.5 MPa,
f_{ctd} = $f_{ctm}/ (\gamma_c$= 1.5) = 2.3 MPa, $f_{ck\ 0.05}$ = 0.7 × f_{ctm} = 2.5 MPa
$f_{ck\ 0.05} / (\gamma_c$= 1.5) = 1.7 MPa
f_{ck} for in-situ slab = 30 MPa, f_{ctm} = 0.3 ×30 $^{0.667}$ = 2.9 MPa,
f_{ctd} = $f_{ctm}/ (\gamma_c$= 1.5) = 1.9 MPa

Step 1: Calculate the shear capacity of concrete alone at an uncracked section
As an approximation, calculate for the composite section, $V_{Rd,c}$ based on the maximum principal stress at the centroidal axis of the composite section.
Stress due to prestress at the centroidal axis of the component beam

$$\sigma_{cp} = -[-\frac{P_s}{A_{Precastbeam}} + \frac{P_s \times e}{I_{Precastbeam}}(y_{bComposite\ beam} - y_{bPrecastbeam})]$$

$$\sigma_{cp} = [\frac{3712 \times 10^3}{4.425 \times 10^5} - \frac{3712 \times 10^3 \times 283}{4.88 \times 10^{10}}(599 - 442)] = 5.0 MPa$$

Take: $f_{ck} = 40$ MPa, $f_{ctd} = 2.3$ MPa and $\alpha_\ell = 075$,

Maximum shear stress τ permitted $= \sqrt{(2.3^2 + 0.75 \times 2.3 \times 5.0)} = 3.7$ MPa.

From Fig. 16.31,

$$S = 750 \times 160 \times (1260 - 599 - 160/2) \text{ for slab} + 300 \times 501^2/2 \text{ for web}$$
$$S = 10.74 \times 10^7$$
$$b_w = 300 \text{ mm}$$

$$V_{Rd,c} = \frac{I \times b_w}{S} \sqrt{\{f_{ctd}^2 + \alpha_\ell \, f_{ctd}\sigma_{cp}\}}$$

$$= \frac{10.07 \times 10^{10} \times 300}{10.74 \times 10^7} \times 3.7 \times 10^{-3} = 1041 \text{kN}$$

(6.4)

Step 2: Check the start of the cracked section

If the flexural stress is greater than $f_{ck, 0.05} / (\gamma_c = 1.5) = 1.7$ MPa, the section should be considered cracked. Assume that the dead load equal to 20 kN/m acts on the precast beam and live load equal to 30 kN/m acts on the composite beam.

$$1.7 = -\frac{P_s}{A} - \frac{P_s \times e}{Z_b} + \frac{q \times 0.5 \times (Lx - x^2)}{Z_b}$$

$$1.7 = -\frac{3712 \times 10^3}{4.425 \times 10^5} - \frac{3712 \times 10^3 \times 283}{101.4 \times 10^6} + \frac{20.0 \times 0.5 \times (24x - x^2) \times 10^6}{101.4 \times 10^6}$$
$$+ \frac{30.0 \times 0.5 \times (24x - x^2) \times 10^6}{168.1 \times 10^6}$$

$$1.7 = -8.39 - 10.36 + 0.188 \times (24x - x^2)$$
$$x^2 - 24x + 108.77 = 0$$

$$x = 6.1 \text{ m and } 17.9 \text{ m}$$

The beam should be considered cracked beyond 6.1 m from the supports. At distances less than 6.1 m from supports, shear capacity of the section is the uncracked value of $V_{Rd, c} = 1041$ kN.

Step 3: Calculate the shear capacity of concrete at a cracked section without shear reinforcement

$$C_{Rd, c} = 0.12$$

From step 1, at the neutral axis, $\sigma_{cp} = 5.0$ MPa

Average stress in stress block, $f_{cd} = 22.7$ MPa, $0.2 \, f_{cd} = 4.5$ MPa

$$\sigma_{cp} = 5.0 > (0.2 \, f_{cd} = 4.5)$$

Take $\sigma_{cp} = 5.0$ MPa in further calculations.

$$k_1 = 0.15$$

Note: Only the steel areas in the tension zone should be included. Ignoring the cables at 1000 mm from the soffit, the centroid of the cables from the soffit is

$$(10 \times 60 + 14 \times 110 + 6 \times 160)/30 = 103 \text{ mm}$$
$$d = 1260 - 103 = 1157 \text{ mm}$$
$$k = 1 + \sqrt{200/d} = 1 + \sqrt{(200/1157)} = 1.42 < 2.0$$
$$v_{min} = 0.035 \, k^{1.5} \sqrt{f_{ck}} = 0.035 \times 1.43^{1.5} \times \sqrt{40} = 0.38 \text{ MPa}$$

(6.3N)

A_{sl} = 30 number 7-wire 15 mm strands each 150 mm^2 = 30 × 150 = 4500 mm^2

$$b_w = 300 \text{ mm}, d = 1157 \text{ mm}$$

$$100\rho_t = (100 \times 4500)/(300 \times 1157) = 1.30 < 2.0$$

Substituting the values into (9.6),

$$V_{Rd,c} = [C_{Rd,c}\, k\{100\rho_1\, f_{ck}\}^{1/3} + k_1\, \sigma_{cp}]b_w d \geq [v_{min} + k_1\, \sigma_{cp}]b_w d \quad (6.2a)$$

$$V_{Rd,c} = [0.12 \times 1.42 \times (1.30 \times 40)^{0.33} + 0.15 \times 5.0] \times 300 \times 1157 \times 10^{-3}$$

$$V_{Rd,c} \geq [0.38 + 0.15 \times 5.0] \times 300 \times 1157 \times 10^{-3}$$

$$V_{Rd,c} = 481 \geq 392.2 \text{ kN}$$

$$V_{Rd,c} = 481 \text{ kN}$$

Step 4: Design necessary shear reinforcement

The shear force VEd at the support is (20 + 30) × 24/2 = 600.0 kN.
The shear capacity of a section cracked in flexure = 481 kN.
The shear capacity of a section uncracked in flexure = 1041 kN.
The beam is cracked beyond 6.1 m from supports.
At 6.1 m, shear force = 600 − 50 × 6.1 = 295 kN < 481kN.
The beam needs only nominal reinforcement.
Maximum s = 0.75d = 0.75 × 975 = 731 mm.
Provide 2-leg H10 links at 700 mm c/c.
Check minimum shear link requirement:

$$A_{sw}/(s \times b_w) > 0.08 \times \sqrt{f_{ck}}/f_{yk} \quad (9.4) \text{ and } (9.5N)$$

$$157/(s \times 300) > 0.08 \times \sqrt{40}/500$$

$$s < 523 \text{ mm}$$

Provide links at500 mm c/c.

16.15 HORIZONTAL SHEAR

In the case of composite beams, it is necessary to ensure that the horizontal shear stress between the precast beam and the cast-in-situ slab as shown in Fig. 16.35 can be safely resisted. If there were to be a shear failure at the slab–beam junction, then composite beam action will be destroyed.

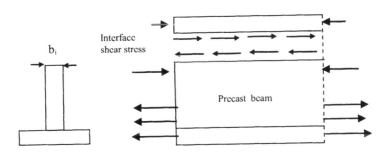

Fig. 16.35 Slab-precast beam interface shear stress.

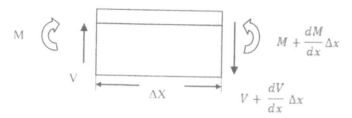

Fig. 16.36 Forces on an element.

Fig. 16.36 shows the bending moment and shear forces acting on element of length Δx. Equilibrium requires that

$$V = \frac{dM}{dx}$$

M = C z, where z is the lever arm and C is the total compressive force due to bending.

The compressive force in the slab is β C, where
β = force in the slab/ total compressive force in the beam

Fig. 16.37 shows the forces acting on an element of slab of length Δx. Equilibrium requires that

$$\tau b_i = \frac{\beta}{z} \times \frac{dM}{dz} = \frac{\beta}{z} V$$

$$\tau = \beta \frac{V}{z \times b_i}$$

where τ is the interface shear stress and b_i is the width of contact between the precast beam and the slab.

Eurocode 2 in clause 6.2.5, equation (6.24) substitutes V_{Ed} for V and v_{Edi} for τ resulting in

$$v_{Edi} = \beta \frac{V_{Ed}}{z \times b_i}$$

(6.24)

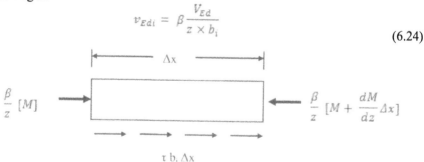

Fig. 16.37 Forces on an element of slab.

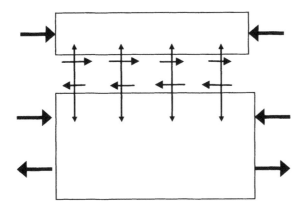

Fig. 16.38 Forces at the interface.

The shear stress v_{Edi} is resisted by a combination of friction and adhesion between the slab concrete and beam concrete. If the shear links are anchored in the slab, it helps to clamp the two elements together and provides a normal force at the interface which helps to resist the interface shear stress as shown in Fig. 16.38. Assuming that shear links are anchored in the slab, and substituting $\alpha = 90^0$, the resistance to shear stress v_{Edi} is given by equation (6.25) of the code as

$$v_{Rdi} = c\ f_{ctd} + \mu\sigma_n + \rho\ f_{yd}\ \mu \le 0.5\ v\ f_{cd} \qquad (6.25)$$

If the interface in the beam is roughened by exposing the aggregates before concreting, $c = 0.45$ and $\mu = 0.7$.

σ_n is the normal external force per unit area across the interface. This can be taken as zero.

f_{yd} = the stress in the shear links.

ρ = Ratio area of link reinforcement crossing the interface/area of the joint.

$v = 0.6[1 - f_{ck}/250]$

16.15.1 Example of Checking for Resistance for Horizontal Shear Stress

Design the shear reinforcement for the beam in section 16.11.1.
From the data in Table 16.8, at ULS

Compression force in slab = 3200 kN
Compression force in beam = 2598 kN
Total compression force = 3200 + 2598 = 5798 kN
$\beta = 3200/5798 = 0.552$
Ultimate moment = 5524 kNm
Lever arm, $z = 5524/5798 = 0.953$ m = 953 mm
Link steel = 157 mm^2 at 500 mm c/c, Area of contact = 500 $\times$ (b_i = 300)
$\rho = 157/ (500 \times 300) = 1.05 \times 10^{-3}$
$f_{yd} = 500/ (\gamma_s = 1.15) = 435$ MPa

Take c= 0.45, μ = 0.7, f_{ck} slab = 30 MPa, v = 0.6[1 – f_{ck}/250] = 0.528
f_{ctm} = 0.3 ×30 $^{0.667}$ = 2.9 MPa, f_{ctd} = f_{ctm}/ (γ_c = 1.5) = 1.9 MPa,

$$f_{cd} = 30/1.5 = 20 \text{ MPa}$$

$$v_{Rdi} = c\, f_{ctd} + \mu\sigma_n + \rho\, f_{yd}\, \mu \le 0.5\, v\, f_{cd}$$
$$v_{Rd,\, i} = 0.45 \times 1.9 + 0.7 \times 0 + 1.05 \times 10^{-3} \times 435 \times 0.7 \le 0.5 \times 0.528 \times 20$$
$$v_{Rd,\, i} = 1.18 \le 5.28$$
$$v_{Rdi} = 1.18 \text{ MPa}$$
$$v_{Edi} = \beta\, \frac{V_{Ed}}{z \times b_1} = 0.552 \times \frac{600 \times 10^3}{953 \times 300} = 1.16 \, MPa$$

$$v_{Rdi} > v_{Edi}$$

Design is satisfactory.

16.16 LOSS OF PRESTRESS IN PRETENSIONED BEAMS

In sections 16.2.1 and 16.3.1 it was stated that although at the time of stressing the cables, the total force is P_{jack}, due to losses that occur during transfer of prestress to concrete and also due to long-term deformation of steel and concrete, there is considerable reduction in prestress at the long-term SLS stage.

16.16.1 Immediate Loss of Prestress at Transfer

The loss at transfer occurs because of the fact that when the force is transferred to concrete, it contracts. Because of the full bond between steel and concrete, steel also suffers the same contraction. Eurocode 2 clause 5.10.4 gives the aspects to be included in calculating the loss. The loss in prestress can be calculated by using a simple model where the prestress and eccentricity are constant over the whole length and all the prestressing steel A_{ps} can be assumed to be concentrated at an eccentricity e. If P_t is the force in the cables, then at the centroid of steel the stress σ_c in concrete is given by

$$\sigma_c = \frac{P_t}{A} + \frac{P_t\, e}{I} e$$

where A and I are respectively the area of cross section and second moment area of pretensioned beam.

The strain ε_c in concrete is given by

$$\varepsilon_c = \sigma_c / E_c$$

where E_c = Young's modulus for concrete considering immediate contraction.

Because of full bond, the strain ε_s in steel is same as strain in concrete.

$$\varepsilon_s = \varepsilon_c = \sigma_c / E_c$$

The stress σ_s in steel corresponding to ε_s is

$$\sigma_s = E_s \, \varepsilon_s = E_s \, \sigma_c / E_c$$

The loss of prestress is given by

$$\text{Loss} = A_{ps} \, \sigma_s = A_{ps} \, E_s \, \sigma_c / E_c$$
$$P_t = P_{jack} - \text{Loss}$$

16.16.1.1 Example of Calculation of Loss at Transfer

Calculate the loss at transfer for the pretensioned beam in section 16.11.1.
The properties of the precast section are:

$$\text{Area, } A_c = 4.425 \times 10^5 \text{ mm}^2$$
$$\text{Second moment of area, } I_c = 4.90 \times 10^{10} \text{ mm}^4$$
$$\text{ybar} = 442 \text{ mm from the soffit}$$
$$z_b = 110.86 \times 10^6 \text{ mm}^3$$

Prestressing force used to stress 32 cables each of area 150 mm² is

$$P_{Jack} = 4700 \text{ kN, eccentricity, } e = 283 \text{ mm.}$$
$$E_p = 195 \text{ GPa}$$

The cables are normally released after one or two days or earlier using steam curing to speed up the gain in strength.   Taking $t = 3$ days,

$$\beta_{cc}(t) = \exp\left\{ s \left[1 - \sqrt{\frac{28}{t}} \right] \right\}$$

(3.2)

Assuming $s = 0.2$ for the type R cement used, $\beta_{cc}(t) = 0.66$

$$f_{ck}(t) = 0.66 \times 30 = 19.9 \text{ MPa, } f_{cm}(t) = f_{ck}(t) + 8 = 27.9 \text{ MPa.}$$
$$E_{cm}(t) = 22 \times (f_{cm}(t)/10)^{0.3} = 29.9 \text{ GPa}$$

A_{ps} = Area of prestressing steel = 32 strands at 150 mm² each = 4800 mm²
Assuming that P_t is in kN, compressive stress σ_c due to prestress at centroid of steel

$$\sigma_c = \frac{P_t \times 10^3}{4.425 \times 10^5} + \frac{P_t \times 10^3 \times 283}{4.90 \times 10^{10}} \times 283 = 3.894 \times 10^{-3} \times P_t$$

$$\text{Loss} = A_{ps} \, E_p \, \sigma_c / E_{cm}(t)$$
$$\text{Loss} = (4800) \times (195/29.9) \times 3.894 \times 10^{-3} \times P_t \times 10^{-3} \text{ kN}$$
$$\text{Loss} = 0.122 \, P_t \text{ kN}$$
$$P_t = P_{jack} - \text{Loss} = 4700 - 0.122 \, P_t$$
$$P_t = 4189 \text{ kN,}$$
$$\text{Transfer loss} = P_{jack} - P_t = 511 \text{ kN}$$

There is 11 % loss of prestress at the time of transfer.

16.16.2 Long-Term Loss of Prestress

After the force has been transferred, concrete continues to contract due to creep. In addition concrete also suffers shrinkage due to loss of moisture. Because the steel is under stress there is reduction in stress due to the relaxation effect. These losses can be calculated using the simple model used in section 16.15.1

(i) Time-dependent losses of prestress for pre and posttensioning
Eurocode 2 gives in clause 5.10.6, equation (5.46) for calculating the loss of stress
due to creep and shrinkage of concrete and relaxation of steel.

$$\Delta P_{c+s+r} = A_p \left[\frac{\epsilon_{cs} E_p + 0.8 \Delta_{pr} + \frac{E_p}{E_{cm}} \times \emptyset(t,t_0) \times \sigma_{c,QP}}{1 + \frac{E_p}{E_{cm}} \times \frac{A_p}{A_c} \times \left(1 + \frac{A_c}{I_c} \times z_{cp}^2\right)\{1 + 0.8 \emptyset(t,t_0)\}} \right] \quad (5.46)$$

Use of this equation is demonstrated for the precast section example in section
16.16.1.1.

A_p = area of prestressing steel = 4800 mm^2

A_c = cross sectional area of concrete = 4.425 × 10^5 mm^2

Second moment of area, I_c = 4.90 × 10^{10} mm^4

z_{cp} = eccentricity of prestress = 283 mm

$1 + (A_c/I_c) \times z_{cp}^2 = 1.723$

E_p = modulus of elasticity for steel = 195 GPa

f_{ck} = 30 MPa

E_{cm} = secant modulus of elasticity = E_{cm} = 22 × [(f_{ck} + 8)/10]$^{0.3}$ = 32.8 GPa

$(E_p/E_{cm}) \times (A_p/A_c) \times 1.723 = 0.11$

Φ (t, t0) = creep coefficient at time t at load application at t_0.

For the section, A_c == 4.425 × 10^5 mm^2

Perimeter, u = 2 × 1100 + 2 × 750 = 3700 mm

h_0 = notional size = 2 × A_c/u = 239 mm

From code Fig. 3.1b, for relative humidity, RH = 80% and the beam is loaded at
t_0 = 3 days, for class S cement, φ (∞, t_0) ≈ 3.0.

$1 + 0.8 \times \varphi$ (∞, t_0) = 3.4

Denominator in code equation (5.46) is equal to $1 + 0.11 \times 3.4 = 1.374$

$\Delta\sigma_{pr}$ = Relaxation loss of stress

Accurate calculations can be done using equations in clause 3.3.2 of Eurocode. An
approximate estimation can be made as follows.
For low relaxation strands (Class 2), 1000 hour loss is 2.5% when tensioned to
0.9 $f_{p0.1k}$.
$f_{p0.1k}$ ≈ 0.88 f_{pk}. f_{pk} = 1860 MPa. 0.9 $f_{p0.1k}$ = 0.9 × 0.88 × 1860 = 1473 MPa.
1000 hour loss = 2.5/100 × 1473 = 37 MPa.
In the long run relaxation loss can be taken as approximately 1.5 times the 1000
hour loss.

$\Delta\sigma_{pr}$ = 1.5 × 37 = 55 MPa

$0.8 \times \Delta\sigma_{pr}$ = 44 MPa

ϵ_{cs} = Estimated shrinkage strain using the data in clause 3.1.4 of Eurocode 2.
An approximate value is used in the following.
Assume, ϵ_{cs} ≈ 0.36 ×10^{-3}. $\epsilon_{cs} \times E_p$ = 70 MPa.
$\sigma_{c, QP}$ = stress in the concrete adjacent to the tendons due to self weight, initial
prestress and other quasi-permanent loads.

(a) Stress due to prestress: Assuming that $P_t = 4189$ kN, compressive stress σ_c due to prestress at centroid of steel

$$\sigma_c = -\frac{4189 \times 10^3}{4.425 \times 10^5} - \frac{4189 \times 10^3 \times 283}{4.90 \times 10^{10}} \times 283 = -16.3 \, \text{MPa}$$

(b) Stress due to uniformly distributed dead load of 12.5 kN/m and live load of 35.0 kN/m over a simply supported span of 20:

Moment at mid-span = $(12.5 + 37.0) \times 20^2/8 = 2475$ kNm.

Stress at the centroid of steel = $2475 \times 10^6 \times 283/ (4.90 \times 10^{10}) = 14.3$ MPa.

$$\sigma_{c, \, QP} = -16.3 + 14.3 = 2.0 \, \text{MPa}$$

The stress due to dead and live loads is maximum at mid-span and zero at the supports and the distribution is parabolic over the span. An average stress due to dead and live loads $\approx (2/3) \times 14.3 = 9.5$ MPa.

$$\sigma_{c, \, QP} = -16.3 + 9.5 = -6.8 \, \text{MPa}$$

$$\frac{E_p}{E_{cm}} \times \varphi(t, t_0) \times \sigma_{c.QP} = \frac{195}{32.8} \times 3.0 \times 6.8 = 121 MPa$$

Substituting in the expression for ΔP_{c+s+r}:

$$\Delta P_{c+s+r} = 4800 \times (70 + 44 + 121) \times 10^{-3}/1.374 = 821 \, \text{kN}$$

As can be seen, the greatest part, nearly 52%, of the long-term loss, is due to creep. Creep can also substantially increase long-term deformation leading to unacceptable deflection. It is very important to make realistic estimation of the effects of creep.

Final prestress remaining = P_{Jack} – transfer loss – long-term loss

$$= 4700 - 511 - 821 = 3368 \, \text{kN}$$

Final loss % = $(1 - 3368/4700) \times 100 = 28\%$.

16.17 LOSS OF PRESTRESS IN POSTTENSIONED BEAMS

The difference between the losses in pretensioned and posttensioned occurs only due to losses during jacking and transfer. The long-term loss calculations are identical.

(i) Transfer loss

In posttensioned beams, because concrete contracts while the cables are being stressed, any loss due to elastic contraction of concrete can be compensated to a certain extent. However it is rare for all tendons to be stressed at the same time. At each stage of stressing, elastic loss takes place in all the tendons previously stressed. Maximum loss is in the tendons first stressed and minimum loss in the last but one tendon stressed. In clause 5.10.5.1 of the Eurocode 2, it is suggested that as an approximation, the transfer loss is approximately 50% of that in a corresponding pretensioned beam.

Apart from the loss due to elastic contraction, there is an additional loss to consider. After the jacking is done, wedges are driven into the anchors to retain the tension in the tendons. Unfortunately, a certain amount of slip of the cables takes place before the wedges 'bite in'. The tendon slips near the anchor but friction between the tendon and the duct prevents the slip affecting the entire length. The loss in slip at the anchorage is known as draw-in loss. As the amount of slip is same whatever the length of the member, the loss of prestress due to draw-in is particularly important in short members. The reader is referred to references at the end of this chapter for details of calculation.

(ii) Loss due to friction between the cable and the duct and curvature of the tendons

Friction between the duct and the cable reduces the force in the cable away from the jacking end. Loss also occurs because of the curvature of the cables. The loss may be minimized by jacking from both ends of the beam. If P_o is the prestress at the jacking end, then prestress P_x at a distance x from the anchorage is given by the equation

$$P_x = P_0 \, e^{-\mu(\theta+kx)}$$

where k = unintentional angular displacement for internal tendons. k is dependent on many factors such as the type of duct, how well it is supported while concrete is being cast, degree of vibration used. In clause 5.10.5.2(3), the code suggests $0.005 < k < 0.01$ per meter length.

μ = friction coefficient. In Table 5.1 of the code, for internal tendons which fill roughly half the duct, $\mu = 0.19$.

θ = Sum of angular displacement over a distance, x.

As an example, in a simply supported posttensioned beam of span 20 m with a mid-span dip of 200 mm, assuming

$$k = 75 \times 10^{-4} \, /m, \, \mu = 0.19$$
$$\Delta = 200mm$$

Equation for the cable profile is

$$y = \frac{4 \times \Delta}{L^2} x(L - x)$$

$$\frac{dy}{dx} = \tan \theta = \frac{4 \times \Delta}{L^2} (L - 2x)$$

Since θ is small, $\tan \theta = dy/dx = \theta$

At x = 0, dy/dx = $4\Delta/L$ = 0.04 radians

At x = L/2, dy/dx = 0.

At x = L, dy/dx = $-4\Delta/L$ = -0.04 radians

From x = 0 to x = L/2, angular displacement = 0.04 – 0 = 0.04.

$$\mu \, (\theta + kx) = 0.022$$
$$P_{10} = 0.978 \times P_0$$

From x = 0 to x = L, angular displacement = 0.04 – (– 0.04) = 0.08.

$$\mu \, (\theta + kx) = 0.022$$
$$P_{20} = 0.957 \times P_0$$

Due to friction and wobble, the loss of prestress from starting anchorage to mid-span is 2.2% loss and from support to support is 4.3% loss.

It is important to minimize this loss. One way of doing this is to tension the cable from both ends. An effort also must be made to keep the profile of the cable as flat as possible.

16.18 DESIGN OF END BLOCK IN POSTTENSIONED BEAMS

In posttensioned members, after stressing the cables which are inside ducts fixed to the reinforcement cage, cables are anchored at the ends using proprietary anchorages. After anchoring, the ducts are grouted to prevent corrosion of the cables and also to bond the cables to concrete.

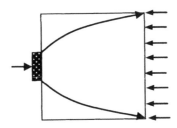

Fig. 16.39 Diffusion of compressive forces in an end block.

When the cables are anchored, a very high force is transferred to the concrete over a small area. As shown in Fig. 16.39, if an axial force representing the force applied to the anchor acts at the end face, the load gradually diffuses into concrete along curved paths and after a certain distance from end, the stresses normal to the cross section become uniform.

Elastic stress studies show that because of the fact that that compressive stresses are inclined to the axis, in order to maintain equilibrium, tensile stresses are required and they vary along the axis as shown in Fig. 16.40.

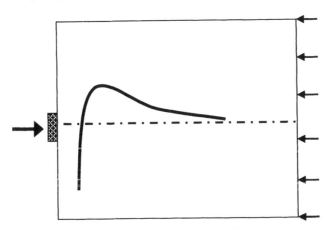

Fig. 16.40 Bursting tensile stress distribution.

A simplified force system can be visualized by replacing the curved stress path by straight inclined struts as shown in Fig. 16.41. In order to maintain equilibrium, a vertical tie is needed. This represents the bursting or splitting force which can cause tensile failure of the end block.

As shown in Fig. 16.42, a more elaborate truss system consisting of concrete struts and steel ties can be visualized which reflects better the gradual diffusion of concentrated force. In clause 8.10.3(5), Eurocode 2 recommends that the dispersion angle can be taken as β = arc tan (2/3) $\approx$ 34°.

In clause 8.10.3(4), Eurocode 2 recommends that if the stress in the ties is limited to 300 MPa, there is no need to check for crack widths. The bearing stress behind the anchorage plates should be checked for limiting bearing stress in concrete.

The bursting stresses are local to the anchorage and generally there is little of an interference effect from neighbouring anchorages. The reinforcement to resist bursting force is designed separately for each anchorage. The reinforcement can be in the form of links or as a spiral.

Design using strut–tie method is discussed in detail in Chapter 18.

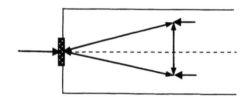

Fig. 16.41 Simple strut–tie model.

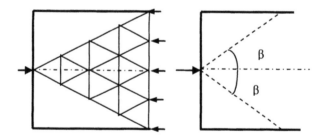

Fig. 16.42 Strut–tie system for an end block.

16.19 REFERENCES

Allen, A.H. (1992). *An Introduction to Prestressed Concrete*. British Cement Association.

Bhatt, P. (2011). *Prestressed Concrete Design to Eurocodes*. Spon Press.

Hurst, M.K. (1998). *Prestressed Concrete Design*, 2nd ed. E & F.N. Spon.

Nawy, Edward G. (2000). *Prestressed Concrete*, 3rd ed. Prentice Hall.

Nicholson, B.A. (1997). *Simple Bridge Design Using Prestressed Beams*. Prestressed Concrete Association.

CHAPTER 17

DEFLECTION AND CRACKING

In normal design practice, reinforced concrete structures are designed for the ultimate limit state. As explained in Chapter 6, the serviceability limit state criteria such as maximum deflection, maximum crack widths are not checked by detailed calculations but by meeting deemed-to-satisfy rules such as

    (a) Ensure satisfy deflection criteria at serviceability limit state by using minimum ratios of span to depth given in Table 7.4N and equations (7.16a) and (7.16b) of Eurocode 2.

    (b) Satisfy crack width criteria at serviceability limit state by restricting maximum diameter of the bar as given in Table 7.2N or by restricting the spacing of tension reinforcement to values given in Table 7.3N of the Eurocode 2.

Only in rare cases is detailed calculation of deflection and crack widths required, the exception being design of liquid retaining structures which are governed by crack width considerations (see Chapter 19). The object of this chapter is to discuss these detailed calculations.

17.1 DEFLECTION CALCULATION

17.1.1 Loads on Structure

The design loads for the serviceability limit state are loads which are sustained loads i.e. load which act for a long time. Therefore the loads to be used in calculating the deflection are:

    (a) Characteristic value of dead loads

    (b) Quasi-permanent imposed loads

In apartments and office buildings only about 25% of the characteristic imposed load is taken as permanently applied.

17.1.2 Analysis of Structure

An elastic analysis based on the gross concrete section may be used to obtain moments for calculating deflections using loads are as set out in 17.1.1.

17.1.3 Method for Calculating Deflection

The method for calculating deflection is set out in clause 7.4.3 of Eurocode 2. It is unrealistic to expect great accuracy in deflection calculations because of a number of factors are difficult to assess which can seriously affect results. Some of the factors are:

    (a) Inaccurate assumptions regarding support restraints
    (b) The actual loading and the amount that is of long-term duration which causes creep cannot be precisely estimated
    (c) Whether the member has or has not cracked
    (d) The difficulty in assessing the effects of finishes and partitions

The method given is to assess curvatures of sections due to moment and to use these values to calculate deflections.

17.1.4 Calculation of Curvatures

The curvature at a section can be calculated using assumptions set out for a cracked or uncracked section. Elastic theory is used for the section analysis.

17.1.5 Cracked Section Analysis

The assumptions used in the analysis of cracked section are as follows:
Strains are calculated on the basis that plane sections remain plane. The reinforcement is elastic with a modulus of elasticity of 200 GPa and concrete in compression is elastic with a secant modulus of elasticity E_{cm}.
Clause 3.1.3 states that the secant modulus of elasticity E_{cm} of the concrete is given by

$$E_{cm} = 22 \left[(f_{ck} + 8)/10 \right]^{0.3} \text{ GPa}$$

E_{cm} value valid between the compressive stress σ_c in concrete in the range

$$0 \le \sigma_c \le 0.4(f_{ck} + 8) \text{ MPa}$$

This value is applicable to concrete made with quartzite aggregates. For limestone and sandstone aggregates the value should be *reduced* by 10% and 30% respectively. For basalt aggregates the value should be *increased* by 20%. The tangent modulus $E_c = 1.05 \ E_{cm}$.
The final creep strain $\varepsilon_{cc}(\infty, t_0)$ at time $t = \infty$ as a multiple of elastic strain (σ_c/E_c) for a concrete loaded at age t_0 is given by Eurocode 2 equation (3.6) as

$$\varepsilon_{cc} (\infty, t0) = \varphi (\infty, t0) (\sigma_c/E_c) \qquad (3.6)$$

The effect of creep due to long-term loads is taken into account by using an effective modulus of elasticity with a value of $E_c/ (1 + \phi(\infty, t_0))$.

To show the method for calculating curvature, consider the doubly reinforced rectangular beam section shown in Fig. 17.1(a). The strain diagram and stresses and internal forces in the section are shown in Fig. 17.1(b) and Fig. 17.1(c) respectively.

The terms used in the figure are defined as follows:

f_c = stress in the concrete in compression
f_{sc} = stress in the compression steel
f_{st} = stress in the tension steel
A_s = area of steel in tension
A_s' = area of steel in compression
x = depth to the neutral axis
h = depth of the beam
d = effective depth
d' = inset of the compression steel
C_c = force in the concrete in compression
C_s = force in the steel in compression
T_s = force in the steel in tension

The following further definitions are required:

E_c = modulus of elasticity of the concrete
E_s = modulus of elasticity of the steel
α_e = modular ratio, E_s / E_c

Note that for quasi-permanent loads the effective value of E_c is used.

E_{eff} = effective modulus of elasticity of the concrete for long term

$$E_{eff} = E_c / \phi(\infty, t_0)$$
$$\phi(\infty, t_0) = \text{creep coefficient}$$

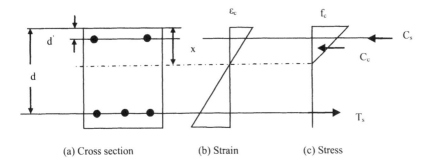

(a) Cross section   (b) Strain   (c) Stress

Fig. 17.1 Stress and strain distribution in a cracked reinforced concrete beam.

Note that although concrete is capable of sustaining a small amount of tensile stress, because of the low tensile strength of concrete, it is common to ignore this altogether.

If the maximum compressive stress in concrete is f_c, the corresponding strain ε_c in concrete is

$$\varepsilon_c = f_c/E_c$$

Assuming full bond, the strains in compression and tension steels are

$$\varepsilon_{sc} = \varepsilon_c (x - d')/x$$
$$\varepsilon_s = \varepsilon_c (d - x)/x$$

The stresses in compression and tension steels are

$$f_{sc} = E_s \, \varepsilon_{sc}$$
$$f_s = E_s \, \varepsilon_s$$

Substituting for strains in steel in terms of concrete strain,

$$f_{sc} = E_s \, \varepsilon_{sc} = E_s \, [\varepsilon_c \, (x - d')/x]$$
$$= E_s \, (f_c/E_c) \, (x - d')/x$$
$$f_{sc} = \alpha_e \, f_c \, (x - d')/x$$
$$\alpha_e = \text{modular ratio}, \; E_s \, / \, E_c$$

Similarly

$$f_s = E_s \, \varepsilon_s = E_s \, [\varepsilon_c \, (d - x)/x]$$
$$= E_s \, (f_c/E_c) \, (d - x)/x$$
$$f_s = \alpha_e \, f_c \, (d - x)/x$$

The internal forces due to compression in concrete, compression steel, tension steel and tension in concrete are given by

$$C_c = 0.5 \, f_c \, b \, x$$
$$C_s = \alpha_e \, f_c \, A_s' \, (x - d')/x$$
$$T_s = \alpha_e \, f_c \, A_s \, (d - x)/x$$

For equilibrium, the sum of the internal forces is zero:

$$C_c + C_s = T_s$$

Substituting for the forces in terms of stresses

$$0.5 f_c \, b \, x + \alpha_e \, f_c \, A_s' \, (x - d')/x = \alpha_e \, f_c \, A_s \, (d - x)/x$$

Multiplying throughout by x

$$0.5 f_c \, b \, x^2 + \alpha_e \, f_c \, A_s' \, (x - d') = \alpha_e \, f_c \, A_s \, (d - x)$$

The value of x can be determined by solving the above quadratic equation.
The sum of the moments of the internal forces about the neutral axis is equal to the applied moment M

$$M = 0.67 \, C_c \, x + C_s \, (x - d') + T_s \, (d - x)$$
$$M = [0.33 \, f_c \, b \, x^3 + \alpha_e \, f_c \, A_s' \, (x - d')^2 + \alpha_e \, f_c \, A_s \, (d - x)^2]/x$$

Note that the area of concrete occupied by the reinforcement has not been deducted in the expressions given above.
The moment M can be expressed as

$$M = f_c \, I_{cr}/ x$$

where

$$I_{cr} = 0.33 \, b \, x^3 + \alpha_e \, A_s'(x - d')^2 + \alpha_e \, A_s \, (d - x)^2$$

I_{cr} is called second moment of area of cracked transformed section.
The compressive strain ε_c in concrete is

$$\varepsilon_c = f_c/E_c$$

The curvature 1/r is

$$1/r = M / (E_c \, I_{cr})$$

17.1.6 Uncracked Section Analysis

For an uncracked section, concrete and steel are both considered to be elastic in tension and in compression. The analysis is similar to the cracked section analysis except that the area of concrete below the neutral axis is uncracked. The stress at the bottom face of the beam is $f_c (h - x)/x$ instead of being zero in a cracked section.

The internal forces are given by
$$C_c = 0.5 f_c\, b\, x$$
$$C_s = \alpha_e\, f_c\, A_s' \,(x - d')/x$$
$$T_s = \alpha_e\, f_c\, A_s \,(d - x)/x$$
$$T_c = 0.5 f_c\, b\, (h - x)^2/x$$

For equilibrium, the sum of the internal forces is zero:
$$C_c + C_s = T_s + T_c$$
$$0.5 f_c\, b\, x + \alpha_e\, f_c\, A_s' \,(x - d')/x = \alpha_e\, f_c\, A_s \,(d - x)/x + 0.5 f_c\, b\, (h - x)^2/x$$

Multiplying throughout by x/f_c
$$0.5\, b\, x^2 + \alpha_e\, A_s' \,(x - d') = \alpha_e\, A_s \,(d - x) + 0.5\, b\, (h - x)^2$$

Simplifying
$$\alpha_e\, A_s'(x - d') = \alpha_e\, A_s \,(d - x) + 0.5\, b\, h^2 - b\, h\, x$$

Solving for x
$$x = \{0.5\, b\, h^2 + \alpha_e \,(A_s\, d + A_s'\, d')\}/\, [b\, h + \alpha_e \,(A_s + A_s')]$$

Note that if the influence of reinforcement is ignored, then $x = 0.5h$.

The sum of the moments of the internal forces about the neutral axis is equal to the external moment M.

$$M = 0.67\, C_c\, x + C_s \,(x - d') + T_s \,(d - x) + 0.67\, T_c \,(h - x)$$
$$M = 0.33\, f_c\, b\, x^2 + \alpha_e\, f_c\, A_s' \,(x - d')^2 /x + \alpha_e\, f_c\, A_s \,(d - x)^2/x$$
$$+\, 0.33\, f_c\, b\, (h - x)^3/x$$
$$M = [0.33\, b\, x^3 + \alpha_e\, A_s' \,(x - d')^2 + \alpha_e\, A_s \,(d - x)^2$$
$$+\, 0.33\, b\, (h - x)^3]\,(f_c/x)$$

Simplifying
$$M = [0.33\, b\, h^3 + \alpha_e\, A_s' \,(x - d')^2 + \alpha_e\, A_s \,(d - x)^2 - b\, h\, x\, (h - x)]\,(f_c/x)$$

The moment M can be expressed as
$$M = I\,(f_c/x)$$
$$I = [0.33\, b\, h^3 + \alpha_e\, A_s' \,(x - d')^2 + \alpha_e\, A_s \,(d - x)^2 - b\, h\, x\, (h - x)]$$

I is called second moment of area of uncracked transformed section.
Very often the influence of reinforcement is ignored and substituting $x = 0.5h$, for an uncracked section, $I = bh^3/12$.

17.1.7 Long-Term Loads: Creep

The effect of creep must be considered for quasi-permanent loads. Load on concrete causes an immediate elastic strain (σ_c/E_c) and a long-term time-dependent strain known as creep strain. The strain due to creep may be much larger than that due to elastic deformation. On removal of the load, most of the strain due to creep is **not fully** recovered.

Creep is discussed in clause 3.1.4 and also in detail in Annex B of Eurocode 2. The creep coefficient $\phi(\infty, t_0)$ is used to evaluate the effect of creep. Values of $\phi(\infty, t_0)$ depend on the age of concrete t_0 at loading, notional size h_0 of the member and ambient relative humidity RH. The notional size h_0 for uniform sections is *twice* the cross sectional area A_c divided by the exposed perimeter u.

In deflection calculations, for calculating the curvature due to the long-term loads, creep is taken into account by using an effective value for the modulus of elasticity of the concrete equal to

$$E_{eff} = E_c/ [1 + \phi(\infty, t_0)]$$

17.1.7.1 Calculation of $\phi(\infty, t_0)$

As an approximation, Fig. 3.1 of Eurocode 2 can be used to calculate the final creep coefficient $\phi(\infty, t_0)$. However, Annex B gives the necessary equations for calculating $\phi(\infty, t_0)$. These are given below along with a numerical example.

$$\phi(\infty, t_0) = \phi_{RH} \times \beta(f_{cm}) \times \beta (t_0)$$

$$\varphi_{RH} = 1 + \frac{(1 - RH)}{0.1 \times h_0^{0.33}}, \quad f_{cm} \leq 35\, MPa$$

$$\varphi_{RH} = \left[1 + \frac{(1 - RH)}{0.1 \times h_0^{0.33}} \times \alpha_1 \right] \times \alpha_2, \quad f_{cm} > 35\, MPa$$

$$\alpha_1 = [\frac{35}{f_{cm}}]^{0.7}, \alpha_2 = [\frac{35}{f_{cm}}]^{0.2}$$

$$\beta(f_{cm}) = 16.8/\sqrt{f_{cm}}$$
$$f_{cm} = f_{ck} + 8$$

Depending on the type of cement, the gain in strength with age varies. This fact can be taken into account by calculating a modified value of t_0 which will be more or less than the actual value of t_0:

$$t_{0,modified} = t_{0,actual} \times [1 + \frac{9}{2 + t_{0,actual}^{1.2}}]^{\alpha}$$

$$\alpha = -1 \text{ for cement Class S}$$
$$\alpha = 0 \text{ for cement Class N}$$
$$\alpha = 1 \text{ for cement Class R}$$

$$\beta(t_0) = \frac{1}{(0.1 + t_{0,modified}^{0.2})}$$

$$h_0 = 2\, A_c/u$$

17.1.7.2 Example of Calculation of $\phi(\infty, t_0)$

Calculate the value of $\phi(\infty, t_0)$ using the following data.
Cement: Class S or N or R.
Concrete: $f_{ck} = 30$ MPa, $f_{cm} = f_{ck} + 8 = 38$ MPa > 35 MPa.
T-beam: web width $b_w = 250$ mm, flange width $b = 1450$ mm, total depth $h = 350$ mm, flange thickness $h_f = 100$ mm.
Concrete is loaded at $t_0 = 7$ days.
Relative humidity RH $= 50\%$.

Step 1: Calculate h_0:
Assume all surfaces exposed to atmosphere:
$$A_c = 1450 \times 100 + 250 \times (350 - 100) = 20.75 \times 10^4 \text{ mm}^2$$
$$u = 2 \times (1450 + 350) = 3600 \text{ mm}$$
$$h_0 = 2 A_c/u = 115 \text{ mm}$$

Step 2: Calculate f_{cm}:
$$f_{cm} = f_{ck} + 8 = 38 \text{ MPa} > 35 \text{ MPa}$$

Step 3: Calculate parameters associated with f_{cm}:
$$\beta(f_{cm}) = 16.8/\sqrt{f_{cm}} = 2.725$$
$$\alpha_1 = [\frac{35}{f_{cm}}]^{0.7} = 0.944$$
$$\alpha_2 = [\frac{35}{f_{cm}}]^{0.2} = 0.984$$

Step 4: Calculate ϕ_{RH}:
RH $= 0.5$, $h_0 = 115$ mm, $\alpha_1 = 0.944$, $\alpha_2 = 0.984$, $f_{cm} > 35$ MPa
$$\phi_{RH} = [1 + (1 - 0.5) \times 0.944/ (0.1 \times 115^{0.333})] \times 0.984 = 1.939$$

Step 5: Calculate $t_{0, \text{ modified}}$:
$$t_{0, \text{ actual}} = 7 \text{ days}$$

$$t_{0, \text{modified}} = t_{0, \text{actual}} \times [1 + \frac{9}{2 + t_{0, \text{actual}}^{1.2}}]^{\alpha}$$

Note: Class S, N and R stand for *Slow* early strength gain, *Normal* early strength gain and *Rapid* early strength gain cements respectively.

Class S cement, $\alpha = -1$, $t_{0, \text{ modified}} = 4$ days
Class N cement, $\alpha = 0$, $t_{0, \text{ modified}} = 7$ days
Class R cement, $\alpha = 1$, $t_{0, \text{ modified}} = 12$ days

Step 6: Calculate $\beta(t_0)$:
$$\beta(t_0) = \frac{1}{(0.1 + t_{0, \text{modified}}^{0.2})}$$

Class S cement, $\beta(t_0) = 1/ (0.1 + 4^{0.2}) = 0.704$
Class N cement, $\beta(t_0) = 1/ (0.1 + 7^{0.2}) = 0.635$
Class R cement, $\beta(t_0) = 1/ (0.1 + 12^{0.2}) = 0.573$

Step 7: Calculate $\phi(\infty, t_0)$:
$\phi(\infty, t_0)$ = $[\phi_{RH} = 1.939] \times [\beta(f_{cm}) = 2.725] \times [\beta (t_0) = 0.704] = 3.7$, class S
$\phi(\infty, t_0)$ = $[\phi_{RH} = 1.939] \times [\beta(f_{cm}) = 2.725] \times [\beta (t_0) = 0.635] = 3.4$, class N
$\phi(\infty, t_0)$ = $[\phi_{RH} = 1.939] \times [\beta(f_{cm}) = 2.725] \times [\beta (t_0) = 0.573] = 3.0$, class R

Step 8: Calculate short term E_{cm}:
$$E_{cm} = 22 \, [(30 + 8)/10]^{0.3} = 32.84 \text{ GPa}$$

Step 9: Calculate long term E_c:
Long term $E_c = 34.5/ [1 + \phi(\infty, t_0)] = 7.0$ GPa, Class S
Long term $E_c = 34.5/ [1 + \phi(\infty, t_0)] = 7.5$ GPa, Class N
Long term $E_c = 34.5/ [1 + \phi(\infty, t_0)] = 8.2$ GPa, Class R

17.1.8 Shrinkage

Concrete shrinks slowly as the water migrates through the pores of concrete. This is termed drying shrinkage ε_{cd}. There is another type of shrinkage called autogenous shrinkage ε_{ca} which is due hardening of the concrete and this develops fairly rapidly. The total shrinkage ε_{cs} is the sum of these two components of shrinkage. Shrinkage is mainly dependent on the ambient relative humidity, the surface area from which moisture can be lost relative to the volume of concrete, and the mix proportions. It is noted that certain aggregates produce concrete with a higher initial drying shrinkage than normal.

17.1.8.1 Calculation of Final Shrinkage Strain $\varepsilon_{cd, \infty}$

Values of basic drying shrinkage strain $\varepsilon_{cd, 0}$ are given by equations (B.11) and (B.12) of Annex B of Eurocode 2.
$$\varepsilon_{cd, 0} = 0.85 \, [(220 + 110 \, \alpha_{ds1}) \times \exp \{-0.1 \times \alpha_{ds2} \times f_{cm}\}] \times 10^{-6} \times \beta_{RH}$$
$$\beta_{RH} = 1.55 \times (1 - RH^3)$$
$$\alpha_{ds1} = 3 \text{ and } \alpha_{ds2} = 0.13 \text{ for Class S cement}$$
$$\alpha_{ds1} = 4 \text{ and } \alpha_{ds2} = 0.12 \text{ for Class N cement}$$
$$\alpha_{ds1} = 6 \text{ and } \alpha_{ds2} = 0.11 \text{ for Class R cement}$$

Example: Using the data in the example in section 17.1.7.1, calculate the value of $\varepsilon_{cd, 0}$.
$f_{ck} = 30$ MPa, $f_{cm} = 38$ MPa, RH = 0.5 (i.e. 50% relative humidity)
$$\beta_{RH} = 1.55 \times (1 - RH^3) = 1.55 \times (1 - 0.5^3) = 1.356$$
$\varepsilon_{cd, 0} = 0.85 \times (220 + 110 \times 3) \times \exp (-0.13 \times 3.8) \times 10^{-6} \times 1.356 = 387 \times 10^{-6}$, S
$\varepsilon_{cd, 0} = 0.85 \times (220 + 110 \times 4) \times \exp (-0.12 \times 3.8) \times 10^{-6} \times 1.356 = 482 \times 10^{-6}$, N
$\varepsilon_{cd, 0} = 0.85 \times (220 + 110 \times 6) \times \exp (-0.11 \times 3.8) \times 10^{-6} \times 1.356 = 668 \times 10^{-6}$, R
The final shrinkage strain $\varepsilon_{cd, \infty} = \varepsilon_{cd, 0} \times k_h$.

Table 17.1 Values of k_h versus h_0

h_o	k_h
100	1.0
200	0.85
300	0.75
≥ 500	0.70

Values of k_h are given in code Table 3.3 reproduced as Table 17.1.

In this example $h_0 = 115$ mm,
$$k_h = 1.0 - (1.0 - 0.85) \times (115 - 100)/(200 - 100) = 0.98$$
$\varepsilon_{cd, \infty} = \varepsilon_{cd, 0} \times k_h = (378, 471, 653) \times 10^{-6}$, for S, N and R cements respectively.

17.1.8.2 Calculation of Final Autogenous Shrinkage Strain $\varepsilon_{ca, \infty}$

$$\varepsilon_{ca, \infty} = 2.5 \times (f_{ck} - 10) \times 10^{-6} = 50 \times 10^{-6}$$

17.1.8.3 Calculation of Final Total Shrinkage Strain $\varepsilon_{cs, \infty}$

$\varepsilon_{cs, \infty} = \varepsilon_{cd, \infty} + \varepsilon_{ca, \infty} = (428, 521, 703) \times 10^{-6}$, for S, N and R class cements.

17.1.8.4 Curvature Due to Shrinkage

Curvature caused by shrinkage strain can be calculated as follows. Fig. 17.2 shows a reinforced concrete beam subjected to strain due to shrinkage of $\varepsilon_{cs, \infty}$.

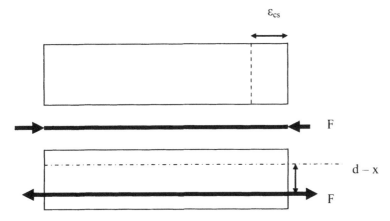

Fig. 17.2 Free and restrained shrinkage of a reinforced concrete beam.

Detaching the reinforcement from the beam, the concrete alone is allowed to shrink freely and does not deflect laterally. However in order to ensure compatibility of deformation, a compressive force F has to be applied to reinforcement alone equal to $\varepsilon_{cs, \infty} \times E_s \times A_s$, where E_s = young's modulus for reinforcement and A_s = area of reinforcement.

At this stage a compressive stress exists in reinforcement only and the concrete is stress free as it has been allowed to shrink freely. Actually this external force does not exist when a reinforced concrete beam shrinks. A force F equal and opposite to the one used for restraining the reinforcement alone is now applied to the *composite reinforced concrete section*. This force which acts at an eccentricity of $(d - x)$ is resisted by both concrete and steel. This force induces a bending moment $M_{Shrinkage}$ in the beam equal to $F \times (d - x)$. The curvature produced $1/r_{cs}$ is given by

$$1/r_{cs} = M_{Shrinkage} / (E_{eff} I) = \varepsilon_{cs, \infty} \times E_s \times A_s \times (d - x) / (E_{c, \, eff} I)$$

Setting

$$\alpha_e = \text{effective modular ratio}, E_s/E_{c, \, eff}$$

$S = A_s (d - x)$, the first moment of area of the reinforcement A_s about the centroidal axis.

$$\varepsilon_{cs, \infty} = \varepsilon_{cs}$$

The expression for curvature due to shrinkage is given by Eurocode 2 equation (7.21) as

$$1/r_{cs} = \varepsilon_{cs} \times \alpha_e \times S / I \qquad\qquad (7.21)$$

In clause 7.4.3(6), it is stated that S and I should be calculated for uncracked and cracked sections in order to calculate the shrinkage curvature for uncracked and cracked sections respectively.

$$E_{c, \, eff} = E_{cm} / [1 + \phi(\infty, t_0)], E_{cm} = 22 (0.8 + 0.1 f_{ck})^{0.3}$$

If there is compression reinforcement, then an additional force F equal to $\varepsilon_{cs, \infty} \times E_s \times A'_s$, and a corresponding moment equal to $-\varepsilon_{cs, \infty} \times E_s \times A'_s (x - d')$ has to be included when calculating the curvature due to shrinkage.

Note that if the reinforcement in the beam is symmetrically placed, there is no curvature due to shrinkage.

17.1.9 Curvature Due to External Loading

Curvature $1/r_b$ is equal to d^2y/dx^2, the second derivative of deflection with respect to distance along the span. Deflection can be calculated directly by integrating either analytically or by numerical integration techniques the differential equation

$$d^2y/dx^2 = M/ (EI).$$

The deflection Δ is calculated from

$$\Delta = KL^2 (1/r_b)$$

where L is the effective span of the member, l/r_b is the curvature at the point of maximum moment which is near mid-span in the case of beams and at support in the case of cantilevers.

$$l/r_b = M/ (EI)$$

K is a constant which depends on the shape of the bending moment diagram

17.1.9.1 Evaluation of Constant K

The method of calculating K is illustrated by a few examples. Expressions for deflection of an elastic beam are given in books on structural analysis. See Bhatt (1999).

Example 1: Simply supported beam carrying uniformly distributed total load W over a span L.
The deflection Δ at mid-span is given by

$$\Delta = 5WL^3/ (384\ EI)$$

Bending moment M at mid-span is

$$M = W\ L/8$$

Replacing the load W by moment M

$$\Delta = 5ML^2/ (48\ EI) = K\ L^2\ (1/r_b),\ K = 5/48$$

Example 2: Simply supported beam carrying a concentrated load W at mid-span over a span L.
The deflection Δ at mid-span is given by

$$\Delta = WL^3/ (48\ EI)$$

Bending moment M at mid-span is

$$M = W\ L/4$$

Replacing the load W by moment M

$$\Delta = 4ML^2/ (48\ EI) = K\ L^2\ (1/r_b),\ K = 1/12$$

Example 3: An intermediate span beam carrying uniformly distributed total load W over a span L with support moments of M_A and M_B.
The deflection Δ at mid-span is given by

$$\Delta = 5WL^3/ (384\ EI) - (M_A + M_B) \times L^2/ (16\ EI)$$

Bending moment M at mid-span is

$$M = W\ L/8 - (M_A + M_B)/2$$

Replacing the load W by moment M

$$\Delta = 5\{M + (M_A + M_B)/2\}\ L^2/ (48\ EI) - (M_A + M_B)\ L2/ (16\ EI)$$
$$\Delta = \{5\ M/48 - (M_A + M_B)/96\}\ L^2/ (EI)$$
$$\Delta = K\ L^2\ (1/r_b),\ K = 5\ /48 - (M_A + M_B)/ (96\ M)$$

Example 4: Cantilever carrying uniformly distributed total load W over a span L.
The deflection Δ at the tip is given by

$$\Delta = WL^3/ (8\ EI)$$

Bending moment M at support is
$$M = W\, L/2$$
Replacing the load W by moment M
$$\Delta = ML^2/\,(4\ EI) = K\ L^2\ (1/r_b),\ K = \tfrac{1}{4}$$

Example 5: Cantilever carrying a concentrated load W at the tip over a span L. The deflection Δ at the tip is given by
$$\Delta = WL^3/\,(3\ EI)$$
Bending moment M at support is
$$M = W\, L$$
Replacing the load W by moment M
$$\Delta = ML^2/\,(3\ EI) = K\ L^2\ (1/r_b),\ K = 1/3$$

17.2 CHECKING DEFLECTION BY CALCULATION

When calculating the curvature in beams, it has to be appreciated that part of the beam will be fully cracked and part of the beam will be uncracked. In order to allow for this, Eurocode 2 in clause 7.4.3 equation (7.18), gives the following equation for calculating the curvature, $1/r$ of members subjected mainly to flexure.

$$1/r = \zeta\ (1/r_{\text{ cracked}}) + (1 - \zeta)\ (1/r_{\text{ uncracked}}) \qquad\qquad (7.18)$$
$$\zeta = 1 - \beta\ (M_{cr}/M)^{\,2} \qquad\qquad (7.19)$$

M_{cr} = Moment to cause cracking = $z_b \times f_{ctm}$.
z_b = section modulus for the bottom fiber.
For a rectangular beam b × h, $z_b = bh^2/6$.
For other sections, $z_b = I/y_b$.
I = second moment of area about the centroidal axis and y_b = distance from the soffit to the centroidal axis.

$$f_{ctm} = 0.30\ f_{ck}^{\ 0.667}\ \text{for}\ f_{ck} \le 50\ \text{MPa}$$
$$= 2.12\ \ell n\ (1.8 + 0.1\ f_{ck})\ \text{for}\ f_{ck} > 50\ \text{MPa}$$
$$\beta = 1.0\ \text{for a single short-term loading}$$
$$= 0.5\ \text{for sustained loads}$$

17.2.1 Example of Deflection Calculation for T-Beam

A simply supported T-beam of 6 m span carries a dead load including self-weight of 14.8kN/m and an imposed load of 10 kN/m. The T-beam section has the tension reinforcement designed for the ultimate limit state and the bars in the top to support the links. The dimensions of the beam shown in Fig. 17.3 are: web width b_w = 250 mm, flange width b = 1450 mm, total depth h = 350 mm, flange thickness h_f = 100 mm, effective depth d = 300 mm, inset of compression steel d' = 45 mm, compression steel 2H16, A_s' = 402 mm^2, tension steel 3H25, A_s = 1473 mm^2. The materials are f_{ck} = 30 MPa concrete and f_{yk} = 500 MPa reinforcement. Calculate the deflection of the beam at mid-span.

(a) Moments

The deflection calculation will be made for characteristic dead and imposed loads to comply with serviceability limit state requirements. The quasi-permanent load is taken as the dead load plus 25% of the imposed load.

$$\text{Quasi-permanent load} = 14.8 + 0.25 \times 10 = 17.3 \text{ kN/m}$$
$$\text{Mid-span moment } M_{Qp} = 17.3 \times 6^2/8 = 77.85 \text{ kNm}$$

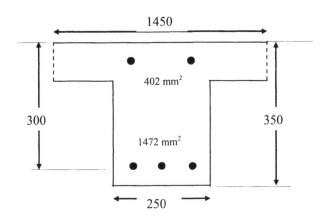

Fig. 17.3 T-beam.

(b) Material properties

$f_{ck} = 30$ MPa, $f_{ctm} = 0.30 \, f_{ck}^{0.667} = 2.9$ MPa, $E_{cm} = 22 \, (0.8 + 0.1 \, f_{ck})^{0.3} = 32.84$ GPa.
From section 17.1.7.1, $t_0 = 7$ days, Class N cement, $\phi(\infty, t_0)] = 3.4$.
$E_c = 34.5/ \, [1 + \phi(\infty, t_0)] = 7.5$ GPa.
$E_s = 200$ GPa, Effective modular ratio, $\alpha_e = 200.0/7.5 = 26.67$.

(c) Section properties

(i) Uncracked section analysis ignoring steel areas

$$\text{Area of cross section, A} = 350 \times 250 + (1450 - 250) \times 100$$
$$= (0.875 + 1.20) \times 10^5 = 2.075 \times 10^5 \text{ mm}^2$$
$$A \times y_b = [0.875 \times 350/2 + 1.20 \times (350 - 100/2)] \times 10^5 = 51.31 \times 10^6 \text{ mm}^3$$
$$y_b = 247 \text{ mm}$$
$$x = 350 - 247 = 103 \text{ mm}$$
$$I = 250 \times 350^3/12 + 250 \times 350 \times (247 - 350/2)^2 + (1450 - 250) \times 100^3/12$$
$$+ (1450 - 250) \times 100 \times (350 - 100/2 - 247)^2$$
$$I = 0.178 \times 10^{10} \text{ mm}^4$$
$$z_b = I/y_b = 7.2 \times 10^6 \text{ mm}^3$$
$$M_{Cracking} = (f_{ctm} = 2.9) \times z_b = 20.88 \text{ kNm}$$

(ii) Uncracked section analysis including steel areas

$$A_s = 1472 \text{ mm}^2, A_s' = 402 \text{ mm}^2, d' = 45 \text{ mm}, h - d = 50 \text{ mm}$$
$$\text{Area of cross section, A} = 2.075 \times 10^5 \text{ mm}^2 + \alpha_e \times (A_s' + A_s) = 2.575 \times 10^5 \text{ mm}^2$$

$$A \times y_b = 51.31 \times 10^6 \text{ mm}^3 + \alpha_e \times [A_s' \times (350 - 45) + A_s \times 50] = 56.55 \times 10^6 \text{ mm}^3$$
$$y_b = 220 \text{ mm}$$
$$x = 350 - 220 = 130 \text{ mm}$$
$$I = 250 \times 350^3/12 + 250 \times 350 \times (220 - 350/2)^2 + (1450 - 250) \times 100^3/12$$
$$+ (1450 - 250) \times 100 \times (350 - 100/2 - 220)^2$$
$$+ \alpha_e \times [A_s' \times (350 - 45 - 220)^2 + A_s \times (220 - 50)^2] = 0.315 \times 10^{10} \text{ mm}^4$$
$$z_b = I/y_b = 14.32 \times 10^6 \text{ mm}^3$$
$$M_{Cracking} = (f_{ctm} = 2.9) \times z_b = 41.53 \text{ kNm}$$

Including the steel areas increases the value of $M_{cracking}$ by 2.0 times!

(iii) Cracked section properties

Assume $x \leq$ depth of flange, 100 mm (see Fig. 17.4)
$$1450 \times x^2/2 + \alpha_e \times A_s' \times (x - 45) = \alpha_e \times A_s \times (d - x)$$
$$x^2 + 68.94 \, x - 16910.3 = 0$$
$$x = 100 \text{ mm}$$
$$I_{cracked} = 1450 \times 100^3/3 + \alpha_e \times [A_s' \times (x - 45)^2 + A_s \times (300 - x)^2]$$
$$= 0.209 \times 10^{10} \text{ mm}^4$$

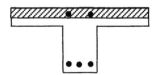

Fig. 17.4 Neutral axis depth in T-beam.

(d) Curvature due to shrinkage

From section 17.1.8.1, shrinkage strain ε_{cs} for type N cement is 521×10^{-6}
$$1/r_{cs} = \varepsilon_{cs} \times \alpha_e \times S \, / \, I$$

(i) Uncracked section ignoring steel areas:
$$I = 0.178 \times 10^{10} \text{ mm}^4, x = 103 \text{ mm}$$
$$S = A_s \times (d - x) - A_s' \times (x - 45) = 0.267 \times 10^6 \text{ mm}^3$$
$$1/r_{cs} = 521 \times 10^{-6} \times 26.67 \times 0.267 \times 10^6/ (0.178 \times 10^{10}) = 2.084 \times 10^{-6} \text{ mm}^{-1}$$

(ii) Uncracked section including steel areas:
$$I = 0.315 \times 10^{10} \text{ mm}^4, x = 130 \text{ mm},$$
$$S = A_s \times (d - x) - A_s' \times (x - 45) = 0.216 \times 10^6 \text{ mm}^3$$
$$1/r_{cs} = 521 \times 10^{-6} \times 26.67 \times 0.216 \times 10^6/ (0.315 \times 10^{10}) = 0.953 \times 10^{-6} \text{ mm}^{-1}$$

(iii) Cracked section including steel areas:
$$I = 0.209 \times 10^{10} \text{ mm}^4, x = 100 \text{ mm},$$
$$S = A_s \times (d - x) - A_s' \times (x - 45) = 0.272 \times 10^6 \text{ mm}^3$$
$$1/r_{cs} = 521 \times 10^{-6} \times 26.67 \times 0.272 \times 10^6/ (0.209 \times 10^{10}) = 1.808 \times 10^{-6} \text{ mm}^{-1}$$

(e) Curvature due to load
Mid-span moment = 77.85 kNm

$$\text{Curvature} = M/(EI)$$
$$E_{c,\,eff} = 7.5 \text{ GPa}$$

(i) Uncracked section ignoring steel areas: $I = 0.178 \times 10^{10} \text{ mm}^4$
$$\text{Curvature, } 1/r = 5.832 \times 10^{-6} \text{ mm}^{-1}$$

(ii) Uncracked section including steel areas: $I = 0.315 \times 10^{10} \text{ mm}^4$
$$\text{Curvature, } 1/r = 3.295 \times 10^{-6} \text{ mm}^{-1}$$

(iii) Cracked section including steel areas: $I = 0.209 \times 10^{10} \text{ mm}^4$
$$\text{Curvature, } 1/r = 4.967 \times 10^{-6} \text{ mm}^{-1}$$

(f) Sum of curvatures

The final curvature l/r_b is the curvature under the quasi-permanent load plus shrinkage curvature.

(i) Uncracked section excluding steel areas
$$\text{Curvature due to shrinkage} = 2.084 \times 10^{-6} \text{ mm}^{-1}$$
$$\text{Curvature due to load} = 5.832 \times 10^{-6} \text{ mm}^{-1}$$
$$\text{Total curvature} = 7.916 \times 10^{-6} \text{ mm}^{-1}$$

(ii) Uncracked section including steel areas
$$\text{Curvature due to shrinkage} = 0.953 \times 10^{-6} \text{ mm}^{-1}$$
$$\text{Curvature due to load} = 3.295 \times 10^{-6} \text{ mm}^{-1}$$
$$\text{Total curvature} = 4.248 \times 10^{-6} \text{ mm}^{-1}$$

(iii) Cracked section including steel areas
$$\text{Curvature due to shrinkage} = 1.808 \times 10^{-6} \text{ mm}^{-1}$$
$$\text{Curvature due to load} = 4.967 \times 10^{-6} \text{ mm}^{-1}$$
$$\text{Total curvature} = 6.775 \times 10^{-6} \text{ mm}^{-1}$$

(g) Final curvature and deflection

Calculate curvature using the Eurocode 2 equation (7.18).
$$1/r = \zeta\,(1/r_{\text{cracked}}) + (1 - \zeta)\,(1/r_{\text{uncracked}}) \tag{7.18}$$

(i) Uncracked section excluding steel areas
$$M_{\text{Cracking}} = (f_{\text{ctm}} = 2.9) \times z_b = 20.88 \text{ kNm}$$
$$M = 77.85 \text{ kNm}$$
$$M_{\text{cr}}/M = 20.88/77.85 = 0.268$$

Calculate ζ using Eurocode 2 equation (7.19).
$$\zeta = 1 - (\beta = 0.5) \times 0.268^2 = 0.964 \tag{7.19}$$
$$\text{Curvature, } 1/r = 0.964 \times 6.775 \times 10^{-6} + (1 - 0.964) \times 7.916 \times 10^{-6}$$
$$= 6.816 \times 10^{-6} \text{ mm}^{-1}$$

For a simply supported beam carrying a uniform load, $K = 5/48$.
$$\text{Deflection } a = K\,L^2\,(1/r_b), \quad L = 6000 \text{ mm}$$
$$\text{Mid-span deflection, } \Delta = (5/48) \times L^2 \times (1/r) = 26 \text{ mm}$$
$$\text{Permissible, } \Delta = \text{Span}/250 = 6000/250 = 24 \text{ mm}$$

(ii) Uncracked section including steel areas
$$M_{Cracking} = (f_{ctm} = 2.9) \times z_b = 41.53 \text{ kNm}$$
$$M_{cr}/M = 41.53/77.85 = 0.533$$
$$\zeta = 1 - (\beta = 0.5) \times 0.533^2 = 0.858 \quad\quad (7.19)$$
$$\text{Curvature, } 1/r = 0.858 \times 6.775 \times 10^{-6} + (1 - 0.858) \times 4.248 \times 10^{-6}$$
$$= 6.416 \times 10^{-6} \text{ mm}^{-1}$$
$$\text{Mid-span deflection, } \Delta = (5/48) \times L^2 \times (1/r) = 24 \text{ mm}$$
$$\text{Permissible, } \Delta = \text{Span}/250 = 6000/250 = 24 \text{ mm}$$
Note that the two sets of results are reasonably close.

17.3 CALCULATION OF CRACK WIDTHS

Calculation of crack widths in connection with the design of structures retaining aqueous liquids was discussed in Chapter 17. In this section determination of crack widths in connection with beams will be discussed.

17.3.1 Cracking in Reinforced Concrete Beams

Concrete is weak in tension. A reinforced concrete beam cracks in flexure on the tension face when the tensile strength of the concrete is exceeded. Tension cracks due to bending (not shear cracks) form at fairly small values of bending moment. These cracks exist even at serviceability limit state loads. As long as these cracks are not unacceptably wide and are well spaced, they are harmless and do not distract from appearance or encourage corrosion of reinforcement. It has to be appreciated that cracking is a semi-random phenomenon and that it is not possible to predict an absolute maximum crack width or the exact spacing of cracks. Generally the closer the spacing, lesser will be the width of an average crack.

Strain steel is an approximate measure of the crack width. For example if σ_{sm} is the mean stress in steel, the corresponding mean strain ε_{sm} is equal to σ_{sm}/E_s, where E_s is Young's modulus for steel.

If $S_{r, max}$ is the maximum crack spacing, then
$$\text{Average crack width}/ S_{r, max} \approx \varepsilon_{sm}$$
However it has to be recognized that concrete losses its tensile strength only at cracks. Between cracks, concrete is still capable of resisting tensile stress. This is known as tension stiffening. In order to get a more realistic measure of the crack width, it is necessary to modify the relationship between crack spacing, average crack width and strain in steel by including the average strain in concrete as well. In other words,
$$w_k/ S_{r, max} \approx (\varepsilon_{sm} - \varepsilon_{cm})$$
where w_k = average crack width, ε_{cm} = mean strain in concrete between cracks.

This is the basic concept used in the code calculation of crack widths and spacing as given in section 7.3.4 of Eurocode 2. The following equations may be used for calculating crack widths.

(a) Crack width w_k may be calculated using code equation (7.8).

$$w_k = S_{r, max} (\varepsilon_{sm} - \varepsilon_{cm}) \qquad (7.8)$$

where
$S_{r, max}$ = maximum crack spacing.
ε_{sm} = mean strain in reinforcement.
ε_{cm} = mean strain in concrete between cracks.

(b) $(\varepsilon_{sm} - \varepsilon_{cm})$ may be calculated from the code equation (7.9).

$$(\varepsilon_{sm} - \varepsilon_{cm}) = \frac{\sigma_s - k_t \frac{f_{ct.eff}}{\rho_{p.eff}} (1 + \alpha_e \, \rho_{p.eff})}{E_s} \geq 0.6 \frac{\sigma_s}{E_s} \qquad (7.9)$$

where
σ_s = stress in tension reinforcement assuming cracked section.
α_e = modular ratio = E_s/E_{cm}.
$\rho_{p, eff} = A_s/A_{c, eff}$.
k_t = 0.6 for short term loading.
    = 0.4 for long term loading.
$A_{c, eff}$ = Effective area of concrete surrounding the reinforcement as shown in Fig. 17.5.

$$h_{c, eff} = min [2.5 (h - d); (h - x); 0.5h]$$

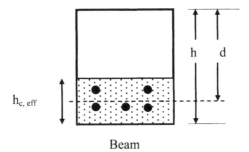

Beam

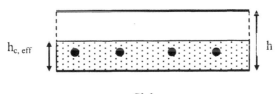

Slab

Fig. 17.5 $h_{c, eff}$ for beam and slab.

(c) $S_{r, max}$ may be calculated from code equation (7.31)

In the case of bending where high bond bars act as tension reinforcement,
$$S_{r,\,max} = k_3\,c + k_1\,k_2\,k_4\,\varphi/\,\rho_{p,\,eff} \qquad (7.11)$$
where
$k_1 = 0.8$ for high bond bars, $k_2 = 0.5$ for bending, $k_3 = 3.4$, $k_4 = 0.425$
$$S_{r,\,max} = 3.4\,c + 0.128\,\varphi/\,\rho_{p,\,eff}$$
c = cover to longitudinal reinforcement.
φ = bar diameter.
$$\rho_{p,\,eff} = A_s/A_{c,\,eff} \qquad (7.10)$$
If there are n_1 bars of diameter φ_1 etc., an equivalent diameter φ_{eq} is given by
$$\varphi_{eq} = \Sigma n_i\,\varphi_i^2/\,\Sigma n_i\,\varphi_i$$

17.4 EXAMPLE OF CRACK WIDTH CALCULATION FOR T-BEAM

The beam chosen is shown in Fig. 17.2. The total moment at the section due to service loads is 111.6 kNm.
The materials are concrete, $f_{ck} = 30$ MPa and reinforcement, $f_{yk} = 500$ MPa.
Determine the spacing and width of cracks.

Solution: The properties of the cracked section are computed first. The values for the moduli of elasticity are as follows:
Reinforcement = $E_s = 200$ GPa
$$\text{Concrete} = E_{cm} = 22 \times (0.8 + 0.1f_{ck})^{0.3} = 32.84 \text{ GPa}$$
$$\text{Modular ratio} = \alpha_e = 200/32.84 = 6.1$$
Assume neutral axis depth $x < h_f$, depth of flange. Ignore tensile strength of concrete.
$$\text{Compression force in concrete, } C_c = 0.5\,f_c \times 1450\,x$$
Compression force in compression steel, $C_s = A_s' \times f_{sc} = 402 \times \alpha_e \times f_c\,(x-45)/x$
Tensile force in steel, $T_s = A_s \times f_{st} = 1472 \times \alpha_e \times f_c\,(300-x)/x$
For equilibrium in the axial direction:
$$C_c + C_s - T_s = 0$$
Multiplying throughout by x/f_c and simplifying
$$0.5 \times 1450\,x^2 + 402 \times \alpha_e \times (x-45) - 1472 \times \alpha_e \times (300-x) = 0$$
Simplifying
$$x^2 + 15.77\,x - 3867.7 = 0$$
$$x = 55 \text{ mm}$$
$$M = C_c \times 0.67\,x + C_s \times (x-45) + T_s \times (300-x)$$
$$M = 111.6 \text{ kNm}$$
$$M = C_c \times 0.67\,x + C_s \times (x-45) + T_s \times (300-x)$$
$$M = (f_c/x)\,[0.33 \times 1450 \times x^3 + 402 \times \alpha_e \times (x-45)^2 + 1472 \times \alpha_e \times (300-x)^2]$$
$$f_c = 9.8 \text{ MPa}$$
The compressive strain εc in the concrete is
$$\varepsilon_c = f_c/E_c = 9.8/\,(32.84 \times 10^3) = 0.3 \times 10^{-3}$$
Strain ε_s at steel level
$$\varepsilon_s = (\varepsilon_c\,/x)\,(d-x)$$

$$\varepsilon_s = 0.3 \times 10^{-3} \times (300 - 55)/55 = 1.336 \times 10^{-3}$$
$$\sigma_s = \varepsilon_s \times E_s = 267 \text{ MPa}$$
$$x = 55 \text{ mm}, h = 350 \text{ mm}, d = 300 \text{ mm}$$
$$h_{c,\,ef} = \min [(2.5 \times (350 - 300); (350 - 55)/3; 350/2]$$
$$h_{c,\,ef} = \min [125; 98; 175] = 98 \text{ mm}$$
$$A_{c,\,ef} = b_w \times h_{c,\,ef} = 24583 \text{ mm}^2$$
$$\rho_{p,\,eff} = As/A_{c,\,ef} = 1472/24583 = 0.06$$
$$f_{ct,\,eff} = f_{ctm} = 0.3 \times f_{ck}^{\,0.67} = 2.9 \text{ MPa}$$
$$(k_t = 0.4) \times (2.9/\,0.06) \times (1 + 6.1 \times 0.06) = 26 \text{ MPa}$$
$$(\sigma_s = 267) - 26 = 241 \text{ MPa} > 0.6\,\sigma_s$$
$$(\varepsilon_{sm} - \varepsilon_{cm}) = 241/E_s = 1.21 \times 10^{-3} \quad\quad (7.9)$$
$$S_{r,\,max} = 3.4\,c + 0.128\,\varphi/\,\rho_{p,\,eff}$$
$$\varphi = 25 \text{ mm}$$
$$\text{Cover}, c = h - d - \varphi/2 = 37.5 \text{ mm}$$
$$S_{r,\,max} = 3.4\,c + 0.128\,\varphi/\,\rho_{p,\,eff} = 3.4 \times 37.5 + 0.128 \times 25/\,0.06 = 181 \text{ mm} \quad (7.11)$$
$$w_k = S_{r,\,max}\,(\varepsilon_{sm} - \varepsilon_{cm}) = 181 \times 1.21 \times 10^{-3} = 0.22 \text{ mm} \quad\quad (7.8)$$

Note: In the above calculation, E_{cm} for concrete was taken as the short term value. However as the load is sustained load, then it is more meaningful to take the value of E_{cm} as affected by creep strains. In that case taking $\phi(\infty, t_0)] = 3.4$,
$$E_c = 34.5/\,[1 + \phi(\infty, t_0)] = 7.5 \text{ GPa}$$
Modular ratio $= \alpha_e = 200/7.5 = 26.7$
$$x^2 + 69.03\,x - 16929.1 = 0$$
$$x = 100 \text{ mm}$$
$$M = C_c \times 0.67\,x + C_s \times (x - 45) + T_s \times (300 - x)$$
$$M = 111.6 \text{ kNm}$$
$$M = C_c \times 0.67\,x + C_s \times (x - 45) + T_s \times (300 - x)$$
$$= (f_c/x)\,[0.33 \times 1450 \times x^3 + 402 \times \alpha_e \times (x - 45)^2 + 1472 \times \alpha_e \times (300 - x)^2]$$
$$f_c = 5.4 \text{ MPa}$$
The compressive strain εc in the concrete is
$$\varepsilon_c = f_c/E_c = 5.4/\,(7.5 \times 10^3) = 0.713 \times 10^{-3}$$
Strain ε_s at steel level
$$\varepsilon_s = (\varepsilon_c/x)\,(d - x)$$
$$\varepsilon_s = 0.713 \times 10^{-3} \times (300 - 100)/100 = 1.425 \times 10^{-3}$$
$$\sigma_s = \varepsilon_s \times E_s = 285 \text{ MPa}$$
$$x = 100 \text{ mm}, h = 350 \text{ mm}, d = 300 \text{ mm}$$
$$h_{c,\,ef} = \min [(2.5 \times (350 - 300); (350 - 100)/3; 350/2]$$
$$h_{c,\,ef} = \min [125; 83; 175] = 83 \text{ mm}$$
$$A_{c,\,ef} = b_w \times h_{c,\,ef} = 20833 \text{ mm}^2$$
$$\rho_{p,\,eff} = As/A_{c,\,ef} = 1472/20833 = 0.071$$
$$f_{ct,eff} = f_{ctm} = 0.3 \times f_{ck}^{\,0.67} = 2.9 \text{ MPa}$$
$$(k_t = 0.4) \times (2.9/\,0.071) \times (1 + 26.7 \times 0.071) = 47 \text{ MPa}$$
$$(\sigma_s = 285) - 47 = 238 \text{ MPa} > 0.6\,\sigma_s$$
$$(\varepsilon_{sm} - \varepsilon_{cm}) = 238/E_s = 1.19 \times 10^{-3}$$
$$S_{r,\,max} = 3.4\,c + 0.128\,\varphi/\,\rho_{p,\,eff}$$

$$\varphi = 25 \text{ mm}$$
$$\text{Cover, } c = h - d - \varphi/2 = 37.5 \text{ mm}$$
$$S_{r, \max} = 3.4 \, c + 0.128 \, \varphi/ \, \rho_{p, eff} = 3.4 \times 37.5 + 0.128 \times 25/ \, 0.071 = 173 \text{ mm}$$
$$w_k = Sr_{, \max} \, (\varepsilon_{sm} - \varepsilon_{cm}) = 173 \times 1.19 \times 10^{-3} = 0.21 \text{ mm}$$

Quite evidently, the effect of using a creep affected value of E_{cm} does not make a great difference to the calculated crack width.

17.5 REFERENCES

Bhatt, P. (1999). *Structures*. Longman.

Ghali, A., Favre, R. and Elbadry, M. (2002). *Concrete Structures: Stresses and Deformations*. E & FN Spon.

Gilbert, I.G. and Ranzi, G. (2011). *Time Dependent Behaviour of Concrete Structures*. Spon Press.

CHAPTER 18

A GENERAL METHOD OF DESIGN AT ULTIMATE LIMIT STATE

18.1 INTRODUCTION

In the previous chapters, designs of different structural elements such as beams, columns, slabs were discussed. The design equations developed were largely based on tests on elements similar to the ones commonly met in practice. This approach has the obvious disadvantage that the methods cannot be safely used outside the range of parameters tested. In practice one has to design structures which are outside the norm of day-to-day structures. The design of such structures needs a design approach which has wide validity. Such an approach needs to be based on

- Sound theoretical principles such as theory of elasticity and plasticity
- Material behaviour

The object of this chapter is to present such an approach. Nowadays analysis of complex structures to determine the elastic stress distribution using finite element programs has become commonplace. Unfortunately design at ultimate limit state cannot be based on the assumption that the material will remain elastic as this makes the designs extremely expensive. This means it is necessary to allow yielding of the sections at ultimate limit state. In other words design method needs to be based on theory of plasticity. However application of traditional theory of plasticity to the design of reinforced concrete structures suffers from two fundamental limitations. They are

- It assumes large ductility of material such as steel. Unfortunately reinforced concrete is a material which has limited ductility even when yielding of steel precedes the crushing of concrete.
- It can be used to predict the collapse load of a structure which is already fully designed!

Fortunately a close examination of the limit theorems of the theory of plasticity points to a way of overcoming both of the limitations stated above.

18.2 LIMIT THEOREMS OF THE THEORY OF PLASTICITY

Theory of plasticity is based on three fundamental assumptions:
- The material has unlimited ductility.
- Failure by buckling is prevented.
- At collapse sufficient areas yield to convert the structure into a mechanism.

These concepts lead to the following three conditions.

1. Equilibrium condition: The state of stress and the applied load must be in equilibrium.
2. Yield condition: The state of stress at no point should violate the yield criterion for the material.
3. Mechanism condition: At ultimate load sufficient parts of the structure must yield in order to convert the structure into a mechanism indicating it cannot support additional load.

If the calculation of the maximum load carried is based on satisfying only equilibrium and yield conditions, it is possible that insufficient areas have yielded to convert the structure into a mechanism. This indicates that the load calculated is equal to or less than the true collapse load. The load is called a lower bound to the true collapse load. For example Hillerborg's strip method of design of slabs discussed in section 8.11 of Chapter 8 is an example of design using the lower bound approach as the loading on the strips is based on simply satisfying equilibrium and the corresponding bending moments and the reinforcement satisfy the yield condition.

On the other hand, if the collapse load is calculated for an assumed collapse mechanism as was done by the yield line analysis method discussed in Chapter 8, then the collapse load calculated is equal to or greater than the true collapse load as there might be other mechanisms which yield a smaller collapse load. The maximum load calculated by this method is called an upper bound to the true collapse load.

It is worth pointing out that the calculations are directed at determining the collapse load. There is no attention paid to the behaviour of the structure at working or serviceability limit state.

18.3 REINFORCED CONCRETE AND LIMIT THEOREMS OF THE THEORY OF PLASTICITY

Two aspects of a steel structure compared with a reinforced concrete structure need to be focused on before considering how to apply principles of the theory of plasticity to the design of reinforced concrete structures.

*   Reinforced concrete has very limited plasticity.
*   Provided buckling is prevented for a given cross section and yield stress, the maximum capacity of a steel section is fixed. In the case of reinforced concrete, the maximum capacity can be changed by varying the amount of reinforcement provided.

Taking note of the above observations, the following procedure can be adopted for designing a structure to carry a given ultimate load.

Equilibrium condition: Based on experience and preliminary design, assume cross sections dimensions and carry out an elastic analysis to determine the stress distribution when the structure is carrying the ultimate load. The object of this step is to simply obtain a set of stresses in equilibrium with the applied load. It is appreciated that this might not be the actual stress distribution at ultimate load as certain sections of the structure might yield, disturbing the elastic stress distribution. It is not always necessary to use elastic stress distribution. However, especially for complex structures, elastic stress distribution is easily calculated using elastic finite element analysis packages. The main aim of this step is to satisfy the equilibrium condition.

Yield condition: Using the elastic stresses and a known yield criterion for the combination of stresses considered, determine the necessary reinforcement. Because the reinforcement is determined so as not to violate the yield condition, this step satisfies the yield condition.

Mechanism condition: If the reinforcement is tailored to satisfy the yield condition at all points of the structure, then the structure will yield at all points and the structure will automatically become a mechanism. This step will be explained in a later section and illustrated with examples.

However in practice, it will not be practicable to vary reinforcement from point to point. In other words the structure might not become a mechanism and the true ultimate load will be larger than the design ultimate load. The design ultimate load becomes a lower bound to the true ultimate load.

Ductility demand: One advantage of this approach is that the load at which yielding occurs first at a section and the load at which collapse occurs will be fairly small and will be even smaller when compared with the design ultimate load. This greatly reduces the need for large ductility and hence recognizes the limitations of reinforced concrete as a ductile material.

18.4 DESIGN OF REINFORCEMENT FOR IN-PLANE STRESSES

Fig. 18.1 shows a membrane element in a state of plane stress. The normal forces per unit length in the x- and y-directions are respectively N_x and N_y. The shear force per unit length is N_{xy}. The sign convention for the normal stresses is tensile stress is positive. Shear stress is positive as shown in Fig. 18.1. If the thickness of the element is t, the stresses are related to forces per unit length as follows.

$$\sigma_x = N_x/t, \ \sigma_y = N_y/t, \ \tau_{xy} = N_{xy}/t$$

The applied forces per unit length (N_x, N_y and N_{xy}) are resisted by stresses in concrete and stresses in steel.

If the principal stresses in concrete are σ_1 and σ_2 with σ_1 inclined at an angle θ to the x-axis, the corresponding forces in concrete per unit length in the cartesian coordinate system are as shown in Fig. 18.2, where $N_1 = \sigma_1 t$ and $N_2 = \sigma_2 t$.

It is assumed here that on a principal plane, any shear resistance due to intergranular friction is ignored.

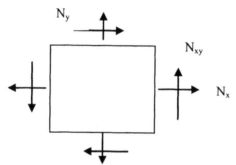

Fig. 18.1 In-plane stresses.

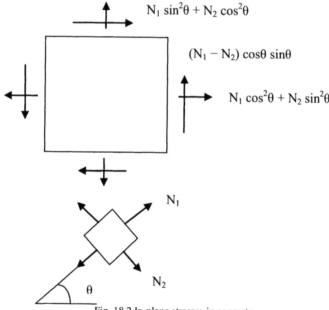

Fig. 18.2 In-plane stresses in concrete.

If the areas of steel in x- and y-directions are respectively A_x and A_y and the stress in steel f_x and f_y, the resistance due to reinforcement is as shown in Fig. 18.3.

It is assumed here that any resistance to shear due to dowel action of steel is ignored as this comes into play only after the formation of large width of cracks at the cracked plane.

Equating the applied force to the combined resistance of steel and concrete,

$$N_x = N_1 \cos^2\theta + N_2 \sin^2\theta + A_x\, f_x$$
$$N_y = N_2 \cos^2\theta + N_1 \sin^2\theta + A_y\, f_y$$
$$N_{xy} = (N_1 - N_2) \cos\theta \sin\theta = 0.5(N_1 - N_2) \sin 2\theta \quad (C18.1)$$

Note that the maximum value of $N_{xy} = 0.5(N_1 - N_2)$.

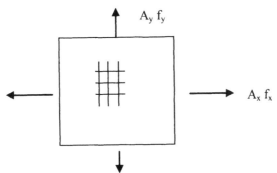

Fig. 18.3 Uniaxial stresses in steel.

The object of design is to match resistance to applied force with minimum steel consumption and without exceeding the strength of concrete in compression. It is assumed that the tensile strength of steel can be ignored.

Eliminating N_2 and θ from equation (C18.1) by equating the product of first two equations to the square of the third equation,

$$(N_x - A_x\, f_x)\, (N_y - A_y\, f_y) - N^2_{xy} = 0 \qquad (C18.2)$$

This is known as the yield criterion for orthogonally reinforced concrete for in-plane stress conditions.

Adding the first two equations of (C18.1),

$$N_x + N_y = N_1 + N_2 + A_x\, f_x + A_y\, f_y \qquad (C18.3)$$

In order to determine he reinforcement areas for a given state of stress, four possible scenarios need to be considered as follows.

Case 1: The principal forces in concrete N_1 and N_2 are both compressive and do not exceed the permissible compressive stress for uncracked concrete equal to $f_{cd} = f_{ck}/\gamma_c$. In such a case there is no need for reinforcement. If the compressive stresses are larger than permissible, an increase in the thickness t instead of compressive reinforcement gives the optimal solution.

The principal stresses are:

$$N_1 = 0.5\{(N_x + N_y) + \sqrt{[(N_x - N_y)^2 + 4\, N^2_{xy}]} \geq -f_{cd}\, t$$
$$N_2 = 0.5\{(N_x + N_y) - \sqrt{[(N_x - N_y)^2 + 4\, N^2_{xy}]} \geq -f_{cd}\, t$$

Case 2: If the major principal force N_1 is tensile, it is set equal to zero. Equation (C18.3) simplifies to

$$N_x + N_y = N_2 + A_x\, f_x + A_y\, f_y \qquad (C18.4)$$

Equation (C18.4) together with equation (C18.2) can be used to derive the reinforcement for two cases as follows.

Case 2a: No reinforcement is needed in the x-direction and steel in the y-direction reaches the permissible tensile stress equal to $f_{yd} = f_{yk}/\gamma_s$. The compressive stress in cracked concrete is equal to or less than $v f_{cd} = (f_{ck}/\gamma_c) \times 0.6 \times (1 - f_{ck}/250)$.
Equation (C18.2) reduces to

$$N_x\, (N_y - A_y\, f_{yd}) - N^2_{xy} = 0$$

$$A_y \, f_{yd} = N_y - N^2_{xy}/N_x \tag{C18.5}$$
$$A_y = 0 \text{ if } N_x \, N_y/N^2_{xy} = 1$$

From equation (C18.4),

$$N_2 = N_x + N_y - A_y \, f_y = N_x + N^2_{xy}/N_x \tag{C18.6}$$

Case 2b: No reinforcement is needed in the y-direction and steel in the x-direction reaches the permissible tensile stress equal to $f_{yd} = f_{yk}/\gamma_s$. The compressive stress in cracked concrete is equal to or less than $vf_{cd} = (f_{ck}/\gamma_c) \times 0.6 \times (1 - f_{ck}/250)$
Equation (C18.2) reduces to

$$N_y \, (N_x - A_x \, f_{yd}) - N^2_{xy} = 0$$
$$A_x \, f_{yd} = N_x - N^2_{xy}/N_y \tag{C18.7}$$
$$A_x = 0 \text{ if } N_x \, N_y/N^2_{xy} = 1$$

From equation (C18.4),

$$N_2 = N_x + N_y - A_x \, f_x = N_y + N^2_{xy}/N_y \tag{C18.8}$$

Case 3: In this case both A_x and A_y are greater than zero and $f_x = f_y = f_{yd}$. The object is to minimize the total quantity of steel $A_{sTotal} = A_x + A_y$.

$$A_{sTotal} = (A_x + A_y) \, f_{yd}$$

Substituting for A_y from equation (C18.2),

$$A_{sTotal} = A_x \, f_{yd} + N_y - N^2_{xy}/ (A_x \, f_{yd} - N_x)$$

To obtain the minimum value of A_{sTotal}, differentiating S with respect to A_x, and setting d A_{sTotal} /dA_x to zero

$$1 = N^2_{xy}/ (A_x \, f_{yd} - N_x)^2 = 0$$

Since $(A_x f_{yd} - N_x) > 0$,

$$A_x f_{yd} - N_x = \left| N_{yx} \right|$$

$\left| N_{yx} \right|$ is the positive numerical value of N_{xy}.

$$A_x \, f_{yd} = N_x + \left| N_{yx} \right|$$
$$A_y \, f_{yd} = N_y + \left| N_{yx} \right| \tag{C18.9}$$
$$N_2 = -2 \left| N_{yx} \right| \tag{C18.10}$$

The compressive stress in cracked concrete is equal to or less than

$$vf_{cd} = (f_{ck}/\gamma_c) \times 0.6 \times (1 - f_{ck}/250).$$

Note that

$$A_x \, f_{yd} = 0 \text{ if } N_x + \left| N_{yx} \right| = 0$$
$$A_y \, f_{yd} = 0 \text{ if } N_y + \left| N_{yx} \right| = 0$$

Summary: From the consideration of the four cases discussed above, the regions of application to the individual cases can be summarised as follows.
Case 1: $N_x \, N_y/ N^2_{xy} > 1, \; A_x = A_y = 0$

Case 2a: $N_x/\left| N_{yx} \right| < -1 \text{ and } Nx \, N_y/ N^2_{xy} \le 1$
$$A_x = 0, \, A_y \, f_{yd} = N_y - N^2_{xy}/N_x$$
$$N_2 = N_x + N_y - A_y \, f_y = N_x + N^2_{xy}/N_x$$

Case 2b: $N_y/\left| N_{yx} \right| < -1 \text{ and } N_x \, N_y/ N^2_{xy} \le 1$
$$A_y = 0, \, A_x \, f_{yd} = N_x - N^2_{xy}/N_y$$
$$N_2 = N_y + N^2_{xy}/N_y$$

Case 3:

$$N_x/|N_{yx}| \geq -1 \text{ and } N_y/|N_{yx}| \geq -1$$
$$A_x f_{yd} = N_x + |N_{yx}|$$
$$A_y f_{yd} = N_y + |N_{yx}|$$
$$N_2 = -2|N_{yx}|$$

Fig. 18.4 summarises in a graphical form the equations for the design of reinforcement.

The equations in this section are same as the equations in Annex F of Eurocode 2 except for a change in notation such as σ_{Edx} for σ_x, σ_{Edy} for σ_y, τ_{Edxy} for τ_{xy}, f_{tdx} for $A_x f_{yd}/t$ and f_{tdy} for $A_y f_{yd}/t$. The Annex assumes that compressive stress is positive.

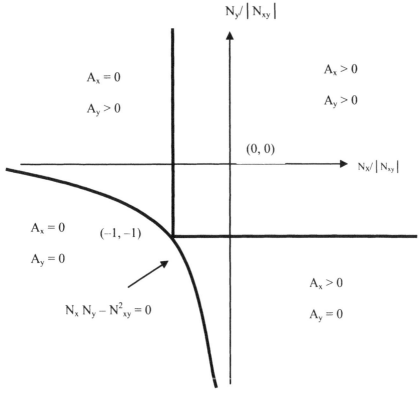

Fig. 18.4 Design zones.

18.4.1 Examples of Reinforcement Calculations

In the following examples, assume:
Thickness, $t = 350$ mm, $f_{yk} = 500$ MPa, $\gamma_s = 1.15$, $f_{yd} = 435$ MPa, $f_{ck} = 30$ MPa, $\gamma_c = 1.50$, $f_{cd} = 20$ MPa, $v = 0.6 (1 - f_{ck}/250) = 0.528$, $vf_{cd} = 10.56$ MPa.
Maximum shear stress $= vf_{cd}/2 = 5.28$ MPa.

Reinforcement is required for the given values of (N_x, N_y and N_{xy}).

Example 1a: $N_x = 4000$N/mm, $N_y = -2000$ N/mm, $N_{xy} = -1200$ N/mm.
$N_x/|N_{yx}| = 3.3 > -1$, $N_y/|N_{yx}| = -1.7 < -1$, $N_x N_y/N^2_{xy} = -5.6 < 1$
As $N_x N_y/N^2_{xy} \leq 1$ and $N_y/|N_{yx}| < -1$, this combination falls into case 2b.
$$A_y = 0, \ Ax \times 435 = N_x - N^2_{xy}/N_y, \quad A_x = 10.0 \text{ mm}^2/\text{mm}$$
Provide H25 bars at 45 mm centres giving 10.9 mm^2/mm.
$$N_2 = N_y + N^2_{xy}/N_y = -2720 \text{ N/mm}, \ \sigma_2 = N_2/t = -7.8 \text{ MPa}$$
Compressive stress is less than vf_{cd}.
$$[\,|N_{yx}| = 1200] < [0.5 \ \sigma_2 \ t = 1365]$$

Example 1b: $N_x = -2400$N/mm, $N_y = 3000$ N/mm, $N_{xy} = 1600$ N/mm.
$N_x/|N_{yx}| = -1.5 < -1$, $N_y/|N_{yx}| = 1.93 > -1$, $N_x N_y/N^2_{xy} = -2.8 < 1$
As $N_x N_y/N^2_{xy} \leq 1$ and $N_x/|N_{yx}| < -1$, this combination falls into case 2a.
$$A_x = 0, \ A_y \times 435 = N_y - N^2_{xy}/N_x, \quad A_y = 9.35 \text{ mm}^2/\text{mm}$$
Provide H25 bars at 40 mm centres giving 12.3 mm^2/mm.
$$N_2 = N_x + N^2_{xy}/N_x = -3467 \text{ N/mm}, \ \sigma_2 = N_2/t = -9.9 \text{ MPa}.$$
Compressive stress is less than vf_{cd}.
$$[\,|N_{yx}| = 1600] < [0.5 \ \sigma_2 \ t = 1732]$$

Example 1c: $N_x = 2000$ N/mm, $N_y = 3000$ N/mm, $N_{xy} = 1800$ N/mm.
As $N_x/|N_{yx}| \geq -1$ and $N_y/|N_{yx}| \geq -1$, this combination falls into case 3.
$$A_x \times 435 = N_x + |N_{yx}|, \quad A_x = 8.74 \text{ mm}^2/\text{mm}$$
Provide H25 bars at 50 mm centres giving 9.82 mm^2/mm
$$A_y \times 435 = N_y + |N_{yx}|, \quad A_y = 11.0 \text{ mm}^2/\text{mm}$$
Provide H25 bars at 40 mm centres giving 12.3 mm^2/mm.
$$N_2 = -2\,|N_{yx}| = -3600 \text{ N/mm}, \ \sigma_2 = N_2/t = -10.3 \text{ MPa}$$
Compressive stress is less than vf_{cd}.
$$[\,|N_{yx}| = 1800] \approx [0.5 \ \sigma_2 \ t = 1800]$$

Example 1d: $N_x = -2400$ N/mm, $N_y = -3000$ N/mm, $N_{xy} = 2000$ N/mm.
$$N_x/|N_{yx}| = -1.2, N_x/|N_{yx}| = -1.5, N_x N_y/N^2_{xy} = 1.8$$
As $N_x N_y/N^2_{xy} > 1$, this combination falls into case 1.
$$A_x = A_y = 0$$
$N_1 = 0.5\{(N_x + N_y) + \sqrt{[(N_x - N_y)^2 + 4 N^2_{xy}]} = -678 \text{ N/m}, \ \sigma_1 = N_1/t = -1.9 \text{ MPa}$
$N_2 = 0.5\{(N_x + N_y) - \sqrt{[(N_x - N_y)^2 + 4 N^2_{xy}]} = -4722 \text{ N/mm},$
$$\sigma_2 = N_2/t = -13.5 \text{ MPa}. \geq - (f_{cd} = 20 \text{ MPa})$$

18.4.2 An Example of Application of Design Equations

Fig. 18.5 shows a box girder 1.2 m wide × 1.7 m deep with 200 mm thick walls. At a cross section at ultimate limit state, it is subjected to the following load combinations: twisting moment, T = 10000 kNm, bending moment, M = 1500 kNm and shear force, V = 900 kN. Design the necessary reinforcement assuming that $f_{yd} = 435$ MPa, $f_{cd} = 20$ MPa, $vf_{cd} = 10.56$ MPa.

Step 1: Carry out an elastic stress analysis
(i) Torsion:
Centre line dimensions of the box:

Width = 1.2 – 0.2 = 1.0 m, depth = 1.7 – 0.2 = 1.5 m

Area enclosed by centre line, $A_k = 1.0 \times 1.5 = 1.5$ m^2
From equation 6.26 of Euroode 2, shear flow in the sides of the box, N_{xy}

$$N_{xy} = T/ (2 \times A_k) = 333 \text{ kN/m} = 333 \text{ N/mm}$$

(ii) Shear force:

Second moment of area, $I = (1.2 \times 1.7^3 - 1.0 \times 1.5^3)/12 = 0.345$ m^4

Shear flow N_{xy} is calculated at three levels in the box as follows.

Top flange-web junction and bottom flange-web junction:
First moment S of area about the centroidal axis of half the area above the junction,

$$S = 0.5 \times (1.2 \times 0.2) \times (1.7/2 - 0.2/2) = 0.09 \text{ m}^2$$
$$N_{xy} = V \times S/I = 900 \times 0.09/ 0.345 = 235 \text{ kN/m} = 235 \text{ N/mm}$$

Shear flow in the flanges will be distributed in a triangular fashion as shown in
Fig. 18.5.   The average value is $0.5 \times 235 = 118$ N/mm.   It acts in opposite
directions on the two halves of the flange.   It is additive to the shear flow from
torsion on the left half of the flange and subtractive on the right half of the flange.

Centroidal axis:
First moment of area about the centroidal axis of half the area above the centroidal
axis,

$$S = 0.5 \times [(\{1.2 \times 1.7/2^2\}/ 2) - (\{0.8 \times 1.3/2^2\}/ 2)] = 0.13225 \text{ m}^2$$
$$N_{xy} = V \times S/I = 900 \times 0.13225 / 0.345 = 345 \text{ kN/m} = 345 \text{ N/mm}$$

An average N_{xy} in the web $\approx 235 + (2/3) (345 - 235) = 308$ N/mm.
Note the factor 2/3 comes from the fact that in a parabola, the average width is 2/3
the maximum width.

(iii) Bending moment:
N_x in the top flange: $- (M/I) \times y \times t = (1500/0.345) \times (1.5/2) \times 0.2 = -652$ N/mm.
N_x in the bottom flange: $(M/I) \times y \times t = (1500/0.345) \times (1.5/2) \times 0.2 = 652$ N/mm

Average N_x in the top half of the web $= -652/2 = -326$ N/mm.
Average N_x in the bottom half of the web $= 652/2 = 326$ N/mm.

Step 2: Calculate the total forces at different sections of the box
Top flange:

$$N_x = -652 \text{ N/mm}, N_y = 0$$
$$N_{xy} = 333 \text{ (torsion)} + 118 \text{ (shear)} = 451 \text{ N/mm}$$
$$N_x/|N_{yx}| = -1.45, N_y/|N_{yx}| = 0, N_x N_y/N^2_{xy} = 0$$

As $N_x/|N_{yx}| < -1$ and $N_x N_y/ N^2_{xy} \le 1$, this falls into case 2a.

$$A_x = 0$$
$$A_y f_{yd} = N_y - N^2_{xy}/N_x = 312, A_y = 0.72 \text{ mm}^2/\text{mm}$$

Provide H16 bars at 275 mm centres which gives $A_y = 0.73$ mm^2/mm.
$$N_2 = N_x + N^2_{xy}/N_x = -964 \text{ N/mm}, \sigma_2 = 4.8 \text{ MPa} < v \text{ } f_{cd}$$

Bottom flange:

$$N_x = 652 \text{ N/mm}, \, N_y = 0$$
$$N_{xy} = 333 \text{ (torsion)} + 118 \text{ (shear)} = 451 \text{ N/mm}$$
$$N_x / |N_{xy}| = 1.45, \, N_y / |N_{yx}| = 0, \, N_x N_y / N^2_{xy} = 0$$

As $N_x / |N_{yx}| \geq -1$ and $N_y / |N_{yx}| \geq -1$, this falls into case 3.

$$N_x + |N_{yx}| = 1103 \text{ N/mm}, \, N_y + |N_{yx}| = 451 \text{ N/mm}$$
$$A_x = 2.54 \text{ mm}^2/\text{mm}$$

Over a total width of flange equal to 1200 mm, $A_x = 3048 \text{ mm}^2$.

Provide 7H25 bars which gives a total area of 3436 mm^2.

$$N_y + |N_{yx}| = 451 \text{ N/mm}$$
$$A_y = 1.04 \text{ mm}^2/\text{mm}$$

H16 bars at 190 mm centres gives $A_y = 1.06 \text{ mm}^2/\text{mm}$. This will be part of shear links.

$$N_2 = -2|N_{yx}| = -902 \text{ N/mm}, \, \sigma_2 = 4.8 \text{ MPa} < v \, f_{cd}$$

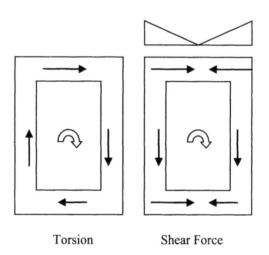

Torsion   Shear Force

Fig. 18.5 Shear flow due to torsion and shear force.

Top half of web:

$$N_x = -326 \text{ N/mm}, \, N_y = 0$$
$$N_{xy} = 333 \text{ (torsion)} + 308 \text{ (shear)} = 641 \text{ N/mm}$$
$$N_x / |N_{yx}| = -0.51, \, N_y / |N_{yx}| = 0, \, N_x N_y / N^2_{xy} = 0$$

As $N_x / |N_{yx}| > -1$ and $Nx \, N_y / N^2_{xy} \leq 1$, this falls into case 3.

$$A_x f_{yd} = N_x + |N_{yx}| = 314, \, A_x = 0.72 \text{ mm}^2/\text{mm}$$

Over a depth of $1.7/2 = 0.85$ m, $A_x = 616 \text{ mm}^2$.

Provide 4H16 giving a total area of 804 mm^2.

$$A_y \, f_{yd} = N_y + |N_{yx}| = 641, \, A_y = 1.47 \text{ mm}^2/\text{mm}$$

Provide H16 at 130 mm centres giving $A_y = 1.55 \text{ mm}^2/\text{mm}$. This will be part of shear links.

$$N_2 = -2|N_{yx}| = -1282 \text{ N/mm}, \, \sigma_2 = -6.4 \text{ MPa} < v \, f_{cd}$$

Bottom half of web:
$$N_x = 326 \text{ N/mm}, N_y = 0$$
$$N_{xy} = 333 \text{ (torsion)} + 308 \text{ (shear)} = 641 \text{ N/mm}$$
$$N_x/|N_{yx}| = 0.51, N_y/|N_{yx}| = 0, N_x N_y/N^2_{xy} = 0$$
$A_s N_x/|N_{yx}| > -1$ and $N_y/|N_{yx}| > -1$, this falls into case 3.
$$A_x f_{yd} = N_x + |N_{yx}| = 967, A_x = 2.22 \text{ mm}^2/\text{mm}$$
Over a depth of $1.7/2 = 0.85$ m, $A_x = 1890 \text{ mm}^2$.
Provide 4H25 giving a total area of 1964 mm^2.
$$A_y f_{yd} = N_y + |N_{yx}| = 641, A_y = 1.47 \text{ mm}^2/\text{mm}$$
Provide H16 at 130 mm centres giving $A_y = 1.55 \text{ mm}^2/\text{mm}$. This will be part of shear links.
$$N_2 = -2|N_{yx}| = -1282 \text{ N/mm}, \sigma_2 = -6.4 \text{ MPa} < v \, f_{cd}$$

Summary

Longitudinal reinforcement:
Bottom flange: 7H25, Top flange = 0.
Top half of web: 4H16, Bottom half of web: 4H25.
Shear Link reinforcement: Provide H16 at 130 mm centres. Note that the reinforcement in the web (H16 at 130 mm centres) overrides the calculated reinforcement in the top flange (H16 at 275 centres) and bottom flange (H16 at 190 mm centres).

This example demonstrates a few of the key concepts of the design principles based on the lower bound plasticity limit theorem. Although it is best to use the elastic stress distribution in design to limit ductility demand, it is not always convenient to do so because elastic stress distribution leads to highly variable distribution of reinforcement. Averaging out the stresses while maintaining equilibrium leads to a more convenient distribution of reinforcement.

In a similar way, although it is best for the direction of reinforcement to coincide with the direction of principal stresses calculated from elastic stress analysis, practical limitations might prevent this. Orthogonal layout of reinforcement is most common, although occasionally skew layout of reinforcement can be adopted as well.

18.4.3 Presence of Prestressing Strands

Prestressing is used to apply external compression to prevent the formation cracks at serviceability limit state. Fig. 18.6 shows an idealized stress–strain relationship for a prestressing strand. As shown in Fig. 18.6, although the design ultimate stress is f_{pd}, after all the losses have taken place, at the serviceability limit state, the prestress in the strands will be f_{pe}. From the serviceability limit state to ultimate limit state, the prestressing strands will act as ordinary reinforcement with a design stress f_{yd} equal to $(f_{pd} - f_{pe})$.

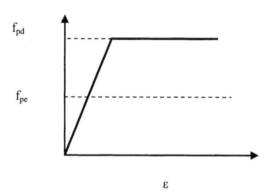

Fig. 18.6 Stress–strain relationship for prestressing strand.

The design equations from 18.3 can be used by replacing
$A_x f_{yd}$ by $[A_{xu} f_{yd} + A_{xp} (f_{pd} - f_{pex})]$ and $A_y f_{yd}$ by $[A_{yu} f_{yd} + A_{yp} (f_{pd} - f_{pey})]$
where:
A_{xu} and A_{yu} are respectively the areas of unstressed (or ordinary) reinforcement areas in x- and y-directions.
A_{xp} and A_{yp} are respectively the areas of stressed (or prestressed) strand areas in x- and y-directions.
f_{pex} and f_{pey} are respectively the prestress in the strands in x- and y-directions.

Example: Design the necessary reinforcement for the following conditions.
Forces per unit length from loads at ultimate limit state:
$$N_x = 1390 \text{ N/mm}, N_y = 2080 \text{ N/mm}, N_{xy} = 1250 \text{ N/mm}$$
The prestress is applied in the x-direction only.
Thickness $t = 250$ mm, $f_{pd} = 1617$ MPa, $f_{pex} = 1068$ MPa, $f_{yd} = 435$ MPa, $vf_{cd} = 10.56$ MPa, $A_{xp} = 0.47$ mm²/mm
Prestress in N/mm = $-A_{xp} \times f_{pex} = -0.47 \times 1068 = -502$ N/mm.
Net $N_x = 1390 - 502 = 888$ N/mm, $N_y = 2080$ N/mm, $N_{xy} = 1250$ N/mm.
$$N_x/|N_{xy}| = 0.71, N_y/|N_{xy}| = 1.66, N_x N_y/N_{xy}^2 = 1.33$$
$$N_x/|N_{yx}| \geq -1 \text{ and } N_y/|N_{yx}| \geq -1$$
This falls into case 3 zone.
$$A_x f_{yd} + A_{xp} \times (f_{pd} - f_{pex}) = N_x + |N_{yx}| = 888 + 1250 = 2138 \text{ N/mm}$$
$$A_x \times 435 + 0.47 \times (1617 - 1068) = 2138$$
$$A_x = 4.33 \text{ mm}^2/\text{mm}.$$
Provide H25 at 100 mm centres. $A_x = 4.91$ mm²/mm.
$$A_y f_{yd} = N_y + |N_{yx}| = 2080 + 1250 = 3330 \text{ N/mm}$$
$$A_x = 7.66 \text{ mm}^2/\text{mm}.$$
Provide H25 at 60 mm centres. $A_x = 8.18$ mm²/mm
$$N_2 = -2|N_{yx}| = -2500 \text{ N/mm}, \sigma_2 = -10.0 \text{ MPa} > -vf_{cd}$$

18.5 REINFORCEMENT DESIGN FOR FLEXURAL FORCES

When only flexural forces act as on a slab as shown in Fig. 18.7, the equations developed in section 8.11, Chapter 8 can be used to design the necessary reinforcement at top and bottom faces in the x- and y-directions. Although these equations are not included in Eurocode 2, they do not conflict with clause 5.6.1(3)P. The only requirement is to adhere to the requirements of clause 5.6.2(2):

i. The area of tensile reinforcement is limited such that at any section,
$$x_u/d \leq 0.25, \ f_{ck} \leq 50 \ \text{MPa}, \quad x_u/d \leq 0.15, \ f_{ck} \geq 55 \ \text{MPa}$$

ii. Reinforcing steel is class B or C which have higher ductility compared to Class A steel.

iii. The ratio of moments at intermediate supports to the moments in the span should be between 0.5 and 2.0.

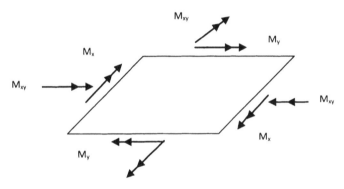

Fig. 18.7 Flexural forces.

M_x and M_y respectively are bending moments about the y- and x-axes and M_{xy} is the twisting moment. The convention used in Fig. 18.8 is the right hand cork screw rule which states that if the right hand thumb points in the direction of the vector, the moment acts in the direction the fingers curl.

The design equations for flexural reinforcement from Chapter 8, section 8.11.1 and section 8.11.2 are repeated here for convenience. In the following, positive bending moments are sagging moments which cause tension on the bottom face.

The rules for calculating the moment of resistance required for flexural steel at bottom M_{yu}^b and M_{xu}^b are as follows.

(a) If $\dfrac{M_x}{|M_{xy}|} \geq -1.0$ and $\dfrac{M_y}{|M_{xy}|} \geq -1.0$, then $M_{xu}^b = M_x + |M_{xy}|$, $M_{yu}^b = M_y + |M_{xy}|$.

(b) If $\dfrac{M_x}{|M_{xy}|} < -1.0$ and $M_y - \dfrac{M_{xy}^2}{M_x} > 1.0$ then $M_{xu}^b = 0$, $M_{yu}^b = M_y - \dfrac{M_{xy}^2}{M_x}$.

(c) If $\dfrac{M_y}{|M_{xy}|} < -1.0$ and $M_x - \dfrac{M_{xy}^2}{M_y} > 1.0$ then $M_{yu}^b = 0$, $M_{xu}^b = M_x - \dfrac{M_{xy}^2}{M_y}$.

(d) If none of the above conditions are valid, then $M_{yu}^b = M_{xu}^b = 0$.

In a manner similar to the determination of sagging moment of resistance, if the ultimate *hogging* moments of resistance provided by steel in x- and y-directions are M_{xu}^t and M_{yu}^t respectively, then the rules for calculating the moment of resistance required for flexural steel at top are as follows. Note that the values of M_{xu}^t and M_{yu}^t are both negative, indicating that they correspond to hogging bending moments requiring steel at the top of the slab.

(a) If $\dfrac{M_x}{|M_{xy}|} \le 1.0$ and $\dfrac{M_y}{|M_{xy}|} \le 1.0$, then $M_{xu}^t = M_x - |M_{xy}|$, $M_{yu}^t = M_y - |M_{xy}|$.

(b) If $\dfrac{M_y}{|M_{xy}|} > 1.0$ and $M_x - \dfrac{M_{xy}^2}{M_y} < 0$ then $M_{yu}^t = 0$, $M_{xu}^t = M_x - \dfrac{M_{xy}^2}{M_y}$.

(c) If $\dfrac{M_x}{|M_{xy}|} > 1.0$ and $M_y - \dfrac{M_{xy}^2}{M_x} < 0$ then $M_{xu}^t = 0$, $M_{yu}^t = M_y - \dfrac{M_{xy}^2}{M_x}$.

(d) If none of the above conditions are true, then $M_{yu}^t = M_{xu}^t = 0$.

18.6 REINFORCEMENT DESIGN FOR COMBINED IN-PLANE AND FLEXURAL FORCES

Very often in shell structures, an element can be subjected to a combination of in-plane forces (N_x, N_y and N_{xy}) as shown in Fig. 18.1 and flexural forces (M_x, M_y and M_{xy}) as shown in Fig. 18.8. The design of reinforcement for such a combination is best carried out using the sandwich model as shown in Fig. 18.8.

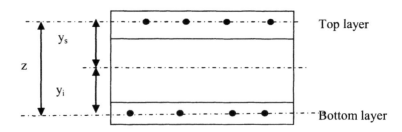

Fig. 18.8 Sandwich model.

The thickness is divided into a top layer and a bottom layer. The thickness of each layer is twice the distance from the outer layer to the centre of steel.

Using notation, y_s and y_i are the distances to the superior (or top) and inferior (or bottom) layers from the centroidal axis,
$$z = y_s + y_i$$
A force N can be distributed to top and bottom layers as N_s and N_i respectively.
$$N = N_s + N_i$$
For the resultant to coincide with the centroidal axis,
$$N_s \times y_s = N_i \times y_i$$
Solving for N_s and N_i
$$N_s = N (1 - y_s/z), N_i = N (1 - y_i/z)$$
Similarly a moment M can be replaced by a couple of forces $\pm M/z$.
The total force in the top and bottom layers are
$$N_s = N (1 - y_s/z) - M/z$$
$$N_i = N (1 - y_i/z) + M/z$$
As the reinforcements in the x- and y-directions cannot lie in the same plane, y_s and y_i on the face normal to x- and y-axes will be different. In order to keep the calculations simple it is convenient to assume that y_s and y_i values are the average of corresponding values on two perpendicular faces.
Sandwich model is given in *Eurocode 2- Part 2: Concrete Bridges-Design and detailing, Annex LL.* Equations (LL.128) to (LL.142) correspond to the equations given above.

18.6.1 Example of Design for Combined In-Plane and Flexural Forces

Design the necessary reinforcement for a slab 500 mm thick subjected to the following combination of forces at the ultimate limit state. Assume $f_{yd} = 435$ MPa, $f_{cd} = 20$ MPa and $vf_{cd} = 10.56$ MPa:
$$N_x = 1000 \text{N/mm}, N_y = 1400 \text{ N/mm}, N_{xy} = 900 \text{ N/mm}$$
$$M_x = 60 \text{ kNm/m}, M_y = 100 \text{ kNm/m}, M_{xy} = 60 \text{ kNm/m}$$
Assume cover to steel = 30mm, reinforcement H20.

Solution:
(i) Take average values only.
$$\text{Thickness of top and bottom layers} = 2 \times 30 + 20 = 80 \text{ mm}$$
$$z = 500 - 80 = 420 \text{ mm}$$
$$y_s = y_i = z/2 = 210 \text{ mm}$$

(ii) Determine the in-plane forces in the top and bottom layers due to applied force and moments.
$$M_x/z = 60 \times 10^6/ (10^3 \times z) = 143 \text{ N/mm},$$
$$M_y/z = 100 \times 10^6/ (10^3 \times z) = 238 \text{ N/mm},$$
$$M_x/z = 100 \times 10^6/ (10^3 \times z) = 238 \text{ N/mm},$$
$$N_x (1 - y_s/z) = N_x (1 - y_i/z) = 500 \text{ N/mm}$$
$$N_y (1 - y_s/z) = N_y (1 - y_i/z) = 700 \text{ N/mm}$$
$$N_{xy} (1 - y_s/z) = N_{xy} (1 - y_i/z) = 450 \text{ N/mm}$$

(iii) Design of reinforcement

Top layer:
$$N_x = 500 - 143 = 357 \text{ N/mm},$$
$$N_y = 700 - 238 = 462 \text{ N/mm},$$
$$N_{xy} = 450 - 238 = 212 \text{ N/mm}$$
$$N_x/|N_{xy}| = 1.68, \ N_y/|N_{xy}| = 2.18, \ N_x N_y/N^2_{xy} = 3.66$$
As $N_x/|N_{yx}| > -1$ and $Ny/|N_{yx}| > -1$, this falls into case 3.
$$A_x f_{yd} = N_x + |N_{xy}| = 569, \ A_x = 1.31 \text{ mm}^2/\text{mm}$$
Provide H20 at 225 mm c/c giving 1.40 mm^2/mm.
$$A_y \ f_{yd} = N_y + |N_{xy}| = 674, \ A_x = 1.55 \text{ mm}^2/\text{mm}$$
Provide H20 at 200 mm c/c giving 1.57 mm^2/mm.
$$N_2 = -2|N_{xy}| = -424 \text{ N/mm}, \ t = 80 \text{ mm}, \ \sigma_2 = N_2/t = -5.3 \text{ MPa}.$$
Maximum compressive stress is less than vf_{cd}.

Bottom layer:
$$N_x = 500 + 143 = 643 \text{ N/mm},$$
$$N_y = 700 + 238 = 938 \text{ N/mm},$$
$$N_{xy} = 450 + 238 = 688 \text{ N/mm}$$
$$N_x/|N_{xy}| = 0.93, \ N_y/|N_{xy}| = 1.36, \ N_x N_y/N^2_{xy} = 1.27$$
As $N_x/|N_{yx}| > -1$ and $Ny/|N_{yx}| > -1$, this falls into case 3.
$$A_x f_{yd} = N_x + |N_{xy}| = 1331, \ A_x = 3.06 \text{ mm}^2/\text{mm}$$
Provide H20 at 100 mm c/c giving 3.14 mm^2/mm.
$$A_y \ f_{yd} = N_y + |N_{xy}| = 1626, \ A_x = 3.74 \text{ mm}^2/\text{mm}$$
Provide H20 at 80 mm c/c giving 3.93 mm^2/mm.
$$N_2 = -2|N_{xy}| = -1376 \text{ N/mm}, \ t = 80 \text{ mm}, \ \sigma_2 = N_2/t = -17.2 \text{ MPa}$$
Maximum compressive stress is greater than vf_{cd}. This stress is not a real stress. It is more a reflection of the modelling used.
Design is satisfactory.

18.7 OUT-OF-PLANE SHEAR

Eurocode 2- Part 2: Concrete bridges-Design and detailing, Annex LL, equations (LL.121) to (LL.123) give the procedure for designing for out-of-plane shear forces. The procedure is illustrated by a simple example.

Example: The slab in section 18.5 is subjected to the following out-of-plane shear forces. Check the overall shear capacity of the slab.
$V_x = 120$ N/mm, $V_y = 180$ N/mm
Solution:
$$\text{Resultant shear force } V_{Ed} = \sqrt{(V^2_x + V^2_y)} = 216.3 \text{ N/mm}$$
$$\tan \varphi = V_y/V_x = 1.5, \ \cos \varphi = 0.56, \ \sin \varphi = 0.83$$
From section 18.5, reinforcements at the bottom of the slab are:
$$A_x = \text{H20 at 100 mm c/c giving 3.14 mm}^2/\text{mm}$$
$$A_y = \text{H20 at 80 mm c/c giving 3.93 mm}^2/\text{mm}$$

$$A_s = A_x \cos^2\varphi + A_y \sin^2\varphi = 3.14 \times (0.56)^2 + 3.93 \times (0.83)^2 = 3.69 \text{ mm}^2/\text{mm}$$
$$d = \text{effective depth} = 500 - 3 - 20/2 = 460 \text{ mm}$$
$$100\rho = 100 \times 3.69/ (d = 460) = 0.80$$

Shear capacity of sections with no shear reinforcement and cracked in flexure is given by equation (6.2a) and (6.2b) of Eurocode 2 in clause 6.2.2 as

$$V_{Rd, c} = [C_{Rd, c} \, k \, (100 \, \rho \, f_{ck})^{0.33}] \, b_w \, d \geq v_{min} \, b_w \, d \qquad (6.2a)$$
$$v_{min} = 0.035 \times k^{1.5} \, \sqrt{f_{ck}} \qquad (6.3N)$$
$$C_{Rd, c} = 0.12, \, k = 1 + \sqrt{(200/d)} = 1.66 < 2.0, \, 100 \, \rho = 0.80 < 2.0, \, f_{ck} = 30 \text{ MPa}$$
$$b_w = 1 \text{ mm}, \, d = 460 \text{ mm}, \, v_{min} = 0.035 \times k^{1.5} \times \sqrt{f_{ck}} = 0.41$$
$$(V_{Rd, c} = 264 \text{N/mm}) > (V_{Ed} = 216.3 \text{ N/mm})$$

Design is satisfactory.

18.8 STRUT–TIE METHOD OF DESIGN

In a structure subjected to in-plane forces, the stress distribution will consist of normal stresses (σ_x and σ_y) and shear stress τ_{xy}. These stresses will result in principal stresses σ_1 and σ_2. If for example σ_1 is tensile and σ_2 is compressive, then one can visualize the external loads being resisted by a combination of tensile and compressive stresses. Compressive stresses can be resisted by concrete and because concrete is weak in tension, tensile stresses can be resisted by steel reinforcement. The external load is thus resisted by a series of concrete struts and steel ties. This is the basic principle behind the very popular strut–tie method of design. This method is very attractive to designers as they can visualize the resistance to a complex stress state provided by a simple system of struts and ties. The method is obviously based on the lower bound limit theorem of plasticity as both equilibrium and yield criteria requirements are satisfied. One word of caution which applies to all such methods is that the method concentrates on providing safety at the ultimate limit state only. There is no attention paid to aspects of serviceability limit state. In order to prevent problems at the serviceability state one should choose a stress state which does not depart drastically from the elastic stress distribution. Eurocode 2 in clause 6.5 gives recommendations for the use of strut–tie method of design.

18.8.1 B and D Regions

Most designs of beams and columns are based on the assumption that strain distribution is linear across the depth or 'plane sections remain plane'. This is known as the Bernoulli assumption and regions where this assumption is valid are known as B regions. Such regions exist away from areas with concentrated loads or reactions, abrupt changes of cross section, etc.

Regions which are not B regions are called D regions where D stands for disturbed. In these regions, assumption of linear distribution of strains is not valid. Fig. 18.9 shows some typical cases of D regions and the associated possible strut–tie models. It is in the D regions the strut–tie method of design finds wide application.

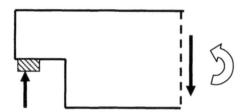

Fig. 18.9(a) D regions at a 'half' joint.

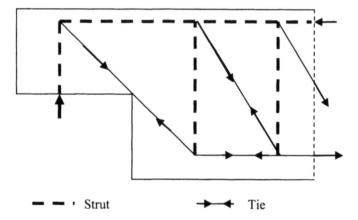

— — · Strut   ►——◄ Tie

Fig. 18.9(b) A possible strut–tie model for a 'half' joint.

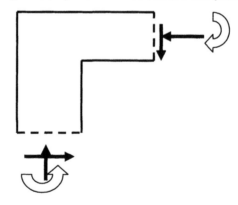

Fig. 18.9(c) D regions at a joint in a frame.

Fig. 18.9(c) shows a joint in a frame subjected to 'closing' bending moments. A possible strut–tie model can be envisaged as shown in Fig. 18.9(d). It clearly shows that an inclined 'strut' is needed to maintain equilibrium at the joint. If the moments are reversed so that the joint is an 'opening' joint, the strut is transformed into a tie.

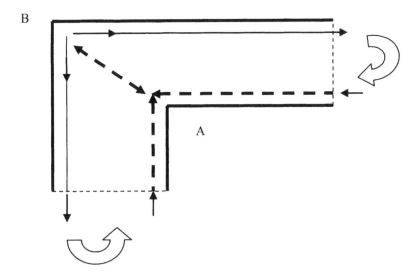

Fig. 18.9(d) A possible strut–tie model: closing moments.

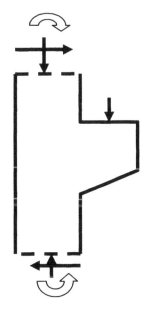

Fig. 18.9(e) D regions at a corbel.

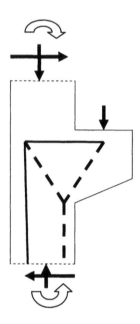

Fig. 18.9(f) Strut–tie model for a corbel.

Fig. 18.9(e) shows a corbel. Corbels are very short cantilevers commonly encountered in industrial structures and also in bridge structures where the corbel is used to support a simply supported beam at a 'half' joint.

Fig. 18.9(f) shows a possible strut–tie model.

18.8.1.1 Saint Venant's Principle

Saint Venant's principle gives an indication of the size of the D region. The principle states that the localized effect of any type of disturbance dies away at a distance of about one member depth. If the structure or the member is larger than the D region, then only these areas need to be designed by the strut–tie method. Away from the disturbance the normal design procedures can be adopted. However if the structure is small enough, then the D region can cover the whole structure. An example of such a structure is a deep beam whose depth is comparable to the span as shown in Fig. 18.10.

18.8.2 Design of Struts

It is well known that if concrete is subjected to uniaxial compression, then the design strength is equal to f_{cd}. If the state of stress is biaxial compression, the design strength will be approximately greater by about 10%. Eurocode 2 in clause 6.5.2, equation (6.55) suggests using a maximum compressive strength of f_{cd}.

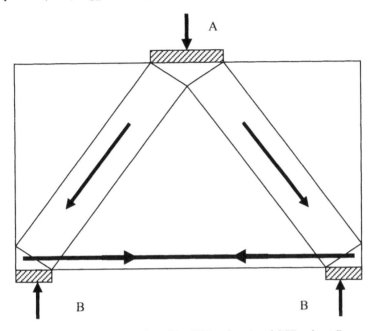

Fig. 18.10 Deep beam modelling with a CCC node at A and CCT nodes at B.

If, however, axial compression is accompanied by tension at right angles, the strength of the strut is greatly reduced. Eurocode 2 in clause 6.5.2, equation (6.56) suggests the design strength as equal to $0.6(1 - f_{ck}/250) f_{cd}$. In addition to limiting the design compressive strength, it is also necessary to design reinforcement to resist the tension developed.

Fig. 18.11 shows two common situations where compression in the strut is accompanied by lateral tension.
If the total width b is less than or equal to 0.5H, where H is the length of the strut, tension reinforcement required can be spread over a height h equal to width b of the strut at both ends. Code equation (6.58) gives the total tension T to be resisted as $T = 0.25 (1 - a/b) F$, where F is the compressive force in the strut and a is the width of the loaded area.
If on the other hand, the total width b is greater than 0.5H, where H is the length of the strut, tension reinforcement required can be smeared over a height h equal to 025H. Code equation (6.59) gives the total tension T to be resisted as
$$T = 0.25 (1 - 1.4 a/H) F.$$

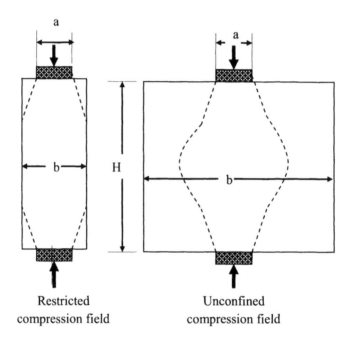

Fig. 18.11 Different compression fields needing transverse tensile reinforcement.

18.8.3 Types of Nodes and Nodal Zones

When using the strut–tie method of design, because the given structure is replaced by a pin-jointed truss, different types of nodes occur where members meet. These nodal zones need to be carefully designed and detailed.

(a) CCC Node: If three compressive forces meet at anode as shown in Fig. 18.12, it is called a CCC node. According to Eurocode 2 equation (6.60), the compression stress in each strut is restricted to a maximum value of $0.6(1 - f_{ck}/250)f_{cd}$. Examples of CCC nodes are nodes A in a frame corner shown in Fig. 18.9(d) and in deep beam shown in Fig. 18.10.

(b) CCT node: If two compressive forces and a tie force anchored in the node through bond meet as shown in Fig. 18.13, it is called a CCT node. According to Eurocode 2 equation (6.61), the compression stress in each strut is restricted to a maximum value of $0.51(1 - f_{ck}/250) f_{cd}$. An example of such a node occurs in the design of a deep beam as shown in Fig. 18.10 at B. Such nodes also occur at the top nodes in 'half' joint shown in Fig. 18.9(b).

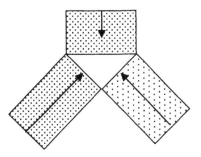

Fig. 18.12 CCC node.

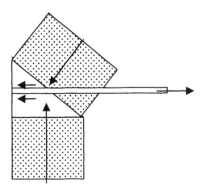

Fig. 18.13 CCT node.

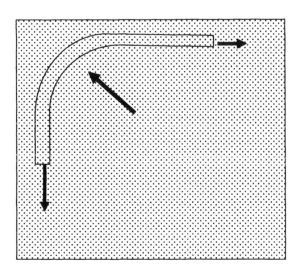

Fig. 18.14 CTT node.

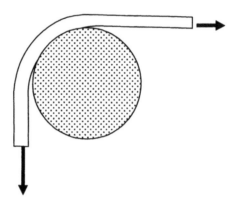

Fig. 18.15 A cross bar inside the bend.

(c) CTT node: If two tensile forces at a compressive force meet at a node it is called a CTT node. Examples of such nodes are B in the frame joint shown in Fig. 18.9(c) and in the bottom nodes in the 'half' joint shown in Fig. 18.9(b).

In the case of the frame joint in Fig. 18.9(d), the tension reinforcement in the beam can be bent over and used as tension reinforcement in the column as shown in Fig. 18.14. The equilibrating compression force is provided by concrete. It is important to ensure that the mandrel is large enough to prevent bearing failure in concrete (see clause 8.3(3) in Eurocode 2). It is desirable to provide a cross bar inside the bend as shown in Fig. 18.15. According to Eurocode 2 equation (6.62), the compression stress in each strut is restricted to a maximum value of $0.45(1 - f_{ck}/250) f_{cd}$.

18.8.4 Elastic Analysis and Correct Strut–Tie Model

Elastic finite element analysis of part of the structure which lies in the B region can be very helpful in correctly modelling the strut–tie model for the given problem.

The idea is illustrated by some typical examples.

Corbel: Fig. 18.9(f) shows a strut–tie model for a corbel. Fig. 18.16(a) and 18.16(b) show the vectorial representation of the major principal stress σ_1 and minor principal stress σ_2 respectively. The major principal stress plot clearly shows the need for horizontal reinforcement at the top of the corbel which is finally anchored in the column. The minor principal plot shows the compressive stress converging from the left and from the right towards the re-entrant corbel indicating that it is a CCC node as shown in Fig. 18.20(c).

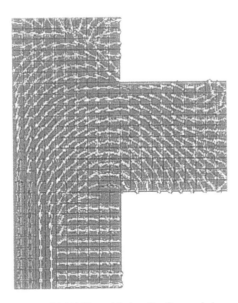

Fig. 18.16(a) Vectorial plot of σ_1 for a corbel.

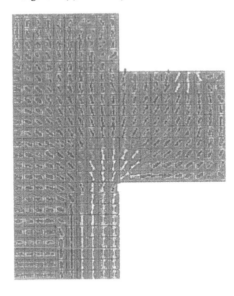

Fig. 18.16(b) Vectorial plot of σ_2 for a corbel.

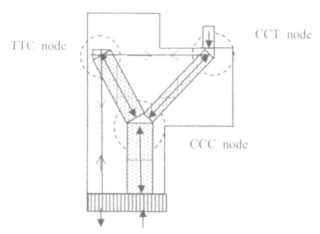

Fig. 18.16(c) Strut−tie model for a corbel with different types of nodes.

Eurocode 2 in Annex J3 gives recommendations for reinforcing corbels (also called brackets). Calavera (2012) also gives detailed drawings for reinforcement (see pages 390 to 395).

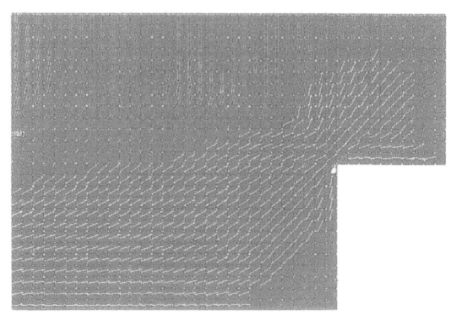

Fig. 18.17(a) Vectorial plot of σ_1 for a 'half joint'.

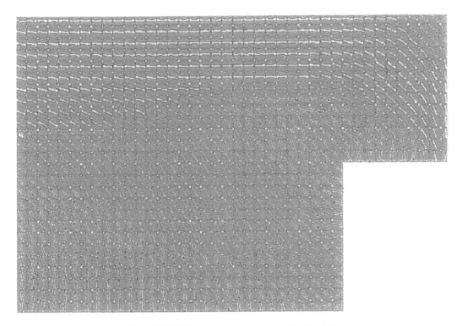

Fig. 18.17(b) Vectorial plot of σ_2 for a 'half joint'.

Half joint: Fig. 18.17(a) and Fig. 18.17(b) show the vectorial representation of the major principal stress σ_1 and minor principal stress σ_2 respectively. The major principal stress plot clearly shows the need for a diagonal reinforcement crossing the re-entrant corner and also reinforcement at the bottom face of the beam. The minor principal plot shows the compressive stress parallel to the top face and also along in the diagonal direction. The plots validate the strut–tie model shown in Fig. 18.9(b).

Calavera (2012) also gives detailed drawings for reinforcement in 'half joints' also called dapped-end beams (see pages 396 to 397).

18.8.5 Example of Design of a Deep Beam Using Strut–Tie Model

Fig. 18.18 shows deep beam. The beam is 5400 mm long × 3000 mm high and 250 mm thick. It is loaded by a uniformly distributed load of 350 kN/m. It is supported on two 400 mm × 250 mm columns. Design the necessary reinforcement assuming $f_{yk} = 500$ MPa, $f_{ck} = 30$ MPa.

Fig. 18.19 shows the vectorial representation of the principal stresses σ_1 and σ_2 from an elastic finite element analysis. It is clear from the diagram that near the bottom surface the stress is horizontal tensile and the compressive stresses flow towards the supports. Guided by the elastic stress distribution, the strut–tie model shown in Fig. 18.20 can be constructed.

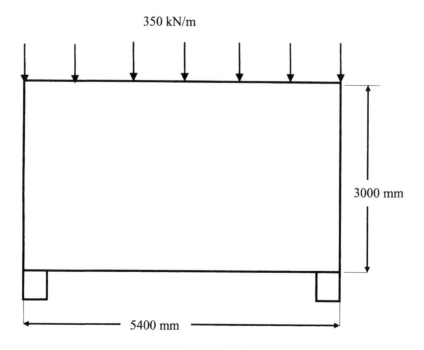

350 kN/m

3000 mm

5400 mm

Fig. 18.18 A deep beam.

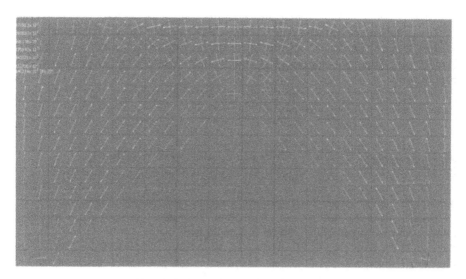

Fig. 18.19 Elastic principal stress distribution.

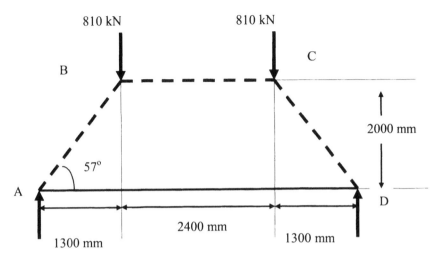

Fig. 18.20 Strut–tie model for the deep beam.

Calculation of the forces in the members:

Strut AB is inclined to the horizontal at $\alpha = \tan^{-1} (2000/1300) = 57°$.
Vertical reaction at A = $300 \times 5.4/2 = 810$ kN.
Force F_{AB} in the strut = $810/\sin 57° = 966$ kN.
Force F_{AD} in the tie = $966 \times \cos 57° = 526$ kN.
Provide 3H25 = 1473 mm².
Force F_{BC} in the strut = $966 \times \cos 57° = 526$ kN.

Design checks:

$$\text{Force } F_{AD} \text{ in the tie} = 526 \text{ kN}$$
$$f_{yk} = 500 \text{ MPa, } f_{yd} = 500/ (\gamma_s = 1.15) = 435 \text{ MPa}$$
$$\text{Area of steel, } A_s = 526 \times 10^3/ 435 = 1209 \text{ mm}^2$$
Provide 3H25, $A_s = 1473$ mm².

Check stresses in the nodal zones:

Fig. 18.21 shows the CCT and CCC nodes occurring in the strut–tie model.
Node A: Node A is a CCT node. Column at A is 400 × 250 mm.
$$\text{Vertical compressive stress in column} = 810 \times 10^3/ (400 \times 250) = 8.1 \text{ MPa}$$
$$\text{Column width} = 400 \text{ mm}$$
Strut is inclined at 57° to the horizontal.
$$\text{Width of strut} = 400/ \cos (90° - 57°) = 477 \text{ mm}$$
$$\text{Compressive stress in the strut, } F_{AB} = 966 \times 10^3/ (477 \times 250) = 8.1 \text{ MPa}$$
$$\text{Vertical height of the joint at A} = 400 \times \tan (33°) = 260 \text{ mm}$$
$$\text{Horizontal stress in the joint due to } F_{AD} = 526 \times 10^3/ (260 \times 250) = 8.1 \text{ MPa}$$
Node A is a CCT node. In a CCT node, according to Eurocode 2 equation (6.61),
the compressive stress is limited to $\sigma_{Rd, max} = 0.51 \times (1 - f_{ck}/250) f_{cd}$.
$$f_{ck} = 30 \text{ MPa, } f_{cd} = f_{ck}/ (\gamma_c = 1.5) = 20 \text{ MPa,}$$
$$\sigma_{Rd, max} = 0.51 \times (1 - 30/250) 20 = 9 \text{ MPa} > 8.1 \text{ MPa.}$$

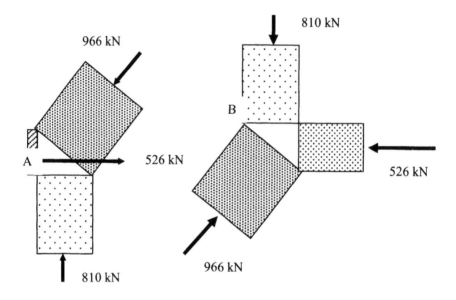

Fig. 18.21 Nodal zones for the deep beam.

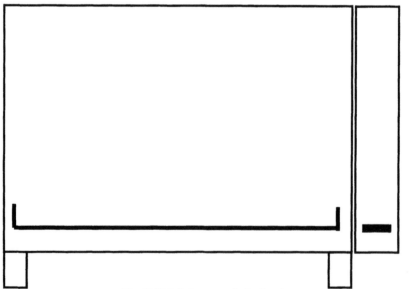

Fig. 18.22 Reinforcement in the deep beam.

Node B: Node B is a CCC node.

   Compressive stress in the strut, $F_{AB} = 966 \times 10^3 / (477 \times 250) = 8.1$ MPa

   Vertical height of the joint at B = $477 \times \sin (90^\circ - 57^\circ) = 260$ mm

Horizontal compressive stress in the joint due to $F_{CB} = 526 \times 10^3 / (260 \times 250)$
$$= 8.1 \text{ MPa}$$

'Column' at B is 400×250 mm.

Vertical compressive stress in column $= 810 \times 10^3 / (400 \times 250) = 8.1$ MPa

The node area at B is subjected to a hydrostatic compressive stress of 8.1 MPa.

In a CCC node, according to Eurocode 2 equation (6.60), the compressive stress is limited to $\sigma_{Rd, max} = 0.6 \times (1 - f_{ck}/250) f_{cd}$.

$$f_{ck} = 30 \text{ MPa}, f_{cd} = f_{ck}/ (\gamma_c = 1.5) = 20 \text{ MPa}$$
$$\sigma_{Rd, max} = 0.6(1 - 30/250) 20 = 10.6 \text{ MPa} > 8.1 \text{ MPa}.$$

The stress state in the nodes is safe.

Fig. 18.22 shows the reinforcement in the deep beam.

18.8.6 Example of Design of Half Joint Using Strut–Tie Model

Fig. 18.23 shows a 'half' beam joint also called a draped end. The beam is 8.0 m simply supported. It is 1500 mm × 800 mm broad. The depth is reduced to 700 mm over a length of 500 mm at each end. The beam supports a uniformly distributed load of 250 kN/m. Assume f_{ck}= 30 MPa and f_{yk} = 500 MPa.

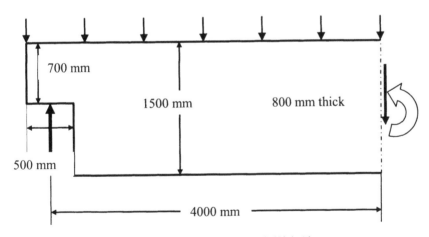

Fig. 18.23 A symmetrical half of a 'half joint' beam.

Fig. 18.24 and Fig. 18.25 show, for the part of the beam near the half joint, from an elastic stress analysis the vector plots of the major principal stress σ_1 and both the principal stresses respectively. Guided by the plot the strut–tie model shown in Fig. 18.26 is adopted.

Analysis:
Reaction at the support $= 250 \times 4.25 = 1062.5$ kN.
Bending moment M at mid-span $= 1062.5 \times 4.0 - 250 \times 4.25^2/2 = 1992.2$ kNm.
$$d = 1500 - 40 \text{ (cover)} - 25/2 = 1436 \text{ mm}.$$

$$k = M/ (b \ d^2 \ f_{ck}) = 1992.2 \times 106/ (800 \times 14362 \times 30) = 0.04.$$
$$z/d = 0.5[1.0 + \sqrt{(1 - 3 \times 0.4)}] = 0.97.$$
$$A_s = 1992.2 \times 10^6/ (0.97 \times 1436 \times 0.87 \times 500) = 3289 \ mm^2.$$

Provide 7H25 = 3437 mm².
Depth x of stress block = 87 mm.

Fig. 18.24 Vector plot of principal stress σ_1 near the 'half joint'.

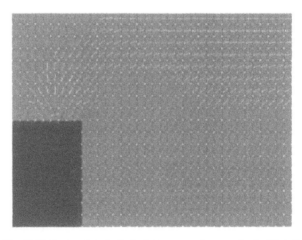

Fig. 18.25 Vector plot of principal stress σ_1 and σ_2 near the 'half joint'.

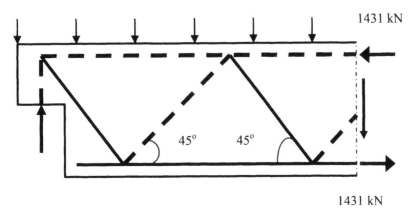

1431 kN

1431 kN

Fig. 18.26 Strut–tie model near the 'half joint'.

The horizontal compressive strut can be located at about $87/2 \approx 44$ mm from the top face.

The horizontal tie can be located at approximately

40(cover) $+ 10$ (links) $+ 25/2 = 63$ mm from the bottom face.

Assuming that the inclined tie is at $45°$ to the horizontal,

Force in the tie = reaction/ $\sin 45° = 1062.5 / (0.707) = 1503$ kN

$$A_s = 1503 \times 10^3 / (0.87 \times 500) = 3454 \text{ mm}^2$$

Provide 7H25 = 3437 mm².

Assuming that the tie has a cover of $30 + 25/2 = 43$ mm, the bar is located at $250 + 43 \sin 45° = 280$ mm from the centre of reaction.

Check the stress state at the node: Fig. 18.27 shows a CCT node. In a CCT node, according to Eurocode 2 equation (6.61), the compressive stress in the struts is limited to

$$\sigma_{Rd, max} = 0.51 \times (1 - f_{ck}/250) f_{cd}$$
$$f_{ck} = 30 \text{ MPa}, f_{cd} = f_{ck}/ (\gamma_c = 1.5) = 20 \text{ MPa}$$
$$\sigma_{Rd, max} = 0.51 \times (1 - 30/250)\, 20 = 9 \text{ MPa}$$

Assuming that the reaction acts over a width of, say, 150 mm, compressive stress σ in the horizontal and vertical strut is

$$\sigma = 1063 \times 10^3 / (800 \times 150) = 8.9 \text{ MPa} < 9 \text{ MPa}$$

Note: Clause 10.9.4.6, Eurocode 2 suggests an alternative model for reinforcement of 'half joints' as shown in Fig. 18.28.

Calavera (2012) suggests the reinforcement details shown in Fig. 18.29 for the strut–tie model in Fig. 18.23.

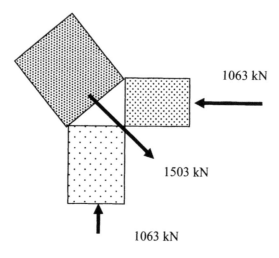

Fig. 18.27 CCT node.

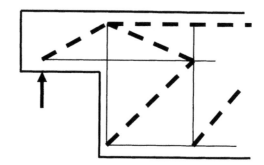

Fig. 18.28 An alternative strut–tie model.

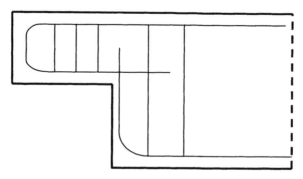

Fig. 18.29 Reinforcement for strut–tie model in Fig. 18.23.

18.9 REFERENCES

Bhatt, P. (2012). *Prestressed Concrete Design to Eurocodes*. Spon Press.

Calavera, Jose. (2012). *Manual for Detailing Reinforced Concrete Structures to EC2*. Spon Press.

Nielsen, M.P. and Hoang, L.C. (2011). *Limit Analysis and Concrete Plasticity*. CRC Press.

CHAPTER 19

DESIGN OF STRUCTURES RETAINING AQUEOUS LIQUIDS

19.1 INTRODUCTION

Structures such as tanks for retaining water or effluents in sewage treatment works are designed using the following codes:

- *BS EN 1992-3:2006 Eurocode 2-Design of concrete structures: Part 3: Liquid retaining and containment structures*
- *U.K. National Annex to BS EN 1992-3:2006 Eurocode 2*
- *BS EN 1991-1-4:2006 Eurocode 1-Actions on structures: Part 4: Silos and Tanks*

The codes generally adopts the relevant clauses of *BS EN 1992-1-1:2004 Eurocode 2-.Design of concrete structures: Part 1-1: General rules and rules for buildings* along with additional clauses as required.

The following is a brief summary of the relevant clauses. The reader should always refer to the complete texts in the codes.

The design is normally carried out according to the limit state principles. However, unlike normal reinforced concrete structures, designs for such structures are often governed by the serviceability limit state considerations of limiting the crack width rather than by ultimate limit state considerations.

19.1.1 Load Factors

When designing a tank for containment of fluids, normally the load to which the walls are designed is the internal pressure. The corresponding partial factor for action is $\gamma_F = 1.2$. At ultimate limit state, the self weight is considered as permanent action with a load factor of 1.35 and maximum height of liquid retained as leading variable action with a load factor of 1.2. In the case of tanks located below ground, the possibility of flotation of the tank when empty due to ground water pressure should be considered. The uplift is normally resisted by the dead weight of the structure.

19.1.2 Crack Width

According to clause 7.3.1, at serviceability limit state in a reinforced concrete structure, the maximum crack width due to direct tension and flexure or restrained temperature and moisture effects should be limited to values dictated by tightness class as defined in code Table 7.105 and given below as Table 19.1.

Table 19.1 Classification of tightness

Tightness class	Requirement for leakage
0	Some degree of leakage acceptable, or leakage of liquids is irrelevant
1	Leakage to be limited to a small amount. Some surface staining or damp patches acceptable.
2	Leakage to be minimal. Appearance not to be impaired by staining.
3	No leakage permitted.

Tightness class 0: Provisions of Eurocode 2, clause 7.3.1, values for maximum crack width as given in code Table 7.1N may be adopted. The clause states that under quasi-permanent load combination, for X0 and XC1 class exposures, the crack width is limited to 0.4 mm and for exposure classes XC2−XC4, XD1−XD2, XS1−XS3, crack width is limited to 0.3 mm.

Tightness class 1: Any cracks which can be expected to pass through the full thickness of the section should be limited to 0.2 mm if $h_0/h \leq 5$ and to 0.05 mm if $h_0/h \geq 35$. Linear interpolation is permitted for intermediate values of h_0/h. Here h = element thickness, h_0 = hydrostatic pressure head. If under the action of quasi-permanent combination of loads the full thickness of the section is not cracked as measured by the compression zone $x_{min} > $ min (50 mm; 0.2 h), then provisions as for tightness class 0 may be adopted.

Tightness class 2: Cracks which can be expected to pass through the full thickness of the section should be avoided.

Tightness class 3: Generally special measures such as liners or prestressing will be required to ensure water tightness.

19.1.2.1 Crack Width Control without Direct Calculation

Where minimum reinforcement as given by code equation (7.1) (see section 6.3.3, Chapter 6), for sections totally in tension, the maximum bar diameter and maximum bar spacing for maximum crack width over the range 0.05 to 0.3 mm are given in Fig. 7.103N and Fig. 7.104N of the code *BS EN 1992-3:2006, Eurocode 2- Design of concrete structures- Part 3: Liquid retaining and containment structures.* The maximum bar diameter can be modified using code equation (7.122) rather than code equation (7.7N) as follows:

$$\varphi_s = \varphi_s' \times \frac{f_{ct\,eff}}{2.9} \times \frac{h}{10(h-d)} \quad \text{for cases of uniform axial tension} \quad (7.122)$$

19.2 BENDING ANALYSIS FOR SERVICEABILITY LIMIT STATE

In Chapter 17, section 17.1.5 the formulae were derived for the stresses in concrete and steel at serviceability limit state when tension and compression steel are present and tensile stress in concrete is ignored. The equation for calculating the neutral axis depth is repeated here for convenience.

$$0.5 f_c \, b \, x^2 + \alpha_e \, f_c \, A_s' \, (x - d') = \alpha_e \, f_c \, A_s \, (d - x)$$

In the absence of compression steel, the lever arm z is

$$z = d - x/3$$

For a given moment M, the stresses in steel and concrete can be determined as follows.

$$M = T\,z = A_s \, f_s \, z = C\,z = 0.5 \, f_c \, b \, x \, z$$

$$f_s = \frac{M}{A_s \, z}, \quad f_c = \frac{2M}{b\,x\,z}$$

19.2.1 Example of Stress Calculation at SLS

Fig 19.1 shows a cantilever wall which is part of a water tank. The tank retains water to a depth of 3.5 m. The base is 400 mm thick overall and is reinforced with H12 bars at 100 mm c/c. Calculate the stresses in concrete and steel at serviceability limit state. Also check whether the moment of resistance is sufficient at ultimate limit state. Assume $f_{ck} = 30$ MPa, $f_{yk} = 500$ MPa.

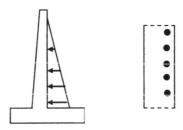

Fig. 19.1 A cantilever wall.

$E_{cm} = 22(0.8 + 0.1 f_{ck})^{0.3} = 32.84$ GPa.
Assume creep coefficient, $\varphi\,(\infty, t_0) = 1.5$.
E_c, long term $= E_{cm}/\,[1 + \varphi\,(\infty, t_0)] = 32.84/2.5 = 13.2$ GPa.

$E_s = 200$ GPa.

$\alpha_e = 200.0/13.2 = 15.2$.

Consider 1 m length of wall. The steel area $A_s = 1131$ mm^2/m, b = 1000 mm

h = 400 mm. Assuming cover = 40 mm,

$$d = 400 - 40 - 12/2 = 354 \text{ mm}$$

(a) Calculate the neutral axis depth:

$$\frac{1}{2}bx^2 = \alpha_e A_s (d-x)$$

$$\frac{1}{2}1000\,x^2 = 15.2 \times 1131 \times (354-x)$$

Simplifying:

$$x^2 + 34.38\,x - 12171.4 = 0, \text{ giving } x = 95 \text{ mm}$$
$$z = d - x/3 = 354 - 95/3 = 323 \text{ mm}$$

(b) Moment at base at serviceability limit state:

Assuming γ = unit weight of water = 10 kN/m^3,

$$M = \gamma H^3/6$$
$$M = 10 \times 3.5^3/6 = 71.46 \text{ kNm/m}$$

(c) Stresses in steel and concrete:

$$f_s = \frac{M}{A_s\,z} = \frac{71.46 \times 10^6}{1131 \times 323} = 196\text{MPa}$$

$$f_c = \frac{2M}{b \times z} = \frac{2 \times 71.46 \times 10^6}{1000 \times 94 \times 323} = 4.7\text{MPa}$$

(d) Moment at ultimate limit state:

Using a load factor of 1.2,

$$M = 1.2 \times 71.46 = 85.75 \text{ kNm/m}$$

(e) Neutral axis depth at ULS:

Assuming that steel yields, equating total tension and total compression,

$$f_{cd}\, b\, 0.8x = 0.87\, f_{yk}\, A_s$$
$$(30/1.5) \times 1000 \times 0.8x = 0.87 \times 500 \times 1131, \text{ giving } x = 31.0 \text{ mm}$$

Check the strain in tension steel:

$$\varepsilon_s = \frac{0.0035}{x}(d-x)$$

$$\varepsilon_s = \frac{0.0035}{31.0}(354-31) = 0.036 > (\frac{0.87 \times 500}{200 \times 10^3} = 0.0022)$$

Steel yields and the assumption is justified.

(f) Moment of resistance at ULS:

$$M_u = 0.87\, f_{yk}\, A_s\, (d - 0.4x)$$
$$= 0.87 \times 500 \times 1131 \times (354 - 0.4 \times 31.0) \times 10^{-6}$$

$$= 168.\ 1\ \text{kNm/m} > 85.75\ \text{kNm/m}$$

The design is satisfactory at ULS.

19.2.2 Crack Width Calculation in a Section Subjected to Flexure Only

Calculation of crack widths and spacing was explained in detail in section 17.3, Chapter 17. In the following, the necessary equations from clause 7.3.1 of the code are summarized. The use of the equations is demonstrated by a numerical example.

(i) Crack width w_k may be calculated using code equation (7.8).

$$w_k = S_{r,\ max}\ (\varepsilon_{sm} - \varepsilon_{cm}) \tag{7.8}$$

(ii) $(\varepsilon_{sm} - \varepsilon_{cm})$ may be calculated from the code equation (7.9).

$$(\varepsilon_{sm} - \varepsilon_{cm}) = \frac{\sigma_s - k_t \dfrac{f_{ct.eff}}{\rho_{p.eff}}\ (1 + \alpha_e\ \rho_{p.eff})}{E_s} \geq 0.6\ \frac{\sigma_s}{E_s} \tag{7.9}$$

where
σ_s = stress in tension reinforcement assuming cracked section
α_e = modular ratio = E_s/E_{cm}
$\rho_{p,\ eff} = A_s/A_{c,\ eff}$
$k_t = 0.4$ for long term loading
$A_{c,\ eff} = b \times h_{c,\ eff}$
$h_{c,\ eff} = \min\ [2.5\ (h - d);\ (h - x);\ 0.5h]$

(iii) $S_{r,\ max}$ may be calculated from code equation (7.11)

$$S_{r,\ max} = 3.4\ c + 0.128\ \varphi/\ \rho_{p,\ eff} \tag{7.11}$$

c = cover to longitudinal reinforcement. φ = bar diameter.
If there are n_i bars of diameter φ_i etc., an equivalent diameter φ_{eq} is given by
$\varphi_{eq} = \Sigma n_i\ \varphi_i^2/\ \Sigma n_i\ \varphi_i$.

Example: Using the data from section 19.2.1,
x = 95 mm, σ_s = 196 MPa, h = 400 mm, c = 40 mm, d = 354 mm
$h_{c,\ ef}$ = min [(2.5× (400 – 354); (400 – 95)/3; 400/2]
$h_{c,\ ef}$ = min [115; 102; 200] = 102 mm
$A_{c,\ ef} = b_w \times h_{c,\ ef} = 1000 \times 102 = 10.2 \times 10^4\ \text{mm}^2$
A_s = H12 at 100 mm = 1131 mm^2/m, $\rho_{p,\ eff} = A_s/A_{c,\ ef} = 1131/10.2 \times 10^4 = 0.011$
$f_{ct,\ eff} = f_{ctm} = 0.3 \times f_{ck}^{0.67} = 2.9$ MPa
Taking creep into account, α_e = 200.0/13.2 = 15.2
$(k_t = 0.4) \times (2.9/0.011) \times (1 + 15.2 \times 0.011) = 123$ MPa
$(\sigma_s = 196) – 123 = 73$ MPa $< (0.6\ \sigma_s = 118)$
Take 0.6 σ_s = 118 MPa
$(\varepsilon_{sm} - \varepsilon_{cm}) = 118/\ (E_s = 200 \times 10^3) = 0.59 \times 10^{-3}$

$$S_{r, max} = 3.4\ c + 0.128\ \varphi/\ \rho_{p,\ eff}$$
$$\varphi = 12\ mm$$
$$\text{Cover, } c = 40\ mm$$
$$S_{r, max} = 3.4\ c + 0.128\ \varphi/\ \rho_{p,\ eff} = 3.4 \times 40 + 0.128 \times 12/\ 0.011 = 276\ mm$$
$$w_k = S_{r, max}\ (\varepsilon_{sm} - \varepsilon_{cm}) = 276 \times 0.59 \times 10^{-3} = 0.15\ mm$$

Choose tightness class 1.

Hydraulic head, $h_0 = 3.5$ m, wall thickness, $h = 400$ mm, $h_0/h = 8.75$. Interpolating between $w_k = 0.2$ mm for $h_0/h \leq 5.0$ and $w_k = 0.0.05$ mm for $h_0/h \geq 35.0$, for $h_0/h = 8.75$, permissible crack width w_k is

$$w_k = 0.05 + (0.2 - 0.05) \times (35 - 8.75)/\ (35.0 - 5.0) = 0.18\ mm$$

Actual crack width of 0.15 mm is less than permissible crack width of 0.18 mm.

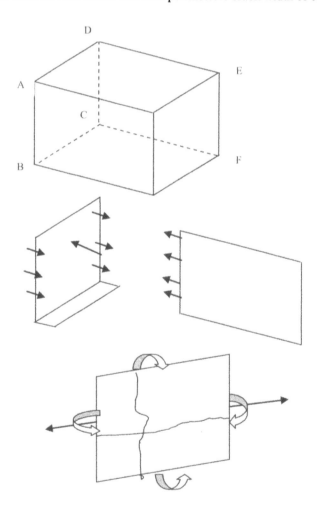

Fig. 19.2 Rectangular tank wall subjected to two-way bending moments and axial tension.

Note that in the case of large tanks under hydrostatic pressure, it is possible to reduce crack width limits in the upper portions of the walls and thus reduce the required steel area.

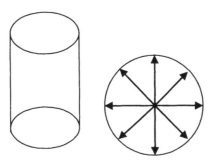

Fig. 19.3 Circular tank wall subjected to bending moment and axial tension at right angles.

19.3 WALL SUBJECTED TO TWO-WAY BENDING MOMENTS AND TENSILE FORCE

Fig. 19.2 shows a rectangular water tank. Normal load acting on the side wall ABCD of the rectangular tank is resisted by two-way bending action. In the vertical direction, the slab acts as a cantilever, fixed at the base. In the horozontal direction, the load is resisted by the wall acting as a beam supported by the front and back walls. The shear forces on the side wall ABCD act as tensile forces on the back wall CDEF and vice versa. The cantilever bending action about the horizontal axis causes horizontal cracks parallel to the base. The bending moment about a vertical axis together with the tensile force on the sides causes vertical cracks paralel to the height.

A similar situation occurs in the case of the circular water tank shown in Fig. 19.3. Radial pressure causes circumferential tension. Depending on the fixity at the base of the wall, fluid pressure may cause bending moment and shear force in the wall as in the case of rectangular tank.

Circular tanks are often prestressed to prevent the formation of cracks due to radial pressure.

19.3.1 Analysis of a Section Subjected to Bending Moment and Direct Tensile Force for Serviceability Limit State

From the equations in section 17.1.5, the total compressive force C is

$$C = bx\frac{f_c}{2} + A'_s \, \alpha_e f_c \, \frac{(x-d')}{x}$$

where A_s is the area per unit length of compression steel.
The total tensile force T is

$$T = A_s f_s = A_s \alpha_e f_c \frac{d-x}{x}$$

For equilibrium in the axial direction,

$$T - C = \text{Applied tensile force N}$$

$$A_s \alpha_e f_c \frac{d-x}{x} - bx\frac{f_c}{2} - A'_s \alpha_e f_c \frac{(x-d')}{x} = N$$

Taking moments about the tension steel,

$$bx\frac{f_c}{2}(d-\frac{x}{3}) + A'_s \alpha_e f_c \frac{(x-d')}{x}(d-d') = M - N\frac{h}{2}$$

where M = applied moment, h = overall depth of the section.
There are two unknowns ,viz., f_c and x in the two equations. Eliminating f_c from the two equations,

$$(M-N\frac{h}{2})A_s\alpha_e(d-x) - A'_s\alpha_e(x-d')\{M+N(d-d'-\frac{h}{2})\} = b\frac{x^2}{2}\{N(d-\frac{x}{3}-\frac{h}{2})+M\}$$

$$(\text{C19.1})$$

Solution of the above cubic equation gives the value of neutral axis depth, *x*. Compressive stress in concrete f_c can be obtained from

$$bx\frac{f_c}{2}(d-\frac{x}{3}) + A'_s\alpha_e f_c \frac{(x-d')}{x}(d-d') = M - N\frac{h}{2} \qquad (\text{C19.2})$$

The equations are valid only if x ≥ d' because it is assumed that the stress in compression steel is actually compressive.

19.3.1.1 Example of Calculation of Stresses Under Bending Moment and Axial Tension

Calculate the stresses using the following data.
$$h = 400 \text{ mm, cover} = 40 \text{ mm}$$
Tension steel: 16 mm diameter bars at 100 mm c/c giving
$$A_s = 2011 \text{ mm}^2/\text{m}, A_s' = 0$$
$$f_{ck} = 30 \text{ MPa}, f_{yk} = 500 \text{ MPa}, \alpha_e = 15$$
Applied actions at serviceability limit state:
$$M = 100 \text{ kNm/m, N} = 60 \text{ kN/m}$$
$$b = 1000 \text{ mm,}$$
$$d = 400 - 40 - 16/2 = 352 \text{ mm}$$

(a) Neutral axis depth:
Substituting A_s' in equation (C19.1),

$$(M-N\frac{h}{2})A_s\alpha_e(d-x) = b\frac{x^2}{2}\{N(d-\frac{x}{3}-\frac{h}{2})+M\}$$

$$(100 \times 10^6 - 60 \times 10^3 \frac{400}{2}) \times 2011 \times 15 \times (352 - x)$$

$$= 1000 \frac{x^2}{2} \{60 \times 10^3 (352 - \frac{x}{3} - \frac{400}{2}) + 100 \times 10^6\}$$

Simplifying:

$$x^3 - 5456 \, x^2 - 265452 \, x + 93.4391 \times 10^6 = 0$$

Solving the cubic equation by trial and error gives $x = 110$ mm.

(b) Compressive stress in concrete:
Substituting A_s' in equation (C19.2),

$$bx \frac{f_c}{2} (d - \frac{x}{3}) = M - N \frac{h}{2}$$

$$1000 \times 109.7 \times \frac{f_c}{2} (352 - \frac{109.7}{3}) = 100 \times 10^6 - 60 \times 10^3 \frac{400}{2}$$

$$f_c = 5.1 \text{ MPa}$$

(c) Tensile stress in steel:
Calculate tensile stress in steel:

$$f_s = \alpha_e f_c \frac{d - x}{x} = 15 \times 5.09 \times \frac{352 - 109.7}{109.7} = 169 \text{MPa}$$

19.3.2 Crack Width Calculation in a Section Subjected to Direct Tension

Fig. 19.4 shows a circular water tank with 350 mm thick walls and 10 m internal diameter and 6.5 m high retains water to a depth of 6.0 m. The base of the tank is designed to be free sliding. Design the reinforcement at the base and calculate the crack width. Assume $f_{ck} = 30$ MPa and $f_{yk} = 500$ MPa.

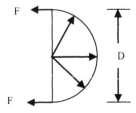

Fig. 19.4 Ring tension.

The pressure p at the base of the tank:
$$p = \gamma H = 10 \text{ kN/m}^3 \times 6 = 60 \text{ kN/m}^2$$
As shown in Fig. 19.4, the circumferential tension at the base T is given by
$$T = p \, D/2 = 60 \times 10/2 = 300 \text{ kN/m}$$
where D = diameter of tank = 10 m.

Using a load factor of $\gamma_F = 1.2$, total tension at ULS is

$$T = 1.2 \times 300 = 360 \text{ kN/m}$$

Total reinforcement required $= 360 \times 10^3/ (0.87 \times f_{yk}) = 828 \text{ mm}^2/\text{m}$.

Assuming 350 mm thick walls, according to clause 7.3.2, equation (7.1) the total minimum steel reinforcement in the wall should not be less than $A_{s, \min}$.

$$A_{s, \min} \, \sigma_s = k_c \, k \, f_{ct, eff} \, A_{ct} \qquad (7.1)$$
$$k_c = 1.0 \text{ for pure tension, } k = 1.0,$$
$$f_{ck} = 30 \text{ MPa}, \, f_{ct, eff} = f_{ctm} = 0.3 \times f_{ck}^{\,0.67} = 2.9 \text{ MPa}$$
$$\sigma_s = f_{yk} = 500 \text{ MPa}$$
$$A_{ct} = 350 \times 1000 = 350 \times 10^3 \text{ mm}^2/\text{m}$$
$$A_{s, \min} = 2030 \text{ mm}^2/\text{m}$$

On each face $A_{s, \min} = 1015 \text{ mm}^2/\text{m}$.

Provide H16 at 175 mm c/c, $A_s = 1149 \text{ mm}^2/\text{m}$ on each face.

Note that because of the sudden release of tension when the walls crack, the total amount of reinforcement is very much greater than that required to resist the tension force in the cracked state.

σ_s = stress in tension reinforcement assuming cracked section. All the tension is resisted by steel.

$$\sigma_s = 300 \times 10^3/ (2 \times 1149) = 131 \text{ MPa}$$

Taking creep into account, $\alpha_e = 200.0/13.2 = 15.2$.

$k_t = 0.4$ for long term loading.

$$\text{Cover} = 40 \text{ mm}, \, d = 350 - 40 - 16/2 = 302 \text{ mm}$$
$$h_{c, eff} = \min [2.5 \, (h - d); \, 0.5h] = \min [2.5 \times (350 - 302); \, 350/2] = 120 \text{ mm}$$
$$A_{c, eff} = b \times h_{c, eff} = 1000 \times 120 = 1.2 \times 10^5 \text{ mm}^2/\text{m}$$
$$f_{ct, eff} = f_{ctm} = 0.3 \times f_{ck}^{\,0.67} = 2.9 \text{ MPa}$$
$$\rho_{p, eff} = A_s/A_{c, eff} = 1149/1.2 \times 10^5 = 0.01$$

$$(\varepsilon_{sm} - \varepsilon_{cm}) = \frac{\sigma_s - k_t \dfrac{f_{ct,eff}}{\rho_{p,eff}} (1 + \alpha_e \, \rho_{p,eff})}{E_s} \geq 0.6 \frac{\sigma_s}{E_s} \qquad (7.9)$$

$$(\varepsilon_{sm} - \varepsilon_{cm}) = (131 - 146)/ E_s < 0.6 \, \sigma_s/ E_s$$
$$\text{Take } (\varepsilon_{sm} - \varepsilon_{cm}) = 0.6\sigma_s/E_s = 131/ (200 \times 10^3) = 0.655 \times 10^{-3}$$

$$S_{r, \max} = k_3 \, c + k_1 \, k_2 \, k_4 \, \varphi/ \rho_{p, eff} \qquad (7.11)$$
$$c = \text{cover to longitudinal reinforcement} = 40 \text{ mm}$$
$$\varphi = \text{bar diameter} = 16 \text{ mm}$$
$$\rho_{p, eff} = 0.01$$
$$S_{r, \max} = 3.4 \, c + 0.128 \, \varphi/ \rho_{p, eff} = 341 \text{ mm}$$

$$S_{r, \max} = k_3 \, c + k_1 \, k_2 \, k_4 \, \varphi/ \rho_{p, eff} \qquad (7.11)$$

where

$k_1 = 0.8$ for high bond bars, $k_2 = 1.0$ for pure tension, $k_3 = 3.4$, $k_4 = 0.425$.
c = cover to longitudinal reinforcement. φ = bar diameter.

19.3.3 Control of Cracking without Direct Calculation

Taking σ_s = 131 MPa and using code Fig. 7.103N for maximum bar diameter and code Fig. 7.104N for maximum spacing in order not to exceed a crack width of w_k = 0.2 mm, maximum bar diameter = 50 mm and maximum bar spacing is 300 mm.
Using code equation (7.122) and adjusting the maximum bar diameter

$$\varnothing_s = \varnothing_s^* \left[\frac{f_{ct,eff}}{2.9}\right] \frac{h}{10(h-d)} = 50 \left[\frac{2.9}{2.9}\right] \frac{350}{10 \times (350-302)} = 37 \, mm$$

Clearly the bar diameter of 16 mm and a spacing of 175 mm are both well below the maximum permitted values.

19.4 CONTROL OF RESTRAINED SHRINKAGE AND THERMAL MOVEMENT CRACKING

Changes in temperature and moisture content of the concrete cause movements. If these movements are restrained they lead to tensile stresses in concrete and possibility of cracks. During hydration of cement, heat is generated and as the concrete cools, it contracts. Similarly loss of moisture leads to drying shrinkage. In most structures these effects are of no significance compared with the stresses due to external loads. However in thin sections such as walls, these effects are important and must be taken into account if the structure is not to be rendered unserviceable due to wide cracks. It is necessary to reinforce the structures to ensure that a number of well distributed cracks of acceptable width occur rather than a few wide cracks.

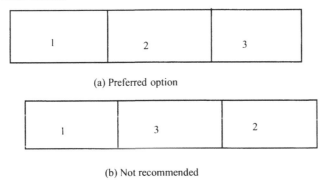

(a) Preferred option

(b) Not recommended

Fig. 19.5 Sequence of construction.

The restraint to movement can be reduced by proper sequence of construction. Fig. 19.5(a) shows the preferred sequence because after each bay is cast, the slab is unrestrained at one edge and can contract during cooling. On the other hand a sequence of construction shown in Fig. 19.5(b) is not recommended because the middle slab is restrained on both sides.

The restraint to movement due to friction between the ground and the slab can be reduced by laying a polythene sheet on a layer of smooth concrete.

19.4.1 Design Options for Control of Thermal Contraction and Restrained Shrinkage

Stresses due to shrinkage and thermal movements are controlled by the provision of movement joints. In Annex N of *BS EN 1992-3:2006 Eurocode 2-Design of concrete structures: Part 3: Liquid retaining and containment structures*, the code allows for the following two options.

1. Design for full restraint. In this case no movement joints are provided and the crack widths and spacings are controlled by the provision of appropriate reinforcement as detailed in sections 19.2.2 and 19.3.3 [see Eurocode 2 clause 7.3.4 and equations (7.8) to (7.12)].

2. Design for free movement. This is achieved by providing close movement joints at greater than 5 m or 1.5 times the wall height. The minimum reinforcement in the wall should be as detailed in Chapter 10 [see Eurocode 2 clauses 9.6.2 to 9.6.4].

Fig. 19.6 shows a typical expansion joint. This has no restraint to movement. There is no continuity of steel or concrete across the joint. An initial gap is provided for expansion and leakage of water is prevented by using a water stop made from rubber or similar materials. If in an expansion joint the initial gap is eliminated, it becomes a complete contraction joint.

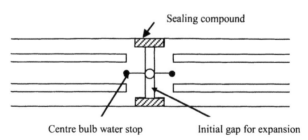

Fig. 19.6 Expansion joint.

19.4.2 Reinforcement Calculation to Control Early-Age Cracking and Thermal Contraction and Restrained Shrinkage

It is well known that increasing any of cement content, placing temperature, pour thickness and restraint contributes to increased risk of early-age cracking. Restraint to contraction occurs when walls are cast onto rigid foundations or large slabs cast as a series of bays. It has to be noted that elastic thermal restraint stresses are relieved to a certain extent by creep.

Designing reinforcement to control crack spacing and width is accomplished in basically two stages:

Stage 1: Ensure minimum steel reinforcement $A_{s,\,min}$ as given in Eurocode 2, clause 7.3.2, equation (7.1) is provided.

$$A_{s,\,min}\,\sigma_s = k_c\,k\,f_{ct,\,eff}\,A_{ct} \tag{7.1}$$
$$k_c = 1.0 \text{ for pure tension, } k = 1.0$$
$$\sigma_s = f_{yk} = 500 \text{ MPa}$$

$f_{ct,\,eff}$ = tensile strength of concrete at the age t when cracks are expected.
The age t when cracks are expected is normally taken as 3 days.

$$f_{ctm} = 0.3 \times f_{ck}\,(t)^{\,0.67},\; f_{ck}\,(t) = f_{cm}\,(t) - 8 \text{ MPa}$$
$$f_{ct,\,eff}\,(t) = f_{cm}\,(t) = \beta_{cc}\,(t) \times (f_{ck} + 8) \tag{3.1}$$
$$\beta_{cc}\,(t) = \text{Exp}\{s\,[1 - \sqrt{(28/t)}]\} \tag{3.2}$$

$s = 0.20, 0.25$ and 0.38 for cements of class R, class N and class S respectively.

Stage 2: Check that the provided steel achieves the specified crack width as specified in Table 7.105 of *EN 1992-3:2006 Eurocode 2 Design of concrete structures: Part 3: Liquid retaining and containment structures* and (repeated as Table 19.1) using the Eurocode 2, Clause 7.3.4, equations (7.8) to (7.12) as given below. See Chapter 17, section 17.3.1 for more details.

19.4.3 Reinforcement Calculation to Control Early-Age Cracking for a Member Restrained at One End

In the case of a member restrained at its end such as a floor slab at the end of a bay, $(\varepsilon_{sm} - \varepsilon_{cm})$ may be calculated from the code equation (M.1) in Annex M of *BS EN 1992-3:2006 Eurocode 2-.Design of concrete structures: Part 3: Liquid retaining and containment structures*.

(i) Calculate $(\varepsilon_{sm} - \varepsilon_{cm})$

$$(\varepsilon_{sm} - \varepsilon_{cm}) = 0.5\,\alpha_e\,k_c\,k\,f_{ct,\,eff}\,[1 + 1/\,(\alpha_e\,\rho)]/\,E_s \tag{M.1}$$

Table 19.2 gives the values of the parameters.
$f_{ct,\,eff}$ = tensile strength of concrete at time t as shown in section 19.4.2.
t can be taken as equal to 3 days for early-age cracking.

$$\alpha_e = \text{modular ratio} = E_s/E_{cm}\,(t),\; \rho = A_s/A_{ct}$$

(ii) Crack spacing $S_{r, max}$ may be calculated from Eurocode 2 equation (7.11):
$$S_{r, max} = k_3 \, c + k_1 \, k_2 \, k_4 \, \varphi / \, \rho_{p, eff} \qquad (7.11)$$
where
$k_1 = 0.8$ for high bond bars, $k_2 = 1.0$ for pure tension, $k_3 = 3.4$, $k_4 = 0.425$.
c = cover to longitudinal reinforcement. φ = bar diameter.

If there are n_i bars of diameter φ_i etc., an equivalent diameter φ_{eq} is given by code equation (7.12).
$$\varphi_{eq} = \Sigma n_i \, \varphi_i^2 / \, \Sigma n_i \, \varphi_i \qquad (7.12)$$

Table 19.2 Values of parameters in equation (M.1)

Parameter	External restraint dominant	Internal restraint dominant
k_c See Eurocode 2, clause 7.3.2	1.0 for pure tension	0.5
k See Eurocode 2, clause 7.3.2	k = 1.0, h ≤ 300 mm k = 0.65, h ≥ 800 mm Intermediate values, interpolate	1.0
A_{ct}, surface area in tension	Full section thickness	20% of section thickness

Note: Table 3.1, CIRIA Manual C660, recommends k = 0.75 for h ≥ 800 mm.

(iii) Crack width w_k may be calculated using code equation (7.8).
$$w_k = S_{r, max} \, (\varepsilon_{sm} - \varepsilon_{cm}) \qquad (7.8)$$
where
$S_{r, max}$ = maximum crack spacing. ε_{sm} = mean strain in reinforcement.
ε_{cm} = mean strain in concrete between cracks.
Note that in between cracks, concrete retains its tensile strength. This is known as tension stiffening.

(iv) Without directly checking the crack width and spacing, the suitability of reinforcement can be checked by calculating σ_s from code equation (M.2),
$$\sigma_s = k_c \, k \, f_{ct, eff} / \, \rho \qquad (M.2)$$
and use Fig. 7.103 to obtain maximum bar diameter and Fig. 7.104N to obtain the maximum spacing for various crack widths.

19.4.4 Example of Reinforcement Calculation to Control Early-Age Cracking in a Slab Restrained at One End

Determine minimum steel and the expected crack width for a 400 mm thick slab with a cover to steel of 40 mm. Assume $f_{ck} = 30$ MPa and $f_{yk} = 500$ MPa. Also check using equation (M.2) and Fig. 7.103N and Fig. 7.104N if the provided spacing is acceptable.

Step 1: Calculate the tensile strength at $t = 3$ days.

$$\text{Take } s = 0.25 \text{ for Class N cement}$$

$$t = 3 \text{ days}, \beta_{cc}(t) = \text{Exp}\{s\,[1 - \sqrt{(28/t)}]\} = 0.6 \tag{3.2}$$

$$f_{cm} = f_{ck} + 8 = 38 \text{ MPa}$$

$$f_{cm}(t) = \beta_{cc}(t) \times (f_{ck} + 8) = 0.6 \times (30 + 8) = 22.7 \text{ MPa} \tag{3.1}$$

$$f_{ck}(t) = f_{cm}(t) - 8 \text{ MPa} = 14.7 \text{ MPa}$$

$$f_{ct,\,eff} = f_{ctm} = 0.3 \times f_{ck}(t)^{0.67} = 1.8 \text{ MPa}$$

Step 2: Calculate the modular ratio at $t = 3$ days.

$$E_{cm} = 22 \times [(f_{ck} + 8)/10]^{0.3} = 22 \times [(30 + 8)/10]^{0.3} = 32.84 \text{ GPa}$$

$$E_{cm}(t) = [f_{cm}(t)/f_{cm}]^{0.3} \times E_{cm} \tag{3.5}$$

$$E_{cm}(t) = [22.7/38.0]^{0.3} \times 32.84 = 4.38 \text{ GPa}$$

$$E_s = 200 \text{ GPa}$$

$$\alpha_e = 200/4.38 = 45.66$$

Step 3: Calculate the minimum steel required.

$$h = 400 \text{ mm}, c = 40 \text{ mm},$$

$$A_{ct} = b_w \times h = 1000 \times 400 = 40 \times 10^4 \text{ mm}^2$$

Take external restraint as dominant, $k_c = 1.0$.

From the data in Table 19.2, interpolate the value of k.

$$k = 1.0 - (1.0 - 0.65) \times (400 - 300)/(800 - 300) = 0.93$$

$$\sigma_s = f_{yk} = 500 \text{ MPa}$$

$$A_{s,\,min} = k_c\, k\, f_{ct,\,eff}\, A_{ct}/\sigma_s = 1.0 \times 0.93 \times 1.8 \times 40 \times 10^4 /500 = 1339 \text{ mm}^2 \tag{7.1}$$

Provide steel at a maximum spacing of say 100 mm to aid proper compaction.
Provide H16 at 100 mm on each face. $A_s = 2 \times 2011 = 4022 \text{ mm}^2/\text{m}$.

$$\rho = A_s/A_{ct} = 4022/(40.0 \times 10^4) = 10.05 \times 10^{-3}$$

Step 4: Calculate the maximum crack spacing $S_{r,\,max}$ and crack width, w_k.

$$(\varepsilon_{sm} - \varepsilon_{cm}) = 0.5\, \alpha_e\, k_c\, k\, f_{ct,\,eff}\, [1 + 1/(\alpha_e\, \rho)]/ E_s \tag{M.1}$$

$$(\varepsilon_{sm} - \varepsilon_{cm}) = 0.5 \times 45.7 \times 1.0 \times 0.93 \times 1.8$$

$$\times [1 + 1/(45.7 \times 10.05 \times 10^{-3})]/(200 \times 10^3) = 0.608 \times 10^{-3}$$

$$c = \text{cover to longitudinal reinforcement} = 40 \text{ mm}$$

$$\varphi = \text{bar diameter} = 16 \text{ mm}.$$

$A_{c,\,eff}$ = Effective area of concrete in tension surrounding reinforcement to a depth of $h_{c,\,eff} = \min\,[h/2;\ 2.5(c + \varphi/2)]$.

Assuming 50% of reinforcement is on each face, $A_s = 4022/2 = 2011 \text{ mm}^2/\text{m}$.

$$h_{c,eff} = \min\,[h/2;\ 2.5(c + \varphi/2)] = \min\,[400/2\,;\ 2.5(40 + 16/2)] = 120 \text{ mm}.$$

$$\rho_{p,\,eff} = A_s/A_{c,eff} = 2011/(120 \times 1000) = 16.76 \times 10^{-3}$$

Note that ρ and ρ_{eff} are two different values. ρ is the ratio of steel to the whole cross section. On the other hand, peff is the ratio of steel area to the concrete area in tension.

$$S_{r,\,max} = k_3\, c + k_1\, k_2\, k_4\, \varphi/\rho_{p,\,eff} \tag{7.11}$$

Take $k_1 = 0.8$ for high bond bars, $k_2 = 1.0$ for pure tension, $k_3 = 3.4$, $k_4 = 0.425$.

$$S_{r, max} = 3.4 \times 40 + 0.8 \times 1.0 \times 0.425 \times 16/ (16.76 \times 10^{-3}) = 461 \text{ mm}$$
$$w_k = S_{r, max} (\varepsilon_{sm} - \varepsilon_{cm}) \tag{7.8}$$
$$w_k = 461 \times 0.608 \times 10^{-3} = 0.28 \text{ mm}$$

Crack width is too large.

If the steel area is increased by providing H20 at 100 mm on each face,
Total $A_s = 6283$ mm^2/m

$$\rho = A_s/A_c = 6283/ (40.0 \times 10^4) = 15.71 \times 10^{-3} = 1.57\%$$
$$(\varepsilon_{sm} - \varepsilon_{cm}) = 0.5 \, \alpha_e \, k_c \, k \, f_{ct, eff} [1 + 1/ (\alpha_e \, \rho)]/ E_s \tag{M.1}$$
$$(\varepsilon_{sm} - \varepsilon_{cm}) = 0.5 \times 45.7 \times 1.0 \times 0.93 \times 1.8$$
$$\times [1 + 1/ (45.7 \times 15.71 \times 10^{-3})]/ (200 \times 10^3) = 0.46 \times 10^{-3}$$
$$\varphi = 20 \text{ mm}$$
$$\rho_{p, eff} = A_s/A_{c,eff}$$

$A_{c, eff}$ = Effective area of concrete in tension surrounding reinforcement to a depth of $h_{c, eff}$ = min [(h/2; 2.5(c + φ/2)].
Assuming 50% of reinforcement is on each face, $A_s = 6283/2 = 3142$ mm^2/m.

$$h_{c, eff} = \text{min } [h/2; \ 2.5(c + \varphi/2)] = \text{min } [400/2; \ 2.5(40 + 20/2)] = 125 \text{ mm}.$$
$$\rho_{p, eff} = A_s/A_{c, eff} = 3142/ (125 \times 1000) = 25.132 \times 10^{-3}$$

$$S_{r, max} = k_3 \, c + k_1 \, k_2 \, k_4 \, \varphi/ \rho_{p, eff} \tag{7.11}$$

$$S_{r, max} = 3.4 \times 40 + 0.8 \times 1.0 \times 0.425 \times 20/ (25.32 \times 10^{-3}) = 405 \text{mm}$$

$$w_k = S_{r, max} (\varepsilon_{sm} - \varepsilon_{cm}) \tag{7.8}$$

$$w_k = 405 \times 0.46 \times 10^{-3} = 0.19 \text{ mm}$$

This is an acceptable crack width.

Step 5: Check the maximum bar diameter and spacing for acceptable crack width.
$$\sigma_s = k_c \, k \, f_{ct, eff} / \rho \tag{M.2}$$
$$\sigma_s = 1.0 \times 0.93 \times 1.8 / (10.05 \times 10^{-3}) = 167 \text{ MPa}$$
From Fig. 7.103N, maximum bar diameter φ_s^* for 0.2 mm wide crack is approximately 40 mm.
Calculate the adjusted the maximum bar diameter φ_s from code equation (7.122)

$$\varphi_s = \varphi_s^* [f_{ct, eff}/2.9] \times \{h/[10 \times (h - d)]\} \tag{7.122}$$
$$h = 400 \text{ mm}, \ d = 400 - 40 - 20/2 = 350 \text{ mm}, \ f_{ct, eff} = 1.8 \text{ MPa}$$
$$\varphi_s = \varphi_s^* [1.8/2.9] \times \{400/[10 \times (400 - 350)]\} = 0.50 \, \varphi_s^* \approx 20 \text{ mm}$$
From Fig. 7.104N, maximum bar spacing for 0.2 mm wide crack is approximately 250 mm.
The provided values are well below acceptable approximate values.

19.4.5 Reinforcement Calculation to Control Early-Age Cracking in a Wall Restrained at One Edge

In the case of a long wall restrained along one end, $(\varepsilon_{sm} - \varepsilon_{cm})$ may be calculated from the code equation (M.3) in Annex M of *BS EN 1992-3:2006 Eurocode 2-Design of concrete structures: Part 3: Liquid retaining and containment structures*:

$$(\varepsilon_{sm} - \varepsilon_{cm}) = R_{ax}\,\varepsilon_{free} \qquad (M.3)$$

where

R_{ax} = the restraint factor

ε_{free} = Strain which would occur if the member was completely unrestrained

The Eurocode 2, Part 3 provides little information on early magnitude of thermal strain. However, using equations (B.11) and (B.12) of Annex B of Eurocode 2, basic drying shrinkage strain $\varepsilon_{cd,\,0}$ can be calculated. Autogenous shrinkage strain ε_{ca} can be calculated using Eurocode 2 equation (3.12). Bamforth (2007) will be found very useful for understanding and calculating the reinforcement to prevent early-age thermal cracking.

Bamforth (2007) gives the following equation for $(\varepsilon_{sm} - \varepsilon_{cm})$

$$(\varepsilon_{sm} - \varepsilon_{cm}) = K_1 \{[\alpha_c\,T_1 + \varepsilon_{ca}]\,R1 + \alpha_c\,T_2\,R_2 + \varepsilon_{cd}\,R_3\} \qquad (CIRIA\ 3.2)$$

and the data required for calculating the possibility of thermal cracking are quite detailed. The manual provides sufficient information to calculate the values of these parameters. The parameters are:

(i) K_1: Coefficient for the stress relaxation due to creep under sustained loading. Take $K_1 = 0.65$.

(ii) T_1, drop in temperature from the maximum to the ambient.
Fig. 19.7 shows qualitative change in temperature with time in days. This is a function of several variables. The main variables are:

(a) Cement type, amount of additives such as fly ash, ggbs (ground granulated blast furnace slag) and total cement content in kg/m³. For a concrete of $f_{ck} = 30$ MPa, total cement content can vary from 350 to 450 kg/m³.

(b).Section thickness: Larger the thickness the greater the rise in temperature above the ambient. This can vary from 20° to 50°C.

(c) Formwork insulation: The temperature rise will be less in steel formwork as compared with plywood formwork.

(d) Placing temperature.

(e) Ambient conditions: Rise in temperature will be less in winter than in summer.

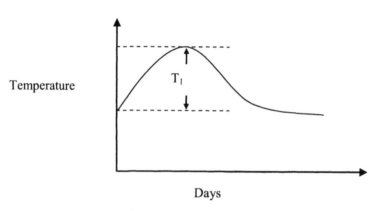

Fig. 19.7 Rise in temperature of concrete after casting.

(iii) α_c = coefficient of thermal expansion: Eurocode 2 in clause 3.1.3(5) gives
$$\alpha_c = 10 \times 10^{-6} / °C$$

(iv) Long term annual ambient temperature change, T_2: In U.K. the recommended values for T_2 are: Casting in winter, $T_2 = 10° C$, casting in summer, $T_2 = 20° C$.

(v) Shrinkage strain ε_{cd} : Values of basic drying shrinkage strain $\varepsilon_{cd, 0}$ is given by equations (B.11) and (B.12) of Annex B of Eurocode 2.
$$\varepsilon_{cd, 0} = 0.85 \, [(220 + 110 \, \alpha_{ds1}) \times exp \, \{-0.1 \times \alpha_{ds2} \times f_{cm}\}] \times 10^{-6} \times \beta_{RH} \quad (B.11)$$
$$\beta_{RH} = 1.55 \times (1 - RH^3), \, RH = \% \text{ ambient relative humidity} \quad (B.12)$$
$$\alpha_{ds1} = 3 \text{ and } \alpha_{ds2} = 0.13 \text{ for Class S cement}$$
$$\alpha_{ds1} = 4 \text{ and } \alpha_{ds2} = 0.12 \text{ for Class N cement}$$
$$\alpha_{ds1} = 6 \text{ and } \alpha_{ds2} = 0.11 \text{ for Class R cement}$$
The final shrinkage strain $\varepsilon_{cd} = \varepsilon_{cd, \infty} = \varepsilon_{cd, 0} \times k_h$.
Values of k_h are given in code Table 3.3 is reproduced in Table 19.3.

Table 19.3 Values of k_h versus h_0

h_0	k_h
100	1.0
200	0.85
300	0.75
≥500	0.70

(vi) Autogenous shrinkage strain ε_{ca}
$$\varepsilon_{ca} = \varepsilon_{ca, \infty} = 2.5 \times (f_{ck} - 10) \times 10^{-6} \quad (3.12)$$

(vii) R_2 and R_3 are restraint factors applying to long-term thermal movement and drying shrinkage respectively.

(viii) Early-age temperature differential ΔT is the temperature difference between the centre and surface of the slab as shown in Fig. 19.8.

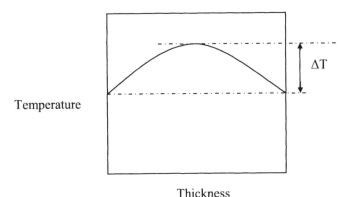

Fig. 19.8 Variation in temperature across the thickness of the slab.

As a simplified approach at the preliminary stage, Bamforth (2007) gives a simplified equivalent to CIRIA equation (3.2) as
$$(\varepsilon_{sm} - \varepsilon_{cm}) = K \left[\alpha_c (T_1 + T_2) + \varepsilon_{cd}\right] \qquad \text{(CIRIA 3.17)}$$
where K = coefficient which takes account of restrain R.
The recommended values are:
$$K = 0.5$$
$\varepsilon_{cd} = 150 \times 10^{-6}$ for U.K. external exposure conditions
$\varepsilon_{cd} = 350 \times 10^{-6}$ for U.K. internal exposure conditions
Recommended values for T_2 in U.K. are:
Casting in winter, $T_2 = 10°$ C and casting in summer, $T_2 = 20°$ C.
T_1: It is a function of f_{ck}, thickness, h and type of formwork. Table 19.4 shows values of T_1 for summer or winter casting for concrete with $f_{ck} = 30$ MPa.

Table 19.4 Recommended values of temperature $T_1°C$

h, mm	Steel formwork					Plywood formwork				
	300	500	700	1000	2000	300	500	700	1000	2000
Summer	18	28	35	43	54	28	36	42	47	55
Winter	12	20	28	37	52	22	30	37	42	55

19.4.6 Example of Reinforcement Calculation to Control Early-Age Cracking in a Wall Restrained at One Edge

Determine minimum steel and the expected crack width for a 400 mm thick wall with a cover to steel of 40 mm. Assume $f_{ck} = 30$ MPa and $f_{yk} = 500$ MPa. The external wall is cast in winter on a strong base.

Step 1: Calculate the minimum steel required.
$$h = 400 \text{ mm}, \; c = 40 \text{ mm},$$
$$A_{ct} = b_w \times h = 1000 \times 400 = 40 \times 10^4 \text{ mm}^2$$
Taking external restrain as dominant, $k_c = 1.0$.

Using the relationship between h and k in Table 19.1, interpolate the value of k
$$k = 1.0 - (1.0 - 0.65) \times (400 - 300)/ (800 - 300) = 0.93$$
$$\sigma_s = f_{yk} = 500 \text{ MPa},$$
From section 19.4.3, $f_{ct, eff} = 1.8$ MPa for f_{ck} (t) at t = 3 days.
$$A_{ct} = 400 \times 1000 = 40 \times 10^4 \text{ mm}^2$$
$A_{s, min} = k_c \ k \ f_{ct, eff} \ A_{ct}/ \sigma_s = 1.0 \times 0.93 \times 1.8 \times 40 \times 10^4 /500 = 1339 \text{ mm}^2$ (7.1)
Provide each face H12 at 160 mm = 707 mm^2/m.
$$\text{Total } A_s = 2 \times 707 = 1414 \text{ mm}^2/\text{m}$$
One half of this value of steel can be placed on each face of the slab.
$$\rho = A_s/A_{ct} = 1414/40.0 \times 10^4 = 3.53 \times 10^{-3}$$

Step 2: Parameters in CIRIA equation (3.17):
$$K = 0.5$$
$\varepsilon_{cd} = 150 \times 10^{-6}$ for U.K. external exposure conditions
Casting in winter, $T_2 = 10° C$
T_1: From Table 19.4, interpolating,
$$T_1 = 12 + (20 - 12) \times (400 - 300)/ (500 - 300) = 16°C$$
Coefficient of thermal expansion, $\alpha_c = 10 \times 10^{-6}/°C$

Step 3: Calculate $(\varepsilon_{sm} - \varepsilon_{cm})$ from CIRIA equation (3.17):
$$(\varepsilon_{sm} - \varepsilon_{cm}) = K \ [\alpha_c \ (T_1 + T_2) + \varepsilon_{cd}] \qquad \text{(CIRIA 3.17)}$$
$$(\varepsilon_{sm} - \varepsilon_{cm}) = 0.5 \times [10 \times 10^{-6} \times (16 + 10) + 150 \times 10^{-6}] = 205 \times 10^{-6}$$

Step 4: Calculate the maximum crack spacing $S_{r, max}$ and crack width, w_k.
$$c = \text{cover to longitudinal reinforcement} = 40 \text{ mm}$$
$$\varphi = \text{bar diameter} = 12 \text{ mm}$$
$A_{c, eff}$ = Effective area of concrete in tension surrounding reinforcement to a
depth of $h_{c, eff} = \min [h/2; 2.5(c + \varphi/2)]$.
Assuming 50% of reinforcement is on each face, $A_s = 1413/2 = 707$ mm^2/m.
$h_{c,eff} = \min [h/2; \ 2.5(c + \varphi/2)] = \min [400/2 ; 2.5(40 + 12/2)] = 115$ mm.
$$\rho_{p, eff} = A_s/A_{c,eff} = 707/(115 \times 1000) = 6.15 \times 10^{-3}$$

*Note that ρ and ρ_{eff} are two different values. ρ is the ratio of steel to the whole
cross section. On the other hand, peff is the ratio of steel area to the concrete
area in tension.*
$$S_{r, max} = k_3 \ c + k_1 \ k_2 \ k_4 \ \varphi/ \rho_{p, eff} \qquad (7.11)$$

Taking $k_1 = 0.8$ for high bond bars, $k_2 = 1.0$ for pure tension, $k_3 = 3.4$, $k_4 = 0.425$.

$$S_{r, max} = 3.4 \times 40 + 0.8 \times 1.0 \times 0.425 \times 12/ (6.15 \times 10^{-3}) = 799 \text{ mm}$$
$$w_k = S_{r, max} \ (\varepsilon_{sm} - \varepsilon_{cm}) \qquad (7.8)$$

$$w_k = 799 \times 205 \times 10^{-6} = 0.16 \text{ mm}$$

19.5 DESIGN OF A RECTANGULAR COVERED TOP UNDERGROUND WATER TANK

Specification

Design a rectangular water tank with two equal compartments as shown in Fig. 19.9.

$$\text{Soil: Unit weight } \gamma = 18 \text{ kN/m}^3$$
$$\text{Concrete: Unit weight } \gamma = 25 \text{ kN/m}^3$$
$$\text{Unit weight of water } \gamma_w = 10 \text{ kN/m}^3$$
$$\text{Soil: Submerged unit weight } \gamma = (18 - \gamma_w) = 8 \text{ kN/m}^3$$
$$\text{Coefficient of friction } \varphi = 30^\circ$$
$$\text{Surcharge} = 12 \text{ kN/m}^2$$

Consider the possibility of water logging up to 1 m below the ground level.
Design crack width = 0.2 mm.
Use f_{ck} = 30 MPa concrete and f_{yk} = 500 MPa steel.
Assume walls and slabs are 400 mm thick. The roof is *not* integrally connected to the walls and is simply supported on the external walls but continuous over the central dividing wall.

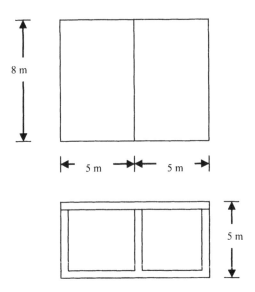

Fig. 19.9 Rectangular water tank.

19.5.1 Check Uplift

Total weight W of the tank when empty:
$$W = \{5 \times 10 - (5 - 0.4 - 0.4)(10 - 0.4 - 0.4 - 0.4)\} \times 8 \times 25$$
$$W = 2608 \text{ kN}$$

As this is a favourable load, using a load factor of $\gamma_{G,\,inf} = 0.90$,
$$W \times \gamma_{G,\,inf} = 2347 \text{ kN}$$

Uplift Pressure of water under the floor due to 4 m head of water
$$\text{Uplift pressure} = 10 \times 4 = 40 \text{ kN/m}^2$$
$$\text{Uplift force} = 8 \times 10 \times 40 = 3200 \text{ kN}$$

Additional weight required to have a factor of safety against floatation of 1.1 is $3200 \times 1.1 - 2347 = 1173$ kN.
This can be provided by extending the base as shown in Fig. 19.10.

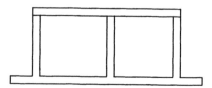

Fig. 19.10 New design of base to increase total weight of tank.

The submerged unit weight of the soil $= 18 - 10 = 8$ kN/m^3.
Pressure due to 1 m high dry soil plus 3.6 m of submerged soil
$$= 1 \times 18 + 3.6 \times 8 = 46.8 \text{ kN/m}^2$$
Submerged weight of slab $= (25 - 10) \times 0.4 = 6.0$ kN/m^2.
If b = width of the projecting base slab, then
$$\{[(10 + 2b) \times (8 + 2b) - 10 \times 8] \times (46.8 + 6.0)\} \times \gamma_{G,\,inf} = 1173$$
If b = 0.65 m, the additional weight is 1192 kN.

19.5.2 Pressure Calculation on Longitudinal Walls

Case 1: Tank empty
Coefficient of active earth pressure:
$$k_a = (1 - \sin\varphi)/ (1 + \sin\varphi) = (1 - 0.5)/ (1 + 0.5) = 0.33$$
$$\text{Pressure due to surcharge} = k_a \times 12 = 4 \text{ kN/m}^2$$
The wall is $5000 - 400 - 400 = 4200$ mm high.
For the top $(1000 - 400) = 600$ mm, unit weight of soil $= 18$ kN/m^3.
Below this level, submerged unit weight of soil $= 8$ kN/m^3.
In addition to the soil pressure there is also the pressure due to ground water.
The pressures at different levels shown in Fig. 19.11 are:

(i) At 400 mm below ground
$$p = 4 \text{ kN/m}^2 \text{ due to surcharge} + k_a \times 18 \times 0.4 = 6.4 \text{ kN/m}^2$$

(ii) at 1000 mm below ground
$$p = 4 \text{ kN/m}^2 \text{ due to surcharge} + k_a \times 18 \times 1.0 = 10.0 \text{ kN/m}^2$$

(iii) at 4600 mm below ground

$$p = 10 + k_a \times 8 \times (4.6 - 1.0) + 10 \times (4.6 - 1.0) \text{ due to ground water}$$
$$p = 55.6 \text{ kN/m}^2$$

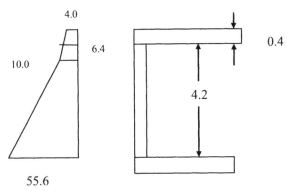

Fig. 19.11 Pressure distribution on the wall.

Case 2: Tank full
Ignore any passive pressure due to soil and assume that the ground is dry.

(i) At 400 mm below ground
$$p = 10 \times 0.4 = 4.0 \text{ kN/m}^2$$

(ii) At 4600 mm below ground
$$p = 10 \times 4.6 = 46 \text{ kN/m}^2$$

19.5.3 Check Shear Capacity

Effective depth: $d = 400 - 40$ mm cover $- 12$ mm bar $/2 = 354$ mm

Case 1: Tank empty
Although part of the lateral pressure is transferred to the sides of the wall, towards the middle of the wall, most of the load is transferred to the base. Using a load factor of $\gamma_F = 1.2$, total shear force at base is approximately
$$V_{Ed} = (\gamma_F = 1.2) \times [0.5 \times (6.4 + 10.0) \times 0.6 + 0.5 \times (10.0 + 55.6) \times 3.6]$$
$$= 147.6 \text{ kN/m}$$

Case 2: Tank full
Total shear force at base is approximately
$$V = (\gamma_F = 1.2) \times [0.5 \times (4.0 + 46.0) \times 4.2] = 126.0 \text{ kN/m}$$

Check whether thickness is sufficient
Use Eurocode 2 equation (6.9) check $V_{Ed} < V_{Rd, max}$.

$$V_{Rd,max} = \alpha_{cw} \, b_w \, z \, v_1 \, f_{cd} \, \frac{1}{(\cot\theta + \tan\theta)} \qquad (6.9)$$

$\alpha_{cw} = 1.0$, $b_w = 1000$ mm, $z = 0.9d = 319$ mm, $f_{cd} = (f_{ck} = 30)/ (\gamma_c = 1.5) = 20$ MPa
$$v_1 = 0.6 \, (1 - f_{ck}/ 250) = 0.528 \qquad (6.6N)$$

Section 6 of clause 6.2.3(109) *Eurocode 2-Design of concrete structures: Part 3: Liquid retaining and containment structures*, recommends that cot θ can be conservatively assumed to be 1.0.

$$V_{Rd,\,max} = [1.0 \times 1000 \times 319 \times 0.528 \times 20/(1 + 1)] \times 10^{-3} = 1684 \text{ kN/m} \qquad (6.9)$$

Section thickness is adequate.

19.5.4 Minimum Steel Area

Calculate the minimum steel required to control early-age thermal cracking
$$h = 400 \text{ mm}, c = 40 \text{ mm},$$
$$A_{ct} = b_w \times h = 1000 \times 400 = 40 \times 10^4 \text{ mm}^2$$
Taking external restraint as dominant, $k_c = 1.0$.
Using the values for h and k in Table 19.1, interpolate the value of k
$$k = 1.0 - (1.0 - 0.65) \times (400 - 300)/ (800 - 300) = 0.93$$
$$\sigma_s = f_{yk} = 500 \text{ MPa} ,$$
From section 19.4.3, $f_{ct,\,eff} = 1.8$ MPa for f_{ck} (t) at t = 3 days
$$A_{ct} = 400 \times 1000 = 40 \times 10^4 \text{ mm}^2$$
$$A_{s,\,min} = k_c \, k \, f_{ct,\,eff} \, A_{ct}/ \sigma_s = 1.0 \times 0.93 \times 1.8 \times 40 \times 10^4 /500 = 1339 \text{ mm}^2 \quad (7.1)$$
Provide on each face H12 at 160 mm = 707 mm²/m.
Total $A_s = 2 \times 707 = 1414$ mm²/m.
$$\rho = A_s/A_{ct} = 1413/ (40.0 \times 10^4) = 3.53 \times 10^{-3}$$
From section 19.4.6, the expected crack width due to wall restraint at the base is 0.16 mm.

Check whether shear reinforcement can be avoided
The shear capacity of a member without any shear reinforcement is given by code equations (6.2a), (6.2b) and (6.3N). Assuming no prestressing,
$$V_{Rd,c} = [C_{Rd,c} \, k \{100 \, \rho_1 \, f_{ck}\}^{1/3}] b_w \, d \geq [v_{min}] b_w d \qquad (6.2a)$$
$$C_{Rd,c} = \frac{0.18}{(\gamma_c = 1.5)} = 0.12 , \quad k = 1 + \sqrt{\frac{200}{d}} = 1 + \sqrt{\frac{200}{354}} = 1.75 \leq 2.0$$
$$100\rho_1 = 100 \frac{A_{sl}}{b_w \, d} = 100 \frac{707}{(1000 \times 354)} = 0.2 \leq 2$$
$$v_{min} = 0.035 k^{1.5} \sqrt{f_{ck}} = 0.15 \qquad (6.3N)$$
$$V_{Rd,\,c} = 135 \text{ kN/m} < (V_{Ed} = 148 \text{ kN/m})$$

Increase reinforcement provided to increase shear resistance and avoid the need for shear reinforcement.

Provide H12 at 90 mm centres on each face.

$$A_s = 1257 \text{ mm}^2/\text{m}$$

$$100\rho_1 = 100\frac{A_{sl}}{b_w d} = 100\frac{1257}{(1000 \times 354)} = 0.355 \le 2$$

$$V_{Rd,c} = 151 \text{ kN/m} > (V_{Ed} = 148 \text{ kN/m})$$

There is no need for shear reinforcement.

Instead of increasing the area of steel, the other option is to increase the thickness of the wall which might be cheaper!

19.5.5 Design of Walls for Bending at Ultimate Limit State

For calculating moments in the walls of the tank, ready-made tables of moment coefficients are available. These coefficients have been obtained from elastic analysis of thin plates using analytical methods based on multiple Fourier series or using the finite element method. Typical results are shown in Table 19.5 for the case of side and bottom edges being clamped and the top edge being free as shown in Fig. 19.12.

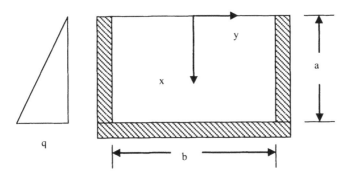

Fig. 19.12 Notation for Table 19.5.

19.5.5.1 Design of Transverse/Side Walls

Although the pressure on the wall is not strictly hydrostatic, the wall can be designed as a 7.2 m wide × 4.2 m high slab clamped on three sides and free at top and subjected to a hydrostatic loading giving base pressures of 55.6 kN/m² for case 1 and 46.0 kN/m² for case 2. Since the pressure difference is not large, design for case 1 and use the same steel area for case 2. Near the base the wall will bend about a horizontal axis and away from the base, and the wall bends as a horizontal clamped beam spanning between the end walls.

Table 19.5 Moment coefficients

b/a	x/a	$y = 0$		$y = b/4$		$y = b/2$
		M_x	M_y	M_x	M_y	M_y
2.0	0	0	0.027	0	0.010	−0.064
	0.25	0.012	0.024	0.006	0.010	−0.060
	0.50	0.016	0.017	0.011	0.010	−0.048
	0.75	−0.007	0.003	−0.002	0.003	−0.024
	1.0	−0.084		−0.058		-
1.5	0	0	0.021	0	0.006	−0.039
	0.25	0.009	0.020	0.004	0.007	−0.044
	0.50	0.015	0.017	0.009	0.008	−0.041
	0.75	0.003	0.006	0.004	0.004	−0.023
	1.0	−0.058		−0.039		-
1.0	0	0	0.010	0	0.002	−0.014
	0.25	0.003	0.012	0.001	0.003	−0.023
	0.50	0.009	0.013	0.005	0.005	−0.028
	0.75	0.008	0.008	0.005	0.004	−0.020
	1.0	−0.032		−0.021		

$$\text{Moment} = \text{Coefficient} \times q \times a^2$$

(a) Vertical bending moment at base

$b = 7.2$ m, $a = 4.2$ m, $b/a = 1.71$. From Table 19.5, interpolating between b/a of 1.5 and 2.0,

$$\text{Bending moment coefficient} \approx (0.084 + 0.058)/2 = 0.071$$

Vertical bending moment M at SLS:

$$M = 0.071 \times 55.6 \times 4.2^2 = 69.64 \text{ kNm/m (SLS)}$$

Vertical bending moment at base at ULS

$$M = (\gamma_F = 1.2) \times 69.64 = 83.57 \text{ kNm/m (ULS)}$$

$$k = M/(b \, d^2 \, f_{ck}) = 83.57 \times 10^6/(1000 \times 354^2 \times 30) = 0.022 < 0.196$$

$$\frac{z}{d} = 0.5[1.0 + \sqrt{(1 - 3\frac{k}{\eta})}] = 0.983$$

$A_s = M/(0.87 \, f_{yk} \, z) = 83.57 \times 10^6/(0.87 \times 500 \times 0.98 \times 354) = 554$ mm²/m.
Provided steel of H12 at 90 mm c/c = 1257 mm²/m on each face is more than adequate.

(b) Horizontal bending moment at fixed vertical edges

From data in Table 19.5, interpolating between b/a of 1.5 and 2.0,

$$\text{Bending moment coefficient} = (0.064 + 0.039)/2 = 0.052$$

$$M \text{ at SLS} = 0.052 \times 55.6 \times 4.2^2 = 51.0 \text{ kNm/m}$$

$$M = (\gamma_F = 1.2) \times 51.0 = 61.2 \text{ kNm/m (ULS)}$$

$$k = M/(b \, d^2 \, f_{ck}) = 61.2 \times 10^6/(1000 \times 354^2 \times 30) = 0.016 < 0.196$$

$$\frac{z}{d} = 0.5[1.0 + \sqrt{(1 - 3k)}] = 0.988$$

$A_s = M/ (0.87\ f_{yk}\ z) = 61.2 \times 10^6/ (0.87 \times 500 \times 0.99 \times 354) = 401\ mm^2/m$

(c) Direct tension in walls

In case 2 there is also direct tension in the horizontal direction in the wall due to water pressure inside he tank on the 10 m long walls. Average pressure p is approximately

$$p = 0.5 \times 46.0 = 23\ kN/m^2$$

Ignoring the resistance provided by the base, tensile force N per meter is

$$N = (0.5 \times 5.0 \times 23) = 57.5\ kN/m\ at\ SLS.$$
$$N = (\gamma_F = 1.2) \times 57.5 = 69\ kN/m.at\ ULS$$

Steel area required to resist this force is approximately

$$A_s = N/ (0.87\ f_{yk}) = 69 \times 10^3/ (0.87 \times 500) = 158\ mm^2/m$$

On each face this is equal to $158/2 = 79\ mm^2/m$.
Adding this steel area to the moment steel needed for horizontal bending,

$$A_s = 401\ (bending) + 79\ (direct\ tension) = 480\ mm^2/m$$

Provided steel of H12 at 90 mm c/c = 1257 mm^2/m on each face is more than adequate.

Note: In the calculations for design of a section subjected to bending and direct tension that follow, the procedure that has been adopted is to calculate the steel area required separately for bending and direct tension and add the two values to obtain the total steel area required. This procedure is approximate. The correct method is to consider moment-axial force interaction as in the case of columns and use a column design chart as explained in Chapter 9 to check the adequacy of reinforcement provided.

(d) Horizontal bending moment at mid-span

From data in Table 19.5, interpolating between b/a of 1.5 and 2.0,

$$Bending\ moment\ coefficient = (0.027 + 0.021)/2 = 0.024$$
$$M\ at\ SLS = 0.024 \times 55.6 \times 4.2^2 = 23.54\ kNm/m$$
$$M = (\gamma_F = 1.2) \times 23.54 = 28.3\ kNm/m\ (ULS)$$
$$k = M/ (b\ d^2\ f_{ck}) = 28.3 \times 10^6/ (1000 \times 354^2 \times 30) = 0.008 < 0.196$$

$$\frac{z}{d} = 0.5[1.0 + \sqrt{(1 - 3\frac{k}{\eta})}] = 0.99$$

$$A_s = M/ (0.87\ f_{yk}\ z) = 28.3 \times 10^6/ (0.87 \times 500 \times 0.99 \times 354) = 186\ mm^2/m$$
$$A_s = 186\ (Bending) + 79\ (Direct\ tension) = 265\ mm^2/m$$

These values are smaller than the bending moment of at support.
Provide on both faces H12 at 90 mm c/c in the horizontal direction.
A_s on each face = 1257 mm^2/m.

19.5.5.2 Crack Width Calculation in Transverse Walls

h = 400 mm, cover = 40 mm, Steel: H12 at 90 mm c/c.
Applied forces at serviceability limit state:

$$M = 51.0 \text{ kNm/m}, N = 57.5 \text{ kN/m (tension)}$$
$$f_{ck} = 30 \text{ MPa}, f_{yk} = 500 \text{ MPa}, \alpha_e = 15.6 \text{ allowing for creep}$$
$$b = 1000 \text{ mm}, d = 354 \text{ mm}, A_s = A_s' = 1257 \text{ mm}^2/\text{m}$$

(a) Calculate the neutral axis depth including compression steel area

Substituting the values in equation (C19.1),

$$(M - N\frac{h}{2})A_s\alpha_e(d - x) - A_s'\alpha_e(x - d')\{M + N(d - d' - \frac{h}{2})\}$$

$$= b\frac{x^2}{2}\{N(d - \frac{x}{3} - \frac{h}{2}) + M\}$$

$$(51.0 \times 10^6 - 57.5 \times 10^3 \times \frac{400}{2}) \times 1257 \times 15.6 \times (354 - x)$$

$$- \{51.0 \times 10^6 + 57.5 \times 10^3 (354 - 46 - \frac{400}{2})\} \times 1257 \times 15.6 \times (x - 46)$$

$$= 1000\frac{x^2}{2}\{57.5 \times 10^3 (354 - \frac{x}{3} - \frac{400}{2}) + 51.0 \times 10^6\}$$

Simplifying:

$$x^3 - 29927.5 \, x^2 - 1.8964 \times 10^6 \times x + 3.258 \times 10^8 = 0$$

Solving by trial and error, $x = 78$ mm.

(b) Calculate the compressive stress in concrete

Substituting the values in equation (C19.2),

$$bx\frac{f_c}{2}(d - \frac{x}{3}) + A_s'\alpha_e f_c \frac{(x - d')}{x}(d - d') = M - N\frac{h}{2}$$

$$\{1000 \times 78 \times \frac{1}{2}(354 - \frac{78}{3}) + 1257 \times 15.6 \times \frac{(78 - 46)}{78} \times (354 - 46)\}f_c$$

$$= 51.0 \times 10^6 - 57.5 \times 10^3 \times \frac{400}{2}$$

Solving, $f_c = 2.6$ MPa.

(c) Calculate the tensile stress in steel

$$f_s = \alpha_e f_c \frac{d - x}{x} = 15.6 \times 2.6 \times \frac{354 - 78}{78} = 143 \text{ MPa}$$

(d) Check maximum bar diameter and maximum spacing using Tables 7.2N and 7.3N respectively of Eurocode 2

Taking steel stress as approximately 160 MPa, for a crack width of 0.2 mm, the maximum bar spacing is 200 mm.

The maximum bar diameter from Table 7.2N is 25 mm. The bar diameter is modified by using Eurocode 2 equation (7.6N) as follows.

$$f_{ct, eff} = 0.3 \times f_{ck}^{067} = 2.9 \text{ MPa}$$

$$\sigma_c = \text{axial stress} = N/(b \, h) = 57.5 \times 10^3/(1000 \times 400) = -0.14 \text{ MPa (tensile)}$$

$$h = 400 \text{ mm} < 1000 \text{ mm}, h^*/h = 1.0$$

$$k_1 = 2/3 \text{ as N is tensile}$$
$$k_c = 0.4 \times \{1 - \sigma_c/[k_1 \times f_{ct,eff}] \times (h/h^*)\} \leq 1.0 \qquad (7.2N)$$
$$k_c = 0.4 \times \{1 - (-0.14)/[0.67 \times 2.9 \times 1.0]\} = 0.43 \leq 1.0$$

$$\phi_s = \phi_s^* \frac{f_{ct,eff}}{2.9} k_c \, 0.5 \frac{h_{cr}}{(h-d)} \qquad (7.6N)$$

h_{cr} = depth of tensile zone immediately prior to cracking.
As the tensile force is very small, cracking is governed by bending moment. h_{cr} can be taken as approximately $h/3 = 400/3 = 133$ mm.
$\Phi_s^* = 25$ mm from Eurocode 2 Table 7.2N.

$$\phi_s = 25 \times \frac{2.9}{2.9} \times 0.43 \times 0.5 \times \frac{155}{(400 - 354)} = 16 \, \text{mm}$$

Both spacing and maximum bar diameter satisfy the requirements for 0, 2 mm crack width

19.5.5.3 Design of Longitudinal Walls

The wall is designed as a 4.4 m × 4.2 m slab clamped on three sides and free at top and subjected to a hydrostatic loading giving at base pressures of 55.6 kN/m² for case 1 and 46.0 kN/m² for case 2. Since the pressure difference is not large, design for case 1 and use the same steel area for case 2 as well.

(a) Vertical bending moment at base
b = 4.4, a = 4.2, b/a ≈ 1.0. From Table 17.9, using the coefficient for b/a = 1.0,
$$\text{Moment at SLS} = 0.032 \times 55.6 \times 4.2^2 = 31.4 \text{ kNm/m (SLS)}$$
$$\text{Moment at ULS} = (\gamma_F = 1.2) \times 31.4 = 37.7 \text{ kNm/m (ULS)}$$
$$k = M/ (b \, d^2 \, f_{ck}) = 37.7 \times 10^6/ (1000 \times 354^2 \times 30) = 0.01 < 0.196$$

$$\frac{z}{d} = 0.5[1.0 + \sqrt{(1 - 3\frac{k}{\eta})}] = 0.99$$

$$A_s = M/ (0.87 \, f_{yk} \, z) = 37.7 \times 10^6/ (0.87 \times 500 \times 0.99 \times 354) = 247 \text{ mm}^2/\text{m}$$

(b) Horizontal bending moment at fixed vertical edges
From Table 17.9, using the coefficient for b/a = 1.0,
$$\text{Moment at SLS} = 0.028 \times 55.6 \times 4.2^2 = 27.5 \text{ kNm/m}$$
$$\text{Moment at ULS} = (\gamma_F = 1.2) \times 27.5 = 33.0 \text{ kNm/m (ULS)}$$
$$k = M/ (b \, d^2 \, f_{ck}) = 33.0 \times 10^6/ (1000 \times 354^2 \times 30) = 0.009 < 0.196$$

$$\frac{z}{d} = 0.5[1.0 + \sqrt{(1 - 3\frac{k}{\eta})}] = 0.99$$

$$A_s = M/ (0.87 \, f_{yk} \, z) = 33.0 \times 10^6/ (0.87 \times 500 \times 0.99 \times 354) = 216 \text{ mm}^2/\text{m}$$

(c) Direct tension in walls
In case 2 there is also direct tension in the horizontal direction in the wall due to water pressure on the 8 m long walls. Average pressure p is approximately

$$p = 0.5 \times 46.0 = 23 \text{ kN/m}^2$$

Ignoring the resistance provided by the base, tensile force N per meter at SLS is

$$N = 0.5 \times 8.0 \times 23 = 92.0 \text{ kN/m}$$

Tensile force N per meter at ULS is

$$N = (\gamma_F = 1.2) \times 92.0 = 110.4 \text{ kN/m}$$

Area of steel required $A_s = 110.4 \times 10^3 / (0.87 \times 500) = 254 \text{ mm}^2/\text{m}$.
Area of steel per face $= 254/2 = 127 \text{ mm}^2/\text{m}$.
Adding this steel area to the moment steel needed for horizontal bending, total area of steel needed: 216 (bending) + 254 (axial tension) $= 470 \text{ mm}^2/\text{m}$.
Provided steel of H12 at 90 mm c/c $= 1257 \text{ mm}^2/\text{m}$ on each face is more than adequate.

(d) Horizontal bending moment at mid-span
From Table 17.9, using the coefficient for b/a = 1.0,

Moment at SLS $= 0.013 \times 55.6 \times 4.2^2 = 12.8 \text{ kNm/m}$.

Moment at ULS $= (\gamma_F = 1.2) \times 12.8 = 15.4 \text{ kNm/m (ULS)}$

$$k = M/ (b \, d^2 \, f_{ck}) = 15.4 \times 10^6 / (1000 \times 354^2 \times 30) = 0.004 < 0.196$$

$$\frac{z}{d} = 0.5[1.0 + \sqrt{(1 - 3\frac{k}{\eta})}] = 0.99$$

$A_s = M/ (0.87 \, f_{yk} \, z) = 15.4 \times 10^6 / (0.87 \times 500 \times 0.99 \times 354) = 101 \text{ mm}^2/\text{m}$.
Adding this steel area to the steel needed for axial tension, total area of steel needed: 101 (bending) + 254 (axial tension) $= 355 \text{ mm}^2/\text{m}$.
Provided steel of H12 at 90 mm c/c $= 1257 \text{ mm}^2/\text{m}$ on each face is more than adequate.

19.5.5.4 Crack Width Calculation in Longitudinal Walls

h = 400 mm, cover = 40 mm, Steel: H12 at 90 mm c/c.
Applied forces at serviceability limit state: M = 27.5 kNm/m, N = 92.0 kN/m.
$f_{ck} = 30$ MPa, $f_{yk} = 500$ MPa, $\alpha_e = 15.6$ allowing for creep.
b = 1000 mm, d = 354 mm, $A_s = A_s' = 1257 \text{ mm}^2/\text{m}$.

(a) Calculate the neutral axis depth including compression steel area

$$(M - N\frac{h}{2})A_s \alpha_e (d - x) - A_s' \alpha_e (x - d')\{M + N(d - d' - \frac{h}{2})\}$$

$$= b\frac{x^2}{2}\{N(d - \frac{x}{3} - \frac{h}{2}) + M\}$$

$$(27.5 \times 10^6 - 92.0 \times 10^3 \times \frac{400}{2}) \times 1257 \times 15.6 \times (354 - x)$$

$$- \{27.5 \times 10^6 + 92 \times 10^3 (354 - 46 - \frac{400}{2})\} \times 1257 \times 15.6 \times (x - 46)$$

$$= 1000 \frac{x^2}{2} \{92.0 \times 10^3 (354 - \frac{x}{3} - \frac{400}{2}) + 27.5 \times 10^6\}$$

Simplifying:
$$x^3 - 29927.5\, x^2 - 1.8964 \times 10^6 \times x + 3.258 \times 10^8 = 0$$
Solving by trial and error, $x = 50$ mm.

(b) Calculate the compressive stress in concrete
$$bx \frac{f_c}{2}(d - \frac{x}{3}) = M - N\frac{h}{2}$$
$$1000 \times 50 \times \frac{f_c}{2}(354 - \frac{50}{3}) = 27.5 \times 10^6 - 92.0 \times 10^3 \times \frac{400}{2}$$

Solving,
$$f_c = 1.1 \text{ MPa}$$

(c) Calculate the tensile stress in steel
$$f_s = \alpha_e f_c \frac{d - x}{x} = 15.6 \times 1.1 \times \frac{354 - 50}{50} = 104 \text{ MPa}$$

(d) Check maximum bar diameter and maximum spacing using Tables 7.2N and 7.3N respectively of Eurocode 2
The calculations can be done as for transverse walls. The steel stress is very low. The crack width can be expected to be smaller than 0.2 mm.

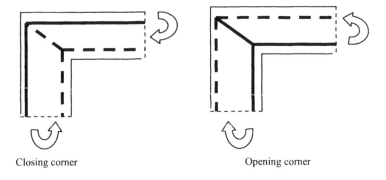

Closing corner   Opening corner

Fig. 19.13 Reinforcing corner for closing and opening joints.

19.5.5.5 Detailing at Corners

Proper detailing of steel at corners is extremely important to realize the full strength of the sections. Section 6 of clause 6.2.3(109) of *Eurocode 2-Design of concrete structures: Part 3: Liquid retaining and containment structures,* recommends using the strut–tie method to determine the required reinforcement for opening and closing corners. Fig. 19.13 shows a schematic reinforcement layout for closing and opening joints, where full lines represent reinforcement and dashed lines concrete struts. Note that in the case of opening joint, the diagonal reinforcement is in tension. Refer to *Annex.2: Frame corners in Eurocode 2* where design of frame corners using strut–tie method is given in detail.

19.5.6 Design of Base Slab at Ultimate Limit State

The slab is subjected to concentrated loads from the walls and bending moments at the ends from the walls. There is also a small amount of direct tension from the internal pressure in the tanks but this has been ignored in the following design.

(i) Longitudinal direction

(1) Tank empty

(a) Load on end walls:
Vertical load from roof slab: $= (5.0/2) \times 0.4 \times 25 = 25.0$ kN/m
Surcharge: $(5/2) \times 12 = 30.0$ kN/m
Weight of wall: $4.2 \times 0.4 \times 25 = 42.0$ kN/m
Weight of soil on the 0.65 m projection is equal to:
$0.65 \times (18.0 \times 1.0$ dry soil at top $+ 8.0 \times 3.2$ submerged soil) $= 28.3$ kN/m
Total $= 25.0 + 30.0 + 42.0 + 28.3 = 125.3$ kN/m
Using a load factor $\gamma_G = 1.35$, concentrated load from the end walls is
$= 1.35 \times 125.3 = 169.2$ kN/m
From previous calculation of wall design, moment from the external pressure
$= 0.071 \times 55.6 \times 4.2^2 = 69.64$ kNm/m at SLS
Using a load factor $\gamma_F = 1.20$, moment from external pressure
$= 1.2 \times 69.64 = 83.57$ kNm/m at ULS

(b) Load on central wall:
Vertical load from roof slab: $= 5.0 \times 0.4 \times 25 = 50.0$ kN/m
Surcharge: $5 \times 12 = 60.0$ kN/m
Weight of wall: $4.2 \times 0.4 \times 25 = 42.0$ kN/m
Total $= 50.0 + 60.0 + 42.0 = 152.0$ kN/m
Using a load factor $\gamma_G = 1.35$, concentrated load from the central walls is
$= 1.35 \times 152.0 = 205.2$ kN/m
Moment from the external pressure $= 0$.

(c) Uplift pressure: There is an uplift pressure of $10 \times 4.0 = 40$ kN/m^2.
Using a load factor $\gamma_F = 1.20$, the uplift pressure is
$$1.20 \times 40.0 = 48.0 \ \text{kN/m}^2$$

(d) Net pressure p on the ground:
$$p = (2 \times 169.2 + 205.2)/9.6 - 48.0 = 8.63 \ \text{kN/m}^2$$
Fig. 19.14 and Fig. 19.15 show respectively the forces on the base slab and the corresponding bending moment distribution.

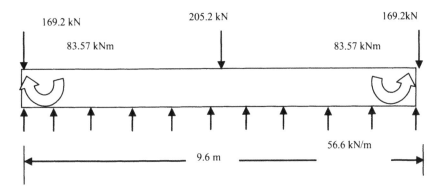

Fig. 19.14 ULS forces on the base slab in the longitudinal direction, tank empty.

Maximum bending moment: Tension at bottom = 83.57 kNm/m at the ends.
Maximum bending moment: Tension at top = 169.2 kNm/m at 2.92 m from the ends.
Maximum shear force = 169.2 kN/m at the ends.

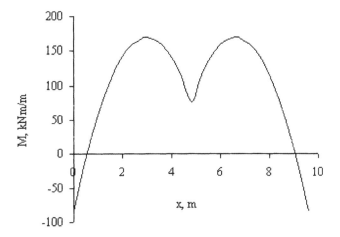

Fig. 19.15 Bending moment distribution in base slab in the longitudinal direction, tank empty.

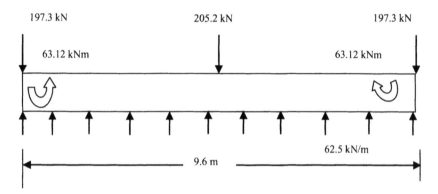

197.3 kN   205.2 kN   197.3 kN

63.12 kNm   63.12 kNm

62.5 kN/m

9.6 m

Fig. 19.16 ULS forces on the base slab in the longitudinal direction, tank full.

(2) Both tanks full and no ground water

(a) Load on end walls:
Vertical load from roof slab: $= (5.0/2) \times 0.4 \times 25 = 25.0$ kN/m
Surcharge: $(5/2) \times 12 = 30.0$ kN/m
Weight of wall: $4.2 \times 0.4 \times 25 = 42.0$ kN/m
Weight of soil on the 0.65 m projection $= 0.65 \times 18.0 \times 4.2 = 49.1$ kN/m
Total $= 25.0 + 30.0 + 42.0 + 49.1 = 146.1$ kN/m
Using a load factor $\gamma_G = 1.35$, concentrated load from the end walls is
$$= 1.35 \times 134.8 = 181.98 \text{ kN/m}$$
From previous calculation of wall design, moment from the external pressure, vertical bending moment at base (SLS):
$$M = 0.071 \times (10 \times 4.2) \times 4.2^2 = 52.60 \text{ kNm/m}$$
Using a load factor $\gamma_F = 1.20$, moment from external pressure at ULS
$$= 1.2 \times 52.60 = 63.12 \text{ kNm/m}$$

(b) Load on central wall:
205.2 kN/m as in Case 1.
Moment from the external pressure $= 0$.

(c) Load from water in the tank:
Using a load factor $\gamma_F = 1.20$, load due to 4.2 m depth of water is
$$= 1.2 \times 4.2 \times 10 = 50.4 \text{ kNm/m}$$
(d) No uplift pressure

(e) Net pressure p on the ground:
$$p = (2 \times 197.3 + 205.2)/9.6 + 50.4 = 112.9 \text{ kN/m}^2$$

(f) Net uniformly distributed load on base slab:
As the load due to water inside the tanks and the base pressure are both uniformly distributed on the base slab, one need to consider only the base pressure due to walls for determining the bending moment in the base slab.

Net uniformly distributed load on the base slab is $112.9 - 50.4 = 62.5$ kN/m^2.

Fig. 19.16 and Fig. 19.17 show respectively the forces on the base slab and the corresponding bending moment distribution.

Maximum bending moment: Tension at top = 374.6 kNm/m at 3.2 m from the ends. Maximum shear force = 197.3 kN/m at the ends.

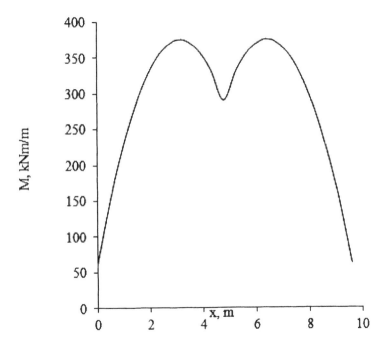

Fig. 19.17 Bending moment distribution in base slab in the longitudinal direction, tank full.

Design of reinforcement:
The maximum bending moment causing tension at top is 374.6 kNm/m from tank full and the maximum bending moment causing tension at bottom is 83.57 kNm/m from tank empty case.

Steel at top:
$M = 374.6$ kNm/m, $b = 1000$ mm, cover = 40 mm, $\varphi = $ H25,
$d = 400 - 40 - 25/2 = 347$ mm, $f_{ck} = 30$ MPa
$k = 374.6 \times 10^6 / (1000 \times 347^2 \times 30) = 0.104 < 0.196$

$$\frac{z}{d} = 0.5[1.0 + \sqrt{(1 - 3\frac{0.104}{1.0})}] = 0.915$$

$$A_s = \frac{374.6 \times 10^6}{0.915 \times 347 \times 0.87 \times 500} = 2712 \text{mm}^2 / \text{m}$$

Provide H25 at 180 c/c = 2727 mm²/m.

Steel at bottom:
M = 83.57 kNm/m, b = 1000 mm, cover = 40 mm, φ = H12,
d = 400 – 40 – 12/2 = 354 mm, f_{ck} = 30 MPa
k = 83.57 × 10⁶/ (1000 × 354² × 30) = 0.02 < 0.196

$$\frac{z}{d} = 0.5[1.0 + \sqrt{(1 - 3\frac{0.02}{1.0})}] = 0.98$$

$$A_s = \frac{83.57 \times 10^6}{0.98 \times 354 \times 0.87 \times 500} = 554 \text{mm}^2 / \text{m}$$

Provide H12 at 200 c/c= 566 mm²/m.

(ii) Transverse direction
The slab is subjected to concentrated loads from the walls and bending moment at the ends from the walls.

(1) Tank empty

(a) Load on end walls:
Vertical load = (γ_G = 1.35) × 123.1 = 166.2 kN/m (from previous calculation for longitudinal wall).
Moment from the external pressure = (γ_F = 1.2) × 0.032 × 55.6 × 4.2²
= 37.7 kNm/m.

(b) Uplift pressure:
Uplift pressure = (γ_F = 1.2) × 10 × 4.0 = 48 kN/m².

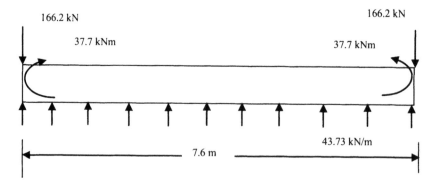

166.2 kN   166.2 kN

37.7 kNm   37.7 kNm

43.73 kN/m

7.6 m

Fig. 19.18 Forces on the base slab in the transverse direction, tank empty.

(c) Net pressure p on the ground:
$$p = 2 \times 166.2/7.6 - 48.0 = -4.3 \text{ kN/m}^2$$
Although calculation indicates that the slab will not be in equilibrium, this is not strictly true because of the presence of the loads on the central wall. Since the overall stability against flotation of the structure has been established, calculation will be continued assuming equilibrium is maintained. Calculation will be somewhat approximate but not seriously so. Fig. 19.18 shows the forces on the base slab. Fig. 19.19 shows the bending moment distribution on the base slab.

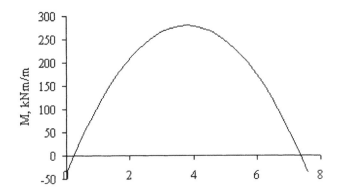

Fig. 19.19 Bending moment on the base slab in the transverse direction, tank empty.

Maximum bending moment causing tension at bottom = 37.7 kNm/m.
Maximum moment causing tension at top = 280.03 kNm/m.
Maximum shear force = 166.2 kN/m at the ends.

(2) Both tanks full and no ground water

(a) Load on end walls:
Vertical load = ($\gamma_G = 1.35$) ×144.2 = 194.7 kN/m (from previous calculation for longitudinal wall).
Moment from the external pressure = ($\gamma_F = 1.2$) × 0.032 × (10 × 4.2) × 4.2^2
$$= 28.4 \text{ kNm/m.}$$
(b) Uplift pressure:
There is no uplift pressure.

(c) Net pressure p on the ground:
$$p = 2 \times 194.7/7.6 = 51.23 \text{ kN/m}^2$$
Fig. 19.20 shows the forces on the base slab. Fig. 19.21 shows the bending moment distribution on the base slab.
Maximum moment causing tension at top = 298.9 kNm/m.

Maximum shear force = 194.7 kN/m at the ends.

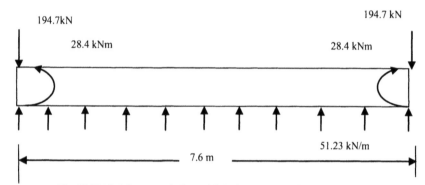

Fig. 19.20 ULS forces on the base slab in the transverse direction, tank full.

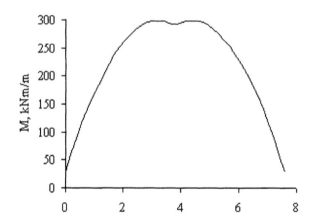

Fig. 19.21 Bending moment distribution on the base slab in the transverse direction, tank full.

Design of reinforcement:
The maximum bending moment causing tension at top is 298.9 kNm/m from tank full case and the maximum bending moment causing tension at bottom is 37.7 kNm/m from tank empty case.

Steel at top:
M = 298.9 kNm/m, b = 1000 mm, cover = 40 mm, φ = H25,
d = 400 – 40 – 25/2 = 347 mm, f_{ck} = 30 MPa
k = 298.9 × 10^6/ (1000 × 347^2 × 30) = 0.083 < 0.196

$$\frac{z}{d} = 0.5[1.0 + \sqrt{(1 - 3\frac{0.083}{1.0})}] = 0.93$$

$$A_s = \frac{298.9 \times 10^6}{0.93 \times 347 \times 0.87 \times 500} = 2200\, mm^2 / m$$

Provide H25 at 220 c/c = 2231 mm^2/m.

Steel at bottom:
M = 37.7 kNm/m, b = 1000 mm, cover = 40 mm, φ = H12.
d = 400 − 40 − 12/2 = 354 mm, f_{ck} = 30 MPa.
k = 37.7 × 10^6/ (1000 × 354^2 × 30) = 0.01 < 0.196

$$\frac{z}{d} = 0.5[1.0 + \sqrt{(1 - 3\frac{0.01}{1.0})}] = 0.99$$

$$A_s = \frac{37.7 \times 10^6}{0.99 \times 354 \times 0.87 \times 500} = 247 \, mm^2 / m$$

Provide H10 at 300 c/c= 262 mm^2/m.

Shear strength check:
Maximum shear force is 194.7 for the transverse direction with the tank being full.

(i) Check the adequacy of the thickness i.e. $V_{Ed} < V_{Rd, max}$ using Eurocode 2 equation (6.9)

$$V_{Rd,max} = \alpha_{cw} \, b_w \, z \, v_1 \, f_{cd} \frac{1}{(\cot \theta + \tan \theta)} \qquad (6.9)$$

α_{cw} = 1.0, b_w = 1000 mm, d = 347 mm, z = 0.9d = 312 mm, f_{cd} = (f_{ck} = 30)/ (γ_c = 1.5) = 20 MPa

$$v_1 = 0.6 \, (1 - f_{ck}/ \, 250) = 0.528 \qquad (6.6N)$$

In section 6 of clause 6.2.3(109), *Eurocode 2-Design of concrete structures: Part 3: Liquid retaining and containment structures,* recommends that cot θ can be conservatively assumed to be 1.0.

$$V_{Rd, \, max} = [1.0 \times 1000 \times 312 \times 0.528 \times 20/(1 + 1)] \times 10^{-3} = 1647 \, kN/m \qquad (6.9)$$

Section thickness is adequate.

(ii) Check whether shear reinforcement can be avoided
The shear capacity of a member without any shear reinforcement is given by code equations (6.2a), (6.2b) and (6.3N). Assuming no prestressing,

$$V_{Rd,c} = [C_{Rd,c} \, k\{100 \, \rho_1 \, f_{ck}\}^{1/3}]b_w \, d \geq [v_{min}]b_w \, d \qquad (6.2a)$$

$$C_{Rd,c} = \frac{0.18}{(\gamma_c = 1.5)} = 0.12 , \; k = 1 + \sqrt{\frac{200}{d}} = 1 + \sqrt{\frac{200}{347}} = 1.76 \leq 2.0$$

Steel at top = A_s = H25 at 180 = 2727 mm^2/m

$$100 \rho_1 = 100 \frac{A_{sl}}{b_w \, d} = 100 \frac{2727}{(1000 \times 347)} = 0.79 \leq 2$$

$$v_{min} = 0.035 k^{1.5} \sqrt{f_{ck}} = 0.15 \qquad (6.3N)$$

$$V_{Rd,c} = 222.8 \text{ kN/m} > (V_{Ed} = 194.7 \text{ kN/m})$$

Final reinforcement:

In longitudinal and transverse directions steel at top is H25 at 180 c/c.

In the longitudinal direction, steel at bottom is H12 at 200 c/c and in the transverse direction steel at bottom is H10 at 300 c/c.

19.6 DESIGN OF CIRCULAR WATER TANKS

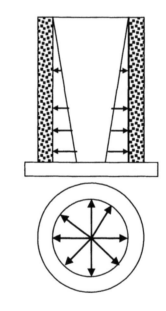

Fig. 19.22 Circular tank.

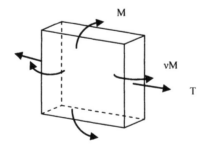

Fig. 19.23 Forces on an element of the wall.

Circular water tanks are commonly employed in reinforced concrete and also in circularly prestressed posttensioned prestressed concrete. Fig. 19.22 shows a circular tank subjected to an internal pressure which can be constant as in the case of gas tanks or increase towards the base as in the case of liquid retaining tanks.

If the tank is not restrained in the radial direction at top and bottom, then considering the tank as a thin-walled cylinder, under a constant internal pressure p, the circumferential tension T in the wall is given by

$$T = p R$$

where R = internal radius of the tank.
The displacement w in the radial direction is given by

$$w = \frac{pR^2}{Et}$$

where E = Young's modulus, t = thickness of the wall.

If the pressure variation is hydrostatic and at p any depth y from the top is

$$p = \gamma y$$

γ = unit weight of the liquid retained.
The circumferential tension T in the wall is given by

$$T = \gamma y \, R.$$

The displacement w in the radial direction at a depth y from the top is given by

$$w = \frac{\gamma R^2}{Et} y$$

If the displacement is constrained at the bottom, then the total pressure p is resisted partly by circumferential tension and partly by bending action in the vertical direction as shown in Fig. 19.23. In addition to the bending moment in the vertical direction, there is also a bending moment in the circumferential direction given by vM, where v is the Poisson's ratio.
The pressure p_t resisted by tension causes a radial displacement w given by

$$w = \frac{p_t R^2}{Et}, \quad \therefore \ p_t = \frac{Et}{R^2} w$$

The pressure p_b resisted by bending action is given by

$$EI \frac{d^4 w}{dy^4} = p_b$$

$I = t^3/12$ per unit length, t = thickness of the wall.
Because of the Poisson effect,

$$EI = \frac{Et^3}{12(1 - v^2)}$$

If p is the internal pressure,

$$p = p_b + p_t = EI \frac{d^4 w}{dy^4} + \frac{Et}{R^2} w$$

The bending moment M and shear force V and circumferential tension T are given by

$$M = -EI\frac{d^2w}{dy^2} \,, V = -EI\frac{d^3w}{dy^3} \,, T = \frac{Et}{R}w$$

The differential equation is known as the beam on elastic foundation equation and can be solved for given boundary conditions.

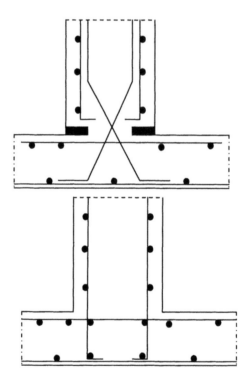

Fig. 19.24 Base details for pinned and fixed joints.

Ready-made tables are available for calculating the circumferential tension T and bending moment M for the two cases of the base of the tank being either fully fixed or pinned. Fig. 19.24 shows the base reinforcement details for achieving pinned and fixed joints. Table 19.6 shows typical values for a specific tank of dimensions $h^2/(Rt) = 8.0$.

19.6.1 Example of Design of a Circular Water Tank

Design an above-ground fixed base water tank for the following specification.

(a) Specification

Internal radius R = 15 m, height h = 6 m, wall thickness t = 300 mm.
Unit weight of water $\gamma = 10 \text{ kN/m}^3$.
$f_{ck} = 30$ MPa, $f_{yk} = 500$ MPa, Design crack width = 0.2 mm.

(b) Calculation of forces

$$\text{Parameter } (h^2/Rt) = 6^2/(15 \times 0.3) = 8.0$$
$$q = \gamma\, h = 10 \times 6 = 60 \text{ kN/m}^2$$

Fig. 19.25, Fig. 19.26 and Fig. 19.27 show the distribution of vertical bending moment, shear force and circumferential tension.
Maximum shear force V at base at ULS:

$$V = (\gamma_F = 1.2) \times 0.063 \times 60 \times 6 = 27.5 \text{ kN/m}$$

Maximum bending moment causing tension on inner face at base at ULS:

$$M = (\gamma_F = 1.2) \times 0.0267 \times 60 \times 6^2 = 69.2 \text{ kNm/m}$$

Maximum bending moment causing tension on the outer face at 0.4 h at ULS:

$$M = (\gamma_F = 1.2) \times 0.0077 \times 60 \text{ x } 6^2 = 19.96 \text{ kNm/m}$$

Maximum ring tension T occurs at mid-height

$$T = (\gamma_F = 1.2) \times 0.43 \times 60 \times 15 = 464 \text{ kN/m}$$

Corresponding moment:

$$M = (\gamma_F = 1.2) \times 0.0066 \times 60 \times 6^2 = 17.1 \text{ kNm/m}$$

Circumferential moment:

$$= \nu\, M = 0.2 \times 17.1 = 3.42 \text{ kNm/m}$$

Table 19.6 Vertical bending moment and ring tension coefficients for cylindrical tanks

$h^2/(Rt) = 8.0$, $\nu = 0.2$				
y/h	Fixed base		Pinned base	
	M	T	M	T
0: Top	0	0.067	0	0.017
0.1	0.0003	0.163	0.0001	0.136
0.2	0.0013	0.256	0.0006	0.254
0.3	0.0028	0.339	0.0016	0.367
0.4	0.0047	0.402	0.0033	0.468
0.5	0.0066	0.430	0.0056	0.545
0.6	0.0077	0.410	0.0084	0.579
0.7	0.0069	0.334	0.0109	0.552
0.8	0.0023	0.210	0.0118	0.446
0.9	–0.0081	0.073	0.0092	0.255
1.0: Base	–0.0267	0	0	0

Moment M = coefficient × $(\gamma\, h^3)$ kNm/m.
Positive moment causes tension on the outer face.
Tension T = coefficient × $(\gamma\, h\, R)$ kN/m.

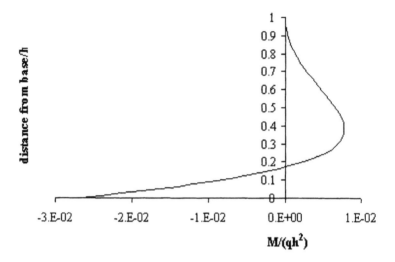

Fig. 19.25 Vertical bending moment in the wall.

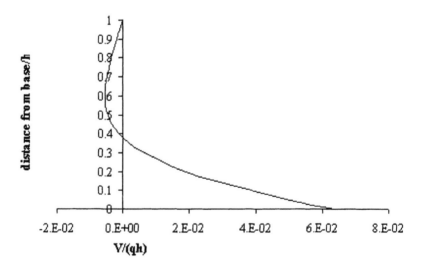

Fig. 19.26 Shear force in the wall.

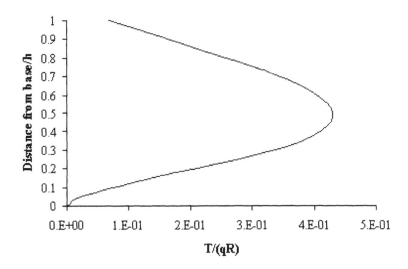

Fig. 19.27 Circumferential tension in the wall.

(c) Design
(i) Check shear capacity
Effective depth:

$$d = 300 - 40 \text{ cover} - 16/2 = 252 \text{ mm}$$
$$V_{Ed} = 27.5 \text{ kN/m}$$

$b_w = 1000\text{mm}, \ z \approx 0.9d = 227 \text{ mm}$

$$v_1 = 0.6 \ (1 - f_{ck}/250) = 0.528 \qquad (6.6N)$$
$$\alpha_{cw} = 1 \text{ for non-prestressed structures.} \qquad (6.10aN)$$

Take $\cot \theta = 1$.
Check adequacy of depth: Use Eurocode 2 equation (6.9)

$$V_{RD,max} = \alpha_{cw} \ b_w \ z \ v_1 \ f_{cd} \frac{1}{(\cot \theta + \tan \theta)} \qquad (6.9)$$
$$V_{Rd} = 1199 \text{ kN}$$

Depth is adequate.

(ii) Steel to control thermal cracking

Step 1: Calculate the tensile strength at t = 3 days.
Take s = 0.25 for Class N cement.

$$t = 3 \text{ days}, \ \beta_{cc} \ (t) = Exp\{s \ [1 - \sqrt{(28/t)}]\} = 0.6 \qquad (3.2)$$
$$f_{cm} = f_{ck} + 8 = 38 \text{ MPa}$$
$$f_{cm} \ (t) = \beta_{cc} \ (t) \times (f_{ck} + 8) = 0.6 \times (30 + 8) = 22.7 \text{ MPa} \qquad (3.1)$$
$$f_{ck} \ (t) = f_{cm} \ (t) - 8 \text{ MPa} = 14.7 \text{ MPa}$$
$$f_{ct, \ eff} = f_{ctm} = 0.3 \times f_{ck} \ (t)^{0.67} = 1.8 \text{ MPa}$$

Step 2: Calculate the modular ratio at t = 3 days.

$$E_{cm} = 22 \times [(f_{ck}+ 8)/10]^{0.3} = 22 \times [(30+ 8)/10]^{0.3} = 32.84 \text{ GPa}$$
$$E_{cm}(t) = [f_{cm}(t)/f_{cm}]^{0.3} \times E_{cm} \qquad\qquad (3.5)$$
$$E_{cm}(t) = [22.7/38.0]^{0.3} \times 32.84 = 4.38 \text{ GPa}$$
$$E_s = 200 \text{ GPa}$$
$$\alpha_e = 200/4.38 = 45.66$$

Step 3: Calculate the minimum steel required to control early-age thermal cracking.
$$h = 300 \text{ mm}, c = 40 \text{ mm},$$
$$A_{ct} = b_w \times h = 1000 \times 300 = 30 \times 10^4 \text{ mm}^2$$
Taking external restraint as dominant, $k_c = 1.0$.
From Table 19.1, $k = 1$
$$\sigma_s = f_{yk} = 500 \text{ MPa},$$
$$f_{ct, eff} = 1.8 \text{ MPa for } f_{ck}(t) \text{ at } t = 3 \text{ days}$$
$$A_{s, min} = k_c k f_{ct, eff} A_{ct}/\sigma_s = 1.0 \times 1.0 \times 1.8 \times 30 \times 10^4/500 = 1080 \text{ mm}^2 \quad (7.1)$$
Provide $A_s = $ H12 at 100 mm $= 1131 \text{ mm}^2/\text{m}$.
One half of this value of steel can be placed on each face of the slab. Each face has H12 at 200 mm.
$$\rho = A_s/A_{ct} = 566/30.0 \times 10^4 = 1.89 \times 10^{-3}$$

Step 4: Calculate the maximum crack spacing $S_{r, max}$ and crack width, w_k.

$$(\varepsilon_{sm} - \varepsilon_{cm}) = 0.5 \alpha_e k_c k f_{ct, eff} [1 + 1/(\alpha_e \rho)]/E_s \qquad\qquad (M.1)$$

$$(\varepsilon_{sm} - \varepsilon_{cm}) = 0.5 \times 45.7 \times 1.0 \times 1.0 \times 1.8 \times [1 + 1/(45.7 \times 1.89 \times 10^{-3}]/(200 \times 10^3)$$
$$= 2.59 \times 10^{-3}$$
$$c = \text{cover to longitudinal reinforcement} = 40 \text{ mm}$$
$$\varphi = \text{bar diameter} = 12 \text{ mm}$$
$A_{c, eff} = $ Effective area of concrete in tension surrounding reinforcement to a depth of
$h_{c, eff} = \min [h/2; 2.5(c + \varphi/2)]$.
Assuming 50% of reinforcement is on each face, $A_s = 566 \text{ mm}^2/\text{m}$.
$$h_{c,eff} = \min [h/2; 2.5(c + \varphi/2)] = \min [300/2; 2.5(40 + 12/2)] = 115 \text{ mm}.$$
$$\rho_{p, eff} = A_s/A_{c,eff} = 566/(115 \times 1000) = 4.92 \times 10^{-3}$$

Note that ρ and ρ_{eff} are two different values. ρ is the ratio of steel to the whole cross section. On the other hand, peff is the ratio of steel area to the concrete area in tension.
$$S_{r, max} = k_3 c + k_1 k_2 k_4 \varphi/\rho_{p, eff} \qquad\qquad (7.11)$$

Taking $k_1 = 0.8$ for high bond bars, $k_2 = 1.0$ for pure tension, $k_3 = 3.4$, $k_4 = 0.425$

$$S_{r, max} = 3.4 \times 40 + 0.8 \times 1.0 \times 0.425 \times 12/(4.92 \times 10^{-3}) = 965 \text{ mm}$$
$$w_k = S_{r, max}(\varepsilon_{sm} - \varepsilon_{cm}) \qquad\qquad (7.8)$$

$$w_k = 965 \times 2.59 \times 10^{-3} = 2.5 \text{ mm}$$
Crack width is too large.

If the steel area is increased by providing H12 at 50 mm on each face and both vertically and horizontally,

$$\text{Total } A_s = 4524 \text{ mm}^2/\text{m}$$
$$\rho = A_s/A_c = 4524/ (30.0 \times 10^{4)} = 15.88 \times 10^{-3} = 1.5\%$$
$$(\varepsilon_{sm} - \varepsilon_{cm}) = 0.5 \, \alpha_e \, k_c \, k \, f_{ct, \, eff} \, [1 + 1/ (\alpha_e \, \rho)]/ E_s \qquad (\text{M.1})$$
$$(\varepsilon_{sm} - \varepsilon_{cm}) = 0.5 \times 45.7 \times 1.0 \times 1.0 \times 1.8 \times [1 + 1/ (45.7 \times 15.88 \times 10^{-3}]/ (200 \times 10^3)$$
$$= 0.49 \times 10^{-3}$$

$$\rho_{p, \, eff} = A_s/A_{c,eff}$$

$A_{c, \, eff}$ = Effective area of concrete in tension surrounding reinforcement to a depth of $h_{c, \, eff} = \min (h/2; 2.5(c + \varphi/2)$.

Assuming 50% of reinforcement is on each face, $A_s = 4524/2 = 2262 \text{ mm}^2/\text{m}$.
$$h_{c,eff} = \min [h/2; \ 2.5(c + \varphi/2)] = \min [300/2 \ ; 2.5(40 + 12/2)] = 115 \text{ mm}.$$
$$\rho_{p, \, eff} = A_s/A_{c,eff} = 2262/(115 \times 1000) = 19.67 \times 10^{-3}$$

$$S_{r, \, max} = k_3 \, c + k_1 \, k_2 \, k_4 \, \varphi/ \, \rho_{p, \, eff} \qquad (7.11)$$

$$S_{r, \, max} = 3.4 \times 40 + 0.8 \times 1.0 \times 0.425 \times 12/ \, 19.67 \times 10^{-3} = 343 \text{mm}$$

$$w_k = S_{r, \, max} \, (\varepsilon_{sm} - \varepsilon_{cm}) \qquad (7.8)$$

$$w_k = 343 \times 0.49 \times 10^{-3} = 0.17 \text{ mm}$$
This is an acceptable crack width.

Step 5: Check the maximum bar diameter and spacing for acceptable crack width.
$$\sigma_s = k_c \, k \, f_{ct,eff} \, / \, \rho \qquad (\text{M.2})$$
$$\sigma_s = 1.0 \times 1.0 \times 1.8 \, / \, 15.88 \times 10^{-3} = 113 \text{ MPa}$$
From Fig. 7.103N, maximum bar diameter φ_s^* for 0.2 mm wide crack is approximately 40 mm.
Calculate the adjusted the maximum bar diameter φ_s from code equation (7.122)

$$\varphi_s = \varphi_s^* \, [f_{ct, \, eff}/2.9] \times \{h/[10 \times (h - d)]\} \qquad (7.122)$$

$h = 400$ mm, $d = 300 - 40 - 12/2 = 254$ mm, $f_{ct, \, eff} = 1.8$ MPa.
$$\varphi_s = \varphi_s^* \, [1.8/2.9] \times \{300/[10 \times (300 - 254)]\} = 0.65 \, \varphi_s^* \approx 26 \text{ mm}$$
From Fig. 7.104N, maximum bar spacing for 0.2 mm wide crack is approximately 250 mm.
The provided values are well below acceptable approximate values.

(iii) Design for vertical bending

a. Vertical steel on inner face
Maximum bending moment causing tension on inner face at base at ULS:
$$M = (\gamma_F = 1.2) \times 0.0267 \times 60 \times 6^2 = 69.2 \text{ kNm/m}$$
$M = 69.2$ kNm/m, $b = 1000$ mm, cover = 40 mm, $\varphi = $ H12,
$$d = 300 - 40 - 12/2 = 254 \text{ mm}, \, f_{ck} = 30 \text{ MPa}$$
$$k = 69.2 \times 10^6/ (1000 \times 254^2 \times 30) = 0.036 < 0.196$$

$$\frac{z}{d} = 0.5[1.0 + \sqrt{(1 - 3\frac{0.036}{1.0})}] = 0.97$$

$$A_s = \frac{69.2 \times 10^6}{0.97 \times 254 \times 0.87 \times 500} = 646 \, mm^2 / m$$

Provide H12 at 175 c/c = 646 mm²/m.

This steel is required for only a height of approximately 0.2 h from base. Above this only minimum steel required. Alternate bars can be terminated beyond (0.2h + anchorage length of 38 φ) = 1656 mm, say 1700 mm above base.

However the minimum steel of H12 at 50 = 2262 provided to minimize early-age thermal cracking supersedes the above value.

b. Vertical steel on outer face
Maximum bending moment causing tension on the outer face at 0.4h at ULS:
$$M = (\gamma_F = 1.2) \times 0.0077 \times 60 \times 6^2 = 19.96 \, kNm/m$$
$$k = 19.96 \times 10^6 / (1000 \times 254^2 \times 30) = 0.01 < 0.196$$

$$\frac{z}{d} = 0.5[1.0 + \sqrt{(1 - 3\frac{0.01}{1.0})}] = 0.99$$

$$A_s = \frac{19.96 \times 10^6}{0.99 \times 254 \times 0.87 \times 500} = 183 \, mm^2 / m$$

However the minimum steel of H12 at 50 = 2262 provided to minimize early-age thermal cracking supersedes the above value.

(iv) Design for ring tension

Maximum ring tension T occurs at mid-height,
$$T = 0.43 \times 60 \times 15 = 387 \, kN/m \text{ at SLS}$$
$$T = (\gamma_F = 1.2) \times 387 = 464 \, kN/m \text{ at ULS}$$
Circumferential moment = 3.42 kNm/m
$$A_s = 464 \times 10^3 / (500/1.15) = 1067 \, mm^2/m$$
Use T16 at 300 on each face giving total $A_s = 1340 \, mm^2/m$.

However the minimum steel of H12 at 50 = 2262 provided to minimize early-age thermal cracking supersedes the above value.

(v) Check crack width:
Moment is small and can be ignored. Stress in steel is due to ring tension.
$$\sigma_s = T/A_s = 387 \times 10^3 / (2262) = 170 \, MPa$$
From code *BS EN 1992-3:2006 Eurocode 2-Design of concrete structures: Part 3: Liquid retaining and containment structures,* for a crack width of 0.2 mm,
Fig. 7.103N gives maximum bar diameter = 25 mm.
Fig. 7.104N, gives maximum bar spacing = 175 mm.

From code equation (7.122), the modified bar diameter is

$$\varphi_s = \varphi_s^* \, (f_{ct, \, eff}/ \, 2.9) \times \{h/[10 \, (h - d)]\} \tag{7.122}$$
$$\varphi_s = 25 \times (1.8/ \, 2.9) \times \{300/[10 \times (300 - 254)]\} = 10 \text{ mm} \approx 12 \text{ mm}$$

The provided steel of H12 at 50 is sufficient to limit the crack width to 0.2 mm.

19.7 REFERENCES

Anchor, Robert D. (1992). *Design of Liquid Retaining Concrete Structures*, 2nd ed. Edward Arnold.

Bamforth, P.B. (2007). *Early-Age Thermal Crack Control in Concrete*. CIRIA.

Batty, Ian and Westbrook, Roger. (1991). *Design of Water Retaining Concrete Structures*. Longman Scientific and Technical.

Ghali, Amin. (1979). *Circular Storage Tanks and Silos*. E&FN Spon.

Perkins, Phillip H. (1986). *Repair, Protection and Water Proofing of Concrete Structures*. Elsevier Applied Science.

CHAPTER 20

U.K. NATIONAL ANNEX

20.1 INTRODUCTION

Eurocodes have been written to be applicable to all counties which belong to the European Union. However there is recognition that differences in construction materials and practices as well as climatic conditions leading for example to differences in wind and snow loading exist. In order to accommodate these differences, Eurocode allows each country to adopt 'nationally determined parameters'. In the U.K., the nationally determined parameters are given in U.K. National Annex (UKNA). In the vast majority of cases the recommendations of Eurocode and UKNA are identical. Only in a small number of cases do they differ. The object of this chapter is to highlight some of these important differences. The reader should refer to the full document for complete details.

20.2 BENDING DESIGN

(a) Design compressive strength: $f_{cd} = \alpha_{cc} f_{ck}/\gamma_c$.
In Eurocode $\alpha_{cc} = 1$, $f_{cd} = f_{ck}/\gamma_c$.
In UKNA, $\alpha_{cc} = 0.85$, $f_{cd} = 085 f_{ck}/\gamma_c$ in flexure and axial loading but may be used in all phenomena.

(b) If concrete strength is determined at an age $t > 28$ days, in Eurocode $\alpha_{cc} = 0.85$ and in UKNA $\alpha_{cc} = 1.0$.

20.2.1 Neutral Axis Depth Limitations for Design Using Redistributed Moments

Eurocode uses the following equations:

$$\delta \geq 0.44 + 1.25(0.6 + \frac{0.0014}{\varepsilon_{cu2}}) \times \frac{x_u}{d} \qquad f_{ck} \leq 50 \text{ MPa}$$
(5.10a)

$$\delta \geq 0.54 + 1.25 \times (0.6 + \frac{0.0014}{\varepsilon_{cu2}}) \times \frac{x_u}{d} \qquad f_{ck} > 50 \text{ MPa}$$
(5.10b)

Note in equation (5.10b), $\delta \geq 0.7$ if Class B and Class C reinforcement is used and $\delta \geq 0.8$ if Class A reinforcement is used.

UKNA uses the following equations for $f_{yk} \leq 500$ MPa:

$$\delta \geq 0.40 + (0.6 + \frac{0.0014}{\varepsilon_{cu2}}) \times \frac{x_u}{d} \qquad f_{ck} \leq 50 \text{ MPa}$$

$$(5.10a)$$

$$\delta \geq 0.40 + (0.6 + \frac{0.0014}{\varepsilon_{cu2}}) \times \frac{x_u}{d} \qquad f_{ck} > 50 \text{ MPa}$$

$$(5.10b)$$

20.3 COVER TO REINFORCEMENT

For $c_{min, dur}$, Eurocode gives values in code Table 4.4N. This is summarised in Chapter 2, section 2.9.
UKNA recommends following the values in BS8500-1:2006.

20.4 SHEAR DESIGN

The main recommendation in UKNA is that unless tests show otherwise, the shear strength of concrete for $f_{ck} > 50$ MPa may be limited to that of $f_{ck} = 50$ MPa.
An additional restriction is that while normally $1 \leq \cot \theta \leq 2.5$, if shear co-exists with externally applied tension, then $\cot\theta = 1.25$.
In Eurocode, $v_1 = v = 0.6(1 - f_{ck}/25)$.
In UKNA $v_1 = 0.6(1 - f_{ck}/25) \times (1 - 0.5 \cos \alpha)$ where $\alpha =$ inclination of shear links to the horizontal. If $\alpha = 90^0$, $v_1 = v = 0.6(1 - f_{ck}/25)$.

20.4.1 Punching Shear

Both in Eurocode and in UKNA, the value of the maximum punching shear stress $v_{Rd, max}$ adjacent to the column in is limited to 0.5 v f_{cd}. However in UKNA it is further required that at the first control perimeter, the shear stress is limited to $2v_{Rd, c}$.

20.5 LOADING ARRANGEMENT ON CONTINUOUS BEAMS AND SLABS

UKNA allows for the following three options:

Option 1: Use the Eurocode recommended loading pattern.

Option 2:
(a) **All** spans carrying ($\gamma_G G_k + \gamma_Q Q_k$).
(b) **Alternate** spans carrying ($\gamma_G G_k + \gamma_Q Q_k$) and other spans carrying $\gamma_G G_k$. The same value of γ_G should be used throughout the structure.

Option 3: For slabs use **all** spans carrying $(\gamma_G\, G_k + \gamma_Q\, Q_k)$ provided:
- In one-way spanning slab the area of each bay exceeds 30m². Bay is the area bounded by the full width of the structure and the width between the lines of support on the other two sides.
- Ratio of $Q_k/G_k \leq 1.25$.
- $Q_k \leq 5$ kN/m².

Note that for design, the resulting moments are redistributed by reducing the support moments (except in cantilevers) by 20% with a consequential increase in span moments.

20.6 COLUMN DESIGN

In Eurocode, the minimum value of diameter of longitudinal reinforcement φ_{min} is 8 mm but in UKNA it is 12 mm.
The maximum spacing of transverse reinforcement $s_{d,\,max}$ is:
Eurocode:
$s_{d,\,max}$ = min [20 × diameter of longitudinal bars; lesser dimension of column; 400 mm]

UKNA:
$s_{d,\,max}$ = as per Eurocode for $f_{ck} \leq 50$ MPa, $\alpha_n\, \alpha_s\, \omega_{wd} \geq 0.04$ for $f_{ck} > 50$ MPa
where
$\alpha_n = 1 - \Sigma b_i^2 / (b_0\, h_0 \times 6)$.
b_i = distance between consecutive bars that are laterally restrained as shown in Fig. 20.1.
$\alpha_s = [1 - 0.5\, s/b_0]\,[1 - 0.5\, s/h_0]$.
ω_{wd} = [volume of confining hoops/volume of concrete] × (f_{yd}/f_{ck}).
s = longitudinal spacing of links.
b_0, h_0 = dimensions of centre lines of links.

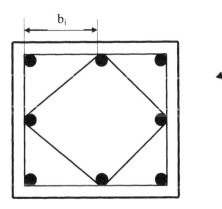

Fig. 20.1 Column cross section.

Example: Fig. 20.2 shows the cross section of a column. H12 links are spaced at 100 mm. Cover to steel = 30 mm. f_{ck} = 60 MPa. Check whether the links are sufficient.

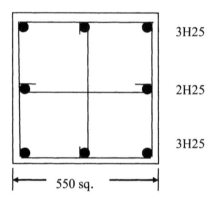

3H25

2H25

3H25

550 sq.

Fig. 20.2 Column cross section.

$$b_0 = h_0 = 550 - 2 \times (30 + 12/2) = 478 \text{ mm}$$
$$b_i = [550 - 2 \times (30 + 12 + 25/2)]/2 = 221 \text{ mm}$$
$$\Sigma b_i^2 = 8 \times 221^2 = 3.89 \times 10^5 \text{ mm}^2$$
$$\alpha_n = 1 - \Sigma b_i^2 / (b_0 \times h_0 \times 6) = 1 - 3.89 \times 10^5 / (478 \times 478 \times 6) = 0.72$$
$$\alpha_s = [1 - 0.5 \text{ s}/b_0] [1 - 0.5 \text{ s}/h_0] = [1 - 0.5 \times 200/478]^2 = 0.625$$
$$\text{Total length of links} = 2 \times [3 \times (550 - 30 - 12] = 3048 \text{ mm}$$
$$\text{Area of H12 link} = 113 \text{ mm}^2$$
$$\text{Volume of link} = 3048 \times 113 = 3.45 \times 10^5 \text{ mm}^3$$
$$\text{Volume of concrete} = b_0 \times h_0 \times s = 22.85 \times 10^6 \text{ mm}^3$$
$$f_{yd} = 500/1.15 = 435 \text{MPa}$$
$$\omega_{wd} = [3.45 \times 10^5 / (22.85 \times 10^6)] \times (435/60) = 0.11$$
$$\alpha_n \times \alpha_s \times \omega_{wd} = 0.72 \times 0.625 \times 0.11 = 0.049 \geq 0.04$$

Link diameter and spacing are satisfactory.

20.7 TIES

(a) Peripheral ties: The forces to be resisted by peripheral ties are as follows.
$$F_{te, per} = \ell_1 q_1 \leq q_2.$$
where ℓ_1 = Length of end span.

Eurocode: q_1= 10 kN/m, q_2 = 70 kN

UKNA: $q_1 = (20 + 4 n_0) \ell_1$, q_2 = 60 kN
where n_0 = number of storeys.

(b) Internal ties: Minimum tensile force $F_{tie, int}$ kN/m width that an internal tie is capable of resisting:

Eurocode: $F_{tie, int} = 20$ kN/m

UKNA: $F_{tie, int} = [(q_k + g_k)/7.5] \times (\ell_x/5) \times F_t \geq F_t$ kN/m

where
ℓ_x = The greater of the distance between the centres of columns, frames or walls supporting two adjacent floor spans in the direction of the tie under consideration.
$F_t = (20 + 4n_0) \leq 60$ kN
Maximum spacing of internal ties = $1.5\ \ell_x$

(c) Horizontal ties:

Eurocode: Horizontal ties to external columns should resist a force $F_{tie, col} = 150$ kN and to the walls should resist a force $F_{tie, fac} = 20$ kN/m of the facade.

UKNA:
$F_{tie, col} = F_{tie, fac}$
$\qquad = \max [2\ F_t \leq \ell_s/ (2.5\ Ft);$ 3% of total design vertical load carried by column or wall at that level]
$F_{tie, col}$ in kN per column.
$F_{tie, col}$ in kN/m of the wall.
ℓ_s = floor to ceiling height in m.

20.8 PLAIN CONCRETE

Design values of compressive and tensile strength of plain concrete are given by:
Eurocode:
$f_{cd} = 0.8\ f_{ck}/ \gamma_c$
$f_{ctd} = 0.8_1\ f_{ctk,\ 0.05}/ \gamma_c$

UKNA:
$f_{cd} = 0.6\ f_{ck}/ \gamma_c$
$f_{ctd} = 0.8\ f_{ctk,\ 0.05}/ \gamma_c$

20.9 ψ FACTORS

Wind loads on buildings:
Eurocode: $\psi_0 = 0.6$.
UKNA: $\psi_0 = 0.5$.

ADDITIONAL REFERENCES

In addition to the references given in the body of the text, the following additional references might be found useful.

Beeby, A.W. and Narayanana, R.S. (2005). *Designers' Guide to EN 1992-1-1 and EN 1992-1-2*. Thomas Telford.

Bennett, D.F.H. and MacDonald, L.A.M. (1992). *Economic Assembly of Reinforcement.* Reinforced Concrete Council and British Cement Association.

Bhatt, P. (1999). *Structures*. Longman.

Booth, Edmund (Ed). (1994). *Concrete Structures in Earthquake Regions.* Longman.

British Standards Institution. (2010). *PD 6887-1:2010, Background Paper to the National Annexes to BS EN 1992-1 and BS EN 1992-3*.

European Concrete Platform. (2008). *Eurocode 2: Commentary*.

European Concrete Platform. (2008). *Eurocode 2: Worked Examples*.

European Concrete Platform. (2007). *How to Design Concrete Structures: Columns*.

Goodchild, C.H. (2009). *Worked Examples to Eurocode 2: Vol 1*. The Concrete Centre.

Institution of Structural Engineers (U.K.). (2006). *Manual for the Design of Concrete Building Structures to Eurocode 2.*

Kotsovos, M.D. and Pavlovic, M.N. (1995). *Finite Element Analysis for Limit State Design*. Thomas Telford.

Kotsovos, Michael D. and Pavlovic, M.N. (1999). *Ultimate Limit State Design of Concrete Structures: A New Approach*. Thomas Telford.

MacGregor, J.G. (1992). *Reinforced Concrete: Mechanics and Design*. Prentice-Hall.

Mosley, W.H., Bungey, J.H. and Hulse, R. (2012). *Reinforced Concrete Design to Eurocode 2*, 6th ed. Palgrave Macmillan.

O'Brien, E.J. and Dixon, A.S and Sheils, E. (2012). *Reinforced and Prestressed Concrete Design*, 2nd ed. Spon Press.

Tubman, J. (1995). *Steel Reinforcement.* CIRIA.

INDEX

Milton Keynes UK
Ingram Content Group UK Ltd.
UKHW030900141024
449569UK00025B/1297